BEAT THE NEET

PHYSICS VOLUME-01 FOR NEET

2000+ Problems with Solutions

BEAT THE NEET

PHYSICS
VOLUME-01
FOR NEET

2000+ Problems with Solutions

DC Pandey
B.Tech, M.Tech, Pantnagar, ID 15722

arihant
ARIHANT PRAKASHAN (Series), MEERUT

arihant

ARIHANT PRAKASHAN (Series), MEERUT

Administrative & Production Offices

Regd. Office
'Ramchhaya' 4577/15, Agarwal Road, Darya Ganj, New Delhi -110002
Tele: 011- 47630600, 43518550

Head Office
Kalindi, TP Nagar, Meerut (UP) - 250002
Tel: 0121-7156203, 7156204

Sales & Support Offices
Agra, Ahmedabad, Bengaluru, Bareilly, Chennai, Delhi, Guwahati, Hyderabad, Jaipur, Jhansi, Kolkata, Lucknow, Nagpur & Pune.

ISBN : 978-93-26199-92-6

PRICE : ₹375.00

PO No : TXT-XX-XXXXXXX-X-XX

Published by Arihant Publications (India) Ltd.

For further information about the books published by Arihant, log on to www.arihantbooks.com or e-mail at info@arihantbooks.com

Follow us on

Preface

Physics is a subject of concepts which needs a lot of practice for NEET aspirants. During the last 30 years of my teaching career, I always felt the need of a problem practice book that can offer quality problems of Physics of all chapters.

To fulfil the need of a good Physics Problem book, I have prepared this book BEAT THE NEET PHYSICS VOLUME - 01. All the problems given in this book are arranged in a systematic manner to enhance the conceptual and analytical ability of the students.

- Each chapter is divided into three levels.
- Previous Years' Questions are given in Level-3. These questions are very important because questions or concepts are repeated every year in many problems.
- Hints/Solutions are also provided at the end of each chapter.

I am very much confident that if you rigorously practice these problems then you will certainly secure good score in Physics. I have taken the utmost care while writing this book but if you find error then please bring it to my notice at the mail id *arihantcorrections@gmail.com*.

Comments and criticism from readers will be highly appreciated and incorporated in the subsequent editions.

I would like to appreciate the efforts done by Mr. Anoop Dhyani in bringing the book in final shape.

DC Pandey
B.Tech, M.Tech, Pantnagar, ID 15722

Content

This book is dedicated to my
honourable grandfather

(Late) Sh. Pitamber Pandey

a Kumaoni poet and a resident
of village Dhaura (Almora), Uttarakhand

Chapter 01

BASIC MATHEMATICS

LEVEL 01 Objective Questions

1 Convert 225° in radian.

(a) $\frac{5\pi}{3}$ (b) $\frac{3\pi}{4}$ (c) $\frac{4\pi}{3}$ (d) $\frac{5\pi}{4}$

2 Value of $\sin\left(14\pi + \frac{\pi}{6}\right) =$

(a) $\frac{1}{2}$ (b) $\frac{1}{\sqrt{2}}$ (c) $\frac{\sqrt{3}}{2}$ (d) $-\frac{1}{2}$

3 $\cos(\alpha+\beta)-\cos(\alpha-\beta)$ is equal to

(a) $2\cos\alpha\cos\beta$ (b) $-2\cos\alpha\cos\beta$ (c) $2\sin\alpha\sin\beta$ (d) $-2\sin\alpha\sin\beta$

4 If $\cos\theta = \frac{5}{13}$, then find the value of $\tan\theta$.

(a) $\frac{5}{12}$ (b) $\frac{12}{13}$ (c) $\frac{12}{5}$ (d) $\frac{13}{5}$

5 Find $\sin^2 120° + \cot^2 120°$.

(a) $\frac{12}{13}$ (b) $\frac{13}{12}$ (c) $\frac{13}{4}$ (d) $\frac{4}{13}$

6 The length of a string between a kite and a point on the ground is 55 m. If the string makes an angle of 53° from ground level and there is no slack in the string, then the height of the kite from ground level is

(a) 55 m (b) 44 m
(c) 33 m (d) 11 m

7 If $\sin\theta = \frac{2}{3}$, find the value of $(\cot\theta \times \sec\theta)$.

(a) $\frac{1}{6}$ (b) $\frac{3}{2}$
(c) $\frac{2}{3}$ (d) $\frac{5}{6}$

8 If $\tan\theta = -\frac{1}{\sqrt{5}}$ and θ lies in the IVth quadrant, then the value of $\cos\theta$ is

(a) $\frac{\sqrt{5}}{\sqrt{6}}$ (b) $\frac{2}{\sqrt{6}}$ (c) $\frac{1}{2}$ (d) $\frac{1}{\sqrt{6}}$

9 If perpendicular of a right-angled triangle is $\frac{1}{4}$ th of its hypotenuse, then find ratio of base to the hypotenuse.

(a) $4B = \sqrt{15}\,H$ (b) $4B = H$
(c) $\sqrt{15}\,H = B$ (d) None of these

10 Distance between two points (8, 4) and (2, a) is 10 unit. Find positive value of a.

(a) 4 (b) 8 (c) 6 (d) 12

11 Which of the following point lies on the line $2x + y + 5 = 0$?

(a) (0, 0) (b) (−2, − 1)
(c) (2, 1) (d) None of these

12 Find slope of a line passing through (2, 3) and (4, − 5).

(a) 4 (b) −4 (c) −2 (d) 2

13 Product of slopes of lines $4y + 2x + 4 = 0$ and $3y - 6x + 10 = 0$, is

(a) −1 (b) 1 (c) 2 (d) −4

14 Intersection point of given lines $y = 4x$ and $y = 2x - 4$ will be

(a) (2, 8) (b) (−2, − 4)
(c) (−2, − 8) (d) (2, − 8)

15 Equation of given line will be

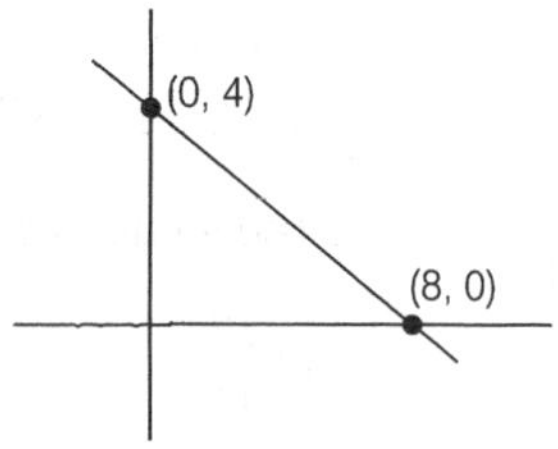

(a) $2y = -x + 8$ (b) $y = \frac{-x}{4} + 2$ (c) $y = -4x + 2$ (d) $y = -2x + 4$

16 If $\sqrt{3}y = 3x + 6$, then angle made by the line with positive X-axis is

(a) 120° (b) 60° (c) 30° (d) 150°

17 Equation for the straight line shown in the figure is

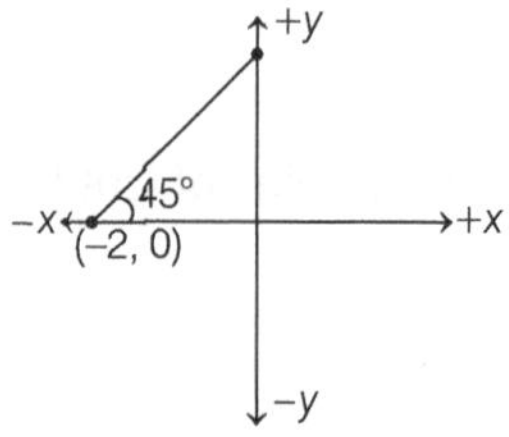

(a) $y = x - 2$ (b) $y = -x + 2$ (c) $y = x + 2$ (d) $y = 2x + 1$

18 If equation of a circle is $(x - 2)^2 + (y - 0)^2 = 16$, find distance of centre from origin.

(a) 2 (b) 4 (c) 8 (d) 16

19 Which of the following match is correct?

(i)	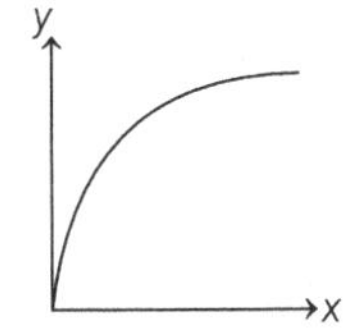	(A)	$y = x^2$
(ii)	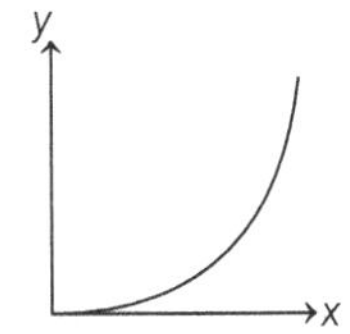	(B)	$y = -x^2$
(iii)	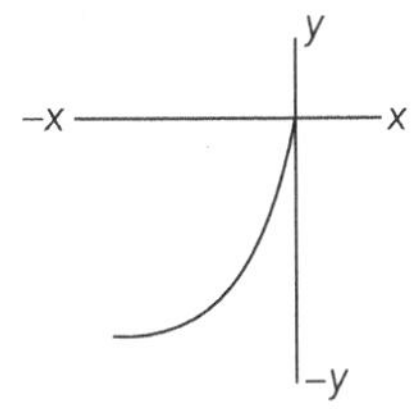	(C)	$x = y^2$
(iv)	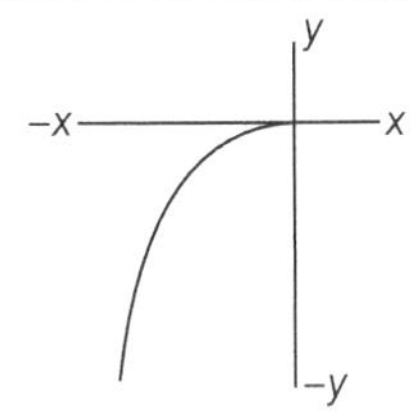	(D)	$x = -y^2$

	(i)	(ii)	(iii)	(iv)
(a)	A	B	C	D
(b)	A	C	B	D
(c)	D	C	B	A
(d)	C	A	D	B

20 Draw graph for equation $4y^2 + 25x^2 = 100$.

(a) 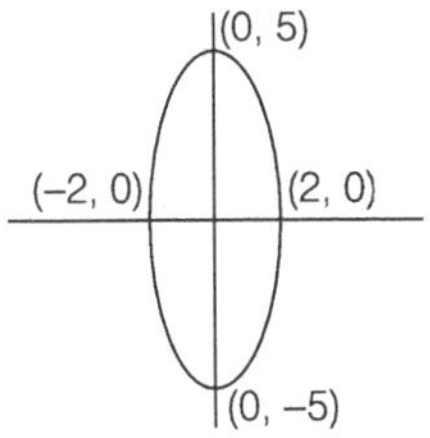

(b) 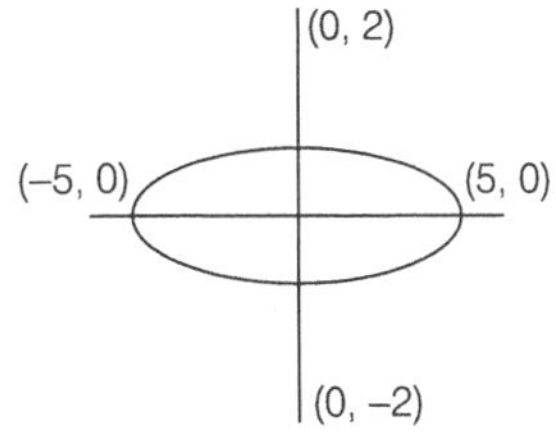

(c) 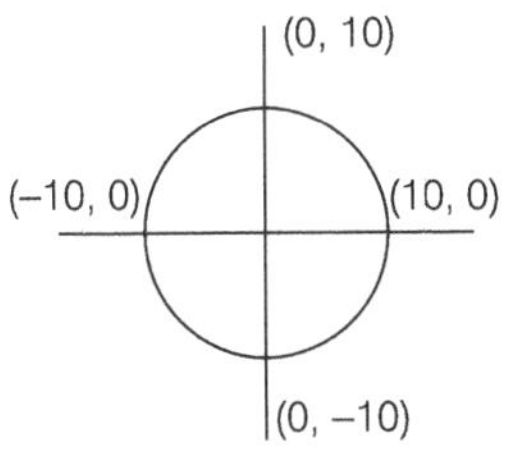

(d) 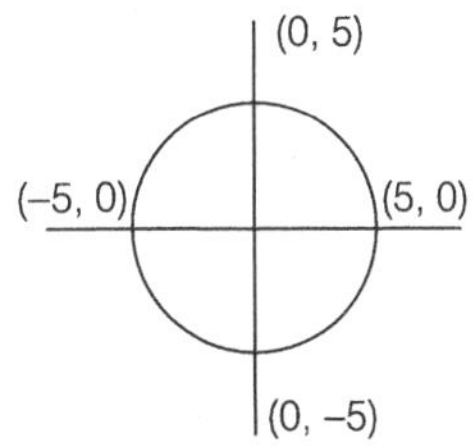

21 Find equation for the given graph.

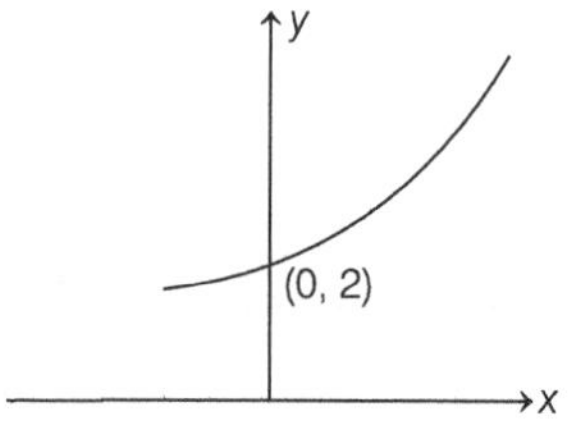

(a) $y = 2x^2$ (b) $x = 2y^2$ (c) $y = e^{2x}$ (d) $y = 2e^{2x}$

22 The equation of a curve is given as $y = x^2 + 2 - 3x$. The curve intersects the X-axis at

(a) (1, 0) (b) (2, 0)
(c) Both (a) and (b) (d) No where

23 If $y = \sin^2 x + \cos^2 x$, then $\frac{dy}{dx}$ is

(a) $4 \sin \cos x$ (b) $2\sin x + 2\cos x$ (c) zero (d) 1

24 Rate of change of y with respect to x will be, if $y = x^4 + 2x^3 + 2$

(a) $4x^3 + 6x^2$ (b) $4x^4 + 5x^2$
(c) $4x^3 + 5x^2$ (d) $4x^3 + 6x^2 + 2$

25 If $y = \sqrt{6x + 5}$, then $\frac{dy}{dx} = ?$

(a) $3\sqrt{6x + 5}$ (b) $\frac{6}{\sqrt{6x + 5}}$ (c) $\frac{3x}{\sqrt{6x + 5}}$ (d) $\frac{3}{\sqrt{6x + 5}}$

26 If current $i = i_0 \sin\left(\omega t + \frac{\pi}{3}\right)$, then $\frac{di}{dt}$ is (where, ω and i_0 are constants)

(a) $i_0 \cos\left(\omega t + \frac{\pi}{3}\right)$ (b) $\frac{i_0}{\omega}\cos\left(\omega t + \frac{\pi}{3}\right)$ (c) $i_0 \omega \sin\left(\omega t + \frac{\pi}{3}\right)$ (d) $i_0 \omega \cos\left(\omega t + \frac{\pi}{3}\right)$

27 If $y = 5x^3 - 4x^2 + 2$, then value of $\frac{d^2y}{dx^2}$ at $x = 1$ is

(a) 22 (b) 7 (c) 24 (d) 3

28 If $y = 4x^2 + 2x$, then slope of y-x graph at $x = 1$ is

(a) 10 (b) 8 (c) 6 (d) 12

29 Find greatest value of y, where $y = 3 \sin\theta - 4\cos\theta$.

(a) 7 (b) 1 (c) 5 (d) 4

30 If $y = 2x^2 - 8x + 5$, then minimum value of y is

(a) 3 (b) 5 (c) -3 (d) -5

31 At $x = 0, y = x^3 - 25$ is

(a) maximum (b) minimum
(c) Neither maximum nor minimum (d) Data insufficient

32 A quantity Q varies with time as $Q = 10 \sin t$, find rate of change of Q with respect to time at $t = \frac{\pi}{4}$.

(a) 5 (b) $\frac{10}{2\sqrt{2}}$ (c) $5\sqrt{2}$ (d) 0

33 Find the maximum value of y, if $y = -x^2 - 4x + 8$.

(a) 12 (b) 20 (c) -4 (d) 8

34 $\int_0^{\pi/2} \sin^2 \theta \, d\theta =$

(a) 1 (b) 0 (c) $\frac{\pi}{2}$ (d) $\frac{\pi}{4}$

35 Evaluate $\int_0^{\pi/4} (\sin x + \cos x) \, dx$.

(a) -1 (b) $\sqrt{2}$ (c) 0 (d) 1

36 The value of $\int_0^1 9x^8 \, dx + \int_0^{\pi/2} \cos x \, dx$ is

(a) 1 (b) 3 (c) 4 (d) 2

37 Find $\int_0^{\frac{\pi}{4}} 4 \sin 2x \, dx$.

(a) 2 (b) -2 (c) 1 (d) -1

38 Value of $\int_0^2 3x^2 \, dx + \int_0^{\pi/2} \sin x \, dx$ is :

(a) 0 (b) 4 (c) 9 (d) None of these

39 $\int_0^{\pi} (\sin x + \cos x) dx$ is

(a) -2 (b) 2 (c) 1 (d) 0

LEVEL 02 Objective Questions

40 If $\cos \theta = \frac{7}{25}$ and θ lies in Ist quadrant, then find $\tan (90° - \theta)$.

(a) $\frac{24}{7}$ (b) $\frac{24}{25}$ (c) $\frac{7}{24}$ (d) $\frac{25}{24}$

41 Value of $\cos^2 22.5°$ can be written as

(a) $\frac{1 + \cos 45°}{2}$ (b) $\frac{1 - \cos 45°}{2}$ (c) $\frac{1 + \sin 45°}{2}$ (d) $\frac{1 - \sin 45°}{2}$

42 Find $\tan\left(\frac{\pi}{4} + \theta\right) \times \tan\left(\frac{\pi}{4} - \theta\right)$ for $\theta = 53°$.

(a) 0 (b) -1 (c) 1 (d) 2

43 If $2 \cos 2\theta = 1$, then $\theta = ?$

(a) 30° (b) 90°
(c) 45° (d) 15°

44 Find the equation of a line which is perpendicular to $y = 3x - 8$ and cutting the Y- axis at (0, 4).

(a) $y = \frac{-x}{3} + 4$ (b) $y = -4x + 3$ (c) $y = -3x + 4$ (d) $y = \frac{-x}{3} - 8$

45 Which pair of line is parallel among the given pairs?

(a) $y = 4x + 8, y = 2x - 6$ (b) $2y + 4x + 8 = 0, \; y + 2x - 8 = 0$
(c) $2y = 4x + 5, \; y = 4x - 5$ (d) $3y = 6x + 8, \; y = -2x + 4$

46 Find equation of a line passing through (−2, 0) and (0, 6).

(a) $y = -3x + 6$ (b) $y = -3x - 2$ (c) $y = 3x + 6$ (d) $y = 3x - 2$

47 Find out y intercept of the given line.

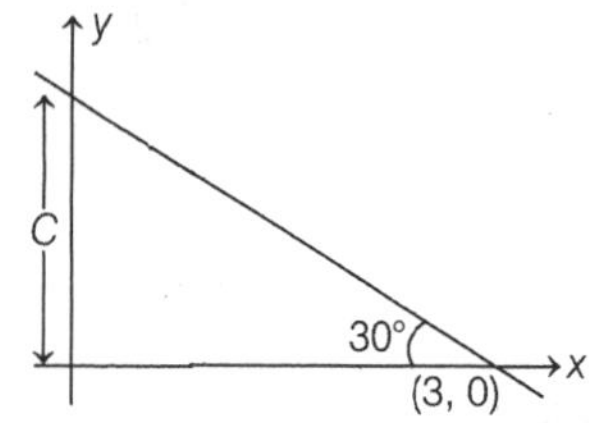

(a) $C = 3$ (b) $C = \sqrt{3}$ (c) $C = 3\sqrt{3}$ (d) $C = -\sqrt{3}$

48 Equation for given curve is

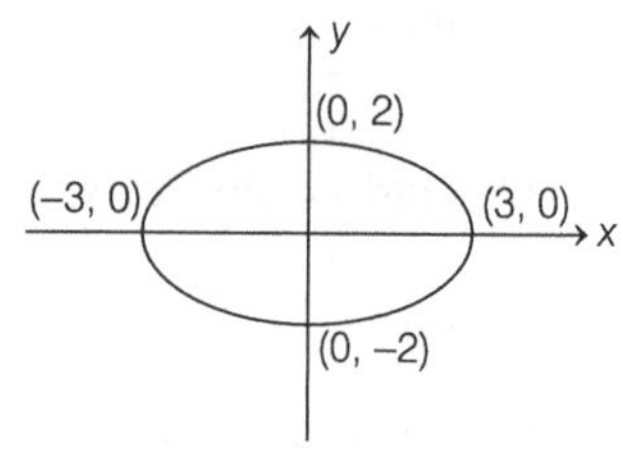

(a) $\frac{x^2}{3} + \frac{y^2}{2} = 1$ (b) $\frac{x^2}{4} + \frac{y^2}{9} = 1$

(c) $\frac{x^2}{9} + \frac{y^2}{4} = 1$ (d) $\frac{x^2}{2} + \frac{y^2}{3} = 1$

49 Equation of a line making an angle of 45° with positive X-axis and cutting Y-axis at (0, 5), is

(a) $y = 2x + 5$ (b) $y = -x + 5$ (c) $y = -x - 5$ (d) $y = x + 5$

50 $\frac{d}{dx}\left[\frac{6x^2 - 2x}{3x - 1}\right]$ equal to

(a) $\frac{6}{(3x-1)^2}$ (b) $\frac{-6}{(3x-1)^2}$ (c) $\frac{6x+1}{(3x-1)^2}$ (d) 2

51 If $y = \sin x + \cos x$, then $\frac{d^2y}{dx^2}$ is equal to

(a) $\sin x - \cos x$ (b) $\cos x - \sin x$ (c) $-(\sin x + \cos x)$ (d) None of these

52 The following graph corresponds to which equation?

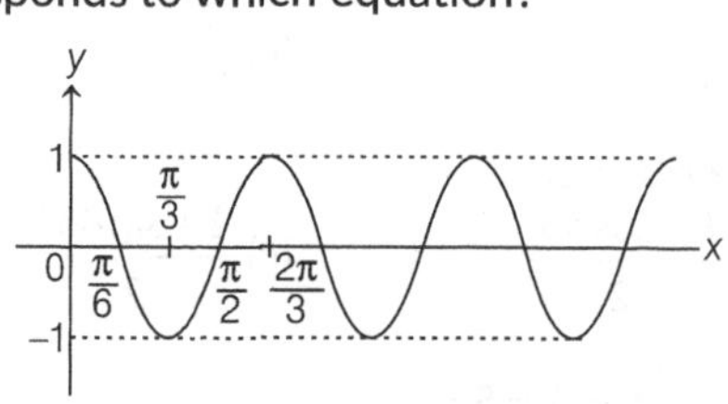

(a) $y = \cos x$ (b) $y = 3\cos x$ (c) $y = 3\cos 3x$ (d) $y = \cos 3x$

53 If rate of change of radius is 2 m/s, then find rate of change of area of circle, when radius is $\frac{1}{\pi}$ m.

(a) $3\ m^2/s$ (b) $4\ m^2/s$

(c) $2\pi\ m^2/s$ (d) $3\pi\ m^2/s$

54 Slope of a curve $y = x - x^2$ at $x = 4$ will be

(a) −7 (b) −9 (c) 8 (d) −5

55 A particle moves along the curve $12y = x^3$, which coordinate changes at faster rate at $x = 10$?

(a) X-coordinate (b) Y-coordinate

(c) Both X and Y-coordinate (d) Data insufficient

56 A tangent is drawn at point P of curve $y = ax(x+b)$. Find the value of ab.

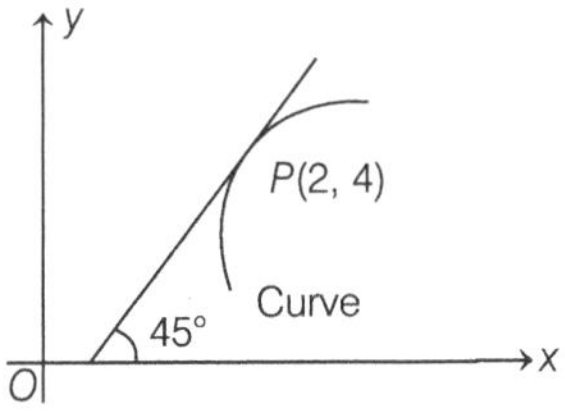

(a) 1 (b) 2 (c) 3 (d) 4

57 If $y = 4x^2 + 2x$, then graph between $\frac{dy}{dx}$ and x is

(a) 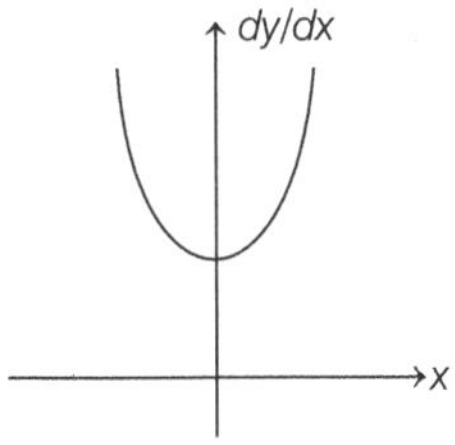
(b) 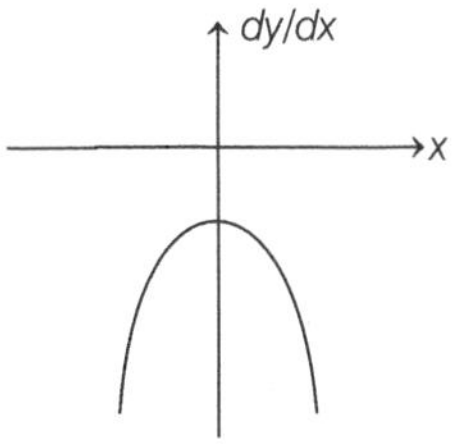
(c) 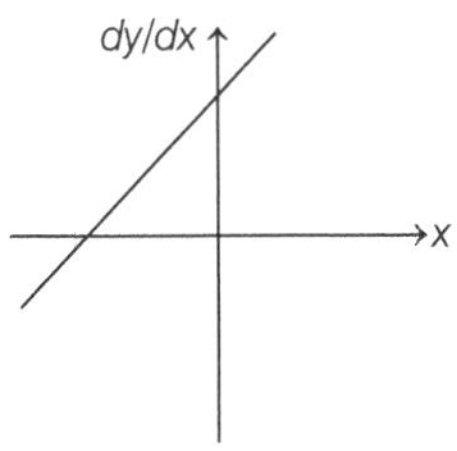
(d) 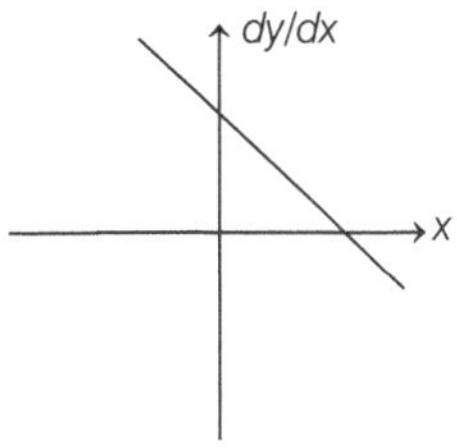

58 Radius of a metallic plate is given as $r = 2t^2$, then rate of increase of area of the plate at $t = 2$ s is

(a) 64π (b) 128π (c) 32π (d) 144π

59 Equation of given curve is $y = \frac{x^2}{2}$, then coordinates of point on curve at which tangent makes an angle 45° with the X-axis are

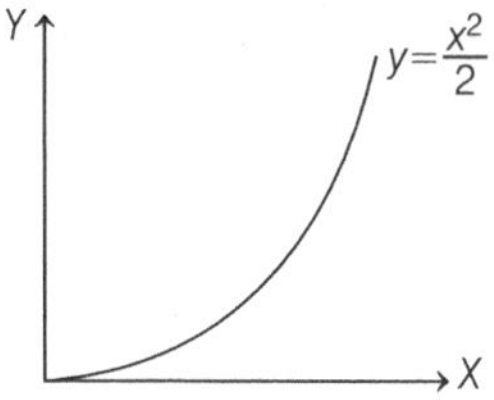

(a) $\left(1, \frac{1}{2}\right)$ (b) (1, 1)

(c) (1, 2) (d) None of these

60 If $y = \sin\theta + \sqrt{3}\cos\theta$, then range of y is

(a) $0 \le y \le 2$ (b) $-2 \le y \le 0$ (c) $-2 \le y \le 2$ (d) $0 \le y \le 4$

61 What is the minimum value of $\frac{2}{4 + \sin\theta + \sqrt{3}\cos\theta}$?

(a) 0 (b) 1 (c) $\frac{1}{3}$ (d) $\frac{1}{2}$

62 For the function $y = \frac{1}{1+\cos^2 x}$, select correct alternatives.

(i) $y_{max} = 1$ (ii) $y_{max} = \frac{1}{2}$ (iii) $y_{min} = \frac{1}{2}$ (iv) $y_{min} = 1$

(a) (i, ii) (b) (i, iii) (c) (ii, iv) (d) (i, ii, iv)

63 The side of a square is increasing at the rate of 0.2 cm/s. The rate of increase of perimeter with respect to time, is

(a) 0.2 cm/s (b) 0.4 cm/s (c) 0.6 cm/s (d) 0.8 cm/s

64 If $x = 2t, y = 4t^2$, then calculate $\int y dx$.

(a) $x^2 + C$ (b) $\frac{x^3}{3} + C$ (c) $x^3 + C$ (d) $2x + C$

65 Area bounded by curve $y = \sin x$, with X-axis when x varies from 0 to $\pi/2$ is

(a) 1 unit (b) 2 unit (c) 3 unit (d) None of these

66 Find the average value of y from graph given as x changes from 0 to 20.

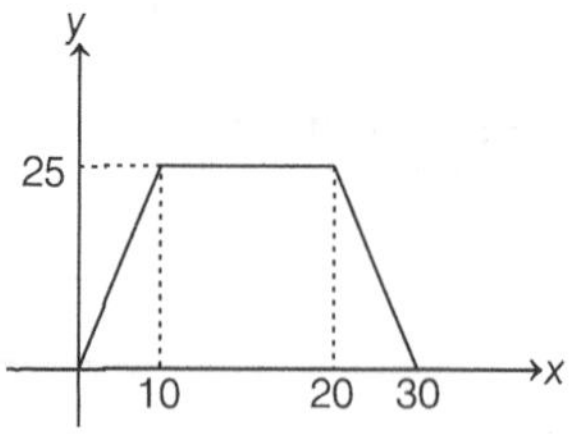

(a) 18 (b) 16 (c) 16.75 (d) 18.75

67 Find the area under the shaded region for curve $y = 3x^2$.

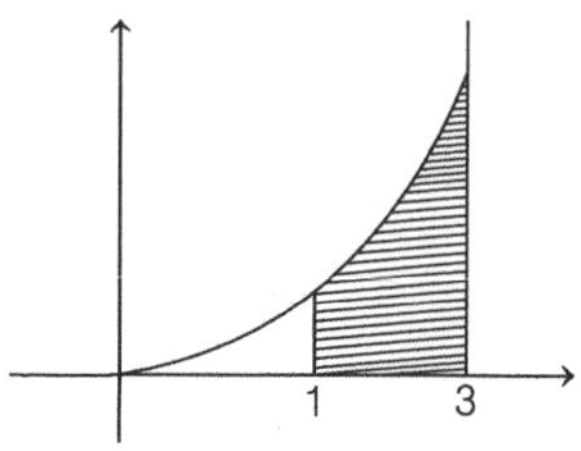

(a) 80 (b) 26 (c) 20 (d) 40

LEVEL 03 Subjective Questions

68 Value of y decreases exponentially from $y = 10$ to $y = 6$. Plot a x-y graph.

69 Value of y increases exponentially from $y = -4$ to $y = +4$. Plot a x-y graph.

70 The graph shown in figure is exponential. Write down the equation corresponding to the graph.

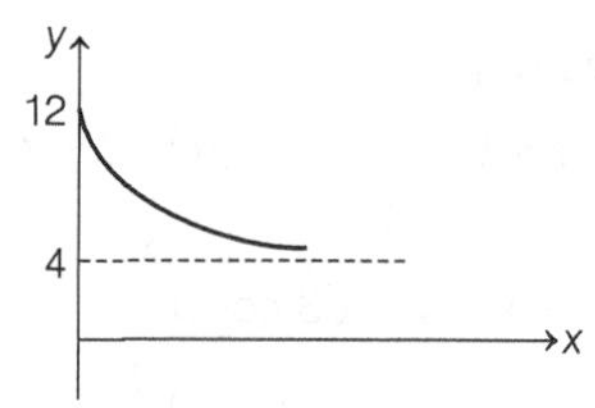

71 The graph shown in figure is exponential. Write down the equation corresponding to the graph.

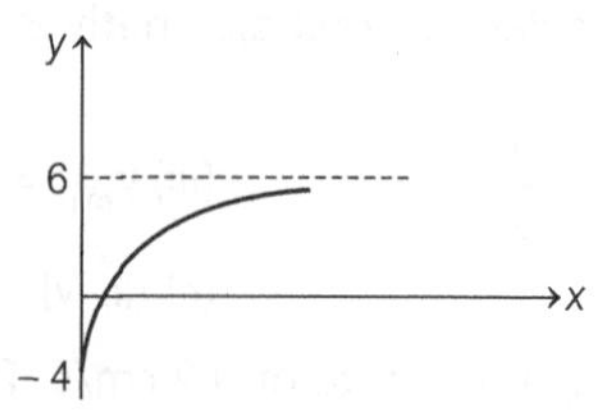

72 In the equation $y = 6 + 4(1 - e^{-kt})$, what is the maximum value of y ?

73 Momentum of a particle is increased by 2% without change in its mass. Find the percentage change in its kinetic energy.

74 Momentum of a particle is decreased by 20% without change in its mass. Find the percentage change in its kinetic energy.

75 Kinetic energy of a particle is decreased by 3% without change in its mass. Find the percentage change in its momentum.

76 Kinetic energy of a particle is increased by 50% without change in its mass. Find the percentage change in its momentum.

77 Height of Kartik is 5 ft and of Atharva 4 ft. Divide ₹1000 in inverse ratio of square of their heights.

ANSWER KEY

Level 1 Objective Questions

1 (d)	**2** (a)	**3** (d)	**4** (c)	**5** (b)	**6** (b)	**7** (b)	**8** (a)	**9** (a)	**10** (d)
11 (b)	**12** (b)	**13** (a)	**14** (c)	**15** (a)	**16** (b)	**17** (c)	**18** (a)	**19** (d)	**20** (a)
21 (d)	**22** (c)	**23** (c)	**24** (a)	**25** (d)	**26** (d)	**27** (a)	**28** (a)	**29** (c)	**30** (c)
31 (c)	**32** (c)	**33** (a)	**34** (d)	**35** (d)	**36** (d)	**37** (a)	**38** (c)	**39** (b)	

Level 2 Objective Questions

40 (c)	**41** (a)	**42** (c)	**43** (a)	**44** (a)	**45** (b)	**46** (c)	**47** (b)	**48** (c)	**49** (d)
50 (d)	**51** (c)	**52** (d)	**53** (b)	**54** (a)	**55** (a)	**56** (c)	**57** (c)	**58** (b)	**59** (a)
60 (c)	**61** (c)	**62** (b)	**63** (d)	**64** (b)	**65** (a)	**66** (d)	**67** (b)		

Level 3 Subjective Questions

68

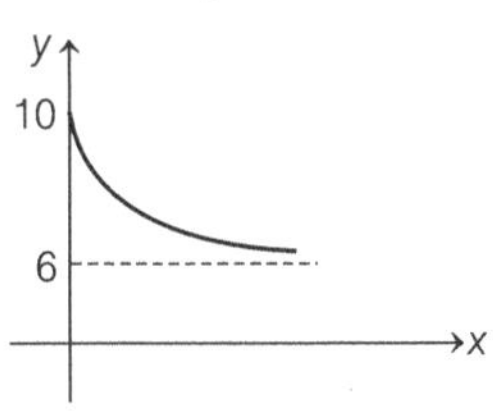

69

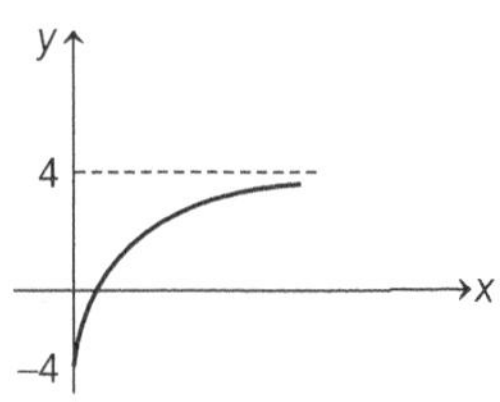

70 $y = 4 + 8e^{-Kx}$. Here, K is a positive constant

71 $y = -4 + 10(1 - e^{-Kx})$. Here, K is positive constant

72 10

73 4%

74 –36%

75 –1.5%

76 Approximately 22%

77 ₹ $\frac{16000}{41}$, ₹ $\frac{25000}{41}$

Solutions

LEVEL 01 Objective Questions

1. $225° = 225 \times \dfrac{\pi}{180} = \dfrac{5\pi}{4}$ rad

2. As we know, sine function repeats after even integer.

$$\therefore \quad \sin(2n\pi + \theta) = \sin\theta$$

$$\Rightarrow \quad \sin\left(14\pi + \frac{\pi}{6}\right) = \sin\frac{\pi}{6}$$

$$= \sin 30° = \frac{1}{2}$$

3. We know,

$$\cos(A + B) = \cos A \cos B - \sin A \sin B$$

$$\cos(A - B) = \cos A \cos B + \sin A \sin B$$

So, $\cos(\alpha + \beta) - \cos(\alpha - \beta)$

$= \cos\alpha\cos\beta - \sin\alpha\sin\beta - (\cos\alpha\cos\beta + \sin\alpha\sin\beta)$

$= \cos\alpha\cos\beta - \sin\alpha\sin\beta - \cos\alpha\cos\beta - \sin\alpha\sin\beta$

$= -2\sin\alpha\sin\beta$

4. We know, $\sin^2\theta + \cos^2\theta = 1$

$$\therefore \quad \sin\theta = \sqrt{1 - \cos^2\theta}$$

$$= \sqrt{1 - \left(\frac{5}{13}\right)^2}$$

$$= \sqrt{\frac{144}{169}} = \frac{12}{13}$$

$$\therefore \quad \tan\theta = \frac{\sin\theta}{\cos\theta} = \frac{\frac{12}{13}}{\frac{5}{13}} = \frac{12}{5}$$

5. We know, $\sin 120° = \sin(90° + 30°)$

$$= \cos 30° = \frac{\sqrt{3}}{2}$$

Similarly, for $\cot 120° = \cot(90° + 30°)$

$$= -\tan 30°$$

$$= \frac{-1}{\sqrt{3}}$$

Substituting the values in $\sin^2 120° + \cot^2 120°$, we get

$$\left(\frac{\sqrt{3}}{2}\right)^2 + \left(\frac{-1}{\sqrt{3}}\right)^2 = \frac{3}{4} + \frac{1}{3}$$

$$= \frac{9 + 4}{12} = \frac{13}{12}$$

6. Given, length of a string between a kite and point is 55 m.

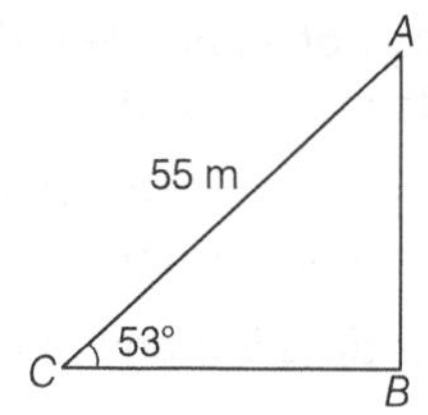

We know, $\quad \sin\theta = \dfrac{AB}{AC}$

$$\sin 53° = \frac{AB}{55}$$

$$\frac{4}{5} \times 55 = AB$$

$$AB = 44 \text{ m}$$

Therefore, the height of the kite from the ground is 44 m.

7.

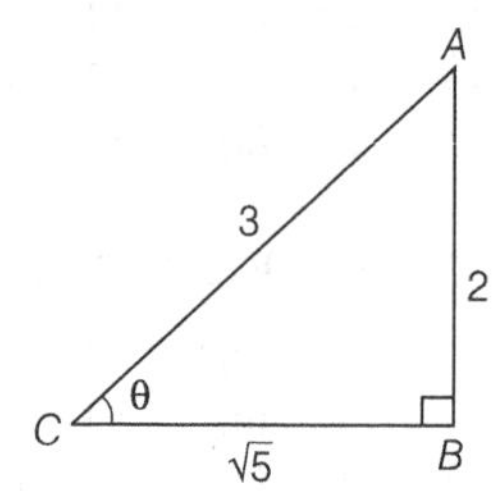

$\cot\theta \times \sec\theta$

$$= \frac{\cos\theta}{\sin\theta} \times \frac{1}{\cos\theta} = \frac{1}{\sin\theta} = \frac{3}{2}$$

8. $-\tan\theta = \dfrac{1}{\sqrt{5}} = \tan(-\theta)$

$$\therefore \quad \cos(-\theta) = \cos\theta = \frac{1}{\sec(-\theta)} = \frac{1}{\sqrt{\sec^2(-\theta)}}$$

$$= \frac{1}{\sqrt{1 + \tan^2(-\theta)}} = \frac{1}{\sqrt{1 + \frac{1}{5}}} = \frac{\sqrt{5}}{\sqrt{6}}$$

9. Let ΔABC be right-angled and its hypotenuse be H.

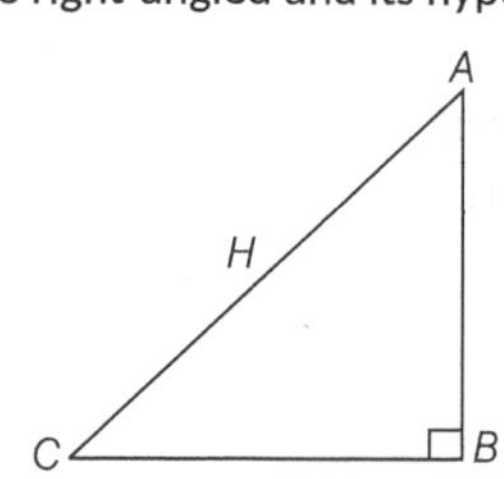

Since, the perpendicular is $\frac{1}{4}$th of H.

$\therefore \qquad AB = \frac{H}{4}$

Using Pythagoras theorem,

$$BC = \sqrt{H^2 - \left(\frac{H}{4}\right)^2} = \sqrt{H^2 - \frac{H^2}{16}} = \sqrt{\frac{15}{16}H^2} = \frac{\sqrt{15}H}{4}$$

Now, $\qquad \frac{B}{H} = \frac{\sqrt{15}}{4}$

$\Rightarrow \qquad 4B = \sqrt{15}H$

10. We have two points, so

$$x_1 = 8,\ y_1 = 4$$

and $\qquad x_2 = 2,\ y_2 = a$

Distance formula $= \sqrt{(x_2 - x_1)^2 + (y_2 - y_1)^2}$

Substituting the values of x and y, we get

$$= \sqrt{(2-8)^2 + (a-4)^2} = 10$$

Squaring on both sides,

$$(-6)^2 + (a-4)^2 = 100$$
$$(a-4)^2 = 100 - 36$$
$$(a-4)^2 = 8^2$$
$$a - 4 = \pm 8$$
$$a = 12$$

or $\qquad a = -4$

So, the positive value of a is 12.

11. For point to be lies on a line, it must satisfy the equation.

$\therefore \qquad 2x + y + 5 = 0$

$$2(-2) + (-1) + 5 = 0$$
$$-5 + 5 = 0$$

Therefore, point $(-2, -1)$ lies on the given line.

12. For slope of a line,

$$m = \frac{y_2 - y_1}{x_2 - x_1} = \frac{-5-3}{4-2} = -4.$$

The slope of a given line is -4

13. Given, $4y + 2x + 4 = 0$ and $3y - 6x + 10 = 0$

Rearranging the first equation,

$$4y = -2x - 4$$

$\Rightarrow \qquad y = -\frac{1}{2}x - 1$

Similarly, for the second equation,

$$3y = 6x - 10 \Rightarrow y = 2x - \frac{10}{3}$$

For the first and second equations slopes are $-\frac{1}{2}$ and 2.

Therefore, product of slope $= -\frac{1}{2} \times 2 = -1$

14. Solving the given equations, we get

$$x = -2 \text{ and } y = -8$$

15. For slope of a line,

$$m = \frac{y_2 - y_1}{x_2 - x_1} = \frac{0-4}{8-0} = -\frac{1}{2}$$

For intercept of a line,

$$y = mx + C$$
$$4 = -\frac{1(0)}{2} + C$$
$$C = 4$$

Therefore, the equation of a line passing through points (0, 4) and (8, 0) is

$$y = -\frac{x}{2} + 4$$

or $\qquad 2y = -x + 8$

16. Given, $\sqrt{3}y = 3x + 6$

$\Rightarrow \qquad y = \sqrt{3}x + \frac{6}{\sqrt{3}}$

We know, $y = mx + C$

Now, $m = \sqrt{3}$

$\Rightarrow \qquad \theta = 60°$

Therefore, the angle made by the line with the positive X-axis is 60°.

17. We know, slope, $m = \tan\theta$

$$m = 1 \text{ as } \theta = 45°$$

For intercept of a line,

$$0 = -2 + C$$

$\Rightarrow \qquad C = 2$

Therefore, equation of a straight line is $y = x + 2$.

18. Centre of circle is (2, 0) and radius is 4. So, distance of centre from origin is 2.

19. The standard terms for parabola as :

(A) $y = x^2$, a parabola will pass through origin and opens upward.

(B) $y = -x^2$, a parabola will pass through origin and opens downward.

(C) $x = y^2$, a parabola will pass through origin and opens rightward.

(D) $x = -y^2$, a parabola passes through origin and opens leftward.

Therefore, option (d) is correct.

20. Given equation is

$$\frac{x^2}{4} + \frac{y^2}{25} = 1 \text{ or } \frac{x^2}{(2)^2} + \frac{y^2}{(5)^2} = 1$$

Comparing this with standard equation of ellipse,

$$\frac{x^2}{a^2} + \frac{y^2}{b^2} = 1$$

We get, $a = 2$ and $b = 5$

21. At $x = 0$ and $y = 2$

Then, y increases as x increases.

Therefore, the best option is (d).

22. $\because$ We know that $F(x) = x^2 - 3x + 2$ is a quadratic equation. Since a quadratic equation has always two roots, so let us assume that the graph of given equation cuts X-axis at points $(x_1, 0)$ and $(x_2, 0)$.

$\because$ It cuts X-axis,

$\therefore \quad x^2 - 3x + 2 = 0 \quad \Rightarrow \quad x^2 - 2x - x + 2 = 0$

$\Rightarrow x(x-2) - 1(x-2) = 0 \quad \Rightarrow \quad (x-1)(x-2) = 0$

$\Rightarrow \quad x = 1, 2$

$\therefore$ The graph cuts X-axis at points (1, 0) and (2, 0).

23. Given, $y = \sin^2 x + \cos^2 x$

The right side of the equation is 1, i.e.

$\Rightarrow \quad y = 1$

$\therefore \quad \frac{dy}{dx} = \frac{d(1)}{dx} = 0$

24. Given, $y = x^4 + 2x^3 + 2$

$$\frac{dy}{dx} = 4x^3 + 6x^2$$

25. $y = (6x + 5)^{\frac{1}{2}}$

$$\therefore \quad \frac{dy}{dx} = \left[\frac{1}{2}(6x+5)^{-\frac{1}{2}}\right][6] = \frac{3}{\sqrt{6x+5}}$$

26. Given, $i = i_0 \sin\left(\omega t + \frac{\pi}{3}\right)$

$$\frac{di}{dt} = i_0\, \omega \cos\left(\omega t + \frac{\pi}{3}\right)$$

27. Given, $y = 5x^3 - 4x^2 + 2$

$$\frac{dy}{dx} = 15x^2 - 8x$$

$$\frac{d^2y}{dx^2} = 30x - 8$$

Substituting the value of $x = 1$

$$= 30(1) - 8 = 22$$

28. Given, $y = 4x^2 + 2x$

Thus, the slope is at $x = 1$ is given as,

$$\frac{dy}{dx} = 8x + 2 = 8(1) + 2 = 10$$

29. $\frac{dy}{d\theta} = 3\cos\theta + 4\sin\theta$

$$\frac{d^2y}{d\theta^2} = -3\sin\theta + 4\cos\theta$$

At maximum value of y, the first derivative $\frac{dy}{d\theta} = 0$.

$\therefore \quad 3\cos\theta + 4\sin\theta = 0$

or $\quad \tan\theta = -\frac{3}{4}$ or $\theta = 143°$ or $(180° - 37°)$

At this angle, $\frac{d^2y}{d\theta^2}$ is negative.

Hence, y is maximum.

$$y_{max} = 3\sin(180° - 37°) - 4\cos(180° - 37°)$$
$$= 5$$

30. For the minimum value,

$$\frac{dy}{dx} = 0 \quad \Rightarrow \quad \frac{dy}{dx} = \frac{d(2x^2 - 8x + 5)}{dx}$$

$$4x - 8 = 0$$
$$x = 2$$

Substituting the value of $x = 2$

$$= 2(2)^2 - 8(2) + 5 = -3$$

Therefore, the minimum value of y is –3.

31. Given, $y = x^3 - 25$

$$\frac{dy}{dx} = 3x^2 - 0 \quad \Rightarrow \quad \frac{d^2y}{dx^2} = 6x$$

We have, $x = 0$

$$\frac{d^2y}{dx^2} = 6(0) = 0$$

Since, $\frac{d^2y}{dx^2}$ is not < 0 and > 0.

Therefore, it is neither maximum nor the minimum.

32. Given, $Q = 10\sin t$

$\Rightarrow \quad \frac{dQ}{dt}$ = rate of change of Q with t

$$= 10 \cdot \frac{d}{dt}\sin t = 10\cos t$$

At $t = \pi/4$

$$\frac{dQ}{dt} = \frac{10}{\sqrt{2}} = 5\sqrt{2}$$

33. Given, $y = -x^2 - 4x + 8$ ...(i)

$$\frac{dy}{dx} = -2x - 4$$

Putting $\frac{dy}{dx} = 0$,

$$-2x - 4 = 0 \quad \Rightarrow \quad x = -2$$

Substituting $x = -2$ in equation (i), we get

$$y_{max} = -4 - 4(-2) + 8 = 12$$

34. $\int_0^{\pi/2} \sin^2\theta \cdot d\theta = \int_0^{\pi/2} \left(\frac{1 - \cos 2\theta}{2}\right) d\theta$

$$= \left[\frac{1}{2}\theta - \frac{1}{4}\sin 2\theta\right]_0^{\pi/2}$$

Substituting the limits, we get the value $\frac{\pi}{4}$.

35. $\int_0^{\pi/4} -\cos x + \sin x$

$= \left(-\cos\frac{\pi}{4} + \sin\frac{\pi}{4} - (-\cos 0) + \sin 0\right) = 1$

36. Given, $\int_0^1 9x^8 dx + \int_0^{\pi/2} \cos x \, dx$

$= [x^9]_0^1 + [\sin x]_0^{\pi/2} = [(1)^9 - 0] + \left[\sin\frac{\pi}{2} - \sin 0\right] = 2$

37. $4\int_0^{\pi/4} \sin 2x \, dx = 4 \times \frac{1}{2}[-\cos 2x]_0^{\pi/4}$

$= 2[-\cos(2x)]_0^{\pi/4} = -2\left(\cos\frac{\pi}{2} - \cos 0°\right)$

$= -2(0-1) = -2 \times -1 = 2$

38. Given, $I = \int_0^2 3x^2 \, dx + \int_0^{\pi/2} \sin x \cdot dx$

$= \left[(2)^3 + \left(-\cos\frac{\pi}{2}\right) - (-\cos 0)\right]$

$= 8 + 1 = 9$

39. Given, $\int_0^{\pi} (\sin x + \cos x) \, dx$

As we know,

$\int (\sin x) \, dx = -\cos x \Rightarrow \int (\cos x) \, dx = \sin x$

Therefore,

$\int_0^{\pi} -\cos x + \sin x = [-\cos\pi + \sin\pi - (-\cos 0) + \sin 0]$

$= [-(-1) + 0 + 1 - 0] = 2$

LEVEL 02 Objective Questions

40. Given, $\cos\theta = \frac{7}{25}$

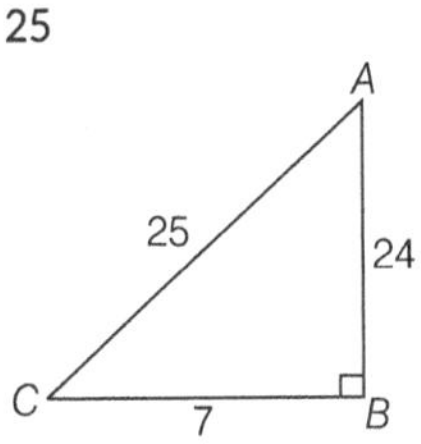

Since, we know, $\cos\theta = \frac{\text{base}}{\text{hypotenuse}} = \frac{7}{25}$

Using Pythagoras theorem,

$AB = \sqrt{(25)^2 - (7)^2}$

$= \sqrt{625 - 49} = \sqrt{576} = 24$

As, $\tan(90° - \theta) = \cot\theta$

Therefore, $\cot\theta = \frac{\text{base}}{\text{perpendicular}} = \frac{BC}{AB} = \frac{7}{24}$

41. Using identity, $2\cos^2\theta = 1 + \cos 2\theta$

$\Rightarrow \cos^2\theta = \frac{1 + \cos 2\theta}{2}$

$\Rightarrow \cos^2 22.5° = \frac{1 + \cos 2(22.5)}{2}$

$\cos^2 22.5° = \frac{1 + \cos 45°}{2}.$

42. Given, $\tan\left[\frac{\pi}{4} + \theta\right] \times \tan\left[\frac{\pi}{4} - \theta\right]$

Using identity, $\tan(A + B) = \frac{\tan A + \tan B}{1 - \tan A \tan B}$

$\tan(A - B) = \frac{\tan A - \tan B}{1 + \tan A \tan B}$

$\frac{\pi}{4}$ is 45° and θ is given 53°.

$\therefore$ Substituting the identity,

$\left(\frac{\tan 45° + \tan 53°}{1 - \tan 45° \tan 53°}\right) \cdot \left(\frac{\tan 45° - \tan 53°}{1 + \tan 45° \tan 53°}\right)$

$\frac{1 + \tan 53°}{1 - \tan 53°} \times \frac{1 - \tan 53°}{1 + \tan 53°} = 1$

43. Given, $2\cos 2\theta = 1$

$\cos 2\theta = \frac{1}{2} \Rightarrow \cos 2\theta = \cos 60°$

$\Rightarrow 2\theta = 60°$ or $\theta = 30°$

44. The negative reciprocal of a slope of a line gives the slope of a line perpendicular to it.

Therefore, slope = 3

slope of $\perp$ line, $= \frac{-1}{3}$

For equation of a line,

Using $y = mx + C$

$y = \frac{-1x}{3} + 4$

Therefore, equation of a straight line is $y = \frac{-x}{3} + 4$.

45. For the two lines to be parallel, their slopes must be equal, but their intercept may or may not be equal.

Therefore, $2y + 4x + 8 = 0$ and $y + 2x - 8 = 0$ lines are parallel as they have the same slope-2.

46. Slope, $m = \frac{y_2 - y_1}{x_2 - x_1} = \frac{6 - 0}{0 - (-2)} = 3$

For y intercept, using $x = -2$ and $y = 0$

We know, the equation of a straight line,

$y = mx + C$

$0 = 3(-2) + C$

$C = 6$

Therefore, equation of a straight line is $y = 3x + 6$.

47. $\frac{C}{3} = \tan 30° = \frac{1}{\sqrt{3}}$

$\therefore \quad C = \sqrt{3}$

48. We know in an ellipse, a is the distance from the centre to the end of the major axis and b is the distance from the centre to the end of the minor axis.

Therefore, $a = 3, b = 2$

Substituting the values in the standard equation of an ellipse, we get

$$\frac{x^2}{a^2} + \frac{y^2}{b^2} = 1 \Rightarrow 1 = \frac{x^2}{(3)^2} + \frac{y^2}{(2)^2} = \frac{x^2}{9} + \frac{y^2}{4}$$

49. Since, the line is making an angle of 45° with the positive X-axis.

Therefore, the slope is 1 and it is positive and intercept is 5. So, line is $y = x + 5$.

50. $\frac{d}{dx}\left[\frac{6x^2 - 2x}{3x - 1}\right] = \frac{d}{dx}\frac{2x\,(3x-1)}{(3x-1)} = \frac{d}{dx}2x = 2$

51. Given, $y = \sin x + \cos x$

$$\frac{dy}{dx} = \cos x - \sin x$$

$$\frac{d^2y}{dx^2} = -\sin x - \cos x = -(\sin x + \cos x)$$

52. If $y = \cos 3x$, then

At $x = 0, y = 1$

At $x = \frac{\pi}{6}, y = 0$

At $x = \frac{\pi}{3}, y = -1$

These values match with the given graph.

53. Given, $\frac{dr}{dt} = 2$ m/s

$\because$ Area of circle $= \pi r^2$

$\therefore$ The rate of change of area of circle,

$$\frac{dA}{dt} = 2\pi r = 2\pi\left(\frac{1}{\pi}\right) \times \frac{dr}{dt} \quad \left(\because \text{radius, } r = \frac{1}{\pi}\right)$$

$$= 2 \times 2 = 4 \text{ m}^2/\text{s}$$

54. Given, $y = x - x^2$

$$\text{Slope} = \frac{dy}{dx} = 1 - 2x$$

At $x = 4$,

$$m = \frac{dy}{dx} = 1 - 2(4) = -7$$

55. Given, $12y = x^3$

$$12\frac{dy}{dt} = 3x^2\frac{dx}{dt}$$

$$4\frac{dy}{dt} = x^2\frac{dx}{dt}$$

At $x = 10$,

$$\frac{4dy}{dt} = (10)^2\frac{dx}{dt} \Rightarrow \frac{dy}{dt} = 25\frac{dx}{dt}$$

So, rate of change of y is 25 times of rate of change of x.

56. $y = ax^2 + abx$

Putting $x = 2$ and $y = 4$, we get

$$4a + 2ab = 4 \quad \text{...(i)}$$

$$\frac{dy}{dx} = 2ax + ab = \text{Slope}$$

At $x = 2$, slope is 1. Hence,

$$4a + ab = 1 \quad \text{...(ii)}$$

Solving Eqs. (i) and (ii), we get

$$ab = 3$$

57. Given, $y = 4x^2 + 2x$

$$\frac{dy}{dx} = 8x + 2$$

Here, $m = 8$ which is positive.

Therefore, option (c) is correct.

58. $A = \pi r^2 = 4\pi t^4 \quad (\because r = 2t^2)$

$$\frac{dA}{dt} = 16\,\pi t^3$$

At $t = 2$ s,

$$\frac{dA}{dt} = (128\,\pi) \text{ units}$$

59. As slope of tangent

$$= \frac{dy}{dx} = \tan\theta$$

Differentiating with respect to x,

$$\tan\theta = \frac{dy}{dx} = \frac{1}{2}\cdot 2x \quad \text{or} \quad \tan 45° = x$$

$$x = 1$$

Now, we have $y = \frac{x^2}{2} = \frac{1}{2}$

Therefore, the coordinates are $\left(1, \frac{1}{2}\right)$.

60. $\sin\theta + \sqrt{3}\cos\theta = 2\left(\frac{1}{2}\sin\theta + \frac{\sqrt{3}}{2}\cos\theta\right)$

$$= 2(\cos 60° \sin\theta + \sin 60° \cos\theta)$$

$$= 2\sin(\theta + 60°)$$

As, $-1 \le \sin(\theta + 60°) \le 1$

$$-2 \le 2\sin(\theta + 60°) \le 2$$

$$\text{Range} = [-2, 2]$$

61. We have obtained maximum value of $\sin\theta + \sqrt{3}\cos\theta$ equal to 2 at $\theta = 30°$. Hence, minimum value of $\frac{2}{4 + \sin\theta + \sqrt{3}\cos\theta}$ will be $\frac{2}{4+2}$ or $\frac{1}{3}$.

62. Given, $y = \dfrac{1}{1+\cos^2 x}$

For y to be maximum, $\cos^2 x$ must be zero.

Substituting $x = 90°$

$$= \frac{1}{1+\cos^2 90°} = \frac{1}{1+0} = 1$$

For y to be minimum, $\cos^2 x$ must be 1.

Substituting $x = 0°$

$$= \frac{1}{1+\cos^2 0°} = \frac{1}{1+1} = \frac{1}{2}$$

63. Let the side of a square be x.

We know, perimeter of square = $4x$

Now, $P = 4x$

$$\frac{dP}{dt} = \frac{4dx}{dt}$$

$$= 4 \times 0.2 = 0.8 \text{ cm/s}$$

Therefore, the rate of increase of the perimeter of square is 0.8 cm/s.

64. Given, $x = 2t, y = 4t^2 \quad \Rightarrow \quad y = x^2$

$$= \int y dx = \int x^2 dx \quad \Rightarrow \quad y = \frac{x^3}{3} + C$$

65. Area means integration.

Hence, area = $\int_0^{\pi/2} \sin x \, dx = [-\cos x]_0^{\pi/2}$

Substituting the limits, we get

Area = 1 unit

66. Average value of y

$$= \frac{\text{Area under the graph}}{\text{Change in } x} = \frac{\frac{1}{2} \times 10 \times 25 + 25 \times 10}{20}$$

$$= \frac{125 + 250}{20} = \frac{375}{20} = 18.75$$

67. Given, $y = 3x^2$

$$\text{Area} = \int_1^3 y dx = \int_1^3 3x^2 dx$$

$$= 3\left[\frac{x^3}{3}\right]_1^3 = 3^3 - 1 = 26$$

LEVEL 03 Subjective Questions

68. y decreases exponentially from $y = 10$ to $y = 6$

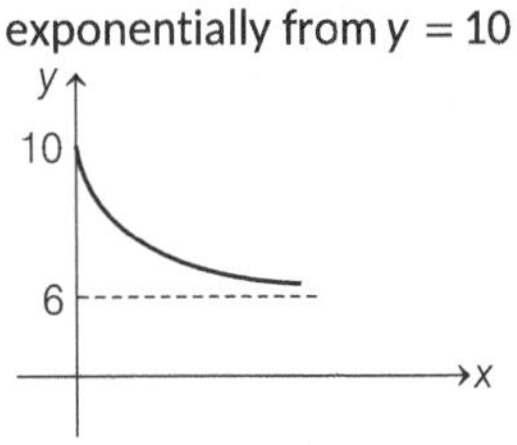

69. y increases exponentially from -4 to $y = 4$

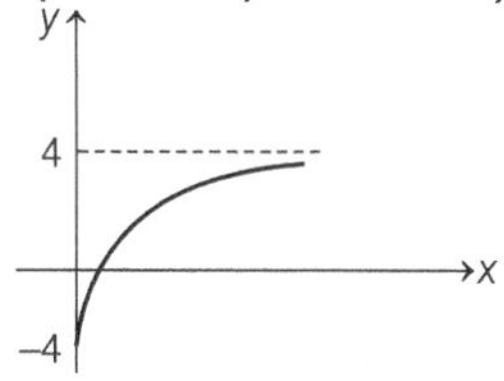

70. The graph is decreasing exponentially in equation will be

$$y = Ke^{-x} + C$$

for $x = 0, y = 12$

$$12 = Ke + 4 \quad \Rightarrow \quad K = 8$$

$\Rightarrow$ required equation is $y = 8e^{-x} + 4$

71. Above -4, the value 10 is increasing exponentially.

72. At $t = 0, y = 6$ and

at $t = \infty, y = 10$

$\therefore \quad y_{max} = 10$

73. $K = \dfrac{p^2}{2m}$ or $K \propto p^2$

$\therefore$ % change in K = 2 (% change in p) = 2 (2%) = 4%

74. $K = \dfrac{p^2}{2m} \quad \Rightarrow \quad p' = 0.8\,p$

$$\therefore \quad K' = \frac{p'^2}{2m} = \frac{(0.8p)^2}{2m} = (0.64)\frac{p^2}{2m} = 0.64K$$

Hence, kinetic energy has been decreased by 36%.

75. $p = \sqrt{2Km}$ or $p \propto K^{\frac{1}{2}}$

$\therefore$ (% change in p) = $\dfrac{1}{2}$ (% change in K) = $\dfrac{1}{2}(-3\%)$

$= -1.5\%$

76. $p = \sqrt{2Km}$ ...(i)

$K' = 1.5\,K$

$\therefore \quad p' = \sqrt{2K'm} = \sqrt{2\,(1.5K)\,m}$ ...(ii)

From Eqs. (i) and (ii), we get

$$p' \approx 1.22\,p$$

Therefore, percentage change in momentum is approximately 22%.

77. Given, height of Kartik = 5 ft

Height of Atharva = 4 ft

Amount to be divided = ₹ 1000

The inverse ratio of their heights = 16 : 25

Let x be the amount that is to be equally divided.

$\therefore$ Kartik will get $16x$ and Atharva will get $25x$.

$\therefore \quad 25x + 16x = 1000$

$\Rightarrow \quad 41x = 1000 \quad \Rightarrow \quad x = ₹\,\dfrac{1000}{41}$

$\therefore$ The amount Kartik will get = $16 \times \dfrac{1000}{41} = ₹\,\dfrac{16000}{41}$

and the amount Atharva will get = $25 \times \dfrac{1000}{41} = ₹\,\dfrac{25000}{41}$

Chapter 02

UNITS, DIMENSIONS AND ERROR ANALYSIS

LEVEL 01 Objective Questions

1 Parallactic second is the unit of

(a) distance (b) time
(c) frequency (d) velocity

2 Which of the following is **not** equal to watt?

(a) Joule/Second (b) Ampere $\times$ Volt
(c) $(\text{Ampere})^2 \times \text{Ohm}$ (d) Ampere/Volt

3 The equation of state for n moles of an ideal gas is $pV = nRT$, where R is constant. The SI unit for R is

(a) JK^{-1} per molecule (b) $JK^{-1} mol^{-1}$
(c) $Jkg^{-1} K^{-1}$ (d) $JK^{-1} g^{-1}$

4 Some physical constants are given in column-I and their dimensional formulae are given in Column-II. Match the correct pairs in the columns.

Column-I	Column-II
A. Planck's constant	p. $[ML^{-1}T^{-2}]$
B. Gravitational constant	q. $[ML^{-1}T^{-1}]$
C. Bulk modulus	r. $[ML^2T^{-1}]$
D. Coefficient of viscosity	s. $[M^{-1}L^3T^{-2}]$

Codes

	A	B	C	D		A	B	C	D
(a)	s	r	q	p	(b)	q	p	r	s
(c)	r	q	p	s	(d)	r	s	p	q

5 Dimensional formula of electrical conductance is

(a) $[M^{-1}L^{-2}T^3A^2]$ (b) $[M^{-1}L^{-2}T^3A^{-2}]$
(c) $[M^{-1}L^{-2}T^{-3}A^2]$ (d) $[M^{-1}L^{-2}T^{-3}A^{-2}]$

6 The quantities A and B are related by the relation $m = A/B$, where m is the linear mass density, A is the force. The dimensions of B are of

(a) pressure (b) work
(c) latent heat (d) None of these

7 Force F and density d are related as $F = \dfrac{\alpha}{\beta + \sqrt{d}}$, then the dimensions of α are

(a) $[M^{3/2} L^{-1/2} T^{-2}]$ (b) $[M^{3/2} L^{-1/2} T^{2}]$
(c) $[M^{3/2} L^{1/2} T^{2}]$ (d) $[M^{3/2} L^{1/2} T^{-2}]$

8 If w, x, y and z are mass, length, time and current respectively, then $\dfrac{x^2 w}{y^3 z}$ has dimensional formula same as

(a) electric potential (b) capacitance
(c) electric field (d) permittivity

9 The formula $P = \dfrac{(x^2 - b)}{at}$ relates power (P), distance (x) and time (t). The dimensional formula of a is

(a) $[ML^0 T^2]$ (b) $[M^{-1}LT^{-2}]$
(c) $[M^{-1}L^0 T^2]$ (d) $[ML^{-2} T^{-2}]$

10 If $\alpha = \dfrac{F}{v^2} \sin \beta t$, then find dimensions of $\dfrac{\alpha}{\beta}$. (Here, F = force, v = velocity, t = time)

(a) $[ML^{-1}T^{-1}]$ (b) $[M^{-1}LT]$ (c) $[ML^{-1}T]$ (d) $[M^{-1}L^{-1}T^{-1}]$

11 The time dependence of a physical quantity p is given by $p = p_0 e^{-\alpha t^2}$, where α is a constant and t is the time, then constant α is

(a) dimensionless (b) dimensions of t^{-2}
(c) dimensions of p (d) dimensions of t^2

12 The velocity v (in cm/s) of a particle is given in terms of time t (in second) by the equation $v = at + \dfrac{b}{t+c}$. The dimensions of a, b and c are

	a	b	c
(a)	$[L^2]$	$[T]$	$[LT^2]$
(b)	$[LT^2]$	$[LT]$	$[L]$
(c)	$[LT^{-2}]$	$[L]$	$[T]$
(d)	$[L]$	$[LT]$	$[T^2]$

13 The equation of a wave is given by $y = A\sin\omega\left[\dfrac{x}{v} - k\right]$, where ω is the angular velocity and v is the linear velocity. Write dimensional formula of k.

(a) $[M^0L^0T^1]$ (b) $[M^1L^0T^{-1}]$
(c) $[M^0L^1T^1]$ (d) $[M^1L^{-1}T^1]$

14 A force F is given by : $F = at + bt^2$, where t represents time. What are dimensions of a and b?

(a) $[MLT^{-3}]$ and $[ML^2T^4]$ (b) $[MLT^{-3}]$ and $[MLT^{-4}]$
(c) $[MLT^{-1}]$ and $[MLT^0]$ (d) $[MLT^{-4}]$ and $[MLT^1]$

15 If A, B and C are three physical quantities, such that A and B are of same dimensions while C has different dimensions, then which of the following may be meaningful?

(a) $\dfrac{(A+C)}{B}$ (b) $\dfrac{(C-2B)}{A}$
(c) $\dfrac{A}{(B+C)}$ (d) $\dfrac{(2A-3B)}{C}$

16 If frequency (f), velocity (v) and density (D) are considered as fundamental units, the dimensional formula for momentum will be

(a) Dvf^2 (b) Dv^2f^{-1}
(c) $D^2v^2f^2$ (d) Dv^4f^{-3}

17 The dimensional formula of product and quotient of two physical quantities A and B are given by $[AB] = [ML^2T^{-2}]$; $\left[\frac{A}{B}\right] = [MT^{-2}]$. The quantities A and B respectively are

(a) force and velocity (b) force and displacement
(c) momentum and displacement (d) work and velocity

18 A quantity A appears in an equation. Pressure $= \frac{Ae^{-Af/kT}}{B}$. Here, T represents temperature, f frequency and k Boltzmann constant. Kinetic energy of a gas molecule is given by $= \frac{3}{2}kT$. What is the dimension of B in length?

(a) 1 (b) 2 (c) 3 (d) 4

19 A spherical body of mass m and radius r is allowed to fall in a medium of viscosity η. The time in which the velocity of the body increases from zero to 0.63 times the terminal velocity (v) is called time constant (τ). Dimensionally τ can be represented by

(a) $\frac{mr^2}{6\pi\eta}$ (b) $\sqrt{\left(\frac{6\pi mr\eta}{g^2}\right)}$ (c) $\frac{m}{6\pi\eta rv}$ (d) None of these

20 Match the following columns.

	Column-I		Column-II
(A)	ε_0	(p)	$[MT^{-2}A^{-1}]$
(B)	μ_0	(q)	$[MLT^{-3}A^{-1}]$
(C)	E	(r)	$[M^{-1}L^{-3}T^4A^2]$
(D)	B	(s)	$[MLT^{-2}A^{-2}]$

Codes

	A	B	C	D
(a)	p	q	r	s
(b)	r	s	q	p
(c)	r	s	p	q
(d)	s	r	p	q

21 The dimensions of $\varepsilon_0 \frac{d\phi_E}{dt}$ are

(a) [AT] (b) [A]
(c) $[ML^2T^{-2}]$ (d) $[ML^{-1}T^{-2}]$

22 If h (Planck's constant), G (universal gravitational constant) and F(force) are fundamental physical quantities, then find out dimensional formula of energy.

(a) $[h^{1/8}G^{5/8}F^{1/2}]$ (b) $[h^{1/2}G^{1/8}F^{5/8}]$
(c) $[h^1G^1F^{1/8}]$ (d) $[F^{1/2}G^{1/8}h^{5/8}]$

23 The expression for centripetal force (F) depends upon mass of body (m), speed (v) of the body and the radius (r) of circular path will be expression for centripetal force

(a) $F = \frac{mv^2}{2r^3}$ (b) $F = \frac{mv^2}{r}$
(c) $F = \frac{mv^2}{r^2}$ (d) $F = \frac{m^2v^2}{2r}$

24 The volume V of water passing at any point of a uniform tube during t seconds is related to the cross-sectional area of the tube (A) and velocity of water (v). The volume is given by $V \propto A^a v^b t^c$, then which of the following will be true?

(a) $a = b = c$
(b) $a \neq b = c$
(c) $a = b \neq c$
(d) $a \neq b \neq c$

25 The standard method of expressing 256.00 in terms of four significant figures is

(a) 2560.0×10^{-1}
(b) 0.2560×10^3
(c) 25.60×10^1
(d) 2.560×10^2

26 The numbers 3.845 and 3.835 on rounding off to 3 significant figures will give

(a) 3.85 and 3.84
(b) 3.84 and 3.83
(c) 3.85 and 3.83
(d) 3.84 and 3.84

27 When 96.54 is divided by 2.40, the correct result is

(a) 40.2250
(b) 40.225
(c) 40.23
(d) 40.2

28 The mass of a box measured by a grocer's balance is 2.3 kg. Two gold pieces of masses 20.15 g and 20.17 g are added to the box. What is the total mass of the box?

(a) 2.34032 kg
(b) 2.34 kg
(c) 2.3 kg
(d) 2.340 kg

29 The value of resistance is 10.845 Ω and the current is 3.23 A. On multiplying, we get the potential difference equal to 35.02935 V. The value of potential difference in terms of significant figures would be

(a) 35 V
(b) 35.0 V
(c) 35.029 V
(d) 35.03 V

30 If in an experiment, observation of a length is 10.28 cm, then actual length may be

(a) 10.2748 cm
(b) 10.2756 cm
(c) 10.2852 cm
(d) 10.2742 cm

31 Measure of two quantities along with the precision of respective measuring instrument is $A = 2.5\ \text{m/s} \pm 0.5\ \text{m/s}$, $B = 0.10\ \text{s} \pm 0.01\ \text{s}$. The value of AB will be

(a) (0.25 ± 0.08) m
(b) (0.25 ± 0.5) m
(c) (0.25 ± 0.05) m
(d) (0.25 ± 0.135) m

32 The length, breadth and thickness of a block are measured as 125.5 cm, 5.6 cm and 0.32 cm, respectively. The measurement is most accurate

(a) of thickness
(b) of length
(c) of breadth
(d) All have the same accuracy

33 The radius of ball is $[5.4 \pm 0.2]$ cm. The percentage error in the volume of the ball is

(a) 11%
(b) 4%
(c) 7%
(d) 9%

34 According to Joule's law of heating, heat produced $H = I^2Rt$, where I is current, R is resistance and t is time. If the errors in the measurement of I,R and t are 3%, 4% and 6% respectively, then error in the measurement of H is

(a) ±17%
(b) ±16%
(c) ±19%
(d) ±25%

35 Percentage error in measuring the radius and mass of a solid sphere 1% and 2%, respectively. Then, error in measurement of moment of inertia about its diameter is

(a) 3%
(b) 6%
(c) 5%
(d) 4%

36 In an experiment, the values of two resistances were measured to be $R_1 = (5.0 \pm 0.2)\Omega$ and $R_2 = (10.0 \pm 0.1)\Omega$. Then, their combined resistance in series will be

(a) $(15 \pm 0.1)\ \Omega$ (b) $(15 \pm 0.2)\ \Omega$
(c) $(15 \pm 2\%)\ \Omega$ (d) $(15 \pm 3\%)\ \Omega$

37 If $P = \frac{A^3}{B^{5/2}}$ and ΔA is absolute error in A and ΔB is absolute error in B, then absolute error ΔP in P is

(a) $\Delta P = \pm\left(3\frac{\Delta A}{A} + \frac{5}{2}\frac{\Delta B}{B}\right)P$ (b) $\Delta P = \pm\left(3\frac{\Delta A}{A} + \frac{5}{2}\frac{\Delta B}{B}\right)$
(c) $\Delta P = \pm\left(3\frac{\Delta A}{A} - \frac{5}{2}\frac{\Delta B}{B}\right)P$ (d) $\Delta P = \pm\left(3\frac{\Delta A}{B} - \frac{5}{2}\frac{\Delta B}{A}\right)P$

38 If displacement of a body is $s = (200 \pm 5)$ m and time taken by it is $t = (20 \pm 0.2)$ s, then percentage error in velocity is :

(a) 2.5 % (b) 3.5 %
(c) 4.5 % (d) None of these

39 If the error in the measurement of radius of circle is 3%, then calculate the error in measurement of area of circle.

(a) 5 % (b) 8 %
(c) 3 % (d) 6 %

40 Dimensional formula of a physical quantity A is $[M^{-1}L^3\ T^{-2}]$. The errors in measuring the quantities M, L and T, respectively are 2% , 3% and 4%. The maximum error that occur in measuring the quantity A is

(a) 9 (b) 10 (c) 14 (d) 19

41 A physical quantity $\rho = \frac{\sqrt{abc^2}}{d^3}$ is determined by measuring a, b, c and d separately with the percentage error of 2%, 3%, 2% and 1%, respectively. Minimum percentage of error is contributed by the measurement of

(a) a (b) b (c) c (d) d

42 A vernier calliper has 10 divisions on the vernier scale which coincide with 9 divisions on the main scale. The least count of the instrument is 0.1 mm. The value of one division on main scale is

(a) 0.5 mm (b) 2 mm
(c) 1 mm (d) 0.1 mm

43 Distance moved by a screw in 5 rotation is 5 mm and total number of circular divisions is 100. The least count of the screw gauge is

(a) 1 mm (b) 0.001 cm
(c) 0.0001 cm (d) 0.001 mm

44 One centimetre on the main scale of vernier callipers is divided into ten equal divisions. If 10 divisions of the vernier scale coincide with 8 divisions of the main scale, the least count of the callipers is

(a) 0.01 cm (b) 0.02 cm
(c) 0.05 cm (d) 0.005 cm

45 The pitch of screw gauge is 0.5 mm and there are 50 divisions on the circular scale. In measuring the thickness of a metal plate, there are five divisions on the main scale and thirty fourth division coincides with the reference line. The thickness of the metal plate is

(a) 2.50 mm (b) 2.34 mm
(c) 2.44 mm (d) 2.84 mm

LEVEL 02 Objective Questions

46 If $K = \frac{d^4x}{dt^4}$, find dimensional formula of K.

(a) $[LT^{-3}]$ (b) $[L^4T^{-4}]$ (c) $[LT^{-4}]$ (d) $[LT^4]$

47 Match the following two columns.

	Column-I		Column-II
(A)	Electrical resistance	(p)	$[ML^3T^{-3}A^{-2}]$
(B)	Electrical potential	(q)	$[ML^2T^{-3}A^{-2}]$
(C)	Specific resistance	(r)	$[ML^2T^{-3}A^{-1}]$
(D)	Specific conductance	(s)	None

Codes

	A	B	C	D		A	B	C	D
(a)	q	s	r	p	(b)	q	r	p	s
(c)	p	q	s	r	(d)	p	r	q	s

48 If the speed of light (c), acceleration due to gravity (g) and pressure (p) are taken as fundamental units, the dimensions of gravitational constant (G) are

(a) c^0gp^{-3} (b) $c^2g^3p^{-2}$

(c) $c^0g^2p^{-1}$ (d) $c^2g^2p^{-2}$

49 If E, m, L and G denote energy, mass, angular momentum and gravitational constant respectively, the quantity $\left(\frac{EL^2}{m^5G^2}\right)$ has the dimensions of

(a) mass (b) length (c) time (d) angle

50 Two numbers $a = 0.92$ and $b = 0.08$ are given. The number of significant figures present in the result after the following operation $a+b$, $a-b$, $a \times b$ and a/b, respectively are

(a) 2, 2, 2, 2 (b) 3, 3, 2, 2

(c) 3, 2, 3, 3 (d) 3, 2, 2, 2

51 If velocity of light c, Planck's constant h and gravitational constant G are taken as fundamental quantities, then the dimensions of length will be

(a) $\sqrt{\frac{ch}{G}}$ (b) $\sqrt{\frac{hG}{c^5}}$ (c) $\sqrt{\frac{hG}{c^3}}$ (d) $\sqrt{\frac{hc^3}{G}}$

52 If A, B and C are physical quantities, having different dimensions, which of the following combinations can never be a meaningful quantity?

(i) $\frac{AB}{C}$ (ii) $\frac{A+B}{C}$ (iii) $AB - C^2$ (iv) $(A-B)/C$

(a) (iii, iv) (b) (i, ii) (c) (ii, iv) (d) (ii, iii)

53 If α kg, β metre and γ second are the fundamental units, 1 calorie can be expressed in new units as

(a) $\alpha^{-1}\beta^2\gamma$ (b) $\alpha^{-1}\beta^{-2}\gamma$ (c) $4.2\,\alpha^{-1}\beta$ (d) $4.2\,\alpha\beta\gamma^{-2}$

54 In a new system of units, unit of mass is 10 kg, unit of length is 1 km and unit of time is 1 min, the value of 1 J in this new hypothetical system is

(a) 3.6×10^{-4} new units (b) 6×10^7 new units

(c) 10^{11} new units (d) 1.67×10^4 new units

55 The period of oscillation of a simple pendulum is given by $T = 2\pi\sqrt{\frac{l}{g}}$, where l is about 100 cm and is known to have 1 mm accuracy. The period is about 2 s. The time of 100 oscillations is measured by a stop watch of least count 0.1 s. The percentage error in g is
(a) 0.1 % (b) 1 % (c) 0.2 % (d) 0.8 %

56 A particle moving on a straight line has initial velocity 2.03 m/s, accelerating uniformly with 1.0 m/s^2. After 0.3 s, the particle will have velocity (in m/s) (upto correct significant figures)
(a) 2.3 (b) 2.0 (c) 2.06 (d) 2.1

57 If $a = 8 \pm 0.08$ and $b = 6 \pm 0.06$. Let $x = a + b, y = a - b, z = a \times b$. The correct order of % error in x, y and z is :
(a) $x = y < z$ (b) $x = y > z$ (c) $x < z < y$ (d) $x > z < y$

58 While measuring the acceleration due to gravity by a simple pendulum, a student makes a positive error of 1% in the length of the pendulum and a negative error of 3% in the value of time period. His percentage error in the measurement of g by the relation $g = 4\pi^2(l/T^2)$ will be
(a) 2% (b) 4% (c) 7% (d) 10%

59 You measure two quantities as $A = (1.0 \pm 0.2)$ m and $B = (2.0 \pm 0.2)$m. You should report correct value for $\sqrt{AB}$ as
(a) (1.4 ± 0.4)m (b) (1.41 ± 0.15)m (c) (1.4 ± 0.3)m (d) (1.4 ± 0.2) m

60 In an experiment, the period of oscillation of a simple pendulum was observed to be 2.63 s, 2.56 s, 2.42 s, 2.71 s and 2.80 s. The mean absolute error is
(a) 0.11 s (b) 0.12 s (c) 0.13 s (d) 0.14 s

61 Resistance of a given wire is obtained by measuring the current flowing in it and the voltage difference applied across it. If the percentage errors in the measurement of the current and the voltage difference are 3% each, then error in the value of resistance of the wire is
(a) 3% (b) 6% (c) zero (d) 1%

62 The refractive index of water measured by the relation $\mu = \frac{\text{real depth}}{\text{apparent depth}}$ is found to have values of 1.34, 1.38, 1.32 and 1.36; the mean value of refractive index with percentage error is:
(a) $1.35 \pm 1.48\%$ (b) $1.35 \pm 0\%$ (c) $1.36 \pm 6\%$ (d) $1.36 \pm 0\%$

63 The accuracy in the measurement of length 50 cm of a simple pendulum is 1 mm. If time period of 2 s is determined by measuring the time of 50 oscillations using a watch of resolution 0.2 s. The percentage error in the determined value of g is
(a) 0.2% (b) 0.3% (c) 0.6% (d) 0.4%

64 The density of a solid ball is to be determined in an experiment. The diameter of the ball is measured with screw gauge whose pitch is 0.5 mm and 50 divisions on circular scale. The reading on the main scale is 2.5 mm and that on circular scale is 20 divisions. If the measured mass of ball has a percentage error of 2%, then the percentage error in density is
(a) 0.9% (b) 2.4% (c) 3.1% (d) 4.2%

65 A student uses a simple pendulum of exactly 1 m length to determine g, the acceleration due to gravity. He uses a stop watch with the least count of 1 s for this and record 40 s for 20 oscillations for this observation, which of the following is true? (T is time period of oscillations)
$\left[\text{Take, } T = 2\pi\sqrt{\frac{L}{g}}\right]$
(a) $\Delta T = 1$ s; $\Delta g = 2.5\%$ (b) $\Delta T = 0.05$ s; $\Delta g = 5\%$
(c) $\Delta T = 0.025$ s; $\Delta g = 5\%$ (d) $\Delta T = 0.025$ s; $\Delta g = 2.5\%$

66 The moment of inertia of a body rotating about a given axis is 6.0 kg-m^2 in the SI system. What is the value of the moment of inertia in a system of units in which the units of length is 5 cm and the unit of mass is 10 g?

(a) 2.4×10^3 g cm^2 (b) 1.5×10^{10} g cm^2
(c) 6.0×10^3 g cm^2 (d) 6.0×10^5 g cm^2

67 If each division on the main scale is 1 mm, which of the following vernier scales gives vernier constant equal to 0.05 mm?

(a) 9 mm divided into 10 divisions (b) 19 mm divided into 20 divisions
(c) 90 mm divided into 100 divisions (d) 99 mm divided into 100 divisions

68 The nth division of main scale coincides $(n+1)$th division of vernier scale. Given one main scale division is equal to a units. Find the least count of the vernier.

(a) $\frac{a}{n+1}$ (b) $(n+1)$
(c) $\frac{(n+1)}{a}$ (d) an

69 In a vernier callipers, 10 divisions of vernier scale coincides with 9 divisions of main scale, the least count of main scale is 0.1 cm. If in the measurement of inner diameter of cylinder, zero of vernier scale lies between 1.3 cm and 1.4 cm of main scale and 2nd division of vernier scale coincides with main scale division, then diameter will be

(a) 1.30 cm (b) 1.34 cm
(c) 1.32 cm (d) 1.36 cm

70 What is the reading of micrometer screw gauge shown in the following figure? Circular scale has 100 divisions.

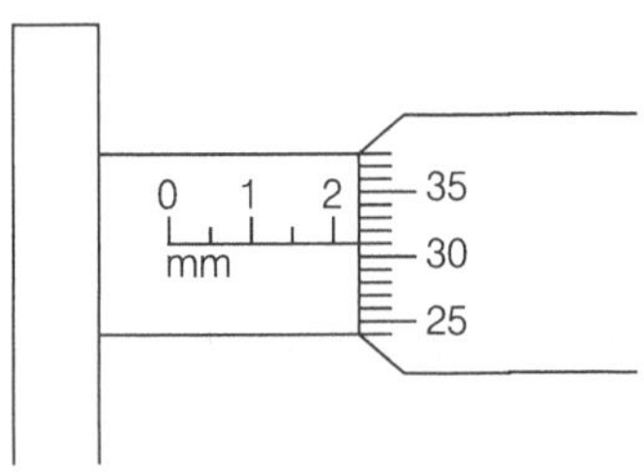

(a) 2.31 mm (b) 2.29 mm
(c) 2.36 mm (d) 2.41 mm

71 The circular scale of a micrometer has 200 divisions and pitch of 2 mm. Find the measured value of thickness of a thin sheet.

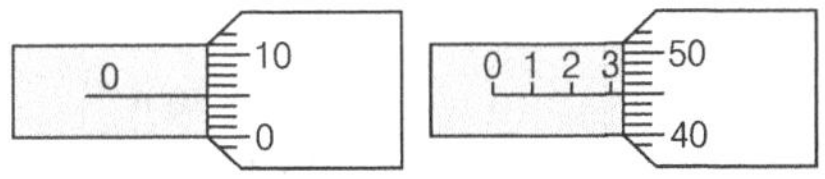

(a) 3.41 mm (b) 6.41 mm
(c) 3.46 mm (d) 3.51 mm

72 A uniform cube is subjected to volume compression. If each side is decreased by 1%, then volumetric strain is

(a) 0.01 (b) 0.06
(c) 0.02 (d) 0.03

73 The values of two resistors are $R_1 = (6 \pm 0.3)$ kΩ and $R_2 = (10 \pm 0.2)$ kΩ. The percentage error in the equivalent resistance when they are connected in parallel is

(a) 3.875 % (b) 2 %
(c) 10.125 % (d) 7 %

LEVEL 03 Previous Years Questions

74 A screw gauge gives the following readings when used to measure the diameter of a wire
Main scale reading : 0 mm, Circular scale reading : 52 divisions.
Given that, 1 mm on main scale corresponds to 100 divisions on the circular scale. The diameter of the wire from the above data is **[NEET 2021]**

(a) 0.52 cm (b) 0.026 cm
(c) 0.26 cm (d) 0.052 cm

75 If force $[F]$, acceleration $[a]$ and time $[T]$ are chosen as the fundamental physical quantities. Find the dimensions of energy. **[NEET 2021]**

(a) $[F][a][T]$ (b) $[F][a][T^2]$ (c) $[F][a][T^{-1}]$ (d) $[F][a^{-1}][T]$

76 If E and G respectively, denote energy and gravitational constant, then $\frac{E}{G}$ has the dimensions of **[NEET 2021]**

(a) $[M^2][L^{-1}][T^0]$ (b) $[M][L^{-1}][T^{-1}]$
(c) $[M][L^0][T^0]$ (d) $[M^2][L^{-2}][T^{-1}]$

77 Time intervals measured by a clock give the following readings 1.25 s, 1.24 s, 1.27 s, 1.21 s and 1.28 s. What is the percentage relative error of the observations? **[NEET 2020]**

(a) 2% (b) 4%
(c) 16% (d) 1.6%

78 A screw gauge has least count of 0.01 mm and there are 50 divisions in its circular scale. The pitch of the screw gauge is **[NEET 2020]**

(a) 0.25 mm (b) 0.5 mm
(c) 1.0 mm (d) 0.01 mm

79 Dimensions of stress are **[NEET 2020]**

(a) $[ML^2T^{-2}]$ (b) $[ML^0T^{-2}]$
(c) $[ML^{-1}T^{-2}]$ (d) $[MLT^{-2}]$

80 Taking into account of the significant figures, what is the value of 9.99 m – 0.0099 m? **[NEET 2020]**

(a) 9.98 m (b) 9.980 m
(c) 9.9 m (d) 9.9801 m

81 The angle of 1′ (minute of arc) in radian is nearly equal to **[NEET 2020]**

(a) 2.91×10^{-4} rad (b) 4.85×10^{-4} rad
(c) 4.80×10^{-6} rad (d) 1.75×10^{-2} rad

82 The unit of thermal conductivity is : **[NEET 2019]**

(a) $J\,m^{-1}\,K^{-1}$ (b) $W\,m\,K^{-1}$
(c) $W\,m^{-1}\,K^{-1}$ (d) $J\,m\,K^{-1}$

83 The main scale of a vernier calliper has n divisions/cm. n divisions of the vernier scale coincide with $(n-1)$ divisions of main scale. The least count of the vernier callipers is **[NEET 2019]**

(a) $\frac{1}{(n+1)(n-1)}$ cm (b) $\frac{1}{n}$ cm
(c) $\frac{1}{n^2}$ cm (d) $\frac{1}{n(n+1)}$ cm

84 A student measured the diameter of a small steel ball using a screw gauge of least count 0.001 cm. The main scale reading is 5 mm and zero of circular scale division coincides with 25 divisions above the reference level. If screw gauge has a zero error of –0.004 cm, the correct diameter of the ball is **[NEET 2018]**

(a) 0.053 cm (b) 0.525 cm (c) 0.521 cm (d) 0.529 cm

85 A physical quantity of the dimensions of length that can be formed out of c, G and $\frac{e^2}{4\pi\varepsilon_0}$ is [c is velocity of light, G is universal constant of gravitation and e is charge] **[NEET 2017]**

(a) $\frac{1}{c^2}\left[G\frac{e^2}{4\pi\varepsilon_0}\right]^{1/2}$ (b) $c^2\left[G\frac{e^2}{4\pi\varepsilon_0}\right]^{1/2}$ (c) $\frac{1}{c^2}\left[\frac{e^2}{G\,4\pi\varepsilon_0}\right]^{1/2}$ (d) $\frac{1}{c}G\frac{e^2}{4\pi\varepsilon_0}$

86 If energy (E), velocity (v) and time (T) are chosen as the fundamental quantities, the dimensional formula of surface tension will be **[CBSE AIPMT 2015]**

(a) $[Ev^{-2}T^{-1}]$ (b) $[Ev^{-1}T^{-2}]$
(c) $[Ev^{-2}T^{-2}]$ (d) $[E^{-2}v^{-1}T^{-3}]$

87 If dimensions of critical velocity v_c of a liquid flowing through a tube are expressed as $[\eta^x\rho^y r^z]$, where η, ρ and r are the coefficient of viscosity of liquid, density of liquid and radius of the tube respectively, then the values of x, y and z are given by **[CBSE AIPMT 2015]**

(a) 1, − 1, − 1 (b) − 1, − 1, 1
(c) − 1, − 1, − 1 (d) 1, 1, 1

88 If force (F), velocity (v) and time (T) are taken as fundamental units, then the dimensions of mass are **[CBSE AIPMT 2014]**

(a) $[FvT^{-1}]$ (b) $[FvT^{-2}]$ (c) $[Fv^{-1}T^{-1}]$ (d) $[Fv^{-1}T]$

89 In an experiment, four quantities a, b, c and d are measured with percentage error 1%, 2%, 3% and 4% respectively. Quantity P is calculated as $P = \frac{a^3b^2}{cd}$%. Error in P is **[NEET 2013]**

(a) 14% (b) 10%
(c) 7% (d) 4%

90 The dimensions of $(\mu_0\varepsilon_0)^{-1/2}$ are **[CBSE AIPMT 2012]**

(a) $[L^{1/2}T^{-1/2}]$ (b) $[L^{-1}T]$
(c) $[LT^{-1}]$ (d) $[L^{1/2}T^{1/2}]$

ANSWER KEY

Level 1 Objective Questions

1 (a)	2 (d)	3 (b)	4 (d)	5 (a)	6 (c)	7 (a)	8 (a)	9 (c)	10 (c)
11 (b)	12 (c)	13 (a)	14 (b)	15 (d)	16 (d)	17 (b)	18 (c)	19 (d)	20 (b)
21 (b)	22 (b)	23 (b)	24 (a)	25 (d)	26 (d)	27 (d)	28 (c)	29 (b)	30 (b)
31 (a)	32 (a)	33 (a)	34 (b)	35 (d)	36 (c)	37 (a)	38 (b)	39 (d)	40 (d)
41 (a)	42 (c)	43 (b)	44 (b)	45 (d)					

Level 2 Objective Questions

46 (c)	47 (b)	48 (c)	49 (d)	50 (c)	51 (c)	52 (c)	53 (d)	54 (a)	55 (c)
56 (a)	57 (c)	58 (c)	59 (d)	60 (a)	61 (b)	62 (b)	63 (c)	64 (c)	65 (b)
66 (b)	67 (b)	68 (a)	69 (c)	70 (a)	71 (b)	72 (d)	73 (a)		

Level 3 Previous Years Questions

74 (d)	75 (b)	76 (a)	77 (d)	78 (b)	79 (c)	80 (a)	81 (a)	82 (c)	83 (c)
84 (d)	85 (a)	86 (c)	87 (a)	88 (d)	89 (a)	90 (c)			

Solutions

LEVEL 01 Objective Questions

1. Parallactic second is the unit of distance.
2. We know, unit of power is watt and it is represented as watt = joule/second, ampere × volt and $(\text{ampere})^2 \times \text{ohm}$
3. We know, SI unit of pressure is Nm^{-2} and for volume is m^3.

   For moles, it is mol and for temperature, it is kelvin.
   $$R = \frac{pV}{nT} = \frac{J}{\text{mol K}} = JK^{-1}\ \text{mol}^{-1}$$
4. The dimensional formula of physical constant is as follows

   Planck's constant $(h) = [ML^2T^{-1}]$

   Gravitational constant $(G) = [M^{-1}L^3T^{-2}]$

   Bulk modulus $(B) = [ML^{-1}T^{-2}]$

   Coefficient of viscosity $(\eta) = [ML^{-1}T^{-1}]$
5. We know, resistance $= \dfrac{1}{G}$
   $$G = \frac{1}{R} = \frac{I}{V} = \frac{A}{[ML^2T^{-3}A^{-1}]}$$
   $$= [M^{-1}L^{-2}T^3A^2]$$
6. Dimensions of quantity $A = [MLT^{-2}]$

   Dimensions of quantity $m = [ML^{-1}]$

   Dimensions of $B = \dfrac{A}{m} = [L^2T^{-2}]$

   Therefore, the dimensions of B is of latent heat.
7. Given, $F = \dfrac{\alpha}{\beta + \sqrt{d}}$

   Dimensions of LHS = Dimensions of RHS

   β and $\sqrt{d}$ must have the same dimensions.
   $$[ML^{-3}]^{\frac{1}{2}} = \beta$$
   $$\beta = M^{\frac{1}{2}}L^{-\frac{3}{2}}$$
   Now, $[MLT^{-2}] = \dfrac{[\alpha]}{\beta}$
   $$\alpha = [MLT^{-2}][M^{\frac{1}{2}}L^{-\frac{3}{2}}]$$
   $$= [M^{\frac{3}{2}}L^{-\frac{1}{2}}T^{-2}]$$
8. Given, $\left[\dfrac{x^3w}{y^3z}\right] = \dfrac{[L^2][M]}{[T^3][A]}$
   $$= [ML^2T^{-3}A^{-1}] = [\text{Electric potential}]$$
9. Given, $P = (x^2 - b)/at$

   Dimension of power $= [ML^2T^{-3}]$
   $$\therefore \quad [ML^2T^{-3}] = \frac{[L^2]}{at} \Rightarrow a = [M^{-1}L^0T^2]$$
10. Since, $\sin \beta t$ is a dimensionless quantity.

    Therefore, $\beta = \dfrac{1}{T}$

    For dimension of α,
    $$\alpha = \frac{[MLT^{-2}]}{[LT^{-1}]^2} = \frac{[MLT^{-2}]}{[L^2T^{-2}]} = [ML^{-1}]$$
    $$\therefore \quad \frac{\alpha}{\beta} = \frac{[ML^{-1}]}{[T^{-1}]} = [ML^{-1}T]$$
11. Given, $p = p_0 e^{-\alpha t^2}$

    As we know, p and p_0 both are pressure. They have the same units.

    Therefore, αt^2 must be dimensionless,
    $$\alpha = \frac{1}{t^2} = [T^{-2}]$$
12. According to principle of homogeneity, the dimension of velocity v, are same.
    $$at \text{ and } \frac{b}{t+c}$$
    For dimension of a,
    $$[LT^{-1}] = at$$
    $$\therefore \quad a = [LT^{-2}]$$
    For dimension of b,
    $$[LT^{-1}] = \frac{b}{t+c}$$
    Here, the dimension of c must be of time
    $$[LT^{-1}] = \frac{b}{t} \Rightarrow b = [L]$$
    Therefore, the dimensions of a, b and c are $[LT^{-2}]$, [L] and [T].
13. $[\omega k] = [M^0L^0T^0]$
    $$\therefore \quad [k] = [\omega]^{-1} = [M^0L^0T^{-1}]^{-1}$$
    $$= [M^0L^0T^1]$$
14. From principle of homogeneity, the dimensions on both sides must be equal.
    $$F = at \Rightarrow a = \frac{F}{t}$$
    $$\frac{[MLT^{-2}]}{[T]} = a$$

$\therefore \quad a = [MLT^{-3}]$

Similarly, $[MLT^{-2}] = bt^2$

$b = [MLT^{-4}]$

The dimensions of a and b are $[MLT^{-3}]$ and $[MLT^{-4}]$.

15. We know that, addition or subtraction can be done only with same physical quantity.

Since, A and B have same dimension, therefore $\frac{(2A-3B)}{C}$ gives the meaningful dimension.

16. Let p be proportional to f, v, D

$\therefore \quad p = [f]^a [v]^b [D]^c$

$[MLT^{-1}] = [T^{-1}]^a [LT^{-1}]^b [ML^{-3}]^c$

Comparing both sides,

$c = 1$

$b - 3c = 1$

$\therefore \quad b = 4$

Also, $\quad a + b = 1$

$\therefore \quad a = -3$

Therefore, $p = [f]^{-3} [v]^4 [D]^1 = Dv^4 f^{-3}$

17. $[AB] \times \left[\frac{A}{B}\right] = [A^2] = [ML^2T^{-2}] \cdot [MT^{-2}]$

$\Rightarrow \quad [A]^2 = [M^2L^2T^{-4}] = [MLT^{-2}]^2$

So, A is force, now $B = \frac{[ML^2T^{-2}]}{A}$

$= \frac{[ML^2T^{-2}]}{[MLT^{-2}]} = [L] = \text{Displacement}$

18. From principle of homogeneity,

Af and kt must have the same dimension.

$[ML^{-1}T^{-2}][B] = [ML^2T^{-1}]$

$B = [M^0L^3T]$

The dimension of B in the length is 3.

19. Since, τ is time constant, so it will have dimension of second (T).

Option (a) has dimensions : $\frac{[ML^2]}{[ML^{-1}T^{-2}]} = [L^3T^2]$

Option (b) has dimensions : $\sqrt{\frac{[ML] \times [ML^{-1}T^{-2}]}{[L^2T^{-4}]}} = [ML^{-1}T]$

Option (c) has dimensions : $\frac{[M]}{[ML^{-1}T^{-2}][L][LT^{-1}]} = \frac{[T^3]}{[L]}$

Hence, option (d) is correct.

20. Dimensions of $\varepsilon_0 = [M^{-1}L^{-3}T^4A^2]$

Dimensions of $\mu_0 = [MLT^{-2}A^{-2}]$

Dimensions of $E = [M^1L^1T^{-3}A^{-1}]$

Dimensions of $B = [M^1T^{-2}A^{-1}]$

21. $\left[\varepsilon_0 \frac{d\phi_E}{dt}\right] = \left[\frac{\varepsilon_0 ES}{t}\right] = \frac{[M^{-1}L^{-3}T^4A^2][MLT^{-3}A^{-1}][L^2]}{[T]}$

$= [A]$

22. $[ML^2T^{-2}] = [h]^x [G]^y [F]^z$

$= [ML^2T^{-1}]^x [M^{-1}L^3T^{-2}]^y [MLT^{-2}]^z$

$= [M^{x-y+z} L^{2x+3y+z} T^{-x-2y-2z}]$

Equating the powers, we get

$x - y + z = 1$ ...(i)

$2x + 3y + z = 2$ ...(ii)

$-x - 2y - 2z = -2$ ...(iii)

Solving these three equations, we get

$x = \frac{1}{2}, y = \frac{1}{8}$ and $z = \frac{5}{8}$

$\therefore \quad [\text{Energy}] = [h^{1/2} G^{1/8} F^{5/8}]$

23. According to question,

$F \propto m^a v^b r^c \Rightarrow F = km^a v^b r^c$

k being a dimensionless constant

From homogeneity of dimensions,

LHS = RHS

$\Rightarrow \quad [MLT^{-2}] = [M]^a [LT^{-1}]^b [L]^c$

$[MLT^{-2}] = [M^a L^{b+c} T^{-b}]$

Comparing powers on both sides, we get

$a = 1$

$b + c = 1$

$-b = -2 \Rightarrow b = 2$

$2 + c = 1 \Rightarrow c = -1$

$\therefore \quad F = kmv^2r^{-1} = \frac{kmv^2}{r}$

The experimental value of k is found to be here

$\therefore \quad F = \frac{mv^2}{r}$

24. $[L^3] = [L^2]^a [LT^{-1}]^b [T]^c$

Equating the powers, we get

$2a + b = 3$ ...(i)

$-b + c = 0$ ...(ii)

$a = b = c = 1$, can satisfy both equations.

25. The powers of ten are not counted as significant figures.

26. In 3.845, the preceeding digit is even, it becomes 3.84 and in 3.835, the preceeding digit is odd, it becomes 3.84.

27. In multiplication or in division, the final result should retain as many significant figures as there are in the original number with the least significant figures.

2.40 contain the least significant figure of 3.

$= \frac{96.54}{2.40} = 40.2$

28. Mass of the box = 2.3 kg

Mass of one gold piece = 0.02015 kg

Mass of other gold piece = 0.02017 kg

Total mass = 2.3 + 0.02015 + 0.02017 = 2.34032 kg

Rounding off = 2.3 kg

29. Result must contain as many significant figures as that in quantities in product with least number of significant figures.

So, $V = 35.0\text{ V}$ (Three significant figures)

30. The actual length must be 10.2756 cm. It is because among the given options, 10.2756 cm is the only length which after rounding off would become 10.28 cm, which is the observation of the length.

31. Here, $A = 2.5\text{ m/s} \pm 0.5\text{ m/s},\ B = 0.10\text{ s} \pm 0.01\text{ s}$

$$AB = 2.5\,(0.10) = 0.25\text{ m}$$

$$\frac{\Delta AB}{AB} = \frac{\Delta A}{A} + \frac{\Delta B}{B} = \frac{0.5}{2.5} + \frac{0.01}{0.10} = 0.3$$

$$\Delta AB = 0.3 \times 0.25 = 0.075 = 0.08\text{ m}$$

The value of $AB = (0.25 \pm 0.08)$ m

32. We have, length = 125.5 cm, breadth = 5.6 cm, thickness = 0.32 cm

In thickness, least count is 0.01 cm. Therefore, it is more accurate.

33. Given, radius of ball = [5.4 ± 0.2] cm

Volume of ball $= \frac{4}{3}\pi R^3$

Percentage error in volume,

$$\frac{\Delta V}{V} = \frac{3\Delta R}{R} \times 100\ \%$$

$$= 3\left(\frac{0.2}{5.4}\right) \times 100\ \%$$

$$= 11\ \%$$

34. Heat produced, $H = I^2RT$

Therefore, $\frac{\Delta H}{H} = \frac{2\Delta I}{I} + \frac{\Delta R}{R} + \frac{\Delta T}{T}$

$$= 2 \times 3 + 4 + 6 = \pm 16\ \%$$

35. Moment of inertia about diameter $= \frac{2}{5}mr^2$

Percentage error in moment of inertia

$$= \left(\frac{\Delta m}{m} + \frac{2\Delta r}{r}\right) \times 100$$

$$= 2\% + 2(1\%) = 4\%$$

36. $R = R_1 + R_2 = 15\ \Omega$

$$\Delta R = (\Delta R_1 + \Delta R_2) = 0.3\ \Omega$$

$$\because \quad \frac{\Delta R}{R} \times 100 = \frac{0.3}{15} \times 100 = 2\%$$

$\therefore$ Correct answer is $(15 \pm 2\%)\ \Omega$

37. Given, $P = \frac{A^3}{B^{5/2}}$

For absolute error,

$$\frac{\Delta P}{P} = \frac{3\Delta A}{A} + \frac{5}{2}\frac{\Delta B}{B}$$

or $$\Delta P = \pm\left(3\frac{\Delta A}{A} + \frac{5}{2}\frac{\Delta B}{B}\right)P$$

38. Displacement of a body = (200 ± 5) m

Time taken = (20 ± 0.2) s

$\therefore$ $s = 200\text{ m},\ \Delta s = 5\text{ m}$

$t = 20\text{ s},\ \Delta t = 0.2\text{ s}$

$$\text{Velocity} = \frac{\text{Displacement}}{\text{Time}}$$

$$\therefore \quad \frac{\Delta v}{v} = \frac{\Delta s}{s} + \frac{\Delta t}{t}$$

$$\therefore \quad = \frac{5}{200} + \frac{0.2}{20} = 0.035$$

Percentage error in velocity = 0.035 × 100

= 3.5 %

39. $A = \pi r^2$

$\therefore$ % error in A = 2 (% error in r)

= 2 (3%) = 6%

40. Given, $A = [M^{-1}L^3T^{-2}]$

$$x = \frac{m^3}{kg\ s^2} = \frac{[L^3]}{[MT^2]}$$

$$\frac{\Delta x}{x} = \frac{\Delta M}{M} + \frac{3\Delta L}{L} + \frac{2\Delta T}{T}$$

$$= 2 + 3(3) + 2(4) = 19\%$$

41. Given, $\rho = \frac{\sqrt{abc^2}}{d^3}$

$$\frac{\Delta\rho}{\rho} = \frac{\Delta a}{a} + \frac{\Delta b}{b} + \frac{2\Delta c}{c} + \frac{3\Delta d}{d}$$

$$= 2 + 3 + 2(2) + 3(1)$$

$$= 12$$

Here, we can see that minimum error contribution is made by a.

42. LC = 1 MSD − 1 VSD

$$0.1\text{ mm} = 1\text{ MSD} - \frac{9}{10}\text{ MSD} = 0.1\text{ MSD}$$

$\therefore$ 1 MSD = 1 mm

43. Least count of screw gauge $= \frac{\text{Pitch}}{\text{Number of divisions}}$

$$\text{Pitch} = \frac{5}{5} = 1\text{ mm}$$

$$\therefore \quad \text{LC} = \frac{1}{100} = 0.01\text{ mm}$$

$$= 0.001\text{ cm}$$

44. $1\text{ MSD} = \frac{1}{10}\text{ cm} = 1\text{ mm}$

10 VSD = 8 MSD

$$1 \text{ VSD} = \frac{8}{10} \text{ MSD}$$

Least count = 1 MSD − 1 VSD

$$= 1 - \frac{8}{10} = \frac{2}{10} \text{ mm}$$

$$= \frac{2}{100} \text{ cm} = 0.02 \text{ cm}$$

45. Pitch = 0.5 mm

Number of divisions = 50

$$\text{Least count} = \frac{\text{Pitch}}{\text{No. of divisions}} = \frac{0.5}{50} = 0.01$$

Main scale reading = 5 (0.5) = 2.5

Thickness of metallic plate = MSR + VSR × LC

$$= 2.5 + 34 \times 0.01 = 2.84 \text{ mm}$$

LEVEL 02 Objective Questions

46. Given, $K = \frac{d^4x}{dt^4} \Rightarrow K = \frac{[L]}{[T^4]} = [LT^{-4}]$

48. Dimensions of $G = [M^{-1}L^3T^{-2}]$

$$[M^{-1}L^3T^{-2}] = [c]\,[g]\,[p]$$

$$[M^{-1}L^3T^{-2}] = [LT^{-1}]^{\alpha}[LT^{-2}]^{\beta}[ML^{-1}T^{-2}]^{\gamma}$$

On equating coefficients, we get

$$\gamma = -1, 3 = \alpha + \beta - \gamma$$

and $\alpha + \beta = 2$ ...(i)

$$-2 = -\alpha - 2\beta - 2\gamma$$

$\alpha + 2\beta = 4$ ...(ii)

On solving Eqs. (i) and (ii), we get

$$\alpha = 0, \beta = 2, \gamma = -1 = [c^0 g^2 p^{-1}]$$

49. Given quantity is $\frac{EL^2}{M^5G^2}$

We know, the dimensions of energy, $E = [MLT^{-2}]$

Angular momentum, $L = [ML^2T^{-1}]$

Gravitational constant, $G = [M^{-1}L^3T^{-2}]$

Substituting each in given equation, gives the dimensionless quantity which is of angle.

50. $a = 0.92$ and $b = 0.08$

$a + b = 0.92 + 0.08 = 1.00$	3 significant figures
$a - b = 0.92 - 0.08 = 0.84$	2 significant figures
$a \times b = 0.92 \times 0.08 = 0.0736$	3 significant figures
$\frac{a}{b} = \frac{0.92}{0.08} = 11.5$	3 significant figures

51. $[L] = [c]^x\,[h]^y[G]^z = [LT^{-1}]^x\,[ML^2T^{-1}]^y[M^{-1}L^3T^{-2}]^z$

Equating the powers, we get

$x + 2y + 3z = 1$ ...(i)

$y - z = 0$ ...(ii)

$-x - y - 2z = 0$ ...(iii)

Solving these three equations, we get

$$x = -\frac{3}{2}, y = z = \frac{1}{2}$$

$$\therefore \quad [L] = \left[\sqrt{\frac{hG}{c^3}}\right]$$

52. The physical quantities having different dimensions cannot be added or subtracted. Therefore, the combinations given in (ii) and (iv) can never be a meaningful quantity.

53. As $1 \text{ cal} = 4.2 \text{ J} = 4.2\left(\frac{\text{kg.m}}{\text{s}^2}\right)$

In new system of units,

$$1 \text{ kg} = \alpha \text{ kg}$$

$$1 \text{ m} = \beta \text{ m}$$

$$1 \text{ s} = \gamma \text{ s}$$

So, $1 \text{ cal} = 4.2\left(\frac{\alpha \text{ kg} \cdot \beta \text{ m}}{(\gamma \text{s})^2}\right)$

$$= 4.2\,\alpha\beta\gamma^{-2} \text{ New units.}$$

54. In new system, mass = 10 kg, length = 1 km, time = 1 min

Unit of joule $[ML^2T^{-2}] = [\text{kg m}^2\text{s}^{-2}]$

$$n_2 = n_1\left[\frac{M_1}{M_2}\right]\left[\frac{L_1}{L_2}\right]^2\left[\frac{T_1}{T_2}\right]^{-2}$$

$$= 1\left[\frac{1\text{g}}{10 \text{ kg}}\right]\left[\frac{1\text{m}}{1 \text{ km}}\right]^2\left[\frac{1\text{s}}{1\text{min}}\right]^{-2}$$

$$= 1\left[\frac{1}{10}\right]\left[\frac{1}{1000}\right]^2\left[\frac{1}{60}\right]^{-2}$$

$$= 3.6 \times 10^{-4} \text{ units}$$

55. We have, $T = 2\pi\sqrt{\frac{L}{g}} \Rightarrow T^2 = 4\pi^2\left(\frac{L}{g}\right)$

$$\Rightarrow \quad g = \frac{4\pi L}{T^2}$$

Percentage error, $\frac{\Delta g}{g} = \frac{\Delta L}{L} + \frac{2\Delta T}{T}$

$$= \frac{0.1}{100} \times 100 + \frac{2 \times 0.1}{2 \times 100} \times 100$$

$$\frac{\Delta g}{g} \times 100 = 0.1 + 2(0.05)$$

$$= 0.2\,\%$$

56. Initial velocity, $u = 2.03$ m/s

Acceleration, $a = 1.0 \text{ m/s}^2$

Time, $t = 0.3$ s

Using equation, $v = u + at$

$$v = 2.03 + 1(0.3)$$

$$= 2.33 \approx 2.3 \text{ m/s}$$

57. We have, $a = 8 \pm 0.08$ and $b = 6 \pm 0.06$

In $x = a + b$, $\quad x = 8 + 6 = 14$

$$\Delta x = 0.08 + 0.06 = 0.14$$

In $y = a - b$, $\quad y = 8 - 6 = 2$

$$\Delta y = 0.08 + 0.06 = 0.14$$

In $z = a \times b$, $\quad z = 8 \times 6 = 48$

$$\therefore \quad \frac{\Delta z}{z} = \frac{\Delta a}{a} + \frac{\Delta b}{b} = \left(\frac{0.08}{8} + \frac{0.06}{6}\right) \times 100 = 2\%$$

$$\frac{\Delta x}{x} \times 100 = \frac{0.14}{14} \times 100 = 1\%$$

$$\frac{\Delta y}{y} \times 100 = \frac{0.14}{2} \times 100 = 7\%$$

$$\therefore \quad x < z < y$$

58. We have, $g = 4\pi^2\left(\frac{1}{T^2}\right)$

$$\therefore \quad \frac{\Delta g}{g} = \frac{\Delta L}{L} + \frac{2\Delta T}{T} = 1 + 2(3) = 7$$

Percentage error = 7%

59. Let $X = \sqrt{AB} = A^{\frac{1}{2}}B^{\frac{1}{2}}$ ...(i)

Now, $\quad X = \sqrt{1.0 \times 2.0} \approx 1.4$ m

Further, from Eq. (i), we get

$$\frac{\Delta X}{X} = \frac{1}{2}\left[\frac{\Delta A}{A} + \frac{\Delta B}{B}\right]$$

or $$\Delta X = \frac{1}{2}\left[\frac{\Delta A}{A} + \frac{\Delta B}{B}\right](X) = \frac{1}{2}\left[\frac{0.2}{1.0} + \frac{0.2}{2.0}\right](1.4) \approx 0.2 \text{ m}$$

Answer is $(X \pm \Delta X)$ or (1.4 ± 0.2) m.

60. The mean period of oscillation of pendulum

$$= \frac{2.63 + 2.56 + 2.42 + 2.71 + 2.80}{5} = 2.62 \text{ s}$$

Now, absolute errors,

$$\Delta T_1 = 2.62 - 2.63 = -0.01$$
$$\Delta T_2 = 2.62 - 2.56 = 0.06$$
$$\Delta T_3 = 2.62 - 2.42 = -0.20$$
$$\Delta T_4 = 2.62 - 2.71 = -0.09$$
$$\Delta T_5 = 2.62 - 2.80 = -0.18$$

The mean of absolute error

$$= \frac{0.01 + 0.06 + 0.20 + 0.09 + 0.18}{5} = \frac{0.54}{5} \simeq 0.11 \text{ s}$$

61. We know, $R = \frac{V}{I}$

$$\frac{\Delta R}{R} = \frac{\Delta V}{V} + \frac{\Delta I}{I} = 3\% + 3\% = 6\%$$

62. The mean value of refractive index,

$$\mu = \frac{\mu_1 + \mu_2 + \mu_3 + \mu_4}{4}$$

$$\mu = \frac{1.34 + 1.38 + 1.32 + 1.36}{4} = 1.35$$

Now, $$\Delta\mu = \frac{(\mu - \mu_1) + (\mu - \mu_2) + (\mu - \mu_3) + (\mu - \mu_4)}{4}$$

$$= \frac{(1.35 - 1.34) + (1.35 - 1.38) + (1.35 - 1.32) + (1.35 - 1.36)}{4}$$

$\Delta\mu = 0$

So, percentage error

$$= 0 \times 100 = 0\%$$

Therefore, the mean value of refractive index with percentage error is $(1.35 \pm 0\ \%)$.

63. We have,

$$\frac{\Delta T}{T} \times 100 = \frac{0.2}{2(50)} \times 100 = 0.2$$

Similarly, $$\frac{\Delta L}{L} \times 100 = \frac{0.1}{50} \times 100 = 0.2$$

Now, $T = 2\pi\sqrt{\frac{L}{g}}$

$$\Rightarrow \quad g = \frac{4\pi^2 L}{T^2} \Rightarrow \frac{\Delta g}{g} = \frac{\Delta L}{L} + \frac{2\Delta T}{T}$$

$$= 0.2 + 2\,(0.2) = 0.6\ \%$$

64. Pitch = 0.5 mm

Circular scale division = 50

Main scale reading = 2.5 mm

Least count $= \frac{0.5}{50} = 0.01$ mm

Circular scale division reading = 20 divisions

Relative error = 2%

Screw gauge reading = Main scale reading

$$+ \left(\frac{\text{Pitch}}{\text{Circular scale division}}\right) \times \text{Circular scale reading}$$

$$= 2.5 + \frac{0.5}{50} \times 20 = 2.7 \text{ mm}$$

The relative percentage error in the density,

$$\frac{\Delta \rho}{\rho} \times 100 = \left(\frac{\Delta M}{M} + \frac{3\Delta D}{D}\right) \times 100$$

$$\therefore \quad \text{Error} = \left(\frac{\Delta M}{M} \times 100\right) + \left(\frac{3\Delta D}{D}\right) \times 100$$

$$= 2 + \left(\frac{3 \times 0.01 \times 100}{2.7}\right) = 2 + 1.1$$

$$= 3.1\ \%$$

65. $T = \frac{t}{n}$ (where, n = number of oscillations)

$\therefore \quad \Delta T = \frac{\Delta t}{n} = \frac{1}{20} = 0.05$ s

Further, $g = \frac{(4\pi^2)L}{T^2}$

4, π and L are exact. Hence,

$$\frac{\Delta g}{g} \times 100 = 2\left(\frac{\Delta T}{T}\right) \times 100$$

$$= 2\left[\frac{\Delta T}{t/n}\right] \times 100$$

$$= \frac{(200)(n)(\Delta T)}{t} = \frac{200 \times 20 \times 0.05}{40}$$

$$= 5\%$$

66. Moment of inertia in SI system = 6.0 kg-m^2

$$T = 6 \times 1000 \times (10^4) \text{ g cm}^2$$

In new system, unit length = 5 cm

Unit mass = 10 g

$\therefore \quad I = 6 \times 1000 \times 10^4 \times (10\text{ g}) \times (5\text{ cm})^2$

$$= 150 \times 10^8 \text{ gcm}^2$$

$$= 1.5 \times 10^{10} \text{ gcm}^2$$

67. Least count = 1 MSD − 1 VSD

Vernier constant = $\frac{1 \text{ MSD}}{\text{No. of divisions on vernier scale}}$

$\frac{1}{\text{No. of divisions}} = 0.05$

No. of divisions = 20 divisions

Let n be the divisions on vernier scale with which 1 mm on main scale reading (MSR) coincides with vernier scale,

$$0.05 = 1 - \frac{n}{100} \Rightarrow n = 95$$

68. $(n+1)$ divisions of vernier scale = n divisions of main scale

$\therefore$ One vernier divisions = $\left(\frac{n}{n+1}\right) \times$ Main scale divisions

Least count = 1 MSD − 1 VSD

$$= \left(1 - \frac{n}{n+1}\right) \text{MSD}$$

$$= \left(\frac{1}{n+1}\right) \text{MSD}$$

$$= \frac{a}{n+1} \qquad (\because 1 \text{ MSD} = a \text{ units}]$$

69. Reading lies from (1.3 − 1.4) cm.

Given, least count = $\frac{1}{n+1}$ (MSD)

$$= \frac{1}{9+1} + \frac{1}{10} \text{MSD}$$

$$= \frac{1}{10} \times 0.1 \text{ cm} = 0.01 \text{ cm}$$

MSR = 1.3 cm

For division lies coincide by second point,

$$\text{VSR} = 2 \times 0.01 = 0.02$$

$\therefore$ Total reading = 1.3 + 0.02 = 1.32 cm

70. Main scale reading = 2

Vernier scale reading = 32

Least count = $\frac{1}{100} = 0.01$

Reading of screw gauge = 2 + 32 × 0.01 ≃ 2.31 mm

71. LC = $\frac{\text{Pitch}}{\text{Number of divisions}} = \frac{2}{200} = 0.01$ mm

Zero error = LC × 5 = 0.05 mm

Thickness = Reading − Zero error

= 3 (Pitch) + 46 (LC) − zero error

= 6 mm + 0.46 mm − 0.05 mm

= 6.41 mm

72. Volume of a cube = L^3

For change in volume, $\frac{\Delta V}{V} = \frac{3\Delta L}{L} = 3(1\%) = 3\%$

Hence, bulk strain = $\frac{\Delta V}{V} = 0.03$

73. $R = \frac{R_1 R_2}{R_1 + R_2} = \frac{6 \times 10}{6 + 10} = 3.75\ \Omega$

Further, $\frac{1}{R} = \frac{1}{R_1} + \frac{1}{R_2}$

Differentiating, we get

$$-\frac{dR}{R^2} = -\frac{dR_1}{R_1^2} - \frac{dR_2}{R_2^2}$$

$$\therefore \quad \Delta R = \left[\frac{\Delta R_1}{R_1^2} + \frac{\Delta R_2}{R_2^2}\right] R^2$$

or $$\frac{\Delta R}{R} \times 100 = \left[\frac{\Delta R_1}{R_1^2} + \frac{\Delta R_2}{R_2^2}\right] \times R \times 100$$

$$= \left[\frac{0.3}{(6)^2} + \frac{0.2}{(10)^2}\right] \times 3.75 \times 100$$

$$= 3.875\%$$

LEVEL 03 Objective Questions

74. Given, the main scale reading, MSR = 0

The circular scale reading, CSR = 52 divisions

Now, we shall determine the least count of the screw gauge,

$$\text{LC} = \frac{p}{n}$$

Here, p is the pitch of the screw,

n is the number of circular divisions in one complete revolution.

$$LC = \frac{1}{100} \text{ mm} \Rightarrow LC = 0.01 \text{ mm}$$

$$\Rightarrow LC = 0.001 \text{ cm}$$

Thus, the least count of the screw gauge is 0.001 cm.

Therefore, diameter of the wire of screw gauge,

$$D = \text{MSR} + (\text{CSR} \times \text{LC})$$

$$\Rightarrow D = 0 + (52 \times 0.001)$$

$$\Rightarrow D = 0.052 \text{ cm}$$

75. Given, fundamental physical quantities are force [F], acceleration [a] and time [T].

Now, we shall determine the dimensions of the energy.

Energy depends on force, acceleration and time as,

$$[E] = [F]^a [a]^b [T]^c$$

$$\Rightarrow [ML^2 T^{-2}] = [MLT^{-2}]^a [LT^{-2}]^b [T]^c$$

$$\Rightarrow [ML^2T^{-2}] = [M]^a [L]^{a+b} [T]^{-2a-2b+c}$$

Comparing the powers of M, L and T on both sides, we get

$$a = 1, a + b = 2$$

and $$-2a - 2b + c = -2$$

$$\Rightarrow 1 + b = 2 \Rightarrow b = 1,$$

$$\Rightarrow -2(1) - 2(1) + c = -2$$

$$\Rightarrow c = 2$$

The dimensions of the energy are $[F]^1[a]^1[T]^2$.

76. The dimensions of energy

$$[E] = [F] \cdot [d]$$

$$\Rightarrow [E] = [MLT^{-2}][L]$$

$$\Rightarrow [E] = [ML^2T^{-2}]$$

As we know that, the expression of gravitational force,

$$F = \frac{GM_1M_2}{r^2} \Rightarrow G = \frac{Fr^2}{M_1M_2}$$

$$\therefore [G] = \frac{[F][r^2]}{[M_1][M_2]} \Rightarrow [G] = \frac{[MLT^{-2}][L]^2}{[M][M]}$$

$$\Rightarrow [G] = [M^{-1}L^3T^{-2}]$$

The dimensions of

$$\frac{E}{G} = \frac{[ML^2T^{-2}]}{[M^{-1}L^3T^{-2}]} \Rightarrow \left[\frac{E}{G}\right] = [M^2L^{-1}T^0]$$

77. Mean time interval

$$\bar{T} = \frac{1.25 + 1.24 + 1.27 + 1.21 + 1.28}{5}$$

$$\Rightarrow = \frac{6.25}{5} = 1.25 \text{ s}$$

Mean absolute error,

$$\Delta\bar{T} = \frac{|\Delta T_1| + |\Delta T_2| + |\Delta T_3| + |\Delta T_4| + |\Delta T_5|}{5}$$

$$\Rightarrow = \frac{|1.25 - 1.25| + |1.25 - 1.24| + |1.25 - 1.27| + |1.25 - 1.21| + |1.25 - 1.28|}{5}$$

$$\Rightarrow = \frac{0 + 0.01 + 0.02 + 0.04 + 0.03}{5} = \frac{0.1}{5} = 0.02 \text{ s}$$

$$\therefore \text{Percentage relative error} = \frac{\Delta\bar{T}}{T} \times 100$$

$$= \frac{0.02}{1.25} \times 100 = 1.6\%$$

78. Given, least count = 0.01 mm

Number of divisions on circular scale = 50

Pitch of the screw gauge = least count × number of divisions on circular scale

$$= 0.01 \times 50 = 0.5 \text{ mm}$$

79. $\because$ Stress $= \dfrac{\text{Force}}{\text{Area}}$

$$\therefore \text{Dimensions of stress} = \frac{[MLT^{-2}]}{[L^2]} = [ML^{-1}T^{-2}]$$

80. The difference between 9.99 m and 0.0099 m is

$$= 9.99 - 0.0099 = 9.9801 \text{ m}$$

Taking significant figures into account, as both the values has two significant figures after decimal.

So, their difference will also have two significant figures after decimal, i.e. 9.98 m.

81. $1 \text{ min} = \dfrac{1}{60} \text{ degree} = \dfrac{1}{60} \times \dfrac{\pi}{180} \text{ rad}$

$$= 2.91 \times 10^{-4} \text{ rad}$$

82. The rate of heat flow through a conductor of length L and area of cross-section A is given by

$$\frac{dQ}{dt} = KA\frac{\Delta T}{L} \text{ J/s or watt}$$

where, K = coefficient of thermal conductivity and ΔT = change in temperature

$$\Rightarrow K = \frac{L}{A\Delta T}\frac{dQ}{dt}$$

$$\therefore \text{Unit of } K = \frac{\text{metre}}{(\text{metre})^2 \times \text{kelvin}} \times \text{watt} = \text{Wm}^{-1}\text{K}^{-1}$$

83. As it is given that n divisions of vernier scale coincide with $(n - 1)$ divisions of main scale, i.e.

$$n(\text{VSD}) = (n - 1)\text{MSD}$$

$$\Rightarrow 1\text{VSD} = \frac{(n-1)}{n}\text{MSD} \quad \text{...(i)}$$

The least count is the difference between one main scale division (MSD) and one vernier scale division (VSD).

$$\therefore \text{Least Count (LC)} = 1\text{MSD} - 1\text{VSD}$$

$$= 1\text{MSD} - \frac{(n-1)}{n}\text{MSD} \quad \text{[from Eq. (i)]}$$

$$= \left(1 - \frac{(n-1)}{n}\right)\text{MSD} = \frac{1}{n}\text{MSD}$$

Here, 1 MSD $= \frac{1}{n}$ cm

$\Rightarrow \quad LC = \frac{1}{n} \times \frac{1}{n}$ cm $= \frac{1}{n^2}$ cm

84. Given, least count of screw gauge, LC = 0.001 cm

Main scale reading,

$$MSR = 5 \text{ mm} = 0.5 \text{ cm}$$

Number of coinciding divisions on the circular scale, i.e. Vernier scale reading, VSR = 25

Here, zero error = −0.004 cm

Final reading obtained from the screw gauge is given as

$$= MSR + VSR \times LC - \text{ zero error}$$

Final reading from the screw gauge

$$= 0.5 + 25 \times 0.001 - (-0.004)$$
$$= 0.5 + 0.025 + 0.004$$
$$= 0.5 + 0.029$$
$$= 0.529 \text{ cm}$$

Thus, the diameter of the ball is 0.529 cm.

85. As force $F = \frac{e^2}{4\pi\varepsilon_0 r^2} \Rightarrow \frac{e^2}{4\pi\varepsilon_0} = r^2 \cdot F$

Putting dimensions of r and F, we get

$$\Rightarrow \left[\frac{e^2}{4\pi\varepsilon_0}\right] = [ML^3 T^{-2}] \quad \text{...(i)}$$

Also, force, $F = \frac{Gm^2}{r^2}$

$$\Rightarrow [G] = \frac{[MLT^{-2}][L^2]}{[M^2]}$$

$$\Rightarrow [G] = [M^{-1} L^3 T^{-2}] \quad \text{...(ii)}$$

and $\left[\frac{1}{c^2}\right] = \frac{1}{[L^2 T^{-2}]} = [L^{-2} T^2] \quad \text{...(iii)}$

Now, checking optionwise,

$$= \frac{1}{c^2}\left(\frac{Ge^2}{4\pi\varepsilon_0}\right)^{1/2} = [L^{-2} T^2][L^6 T^{-4}]^{1/2} = [L]$$

86. We know that,

Surface tension $(S) = \frac{\text{Force [F]}}{\text{Length [L]}}$

So, $[S] = \frac{[MLT^{-2}]}{[L]} = [ML^0 T^{-2}]$

Energy (E) = Force × Displacement

$\Rightarrow \quad [E] = [ML^2 T^{-2}]$

Velocity $(v) = \frac{\text{Displacement}}{\text{Time}}$

$\Rightarrow \quad [v] = [LT^{-1}]$

As, $\quad S \propto E^a v^b T^c$

where, a, b, c are constants.

From the principle of homogeneity,

$$[LHS] = [RHS]$$

$$\Rightarrow [ML^0T^{-2}] = [ML^2 T^{-2}]^a [LT^{-1}]^b [T]^c$$

$$\Rightarrow [ML^0 T^{-2}] = [M^a L^{2a+b} T^{-2a-b+c}]$$

Equating the power on both sides, we get

$$a = 1, 2a + b = 0, b = -2$$

$$\Rightarrow -2a - b + c = -2$$

$$\Rightarrow c = (2a + b) - 2$$

$$= 0 - 2 = -2$$

So $\quad [S] = [Ev^{-2} T^{-2}] = [Ev^{-2} T^{-2}]$

87. Key Concept According to principle of homogeneity of dimension, a physical quantity equation will be dimensionally correct, if the dimensions of all the terms occurring on both sides of the equations are same.

Given critical velocity of liquid flowing through a tube are expressed as

$$v_c \propto \eta^x \rho^y r^z$$

Coefficient of viscosity of liquid,

$$\eta = [ML^{-1}T^{-1}]$$

Density of liquid, $\rho = [ML^{-3}]$

Radius of a tube $r = [L]$

Critical velocity of liquid, $v_c = [ML^0T^{-1}]$

$$\Rightarrow [M^0L^1T^{-1}] = [ML^{-1}T^{-1}]^x [ML^{-3}]^y [L]^z$$

$$[M^0L^1T^{-1}] = [M^{x+y} L^{-x-3y+z} T^{-x}]$$

Comparing exponents of M, L and T, we get

$$x + y = 0, -x - 3y + z = 1, -x = -1$$

$$\Rightarrow z = -1, x = 1, y = -1$$

88. We know that,

$$F = ma \Rightarrow F = \frac{mv}{t} \Rightarrow m = \frac{Ft}{v}$$

$$[M] = \frac{[F][T]}{[v]} = [Fv^{-1} T]$$

89. As given, $P = \frac{a^3 b^2}{cd}$

$$\therefore \frac{\Delta P}{P} \times 100 = \left(\frac{3\Delta a}{a} + \frac{2\Delta b}{b} + \frac{\Delta c}{c} + \frac{\Delta d}{d}\right) \times 100$$

$$= 3\frac{\Delta a}{a} \times 100 + 2\frac{\Delta b}{b} \times 100 + \frac{\Delta c}{c} \times 100 + \frac{\Delta d}{d} \times 100$$

$$= 3 \times 1 + 2 \times 2 + 3 + 4$$
$$= 3 + 4 + 3 + 4 = 14\%$$

90. $(\mu_0\varepsilon_0)^{-1/2}$ is the expression for velocity of light.

As $\quad c = \frac{1}{\sqrt{\mu_0 \varepsilon_0}}$

So, dimension of $c = [LT^{-1}]$

Chapter 03

VECTORS

LEVEL 01 Objective Questions

1 Six vectors, **a** to **f** have the magnitudes and directions indicated in the figures. Which of the following statement is true?

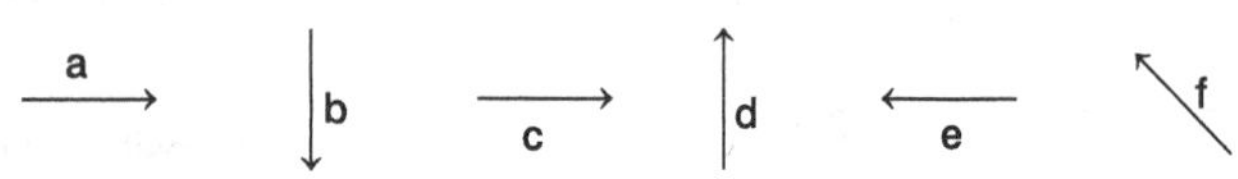

(a) $\mathbf{b}+\mathbf{c}=\mathbf{f}$ (b) $\mathbf{d}+\mathbf{c}=\mathbf{f}$ (c) $\mathbf{d}+\mathbf{e}=\mathbf{f}$ (d) $\mathbf{b}+\mathbf{e}=\mathbf{f}$

2 Two vectors **A** and **B** have components A_x, A_y, A_z and B_x, B_y, B_z, respectively. If $\mathbf{A}+\mathbf{B}=0$, then

(a) $A_x = B_x, A_y = -B_y, A_z = -B_z$ (b) $A_x = B_x, A_y = B_y, A_z = -B_z$
(c) $A_x = B_x, A_y = B_y, A_z = B_z$ (d) $A_x = -B_x, A_y = -B_y, A_z = -B_z$

3 In the following figure, **E** equals

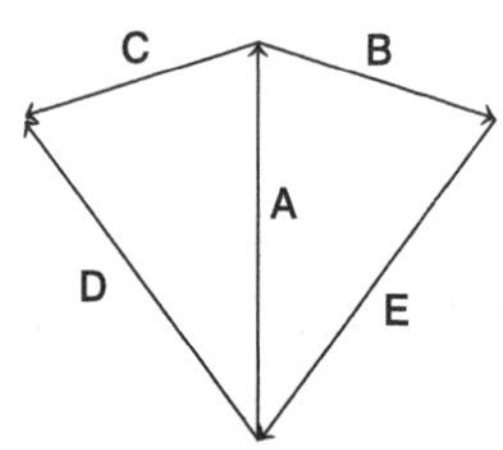

(a) **A** (b) **B** (c) $\mathbf{A}+\mathbf{B}$ (d) $-(\mathbf{A}+\mathbf{B})$

4 What is the angle between following vectors?

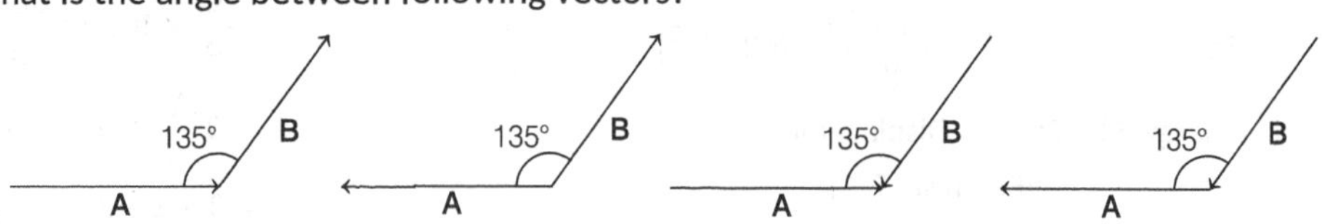

(a) 45°, 135°, 135°, 45° (b) 135°, 135°, 45°, 45°
(c) 135°, 45°, 45°, 135° (d) 135°, 45°, 45°, 45°

5 Two vectors (**P** and **Q**) of same magnitude are arranged in the following manner.

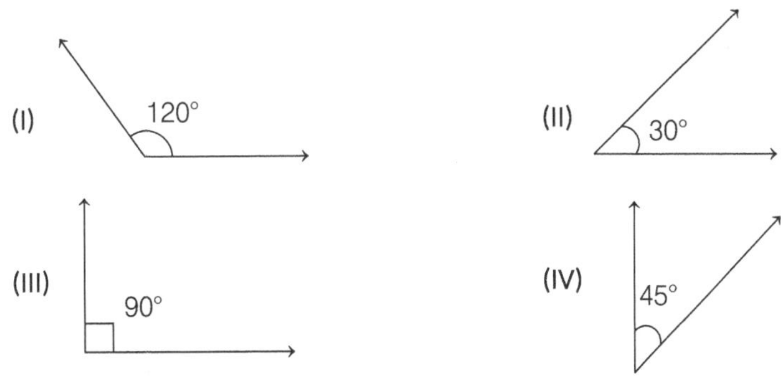

Which of the following are correct regarding magnitude of resultant?

(a) I = II = III = IV (b) I > II > III > IV (c) III > I > II > IV (d) II > IV > III > I

6 For figure, the correct relation is

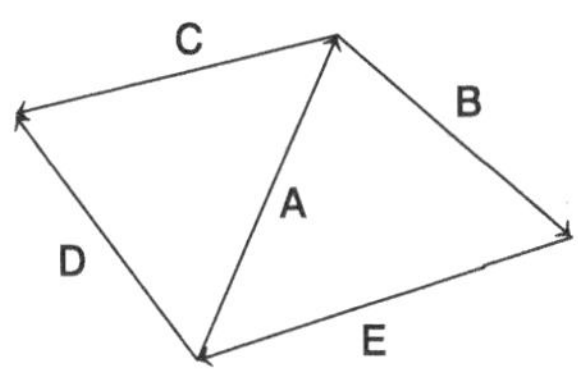

(a) $\mathbf{A}+\mathbf{B}+\mathbf{E}=0$ (b) $\mathbf{C}-\mathbf{D}=\mathbf{A}$ (c) $\mathbf{B}+\mathbf{E}-\mathbf{C}=\mathbf{D}$ (d) All of these

7 Three forces are acting on a particle as shown in the figure. To have the resultant force only along the y-direction, the magnitude of the minimum additional force needed is

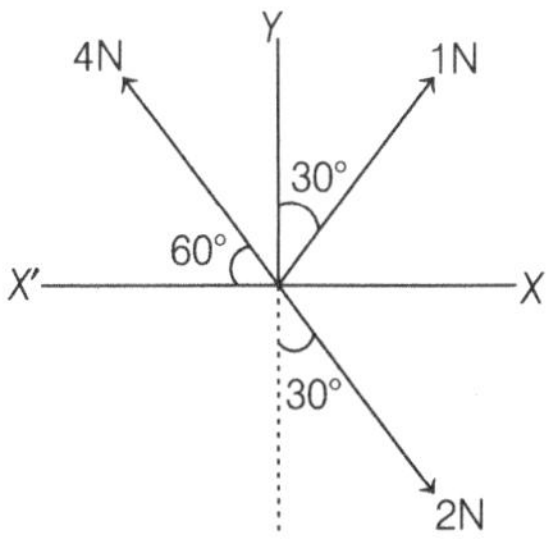

(a) 0.866 N
(b) 1.732 N
(c) 0.5 N
(d) 4 N

8 A particle is being acted upon by four forces of 30 N due East, 20 N due North, 50 N due West and 40 N due South. The resultant force will be

(a) $20\sqrt{2}$, 60° South to West (b) $20\sqrt{2}$, 45° South to West
(c) $20\sqrt{2}$, 45° North to East (d) $20\sqrt{2}$, 45° South to East

9 Two forces $\mathbf{F}_1$ and $\mathbf{F}_2$ are acting on a body. One force is double that of the other force and the resultant is equal to the greater force. Then, the angle between the two forces is

(a) $\cos^{-1}(1/2)$ (b) $\cos^{-1}(-1/2)$
(c) $\cos^{-1}(-1/4)$ (d) $\cos^{-1}(1/4)$

10 Find the magnitude of the resultant of shown forces.

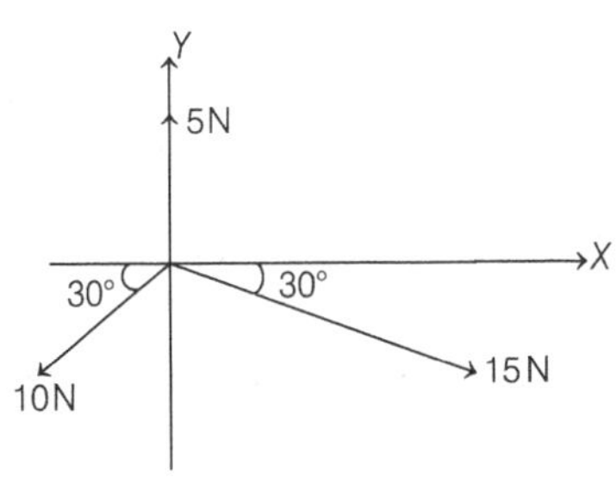

(a) $5\sqrt{3}$ N (b) $3\sqrt{7}$ N (c) $6.5\sqrt{3}$ N (d) $5\sqrt{7}$ N

11 The resultant of forces shown in the following figure is

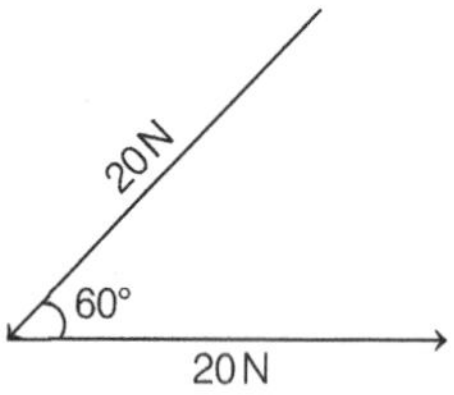

(a) 20 N (b) 10 N (c) $20\sqrt{3}$ N (d) 5 N

12 At what angle should the two force vectors $2F$ and $\sqrt{2}\,F$ act, so that the resultant force is $\sqrt{10}\,F$?

(a) 45° (b) 60° (c) 120° (d) 90°

13 100 coplanar forces each equal to 10 N act on a body. Each force makes angle $\frac{\pi}{50}$ with the preceding force. What is the resultant of the forces?

(a) 1000 N (b) 500 N
(c) 250 N (d) Zero

14 A particle moves along a path $ABCD$ as shown in the following figure. The magnitude of net displacement of the particle from position A to position D is

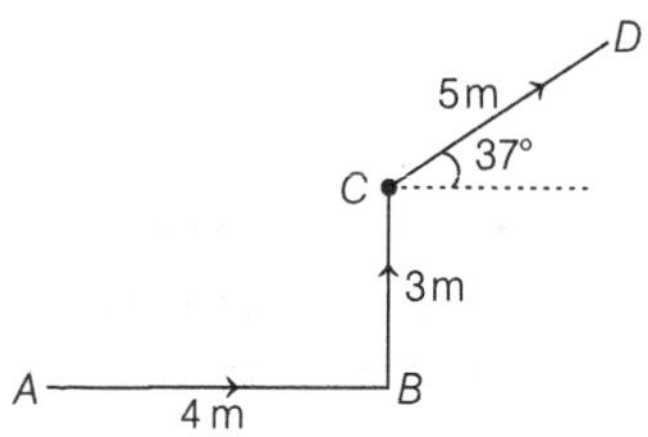

(a) 10 m (b) $5\sqrt{2}$ m (c) 9 m (d) $7\sqrt{2}$ m

15 Find the value of $|-\hat{\mathbf{i}}| + |\hat{\mathbf{j}}| + |-\hat{\mathbf{k}}|$.

(a) 1 (b) 2 (c) 3 (d) None of these

16 The angles which a vector $\hat{\mathbf{i}} + \hat{\mathbf{j}} + \sqrt{2}\hat{\mathbf{k}}$ makes with X, Y and Z-axes respectively are

(a) 60°, 60°, 60° (b) 45°, 45°, 45° (c) 60°, 60°, 45° (d) 45°, 45°, 60°

17 If $\mathbf{a} = 6\sqrt{3}\,\hat{\mathbf{i}} + 8\sqrt{3}\,\hat{\mathbf{j}} + 10\,\hat{\mathbf{k}}$, then what will be $\cos\gamma$? If γ is the angle made by a with positive Z-axis.

(a) 1 (b) $-\frac{1}{2}$ (c) $\frac{\sqrt{3}}{2}$ (d) $\frac{1}{2}$

18 If direction cosines of a vector are $\frac{1}{3}, \frac{1}{3}$ and $\frac{\sqrt{7}}{3}$ respectively, then write the vector in the form of $\hat{\mathbf{i}}, \hat{\mathbf{j}}$ and $\hat{\mathbf{k}}$. Magnitude of vector is 9.

(a) $3\hat{\mathbf{i}} + \hat{\mathbf{j}} + \sqrt{7}\,\hat{\mathbf{k}}$ (b) $\hat{\mathbf{i}} + \hat{\mathbf{j}} + \sqrt{7}\,\hat{\mathbf{k}}$ (c) $9\hat{\mathbf{i}} + 9\hat{\mathbf{j}} + \sqrt{7}\,\hat{\mathbf{k}}$ (d) $3\hat{\mathbf{i}} + 3\hat{\mathbf{j}} + 3\sqrt{7}\,\hat{\mathbf{k}}$

19 If $\mathbf{A} = 2\hat{\mathbf{i}} + 3\hat{\mathbf{j}} - \hat{\mathbf{k}}$ and $\mathbf{B} = -\hat{\mathbf{i}} + 3\hat{\mathbf{j}} + 4\hat{\mathbf{k}}$, then projection of $\mathbf{A}$ on $\mathbf{B}$ will be

(a) $\frac{3}{\sqrt{13}}$ (b) $\frac{3}{\sqrt{26}}$ (c) $\sqrt{\frac{3}{26}}$ (d) $\sqrt{\frac{3}{13}}$

20 The scalar product of two vectors is $2\sqrt{3}$ and the magnitude of their vector product is 2. The angle between them is

(a) 30° (b) 45°
(c) 60° (d) 90°

21 Given $\mathbf{A} = \hat{\mathbf{i}} + 2\hat{\mathbf{j}} - 3\hat{\mathbf{k}}$. When $\mathbf{B}$ is added to $\mathbf{A}$, we get unit vector along X-axis, then $\mathbf{B}$ is
(a) $-2\hat{\mathbf{j}} + 3\hat{\mathbf{k}}$ (b) $-\hat{\mathbf{i}} - 2\hat{\mathbf{j}}$ (c) $-\hat{\mathbf{i}} + 3\hat{\mathbf{k}}$ (d) $2\hat{\mathbf{j}} - 3\hat{\mathbf{k}}$

22 If $\mathbf{A} = 2\hat{\mathbf{i}} + \hat{\mathbf{j}} - \hat{\mathbf{k}}$, $\mathbf{B} = \hat{\mathbf{i}} + 2\hat{\mathbf{j}} + 3\hat{\mathbf{k}}$ and $\mathbf{C} = 6\hat{\mathbf{i}} - 2\hat{\mathbf{j}} - 6\hat{\mathbf{k}}$, then the angle between $(\mathbf{A} + \mathbf{B})$ and $\mathbf{C}$ will be
(a) 30° (b) 45° (c) 60° (d) 90°

23 Maximum and minimum value of two given vectors are 17 and 7, respectively. If these vectors acts perpendicular to each other, then magnitude of resultant will be
(a) 14 (b) 16 (c) 18 (d) 13

24 If the ratio of maximum and minimum magnitudes of the resultant of two vectors **a** and **b** is 3 : 1, then $|\mathbf{a}|$ is equal to
(a) $|\mathbf{a}|$ (b) $2|\mathbf{b}|$ (c) $3|\mathbf{b}|$ (d) $4|\mathbf{b}|$

25 If $\dfrac{|\mathbf{a} + \mathbf{b}|}{|\mathbf{a} - \mathbf{b}|} = 1$, the angle between **a** and **b** is
(a) 0° (b) 45° (c) 90° (d) 60°

26 **A** and **B** are two vectors in a plane at an angle of 60° with each other. **C** is another vector perpendicular to the plane containing vectors **A** and **B**. Which of the following relations is possible?
(a) $\mathbf{A} + \mathbf{B} = \mathbf{C}$ (b) $\mathbf{A} + \mathbf{C} = \mathbf{B}$ (c) $\mathbf{A} \times \mathbf{B} = \mathbf{C}$ (d) $\mathbf{A} \times \mathbf{C} = \mathbf{B}$

27 If $\mathbf{A} \times \mathbf{B} = 0$ and $\mathbf{A} \cdot \mathbf{B} = -AB$, then angle between **A** and **B** is
(a) zero (b) $\pi/4$ (c) $\pi/2$ (d) π

28 If $\mathbf{A} = 3\hat{\mathbf{i}} + 4\hat{\mathbf{j}}$ and $\mathbf{B} = 6\hat{\mathbf{i}} + 8\hat{\mathbf{j}}$, then which of the following is **not** true?
(a) $\mathbf{A} \times \mathbf{B} = 0$ (b) $\dfrac{A}{B} = \dfrac{1}{2}$ (c) $\mathbf{A} \cdot \mathbf{B} = 48$ (d) $A = 5$

29 A vector which must be added to the vectors $\hat{\mathbf{i}} - 3\hat{\mathbf{j}} + 2\hat{\mathbf{k}}$ and $3\hat{\mathbf{i}} + 6\hat{\mathbf{j}} - 7\hat{\mathbf{k}}$, so that the resultant vector is a unit vector along Y-axis is
(a) $4\hat{\mathbf{i}} + 2\hat{\mathbf{j}} + 5\hat{\mathbf{k}}$ (b) $-4\hat{\mathbf{i}} - 2\hat{\mathbf{j}} + 5\hat{\mathbf{k}}$ (c) $3\hat{\mathbf{i}} + 4\hat{\mathbf{j}} + 5\hat{\mathbf{k}}$ (d) Null vector

30 What is the angle between **A** and **B**, if **A** and **B** are the adjacent sides of a parallelogram drawn from a common point and the area of the parallelogram is $\dfrac{AB}{2}$?
(a) $\dfrac{\pi}{3}$ (b) $\dfrac{\pi}{6}$ (c) $\dfrac{\pi}{2}$ (d) π

31 The value of $(\mathbf{A} + \mathbf{B}) \times (\mathbf{A} - \mathbf{B})$ is
(a) 0 (b) $\mathbf{A}^2 - \mathbf{B}^2$ (c) $\mathbf{B} \times \mathbf{A}$ (d) $2(\mathbf{B} \times \mathbf{A})$

32 $|\mathbf{A} \times \mathbf{B}|^2 + |\mathbf{A} \cdot \mathbf{B}|^2 =$
(a) zero (b) $A^2 B^2$ (c) AB (d) $\sqrt{AB}$

33 The angle between the vectors **A** and **B** is θ. Then, value of $\mathbf{A} \cdot (\mathbf{B} \times \mathbf{A})$ is
(a) $\mathbf{A}^2 \mathbf{B}$ (b) zero (c) $\mathbf{A}^2 \mathbf{B} \sin\theta$ (d) $\mathbf{A}^2 \mathbf{B} \cos\theta$

34 If $(\mathbf{A} - \mathbf{B})$ is a unit vector along Y-axis and $\mathbf{A} = 3\hat{\mathbf{i}} + 2\hat{\mathbf{j}} + \hat{\mathbf{k}}$, then what is **B** ?
(a) $-3\hat{\mathbf{i}} - \hat{\mathbf{j}} - \hat{\mathbf{k}}$ (b) $3\hat{\mathbf{i}} + \hat{\mathbf{j}} + \hat{\mathbf{k}}$ (c) $2\hat{\mathbf{i}} + \hat{\mathbf{j}} + \hat{\mathbf{k}}$ (d) $3\hat{\mathbf{i}} - \hat{\mathbf{j}} + \hat{\mathbf{k}}$

35 The x and y-components of vector **A** have numerical values 6 and 6 respectively and that of $(\mathbf{A} + \mathbf{B})$ have numerical values 10 and 9. What is the magnitude of **B**?
(a) 2 (b) 3 (c) 4 (d) 5

LEVEL 02 Objective Questions

36 If $\mathbf{A} = 2\hat{\mathbf{i}} + p\hat{\mathbf{j}} + q\hat{\mathbf{k}}, \mathbf{B} = 5\hat{\mathbf{i}} + 7\hat{\mathbf{j}} + 3\hat{\mathbf{k}}$ and $\mathbf{A} \parallel \mathbf{B}$, then values of p and q are

(a) $\frac{14}{5}$ and $\frac{6}{5}$
(b) $\frac{14}{3}$ and $\frac{6}{5}$
(c) $\frac{6}{5}$ and $\frac{1}{3}$
(d) $\frac{3}{4}$ and $\frac{1}{4}$

37 Five forces 2 N, $\sqrt{3}$ N, 5 N, $\sqrt{3}$ N and 2 N respectively acts simultaneously at a particle P as shown in the following figure.

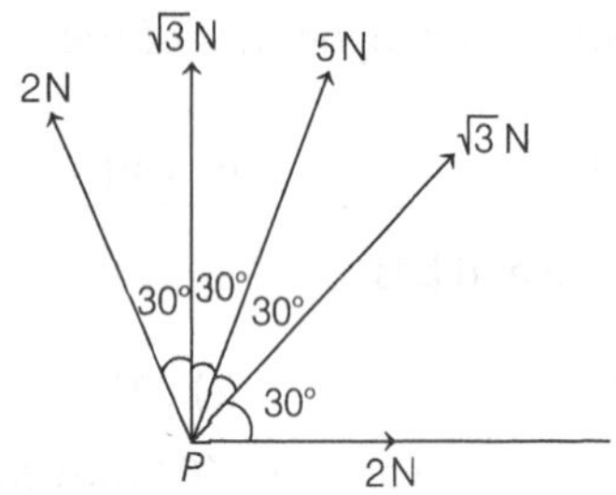

The resultant force on the particles P is

(a) 10 N making an angle 60° with X-axis
(b) 10 N making an angle 60° with Y-axis
(c) 20 N along Y-axis
(d) None of these

38 Vector **A** is of length 2 cm and is 60° above the X-axis in the first quadrant. Vector **B** is of length 2 cm and 60° below the X-axis in the fourth quadrant. The sum $\mathbf{A} + \mathbf{B}$ is a vector of magnitude

(a) 2 along $+Y$-axis
(b) 2 along $+X$-axis
(c) 1 along $-X$-axis
(d) 2 along $-X$-axis

39 Two vectors **A** and **B** have equal magnitudes. If magnitude of $\mathbf{A} + \mathbf{B}$ is equal to n times the magnitude of $\mathbf{A} - \mathbf{B}$, then the angle between **A** and **B** is

(a) $\cos^{-1}\left(\frac{n-1}{n+1}\right)$
(b) $\cos^{-1}\left(\frac{n^2-1}{n^2+1}\right)$
(c) $\sin^{-1}\left(\frac{n-1}{n+1}\right)$
(d) $\sin^{-1}\left(\frac{n^2-1}{n^2+1}\right)$

40 Two vectors **A** and **B** of unknown magnitudes are as shown alongside, respectively.
Then, $(\mathbf{A} - \mathbf{B})$ cannot be

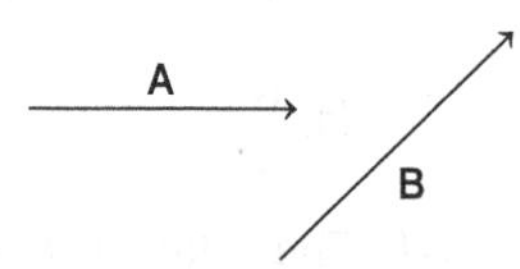

(a) 
(b) 
(c) 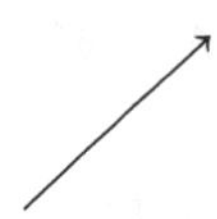
(d) 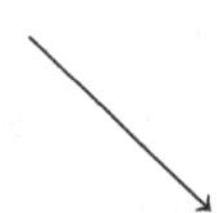

41 Two forces $\mathbf{F}_1$ and $\mathbf{F}_2$ of equal magnitudes produce a resultant of 4 N. When $\mathbf{F}_2$ is reversed in direction, the magnitude of resultant will be $2\sqrt{5}$ N. Find magnitude of $\mathbf{F}_1$.

(a) 2 N
(b) 3 N
(c) 4 N
(d) 5 N

42 Let the angle between two non-zero vectors **A** and **B** be 120° and its resultant be **C**.

(a) C must be equal to $|\mathbf{A} - \mathbf{B}|$
(b) C must be less than $|\mathbf{A} - \mathbf{B}|$
(c) C must be greater than $|\mathbf{A} - \mathbf{B}|$
(d) C may be equal to $|\mathbf{A} - \mathbf{B}|$

43 Vectors **A**, **B**, **C**, **D**, **E** and **F** with magnitudes 3, 2, 3, 5, 6, 4 respectively are shown in the following figure.

Find $\left|\mathbf{A}+\frac{\mathbf{B}}{2}+\frac{\mathbf{C}}{3}+\frac{\mathbf{D}}{5}+\frac{\mathbf{E}}{6}+\frac{\mathbf{F}}{4}\right|$.

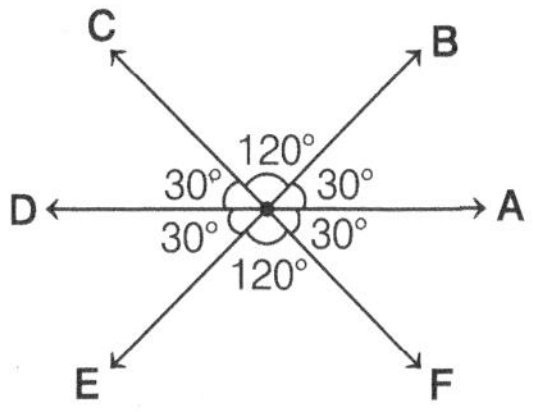

(a) 1 (b) 2 (c) 3 (d) 4

44 For the given figure, if $|\mathbf{A}| = 5, |\mathbf{B}| = 10$ and $|\mathbf{C}| = 4\sqrt{2}$, then $\frac{\mathbf{A}}{5}+\frac{\mathbf{B}}{10}+\frac{\mathbf{C}}{2}$ will be in the direction

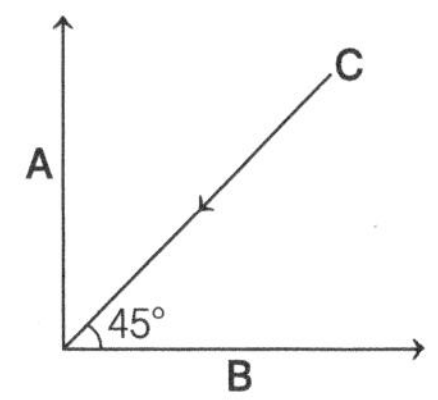

(a)  (b)  (c)  (d) 

45 A string is pulled with a force $F = 100$ N along AB as shown in the following figure. Write F as a factor.

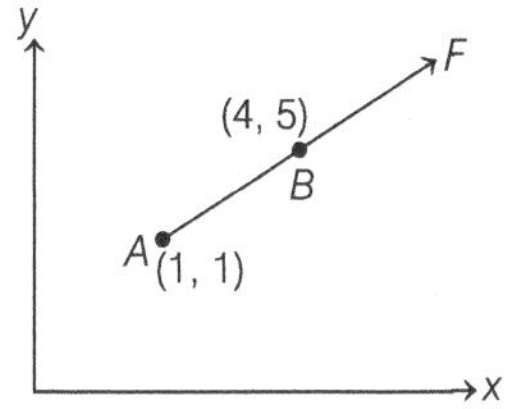

(a) $(3\hat{\mathbf{i}}+4\hat{\mathbf{j}})$ N (b) $(60\hat{\mathbf{i}}+80\hat{\mathbf{j}})$ N (c) $(80\hat{\mathbf{i}}+60\hat{\mathbf{j}})$ N (d) $(40\hat{\mathbf{i}}+50\hat{\mathbf{j}})$ N

46 Three vectors **a**, **b** and **c** bear the following relations $\mathbf{a}-\mathbf{b}+\mathbf{c}=0, b=\sqrt{a^2+c^2}$ and $a=\sqrt{3}c$. Angle between vectors ....

(a) **a** and **b** is 30° (b) **a** and **b** is 60°
(c) **b** and **c** is 30° (d) **a** and **c** is 60°

47 The vector component of vector $\mathbf{A}=3\hat{\mathbf{i}}+4\hat{\mathbf{j}}+5\hat{\mathbf{k}}$ along vector $\mathbf{B}=\hat{\mathbf{i}}+\hat{\mathbf{j}}+\hat{\mathbf{k}}$ is

(a) $2\hat{\mathbf{i}}+2\hat{\mathbf{j}}+2\hat{\mathbf{k}}$ (b) $3\hat{\mathbf{i}}+3\hat{\mathbf{j}}+3\hat{\mathbf{k}}$
(c) $\frac{4}{\sqrt{3}}(\hat{\mathbf{i}}+\hat{\mathbf{j}}+\hat{\mathbf{k}})$ (d) $5\hat{\mathbf{i}}+5\hat{\mathbf{j}}+5\hat{\mathbf{k}}$

48 Given that $\mathbf{A}+\mathbf{B}=\mathbf{C}$ and that **C** is perpendicular to **A**. Further if $|\mathbf{A}|=|\mathbf{C}|$, then what is the angle between **A** and **B**?

(a) $\frac{\pi}{4}$ rad (b) $\frac{\pi}{2}$ rad (c) $\frac{3\pi}{4}$ rad (d) π rad

49 If $\mathbf{a}_1$ and $\mathbf{a}_2$ are two non-collinear unit vectors and if $|\mathbf{a}_1 + \mathbf{a}_2| = \sqrt{3}$, then the value of $(\mathbf{a}_1 - \mathbf{a}_2)\cdot(2\mathbf{a}_1 + \mathbf{a}_2)$ is

(a) 2 (b) $\frac{3}{2}$ (c) $\frac{1}{2}$ (d) 1

50 If **a** and **b** are two unit vectors such that $\mathbf{a} + 2\mathbf{b}$ and $5\mathbf{a} - 4\mathbf{b}$ are perpendicular to each other, then the angle between **a** and **b** is

(a) 45° (b) 60°
(c) $\cos^{-1}\left(\frac{1}{3}\right)$ (d) $\cos^{-1}\left(\frac{2}{7}\right)$

51 If $\mathbf{A} = 2\hat{\mathbf{i}} + \sqrt{7}\,\hat{\mathbf{j}}$ and $\mathbf{B} = 5\hat{\mathbf{i}} + \sqrt{7}\,\hat{\mathbf{j}} - 3\hat{\mathbf{k}}$, then the vector whose magnitude is equal to $\mathbf{A}\cdot\mathbf{B}$ and parallel to $\mathbf{B} - \mathbf{A}$ is

(a) $\frac{17}{\sqrt{2}}(\hat{\mathbf{k}} - \hat{\mathbf{j}})$ (b) $\frac{17}{\sqrt{2}}(\hat{\mathbf{i}} - \hat{\mathbf{k}})$ (c) $3\hat{\mathbf{i}} - 3\hat{\mathbf{k}}$ (d) $3\hat{\mathbf{k}} - 3\hat{\mathbf{i}}$

52 If **a** and **b** are non-collinear unit vectors and $|\mathbf{a} + \mathbf{b}| = \sqrt{3}$, then the value of $(2\mathbf{a} - 5\mathbf{b})\cdot(3\mathbf{a} + \mathbf{b})$ is

(a) $\frac{41}{2}$ (b) $\frac{11}{2}$ (c) $-\frac{11}{2}$ (d) $-\frac{41}{2}$

53 **A** and **B** are two vectors given by $\mathbf{A} = 2\hat{\mathbf{i}} + 3\hat{\mathbf{j}}$ and $\mathbf{B} = \hat{\mathbf{i}} + \hat{\mathbf{j}}$. Then, magnitude of the component of **A** along **B** is

(a) $\frac{5}{\sqrt{2}}$ (b) $\frac{3}{\sqrt{2}}$ (c) $\frac{7}{\sqrt{2}}$ (d) $\frac{1}{\sqrt{2}}$

54 The angle between following two vectors

$$\mathbf{R} = -\hat{\mathbf{i}} + \frac{\hat{\mathbf{j}}}{3} + \hat{\mathbf{k}} \quad \text{and} \quad \mathbf{S} = x\hat{\mathbf{i}} + 3\hat{\mathbf{j}} + (x-1)\hat{\mathbf{k}}$$

(a) is obtuse angle (b) is acute angle (c) is right angle (d) depend on X

55 Two vectors $\mathbf{a} = 3\hat{\mathbf{i}} + 8\hat{\mathbf{j}} - 2\hat{\mathbf{k}}$ and $\mathbf{b} = 6\hat{\mathbf{i}} + 16\hat{\mathbf{j}} + x\hat{\mathbf{k}}$ are such that the component of **b** perpendicular to **a** is zero. Then, the value of x will be

(a) 8 (b) -4 (c) 4 (d) -8

56 If **A** and **B** are two non-zero vectors such that $|\mathbf{A} + \mathbf{B}| = \frac{|\mathbf{A} - \mathbf{B}|}{2}$ and $|\mathbf{A}| = 2|\mathbf{B}|$, then the angle between **A** and **B** is

(a) 37° (b) 53° (c) $\cos^{-1}(-3/4)$ (d) $\cos^{-1}(-4/3)$

57 Given that $\mathbf{A} + \mathbf{B} = \mathbf{R}$ and $\mathbf{A} + 2\mathbf{B}$ is perpendicular to **A**, then

(a) $2B = R$ (b) $B = 2R$
(c) $B = R$ (d) $B^2 = 2R^2$

58 The forces **P** and **Q** have a resultant perpendicular to **P**. The angle between the forces is

(a) $\tan^{-1}(-\mathbf{P}/\mathbf{Q})$ (b) $\tan^{-1}(\mathbf{P}/\mathbf{Q})$
(c) $\sin^{-1}(\mathbf{P}/\mathbf{Q})$ (d) $\cos^{-1}(-\mathbf{P}/\mathbf{Q})$

59 Consider three vectors $\mathbf{A} = \hat{\mathbf{i}} + \hat{\mathbf{j}} - 2\hat{\mathbf{k}}$, $\mathbf{B} = \hat{\mathbf{i}} - \hat{\mathbf{j}} + \hat{\mathbf{k}}$ and $\mathbf{C} = 2\hat{\mathbf{i}} - 3\hat{\mathbf{j}} + 4\hat{\mathbf{k}}$. A vector **X** of the form $\alpha\mathbf{A} + \beta\mathbf{B}$ (α and β are numbers) is perpendicular to **C**. The ratio of α and β is

(a) 1 : 1 (b) 2 : 1
(c) –1 : 1 (d) 3 : 1

60 If vectors $\mathbf{A} = \hat{\mathbf{i}} + 2\hat{\mathbf{j}} + 4\hat{\mathbf{k}}$ and $\mathbf{B} = 5\hat{\mathbf{i}}$ represent the two sides of a triangle, then the third side of the triangle has length equal to

(a) $\sqrt{56}$ (b) $\sqrt{21}$
(c) 5 (d) 6

61 If in the following figure, $\mathbf{AC} = \hat{\mathbf{i}} + 2\hat{\mathbf{j}} + 4\hat{\mathbf{k}}$ and $\mathbf{BD} = \hat{\mathbf{i}} - 3\hat{\mathbf{j}} + \hat{\mathbf{k}}$, then **BC** is

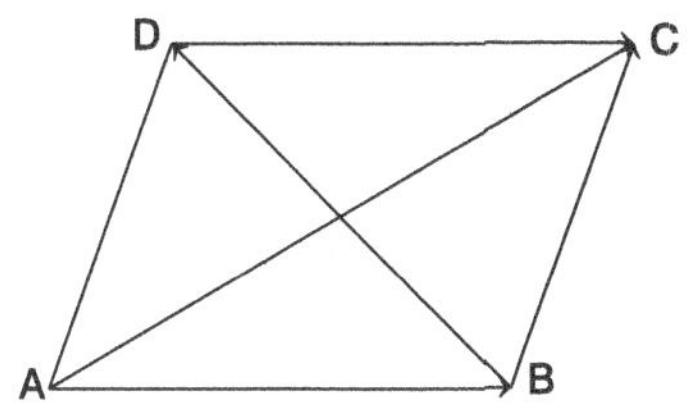

(a) $\frac{3}{2}\hat{\mathbf{i}} - \frac{1}{2}\hat{\mathbf{j}} + 5\hat{\mathbf{k}}$ (b) $\hat{\mathbf{i}} - \frac{1}{2}\hat{\mathbf{j}} + \frac{5}{2}\hat{\mathbf{k}}$ (c) $2\hat{\mathbf{i}} - \hat{\mathbf{j}} + 5\hat{\mathbf{k}}$ (d) $\frac{3}{2}\hat{\mathbf{i}} - 2\hat{\mathbf{j}} + 3\hat{\mathbf{k}}$

62 If diagonals of a parallelogram are $(\hat{\mathbf{i}} - \hat{\mathbf{j}})$ and $(\hat{\mathbf{i}} + \hat{\mathbf{j}})$, then area of the parallelogram will be

(a) 0.5 unit (b) 1 unit (c) 2 unit (d) 4 unit

63 **A**, **B** and **C** are three coplanar vectors, then

(a) $(\mathbf{A} \cdot \mathbf{B}) \times \mathbf{C} = 0$ (b) $(\mathbf{A} \times \mathbf{B}) \cdot \mathbf{C} = 0$ (c) $(\mathbf{A} \times \mathbf{B}) \times \mathbf{C} = 0$ (d) $(\mathbf{A} \cdot \mathbf{B}) \cdot \mathbf{C} = 0$

64 For two unit vectors $\hat{\mathbf{a}}$ and $\hat{\mathbf{b}}$; If $|\hat{\mathbf{a}} \cdot \hat{\mathbf{b}}| = \frac{1}{2}$, then $|\hat{\mathbf{a}} - \hat{\mathbf{b}}|$ may be

(a) 1 (b) $\frac{\sqrt{3}}{2}$ (c) Both (a) and (b) (d) None of these

65 A parallelogram is formed with **a** and **b** as the adjacent sides. Let $\mathbf{d}_1$ and $\mathbf{d}_2$ be the diagonals of the parallelogram, then $a^2 + b^2 = ?$

(a) $\frac{d_1^2 + d_2^2}{2}$ (b) $\frac{d_1^2 - d_2^2}{2}$ (c) $d_1^2 + d_2^2$ (d) $d_1^2 - d_2^2$

LEVEL 03 Previous Years Questions

66 If the magnitude of sum of two vectors is equal to the magnitude of difference of the two vectors, the angle between these vectors is **[NEET 2016, CBSE AIPMT 1991]**

(a) 90° (b) 45° (c) 180° (d) 0°

67 If vectors $\mathbf{A} = \cos \omega t\ \hat{\mathbf{i}} + \sin \omega t\ \hat{\mathbf{j}}$ and $\mathbf{B} = \cos \frac{\omega t}{2}\hat{\mathbf{i}} + \sin\frac{\omega t}{2}\hat{\mathbf{j}}$ are functions of time, then the value of t at which they are orthogonal to each other, is **[CBSE AIPMT 2015]**

(a) $t = \frac{\pi}{4\omega}$ (b) $t = \frac{\pi}{2\omega}$ (c) $t = \frac{\pi}{\omega}$ (d) $t = 0$

68 **A** and **B** are two vectors and θ is the angle between them. If $|\mathbf{A} \times \mathbf{B}| = \sqrt{3}\,(\mathbf{A} \cdot \mathbf{B})$, then the value of θ is **[CBSE AIPMT 2007]**

(a) 60° (b) 45° (c) 30° (d) 90°

69 If a vector $2\hat{\mathbf{i}} + 3\hat{\mathbf{j}} + 8\hat{\mathbf{k}}$ is perpendicular to the vector $4\hat{\mathbf{j}} - 4\hat{\mathbf{i}} + \alpha\hat{\mathbf{k}}$, then the value of α is **[CBSE AIPMT 2005]**

(a) –1 (b) $\frac{1}{2}$ (c) $-\frac{1}{2}$ (d) 1

70 If $|\mathbf{A} \times \mathbf{B}| = \sqrt{3}\,\mathbf{A} \cdot \mathbf{B}$, then the value of $|\mathbf{A} + \mathbf{B}|$ is **[CBSE AIPMT 2004]**

(a) $(A^2 + B^2 + AB)^{1/2}$ (b) $\left(A^2 + B^2 + \frac{AB}{\sqrt{3}}\right)^{1/2}$

(c) $A + B$ (d) $(A^2 + B^2 + \sqrt{3}\,AB)^{1/2}$

71 The vector sum of two forces is perpendicular to their vector differences. In that case, the forces **[CBSE AIPMT 2003]**

(a) are not equal to each other in magnitude
(b) cannot be predicted
(c) are equal to each other
(d) are equal to each other in magnitude

72 If a unit vector is represented by $0.5\hat{\mathbf{i}} + 0.8\hat{\mathbf{j}} + c\hat{\mathbf{k}}$, then the value of c is **[CBSE AIPMT 1999]**

(a) 1
(b) $\sqrt{0.11}$
(c) $\sqrt{0.01}$
(d) 0.39

73 Which of the following is not a vector quantity? **[CBSE AIPMT 1995]**

(a) Speed
(b) Velocity
(c) Torque
(d) Displacement

74 The angle between the two vectors $\mathbf{A} = 3\hat{\mathbf{i}} + 4\hat{\mathbf{j}} + 5\hat{\mathbf{k}}$ and $\mathbf{B} = 3\hat{\mathbf{i}} + 4\hat{\mathbf{j}} - 5\hat{\mathbf{k}}$ will be **[CBSE AIPMT 1994]**

(a) 0°
(b) 45°
(c) 90°
(d) 180°

75 If the magnitude of sum of two vectors is equal to the magnitude of difference of the two vectors, the angle between these vectors is **[NEET 2016, CBSE AIPMT 1991]**

(a) 90°
(b) 45°
(c) 180°
(d) 0°

76 The resultant of $\mathbf{A} \times \mathbf{0}$ will be equal to **[CBSE AIPMT 1992]**

(a) zero
(b) A
(c) zero vector
(d) unit vector

77 The angle between $\mathbf{A}$ and $\mathbf{B}$ is θ. The value of the triple product $\mathbf{A} \cdot (\mathbf{B} \times \mathbf{A})$ is **[CBSE AIPMT 1989]**

(a) A^2B
(b) zero
(c) $A^2B\sin\theta$
(d) $A^2B\cos\theta$

78 The magnitudes of vectors **A**, **B** and **C** are 3, 4 and 5 units respectively. If $\mathbf{A} + \mathbf{B} = \mathbf{C}$, the angle between **A** and **B** is **[CBSE AIPMT 1988]**

(a) $\frac{\pi}{2}$
(b) $\cos^{-1}(0.6)$
(c) $\tan^{-1}\left(\frac{7}{5}\right)$
(d) $\frac{\pi}{4}$

ANSWER KEY

Level 1 Objective Questions

1 (c)	**2** (d)	**3** (d)	**4** (a)	**5** (d)	**6** (a)	**7** (c)	**8** (b)	**9** (c)	**10** (a)
11 (a)	**12** (a)	**13** (d)	**14** (a)	**15** (c)	**16** (c)	**17** (d)	**18** (d)	**19** (b)	**20** (a)
21 (a)	**22** (d)	**23** (d)	**24** (b)	**25** (c)	**26** (c)	**27** (a)	**28** (c)	**29** (b)	**30** (b)
31 (d)	**32** (b)	**33** (b)	**34** (b)	**35** (d)					

Level 2 Objective Questions

36 (a)	**37** (b)	**38** (b)	**39** (b)	**40** (c)	**41** (b)	**42** (b)	**43** (b)	**44** (a)	**45** (b)
46 (a)	**47** (c)	**48** (c)	**49** (c)	**50** (b)	**51** (b)	**52** (c)	**53** (a)	**54** (c)	**55** (b)
56 (c)	**57** (c)	**58** (d)	**59** (a)	**60** (d)	**61** (b)	**62** (b)	**63** (b)	**64** (c)	**65** (a)

Level 3 Previous Years Questions

66 (a)	**67** (c)	**68** (a)	**69** (c)	**70** (a)	**71** (d)	**72** (b)	**73** (a)	**74** (c)	**75** (a)
76 (c)	**77** (b)	**78** (a)							

Solutions

LEVEL 01 Objective Questions

1. The **f** vector can be the resultant of **d** and **e** vectors.
   Therefore, $\mathbf{d} + \mathbf{e} = \mathbf{f}$

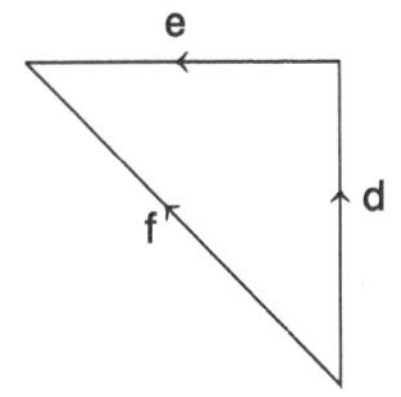

2. Given, $\mathbf{A} + \mathbf{B} = 0 \Rightarrow \mathbf{A} = -\mathbf{B}$
   Therefore, $A_x = -B_x$, $A_y = -B_y$ and $A_z = -B_z$
3. We know, $\mathbf{A} + \mathbf{B} + \mathbf{E} = 0$
   $\Rightarrow \quad \mathbf{E} = -(\mathbf{A} + \mathbf{B})$
4. No solution is required.
5. We know that, $R = \sqrt{A^2 + B^2 + 2AB\cos\theta}$
   R will be maximum when $\theta = 30°$ and minimum for $\theta = 120°$.
   In 45° and 30°, R will be more for 30°.
6. $\mathbf{A} + \mathbf{B} + \mathbf{E}$ makes a closed regular polygon, hence, $\mathbf{A} + \mathbf{B} + \mathbf{E} = 0$.
7. The component of 1 N and 2 N forces along X - axis
   $$= 1\cos 60° + 2\sin 30°$$
   $$= \frac{1}{2} + 1 = 1.5$$
   The component of 4 N along X - axis
   $$= -4\sin 30° = -4 \times \frac{1}{2} = -2\text{ N}$$
   Therefore, the net force along X - axis,
   $$F_x = -2 + 1.5 = -0.5\text{ N}$$
   If a force of 0.5 N is applied along X - axis, the resultant force become zero and the resultant force will be along Y - axis.
8. In horizontal, the net force is 20 N and in vertical, it is 20 N.

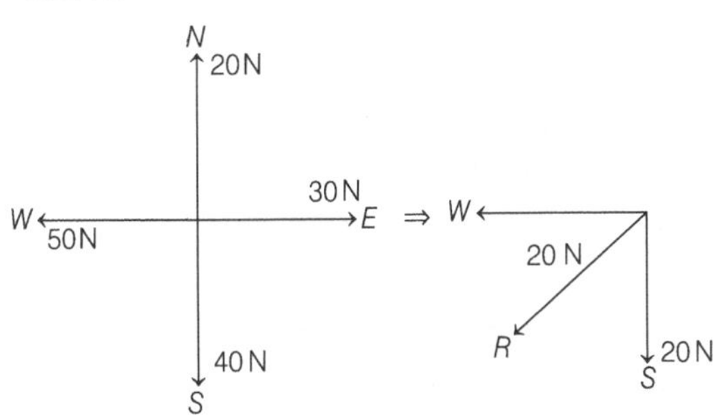

The resultant force, $F_{res} = \sqrt{(20)^2 + (20)^2}$
$= 20\sqrt{2}$ in South-West direction.

9. Given, $\mathbf{F}_2 = 2\mathbf{F}_1$
   Resultant $= \mathbf{F}_2$
   Therefore, $\mathbf{F}_2 = \sqrt{\mathbf{F}_1^2 + \mathbf{F}_2^2 + 2\mathbf{F}_1\mathbf{F}_2\cos\theta}$
   $$2\mathbf{F}_1 = \sqrt{\mathbf{F}_1^2 + (2\mathbf{F}_1)^2 + 2\mathbf{F}_1(2\mathbf{F}_1)\cos\theta}$$
   Squaring on both sides,
   $$4\mathbf{F}_1^2 = \mathbf{F}_1^2 + 4\mathbf{F}_1^2 + 4\mathbf{F}_1^2\cos\theta$$
   $$-\mathbf{F}_1^2 = 4\mathbf{F}_1^2\cos\theta$$
   $$-\frac{1}{4} = \cos\theta$$
   $$\Rightarrow \quad \theta = \cos^{-1}\left(-\frac{1}{4}\right)$$
10.

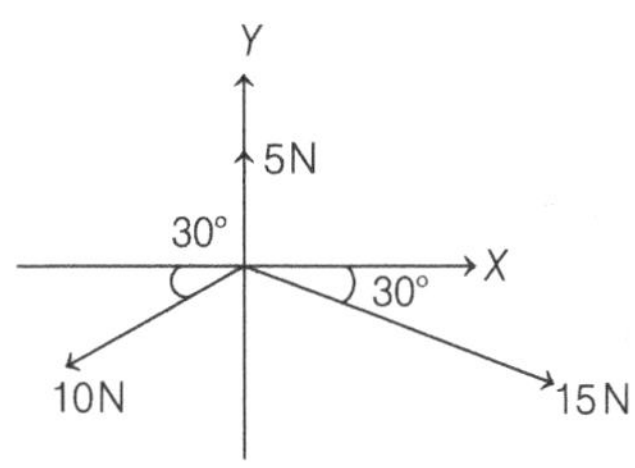

$$F_X = 15\cos 30° - 10\cos 30°$$
$$= 2.5\sqrt{3}\text{ N}$$
$$F_Y = 5 - 10\sin 30° - 15\sin 30°$$
$$= -7.5\text{ N}$$
$$\therefore \quad F = \sqrt{F_X^2 + F_Y^2} = 5\sqrt{3}\text{ N}$$

11. $\theta = 120°$
    $$R = \sqrt{20^2 + 20^2 + 2 \times 20 \times 20 \times \cos 120°}$$
    $$= \sqrt{(20)^2} = 20\text{ N} \qquad \left(\because \cos 120° = -\frac{1}{2}\right)$$
12. Let θ be the angle between the two vectors.
    $$(\sqrt{10})^2 = (2)^2 + (\sqrt{2})^2 + (2) \times 2 \times \sqrt{2}\cos\theta$$
    $$10 = 4 + 2 + 4\sqrt{2}\cos\theta$$
    $$\cos\theta = \frac{1}{\sqrt{2}}$$
    $$\Rightarrow \quad \theta = 45°$$
13. Total angle $= 100 \times \frac{\pi}{50} = 2\pi$
    Since, each force is opposed by an equal force.
    Therefore, the net resultant will be zero.

14. $s = \sqrt{x^2 + y^2} = \sqrt{(4 + 5\cos 37°)^2 + (3 + 5\sin 37°)^2}$
$= 10$ m

15. Given, $|-\hat{\mathbf{i}}| + |\hat{\mathbf{j}}| + |-\hat{\mathbf{k}}| = 1 + 1 + 1 = 3$

16. Given, $\mathbf{R} = \hat{\mathbf{i}} + \hat{\mathbf{j}} + \sqrt{2}\,\hat{\mathbf{k}}$

Comparing with the vector $R = R_x\hat{\mathbf{i}} + R_y\hat{\mathbf{j}} + R_z\hat{\mathbf{k}}$

$\therefore \quad R_x = 1,\ R_y = 1,\ R_z = \sqrt{2}$

$\Rightarrow \quad R = \sqrt{R_x^2 + R_y^2 + R_z^2} = 2$

$$\cos\alpha = \frac{R_x}{R} = \frac{1}{2} \Rightarrow \alpha = 60°$$

$$\cos\beta = \frac{R_y}{R} = \frac{1}{2} \Rightarrow \beta = 60°$$

and $\quad \cos\gamma = \frac{R_z}{R} = \frac{1}{\sqrt{2}} \Rightarrow \gamma = 45°$

Therefore, the angles with X, Y and Z - axes are 60°, 60° and 45°.

17. Given, $\mathbf{a} = 6\sqrt{3}\,\hat{\mathbf{i}} + 8\sqrt{3}\,\hat{\mathbf{j}} + 10\,\hat{\mathbf{k}}$

$$|\mathbf{a}| = \sqrt{(6\sqrt{3})^2 + (8\sqrt{3})^2 + (10)^2} = 20$$

$\because \quad \mathbf{a} = a_x\hat{\mathbf{i}} + a_y\hat{\mathbf{j}} + a_z\hat{\mathbf{k}}$
$= 6\sqrt{3}\hat{\mathbf{i}} + 8\sqrt{3}\,\hat{\mathbf{j}} + 10\hat{\mathbf{k}}$

$\therefore \quad \cos\gamma = \frac{a_z}{a} = \frac{10}{20} = \frac{1}{2}$

18. $A_x = 9\cos\alpha = 9 \times \frac{1}{3} = 3$

$A_y = 9\cos\beta = 9 \times \frac{1}{3} = 3$

$A_z = 9\cos\gamma = 9 \times \frac{\sqrt{7}}{3} = 3\sqrt{7}$

$\therefore \quad \mathbf{A} = A_x\,\hat{\mathbf{i}} + A_y\,\hat{\mathbf{j}} + A_z\hat{\mathbf{k}} = 3\,\hat{\mathbf{i}} + 3\,\hat{\mathbf{j}} + 3\sqrt{7}\,\hat{\mathbf{k}}$

19. We know that, the projection **A** on **B**

$$= \frac{\mathbf{A}\cdot\mathbf{B}}{|\mathbf{B}|} = \frac{(2\hat{\mathbf{i}} + 3\hat{\mathbf{j}} - \hat{\mathbf{k}})\cdot(-\hat{\mathbf{i}} + 3\hat{\mathbf{j}} + 4\hat{\mathbf{k}})}{|(-\hat{\mathbf{i}} + 3\hat{\mathbf{j}} + 4\hat{\mathbf{k}})|} = \frac{3}{\sqrt{26}}$$

20. Let θ be the angle between A and B.

$|\mathbf{A}\cdot\mathbf{B}| = AB\cos\theta$ and $|\mathbf{A}\times\mathbf{B}| = AB\sin\theta$

$$\tan\theta = \frac{|\mathbf{A}\times\mathbf{B}|}{|\mathbf{A}\cdot\mathbf{B}|} = \frac{2}{2\sqrt{3}} = \frac{1}{\sqrt{3}}$$

$\theta = 30°$

21. Given, $\mathbf{A} = \hat{\mathbf{i}} + 2\hat{\mathbf{j}} - 3\hat{\mathbf{k}}$

If we add a vector **B** to **A**, then we get

$\mathbf{A} + \mathbf{B} = \hat{\mathbf{i}} \Rightarrow \mathbf{B} = \hat{\mathbf{i}} - \mathbf{A}$
$= \hat{\mathbf{i}} - (\hat{\mathbf{i}} + 2\hat{\mathbf{j}} - 3\hat{\mathbf{k}}) = -2\hat{\mathbf{j}} + 3\hat{\mathbf{k}}$

22. Given, $\mathbf{A} = 2\hat{\mathbf{i}} + \hat{\mathbf{j}} - \hat{\mathbf{k}}, \mathbf{B} = \hat{\mathbf{i}} + 2\hat{\mathbf{j}} + 3\hat{\mathbf{k}}$
and $\mathbf{C} = 6\hat{\mathbf{i}} - 2\hat{\mathbf{j}} - 6\hat{\mathbf{k}}$

$\therefore \ (\mathbf{A} + \mathbf{B}) = (2\hat{\mathbf{i}} + \hat{\mathbf{j}} - \hat{\mathbf{k}}) + (\hat{\mathbf{i}} + 2\hat{\mathbf{j}} + 3\hat{\mathbf{k}}) = 3\hat{\mathbf{i}} + 3\hat{\mathbf{j}} + 2\hat{\mathbf{k}}$

Now, $(\mathbf{A} + \mathbf{B})\cdot\mathbf{C}$
$= (3\hat{\mathbf{i}} + 3\hat{\mathbf{j}} + 2\hat{\mathbf{k}})\cdot(6\hat{\mathbf{i}} - 2\hat{\mathbf{j}} - 6\hat{\mathbf{k}}) = 18 - 6 - 12 = 0$

As, dot product is zero.

i.e. $\quad \theta = 90°$

23. We know that, $R = \sqrt{A^2 + B^2 + 2AB\cos\theta}$

R will be maximum when $\theta = 0°$ and minimum when $\theta = 180°$

$R_{max} = \mathbf{A} + \mathbf{B}$

$17 = \mathbf{A} + \mathbf{B}$ ...(i)

Similarly, $\quad R_{min} = \mathbf{A} - \mathbf{B}$

$7 = \mathbf{A} - \mathbf{B}$ ...(ii)

On solving Eqs. (i) and (ii), we get

$\mathbf{A} = 12$ and $\mathbf{B} = 5$

$\therefore$ Magnitude of resultant $= \sqrt{144 + 25} = 13$

24. The resultant is maximum, when θ is 0° and minimum when θ is 180°.

$$\frac{|\mathbf{a} + \mathbf{b}|}{|\mathbf{a} - \mathbf{b}|} = \frac{\sqrt{\mathbf{a}^2 + \mathbf{b}^2 + 2\mathbf{ab}}}{\sqrt{\mathbf{a}^2 + \mathbf{b}^2 - 2\mathbf{ab}}} = \frac{3}{1}$$

$\mathbf{a}^2 + \mathbf{b}^2 + 2\mathbf{ab} = 9\mathbf{a}^2 + 9\mathbf{b}^2 - 18\mathbf{ab}$

$\Rightarrow \quad 8\mathbf{a}^2 + 8\mathbf{b}^2 - 20\mathbf{ab} = 0$

$\Rightarrow \quad 2\mathbf{a}^2 + 2\mathbf{b}^2 - 5\mathbf{ab} = 0$

Let $\mathbf{a} = x\cdot\mathbf{b}$

$\Rightarrow \quad 2x^2\mathbf{b}^2 + 2\mathbf{b}^2 - 5x\mathbf{b}^2 = 0$

$\Rightarrow \quad 2x^2 + 2 - 5x = 0$

$\Rightarrow \quad 2x^2 - 4x - x + 2 = 0$

$\Rightarrow \quad 2x(x - 2) - 1(x - 2) = 0$

$\Rightarrow \quad (2x - 1)(x - 2) = 0$

$\Rightarrow \quad x = 2 \text{ or } x = \frac{1}{2}$

$\mathbf{a} = 2\mathbf{b}$

25. Given, $\frac{|\mathbf{a} + \mathbf{b}|}{|\mathbf{a} - \mathbf{b}|} = 1$

$\Rightarrow |\mathbf{a}^2 + \mathbf{b}^2 + 2\mathbf{ab}\cos\theta| = |\mathbf{a}^2 + \mathbf{b}^2 - 2\mathbf{ab}\cos\theta|$

$\Rightarrow \ 4\mathbf{ab}\cos\theta = 0 \Rightarrow \cos\theta = 0$

$\Rightarrow \quad \theta = 90°$

26. Vector **C** is perpendicular to both vectors **A** and **B**, hence it can be equal to their cross product.

27. $\mathbf{A}\times\mathbf{B} = 0 \Rightarrow \mathbf{A} \parallel \mathbf{B}$

28. Given, $\quad \mathbf{A} = 3\hat{\mathbf{i}} + 4\hat{\mathbf{j}}$ and $\mathbf{B} = 6\hat{\mathbf{i}} + 8\hat{\mathbf{j}}$

$\mathbf{A}\cdot\mathbf{B} = 18 + 32 = 50$

29. Given, $\quad \mathbf{A} = \hat{\mathbf{i}} - 3\hat{\mathbf{j}} + 2\hat{\mathbf{k}}$

$\mathbf{B} = 3\hat{\mathbf{i}} + 6\hat{\mathbf{j}} - 7\hat{\mathbf{k}}$

$\mathbf{R} = x\hat{\mathbf{i}} + y\hat{\mathbf{j}} + z\hat{\mathbf{k}}$

Now, $\mathbf{A} + \mathbf{B} + \mathbf{R} = \hat{\mathbf{j}}$

$(\hat{\mathbf{i}} - 3\hat{\mathbf{j}} + 2\hat{\mathbf{k}}) + (3\hat{\mathbf{i}} + 6\hat{\mathbf{j}} - 7\hat{\mathbf{k}}) + x\hat{\mathbf{i}} + y\hat{\mathbf{j}} + z\hat{\mathbf{k}} = \hat{\mathbf{j}}$

Equating coefficients, we get

$$4 + x = 0 \Rightarrow x = -4$$
$$3 + y = 1 \Rightarrow y = -2$$
$$-5 + z = 0 \Rightarrow z = 5$$

So, vector $= (-4\hat{\mathbf{i}} - 2\hat{\mathbf{j}} + 5\hat{\mathbf{k}})$.

30. Given, area of parallelogram $= \dfrac{AB}{2}$

$$|\mathbf{A} \times \mathbf{B}| = \frac{AB}{2}$$

$$\therefore \quad AB \sin\theta = \frac{AB}{2}$$

$$\sin\theta = \frac{1}{2}$$

$$\Rightarrow \quad \theta = 30^\circ \text{ or } \frac{\pi}{6}$$

31. Given, $(\mathbf{A} + \mathbf{B}) \times (\mathbf{A} - \mathbf{B})$

$= (\mathbf{A} \times \mathbf{A}) - (\mathbf{A} \times \mathbf{B}) + (\mathbf{B} \times \mathbf{A}) - (\mathbf{B} \times \mathbf{B})$

$= 0 - (\mathbf{A} \times \mathbf{B}) - (\mathbf{A} \times \mathbf{B}) - 0 \qquad [\because (\mathbf{B} \times \mathbf{A}) = -(\mathbf{A} \times \mathbf{B})]$

$= -2(\mathbf{A} \times \mathbf{B}) = 2(\mathbf{B} \times \mathbf{A})$

32. Given, $|\mathbf{A} \times \mathbf{B}|^2 + |\mathbf{A} \cdot \mathbf{B}|^2$

$$= (AB\sin\theta)^2 + (AB\cos\theta)^2$$
$$= A^2B^2\sin^2\theta + A^2B^2\cos^2\theta$$
$$= A^2B^2(\sin^2\theta + \cos^2\theta)$$
$$= A^2B^2$$

33. The angle between the two vectors **A** and $\mathbf{B} \times \mathbf{A}$ is $\pi/2$,

so $\qquad \mathbf{A} \cdot (\mathbf{B} \times \mathbf{A}) = 0$

Dot product of two vectors is zero when they are perpendicular to each other.

34. As, $\mathbf{A} - \mathbf{B}$ is a unit vector along Y - axis,

$$\mathbf{A} - \mathbf{B} = \hat{\mathbf{j}}$$

$$\Rightarrow \quad \mathbf{B} = \mathbf{A} - \hat{\mathbf{j}}$$
$$= 3\hat{\mathbf{i}} + 2\hat{\mathbf{j}} + \hat{\mathbf{k}} - \hat{\mathbf{j}}$$
$$= 3\hat{\mathbf{i}} + \hat{\mathbf{j}} + \hat{\mathbf{k}}$$

35. $B_x = 10 - 6 = 4$

$B_y = 9 - 6 = 3$

$$\therefore \quad \mathbf{B} = \sqrt{B_x^2 + B_y^2} = 5$$

LEVEL 02 Objective Questions

36. As the vectors **A** and **B** are parallel.

$$\Rightarrow \quad \mathbf{A} \times \mathbf{B} = \mathbf{0}$$

$$(2\hat{\mathbf{i}} + p\hat{\mathbf{j}} + q\hat{\mathbf{k}}) \times (5\hat{\mathbf{i}} + 7\hat{\mathbf{j}} + 3\hat{\mathbf{k}}) = \mathbf{0}$$

$$(3p - 7q)\,\hat{\mathbf{i}} + (5q - 6)\,\hat{\mathbf{j}} + (14 - 5p)\,\hat{\mathbf{k}} = \mathbf{0}$$

So, $\quad 3p - 7q = 0,\ (5q - 6) = 0,\ 14 - 5p = 0$

$$\therefore \quad p = \frac{14}{5},\ q = \frac{6}{5}$$

37. Sum of components of applied forces along X and Y-axis are

$$\Sigma F_x = 2 + \sqrt{3}\cos 30^\circ + 5\cos 60^\circ + \sqrt{3}\cos 90^\circ + 2\cos 120^\circ$$

$$= 2 + \sqrt{3} \times \frac{\sqrt{3}}{2} + 5 \times \frac{1}{2} + \sqrt{3} \times 0 - 2 \times \frac{1}{2}$$

$$= 2 + \frac{3}{2} + \frac{5}{2} - 1 = 5\text{ N}$$

$$\Sigma F_y = 2 \times \sin 0^\circ + \sqrt{3}\sin 30^\circ + 5\sin 60^\circ + \sqrt{3}\sin 90^\circ + 2\sin 120^\circ$$

$$= 0 + \frac{\sqrt{3}}{2} + 5 \times \frac{\sqrt{3}}{2} + \sqrt{3} + 2 \times \frac{\sqrt{3}}{2}$$

$$= 5\sqrt{3}\text{ N}$$

Magnitude of resultant of ΣF_x and ΣF_y is

$$R = \sqrt{(\Sigma F_x)^2 + (\Sigma F_y)^2}$$
$$= \sqrt{25 + 75} = 10\text{ N}$$

Angle made by resultant with X-axis

$$\theta = \tan^{-1}\left(\frac{\Sigma F_y}{\Sigma F_x}\right) = \tan^{-1}\left(\frac{1}{\sqrt{3}}\right)$$

or $\qquad \theta = 30^\circ$

or angle made by R with Y-axis $= 60^\circ$

38.

$$\mathbf{A} = 2\cos 60^\circ\,\hat{\mathbf{i}} + 2\sin 60^\circ\,\hat{\mathbf{j}}$$
$$\mathbf{B} = 2\cos(60^\circ)\,\hat{\mathbf{i}} + 2\sin 60^\circ\,(-\hat{\mathbf{j}})$$
$$= 2\cos 60^\circ\,\hat{\mathbf{i}} - 2\sin 60^\circ\,\hat{\mathbf{j}}$$

$$\mathbf{A} + \mathbf{B} = 2 \times 2\cos 60^\circ = 2 \text{ along } +X\text{-axis}$$

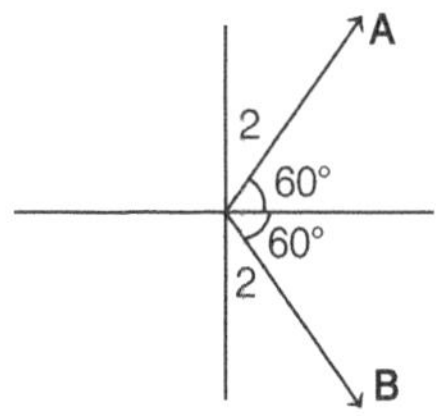

39. We have given, $|\mathbf{A} + \mathbf{B}| = n|\mathbf{A} - \mathbf{B}|$

$$(\mathbf{A}^2 + \mathbf{B}^2 + 2\mathbf{AB}\cos\theta) = n^2(\mathbf{A}^2 + \mathbf{B}^2 - 2\mathbf{AB}\cos\theta)$$

Hence, $\cos\theta = \dfrac{n^2 - 1}{n^2 + 1}$

$$\Rightarrow \quad \theta = \cos^{-1}\left(\frac{n^2 - 1}{n^2 + 1}\right)$$

40.

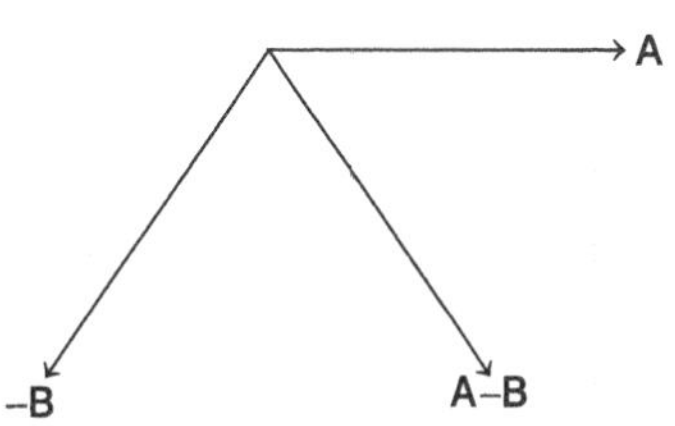

So, direction of $\mathbf{A} - \mathbf{B}$ can't be in upward direction.

41. $2F\cos\frac{\theta}{2} = 4$ ...(i)

$2F\sin\frac{\theta}{2} = 2\sqrt{5}$ ...(ii)

Squaring and adding Eqs. (i) and (ii), we get

$$4F^2 = 36$$

$\therefore$ $F = 3$ N

42. From $\theta = 0° - 90°$, $|\mathbf{A}+\mathbf{B}| > |\mathbf{A}-\mathbf{B}|$
and from $\theta = 90° - 180°$, $|\mathbf{A}+\mathbf{B}| < |\mathbf{A}-\mathbf{B}|$

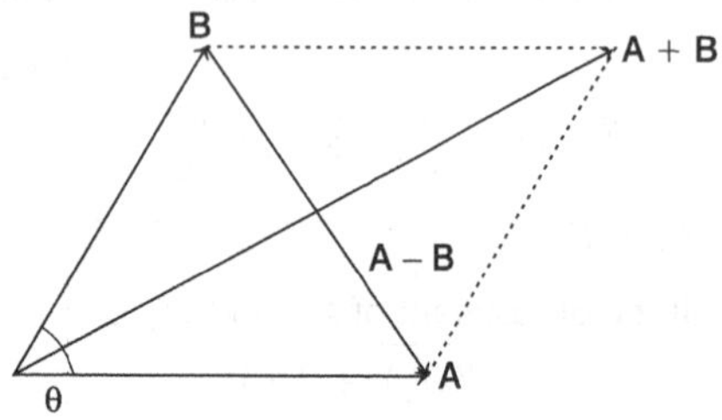

43. $\mathbf{A} + \frac{\mathbf{B}}{2} + \frac{\mathbf{C}}{3} + \frac{\mathbf{D}}{5} + \frac{\mathbf{E}}{6} + \frac{\mathbf{F}}{4}$

$= 3\hat{\mathbf{i}} + (1\cos 30°\,\hat{\mathbf{i}} + 1\sin 30°\,\hat{\mathbf{j}}) + (1\sin 30°\,\hat{\mathbf{j}} - 1\cos 30°\,\hat{\mathbf{i}}) + (-\hat{\mathbf{i}}) + (-1\cos 30°\,\hat{\mathbf{i}} - 1\sin 30°\,\hat{\mathbf{j}}) + (1\cos 30°\,\hat{\mathbf{i}} - 1\sin 30°\,\hat{\mathbf{j}})$

$= 2\hat{\mathbf{i}}$

So, $\left|\mathbf{A} + \frac{\mathbf{B}}{2} + \frac{\mathbf{C}}{3} + \frac{\mathbf{D}}{5} + \frac{\mathbf{E}}{6} + \frac{\mathbf{F}}{4}\right| = 2$

44.

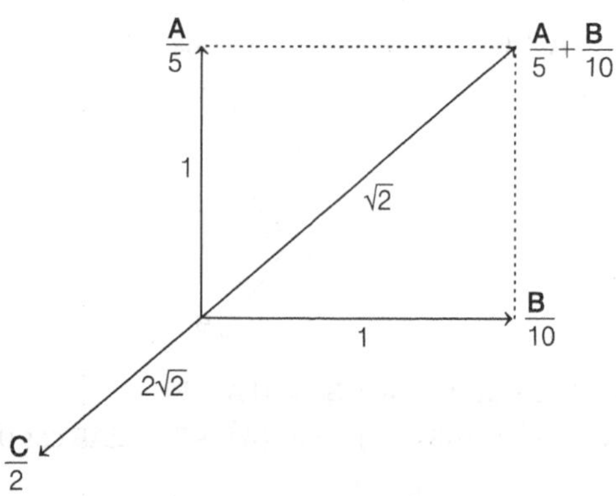

We can see that $\frac{\mathbf{A}}{5} + \frac{\mathbf{B}}{10} + \frac{\mathbf{C}}{2}$ is in the direction of **C**.

45. Vector $\mathbf{AB} = (4-1)\hat{\mathbf{i}} + (5-1)\hat{\mathbf{j}}$

$= 3\hat{\mathbf{i}} + 4\hat{\mathbf{j}}$

So, $\hat{\mathbf{AB}} = \frac{\mathbf{AB}}{|\mathbf{AB}|} = \frac{3\hat{\mathbf{i}} + 4\hat{\mathbf{j}}}{5}$

Force $= 100\,\hat{\mathbf{AB}} = (60\hat{\mathbf{i}} + 80\hat{\mathbf{j}})$ N

46. As, $\mathbf{a} + \mathbf{c} = \mathbf{b}$

$\Rightarrow$ $a^2 + c^2 + 2\mathbf{a}\cdot\mathbf{c} = b^2$

$\Rightarrow$ $\mathbf{a}\cdot\mathbf{c} = 0$ or $\mathbf{a} \perp \mathbf{c}$

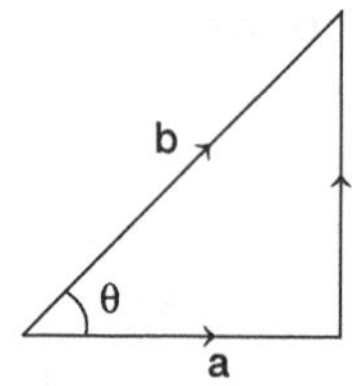

Also, from vector triangle,

$$\tan\theta = \frac{|\mathbf{c}|}{|\mathbf{a}|} = \frac{c}{\sqrt{3}c} = \frac{1}{\sqrt{3}}$$

or $\theta = 30°$

47. Vector component of **A** along **B**

$$= \frac{\mathbf{A}\cdot\mathbf{B}}{|\mathbf{B}|}\cdot\hat{\mathbf{B}} = \frac{12}{\sqrt{3}}\cdot\frac{\hat{\mathbf{i}}+\hat{\mathbf{j}}+\hat{\mathbf{k}}}{(\sqrt{3})^2}$$

$$= \frac{4}{\sqrt{3}}(\hat{\mathbf{i}}+\hat{\mathbf{j}}+\hat{\mathbf{k}})$$

48. Given, $\mathbf{A} + \mathbf{B} = \mathbf{C}$ and $|\mathbf{A}| = |\mathbf{C}|$

$\mathbf{B} = \mathbf{C} - \mathbf{A}$

$\mathbf{B}^2 = C^2 + A^2 - 2\mathbf{C}\cdot\mathbf{A}$

and $|\mathbf{A}| = |\mathbf{C}|$

$\Rightarrow$ $\mathbf{B}^2 = 2A^2$

or $|\mathbf{B}| = \sqrt{2}|\mathbf{A}|$

As $\mathbf{C} \perp \mathbf{A}$

Now, $\mathbf{A} + \mathbf{B} = \mathbf{C}$

$\Rightarrow$ $A^2 + B^2 + 2AB\cos\theta = \mathbf{C}^2$

$A^2 + 2A^2 + 2A\sqrt{2}A\cos\theta = A^2$

$2A^2 + 2A^2\sqrt{2}\cos\theta = 0$

$2A^2(1 + \sqrt{2}\cos\theta) = 0$

$1 + \sqrt{2}\cos\theta = 0$

$\cos\theta = -\frac{1}{\sqrt{2}} = \frac{3\pi}{4}$ rad

49. Given, $\mathbf{a}_1$ and $\mathbf{a}_2$ are two non-collinear unit vectors

$|\mathbf{a}_1| = |\mathbf{a}_2| = 1$ and $|\mathbf{a}_1 + \mathbf{a}_2| = \sqrt{3}$

Squaring on both sides,

$(\mathbf{a}_1 + \mathbf{a}_2)^2 = 3$

$(\mathbf{a}_1 + \mathbf{a}_2)(\mathbf{a}_1 + \mathbf{a}_2) = 3$

$\mathbf{a}_1\cdot\mathbf{a}_1 + \mathbf{a}_2\cdot\mathbf{a}_2 + 2\mathbf{a}_1\cdot\mathbf{a}_2 = 3$

$1 + 1 + 2\mathbf{a}_1\mathbf{a}_2 = 3$

$2\mathbf{a}_1\cdot\mathbf{a}_2 = 1$

$\mathbf{a}_1\cdot\mathbf{a}_2 = \frac{1}{2}$

Now, value of

$(\mathbf{a}_1 - \mathbf{a}_2)(2\mathbf{a}_1 + \mathbf{a}_2) = 2\mathbf{a}_1\cdot\mathbf{a}_1 + \mathbf{a}_1\cdot\mathbf{a}_2 - 2\mathbf{a}_1\mathbf{a}_2 - \mathbf{a}_2\cdot\mathbf{a}_2$

$= 2\mathbf{a}_1^2 - \mathbf{a}_1\cdot\mathbf{a}_2 - \mathbf{a}_2^2$

$= 2\times 1 - \frac{1}{2} - 1 = \frac{1}{2}$

50. $(\mathbf{a} + 2\mathbf{b}) \cdot (5\mathbf{a} - 4\mathbf{b}) = 0$

$\Rightarrow 5a^2 - 4ab\cos\theta + 10ab\cos\theta - 8b^2 = 0$ ...(i)

a and **b** are unit vectors, hence

$$a = b = 1$$

Substituting in Eq. (i), we have

$$\cos\theta = \frac{1}{2} \quad \text{or} \quad \theta = 60^\circ$$

51. $\mathbf{A} \cdot \mathbf{B} = 10 + 7 = 17 =$ Magnitude of vector

$$\mathbf{B} - \mathbf{A} = 3\hat{\mathbf{i}} - 3\hat{\mathbf{k}}$$

Unit vector along $\mathbf{B} - \mathbf{A}$,

$$\hat{\mathbf{n}} = \frac{\mathbf{B} - \mathbf{A}}{|\mathbf{B} - \mathbf{A}|} = \frac{1}{\sqrt{2}}(\hat{\mathbf{i}} - \hat{\mathbf{k}})$$

The desired vector,

$$(\mathbf{A} \cdot \mathbf{B})\,\hat{\mathbf{n}} = \frac{17}{\sqrt{2}}(\hat{\mathbf{i}} - \hat{\mathbf{k}})$$

52. We have, $(2\mathbf{a} - 5\mathbf{b}) \cdot (3\mathbf{a} + \mathbf{b})$

$$= 6\mathbf{a}^2 + 2\mathbf{ab} - 15\,\mathbf{ab} - 5\mathbf{b}^2$$

$$= 6 - 13\,\mathbf{ab} - 5$$

As $(\mathbf{a} + \mathbf{b}) = \sqrt{3}$

$$\Rightarrow \quad \mathbf{ab} = \frac{1}{2}$$

$$= 1 - \frac{13}{2} = -\frac{11}{2}$$

53. Component of vector **A** along vector **B** is

$$\frac{\mathbf{A} \cdot \mathbf{B}}{|\mathbf{B}|^2} \cdot \mathbf{B}$$

and component of **A** perpendicular to **B** is

$$\mathbf{A} - \left(\frac{\mathbf{A} \cdot \mathbf{B}}{|\mathbf{B}|^2}\right) \cdot \mathbf{B}$$

Now in the given question,

$$\mathbf{A} = 2\hat{\mathbf{i}} + 3\hat{\mathbf{j}} \text{ and } \mathbf{B} = \hat{\mathbf{i}} + \hat{\mathbf{j}}$$

$$\Rightarrow \quad \mathbf{A} \cdot \mathbf{B} = 2 + 3 = 5$$

and $\quad |\mathbf{B}| = \sqrt{1^2 + 1^2} = \sqrt{2}$

So, component of **A** along **B**

$$= \left(\frac{\mathbf{A} \cdot \mathbf{B}}{|\mathbf{B}|^2}\right) \cdot \mathbf{B} = \frac{5}{2}(\hat{\mathbf{i}} + \hat{\mathbf{j}})$$

and its magnitude $= \sqrt{\left(\frac{5}{2}\right)^2 + \left(\frac{5}{2}\right)^2}$

$$= \frac{5}{2} \times \sqrt{2} = \frac{5}{\sqrt{2}}$$

54. $\because \quad \mathbf{R} = -\hat{\mathbf{i}} + \frac{\hat{\mathbf{j}}}{3} + \hat{\mathbf{k}}$

$$\mathbf{S} = x\hat{\mathbf{i}} + 3\hat{\mathbf{j}} + (x - 1)\hat{\mathbf{k}}$$

As, $\mathbf{R} \cdot \mathbf{S} = -x + 1 + (x - 1)$

$$= 0$$

As it can be seen that, the dot product of **R** and **S** is 0. It means both the vectors are perpendicular to each other, therefore angle between both the vectors is a right angle.

55. If **a** and **b** are parallel or anti-parallel, then component of **b** perpendicular to be **a** will be zero. For vectors to be parallel or anti-parallel, coefficients of $\hat{\mathbf{i}}$, $\hat{\mathbf{j}}$ and $\hat{\mathbf{k}}$ should bear a constant ratio.

$$\therefore \quad \frac{3}{6} = \frac{8}{16} = -\frac{2}{x} \Rightarrow x = -4$$

56. Given, $|\mathbf{A} + \mathbf{B}| = \frac{|\mathbf{A} - \mathbf{B}|}{2}$ and $|\mathbf{A}| = 2|\mathbf{B}|$

Let θ be the angle between the two vectors,

$$|\mathbf{A} + \mathbf{B}|^2 = \mathbf{A}^2 + \mathbf{B}^2 + 2\,\mathbf{A} \cdot \mathbf{B}$$

$$\frac{|\mathbf{A} - \mathbf{B}|^2}{4} = \mathbf{A}^2 + \mathbf{B}^2 + 2\mathbf{A} \cdot \mathbf{B}$$

$$\mathbf{A}^2 + \mathbf{B}^2 - 2\mathbf{A} \cdot \mathbf{B} = 4(\mathbf{A}^2 + \mathbf{B}^2 + 2\mathbf{A} \cdot \mathbf{B})$$

$$3(\mathbf{A}^2 + \mathbf{B}^2) + 10\mathbf{A} \cdot \mathbf{B} = 0$$

$$3(4\mathbf{B}^2 + \mathbf{B}^2) + 10\,\mathbf{A} \cdot \mathbf{B}\cos\theta = 0$$

$$3(5\mathbf{B}^2) + 10\,\mathbf{AB}\cos\theta = 0$$

$$15\,\mathbf{B}^2 + 10(2\mathbf{B}) \cdot \mathbf{B}\cos\theta = 0$$

$$5\mathbf{B}^2(3 + 4\cos\theta) = 0$$

$$3 + 4\cos\theta = 0$$

$$\Rightarrow \quad \theta = \cos^{-1}\left(-\frac{3}{4}\right)$$

57. Given, $\mathbf{A} + \mathbf{B} = \mathbf{R}$

$$(\mathbf{A} + \mathbf{B}) \cdot (\mathbf{A} + \mathbf{B}) = \mathbf{R} \cdot \mathbf{R}$$

$\therefore \quad A^2 + B^2 + 2\mathbf{A} \cdot \mathbf{B} = R^2$ ...(i)

Now, $(\mathbf{A} + 2\mathbf{B}) \perp \mathbf{A}$

$$(\mathbf{A} + 2\mathbf{B}) \cdot \mathbf{A} = 0$$

$A^2 + 2\mathbf{A} \cdot \mathbf{B} = 0$ ...(ii)

Substituting the value of $(A^2 + 2\mathbf{A} \cdot \mathbf{B})$ from Eq. (ii) to Eq. (i), we get

$$B^2 = R^2$$

$$\Rightarrow \quad |\mathbf{B}| = |\mathbf{R}|$$

58.

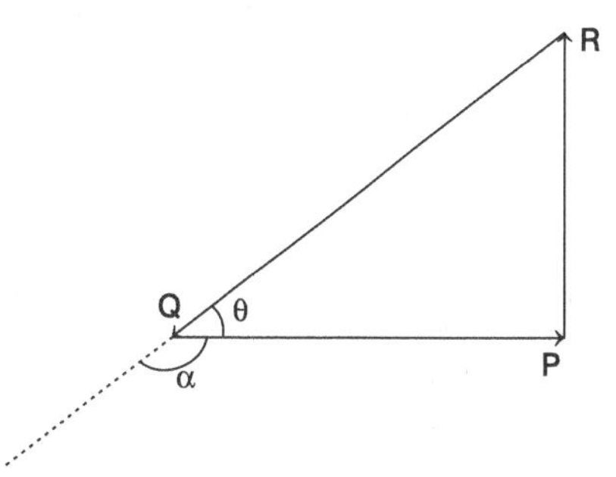

Angle between **P** and **Q** is α.

$$\alpha = 180^\circ - \theta$$

$\therefore \quad \cos\alpha = -\cos\theta = -\dfrac{\mathbf{P}}{\mathbf{Q}}$

$\Rightarrow \quad \alpha = \cos^{-1}\left(-\dfrac{\mathbf{P}}{\mathbf{Q}}\right)$

59. Given, $\mathbf{A} = \hat{\mathbf{i}} + \hat{\mathbf{j}} - 2\hat{\mathbf{k}}, \mathbf{B} = \hat{\mathbf{i}} - \hat{\mathbf{j}} + \hat{\mathbf{k}}$ and $\mathbf{C} = 2\hat{\mathbf{i}} - 3\hat{\mathbf{j}} + 4\hat{\mathbf{k}}$

The vector $\mathbf{X} = \alpha\mathbf{A} + \beta\mathbf{B}$ is perpendicular to $\mathbf{C}$

$= \alpha(\hat{\mathbf{i}} + \hat{\mathbf{j}} - 2\hat{\mathbf{k}}) + \beta(\hat{\mathbf{i}} - \hat{\mathbf{j}} + \hat{\mathbf{k}})$

$= (\alpha + \beta)\hat{\mathbf{i}} + (\alpha - \beta)\hat{\mathbf{j}} + (-2\alpha + \beta)\hat{\mathbf{k}}$

Now, the vector is perpendicular to $\mathbf{C}$

$(\alpha + \beta)\hat{\mathbf{i}} + (\alpha - \beta)\hat{\mathbf{j}} + (-2\alpha + \beta)\hat{\mathbf{k}} \cdot (2\hat{\mathbf{i}} - 3\hat{\mathbf{j}} + 4\hat{\mathbf{k}}) = 0$

$(2\alpha + 2\beta) + (-3\alpha + 3\beta) + (-8\alpha + 4\beta) = 0$

$-9\alpha + 9\beta = 0$

$\Rightarrow \quad \dfrac{\alpha}{\beta} = 1$

60. Given, $\mathbf{A} = (\hat{\mathbf{i}} + 2\hat{\mathbf{j}} + 4\hat{\mathbf{k}})$ and $\mathbf{B} = 5\hat{\mathbf{i}}$

The third side of a triangle $= \mathbf{A} - \mathbf{B}$

$= -4\hat{\mathbf{i}} + 2\hat{\mathbf{j}} + 4\hat{\mathbf{k}}$

Its magnitude $= \sqrt{(-4)^2 + (2)^2 + (4)^2} = 6$

61. Given, $\mathbf{AC} = \hat{\mathbf{i}} + 2\hat{\mathbf{j}} + 4\hat{\mathbf{k}}$ and $\mathbf{BD} = \hat{\mathbf{i}} - 3\hat{\mathbf{j}} + \hat{\mathbf{k}}$

Since, **ABCD** is a parallelogram.

$\therefore \quad \mathbf{AB} = \mathbf{CD} \Rightarrow \mathbf{AD} = \mathbf{BC}$

$\mathbf{AC} = \mathbf{AB} + \mathbf{BC}$ ...(i)

$\mathbf{BD} = -\mathbf{AB} + \mathbf{AD}$ ($\because \mathbf{AD} = \mathbf{BC}$) ...(ii)

Adding Eqs. (i) and (ii), we get

$2\mathbf{BC} = \mathbf{AC} + \mathbf{BD}$

$2\mathbf{BC} = (\hat{\mathbf{i}} + 2\hat{\mathbf{j}} + 4\hat{\mathbf{k}}) + (\hat{\mathbf{i}} - 3\hat{\mathbf{j}} + \hat{\mathbf{k}})$

$2\mathbf{BC} = 2\hat{\mathbf{i}} - \hat{\mathbf{j}} + 5\hat{\mathbf{k}}$

$\mathbf{BC} = \hat{\mathbf{i}} - \dfrac{\hat{\mathbf{j}}}{2} + \dfrac{5\hat{\mathbf{k}}}{2}$

62. Given, the diagonals of a parallelogram are $(\hat{\mathbf{i}} - \hat{\mathbf{j}})$ and $(\hat{\mathbf{i}} + \hat{\mathbf{j}})$.

Area of a parallelogram $= \dfrac{1}{2}(\mathbf{a} \times \mathbf{b})$

$= \dfrac{1}{2}(\hat{\mathbf{i}} - \hat{\mathbf{j}}) \times (\hat{\mathbf{i}} + \hat{\mathbf{j}})$

$= \dfrac{1}{2}(2\hat{\mathbf{k}}) = \dfrac{1}{2}|2\hat{\mathbf{k}}| = \dfrac{1}{2}|\sqrt{4}| = \dfrac{1 \times 2}{2} = 1$ unit

63. If **A**, **B** and **C** are coplanar, then $(\mathbf{A} \times \mathbf{B})$ is perpendicular to **A** and **B** both, or perpendicular to the plane of **A** and **B**. Hence, $(\mathbf{A} \times \mathbf{B})$ is perpendicular to **C** also. Therefore, their dot product is zero.

64. $|\hat{\mathbf{a}} \cdot \hat{\mathbf{b}}| = \dfrac{1}{2}$

$\therefore \quad |1 \times 1 \times \cos\theta| = \dfrac{1}{2}$

$\therefore$ θ is either 60° or 120°.

$\therefore \quad |\mathbf{a} - \mathbf{b}| = 2 \times 1 \times \sin\dfrac{\theta}{2}$

At $\theta = 60°$, value is 1.

At $\theta = 120°$, value is $\dfrac{\sqrt{3}}{2}$.

65. Let θ be the angle between **a** and **b**.

Since we know, the long diagonal is the vector sum of two sides and the short diagonal is the vector difference of two sides.

$\mathbf{d}_1 = \mathbf{a} + \mathbf{b}$ and $\mathbf{d}_2 = \mathbf{a} - \mathbf{b}$

$d_1^2 = a^2 + b^2 + 2ab\cos\theta$

$d_2^2 = a^2 + b^2 - 2ab\cos\theta$

$\therefore \quad d_1^2 + d_2^2 = 2(a^2 + b^2)$

$\Rightarrow \quad (a^2 + b^2) = \dfrac{d_1^2 + d_2^2}{2}$

LEVEL 03 Previous Years Questions

66. Suppose two vectors are **P** and **Q**.

It is given that

$|\mathbf{P} + \mathbf{Q}| = |\mathbf{P} - \mathbf{Q}|$

Let angle between **P** and **Q** is φ.

$\therefore \quad P^2 + Q^2 + 2PQ\cos\phi = P^2 + Q^2 - 2PQ\cos\phi$

$\Rightarrow \quad 4PQ\cos\phi = 0$

$\Rightarrow \quad \cos\phi = 0 \quad [\because P, Q \neq 0]$

$\Rightarrow \quad \phi = \dfrac{\pi}{2} = 90°$

67. For perpendicular vector, we have

$\mathbf{A} \cdot \mathbf{B} = 0$

$[\cos\omega t\hat{\mathbf{i}} + \sin\omega t\hat{\mathbf{j}}]\left[\cos\dfrac{\omega t}{2}\hat{\mathbf{i}} + \dfrac{\sin\omega t}{2}\hat{\mathbf{j}}\right] = 0$

$\Rightarrow \quad \cos\omega t\cos\dfrac{\omega t}{2} + \sin\omega t\sin\dfrac{\omega t}{2} = 0$

$[\because \cos(A - B) = \cos A\cos B + \sin A\sin B]$

$\Rightarrow \quad \cos\left(\omega t - \dfrac{\omega t}{2}\right) = 0 \Rightarrow \cos\dfrac{\omega t}{2} = 0$

$\Rightarrow \quad \dfrac{\omega t}{2} = \dfrac{\pi}{2}$

$\Rightarrow \quad t = \dfrac{\pi}{\omega}$

Thus, time taken by vectors which are orthogonal to each other is $\dfrac{\pi}{\omega}$.

68. Given, $|\mathbf{A} \times \mathbf{B}| = \sqrt{3}(\mathbf{A} \cdot \mathbf{B})$

$\Rightarrow \quad AB\sin\theta = \sqrt{3}\,AB\cos\theta$

$\Rightarrow \quad \tan\theta = \sqrt{3}$

$\therefore \quad \theta = 60°$

69. If two vectors are perpendicular to each other, then their dot product is always equal to zero.

Let, $\mathbf{a} = 2\hat{\mathbf{i}} + 3\hat{\mathbf{j}} + 8\hat{\mathbf{k}}$

$\mathbf{b} = 4\hat{\mathbf{j}} - 4\hat{\mathbf{i}} + \alpha\hat{\mathbf{k}} = -4\hat{\mathbf{i}} + 4\hat{\mathbf{j}} + \alpha\hat{\mathbf{k}}$

According to the above hypothesis,

$$\mathbf{a} \perp \mathbf{b} \Rightarrow \mathbf{a} \cdot \mathbf{b} = 0$$

$$\Rightarrow (2\hat{\mathbf{i}} + 3\hat{\mathbf{j}} + 8\hat{\mathbf{k}}) \cdot (-4\hat{\mathbf{i}} + 4\hat{\mathbf{j}} + \alpha\hat{\mathbf{k}}) = 0$$

$$\Rightarrow -8 + 12 + 8\alpha = 0$$

$$\Rightarrow 8\alpha = -4$$

$$\therefore \alpha = -\frac{4}{8} = -\frac{1}{2}$$

70. Given, $|\mathbf{A} \times \mathbf{B}| = \sqrt{3}\, \mathbf{A} \cdot \mathbf{B}$ ...(i)

But $|\mathbf{A} \times \mathbf{B}| = |\mathbf{A}|\,|\mathbf{B}| \sin\theta = AB \sin\theta$ ...(ii)

and $\mathbf{A} \cdot \mathbf{B} = |\mathbf{A}|\,|\mathbf{B}| \cos\theta = AB \cos\theta$ ...(iii)

From Eqs. (i), (ii) and (iii), we get

$$AB \sin\theta = \sqrt{3}\, AB \cos\theta$$

or $\tan\theta = \sqrt{3}$

$\therefore \theta = 60°$

The addition of vectors **A** and **B** can be given by the law of parallelogram.

$$\therefore |\mathbf{A} + \mathbf{B}| = \sqrt{A^2 + B^2 + 2AB \cos 60°}$$

$$= \sqrt{A^2 + B^2 + 2AB \times \frac{1}{2}}$$

$$= (A^2 + B^2 + AB)^{1/2}$$

71. Let **A** and **B** be two forces. The sum of the two forces,

$\mathbf{F}_1 = \mathbf{A} + \mathbf{B}$ ...(i)

The difference of the two forces,

$\mathbf{F}_2 = \mathbf{A} - \mathbf{B}$ ...(ii)

Since, sum of the two forces is perpendicular to their differences as given, so

$$\mathbf{F}_1 \cdot \mathbf{F}_2 = 0$$

or $(\mathbf{A} + \mathbf{B}) \cdot (\mathbf{A} - \mathbf{B}) = 0$

or $\mathbf{A}^2 - \mathbf{A} \cdot \mathbf{B} + \mathbf{B} \cdot \mathbf{A} - \mathbf{B}^2 = 0$

or $\mathbf{A}^2 = \mathbf{B}^2$

or $|\mathbf{A}| = |\mathbf{B}|$

Thus, the forces are equal to each other in magnitude.

72. Unit vector can be found by dividing a vector with its magnitude, i.e. $\mathbf{A} = \dfrac{\mathbf{A}}{|\mathbf{B}|}$

Let us represent the unit vector by $\hat{\mathbf{n}}$. We also know that the modulus of unit vector is 1, i.e. $|\hat{\mathbf{n}}| = 1$

$\therefore |\hat{\mathbf{n}}| = |0.5\hat{\mathbf{i}} + 0.8\hat{\mathbf{j}} + c\hat{\mathbf{k}}| = 1$

or $\sqrt{(0.5)^2 + (0.8)^2 + c^2} = 1$

or $0.25 + 0.64 + c^2 = 1$

or $0.89 + c^2 = 1$

or $c^2 = 1 - 0.89 = 0.11$

$\Rightarrow c = \sqrt{0.11}$

73. Speed is a scalar quantity.

74. $\mathbf{A} \cdot \mathbf{B} = (3\hat{\mathbf{i}} + 4\hat{\mathbf{j}} + 5\hat{\mathbf{k}}) \cdot (3\hat{\mathbf{i}} + 4\hat{\mathbf{j}} - 5\hat{\mathbf{k}})$

$= 9 + 16 - 25 = 0$

$\Rightarrow$ So, $\theta = 90°$

75. Suppose two vectors are **P** and **Q**.

It is given that

$$|\mathbf{P} + \mathbf{Q}| = |\mathbf{P} - \mathbf{Q}|$$

Let angle between **P** and **Q** be ϕ.

$$\therefore P^2 + Q^2 + 2PQ \cos\phi = P^2 + Q^2 - 2PQ \cos\phi$$

$\Rightarrow 4PQ \cos\phi = 0$

$\Rightarrow \cos\phi = 0$ $[\because P, Q \neq 0]$

$\Rightarrow \phi = \dfrac{\pi}{2} = 90°$

76. The cross product of any vector with a null vector is a null vector or zero vector.

77. In scalar triple product of vectors, the positions of dot and cross can be interchanged, i.e.

$$\mathbf{A} \cdot (\mathbf{B} \times \mathbf{A}) = (\mathbf{A} \times \mathbf{B}) \cdot \mathbf{A}$$
$$= (\mathbf{A} \times \mathbf{A}) \cdot \mathbf{B}$$

But $\mathbf{A} \times \mathbf{A} = 0$

$\therefore \mathbf{A} \cdot (\mathbf{B} \times \mathbf{A}) = 0$

78. In the following figure, $\mathbf{A} + \mathbf{B} = \mathbf{C}$

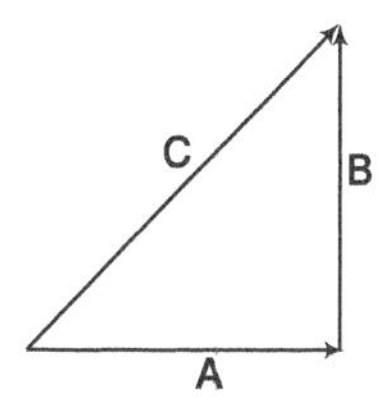

Also, $|\mathbf{A}| = 3,\ |\mathbf{B}| = 4, |\mathbf{C}| = 5$

As $\mathbf{A} + \mathbf{B} = \mathbf{C}$

So, $3^2 + 4^2 + 2 \cdot 4 \cdot 3 \cos\theta = 5^2$

$\cos\theta = 0$

$\Rightarrow \theta = \dfrac{\pi}{2}$

$\Rightarrow$ **A** is perpendicular to **B**.

Chapter 04

MOTION IN ONE DIMENSION

LEVEL 01 Objective Questions

1 A body moves along one quarter of a circle, the ratio of distance to displacement is

(a) $\frac{\pi}{2\sqrt{2}}$ (b) $\frac{2\sqrt{2}}{\pi}$ (c) $\frac{\pi}{\sqrt{2}}$ (d) $\frac{\sqrt{2}}{\pi}$

2 A car covers $\frac{1}{3}$ part of total distance with a speed of 20 kmh^{-1} and second $\frac{1}{3}$ part with a speed of 30 kmh^{-1} and the last $\frac{1}{3}$ part with a speed of 60 kmh^{-1}. The average speed of the car is

(a) 55 kmh^{-1} (b) 30 kmh^{-1}
(c) 45 kmh^{-1} (d) 37.3 kmh^{-1}

3 A person walks 80 m east, then turns right through angle 143°, walks further 50 m and stops. His position relative to the starting point is

(a) 50 m, 53° east of south (b) 50 m, 53° west of south
(c) 50 m, 37° south of east (d) 30 m, 53° south of east

4 An insect starts moving from origin at a constant velocity $(2\hat{\mathbf{i}} + 3\hat{\mathbf{j}})$ m/s for 1 s and then with constant velocity $(2\hat{\mathbf{i}} + 2\hat{\mathbf{j}})$ m/s for $\frac{1}{2}$ s. The distance of the insect from origin after the complete journey is

(a) 4 m (b) 5 m
(c) 7 m (d) 10 m

5 A particle is moving, such that its position coordinates (x, y) are (2m, 3m) at time $t = 0$, (6m, 7m) at time $t = 2$ s and (13m, 14m) at time $t = 5$ s.
Average velocity vector (v_{av}) from $t = 0$ to $t = 5$ s is

(a) $\frac{1}{5}(13\hat{\mathbf{i}} + 14\hat{\mathbf{j}})$ (b) $\frac{7}{3}(\hat{\mathbf{i}} + \hat{\mathbf{j}})$
(c) $2(\hat{\mathbf{i}} + \hat{\mathbf{j}})$ (d) $\frac{11}{5}(\hat{\mathbf{i}} + \hat{\mathbf{j}})$

6 A body is moving with 20 m/s towards east. After 5 s, its velocity becomes 30 m/s towards north and continues. Then, ratio of magnitude of average velocity between (0 to 10 s) and average acceleration between (0 to 10 s) is

(a) $\frac{1}{2}$ (b) $\frac{3}{2}\sqrt{13}$ (c) 5 (d) $5\sqrt{13}$

7 A vehicle travels one-fourth of total distance with speed v_1 and remaining part with speed v_2, then its average speed in the total journey is

(a) $\frac{3v_1 + v_2}{2}$ (b) $\frac{3v_1 v_2}{v_1 + v_2}$ (c) $\frac{4v_1 v_2}{3v_1 + v_2}$ (d) $\frac{4v_1 v_2}{v_1 + v_2}$

8 A particle is moving, such that its position coordinates are varying with time as shown below :

S.No.	Position (x,y) in m	Time (s)
(i)	(0, 2)	0
(ii)	(2, 5)	2
(iii)	(8, 8)	6

The average velocity of the particle from $t = 0$ s to $t = 6$ s is

(a) $(4\hat{i} + 3\hat{j})$ m/s (b) $\left(\frac{4}{3}\hat{i} + \hat{j}\right)$ m/s (c) $(\hat{i} + \hat{j})$ m/s (d) $(3\hat{i} + 4\hat{j})$ m/s

9 Two particles A and B move with constant velocities $\mathbf{v}_1$ and $\mathbf{v}_2$. At the initial moment, their position vectors are $\mathbf{r}_1$ and $\mathbf{r}_2$, respectively. The condition for particles A and B for their collision is

(a) $\mathbf{r}_1 - \mathbf{r}_2 = \mathbf{v}_1 - \mathbf{v}_2$ (b) $\frac{\mathbf{r}_1 - \mathbf{r}_2}{|\mathbf{r}_1 - \mathbf{r}_2|} = \frac{\mathbf{v}_2 - \mathbf{v}_1}{|\mathbf{v}_2 - \mathbf{v}_1|}$ (c) $\mathbf{r}_1 \cdot \mathbf{v}_1 = \mathbf{r}_2 \cdot \mathbf{v}_2$ (d) $\mathbf{r}_1 \times \mathbf{v}_1 = \mathbf{r}_2 \times \mathbf{v}_2$

10 Two particles A and B are moving in xy-plane. Their positions vary with time t according to relation $x_A(t) = 3t$, $x_B(t) = 6$ and $y_A(t) = t$, $y_B(t) = 2 + 3t^2$. The distance between these particles at $t = 1$ s is

(a) 5 m (b) 3 m
(c) 4 m (d) $\sqrt{12}$ m

11 A runner completes one round of a circular path of radius r in 40 s. His displacement after 2 min 20 s will be

(a) zero (b) $2\pi r$
(c) $2r$ (d) $7\pi r$

12 A man walks on a equilateral triangle along path ABC with constant speed. Find ratio of average speed and magnitude of average velocity from A to C.

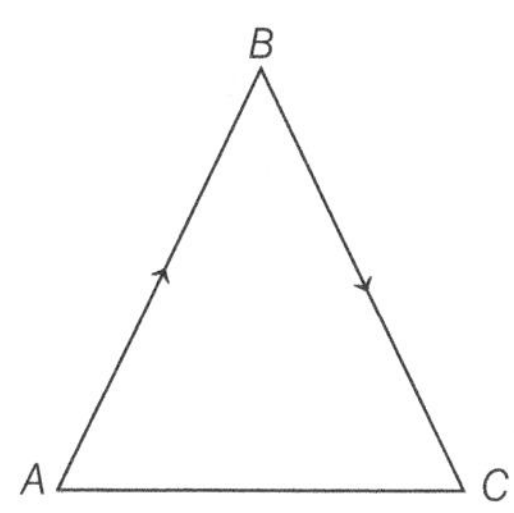

(a) 1
(b) 2
(c) $\frac{1}{2}$
(d) None of the above

13 A motor car is going due north at a speed of 50 km/h, it makes a 90° left turn without changing the speed. The change in velocity of the car is about

(a) 50 km/h (b) $50\sqrt{2}$ km towards south-west
(c) 70 km/h towards north-west (d) None of these

14 A particle moves on a straight line. The product of its acceleration and velocity is constant. The distance moved by the particle in time t is proportional to

(a) t (b) $\sqrt{t}$ (c) $t^{3/2}$ (d) t^2

15 A 150 m long train is moving with a uniform velocity of 45 km/h. The time taken by the train to cross a bridge of length 850 m is

(a) 56 s (b) 68 s (c) 80 s (d) 92 s

16 A particle is projected vertically upwards. It travels equal distance during 5th and 6th second of its motion. Find its projection speed. (Take, $g = 9.8 \text{ m/s}^2$)

(a) 50 m/s (b) 30 m/s (c) 49 m/s (d) 29.4 m/s

17 A particle moves with constant acceleration for 6 s after starting from rest. The distance travelled during the consecutive 2 s interval are in the ratio

(a) 1 : 1 : 1 (b) 1 : 2 : 3 (c) 1 : 3 : 5 (d) 1 : 5 : 9

18 A particle is moving in a straight line with initial velocity u and uniform acceleration a. If the sum of the distances travelled in t th and $(t + 1)$ th s is 100 cm, then its velocity after t s (in cm/s) is

(a) 20 (b) 30 (c) 50 (d) 80

19 A ball is thrown vertically upwards from the ground. It crosses a point at the height of 25 m twice at an interval of 4 s. The ball was thrown with the velocity of

(a) 20 m/s (b) 25 m/s (c) 30 m/s (d) 35 m/s

20 A particle is projected vertically upward with initial velocity 25 m/s. For its motion during third second, which of the following statement is correct?

(a) Displacement of the particle is 10 m.
(b) Distance covered by the particle is 10 m.
(c) Distance covered by the particle is 2.5 m.
(d) None of these

21 A stone thrown upwards with speed u attains maximum height h. Another stone thrown upwards from the same point with speed $2u$ attains maximum height H. What is the relation between h and H?

(a) $2h = H$
(b) $3h = H$
(c) $4h = H$
(d) $5h = H$

22 A truck is moving on the horizontal road with constant speed v. A ball is thrown from the truck vertically up at speed u with respect to truck. What is the distance traversed by the truck when ball returns on the truck?

(a) $\frac{uv}{g}$ (b) $\frac{2uv}{g}$ (c) $\frac{3uv}{g}$ (d) $\frac{uv}{2g}$

23 A body starting from rest covers a distance of 9 m in the fifth second. The constant acceleration of the body is

(a) 2 m/s^2
(b) 0.2 m/s^2
(c) 1.8 m/s^2
(d) 4 m/s^2

24 A body starts from rest and moves with constant acceleration. The ratio of distance covered by the body in nth second to that covered in n second is

(a) $\frac{1}{n}$ (b) $\frac{2n-1}{n^2}$ (c) $\frac{n^2}{2n-1}$ (d) $\frac{2n-1}{2n^2}$

25 A particle is moving along a straight line with constant acceleration. At the end of 10th second, its velocity becomes 20 m/s and in 10th second, it travels a distance of 10 m. Then, the acceleration of the particle will be

(a) 10 m/s^2 (b) 20 m/s^2 (c) $\frac{1}{5} \text{ m/s}^2$ (d) 3.8 m/s^2

26 With what velocity, a ball be projected vertically, so that the distance covered by it in 5th second is twice the distance, it covers in its 6th second? (Take, $g = 10 \text{ m/s}^2$)

(a) 58.8 m/s
(b) 49 m/s
(c) 65 m/s
(d) 19.6 m/s

27 A ball is dropped from the top of a tower, when there is a high speed wind. If the wind exerts a constant force on the ball, then graph of displacement in y- direction *versus* displacement in x- direction is

(a) parabolic (b) straight line (c) elliptical (d) circle

28 Figure shows the path followed by a particle. Its velocity is shown at points A and B.

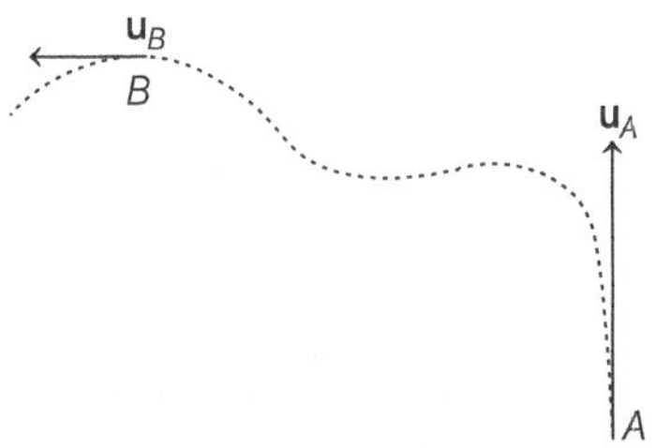

Which of the vectors in the given options best represents the average velocity of the particle for its motion from A to B ?

(a)  (b)  (c)  (d) 

29 A body falling from height h takes t_1 time to reach the ground. The time taken to cover the first half of the height is

(a) $t_2 = \frac{t_1}{\sqrt{2}}$ (b) $t_1 = \frac{t_2}{\sqrt{2}}$ (c) $t_2 = \sqrt{3} t_1$ (d) $t_1 = \sqrt{3}\, t_2$

30 One particle is dropped while another particle is thrown downwards with an initial velocity of 2 m/s, simultaneously. Both these particles are at a separation of 18 m after time t is

(a) 4.5 s (b) 9 s
(c) 9.8 s (d) 18 s

31 The initial velocity of a particle moving along a straight line is 10 m/s and its retardation is 2 m/s^2. The distance moved by the particle in the 5th second of its motion is

(a) 1 m (b) 5 m (c) 12 m (d) 10 m

32 A body is dropped under gravity. The ratio of distances covered by the body in 1st and 6th second of motion is

(a) 1 : 11 (b) 1 : 6
(c) 1 : 5 (d) 5 : 12

33 A body moving with a uniform acceleration crosses a distance of 65 m in the 5th second and 105 m in 9th second. How far will it go in 20 s?

(a) 2040 m (b) 240 m
(c) 2400 m (d) 2004 m

34 Which of the following statements are true ?

(i) A body can have constant speed but varying velocity.
(ii) A body can have constant velocity but varying speed.
(iii) A body can have acceleration without having velocity.
(iv) A body can have velocity without having acceleration.

(a) (ii, iv) (b) (i, iii) (c) (ii, iii) (d) (i, iii, iv)

35 A window is 50 cm long. A stone is falling from a height of 40 cm above the window. It crosses the window in

(a) $\frac{3}{7}$ s (b) $\frac{2}{7}$ s (c) $\frac{1}{7}$ s (d) $\frac{4}{7}$ s

36 A small glass ball is pushed with a speed v from A. It moves on a smooth surface and collides with wall at B. If it loses half of its speed during the collision, what will be the average speed of the ball till it reaches at its initial position?

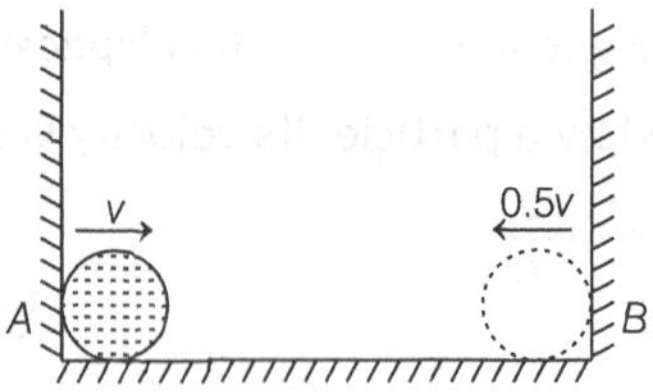

(a) Zero (b) $\frac{v}{3}$ (c) $\frac{2v}{3}$ (d) $\frac{v}{2}$

37 A body A moves with uniform acceleration a and zero initial velocity. Another body B starts at same time from the same point and moves in the same direction with a constant velocity v. The two bodies meet after a time t. The value of t is

(a) $\frac{2v}{a}$ (b) $\frac{v}{a}$ (c) $\frac{v}{2a}$ (d) $\frac{v}{4a}$

38 If a ball is thrown vertically upward with speed u, the distance covered during the last t second of its ascent is

(a) $\frac{1}{2}gt^2$ (b) $ut-\frac{1}{2}gt^2$ (c) $(u-gt)t$ (d) ut

39 A body is falling from a height of 100 m, the time taken for first 50 m is t_1 and remaining 50 m is t_2, then find the ratio of t_1 and t_2.

(a) $\sqrt{2}+1$ (b) $1+\frac{1}{\sqrt{2}}$ (c) $\sqrt{2}-1$ (d) $1-\frac{1}{\sqrt{2}}$

40 A ball A is thrown up vertically with a speed u at the same instant, another ball B is released from a height h. At time t, the speed of A relative to B is (when both are in air)

(a) u (b) $2u$ (c) $u-2gt$ (d) $\sqrt{u^2-gt}$

41 A police jeep wants to chase a thief's car which is moving with speed 40 m/s on a straight track. If the separation between the jeep and car is 400 m, then speed of jeep required to catch the car in 20 s is

(a) 40 m/s (b) 60 m/s (c) 80 m/s (d) 120 m/s

42 A stone is thrown upward from the top of a tower with some initial speed and it reaches the ground in 16 s. Now, it is thrown with the same initial speed downward and it reaches the ground in 9 s. In how much time will it reach the ground, if the stone is allowed to fall freely under gravity from the same place?

(a) 25 s (b) 12.5 s (c) 7 s (d) 12 s

43 For a moving bus, if its displacement-time relation is given as $s=t^3-27t+4$. Then, find its acceleration when its velocity is zero.

(a) 15 m/s^2 (b) 18 m/s^2 (c) 20 m/s^2 (d) 24 m/s^2

44 The acceleration of a particle starting from rest varies with time according to relation $a=\alpha t+\beta$. The velocity of the particle after a time t will be

(a) $\frac{\alpha t^2}{2}+\beta$ (b) $\frac{\alpha t^2}{2}+\beta t$ (c) $\alpha t^2+\frac{1}{2}\beta t$ (d) $\frac{(\alpha t^2+\beta)}{2}$

45 A particle moves along a straight line OX. At a time t (in seconds), the distance x (in metres) of the particle from O is given by $x=40+12t-t^3$. How much did the particle travel before coming to rest?

(a) 24 m (b) 40 m (c) 56 m (d) 16 m

46 A motorcycle is moving with uniform acceleration enters on a bridge with speed u and leaves the bridge with speed $\sqrt{7}u$. Then, find the speed of motorcycle when it was at the mid-point of the bridge. (Assume constant acceleration)
(a) $3u$ (b) $2u$ (c) $\sqrt{5}u$ (d) $\sqrt{3}u$

47 For motion of an object along the x-axis, the velocity v depends on the displacement x as $v = 3x^2 - 2x$, then what is the acceleration at $x = 2$m?
(a) 48 m/s^2 (b) 80 m/s^2
(c) 18 m/s^2 (d) 10 m/s

48 The displacement of a particle moving along x-axis depends on time as $\sqrt{x} = t + 1$. Velocity of particle
(a) increases with time (b) decreases with time
(c) is independent of time (d) None of these

49 A particle is moving with velocity $v = (4t^3 + 3t^2 - 1)$ m/s. The displacement of particle in time $t = 1$ s to $t = 2$ s will be
(a) 21 m (b) 17 m
(c) 13 m (d) 9 m

50 The coordinates of a particle moving in xy-plane vary with time as $x = 4t^2$, $y = 2t$. The path of the particle is
(a) a straight line (b) a circle (c) a parabola (d) an ellipse

51 If velocity of a particle is given by $v = (10 + 2t^2)$ m/s. The average acceleration between 2 s and 5 s is
(a) 2 m/s^2 (b) 4 m/s^2
(c) 12 m/s^2 (d) 14 m/s^2

52 Starting from rest, acceleration of a particle is $a = 2(t - 1)$. The velocity of the particle at $t = 5$ s is
(a) 15 m/s (b) 25 m/s
(c) 5 m/s (d) 0

53 Acceleration-velocity graph of a moving particle is shown in figure. The particle is

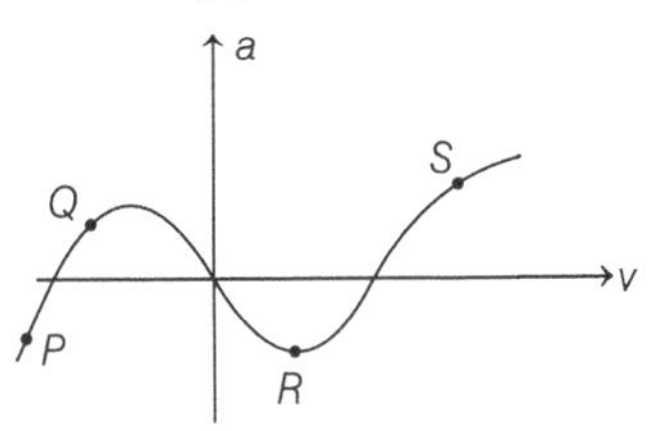

(a) speeding up at P (b) speeding up at Q
(c) speeding down at S (d) speeding up at R

54 The graph shown in figure is a plot of position *versus* time. For which labelled region is the velocity positive and the acceleration negative?

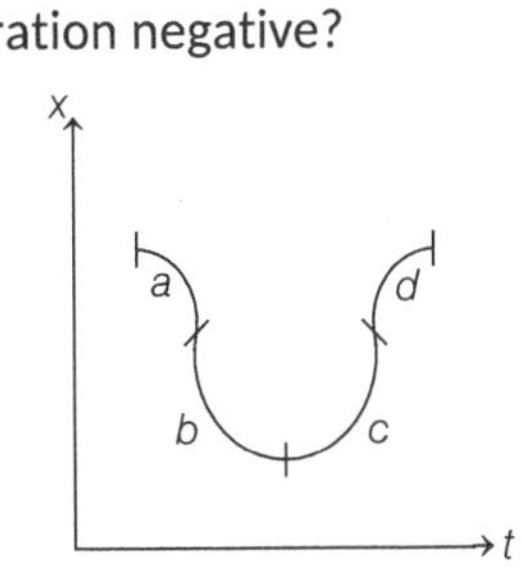

(a) a (b) b (c) c (d) d

55 The graph shows the variation of velocity of a rocket with time. Then, the maximum height attained by the rocket is

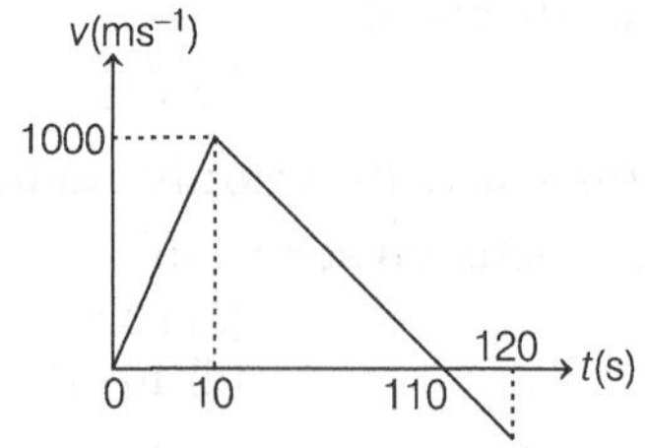

(a) 1.1 km (b) 5 km (c) 55 km (d) None of these

56 The diagram shows the displacement-time graph for a particle moving in a straight line. The distance covered by the particle in the time interval $t = 0$ to $t = 5$s is

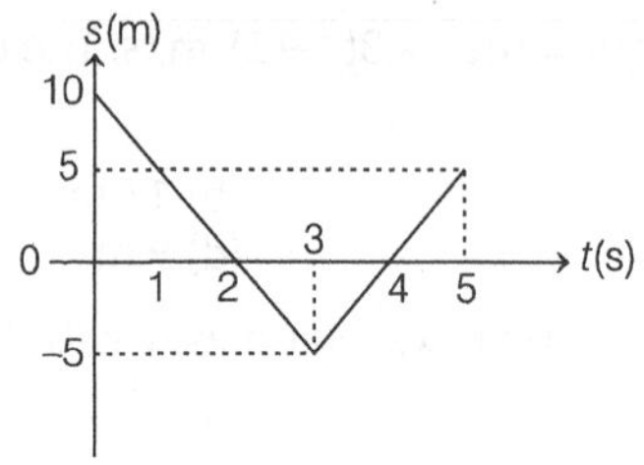

(a) 12.5 m (b) 17.5 m (c) 22.5 m (d) 25.0 m

57 Velocity-time (v-t) graph of a particle moving in a straight line is shown in the given figure.

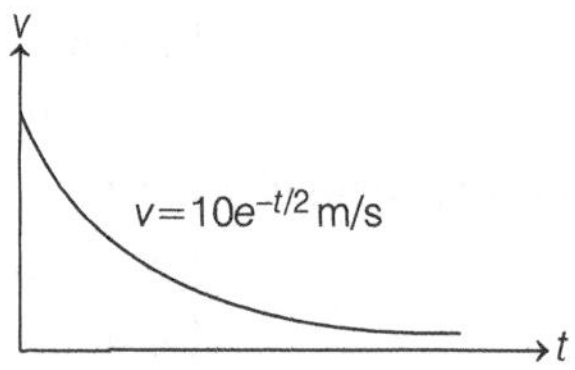

Which of the following graph shows variation in acceleration of the particle with time?

(a)
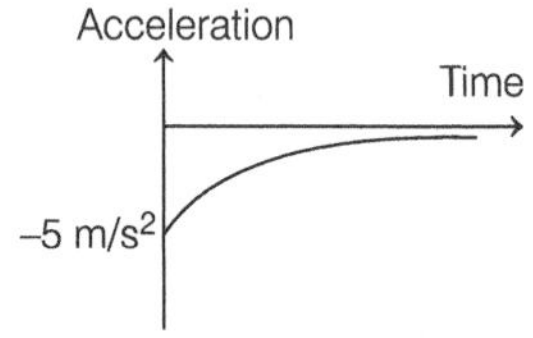

(b)
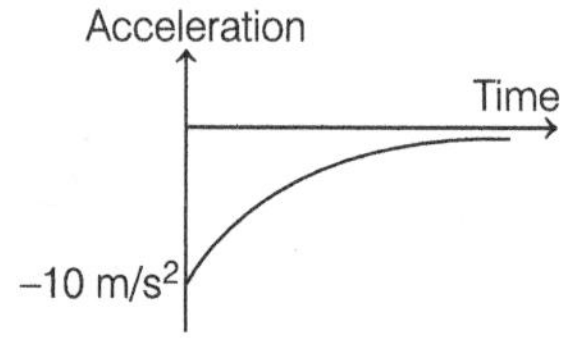

(c)
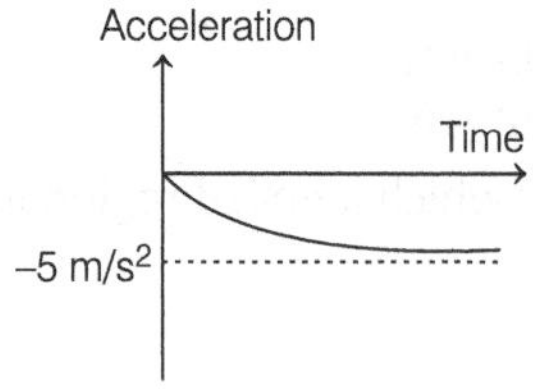

(d)
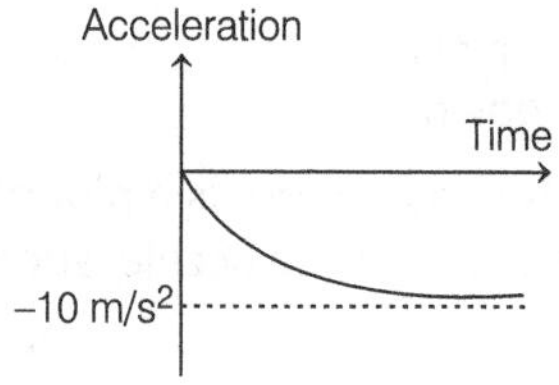

58 A physics student studies rectilinear motion of a body and prepares the following graph

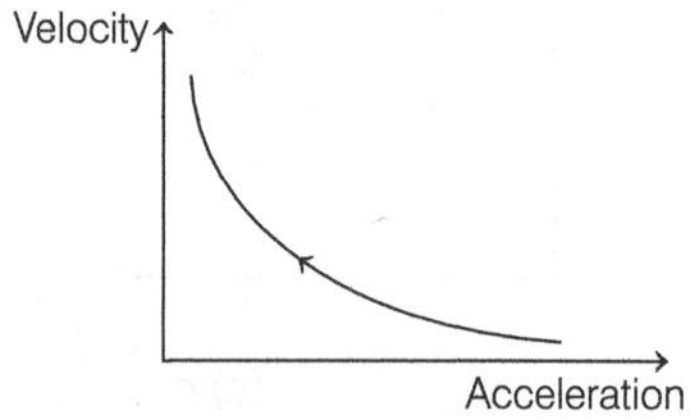

Which of the following conclusions best suits the above graph?
(a) The body is speeding up and its acceleration is decreasing.
(b) The body is slowing down and its acceleration is increasing.
(c) The body is speeding up and its acceleration is increasing.
(d) The given graph cannot describe any physically realisable motion.

59 The graph shows the variation of the velocity v of an object with time t.

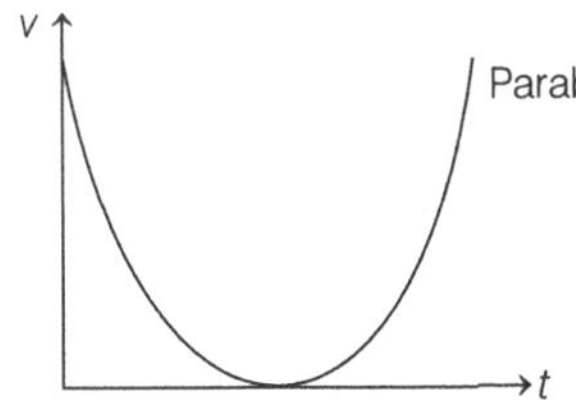

Which one of the following graphs best represents the variation of acceleration a of the object with time t?

(a)
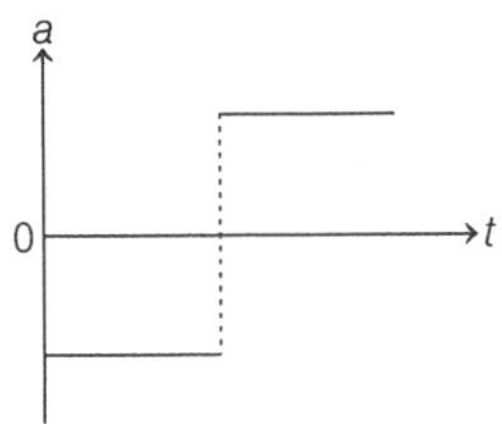

(b)
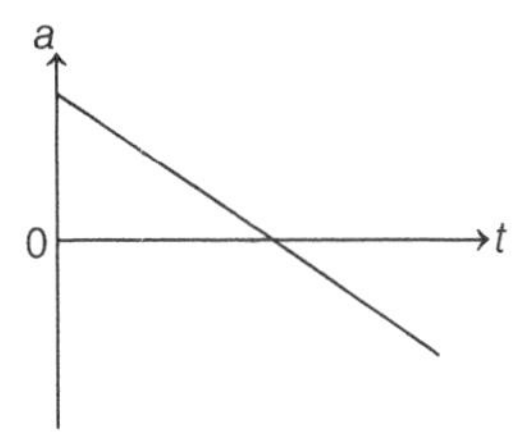

(c)
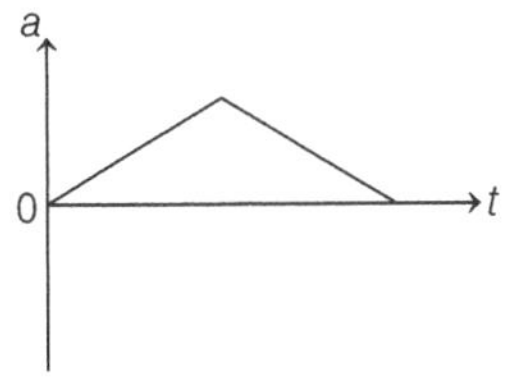

(d)
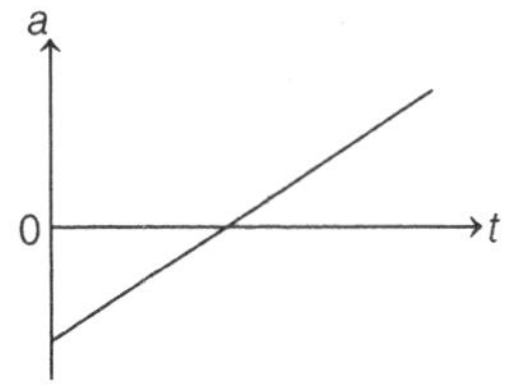

60 Each of the three graphs represents acceleration *versus* time for an object that already has a positive velocity at time t_1. Which graphs show an object whose speed is increasing for the entire time interval between t_1 and t_2?

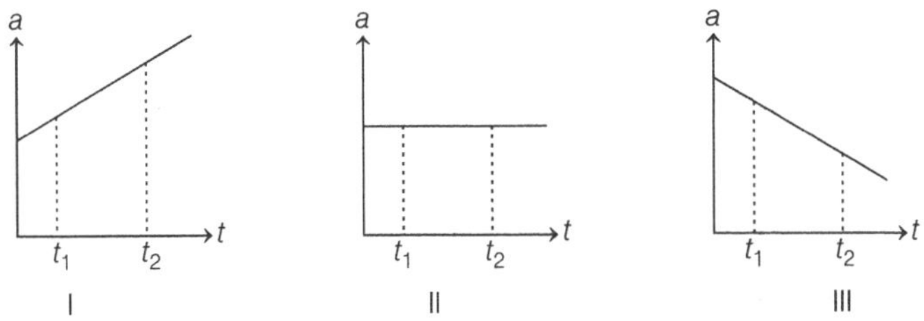

(a) Graph I only (b) Graphs I and II only (c) Graphs I and III only (d) Graphs I, II and III

61 Figure shows two displacement-time graphs of A and B. The relative velocity of particles

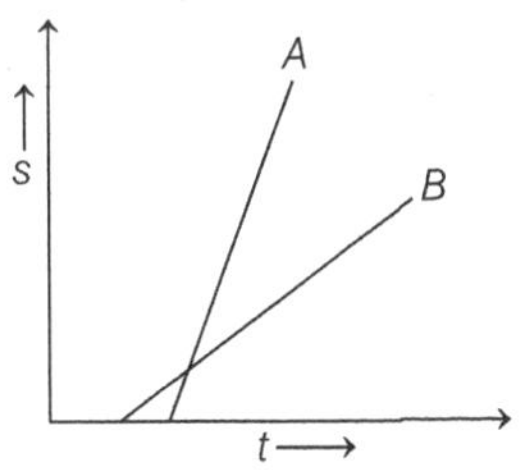

(a) is zero
(b) is (non-zero) constant
(c) continuously increases
(d) continuously decreases

62 From the velocity-time graph, given in figure of a particle moving in a straight line, one can conclude that

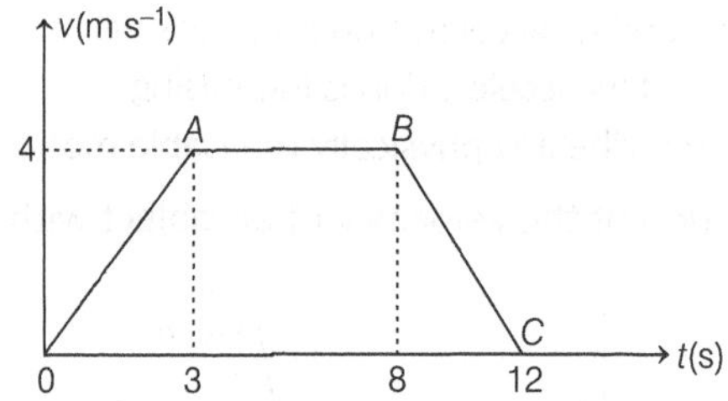

(a) Its average velocity during the 12 s interval is $24/7\ ms^{-1}$.
(b) Its acceleration for the first 3 s is uniform and is equal to $2\ ms^{-2}$.
(c) The body has non-zero acceleration between $t = 3$ s and $t = 8$ s.
(d) The body has a uniform retardation from $t = 8$ s to $t = 12$ s and is equal to $1\ ms^{-2}$.

63 The acceleration *versus* time graph of particle is shown in figure. The respective v-t graph of the particle is

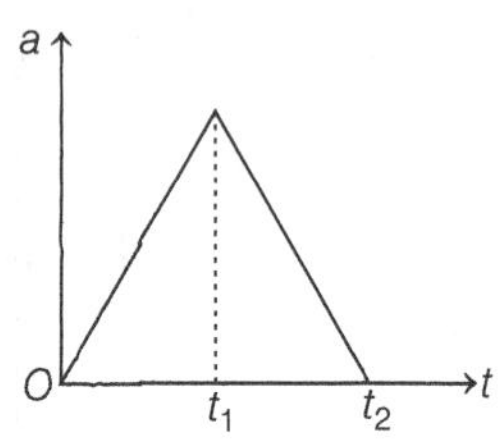

(a) 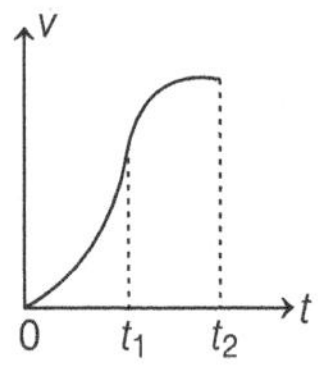

(b) 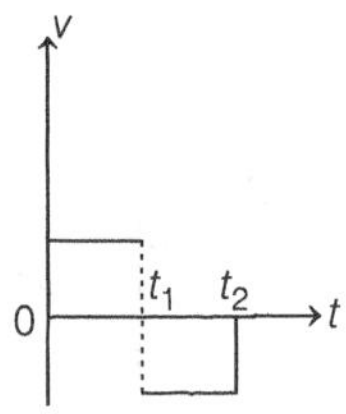

(c) 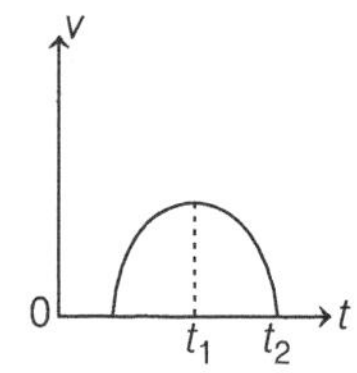

(d) 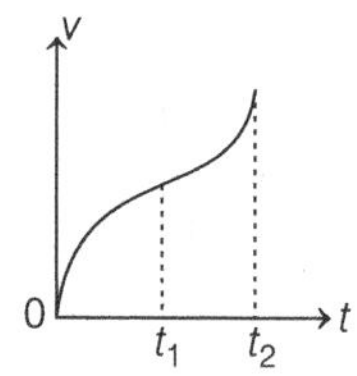

64 The acceleration-time (a-t) graph of a particle moving along a straight line is as shown in figure. If the initial velocity of the particle is 5 m/s, then the velocity at the end of 7 s is

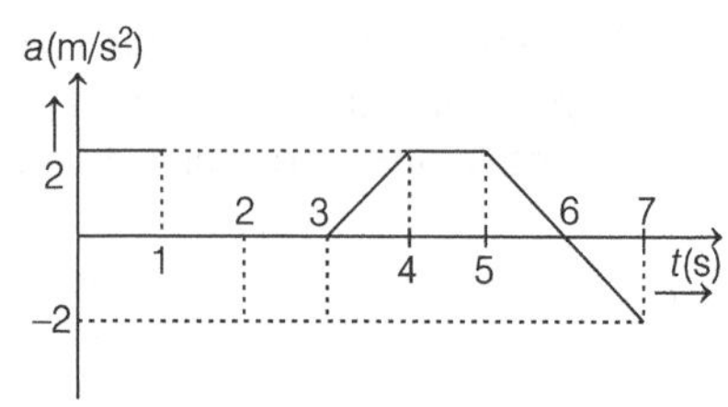

(a) 5 m/s
(b) 6 m/s
(c) 8 m/s
(d) 10 m/s

65 A boy can swim in still water at 1 m/s. He swims across a river flowing at 0.6 m/s which is 336 m wide. If he travels in shortest possible time, then what time he takes to cross the river?
(a) 250 s (b) 420 s (c) 340 s (d) 336 s

66 Car A moves due north at a speed of 40 km/h, while another car B moves due east at a speed of 30 km/h. Find the velocity of car B relative to car A (both in magnitude and direction).
(a) 40 km/h, at an angle $\tan^{-1}\left(\frac{3}{5}\right)$ south of east
(b) 50 km/h, at an angle $\tan^{-1}\left(\frac{3}{5}\right)$ south of east
(c) 40 km/h, at an angle $\tan^{-1}\left(\frac{3}{4}\right)$ south of east
(d) 50 km/h, at an angle $\tan^{-1}\left(\frac{3}{4}\right)$ south of east

67 A person is ordered to swim across a river of width 500 m. At what minimum rate should he swim perpendicular to river flow, so that horizontal drift is 300 m only down the stream. Speed of water current is 3 m/s.

(a) 6 m/s
(b) 3 m/s
(c) 5 m/s
(d) 4 m/s

68 A man crosses a river of width 400 m flowing at 40 m/s in minimum time. If velocity of man in still water is 50 m/s, then the displacement of man while he reaches the opposite bank of the river is

(a) 80 m
(b) $8\sqrt{4}$ m
(c) $80\sqrt{41}$ m
(d) $10\sqrt{41}$ m

69 A bird is flying towards north with a velocity 40 km h^{-1} and a train is moving with velocity 40 km h^{-1} towards east. What is the velocity of the bird noted by a man in the train?

(a) $40\sqrt{2}$ km h^{-1} N - E
(b) $40\sqrt{2}$ km h^{-1} S - E
(c) $40\sqrt{2}$ km h^{-1} N - W
(d) $40\sqrt{2}$ km h^{-1} S - W

70 A boat moves relative to still water with a velocity which is n times the river flow velocity. At what angle to the stream direction must the boat move to cross the river along shortest path?

(a) $\frac{\pi}{2}$
(b) $\sin^{-1}\left(\frac{1}{n}\right)$
(c) $\frac{\pi}{2}+\sin^{-1}\left(\frac{1}{n}\right)$
(d) $\frac{\pi}{2}-\sin^{-1}\left(\frac{1}{n}\right)$

LEVEL 02 Objective Questions

71 A man moves in an open field, such that after moving 10 m in a straight line, he makes a sharp turn of 60° to his left. The total displacement of the man just after 7 such turns is

(a) 10 m
(b) 20 m
(c) 70 m
(d) 30 m

72 A particle moving in a straight line covers half the distance with speed of 3 m/s. The other half of the distance is covered in two equal time intervals with speed of 4.5 m/s and 7.5 m/s, respectively. The average speed of the particle during this motion is

(a) 4.0 m/s
(b) 5.0 m/s
(c) 5.5 m/s
(d) 4.8 m/s

73 Two particles A and B are placed in gravity free space at (0, 0, 0) m and (−30,0,0) m, respectively. Particle A is projected with a velocity $(5\hat{\mathbf{i}}+5\hat{\mathbf{j}}+5\hat{\mathbf{k}})$ ms^{-1}, while particle B is projected with a velocity $(10\hat{\mathbf{i}}+5\hat{\mathbf{j}}+5\hat{\mathbf{k}})$ ms^{-1} simultaneously. Then,

(a) they will collide at (10, 20, 10) m
(b) they will collide at (10, 10, 10) m
(c) they will never collide
(d) they will collide after 6 s

74 Two buses A and B are at positions 50 m and 100 m from the origin at time $t=0$. They start moving in the same direction simultaneously with uniform velocity of 10 m/s and 5 m/s. Determine the position from origin at which A over takes B.

(a) 110 m
(b) 150 m
(c) 100 m
(d) 200 m

75 An aeroplane is to go along straight line from A to B, and back again. The relative speed with respect to wind is V. The wind blows perpendicular to line AB with speed v. The distance between A and B is l. The total time for the round trip is

(a) $\frac{2lV}{\sqrt{V^2-v^2}}$
(b) $\frac{2vl}{V^2-v^2}$
(c) $\frac{2Vl}{V^2-v^2}$
(d) $\frac{2lv}{\sqrt{V^2+v^2}}$

76 A particle's position as a function of time is described as $y(t) = 2t^2 + 3t + 4$. What is the average velocity of the particle from $t = 0$ to $t = 3$ s?

(a) 3 m/s (b) 6 m/s (c) 9 m/s (d) 12 m/s

77 The displacement of a particle is given by $x = (t-2)^2$, where x is in metres and t in second. The distance covered by the particle in first 4 s is

(a) 4 m (b) 8 m (c) 12 m (d) 16 m

78 If $v = (x^2 - 5x + 4)$ m/s, find the acceleration of the particle when velocity of the particle is zero.

(a) Zero (b) 2 (c) 3 (d) None of these

79 The position of a particle varies with time according to the relation $x = (3t^2 - t^3)$ m. The maximum positive velocity of the particle (in m/s) and distance travelled (in m) by the particle upto 4 s respectively, are

(a) 3, 16 (b) 2, 32 (c) 3, 24 (d) 3, 32

80 If velocity of a particle is given by $v = (2t + 3)$ m/s, then average velocity in interval $0 \le t \le 1$ s is

(a) $\frac{7}{2}$ m/s (b) $\frac{9}{2}$ m/s (c) 4 m/s (d) 5 m/s

81 A particle is moving along x-axis whose position is given by $x = 4 - 9t + \frac{t^3}{3}$, then choose the correct statement for this motion.

(a) Direction of motion is not changing at any of the instants.
(b) For $0 < t < 2$ s, the particle is speeding up.
(c) Direction of motion is changing at $t = 3$ s.
(d) For $0 < t < 3$ s, the particle is speeding up.

82 The velocity of the particle at any time t is given by $v = 2t(3-t)$ ms^{-1}. At what time is its velocity maximum ?

(a) 2 s (b) 3 s (c) $\frac{2}{3}$ s (d) $\frac{3}{2}$ s

83 The position of a particle is varying with respect to time as $\mathbf{r} = a(t-1)^2\,\hat{\mathbf{i}} + bt^2\,\hat{\mathbf{j}}$, where a and b are constants, $\mathbf{r}$ is in metre and t is in second. The time after which acceleration becomes normal to the velocity is

(a) $\frac{a}{a+b}$ (b) $\frac{a^2}{(a+b)^2}$ (c) $\frac{a^2}{a+b}$ (d) $\frac{a^2}{a^2+b^2}$

84 Two balloons are moving in air with velocities $\mathbf{v}_1 = \{2t\,\hat{\mathbf{i}} + (t-2)\,\hat{\mathbf{j}}\}$ m/s and $\mathbf{v}_2 = \{(t-4)\,\hat{\mathbf{i}} + t\,\hat{\mathbf{j}}\}$ m/s, then at what t balloons are moving parallel to each other?

(a) 5/4 s (b) 4/5 s
(c) 10/3 s (d) $-3 \pm \sqrt{17}$ s

85 A particle starts from origin at $t = 0$ with velocity $6\,\hat{\mathbf{j}}$ m/s, in xy-plane with constant acceleration $\mathbf{a} = (2\hat{\mathbf{i}} - 2\hat{\mathbf{j}})$ m/s^2. Then, the position coordinates of the particle (in m) when velocity of the particle becomes parallel to x-axis is

(a) (9, 9) (b) (3, 9)
(c) (6, −9) (d) (6, 0)

86 A particle is falling freely under gravity. In first t s, it covers distance x_1 and in the next t s, it covers distance x_2, then t is given by

(a) $\sqrt{\frac{x_2 - x_1}{g}}$ (b) $\sqrt{\frac{x_2 + x_1}{g}}$ (c) $\sqrt{\frac{2(x_2 - x_1)}{g}}$ (d) $\sqrt{\frac{2(x_2 + x_1)}{g}}$

87 A body is thrown vertically upward at $t = 0$. It is at a height 80 m at instants t_1 and t_2. Also, it is at a height 60 m at instants t_1' and t_2' . Then,

(a) $t_1 + t_2 = t_1' + t_2'$ (b) $t_1 + t_2 > t_1' + t_2'$ (c) $t_1 + t_2 < t_1' + t_2'$ (d) None of these

88 A train accelerates from rest at a constant rate α for distance x_1 and time t_1. After that, it retards to rest at constant rate β for a distance x_2 and time t_2. The correct relation among the following is

(a) $\frac{x_1}{x_2} = \frac{\alpha}{\beta} = \frac{t_1}{t_2}$ (b) $\frac{x_1}{x_2} = \frac{\beta}{\alpha} = \frac{t_1}{t_2}$

(c) $\frac{x_1}{x_2} = \frac{\alpha}{\beta} = \frac{t_2}{t_1}$ (d) $\frac{x_1}{x_2} = \frac{\beta}{\alpha} = \frac{t_2}{t_1}$

89 Two bodies move in a straight line towards each other with initial velocities v_1 and v_2 and with constant accelerations a_1 and a_2 directed against their corresponding initial velocities. The maximum initial separation l_{max} between them for which they will meet during the motion is

(a) $\frac{v_1^2}{a_1} + \frac{v_2^2}{a_2}$ (b) $\frac{(v_1 + v_2)^2}{2(a_1 + a_2)}$

(c) $\frac{v_1 v_2}{\sqrt{a_1 a_2}}$ (d) $\frac{v_1^2 - v_2^2}{(a_1 - a_2)}$

90 A particle projected from origin moves in xy-plane with a velocity $\mathbf{v} = 3\hat{\mathbf{i}} + 6x\hat{\mathbf{j}}$, where $\hat{\mathbf{i}}$ and $\hat{\mathbf{j}}$ are the unit vectors along x and y-axes. Find the equation of path followed by the particle.

(a) $y = x^2$ (b) $y = \frac{1}{x^2}$ (c) $y = 2x^2$ (d) $y = \frac{1}{x}$

91 A particle is projected with velocity v_0 along x-axis. The deceleration on the particle is proportional to the square of the distance from the origin, i.e. $|a| = \alpha x^2$. The distance at which the particle stop is

(a) $\sqrt{\frac{3v_0}{2\alpha}}$ (b) $\left(\frac{3v_0}{2\alpha}\right)^{\frac{1}{3}}$ (c) $\sqrt{\frac{2v_0^2}{3\alpha}}$ (d) $\left(\frac{3v_0^2}{2\alpha}\right)^{\frac{1}{3}}$

92 A parachutist after bailing out falls 20 m without friction. When parachute opens, it decelerates at 2 m/s^2. He reached the ground with a speed of 4 m/s. At what height, did he bail out?

(a) 91 m (b) 182 m (c) 293 m (d) 116 m

93 A stationary man observes that the rain is falling vertically downward. When he starts running with a velocity of 12 km/h, he observes that the rain is falling at an angle 60° with the vertical. The actual velocity of rain is

(a) $12\sqrt{3}$ km/h (b) $6\sqrt{3}$ km/h (c) $4\sqrt{3}$ km/h (d) $2\sqrt{3}$ km/h

94 Body A starts from rest with an acceleration a_1. After two sec, another body B starts from rest with an acceleration a_2. If they travel equal distances in fifth sec. The ratio $a_1 : a_2$ will be equal to

(a) 9 : 5 (b) 5 : 9 (c) 5 : 7 (d) 7 : 9

95 A particle moves in a straight line with a constant acceleration. It changes its velocity from 10 ms^{-1} to 20 ms^{-1}, while passing through a distance 135 m in t s. The value of t is (in s)

(a) 12 (b) 9 (c) 10 (d) 1.8

96 An object starts from rest at the origin and moves along the x-axis with constant acceleration of 4 m/s^2. Its average velocity as it goes from $x = 2$ m to $x = 8$ m is

(a) 3 m/s (b) 4 m/s (c) 5 m/s (d) 6 m/s

97 A lift whose cage is 3 m high is moving up with an acceleration of 2 m/s^2. A piece of stone is dropped from the top of the cage of the lift, when its velocity is 8 m/s. If $g = 10$ m/s^2, then the stone will reach the floor of the lift after

(a) 0.3 s (b) 0.5 s (c) 0.4 s (d) 0.7 s

98 Two particles A and B are moving on a straight line as shown in figure. At $t = 0$, A has velocity 36 m/s and has acceleration 2 m/s^2 and B has velocity 30 m/s and has acceleration 4 m/s^2. The separation between A and B is 8 m. Find the time, when B overtakes A eventually.

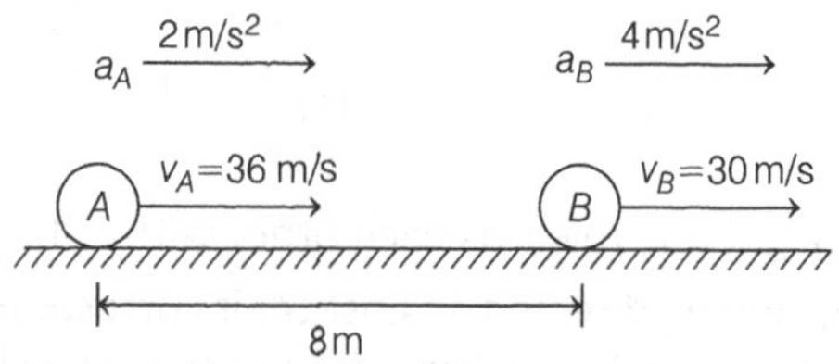

(a) 4 s (b) 2 s (c) 6 s (d) 8 s

99 A body falling freely under gravity passes two points 30 m apart in 1 s. From what point above the upper point, it began to fall? (Take, $g = 9.8$ ms^{-2})

(a) 32.1 m (b) 16.0 m (c) 8.6 m (d) 4.0 m

100 A rocket rises vertically up from the surface of earth, so that it's distance varies from the earth's surface as $l = bt^2$, where b is a constant. After 10 s, the rocket has travelled 2 km. Determine its speed at that moment.

(a) 100 m/s (b) 200 m/s (c) 400 m/s (d) 600 m/s

101 A body is projected vertically upward with a velocity of 10 ms^{-1}. It reaches the maximum vertical height h in time t. In time $\frac{t}{2}$, the height covered is

(a) $\frac{h}{2}$ (b) $\frac{2h}{5}$ (c) $\frac{3h}{4}$ (d) $\frac{5h}{8}$

102 The engine of a vehicle can produce a maximum acceleration of 4 ms^{-2}. Its brakes can produce a maximum retardation of 6 ms^{-2}. The minimum time in which it can cover a distance of 3 km is

(a) 30 s (b) 40 s (c) 50 s (d) 60 s

103 A body is projected vertically upwards from ground. If t_1 and t_2 be the times at which it is at height h above the projection while ascending and descending respectively, then h is

(a) $\frac{1}{2} gt_1t_2$ (b) gt_1t_2 (c) $2\, gt_1t_2$ (d) $2\, hg$

104 A, B, C and D are points in a vertical line, such that $AB = BC = CD$. If a body falls from rest from A, then the times of descend through AB, BC and CD are in the ratio

(a) $1 : \sqrt{2} : \sqrt{3}$ (b) $\sqrt{2} : \sqrt{3} : 1$ (c) $\sqrt{3} : 1 : \sqrt{2}$ (d) $1 : (\sqrt{2} - 1) : (\sqrt{3} - \sqrt{2})$

105 The distance between two trucks moving towards each other is decreasing at the rate of 10 m/s. If these trucks travel in same direction, the separation increases at the rate of 5 m/s. The speeds of the trucks are

(a) $v_1 = 8.5$ m/s, $v_2 = 1.5$ m/s (b) $v_1 = 7.5$ m/s, $v_2 = 2.5$ m/s
(c) $v_1 = 5$ m/s, $v_2 = 5$ m/s (d) None of these

106 Thief's car is moving with a speed of 10 m/s. A police van chasing the car with a speed of 5 m/s fires a bullet at the thief's car with muzzle velocity 72 km/h. Find the speed with which the bullet will hit the car.

(a) 10 m/s (b) 20 m/s (c) 15 m/s (d) 25 m/s

107 Two particles start moving from same position at same time along same straight line. One with constant velocity v and second with constant acceleration a. During the time interval when second particle catches the first particle, the greatest distance between them is

(a) $\frac{v^2}{a}$ (b) $\frac{v^2}{2a}$ (c) $\frac{2v^2}{a}$ (d) $\frac{v^2}{4a}$

108 A stone is thrown upwards from a tower with a velocity 50 ms^{-1}. Another stone is simultaneously thrown downwards from the same location with a velocity 50 ms^{-1}. When the first stone is at the highest point, the relative velocity of the second stone with respect to the first stone is (assume that second stone has not yet reached the ground)

(a) zero (b) 50 ms^{-1}
(c) 100 ms^{-1} (d) 150 ms^{-1}

109 A coin is released inside a lift at a height of 2 m from the floor of the lift. The height of the lift is 10 m. The lift is moving with an acceleration of 9 m/s^2 downwards. The time after which the coin will strike with the floor of lift is (Take, $g = 10\ m/s^2$)

(a) 4 s (b) 2 s (c) $\frac{4}{\sqrt{21}}$ s (d) $\frac{2}{\sqrt{11}}$ s

110 A person throws some balls into the air after every second. The next ball is thrown, when the velocity of the first ball is zero. How high do the balls rise above his hand? (Take, $g = 10\ m/s^2$)

(a) 2 m (b) 5 m
(c) 8 m (d) 10 m

111 A particle on y-axis starts from rest along a straight line $y = 3$ with constant acceleration $2\sqrt{3}\ m/s^2$. Angle made by position vector of the particle with positive x-axis after one second of its motion is

(a) 30° (b) 45°
(c) 60° (d) 90°

112 A particle is moving in free space with some velocity as shown. Forces $\mathbf{F}_1$, $\mathbf{F}_2$ and $\mathbf{F}_3$ of magnitudes 10 N, 10 N and 15 N are applied on the particle,

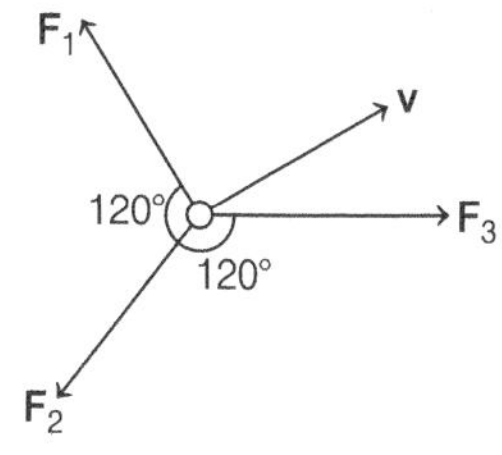

(a) its velocity changes only in magnitude
(b) its velocity changes only in direction
(c) its velocity changes in both magnitude and direction
(d) its velocity remains constant

113 A ball is thrown vertically upwards with a velocity u from the balloon descending with velocity v. The ball will pass by the balloon after time

(a) $\frac{u-v}{2g}$ (b) $\frac{u+v}{2g}$ (c) $\frac{2(u-v)}{g}$ (d) $\frac{2(u+v)}{g}$

114 Two balls are projected simultaneously with the same speed from the top of a tower, one vertically upwards and the other vertically downwards. They reach the ground in 9 s and 4 s, respectively. The height of the tower is (Take, $g = 10\ m/s^2$)

(a) 90 m (b) 180 m
(c) 270 m (d) 360 m

115 A ball is dropped from a long tower at $t = 0$. After 10 s, another ball is thrown in downward direction from the same point with speed v. If these two balls meet at $t = 15$ s, then the value of v is (Take, $g = 10\ m/s^2$)

(a) 50 m/s (b) 100 m/s
(c) 200 m/s (d) 225 m/s

116 The diagram shows the position-time graph for a particle moving in a straight line. Find the average velocity for the interval from $t = 0$ to $t = 5$ s.

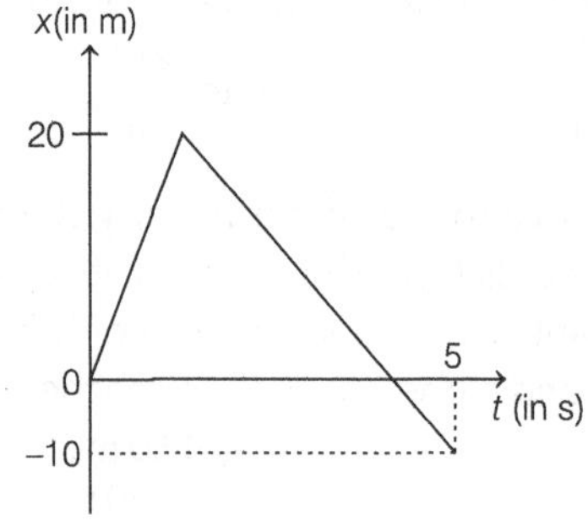

(a) $-2\ \text{ms}^{-1}$ (b) $2\ \text{ms}^{-1}$ (c) $-4\ \text{ms}^{-1}$ (d) $4\ \text{ms}^{-1}$

117 A car is moving with uniform velocity 10 m/s. After $t = 5$ s, brakes are applied to produce constant retardation and car comes to rest at $t = 10$ s. The position-time graph from $t = 0$ to 15 s is

(a)
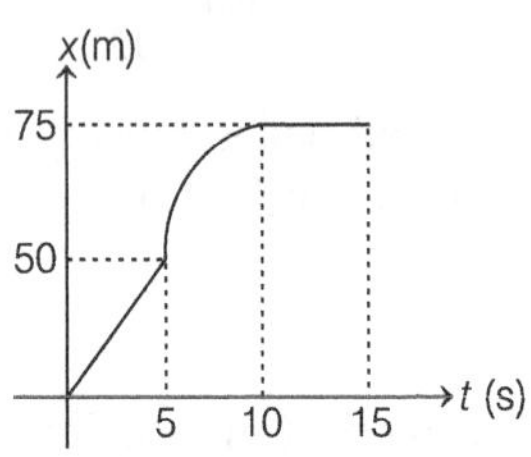

(b)
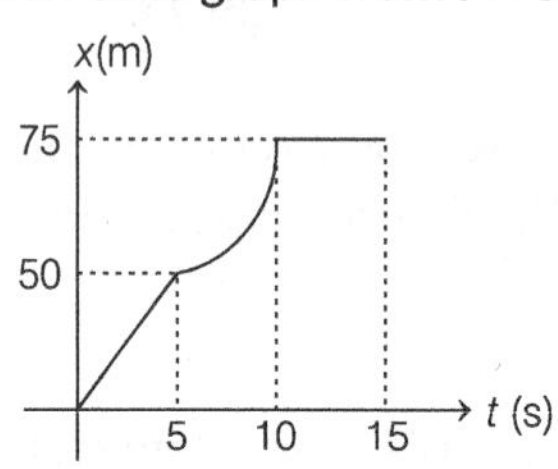

(c)
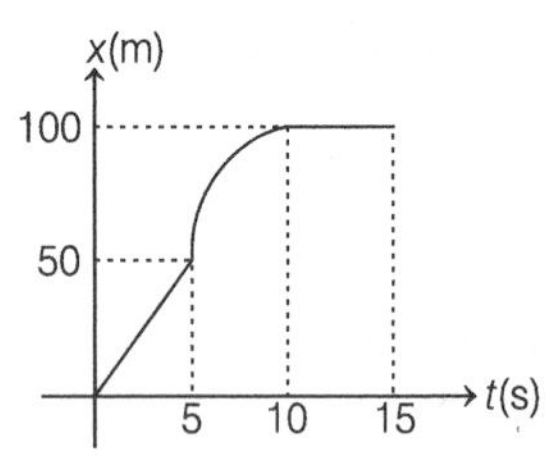

(d)
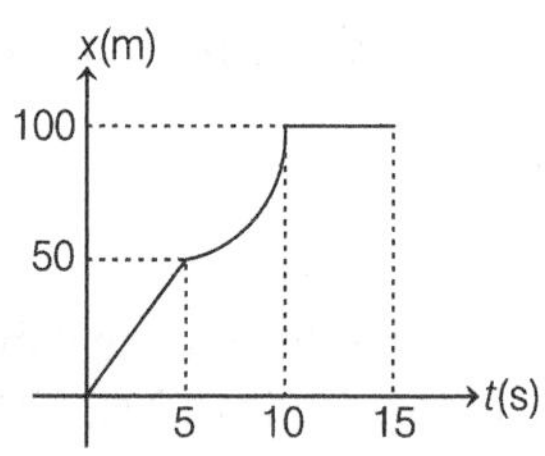

118 Figure shows position *versus* time graph of two rabbits running opposite to each other between two trees. Which of the following statements are true?

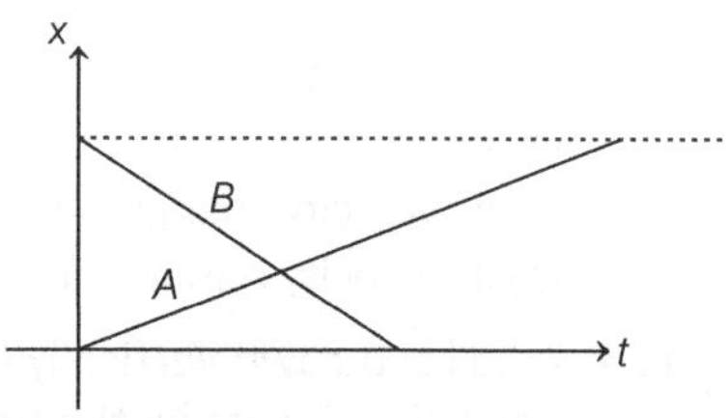

(a) Rabbit *B* has greater magnitude of average velocity.
(b) Rabbit *A* has greater magnitude of average velocity.
(c) Both the rabbits have same displacement.
(d) Both the rabbits have same constant speed.

119 A river flows with a constant speed of 5 kmh^{-1}. The distance-time graph of a person that crosses the river directly is shown in figure. The direction of his velocity relative to water, makes with the direction of river flow, an angle of

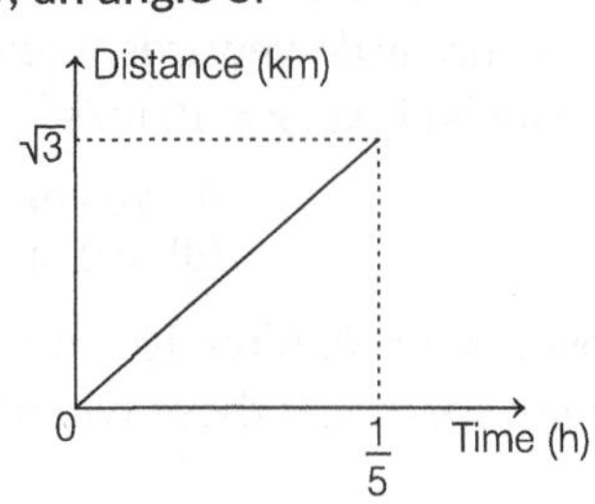

(a) 150° (b) 120°
(c) 135° (d) 127°

120 A stone is thrown upwards from the top of a tower 60 m high at a speed of 20 m/s. The correct position-time graph for the time interval in which it reaches ground, is ($g = 10$ m/s^2 and take origin to be the base of the tower)

(a)
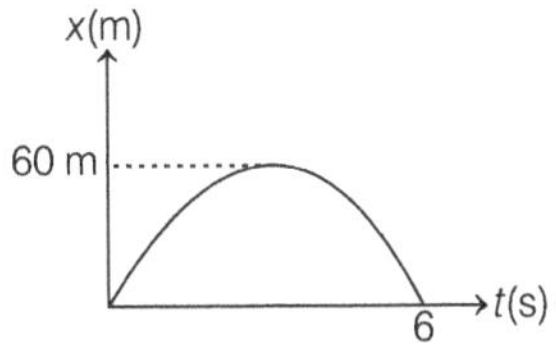

(b)
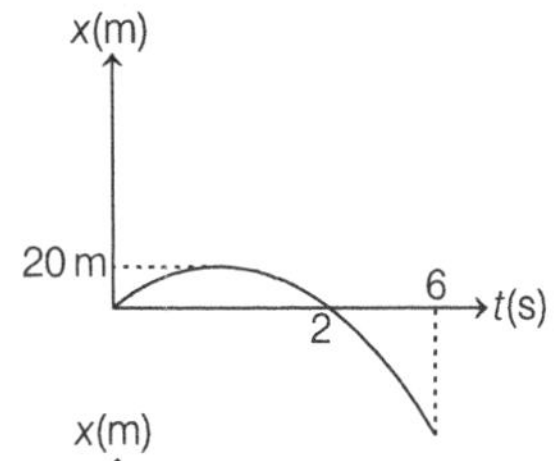

(c)
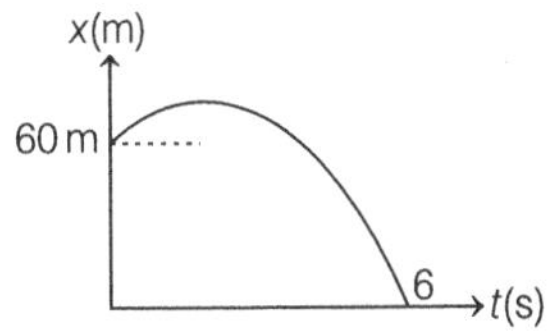

(d)
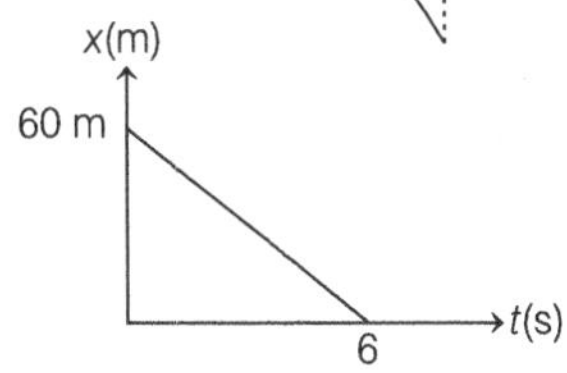

121 A particle moves rectilinearly such that its velocity-time graph is shown in figure. If initially particle is at $x = 5$ m, then position of particle as a function of time t will be

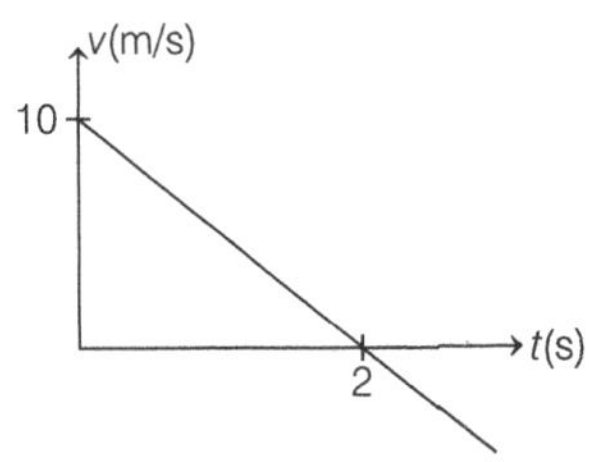

(a) $5 + 10t - 2.5\,t^2$ (b) $10t - 2.5\,t^2$ (c) $5 - 10t + 2.5\,t^2$ (d) $10t + 2.5\,t^2 - 5$

122 Acceleration-time graph of a particle moving along x-axis is as shown. The particle will have the velocity same as its initial velocity at time

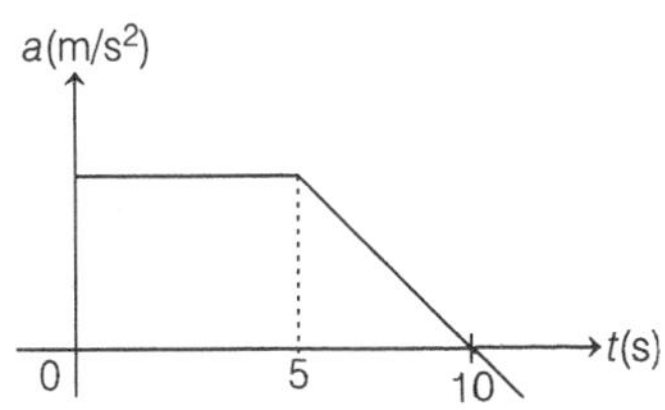

(a) 10 s (b) $(10 + \sqrt{3})$ s (c) $(10 + 5\sqrt{3})$ s (d) $(10 + 2\sqrt{3})$ s

123 Which of the following pairs of graphs does not represent the motion of the same particle in the same interval (curves are parabolic)?

(a)
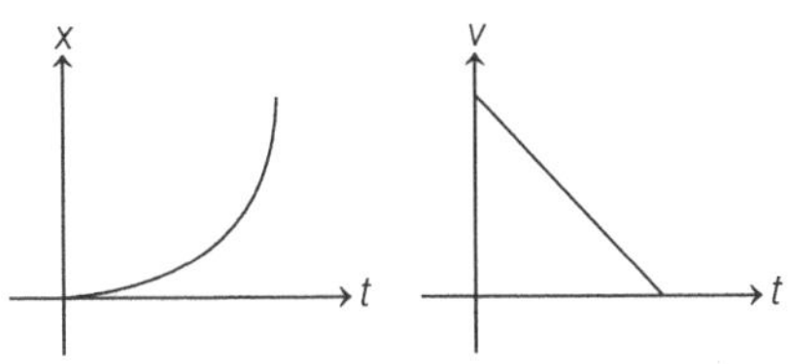

(b)
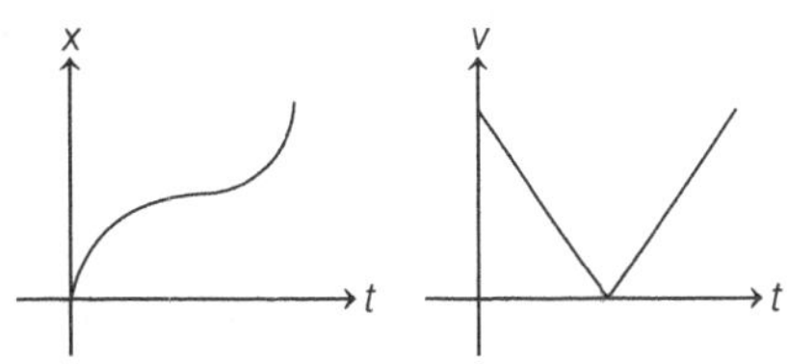

(c)
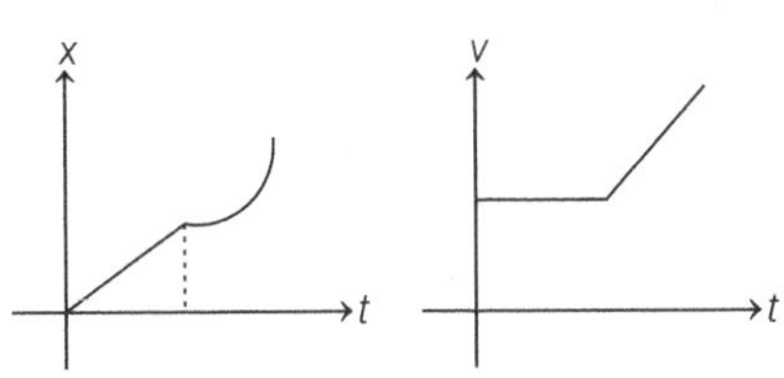

(d)
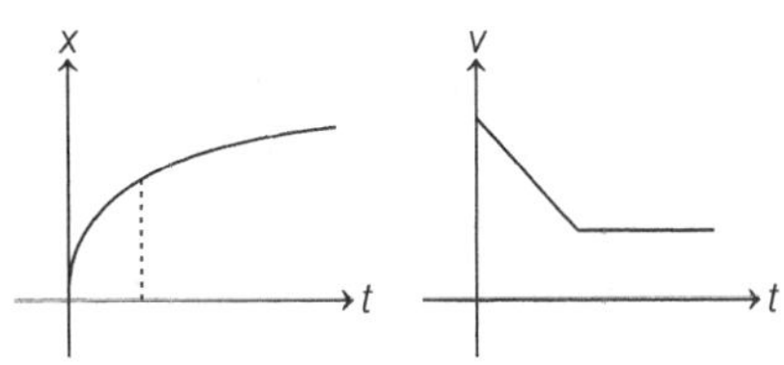

124 Initially car A is 10.5 m ahead of car B. Both start moving at time $t = 0$ in the same direction along a straight line. The velocity-time graph of two cars is shown in figure. The time when the car B will catch the car A, will be

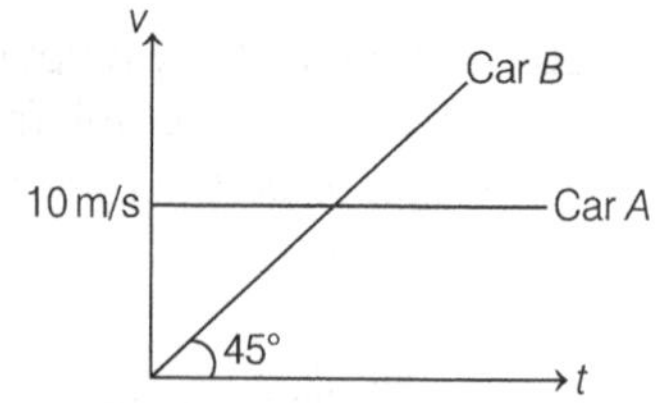

(a) 21 s (b) $2\sqrt{5}$ s
(c) 20 s (d) 15 s

125 The graph given shows the position of two cars A and B as a function of time. The cars move along the x-axis on parallel but separate tracks, so that they can pass each other's position without colliding. Mark the **incorrect** statement.

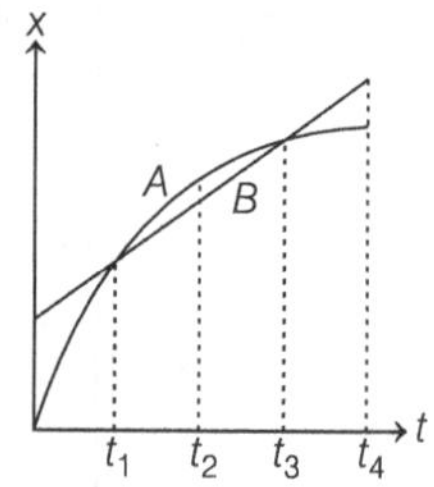

(a) At time t_1, car B overtakes the car A.
(b) At time t_3, car B is moving faster.
(c) At time t_2, the two cars may have the same velocity.
(d) The motion of car A as shown on the graph shows that it is slowing down.

126 The velocity-time graph of a particle moving on a straight line is as shown in figure

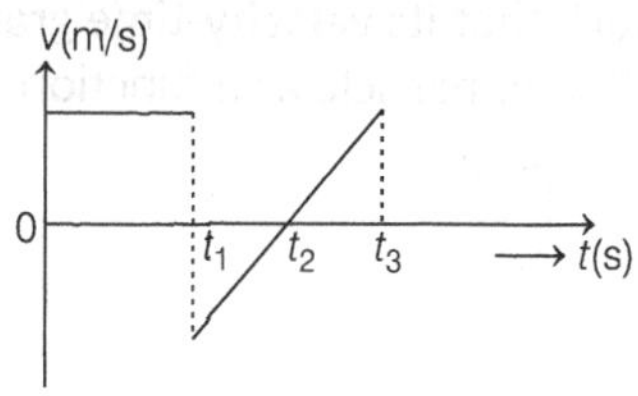

The corresponding displacement-time graph will be

(a)
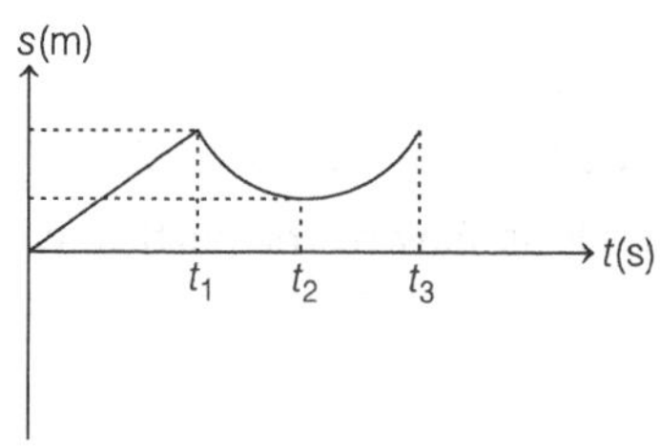

(b)
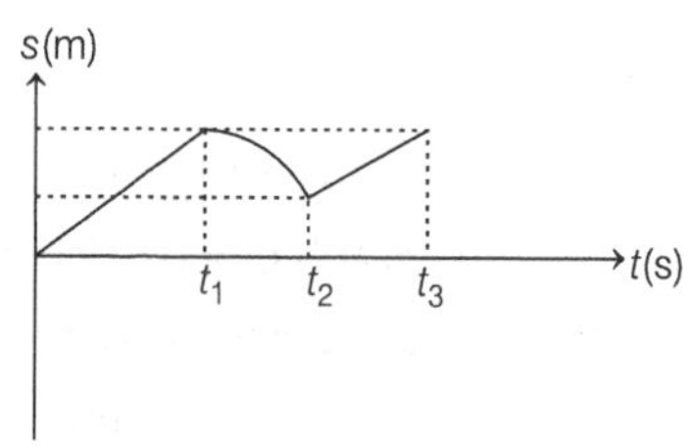

(c)
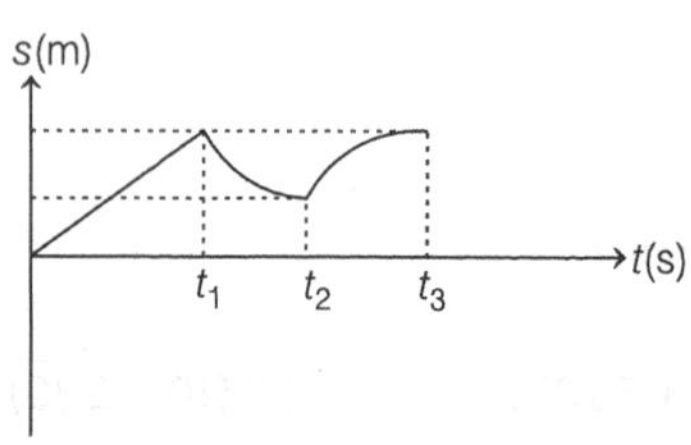

(d)
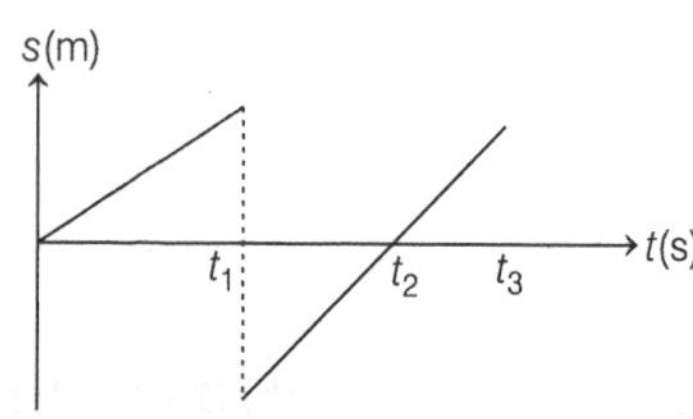

127 A ship A is moving westwards with a speed of 10 km h^{-1} and a ship B, 100 km south of A, is moving northwards with a speed of 10 km h^{-1}. The time after which the distance between them becomes shortest, is

(a) 5 h (b) $5\sqrt{2}$ h
(c) $10\sqrt{2}$ h (d) 3 h

128 A person walks at the rate of 3 km/h. Rain appears to him in vertical direction at the rate of $3\sqrt{3}$ km/h. Find magnitude and direction of true velocity of rain.

(a) 6 km/h, inclined at an angle of 45° to the vertical towards the person's motion
(b) 3 km/h, inclined at an angle of 30° to the vertical towards the person's motion
(c) 6 km/h, inclined at an angle of 30° to the vertical towards the person's motion
(d) 6 km/h, inclined at an angle of 60° to the vertical towards the person's motion

129 To a man running at a speed of 5 m/s, the rain drops appear to be falling at an angle of 45° from the vertical. If the rain drops are actually falling vertically downwards, then velocity of rain (in m/s) is

(a) $5\sqrt{3}$ (b) 5 (c) $5\sqrt{2}$ (d) 4

130 Figure shows two ships moving in xy-plane with velocities v_A and v_B. The ships move, such that B always remains north of A. The ratio $\frac{v_B}{v_A}$ is equal to

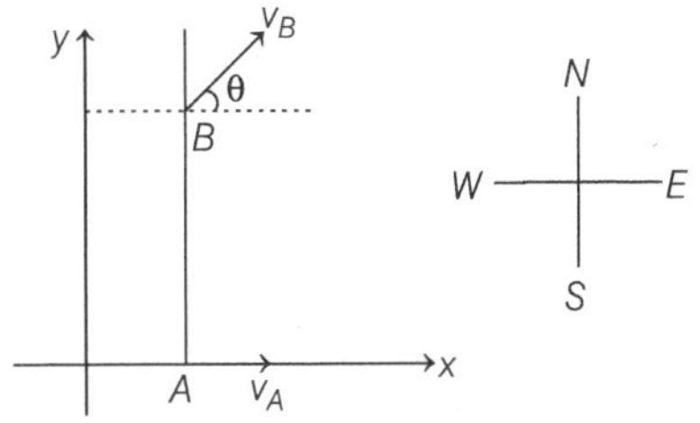

(a) $\cos\theta$ (b) $\sin\theta$ (c) $\sec\theta$ (d) $\text{cosec}\,\theta$

131 A man is crossing a river flowing with velocity of 5 m/s. He reaches at point B directly across at a distance of 60 m in 5 s. His velocity in still water should be

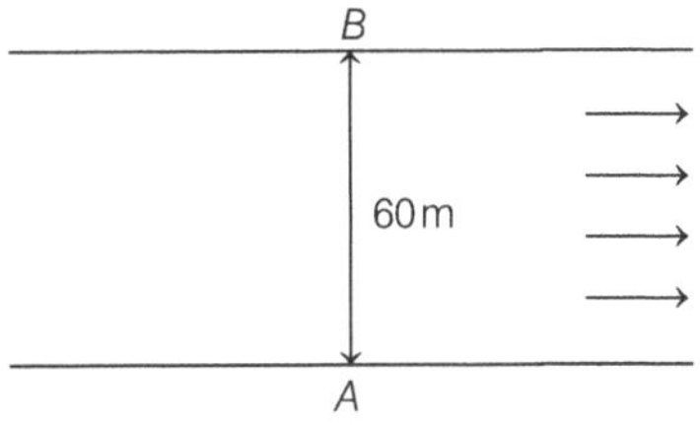

(a) 12 m/s (b) 13 m/s (c) 5 m/s (d) 10 m/s

132 Two particles A and B start moving simultaneously from a point P on straight line paths. Their paths make an angle of 37° with each other as shown in the figure. One of them moves with uniform speed of 10 m/s and the other with 8 m/s. Distance between the particles 40 s after they leave point P is closed to

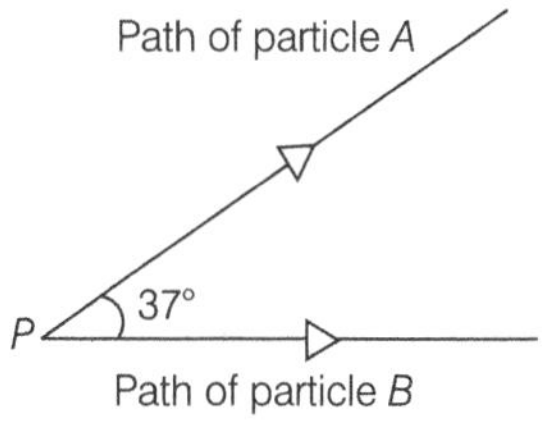

(a) 192 m (b) 240 m
(c) 320 m (d) 426 m

133 An object is moving in xy-plane, its velocity and acceleration at $t = 0$ are represented in figure.

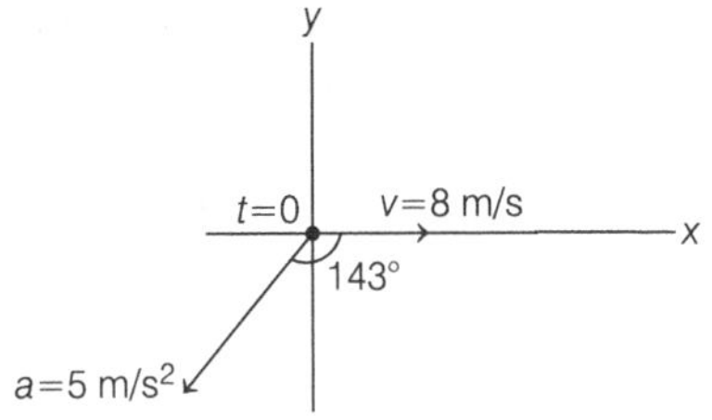

After how much time object crosses the y-axis?

(a) 4 s (b) 2 s (c) 3 s (d) None of these

134 A person in a stationary vehicle on a road has to hold his umbrella at 60° with the vertical to keep the rain away. He throws the umbrella and drives the vehicle at 20 ms^{-1}. He finds the rain drops are falling vertically on him. The speed of the rain drops with respect to the road is

(a) $\frac{40}{\sqrt{3}}$ ms^{-1} (b) $\frac{20}{\sqrt{3}}$ ms^{-1} (c) $\frac{30}{\sqrt{3}}$ ms^{-1} (d) $\frac{25}{\sqrt{3}}$ ms^{-1}

135 A taxi leaves the station X for station Y every 10 min. Simultaneously, a taxi also leaves the station Y for station X every 10 min. The taxis move at the same constant speed and go from X to Y or *vice- versa* in 2 h. How many taxis coming from the other side will meet each taxi enroute from Y to X?

(a) 11 (b) 12 (c) 23 (d) 24

136 The magnitude of velocity of a body is given by $v = |t-6|$ m/s, then average speed of body between 2 s to 6 s will be

(a) 2 m/s (b) 6 m/s (c) 3.5 m/s (d) 4.5 m/s

137 For $v = \sin t$, find average velocity between 0 and t for $0 \le t \le \frac{\pi}{2}$.

(a) $\frac{2}{\pi}$ m/s (b) $\frac{4}{\pi}$ m/s (c) $-\frac{2}{\pi}$ m/s (d) $-\frac{4}{\pi}$ m/s

138 A particle is moving along x-axis, such that its position (x) varies with time (t) as $x = (t \sin t)$ m, then average acceleration of the particle from $t = 0$ s to $t = \frac{\pi}{2}$ s is

(a) $\frac{\pi}{2}$ m/s^2 (b) $\frac{2}{\pi}$ m/s^2 (c) $\frac{\pi}{4}$ m/s^2 (d) $\frac{4}{\pi}$ m/s^2

139 A particle of mass m moves on positive x-axis under the influence of force acting towards the origin given by $-kx^2\hat{\mathbf{i}}$. If the particle starts from rest at $x = a$, the speed it will attain, when it crosses the origin is

(a) $\sqrt{\frac{k}{ma}}$ (b) $\sqrt{\frac{2k}{ma}}$ (c) $\sqrt{\frac{ma}{2k}}$ (d) None of these

140 A particle moves with velocity $\mathbf{v}$ in xy-plane as $\mathbf{v} = C\,(xy^2\hat{\mathbf{i}} + yx^2\hat{\mathbf{j}})$ m/s. Then, the general equation of its path will be

(a) $y = x$ + constant (b) $y = x^2$ + constant
(c) $y^2 = x^2$ + constant (d) $y^3 = x^3$ + constant

141 A particle is released from a certain height $h = 405$ m due to wind, the particle gathers the horizontal velocity component $v_x = ay$, where $a = \frac{1}{5}\,\text{s}^{-1}$ and y is the vertical displacement of the particle from the point of release, then find horizontal drift of the particle when it strikes the ground. (Take, $g = 10\ \text{ms}^{-2}$)

(a) 729 m (b) 243 m
(c) 81 m (d) None of these

142 Two bodies are moving such that their x-coordinates are changing according to the law, $x_1 = -3 + 2t + t^2$ and $x_2 = 7 - 8t + t^2$. The relative speed v of bodies at the time of their meeting will be

(a) 25 m/s (b) 15 m/s (c) 10 m/s (d) 5 m/s

143 Choose the **correct** option(s).

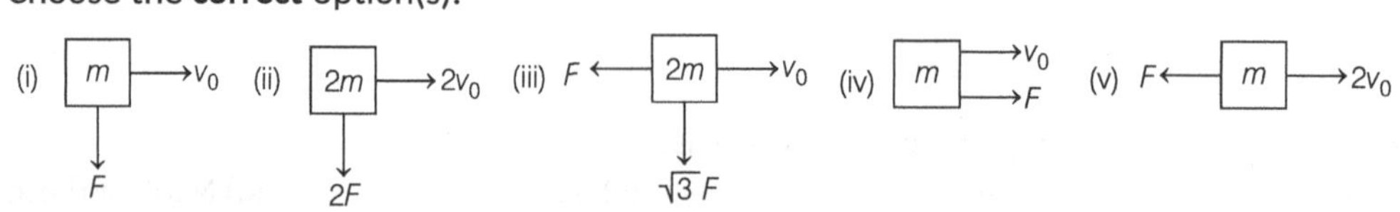

(A) Magnitude of rate of change of velocity is same for all.
(B) Magnitude of rate of change of speed is same for all.
(C) Initially magnitude of rate of change of speed is same for (i) and (ii).
(D) Magnitude of rate of change of speed is same for (iv) and (v).

(a) Only A (b) A, C, D (c) A, B, C, D (d) A, D

144 A bird is tossing (flying to and fro) between two cars moving towards each other on a straight road. One car has a speed of 27 km h^{-1} while the other has the speed of 18 km h^{-1}. The bird starts moving from first car towards the other and is moving with the speed of 36 km h^{-1}, when the two cars were separated by 36 km. The total distance covered by the bird till the cars meet is

(a) 28.8 km (b) 38.8 km (c) 48.8 km (d) 58.8 km

145 A particle moves in the xy-plane with velocity $u_x = 8t - 2$ and $u_y = 2$. If it passes through the point (14, 4) at $t = 2$ s, the equation of its path is

(a) $x^2 = 2 - y$ (b) $x = 2 - y^2$
(c) $x = y^2 - y + 14$ (d) $x = y^2 + y - 2$

146 An aircraft is flying at a height of 3400 m above the ground. If the angle subtended at a ground observation centre by two positions of the aircraft is 10 sec apart is 30°. What is the speed of the aircraft? (The observation point is on the vertical line, which is midway between positions of aircraft.) [Take, tan15° = 0.2679]

(a) 262.4 ms^{-1} (b) 131.2 ms^{-1} (c) 91.1 ms^{-1} (d) 182.2 ms^{-1}

Paragraph for questions from 147 to 149

Two cars travelling on a straight road. The car C_1 is going at a constant speed of 9 m/s. 40 m ahead of C_1 is car C_2 at rest, but starts moving with an acceleration of $1\,m/s^2$.

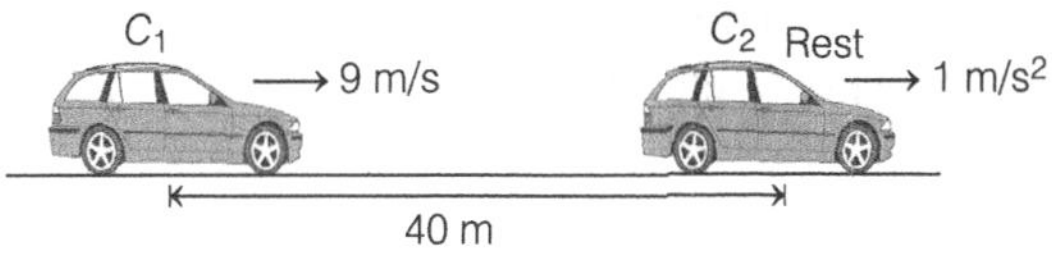

147 The graph correctly representing the position of two cars with time with respect to initial position of C_1

(a)
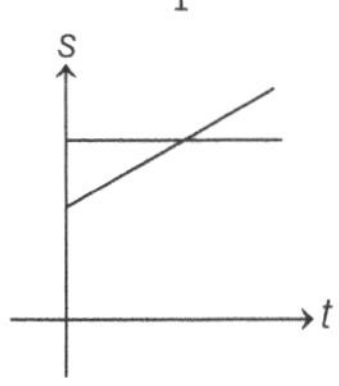

(b)
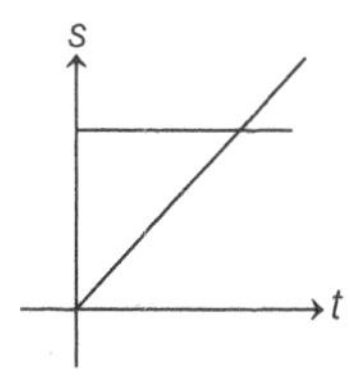

(c)
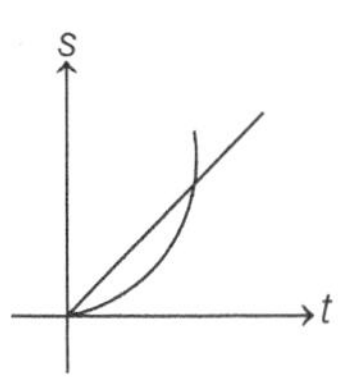

(d)
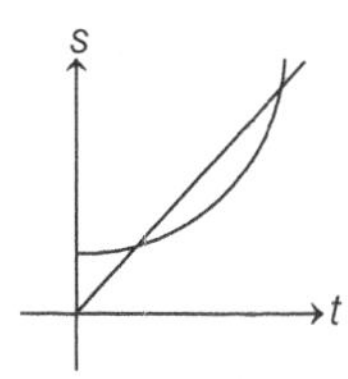

148 The graph correctly representing the velocity of the two cars with time

(a)
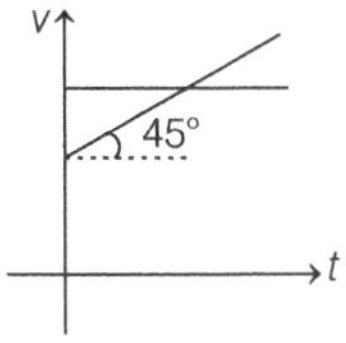

(b)
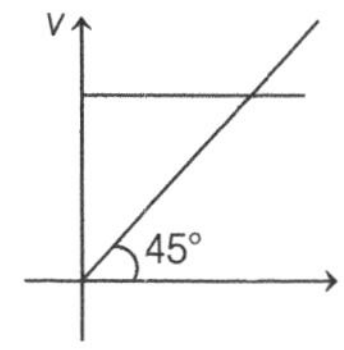

(c)
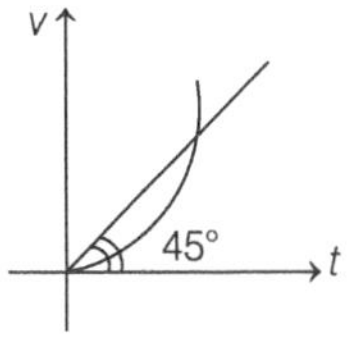

(d)
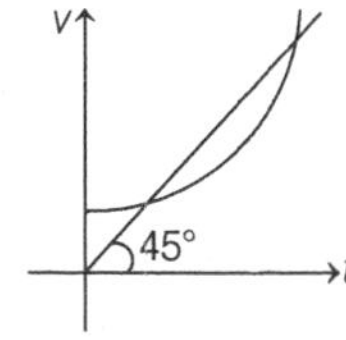

149 During their motion, the car C_1 overtakes the car C_2. But after C_2 overtakes C_1, determine the maximum lead that C_1 can have from C_2.

(a) 0.5 m (b) 9.5 m (c) 81 m (d) 90 m

150 In a car race, car A takes t_0 time less to finish than car B and passes the finishing point with a velocity v_0 more than car B. The car starts from rest and travel with constant acceleration a_1 and a_2. Then, the ratio $\frac{v_0}{t_0}$ is equal to

(a) $\frac{a_1^2}{a_2}$ (b) $\frac{a_1 + a_2}{2}$ (c) $\sqrt{a_1 a_2}$ (d) $\frac{a_2^2}{a_1}$

151 A particle is moving in a straight line and passes through a point O with velocity of 10 ms^{-1}. The particle moves with a constant retardation of 5 ms^{-2} for 3 s and there after moves with constant velocity. How long after leaving O does the particle return to O?

(a) 3 s (b) 8 s
(c) 4.5 s (d) Never

152 A particle is moving in xy-plane. At certain instant of time, the components of its velocity and acceleration are as follows : $v_x = 3$ m/s, $v_y = 4$ m/s, $a_x = 2 \text{ m/s}^2$ and $a_y = 1 \text{ m/s}^2$. The rate of change of speed at this moment is

(a) $\sqrt{10} \text{ m/s}^2$ (b) 4 m/s^2
(c) 10 m/s^2 (d) 2 m/s^2

153 A body travelling along a straight line with a uniform acceleration has velocity 5 m/s at a point A A and 15 m/s at a point B, respectively. If M is the mid-point of AB, then choose **incorrect** statement.

(a) The ratio of time taken by the body to cover distance MB and AM is $\left[\frac{\sqrt{5}-1}{2}\right]$.

(b) The velocity at M is $5\sqrt{5}$ m/s.

(c) Average velocity over AM is $\frac{5(\sqrt{5}+1)}{2}$ m/s.

(d) The product of the acceleration and the distance AB is $200 \text{ m}^2/\text{s}^2$.

154 A ball is thrown vertically upwards with some speed. It reaches two points A and B one after another, such that heights of A and B are one-fourth and three-fourth of the maximum height attained. If the total time of flight is T, the time taken by the ball to travel from A to B, is

(a) $\frac{\sqrt{3}}{4} T$ (b) $\left(\frac{\sqrt{3}-1}{2}\right) T$ (c) $\left(\frac{\sqrt{3}+1}{2}\right) T$ (d) $\frac{T}{\sqrt{2}}$

155 A ball is dropped from a long tower. One second after its fall, air provides horizontal acceleration of $\frac{4g}{3}$. Total displacement covered by the ball after 4 s of its fall will be (Take, $g = 10 \text{ m/s}^2$)

(a) 133.3 m (b) 100 m
(c) 227.8 m (d) 150 m

156 A particle moves a distance x in time t according to the equation $x = \frac{1}{\sqrt{(t+1)}}$. The acceleration of the particle is proportional to

(a) $(\text{velocity})^{\frac{2}{3}}$ (b) $(\text{velocity})^{\frac{5}{3}}$ (c) $(\text{distance})^{\frac{1}{3}}$ (d) $(\text{distance})^{\frac{5}{2}}$

157 A particle starting from rest, travels a distance s with uniform acceleration and then travels a distance of $3s$ with uniform velocity followed by a distance $5s$ with uniform deceleration and comes to rest. Then, the ratio of average speed to the maximum speed of the particle is

(a) $\frac{1}{5}$ (b) $\frac{2}{5}$ (c) $\frac{3}{5}$ (d) $\frac{4}{5}$

158 Between the two stations, a train accelerates uniformly at first, then moves with constant velocity and finally retards uniformly. If the ratio of the time taken be $1:8:1$ and the maximum speed attained be 60 km/h, then what is the average speed over the whole journey?

(a) 48 km/h (b) 52 km/h (c) 54 km/h (d) 56 km/h

159 A particle crossing the origin of the coordinates at time $t = 0$, moves in the xy-plane with a constant acceleration β in the y-direction. If its equation is $y = \alpha x^2$ (α is a constant), its velocity component in the x- direction is

(a) $\sqrt{\frac{\alpha}{\beta}}$ (b) $\sqrt{\frac{\beta}{\alpha}}$ (c) $\sqrt{\frac{\alpha}{2\beta}}$ (d) $\sqrt{\frac{\beta}{2\alpha}}$

160 A man is moving down an incline with a speed v. He meets rain with the same speed but horizontally. The speed of rain is.

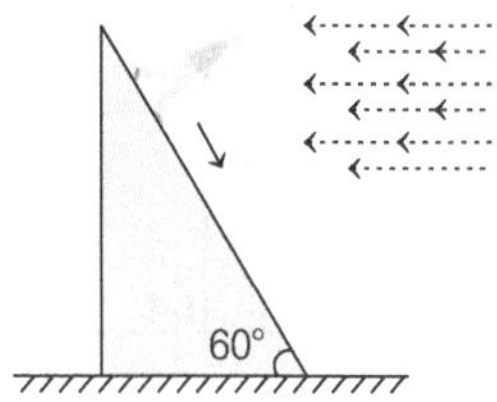

(a) v (b) $\sqrt{3}v$

(c) $\frac{\sqrt{3}}{2}v$ (d) $\frac{1}{2}v$

161 Two particles P and Q are moving in a straight line along positive x-axis. Displacement (s) and time (t) graph is plotted for both the particles.

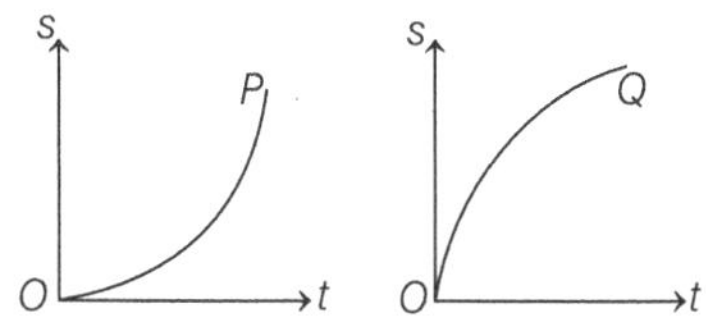

Consider the following statements.

(1) Speed of P is increasing with time.
(2) Speed of Q is increasing with time.
(3) Average speed of P increases with time.
(4) Average speed of Q decreases with time.
(5) Acceleration of P is positive.
(6) Acceleration of Q is negative.

Number of correct statements is

(a) 3 (b) 4 (c) 5 (d) 6

162 The velocity-displacement graph of a particle moving along a straight line is shown in figure

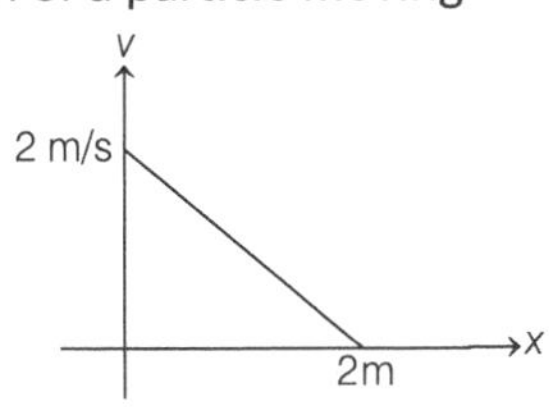

Then, the acceleration-displacement graph is

(a) 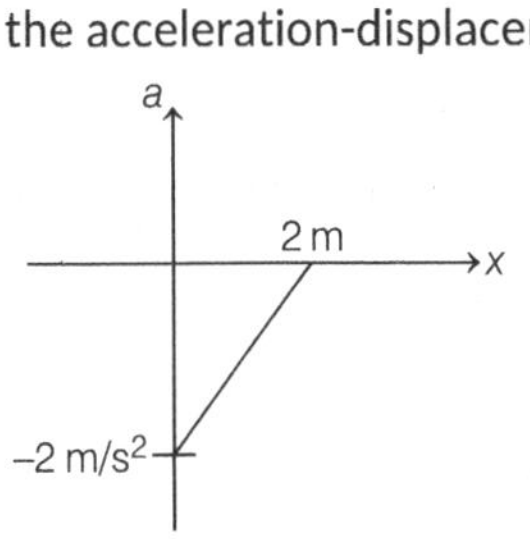

(b) 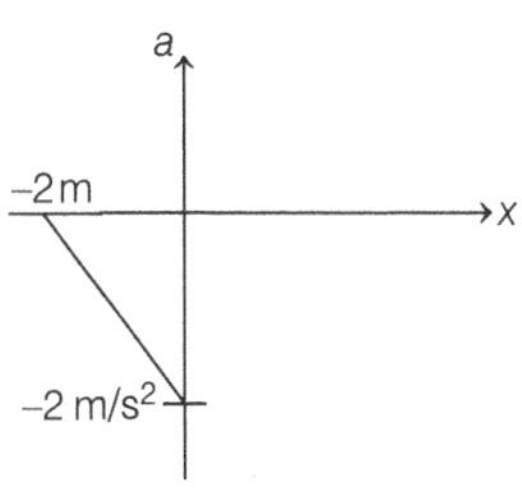

(c) 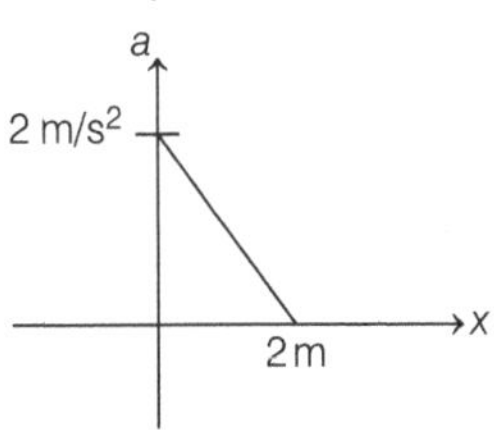

(d) 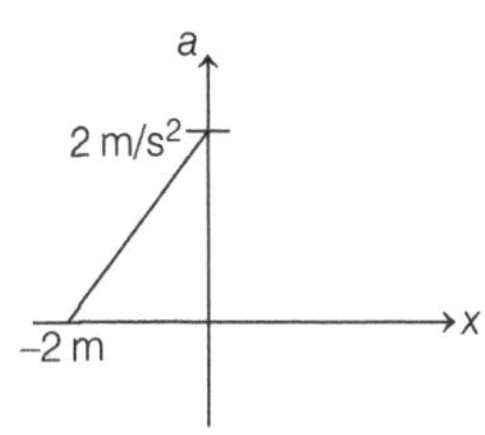

163 The acceleration of a particle which moves along the positive x- axis varies with its position as shown. If the velocity of the particle is 0.8 m/s at $x = 0$, the velocity of the particle at $x = 1.4$ m is (in m/s)

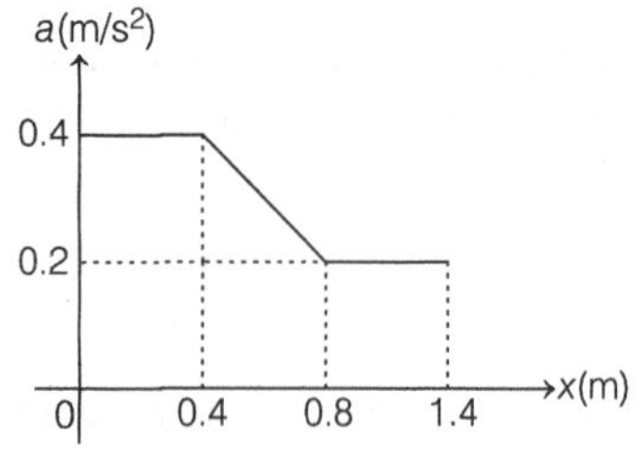

(a) 1.6 (b) 1.2
(c) 1.4 (d) None of these

164 A particle is moving in xy-plane such that the v_x - t and v_y - t graphs are as shown in figure

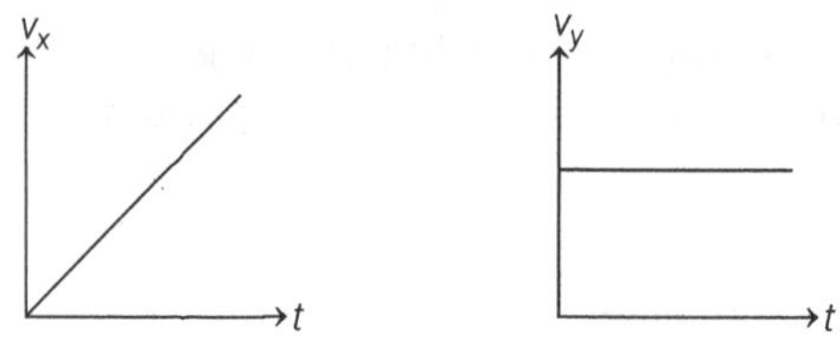

The correct graph representing trajectory on xy-plane is

(a) 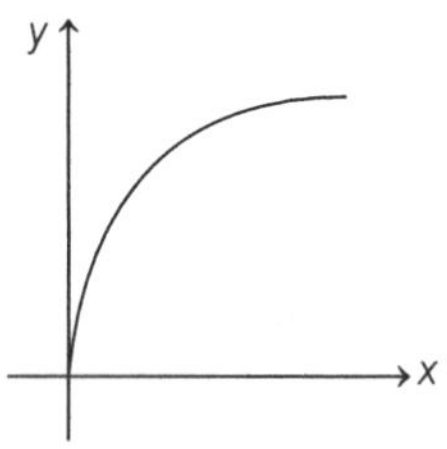

(b) 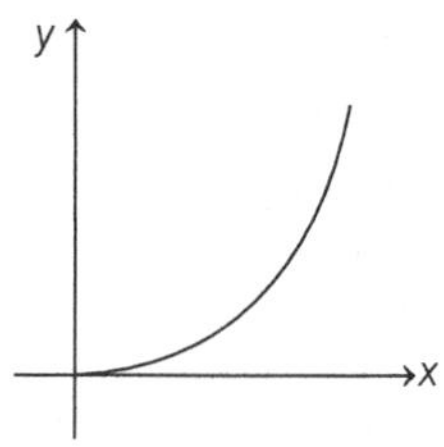

(c) 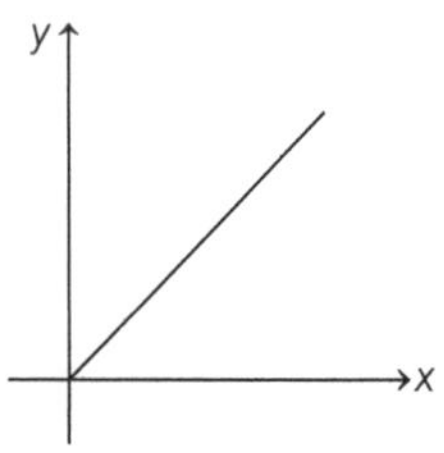

(d) 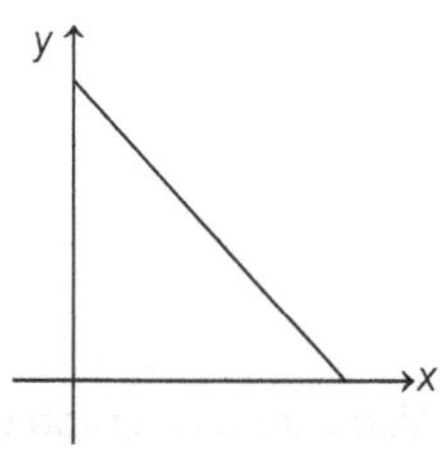

165 In given figure, a smooth square platform $ABCD$ in horizontal plane is moving toward right with a uniform speed $10\sqrt{3}$ m/s. At what angle θ must a particle be projected from B with speed 20 m/s, so that it strikes at the point A?

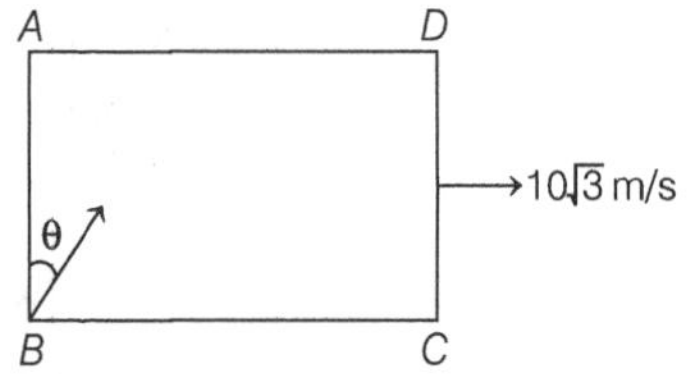

(a) 60° (b) $\cos^{-1}\left(\frac{2}{\sqrt{3}}\right)$

(c) 30° (d) $\sin^{-1}\left(\frac{2}{\sqrt{3}}\right)$

166 A particle starts moving along a straight lines at $t = 0$. Its acceleration-time graph is shown. Correct relation between magnitudes of displacements between time durations $t = 2$ s to $t = 3$ s and $t = 7$ s to $t = 8$ s and represented by s_{23} and s_{78}, respectively is

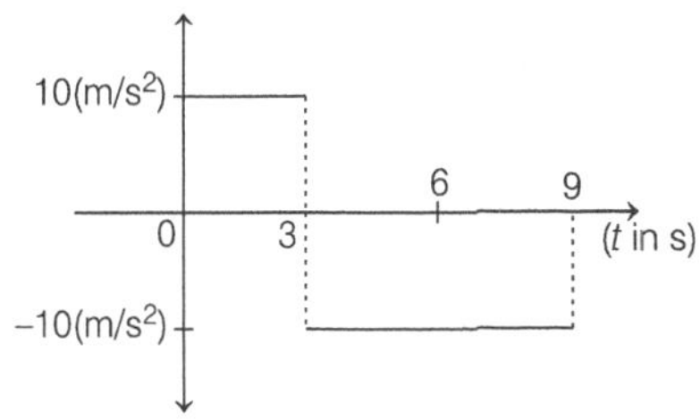

(a) $s_{23} > s_{78}$
(b) $s_{78} > s_{23}$
(c) $s_{23} = s_{78}$
(d) Insufficient information to decide

167 Rain is falling vertically with a speed of 20 m/s relative to air. A person is running in the rain with a velocity of 5 m/s and wind is also blowing with a speed of 15 m/s (both towards east), then velocity of rain with respect to man is
(a) $\sqrt{500}$ m/s, $\tan^{-1}$ (4/2) from vertical
(b) $\sqrt{50}$ m/s, $\tan^{-1}$ (2/3) from vertical
(c) $10\sqrt{5}$ m/s, $\tan^{-1}$ (1/2) from vertical
(d) 20 m/s, $\tan^{-1}$ (4/3) from vertical

168 A swimmer swims in still water at a speed = 5 km/h. He enters a 200 m wide river, having river flow speed = 4 km/h at point A and proceeds to swim at an angle of 127° with the river flow direction. Another point B is located directly across A on the other side. The swimmer lands on the other bank at a point C, from which he walks the distance CB with a speed = 3 km/h. The total time in which he reaches from A to B is
(a) 5 min
(b) 4 min
(c) 3 min
(d) None of these

169 An observer on ground sees a boat crossing a river of width 800 m perpendicular to its stream in 200 s. He also finds a man on a raft floating at speed of 3 m/s with river. The distance travelled by boat as seen by man on the raft in crossing the river is
(a) 800 m
(b) 1000 m
(c) 1200 m
(d) 1600 m

170 A swimmer crosses a flowing stream of breadth b to and fro perpendicular to stream in time T_1. The time taken to cover the same distance up and down the stream is T_2. If T_3 is the time, the swimmer would take to swim a distance $2b$ in still water, then
(a) $T_3 = T_1 + T_2$
(b) $T_1^2 = T_2 T_3$
(c) $T_2^2 = T_1 T_3$
(d) $T_3^2 = T_1 T_2$

LEVEL 03 Previous Years Questions

171 A car starts from rest and accelerates at 5 m/s^2. At $t = 4$ s, a ball is dropped out of a window by a person sitting in the car. What is the velocity and acceleration of the ball at $t = 6$ s? (Take, $g = 10$ m/s^2) **[NEET 2021]**
(a) 20 m/s, 5 m/s^2
(b) 20 m/s, 0
(c) $20\sqrt{2}$ m/s, 0
(d) $20\sqrt{2}$ m/s, 10 m/s^2

172 A small block slides down on a smooth inclined plane, starting from rest at time $t = 0$. Let s_n be the distance travelled by the block in the interval $t = n - 1$ to $t = n$. Then, the ratio $\frac{s_n}{s_{n+1}}$ is **[NEET 2021]**
(a) $\frac{2n-1}{2n}$
(b) $\frac{2n-1}{2n+1}$
(c) $\frac{2n+1}{2n-1}$
(d) $\frac{2n}{2n-1}$

173 A person sitting in the ground floor of a building notices through the window of height 1.5 m, a ball dropped from the roof of the building crosses the window in 0.1 s. What is the velocity of the ball when it is at the topmost point of the window? (Take, $g = 10\ m/s^2$) **[NEET 2020]**

(a) 15.5 m/s (b) 14.5 m/s (c) 4.5 m/s (d) 20 m/s

174 A ball is thrown vertically downward with a velocity of 20 m/s from the top of a tower. It hits the ground after some time with a velocity of 80 m/s. The height of the tower is (Take, $g = 10\ m/s^2$) **[NEET 2020]**

(a) 340 m (b) 320 m (c) 300 m (d) 360 m

175 A person travelling in a straight line moves with a constant velocity v_1 for certain distance 'x' and with a constant velocity v_2 for next equal distance. The average velocity v is given by the relation **[NEET 2019]**

(a) $\frac{1}{v} = \frac{1}{v_1} + \frac{1}{v_2}$ (b) $\frac{2}{v} = \frac{1}{v_1} + \frac{1}{v_2}$ (c) $\frac{v}{2} = \frac{v_1 + v_2}{2}$ (d) $v = \sqrt{v_1 v_2}$

176 A person standing on the floor of an elevator drops a coin. The coin reaches the floor in time t_1 if the elevator is at rest and in time t_2 if the elevator is moving uniformly. The which of the following option is correct? **[NEET 2019]**

(a) $t_1 < t_2$ or $t_1 > t_2$ depending upon whether the lift is going up or down
(b) $t_1 < t_2$
(c) $t_1 > t_2$
(d) $t_1 = t_2$

177 A toy car with charge q moves on a frictionless horizontal plane surface under the influence of a uniform electric field **E**. Due to the force $q\mathbf{E}$, its velocity increases from 0 to 6 m/s in one second duration. At that instant, the direction of the field is reversed. The car continues to move for two more seconds under the influence of this field. The average velocity and the average speed of the toy car between 0 to 3 s are respectively, **[NEET 2018]**

(a) 1 m/s, 3.5 m/s (b) 1 m/s, 3 m/s (c) 2 m/s, 4 m/s (d) 1.5 m/s, 3 m/s

178 Preeti reached the metro station and found that the escalator was not working. She walked up the stationary escalator in time t_1. On other days, if she remains stationary on the moving escalator, then the escalator takes her up in time t_2. The time taken by her to walk up on the moving escalator will be **[NEET 2017]**

(a) $\frac{t_1 + t_2}{2}$ (b) $\frac{t_1 t_2}{t_2 - t_1}$ (c) $\frac{t_1 t_2}{t_2 + t_1}$ (d) $t_1 - t_2$

179 If the velocity of a particle is $v = At + Bt^2$, where A and B are constants, then the distance travelled by it between 1 s and 2 s is **[NEET 2016]**

(a) $3A + 7B$ (b) $\frac{3}{2}A + \frac{7}{3}B$ (c) $\frac{A}{2} + \frac{B}{3}$ (d) $\frac{3}{2}A + 4B3$

180 Two cars P and Q start from a point at the same time in a straight line and their positions are represented by $X_P(t) = at + bt^2$ and $X_Q(t) = ft - t^2$. At what time do the cars have the same velocity? **[NEET 2016]**

(a) $\frac{a - f}{1 + b}$ (b) $\frac{a + f}{2(b - 1)}$ (c) $\frac{a + f}{2(1 + b)}$ (d) $\frac{f - a}{2(1 + b)}$

181 A particle of unit mass undergoes one-dimensional motion, such that its velocity varies according to $v(x) = \beta x^{-2n}$ where, β and n are constants and x is the position of the particle. The acceleration of the particle as a function of x, is given by **[CBSE AIPMT 2015]**

(a) $-2n\beta^2 x^{-2n-1}$ (b) $-2n\beta^2 x^{-4n-1}$
(c) $-2\beta^2 x^{-2n+1}$ (d) $-2n\beta^2 e^{-4n+1}$

182 The motion of a particle along a straight line is described by equation $x = 8 + 12t - t^3$ where, x is in metre and t in s.
The retardation of the particle, when its velocity becomes zero is [CBSE AIPMT 2012]
(a) 24 ms^{-2} (b) zero
(c) 6 ms^{-2} (d) 12 ms^{-2}

183 A body is moving with velocity 30 m/s towards East. After 10s, its velocity becomes 40 m/s towards North. The average acceleration of the body is [CBSE AIPMT 2011]
(a) 7 m/s^2 (b) $\sqrt{7} \text{ m/s}^2$
(c) 5 m/s^2 (d) 1 m/s^2

184 A boy standing at the top of a tower of 20 m height drops a stone. Assuming, $g = 10 \text{ ms}^{-2}$, the velocity with which it hits the ground is [CBSE AIPMT 2011]
(a) 20 m/s (b) 40 m/s
(c) 5 m/s (d) 10 m/s

185 A ball is dropped from a high rise platform at $t = 0$ starting from rest. After 6 s, another ball is thrown downwards from the same platform with a speed v. The two balls meet at $t = 18$ s. What is the value of v? (Take $g = 10 \text{ ms}^{-2}$) [CBSE AIPMT 2010]
(a) 74 ms^{-1} (b) 55 ms^{-1}
(c) 40 ms^{-1} (d) 60 ms^{-1}

ANSWER KEY

Level 1 Objective Questions

1 (a)	2 (b)	3 (b)	4 (b)	5 (d)	6 (c)	7 (c)	8 (b)	9 (b)	10 (a)
11 (c)	12 (b)	13 (b)	14 (c)	15 (c)	16 (c)	17 (c)	18 (c)	19 (c)	20 (c)
21 (c)	22 (b)	23 (a)	24 (b)	25 (b)	26 (c)	27 (b)	28 (b)	29 (a)	30 (b)
31 (a)	32 (a)	33 (c)	34 (d)	35 (c)	36 (c)	37 (a)	38 (a)	39 (a)	40 (a)
41 (b)	42 (d)	43 (b)	44 (b)	45 (d)	46 (b)	47 (b)	48 (a)	49 (a)	50 (c)
51 (d)	52 (a)	53 (a)	54 (d)	55 (c)	56 (d)	57 (a)	58 (a)	59 (d)	60 (d)
61 (b)	62 (d)	63 (a)	64 (d)	65 (d)	66 (d)	67 (c)	68 (c)	69 (c)	70 (c)

Level 2 Objective Questions

71 (a)	72 (a)	73 (d)	74 (b)	75 (a)	76 (c)	77 (b)	78 (a)	79 (c)	80 (c)
81 (c)	82 (d)	83 (d)	84 (d)	85 (a)	86 (a)	87 (a)	88 (b)	89 (b)	90 (a)
91 (d)	92 (d)	93 (c)	94 (b)	95 (b)	96 (d)	97 (d)	98 (a)	99 (a)	100 (c)
101 (c)	102 (c)	103 (a)	104 (d)	105 (b)	106 (c)	107 (b)	108 (c)	109 (b)	110 (b)
111 (c)	112 (c)	113 (d)	114 (b)	115 (c)	116 (a)	117 (a)	118 (a)	119 (b)	120 (c)
121 (a)	122 (c)	123 (a)	124 (a)	125 (a)	126 (a)	127 (a)	128 (c)	129 (b)	130 (c)
131 (b)	132 (b)	133 (a)	134 (a)	135 (c)	136 (a)	137 (a)	138 (b)	139 (d)	140 (c)
141 (b)	142 (c)	143 (b)	144 (a)	145 (c)	146 (d)	147 (d)	148 (b)	149 (a)	150 (c)
151 (c)	152 (d)	153 (d)	154 (a)	155 (a)	156 (b)	157 (c)	158 (c)	159 (d)	160 (a)
161 (c)	162 (a)	163 (b)	164 (a)	165 (a)	166 (a)	167 (c)	168 (b)	169 (b)	170 (b)

Level 3 Previous Years Questions

171 (d)	172 (b)	173 (b)	174 (c)	175 (b)	176 (d)	177 (b)	178 (c)	179 (b)	180 (d)
181 (b)	182 (d)	183 (c)	184 (a)	185 (a)					

Solutions

LEVEL 01 Objective Questions

1. Displacement of a particle, $d_1 = \sqrt{2}r$

Distance of particle, $d_2 = \frac{\pi r}{2}$

So, $\frac{d_2}{d_1} = \frac{\pi r/2}{\sqrt{2}r} = \frac{\pi}{2\sqrt{2}}$

2. The time taken for covering $\frac{1}{3}$ of total distance,

$$t_1 = \frac{s}{3\,(20)} = \frac{s}{60}$$

Similarly, t_2 and t_3, $t_2 = \frac{s}{3\,(30)} = \frac{s}{90}$ and $t_3 = \frac{s}{180}$

Therefore, $v_{average} = \frac{s}{t_1 + t_2 + t_3} = \frac{s}{\frac{s}{60} + \frac{s}{90} + \frac{s}{180}}$

$= 30 \text{ kmh}^{-1}$

3.

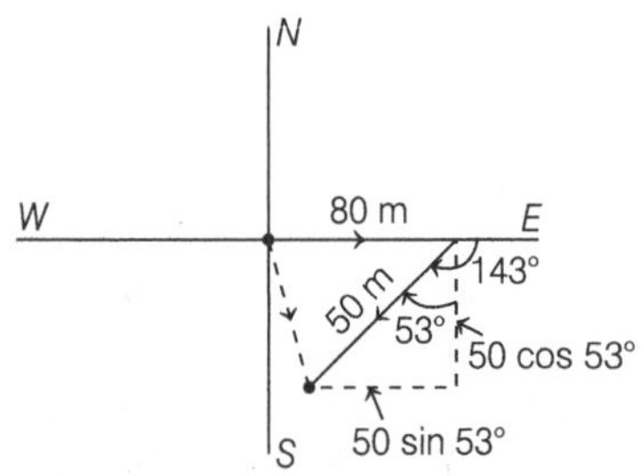

Net displacement,

$$\mathbf{s} = (80 - 50 \sin 53°)\,\hat{\mathbf{i}} + (50 \cos 53°)\,(-\hat{\mathbf{j}})$$

$$= \left(80 - 50 \times \frac{4}{5}\right)\hat{\mathbf{i}} - \left(50 \times \frac{3}{5}\right)\hat{\mathbf{j}}$$

$$= 40\,\hat{\mathbf{i}} - 30\,\hat{\mathbf{j}}$$

$$|\mathbf{s}| = \sqrt{40^2 + 30^2} = 50 \text{ m}$$

Also, **s** is 53° west of south.

4. The coordinates of insect after one second

$$= (2\hat{\mathbf{i}} + 3\hat{\mathbf{j}})\,1 = (2, 3)$$

Similarly, the coordinates after $\frac{1}{2}$ second

$$x = 2 + 2\left(\frac{1}{2}\right),\ y = 3 + 2\left(\frac{1}{2}\right)$$

or $\quad x = 3\text{m},\ y = 4\text{m}$

So, distance from origin $= \sqrt{3^2 + 4^2} = 5$ m

5. Initial position of a particle $= 2\hat{\mathbf{i}} + 3\hat{\mathbf{j}}$

Final position of a particle $= 13\hat{\mathbf{i}} + 14\hat{\mathbf{j}}$

$$r_f - r_i = (13\hat{\mathbf{i}} + 14\hat{\mathbf{j}}) - (2\hat{\mathbf{i}} + 3\hat{\mathbf{j}}) = 11\hat{\mathbf{i}} + 11\hat{\mathbf{j}}$$

Therefore, average velocity vector (v_{av}) from $t = 0$ to 5 s

$$= \frac{11}{5}(\hat{\mathbf{i}} + \hat{\mathbf{j}})$$

6. Average velocity $= \frac{\text{Displacement}}{\text{Total time}}$

$$= \frac{(20 \times 5)\hat{\mathbf{i}} + (30 \times 5)\hat{\mathbf{j}}}{10}$$

$$= \frac{\sqrt{(100)^2 + (150)^2}}{10} = \sqrt{325}$$

Similarly, average acceleration $= \frac{30\hat{\mathbf{j}} - 20\hat{\mathbf{i}}}{10}$

$$= \frac{\sqrt{900 + 400}}{10} = \sqrt{13}$$

$\therefore \quad \text{ratio} = \frac{\sqrt{325}}{\sqrt{13}} = 5$

7.

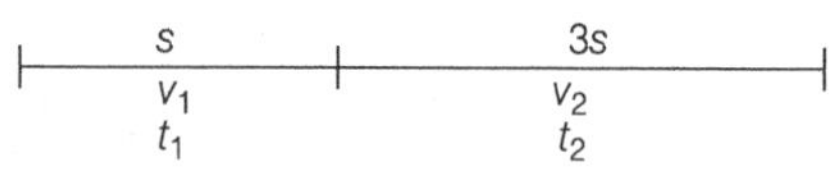

$$t_1 = \frac{s}{v_1},\ t_2 = \frac{3s}{v_2}$$

Average speed $= \frac{4s}{t_1 + t_2} = \frac{4s}{\frac{s}{v_1} + \frac{3s}{v_2}} = \frac{4\,v_1 v_2}{3\,v_1 + v_2}$

8. The position of a particle at time $t = 6$ s and 0 s is

$$r_B = 8\hat{\mathbf{i}} + 8\hat{\mathbf{j}} \text{ and } r_A = 2\hat{\mathbf{j}}$$

Now, average velocity $= \frac{\mathbf{r}_B - \mathbf{r}_A}{\Delta t}$

$$= \frac{8\hat{\mathbf{i}} + 8\hat{\mathbf{j}} - 2\hat{\mathbf{j}}}{6}$$

$$= \frac{8\hat{\mathbf{i}} + 6\hat{\mathbf{j}}}{6} = \left(\frac{4\hat{\mathbf{i}}}{3} + \hat{\mathbf{j}}\right) \text{m/s}$$

9. Let the particle be collide at point A and $\mathbf{r}_3$ be the position vector.

$$\mathbf{r}_3 = \mathbf{r}_1 + \mathbf{v}_1 t = \mathbf{r}_2 + \mathbf{v}_2 t$$

So, $\quad \mathbf{r}_1 - \mathbf{r}_2 = (\mathbf{v}_2 - \mathbf{v}_1)\,t \quad$ ...(i)

$\Rightarrow \quad t = \frac{|\mathbf{r}_1 - \mathbf{r}_2|}{|\mathbf{v}_2 - \mathbf{v}_1|} \quad$ ...(ii)

From Eqs. (i) and (ii), we get

$$\frac{\mathbf{r}_1 - \mathbf{r}_2}{|\mathbf{r}_1 - \mathbf{r}_2|} = \frac{\mathbf{v}_2 - \mathbf{v}_1}{|\mathbf{v}_2 - \mathbf{v}_1|}$$

10. Given, $x_A = 3t,\ x_B = 6$ and $y_A = t,\ y_B = 2 + 3t^2$

At time $t_1 = 1$ s, coordinates are

$$x_A = 3\hat{\mathbf{i}},\ x_B = 6\hat{\mathbf{i}},\ y_A = \hat{\mathbf{j}} \text{ and } y_B = 5\hat{\mathbf{j}}$$

So, distance $AB = \sqrt{(x_A - x_B)^2 + (y_A - y_B)^2}$

$$= \sqrt{(3-6)^2 + (1-5)^2}$$

$$= 5 \text{ m}$$

11. In time 2 min 20 s = 140 s

In $t = 120$ s, as it is multiple of 40 s, it will cover 3 round. Hence, its displacement is zero.

In remaining 20 s, the particle will cover half round.

Hence, displacement = $2r$

12. Average speed = Constant speed

$$= v \text{ (say)}$$

$$\text{Average velocity} = \frac{s}{t} = \frac{AC}{(AB + BC)/v} = \frac{v}{2}$$

$\therefore$ The ratio is 2.

13. Initial velocity, $v_i = 50\,\hat{\mathbf{j}}$

Final velocity, $v_f = -50\,\hat{\mathbf{i}}$

Change in velocity, $\Delta v = -50\,\hat{\mathbf{i}} - 50\,\hat{\mathbf{j}}$

Magnitude of change in velocity

$$= \sqrt{(-50)^2 + (-50)^2} = 50\sqrt{2} \text{ km/h}$$

Hence, resultant velocity is in south-west direction.

14. We have given, $av = k$ ...(i)

Now we know, $a = \frac{dv}{dt}$

$\therefore \quad \frac{k}{v} = \frac{dv}{dt}$ [using Eq. (i)]

$\Rightarrow \quad vdv = kdt$

$$\int vdv = k\int dt$$

$$\frac{v^2}{2} = kt$$

$$v = (2k)^{1/2} t^{1/2}$$

Now, $v = \frac{dx}{dt}$

$$\frac{dx}{dt} = (2k)^{1/2} t^{1/2}$$

$$\int dx = (2k)^{1/2} \int t^{1/2} dt$$

$$x = (2k)^{1/2} \cdot \frac{2}{3} \cdot t^{3/2}$$

$$x \propto t^{3/2}$$

15. The total distance covered by train

$$= (850 + 150) \text{ m} = 1000 \text{ m}$$

$$\text{So, velocity} = \frac{\text{displacement}}{\text{time}}$$

$$t = \frac{1000}{12.5} = 80 \text{ s}$$

16. $s_5 = s_6 = 4.9$ m

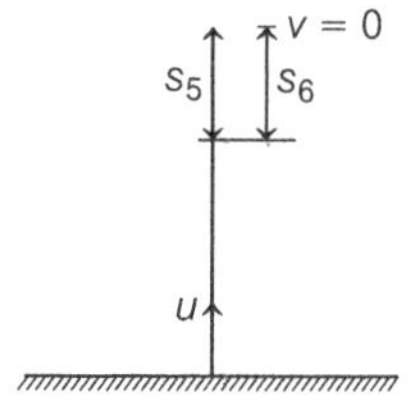

$\therefore \quad t_{up} = \frac{u}{g} = 5$ s

or $\quad u = 5g = 5 \times 9.8$

$$= 49 \text{ m/s}$$

17. We know, the distance travelled,

$$s = ut + \frac{1}{2}at^2$$

Now, interval wise distance travelled are

$$s_1 - 0 = \frac{a(1)^2}{2} = \frac{a}{2}$$

$$s_2 - s_1 = \frac{a(2)^2}{2} - \frac{a(1)^2}{2} = \frac{3a}{2}$$

$$s_3 - s_2 = \frac{a(3)^2}{2} - \frac{a(2)^2}{2} = \frac{5a}{2}$$

Hence, the ratio is 1 : 3 : 5.

18. The distance travelled in nth second,

$$s_n = u + \frac{1}{2} \cdot a(2n-1)$$

So, distance travelled in t th and $(t+1)$ th second,

$$s_t = u + \frac{a}{2}(2t-1) \quad \text{...(i)}$$

$$s_{t+1} = u + \frac{a}{2}(2t+1) \quad \text{...(ii)}$$

According to question,

$$s_t + s_{t+1} = 100$$

Putting the values from Eqs. (i) and (ii), we get

$$2(u + at) = 100$$

$$v = 50 \text{ cm/s}$$

19.

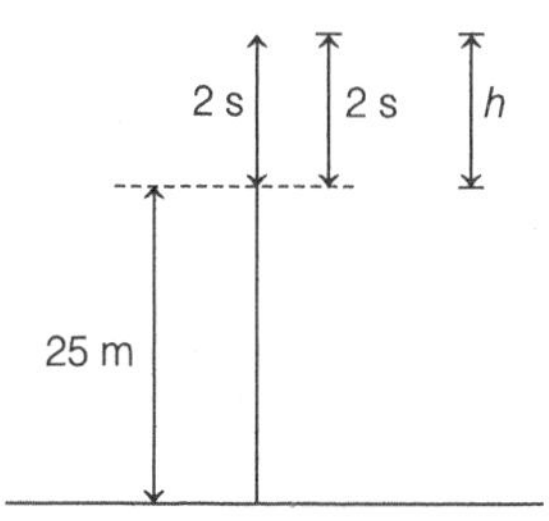

$$h = \frac{1}{2}gt^2 = \frac{1}{2} \times 10 \times (2)^2$$

$$= 20 \text{ m}$$

$$h = 25 + 20 = \frac{u^2}{2g}$$

$$45 = \frac{u^2}{2g} \Rightarrow u^2 = 45 \times 2 \times 10$$

$$\Rightarrow \quad u^2 = 900$$

$$\therefore \quad u = 30 \text{ m/s}$$

20.

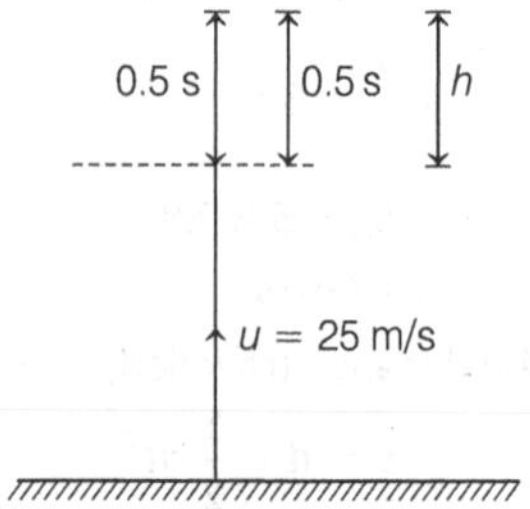

$$t_{up} = \frac{u}{g} = \frac{25}{10} = 2.5 \text{ s}$$

$$h = \frac{1}{2}gt^2 = \frac{1}{2} \times 10 \times (0.5)^2 = 1.25 \text{ m}$$

$$\therefore \quad d = 2h = 2.5 \text{ m}$$

21. We know, maximum height $= \frac{u^2}{2g}$

with velocity $2u = \frac{(2u)^2}{2g}$

$$\therefore \quad H = 4h$$

22. A ball is thrown vertically upwards. Since, acceleration is negative, its velocity changes from u to 0.

Total time taken by ball, $t = \frac{2u}{g}$

Thus, the distance transversed by truck

$$= vt = \frac{2uv}{g}$$

23. We know, the displacement in the nth second is given as

$$s_n = u + \frac{a}{2}(2n - 1)$$

$$9 = 0 + \frac{a}{2}[2(5) - 1]$$

$$9 = \frac{9a}{2} \Rightarrow a = 2 \text{ m/s}^2$$

24. The distance covered in n second,

$$s = un + \frac{1}{2}an^2$$

Similarly, the distance covered in nth second,

$$s_n = u + \frac{a}{2}(2n - 1)$$

Thus, $\frac{s_n}{s} = \frac{2n-1}{n^2}$

25. We have, $v = u + at$

$$20 = u + 10a \quad \text{...(i)}$$

Now, the distance covered in nth second,

$$s_n = u + \frac{a}{2}[2n - 1]$$

$$10 = u + \frac{a}{2}[2(10) - 1] = u + \frac{19a}{2}$$

From 1st equation, $10 = 20 - 10a + \frac{19a}{2}$

$$\Rightarrow \quad a = 20 \text{ m/s}^2$$

26. The distance covered in the nth second,

$$s_n = u + \frac{a}{2}(2n - 1)$$

We have, $s_5 = 2s_6$

$$u - \frac{1}{2}g(2 \times 5 - 1) = 2\left[u - \frac{1}{2}g(2 \times 6 - 1)\right]$$

$$u = 65 \text{ m/s}$$

27. The displacement in y-direction, $y = \frac{1}{2}gt^2$

In x-direction, $x = \frac{1}{2}a_x t^2 = \frac{1}{2}\frac{Ft^2}{m}$

$$\frac{y}{x} = \frac{1/2gt^2}{1/2\,Ft^2/m}$$

$$y = \frac{mg}{F}x$$

Since $y \propto x$, graph is a straight line.

28. We know that, the average velocity of a particle

$$= \frac{\text{net displacement}}{\text{total time}}$$

The direction of displacement vector will be in the direction of velocity vector.

Hence, (b) represents the best resultant of average velocity.

29. We have, $h = \frac{1}{2}gt_1^2$ ...(i)

Let the time taken to cover the first half of height is t_2.

$$\frac{h}{2} = \frac{1}{2}gt_2^2$$

$$h = g\,t_2^2 \quad \text{...(ii)}$$

From Eqs. (i) and (ii), we get

$$\frac{1}{2}gt_1^2 = gt_2^2$$

$$t_2 = \frac{t_1}{\sqrt{2}}$$

30. The velocity of particle, $v_1 = 2$ m/s

$$v_{relative} = 2 \text{ m/s}$$

$$s_{relative} = 18 \text{ m}$$

$$a_{relative} = 0$$

Therefore, $t = \frac{s_{relative}}{v_{relative}} = \frac{18}{2} = 9 \text{ s}$

31. $s_t = u + at - \frac{1}{2}a$

$= 10 - 2 \times 5 - \frac{1}{2} \times (-2) = 1\text{ m}$

32. The distance covered in the nth second is given as

$$s_n = u + \frac{a}{2}(2n-1)$$

Here, $u = 0$

In 1st second,

$$s_n = \frac{a}{2}$$

In 6th second,

$$s_6 = \frac{a}{2}[2(6)-1] = \frac{11a}{2}$$

Therefore, $\frac{s_1}{s_6} = \frac{1}{11}$

33. The distance covered in the nth second is given as,

$$s_n = u + \frac{a}{2}(2n-1)$$

In 5th second,

$$65 = u + \frac{a}{2}[2(5)-1]$$

$$65 = u + \frac{9a}{2}$$

$$130 = 2u + 9a \quad ...(i)$$

Similarly, in 9 th second,

$$105 = u + \frac{a}{2}[2(9)-1]$$

$$105 = u + \frac{a}{2} \times 17$$

$$210 = 2u + 17a \quad ...(ii)$$

On solving Eqs. (i) and (ii), we get

$a = 10\text{ m/s}^2$ and $u = 20\text{ m/s}$

Therefore, the distance covered in first 20 seconds

$= ut + \frac{1}{2}at^2 = 20 \times 20 + \frac{1}{2}(10)(20)^2$

$= 2400\text{ m}$

34. (i) Uniform circular motion.

(iii) When a particle is projected vertically upwards, then at highest point, $\mathbf{v} = 0$ but $\mathbf{a} \neq 0$.

(iv) Constant velocity.

35.

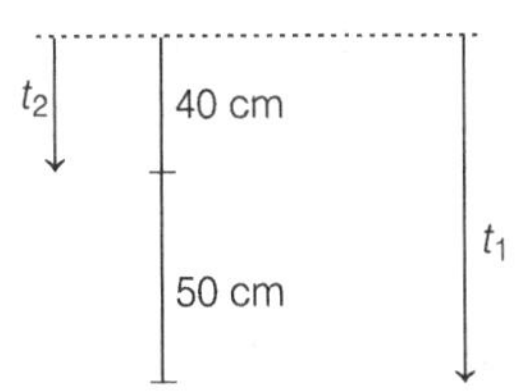

$$t = t_1 - t_2 = \sqrt{\frac{2h_1}{g}} - \sqrt{\frac{2h_2}{g}}$$

$$= \sqrt{\frac{2 \times 0.9}{9.8}} - \sqrt{\frac{2 \times 0.4}{9.8}}$$

$$= \frac{3}{7} - \frac{2}{7} = \frac{1}{7}\text{ s}$$

36. Average speed $= \frac{\text{Distance}}{\text{Time}}$

$$= \frac{2d}{\frac{d}{v} + \frac{d}{0.5v}} = \frac{2v}{3}$$

37. The position of A, $s_A = \frac{1}{2}at^2$

Similarly, the position of B, $s_B = vt$

When they meet, $s_A = s_B$

$$vt = \frac{1}{2}at^2$$

$$\Rightarrow \quad t = \frac{2v}{a}$$

38. Time of ascent, $T = \frac{u}{g}$

Let h be the distance covered in the last t s of its ascent.

Velocity at that point, $v = u - g(T - t)$

$$v = u - g\left(\frac{u}{g} - t\right) = gt$$

So, $h = vt - \frac{1}{2}gt^2 = gt^2 - \frac{1}{2}gt^2 = \frac{1}{2}gt^2$

39. For the first 50 m,

$$50\text{ m} = 5t_1^2$$

$$\Rightarrow \quad t_1 = \sqrt{10}\text{ s}$$

Now, for 100 m,

$$100 = 5t^2 \Rightarrow t = \sqrt{20}\text{ s}$$

So, $t_1 + t_2 = \sqrt{20}$

$$t_2 = \sqrt{20} - \sqrt{10}$$

$$\therefore \quad \frac{t_1}{t_2} = \sqrt{2} + 1$$

40. The velocity of ball A at any point,

$$v_A = u - gt$$

Since, the ball B is released from height at any time,

$$v_B = -gt$$

So, relative speed $= v_A - v_B = u - gt - (-gt) = u$

41. Given, the speed of police jeep = 40 m/s

Separation between jeep and car = 400 m

Now, the distance travelled by the thief in 20 s

$= 40 \times 20 = 800\text{ m}$

Hence, for the police, the total distance

$= 400 + 800 = 1200\text{ km}$

Therefore, $v = \frac{1200}{20} = 60\text{ m/s}$

42. In the first situation,

$$-h = ut_1 - \frac{1}{2}gt_1^2 \quad (\because u_1 = u) \qquad \text{...(i)}$$

Now, in second situation when the stone is thrown downwards,

$$-h = -ut_2 - \frac{1}{2}gt_2^2 \quad (\because u_2 = u) \qquad \text{...(ii)}$$

From Eqs. (i) and (ii), we get

$$u = \frac{1}{2}g(t_1 - t_2)$$

Substituting u in Eq. (ii), we get

$$h = \frac{1}{2}gt_1t_2$$

Now, when the stone is allowed to fall freely under gravity,

$$h = \frac{1}{2}gt^2$$

$$\frac{1}{2}gt_1t_2 = \frac{1}{2}gt^2$$

$$t = \sqrt{t_1t_2} = \sqrt{16 \times 9}$$

$$= 12 \text{ s}$$

43. Given, $s = t^3 - 27t + 4$

Now, $v = \frac{ds}{dt} = 3t^2 - 27$

If velocity is zero, i.e.

$$3t^2 - 27 = 0$$

$$3t^2 = 27$$

$$t^2 = 9 \text{ s} \Rightarrow t = 3 \text{ s}$$

Now, $a = \frac{dv}{dt} = 6t = 6(3) = 18 \text{ m/s}^2$

44. We have, $a = \alpha t + \beta$

$$a = \frac{dv}{dt}$$

$$dv = a\,dt$$

$$dv = (\alpha t + \beta)\,dt$$

$$\int_0^v dv = \int_0^t (\alpha t + \beta)\,dt$$

$$v = \frac{\alpha t^2}{2} + \beta t$$

45. We have, $x = 40 + 12t - t^3$

Position of particle at $t = 0$,

$$x = 40 + 12(0) - (0)^3 = 40 \text{ m}$$

Particle comes to rest, therefore the velocity becomes zero.

$$v = \frac{dx}{dt} = 12 - 3t^2$$

$$12 = 3t^2$$

$$\Rightarrow \quad t = 2 \text{ s}$$

Now, position of particle from origin at time $t = 2$ s,

$$x = 40 + 12(2) - (2)^3 = 56 \text{ m}$$

Therefore, the distance travelled by the particle

$$= 56 \text{ m} - 40 \text{ m} = 16 \text{ m}$$

46. Let the total distance be s and at mid-point $= \frac{s}{2}$

We know, $s = \frac{v^2 - u^2}{2a} = \frac{(\sqrt{7}u)^2 - u^2}{2a} = \frac{7u^2 - u^2}{2a}$

$$= \frac{6u^2}{2a} = \frac{3u^2}{a} \qquad \text{...(i)}$$

Let the speed of motorcycle at the mid-point be v'.

$$v'^2 = u^2 + 2a\left(\frac{s}{2}\right)$$

$$s = \frac{v'^2 - u^2}{a} \qquad \text{...(ii)}$$

On solving Eqs. (i) and (ii), we get

$$v' = 2u$$

47. Acceleration, $a = \frac{dv}{dt} = \frac{dv}{dx}\cdot\frac{dx}{dt} = v\cdot\frac{dv}{dx}$

So, $a = (3x^2 - 2x)(6x - 2)$

At $x = 2$ m,

$$a = (3 \times 2^2 - 2 \times 2)(6 \times 2 - 2)$$

$$= 8 \times 10 = 80 \text{ m/s}^2$$

48. Given, $\sqrt{x} = t + 1$

$$\Rightarrow \quad x = (t + 1)^2$$

$$v = \frac{dx}{dt} = 2(t + 1)$$

As t increases, velocity also increases with time.

49. Given, $v = (4t^3 + 3t^2 - 1)$ m/s

$$v = \frac{ds}{dt}$$

$$\int_0^s ds = \int_1^2 v\,dt$$

$$\int_0^s ds = \int_1^2 (4t^3 + 3t^2 - 1)\,dt$$

$$s = [t^4 + t^3 - t]_1^2$$

$$= [(2)^4 + (2)^3 - 2] - (1)^4 + (1)^3 - (1)$$

$$= 16 + 8 - 2 - 1 = 21 \text{ m}$$

50. Given, $x = 4t^2$, $y = 2t$

Eliminating t, we get

$$x = y^2$$

Hence, it is a parabolic path.

51. Given, $v = (10 + 2t^2)$ m/s

We know, average acceleration $= \frac{v_f - v_i}{\Delta t}$

$$= \frac{v(5) - v(2)}{\Delta t} = \frac{[10 + 2(5)^2] - [10 + 2(2)^2]}{3}$$
$$= 14 \text{ m/s}^2$$

52. Given, $a = 2(t - 1) = 2t - 2$

$$\frac{dv}{dt} = 2t - 2$$
$$\int dv = \int (2t - 2)\, dt$$
$$v = t^2 - 2t + C = (5)^2 - 2(5) + 0$$
$$= 25 - 10 = 15 \text{ m/s}$$

53. At point Q, the acceleration and the velocity both are positive. Hence, the particle is speeding up at P.

54. The slope of a graph position *versus* time gives the velocity.

At point d, slope is positive and decreasing.

Therefore, the velocity is positive and acceleration is negative.

55. The maximum height attained by the rocket will be equal to the positive area under v - t graph

$$= \frac{1}{2} \times 110 \times 1000 = 55 \text{ km}$$

56. The total distance travelled by the particle

$$= 10 + 5 + 5 + 5 = 25 \text{ m}$$

57. Given velocity, $v = 10e^{-\frac{t}{2}}$

Acceleration, $a = \frac{dv}{dt} = 10 \times \frac{-1}{2} \times e^{-\frac{t}{2}}$

$\Rightarrow \qquad a = -5\, e^{-\frac{t}{2}}$

When $t = 0$, $a = -5 \text{ m/s}^2$

Also as $t \to \infty$, $a \to 0$

Hence, correct graph is option (a).

58. From the graph, we can see that velocity is increasing and acceleration is decreasing.

59. Slope of v - t graph $= a$

(i) Slope is negative and decreasing

(ii) Slope is zero

(iii) Slope is positive and increasing

60. In all three graphs, velocity and acceleration both are positive.

61. $|\mathbf{v}_A - \mathbf{v}_B| = \text{slope}_A - \text{slope}_B = C_1 - C_2 = \text{constant}$

$\neq 0 \quad \text{as} \quad C_1 \neq C_2$

62. Average velocity $= \frac{s}{t} = \frac{\text{Area under } v\text{ - }t \text{ graph}}{\text{Time}}$

63. We know, $a = \frac{dv}{dt}$

$$dv = a\,dt$$

From graph, we can say, $a \propto t$

$$a = kt$$
$$\int dv = \int kt\,dt$$
$$v \propto t^2$$

From 0 to t_1, acceleration is positive and increasing; and from t_1 to t_2, it is decreasing.

Therefore, in 0 to t_1, v - t graph parabolic upwards and in t_1 to t_2, parabolic downwards.

64. $v_f - v_i$ = Area under a - t graph.

$\therefore \quad v_f = v_i$ + Area under a - t graph.

65.

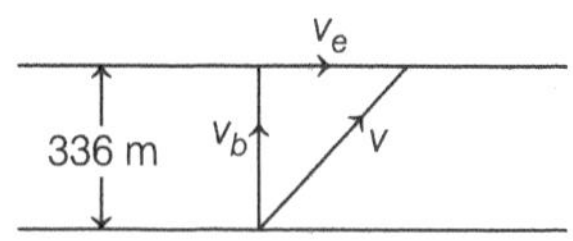

Distance = 336 m

Speed (in direction v_b) = 1 m/s

Time to cross $= \frac{\text{Distance}}{\text{speed}} = \frac{336}{1} = 336 \text{ s}$

66.

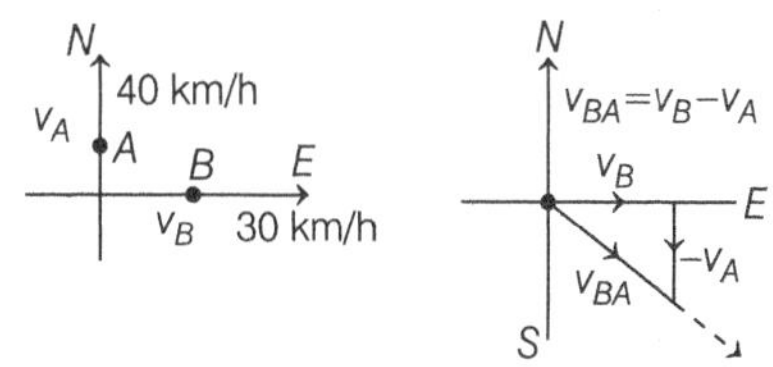

Velocity of car $A = 40\,\hat{\mathbf{j}}$

Velocity of car $B = 30\,\hat{\mathbf{i}}$

$$\mathbf{v}_{BA} = \mathbf{v}_B - \mathbf{v}_A = 30\,\hat{\mathbf{i}} - 40\,\hat{\mathbf{j}}$$
$$|\mathbf{v}_{BA}| = \sqrt{30^2 + 40^2} = 50 \text{ km/h}$$
$$\tan\theta = \frac{30}{40} \Rightarrow \theta = \tan^{-1}\left(\frac{3}{4}\right)$$

67. $t = \frac{300}{3} = 100 \text{ s}$

So, the speed at which swimmer must swim to cross

$$= \frac{500}{100} = 5 \text{ m/s}$$

68.

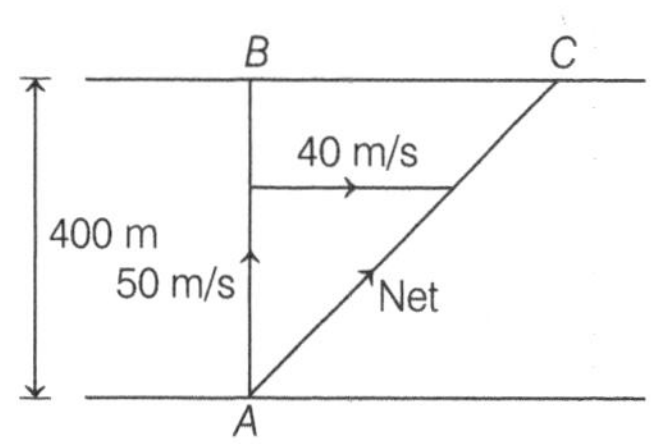

$$t = \frac{400}{50} = 8 \text{ s}$$
$$BC = 40 \times t = 320 \text{ m}$$
$$AC = \sqrt{AB^2 + BC^2}$$
$$= \sqrt{(400)^2 + (320)^2}$$
$$= 80\sqrt{41} \text{ m}$$

69. The relative velocity of bird with respect to train,

$$v_{BT} = \mathbf{v}_B + (-\mathbf{v}_T)$$

$$|v_{BT}| = \sqrt{(\mathbf{v}_B)^2 + (-\mathbf{v}_T)^2}$$

$$= \sqrt{40^2 + 40^2}$$

$$= 40\sqrt{2} \text{ km/h north-west}$$

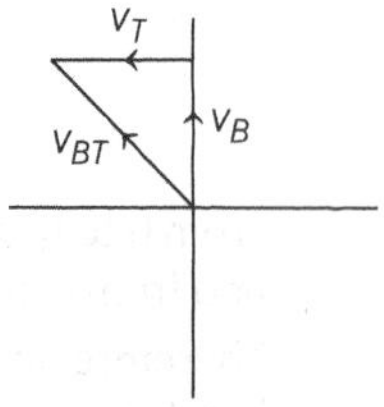

70.

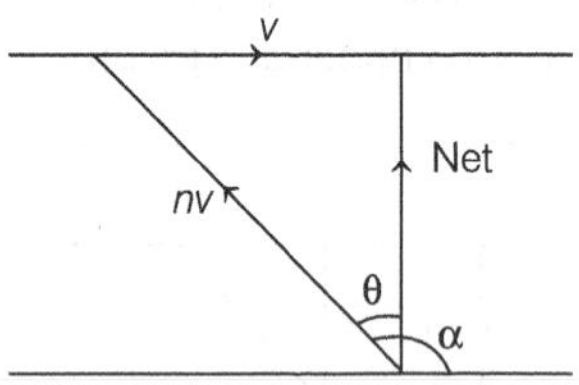

$$\sin\theta = \frac{v}{nv}$$

$$\therefore \quad \theta = \sin^{-1}\left(\frac{1}{n}\right)$$

$$\because \quad \alpha = \frac{\pi}{2} + \theta$$

$$\alpha = \frac{\pi}{2} + \sin^{-1}\left(\frac{1}{n}\right)$$

LEVEL 02 *Objective Questions*

71. After 6 turns, he will return to its original position.

72. In first condition, $t_1 = \frac{s}{2(3)} = \frac{s}{6}$

In second condition, $\frac{s}{2} = 4.5\, t_2 + 7.5\, t_3$

As, $t_2 = t_3 \qquad 12t_2 = \frac{s}{2}$

Therefore, the total time taken,

$$t = \frac{s}{6} + \frac{s}{24} + \frac{s}{24} = \frac{s}{4}$$

$\therefore$ Average speed $= \frac{s}{s/4} = 4$ m/s

73. $\mathbf{v}_{AB} = \mathbf{v}_A - \mathbf{v}_B = (-5\hat{\mathbf{i}})$ m/s

$$\mathbf{r}_{BA} = \mathbf{r}_B - \mathbf{r}_A = (-30\ \hat{\mathbf{i}}) \text{ m}$$

$\mathbf{v}_{AB}$ is along $\mathbf{r}_{BA}$. So, they will definitely collide after time,

$$t = \frac{|\mathbf{r}_{BA}|}{|\mathbf{v}_{AB}|} = \frac{30}{5}$$

$$= 6 \text{ s}$$

74. $50 + v_B t = v_A t$

$$50 + 5t = 10t$$

$$t = 10 \text{ s}$$

So, the position at which bus *A* overtakes bus *B*,

$$x = 50 + v_A t = 50 + 10 \times 10 = 150 \text{ m}$$

75. For no drift, $V\sin\theta = v \Rightarrow \sin\theta = \frac{v}{V}$

$$\therefore \quad \cos\theta = \sqrt{1 - \frac{v^2}{V^2}}$$

Total time for a round trip $= \frac{2l}{\cos\theta} = \frac{2l}{\sqrt{1 - \frac{v^2}{V^2}}}$

76. Given, $y(t) = 2t^2 + 3t + 4$

Now, average velocity

$$= \frac{y(3) - y(0)}{3} = \frac{2(3)^2 + 3(3) + 4 - 2(0)^2 - 3(0) - 4}{3}$$

$$= \frac{31 - 4}{3} = 9 \text{ m/s}$$

77. Given, $x = (t-2)^2$

$$x = t^2 - 4t + 4$$

Now, $v = \frac{dx}{dt} = 2t - 4$

Distance covered $= \int v \cdot dt$

$$= \int_0 t^2 - 4t \text{ for } t = 0 \text{ to } 2 \text{ and } t = 2 \text{ to } 4$$

$$= 8 \text{ m}$$

78. We have given, $v = (x^2 - 5x + 4)$ m/s

$$a = \frac{dv}{dt} = 2x \cdot \frac{dx}{dt} - 5 \cdot \frac{dx}{dt} = 2xv - 5v$$

As $v = 0$, therefore $a = 0$

79. Given, $x = (3t^2 - t^3)$ m

$$v = \frac{dx}{dt} = 6t - 3t^2$$

At $t = 2$ s

$$\therefore \quad v = 6(2) - 3(2)^2 = 0$$

Now, the distance travelled in 2 s

$$= 3(2)^2 - (2)^3 = 4 \text{ m}$$

In next 2 s

$$= x_4 - x_2 = [3(4)^2 - (4)^3 - 4] = |-20|$$

$\therefore$ Total distance $= 4 + 20 = 24$ m

Velocity is maximum where $\frac{dv}{dt}$ or a or $6 - 6t = 0$ or $t = 1$ s

$$\therefore \quad v_{max} = 6(1) - 3(1)^2 = 3 \text{ m/s}$$

80. Given, $v = (2t + 3)$ m/s

Now, the average velocity = in time interval $0 \le t \le 1$

$$= \frac{\int_0^1 v\,dt}{\int_0^1 dt} = \frac{\int_0^1 (2t+3)\,dt}{\int_0^1 dt} = \frac{t^2 + 3t|_0^1}{t|_0^1}$$

$$= \frac{1^2 + 3 \times 1 - (0 - 0)}{1 - 0} = 4 \text{ m/s}$$

81. Given, $x = 4 - 9t + \frac{t^3}{3}$

$$v = \frac{dx}{dt} = -9 + t^2 = 0$$

$$t = 3 \text{ s}$$

At time 3 s, the direction of the motion changes.

82. Given, $v = 2t(3 - t)$ m/s

$$= 6t - 2t^2$$

$$\frac{dv}{dt} = -4t + 6 = 0$$

$$t = \frac{3}{2} \text{ s}$$

83. Given, $\mathbf{r} = a(t-1)^2\,\hat{\mathbf{i}} + bt^2\,\hat{\mathbf{j}}$

$$\mathbf{v} = \frac{d\mathbf{r}}{dt} = 2a(t-1)\,\hat{\mathbf{i}} + 2bt\,\hat{\mathbf{j}}$$

$$\mathbf{a} = \frac{d\mathbf{v}}{dt} = 2a\hat{\mathbf{i}} + 2b\hat{\mathbf{j}}$$

Now, $\mathbf{a} \cdot \mathbf{v} = 0$

$$2a \cdot 2a\,(t-1) + 2bt \cdot 2b = 0$$

$$4a^2t - 4a^2 + 4b^2t = 0$$

$$\Rightarrow \quad t = \frac{a^2}{a^2 + b^2}$$

84. We have given, $v_1 = \{2t\hat{\mathbf{i}} + (t-2)\,\hat{\mathbf{j}}\}$ m/s

$$v_2 = \{(t-4)\,\hat{\mathbf{i}} + t\,\hat{\mathbf{j}}\} \text{ m/s}$$

For balloons to move parallel,

$$\mathbf{v}_1 \times \mathbf{v}_2 = 0$$

$$[2t^2 - (t-2)(t-4)]\,\hat{\mathbf{j}} = 0$$

$$2t^2 - t^2 + 6t - 8 = 0$$

On solving this, we get

$$t = -3 \pm \sqrt{17} \text{ s}$$

85. We have given, $v = 6\hat{\mathbf{j}}$ m/s

Now, $\quad v_y = u_y + at$

$v_y = 0$, when velocity become parallel

$$0 = 6\hat{\mathbf{j}} - 2\hat{\mathbf{j}}\,t$$

$$t = 3 \text{ s}$$

Now, the x-coordinate of the particle

$$= \frac{1}{2}(2)(3)^2 = 9$$

Similarly, y-coordinate of the particle

$$= 6 \times 3 - \frac{1}{2}(2)(3)^2 = 9$$

86. $x_1 = \frac{1}{2}gt^2 \quad$ ...(i)

$$x_1 + x_2 = \frac{1}{2}g\,(2t)^2 \quad \text{...(ii)}$$

Subtracting Eq. (i) from Eq. (ii), we get

$$x_2 - x_1 = gt^2$$

$$\therefore \quad t = \sqrt{\frac{x_2 - x_1}{g}}$$

87. $t_1 + t_2 = t_1' + t_2'$

The sum of time remains same.

88. According to the question,

For the 1st half of accelerated motion,

$$v^2 = 2\alpha x_1 \quad (\text{as } u = 0) \quad \text{...(i)}$$

At the end of first t_1 s, it gains velocity,

$$v = \alpha t_1$$

Substituting v in Eq. (i), we get

$$\alpha^2 t_1^2 = 2\alpha x_1 \Rightarrow x_1 = \frac{\alpha t_1^2}{2}$$

Now, for the rest half journey,

$$0^2 = (\alpha t_1)^2 - 2\beta x_2$$

or $$x_2 = \frac{\alpha^2(t_1^2)}{2\beta}$$

and $$0 = \alpha t_1 - \beta t_2$$

$$\frac{\alpha}{\beta} = \frac{t_2}{t_1}$$

and $$\frac{x_1}{x_2} = \frac{\alpha(t_1)^2}{2}\,\frac{2\beta}{\alpha^2(t_1)^2} = \frac{\beta}{\alpha}$$

Therefore, $\frac{x_1}{x_2} = \frac{\beta}{\alpha} = \frac{t_1}{t_2}$

89. The distance x_1 travelled by the body 1,

$$x_1 = v_1 t - \frac{1}{2}a_1 t^2$$

Similarly, for the body 2,

$$x_2 = v_2 t - \frac{1}{2}a_2 t^2$$

Now, $x_1 + x_2 = x = (v_1 + v_2)\,t - \frac{1}{2}(a_1 + a_2)\,t^2 \quad$ ...(i)

$$\frac{dx}{dt} = (v_1 + v_2) - (a_1 + a_2)\,t = 0$$

$$t = \frac{(v_1 + v_2)}{a_1 + a_2}$$

Substituting the t in Eq. (i), we get

$$= (v_1 + v_2)\left(\frac{v_1 + v_2}{a_1 + a_2}\right) - \frac{1}{2}(a_1 + a_2)\left(\frac{v_1 + v_2}{a_1 + a_2}\right)^2$$

$$= \frac{1}{2}\frac{(v_1 + v_2)^2}{(a_1 + a_2)}$$

90. Given, $\mathbf{v} = 3\hat{\mathbf{i}} + 6x\hat{\mathbf{j}}$

$$\frac{dx}{dt} = 3 \quad \text{and} \quad \frac{dy}{dt} = 6x$$

Now, $\int dx = \int 3dt$

$$x = 3t \Rightarrow t = \frac{x}{3}$$

Now, $dy = 6xdt$

$dy = 6\,(3t)\,dt$

$dy = 18\,t\,dt$

Integrating on both sides,

$$y = 9t^2 = 9\left(\frac{x}{3}\right)^2 = x^2$$

Therefore, $y = x^2$

91. Given, $a = -\alpha x^2$

$$\frac{vdv}{dx} = -\alpha x^2$$

$$\int_{v_0}^{0} vdv = \alpha \int_{0}^{s} x^2 dx$$

$$\left[\frac{v^2}{2}\right]_{v_0}^{0} = \alpha\left[\frac{x^3}{3}\right]_0^s$$

$$\frac{v_0^2}{2} = \frac{\alpha s^3}{3}$$

$$s^3 = \frac{3v_0^2}{2\alpha}$$

$$s = \left(\frac{3v_0^2}{2\alpha}\right)^{1/3}$$

92. The initial velocity of parachutist, $u = \sqrt{2gh}$

$$= \sqrt{2(10)(20)} = 20 \text{ m/s}$$

Therefore, $s = \dfrac{v^2 - u^2}{2a}$

$$= \frac{(4)^2 - (20)^2}{2(-2)} = 96 \text{ m}$$

So, the total height balloon bail out

$= 96 + 20 = 116$ m

93. So, $\mathbf{v}_{RM} = \mathbf{v}_R - \mathbf{v}_M$ or $\mathbf{v}_{RM} = \mathbf{v}_R + (-\mathbf{v}_M)$

Here, $\mathbf{v}_R$ is vertical and $\mathbf{v}_M$ or $-\mathbf{v}_M$ is horizontal.

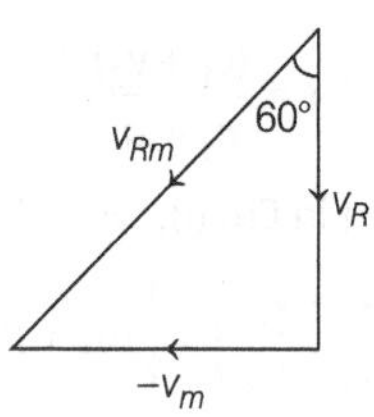

$\therefore$ $\tan 60° = \dfrac{\mathbf{v}_M}{\mathbf{v}_R} \Rightarrow \sqrt{3} = \dfrac{\mathbf{v}_M}{\mathbf{v}_R}$

$\therefore$ $\mathbf{v}_R = 4\sqrt{3}$ km/h (given, $\mathbf{v}_M = 12$ km/h)

94. $s_t = u + at - \dfrac{1}{2}a$

Here, $u = 0$

$\therefore$ $s_t = a\left(t - \dfrac{1}{2}\right)$

s_5 of first particle $= s_3$ of second particle

$\Rightarrow$ $a_1\left(5 - \dfrac{1}{2}\right) = a_2\left(3 - \dfrac{1}{2}\right)$

$\therefore$ $\dfrac{a_1}{a_2} = \dfrac{5}{9}$

95. We have given, initial velocity, $u = 10 \text{ ms}^{-1}$

Final velocity, $v = 20 \text{ ms}^{-1}$

Acceleration, $a = \dfrac{20 - 10}{t} = \dfrac{10}{t}$

Therefore, $s = ut + \dfrac{1}{2}at^2$

$$135 = 10t + \frac{1}{2}\left(\frac{10}{t}\right)t^2$$

$\Rightarrow$ $135 = 15t$

$\Rightarrow$ $t = 9$ s

96. The velocity attained from $x = 0$ to 2m,

$$v^2 = 2\,(4)\,(2) = 4 \text{ m/s}$$

Similarly, the velocity attained from 0 to 8 m,

$$v^2 = 2(4)(8) = 8 \text{ m/s}$$

So, average velocity $= \dfrac{8 + 4}{2} = 6$ m/s

97. Stone relative to cage :

$u_{rel} = 0$

$a_{rel} = 10 + 2 = 12 \text{ m/s}^2$ (downwards)

$s_{rel} = 3$ m (downwards)

$s_{rel} = \dfrac{1}{2}a_{rel}\,t^2$

$3 = \dfrac{1}{2} \times 2 \times t^2$

$3 = t^2 \Rightarrow t = 0.7$ s

98. $s_A = s_B + 8$

$$\left[36\,t + \frac{1}{2} \times 2 \times t^2\right] = \left[30\,t + \frac{1}{2} \times 4 \times t^2\right] + 8$$

$$t^2 - 6t + 8 = 0$$

$t = 2$ s and $t = 4$ s

At $t = 2$ s, A will first overtake B and then at $t = 4$ s, B will again overtake A.

99. Let us suppose the body passes the upper point at t s and lower point $(t + 1)$ s,

$$s_2 - s_1 = \frac{1}{2}g\,(t + 1)^2 - \frac{1}{2}gt^2$$

$$30 \text{ m} = \frac{1}{2} \times 9.8\,(2t + 1)$$

$\Rightarrow$ $t = 2.56$ s

So, $s_1 = \dfrac{1}{2}gt^2 = \dfrac{1}{2} \times 9.8 \times (2.56)^2$

$= 32.1$ m

100. Given, $L = bt^2 = 2000$ m

$$v = \frac{dL}{dt} = 2bt = 2t\left(\frac{L}{t^2}\right) = \frac{2L}{T}$$

At $t = 10$ s,

$$v = \frac{2(2000)}{10} = 400 \text{ m/s}$$

101. At maximum height, $v = 0 = u - gt$

$$t = \frac{u}{g} = 1 \text{ s}$$

Now, $h' = ut + \frac{1}{2}gt^2$

$$= 10 \times \frac{1}{2} - \frac{1(10)}{2}\left(\frac{1}{4}\right) = \frac{3h}{4}$$

102. Given, maximum acceleration, $\alpha = 4 \text{ m/s}^2$

Maximum retardation, $\beta = 6 \text{ m/s}^2$

Distance = 3 km

$$\frac{\alpha\beta \times t^2}{2(\alpha + \beta)} = 3 \times 10^3 \text{ m}$$

On substituting values, we get

$$t = 50 \text{ s}$$

103. Let OA be the height h.

$$h = ut_1 - \frac{1}{2}gt_1^2 \quad \text{...(i)}$$

$$O \text{ to } B = ut_2 - \frac{1}{2}gt_2^2 \quad \text{...(ii)}$$

Solving Eqs. (i) and (ii), we get

$$h = \frac{1}{2}gt_1t_2$$

104. Time of descend through B to $A = \sqrt{\frac{2H}{g}}$

Similarly, through B to $C = \sqrt{\frac{4H}{g}}$

and t_3 from D to $A = \sqrt{\frac{6H}{g}}$

$\therefore$ Time taken along $BC = t_{AC} - t_{AB} = \sqrt{\frac{2H}{g}}(\sqrt{2} - 1)$

Time taken along $CD = t_{AD} - t_{AC} = \sqrt{\frac{2H}{g}}(\sqrt{3} - \sqrt{2})$

$\therefore$ $AB : BC : CD$

$$= 1 : (\sqrt{2} - 1) : (\sqrt{3} - \sqrt{2})$$

105. As the trucks are moving towards each other,

$$v_1 + v_2 = 10 \quad \text{...(i)}$$

and in case of 2nd condition, their separation increases

$$v_1 - v_2 = 5 \quad \text{...(ii)}$$

From equations (i) and (ii), we get

$$v_1 = 7.5 \text{ m/s and } v_2 = 2.5 \text{ m/s}$$

106. Speed of bullet $= 72 \times \frac{5}{18} = 20$ m/s

The net bullet speed = speed of bullet + speed of van

$$= 20 + 5 = 25 \text{ m/s}$$

So, the velocity with which bullet will hit car

$$= 25 - 10 = 15 \text{ m/s}$$

107. The greatest distance between the particles would be when they have the same velocity.

Now, let the second particle has a velocity at time t,

$$v = at \Rightarrow t = \frac{v}{a}$$

Distance covered by first particle, $s_1 = vt$

$$= \frac{vv}{a}$$

$$\Rightarrow \quad s_1 = \frac{v^2}{a}$$

Similarly, the distance covered by second particle,

$$= s_2 = \frac{1}{2}a\left(\frac{v}{a}\right)^2 = \frac{1}{2}\frac{v^2}{a}$$

Therefore, the distance between them

$$= s_1 - s_2 = \frac{v^2}{2a} - \frac{1v^2}{2a} = \frac{v^2}{2a}$$

108. Time taken by stone one,

$$t = \frac{u}{a} = \frac{50}{10} = 5 \text{ s}$$

Now, the velocity attain by second stone in 5 s,

$$v = u + at = 50 + 10 \times 5 = 100 \text{ m/s}$$

So, the relative velocity between A and B is 100 m/s.

109. Given, $u = 0$, $a_{relative} = 10 - 9 = 1 \text{ m/s}^2$,

$$s = 2\text{m}$$

$$2 = \frac{1}{2}(1)t^2 \Rightarrow t = 2 \text{ s}$$

So after 2 s, the coin will strike the lift.

110. The person throws balls into air vertically upward in regular interval of 1 s and next ball is thrown when the velocity of ball thrown earlier becomes zero.

$$0 = u - gt$$

As, $g = 10 \text{ m/s}^2$, $t = 1$ s,

$$u = 10 \text{ m/s}$$

Height to the ball rises, $h = \frac{u^2}{2g} = \frac{(10)^2}{2 \times 10} = 5$ m

111. The displacement of particle $= \frac{1}{2}at^2$

$$= \frac{1}{2}(2\sqrt{3})(1)^2 = \sqrt{3} \text{ m}$$

So, $\tan\theta = \frac{3}{\sqrt{3}} = \sqrt{3}$

$$\Rightarrow \quad \theta = 60°$$

112. As the net resultant of forces will be along $\mathbf{F}_3$, therefore as a result, acceleration is produced.

Hence, the velocity changes in both magnitude and direction.

113. From the frame of reference, the velocity with which ball is thrown $(u+v)$.

Therefore, its relative time, $t = \dfrac{2(u+v)}{g}$

114. The distance covered by ball (I) in 9 s,

$$s = 9u - \frac{1}{2}g(9)^2 \qquad \text{...(i)}$$

Similarly, the distance covered by ball (II) in 4 s,

$$s = 4u + \frac{1}{2}g(4)^2 \qquad \text{...(ii)}$$

On solving Eqs. (i) and (ii), we get

$$s = 180 \text{ m}$$

115. The distance travelled by the 1st ball in 15 s

$$= \frac{1}{2}gt^2 = \frac{1}{2}(10)(15)^2$$

$$= 1125 \text{ m}$$

Now, the second ball has to travel the same distance to meet first ball in $(15-10) = 5$ s,

$$1125 = v(5) + \frac{1}{2} \times 10 \times (5)^2$$

$$\Rightarrow \quad v = \frac{1125 - 125}{5}$$

$$= 200 \text{ m/s}$$

116. Average velocity $= \dfrac{\text{Displacement}}{\text{Total time}}$

$$= -\frac{10}{5} = -2 \text{ m/s}$$

117. For $t < 5$ s, the car is moving with uniform velocity,

$$s_{0\text{-}5} = 10t = (10 \times 5)$$

$$= 50 \text{ m}$$

For $t = 5 < t < 10$,

$$0 = 10 - a(10 - 5)$$

$$a = 2 \text{ m/s}^2$$

$$s_{5-10} = 10 \times 5 - \frac{1}{2} \times 2 \times (5)^2$$

$$= 25 \text{ m}$$

$$\therefore \quad s_{0-10} = 50 + 25 = 75 \text{ m}$$

118. Slopes are constants. Hence, speeds are constant. Therefore, average speeds are constant.

Slope of B is more. Hence, constant speed or average speed of B will be more.

119. Net velocity $= \dfrac{\sqrt{3}}{1/5} = 5\sqrt{3}$ km/h

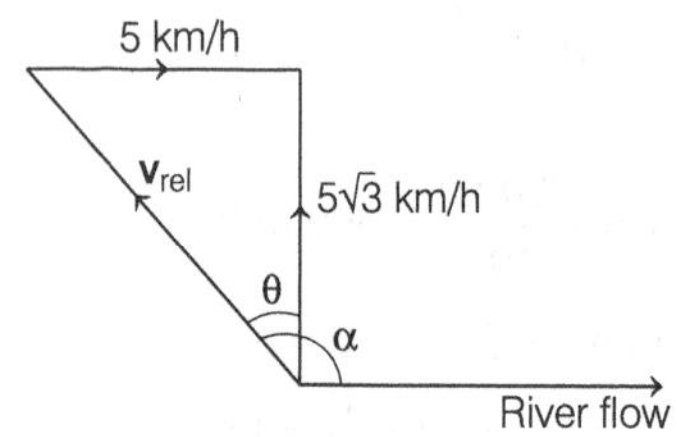

$$\tan\theta = \frac{5}{5\sqrt{3}} = \frac{1}{\sqrt{3}}$$

$$\therefore \quad \theta = 30°$$

$$\therefore \quad \alpha = 120°$$

Relative velocity is suppose $\mathbf{v}_{rel}$.

120. (i) Graph will be parabola.

(ii) At $t = 0$, $x = 60$ m.

(iii) x will first increase and then decrease.

121. $\dfrac{dv}{dt} = -\dfrac{10}{2}$

$$dv = -5\int dt$$

$$v = -5t + C$$

At $t = 0$, $v = 10$, $C = 10$

$$v = -5t + 10$$

$$\frac{dx}{dt} = -5t + 10$$

$$\int dx = \int(-5t + 10)\,dt$$

$$x = -\frac{5t^2}{2} + 10t + C_1$$

$$x = -2.5t^2 + 10t + 5$$

122. If particle will have the same velocity as initial, then area under the curve should be zero.

The slope of a - t graph $= \dfrac{a}{5}$

Now, area from 0 to 10 s

$$= 5a + \frac{5a}{2}$$

On solving this, $t = (10 + 5\sqrt{3})$ s

123.

(a)

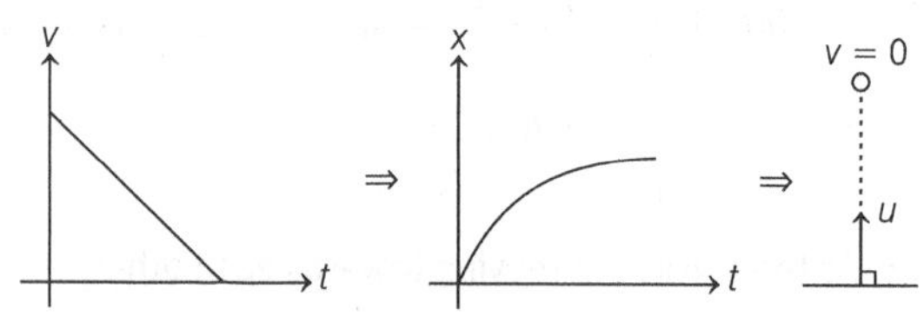

124. The distance travelled by car $A = 10t$

Now, $s_B = \frac{1}{2}at^2 = \frac{1}{2}t^2$

As, $a = \tan 45°$

Since, the car A is initially 10.5 m ahead of car B, thus we get

$$s_A + 10.5 = \frac{1}{2}t^2$$

$$10.5 + 10t = \frac{1}{2}t^2$$

On solving quadratic equation, we get

$$t = 21 \text{ s}$$

125. At time t_1 and t_2, the position of both cars is same.
In time t_1, the position of car A is more than that of car B. Hence, the car A is overtaking the car B.

126. From 0 to t_1, motion is uniform in positive direction.
From t_1 to t_2, motion is retarded in negative direction.
From t_2 to t_3, motion is accelerated in positive direction.

127. Given, velocity of A, $\mathbf{v}_A = -10\,\hat{\mathbf{i}}$ km/h

Velocity of B, $\mathbf{v}_B = 10\,\hat{\mathbf{j}}$ km/h

$$\mathbf{v}_{AB} = \mathbf{v}_A - \mathbf{v}_B = |-10\hat{\mathbf{i}} - 10\hat{\mathbf{j}}|$$

$$|\mathbf{v}_{AB}| = \sqrt{10^2 + 10^2} = 10\sqrt{2} \text{ km/h}$$

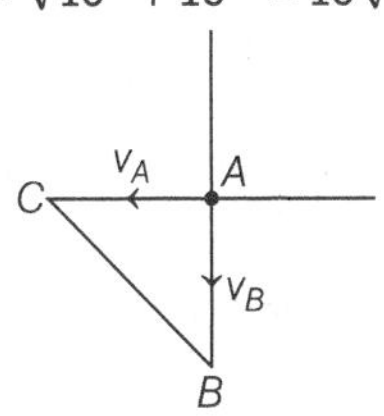

So, the shortest distance between the ships,

$$\sin 45° = \frac{AB}{100} \quad \text{(here, } AC = \text{shortest distance)}$$

$$AB = \frac{100}{\sqrt{2}} \text{ km}$$

$$\text{So, } t = \frac{100/\sqrt{2}}{|\mathbf{v}_{AB}|} = \frac{100/\sqrt{2}}{10\sqrt{2}} = 5 \text{ h}$$

128. The velocity of rain $= \sqrt{(3)^2 + (3\sqrt{3})^2} = 6$ km/h

and direction, $\tan\theta = \frac{3}{3\sqrt{3}} = 30°$

129. $\mathbf{v}_{RM}$ or $\mathbf{v}_R - \mathbf{v}_M$ is at 45° from vertical.
Hence, $|\mathbf{v}_R| = |\mathbf{v}_M| = 5$ m/s

130. x - component of velocities should be same.

Hence, $v_B \cos\theta = v_A$ or $\frac{v_B}{v_A} = \sec\theta$

131. Velocity in x-direction, $v_x = 5$ m/s

Velocity in y-direction, $v_y = \frac{60}{5} = 12$ m/s

So, the velocity in still water

$$= \sqrt{(5)^2 + (12)^2} = 13 \text{ m/s}$$

132. Distance = |Relative velocity| × Time

$$= |\mathbf{v}_A - \mathbf{v}_B| \times \text{Time}$$

133. The components of acceleration along x - axis,

$$a_x = -a\cos 37° = -4 \text{ m/s}^2$$

$$v_x = 8 \text{ m/s}$$

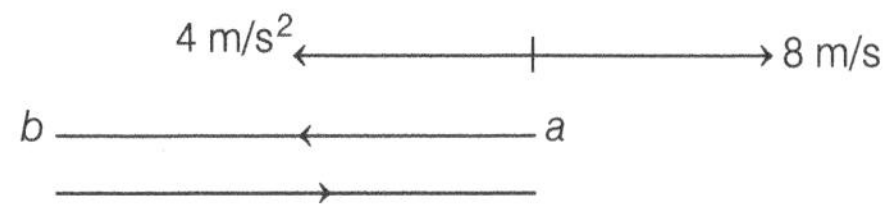

$$t = t_{ab} + t_{ba} = 2\left|\left(\frac{v_x}{a_x}\right)\right|$$

$$= 2 \times \frac{8}{4} = 4 \text{ s}$$

134. Let the velocity of rain $= a\hat{\mathbf{i}} \mp b\hat{\mathbf{j}}$

$$\mathbf{v}_r = |\mathbf{v}_r|\sin 60°\,\hat{\mathbf{j}} \mp v_r \cos 60°\,\hat{\mathbf{j}}$$

$$\mathbf{v}_m = 20\,\hat{\mathbf{i}}$$

So, $\mathbf{v}_{rm} = |-\mathbf{v}_r|\cos 60°\,\hat{\mathbf{j}}$

$$\mathbf{v}_r - \mathbf{v}_m = -\frac{|\mathbf{v}_r|\,\hat{\mathbf{j}}}{2}$$

$$|\mathbf{v}_r|\sin 60°\,\hat{\mathbf{i}} - |\mathbf{v}_r|\cos 60°\,\hat{\mathbf{j}} - 20\,\hat{\mathbf{i}} = -\frac{|\mathbf{v}_r|}{2}\,\hat{\mathbf{j}}$$

$$\mathbf{v}_r \sin 60° - 20 = 0$$

$$\mathbf{v}_r = \frac{20}{\sin 60°} = \frac{40}{\sqrt{3}} \text{ m/s}$$

135. In 2 h, number of taxis leave from X to Y

$$= \frac{2 \times 60}{10} = 12$$

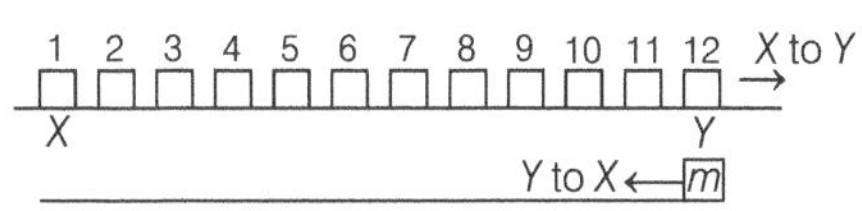

So, any taxi leaving Y will across 11 taxis from X and it also meets 12 taxis which are leaving X(to Y) in 2 h of journey.

Hence,

$$\text{total taxis} = 11 + 12 = 23$$

136. Given, $v = (t - 6)$ m/s

$$\frac{dx}{dt} = t - 6$$

$$\int dx = \int_2^6 (t-6)\,dt$$

$$x = \left[\frac{t^2}{2} - 6t\right]_2^6 = 8 \text{ m}$$

So, average speed $= \frac{8}{4} = 2$ m/s

137. Given, $v = \sin t$

Average value of a function in interval a to b is,

$$f_{avg} = \frac{1}{b-a} \cdot \int_a^b f(x)\, dx$$

$$v_{avg} = \frac{\int v dt}{\int dt} = \frac{\int_0^{\pi/2} \sin t\, dt}{\int_0^{\pi/2} dt} = \frac{2}{\pi} \text{ m/s}$$

138. Given, $x = (t \sin t)$ m

$$\frac{dx}{dt} = t\left(\frac{d(\sin t)}{dt}\right) + \frac{\sin t d(t)}{dt}$$

$$= t(\cos t) + \sin t$$

Now, $v_{i_{t=0}} = 0 \times \cos 0 + \sin 0 = 0$

$$v_{f_{t=\pi/2}} = \frac{\pi}{2}\cos\frac{\pi}{2} + \sin\frac{\pi}{2} = 1$$

Therefore, $a = \frac{v_f - v_i}{t_f - t_i} = \frac{1-0}{\frac{\pi}{2} - 0} = \frac{2}{\pi} \text{ m/s}^2$

139. Given, $F = -kx^2\hat{\mathbf{i}}$

$$a = -\frac{kx^2\hat{\mathbf{i}}}{m}$$

We know, $a = \frac{vdv}{dx}$

$$\frac{vdv}{dx} = -\frac{kx^2\hat{\mathbf{i}}}{m}$$

On integrating, we get

$$\frac{v^2}{2} = -\frac{kx^3}{3m} + C$$

At $x = a, v = 0$

So, $C = \frac{ka^3}{3m}$

$\therefore$ Velocity at origin is, $v = \sqrt{\frac{2ka^3}{3m}}$

140. Given, $\mathbf{v} = C(xy^2\,\hat{\mathbf{i}} + yx^2\,\hat{\mathbf{j}})$

$$v = \frac{dx}{dt} \text{ or } \frac{dy}{dt}$$

$$dx = Cxy^2 dt \text{ and } dy = Cyx^2 dt$$

Now, $\frac{dx}{dy} = \frac{Cxy^2dt}{Cyx^2dt} = \frac{y}{x}$

$$xdx = ydy$$

On integrating both sides, we get

$$\int xdx = \int ydy$$

$$\frac{x^2}{2} = \frac{y^2}{2} + C$$

$$x^2 = y^2 + C$$

or $\quad y^2 = x^2 + C$

141. Time of flight $= \sqrt{\frac{2H}{g}} = \sqrt{\frac{2(405)}{10}} = 9$ s

Now, $v_x = ay$

$$\frac{dx}{dt} = ay = \frac{y}{5} \quad \text{...(i)}$$

$$y = \frac{1}{2}gt^2 \quad \text{...(ii)}$$

From Eqs. (i) and (ii), we get

$$\frac{dx}{dt} = \frac{1}{2}agt^2$$

$$x = \frac{1}{2}ag\int_0^t t^2 dt = \frac{agt^3}{6}$$

So, $x = \frac{1}{5}\frac{(10)(9)^3}{6} = 243$ m

142. Given, $x_1 = -3 + 2t + t^2$ and $x_2 = 7 - 8t + t^2$

So, $v_1 = \frac{dx_1}{dt} = 2t + 2$ and $v_2 = \frac{dx_2}{dt} = 2t - 8$

Therefore, their relative speed,

$$v_{12} = v_1 - v_2 = 2t + 2 - (2t - 8)$$
$$= 10 \text{ m/s}$$

143. Case (i) $a_t = 0 \Rightarrow v = v_0 =$ constant

Case (ii) $a_t = 0 \Rightarrow v = 2v_0 =$ constant

Case (iii), (iv) and (v), $a_t = \frac{F}{m} =$ same

144. Given, speed of one car = 27 km/h

Speed of other car = 18 km/h

So, their relative speed of each car = 18 + 27

= 45 km/h

Distance between two cars = 36 km

So, the time at which cars will meet

$$= \frac{36}{45} = 0.8 \text{ h}$$

Now, speed of bird = 36 km/h

$\therefore$ Distance covered by bird = 36 × 0.8

= 28.8 km

145. We have given, $u_x = 8t - 2$ and $u_y = 2$

So, $\int_4^y dy = \int_2^t 2dt = |y|_2^y = 2|t|_2^t$

$\therefore \quad y - 4 = 2(t - 2)$

$$t = \frac{y}{2}$$

Similarly, velocity in x-direction,

$$\int_{14}^x dx = \int_2^t (8t - 2)\, dt$$

$$x|_{14}^x = 4t^2 - 2t|_2^t$$

$$x - 14 = 4(t^2 - 4) - 2(t - 2)$$

$$x - 14 = 4\left(\left(\frac{y}{2}\right)^2 - 4\right) - 2\left(\frac{y}{2} - 2\right)$$

$$x - 14 = 4\left(\frac{y^2}{4}\right) - 4 - 2\left(\frac{y}{2} - 2\right)$$

$$x = y^2 - y + 14$$

146.

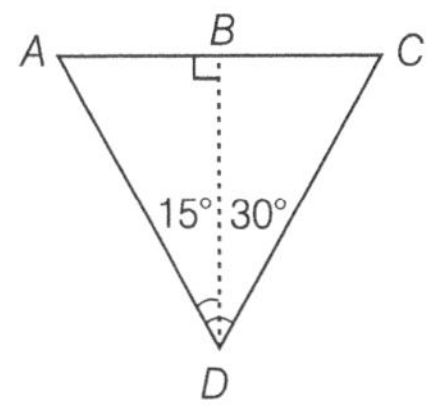

In ΔABD,

$$\tan 15° = \frac{AB}{3400}$$

or $AC = 2AB = 2 \times 3400 \times \tan 15° = 1822.4$ m

Thus, speed of aircraft $= \frac{1822.4}{10} = 182.24$ m/s

147. Initially, C_1 will overtake C_2 because C_1 has some initial velocity. But later on C_2 will further overtake C_1 because C_2 is moving with a constant acceleration and C_1 with constant velocity.

$$s_{C_1} = 9t \qquad (\because s = vt)$$

$\therefore$ s_{C_1} *versus* time is a straight line passing through origin.

$$s_{C_2} = 40 + \frac{1}{2} \times 1 \times t^2 \qquad (\because s = s_0 + \frac{1}{2}at^2)$$

$$= 40 + 0.5\,t^2$$

$\therefore$ s_{C_2} *versus* time graph is a parabola.

At $t = 0$, $s_{C_2} = 40$ m

148. $v_{C_1} = 9$ m/s = constant

$v_{C_2} = at = t \qquad (\because a = 1\text{ m/s}^2)$

149. At maximum lead,

$$v_{C_1} = v_{C_2} \quad \text{or} \quad t = 9\text{ s}$$

$\therefore$ At $t = 9$ s, there will be maximum distance between C_1 and C_2,

$$s_{C_1} = 9 \times 9 = 81\text{ m}$$

$$s_{C_2} = 40 + (0.5)(9^2) = 80.5\text{ m}$$

$\therefore$ Maximum lead $= 81 - 80.5 = 0.5$ m

Note : s_{C_2} is measured from the initial position of C_1.

150. The velocity of car $A = \sqrt{2as}$

Now, the time taken by car A to travel distance,

$$s = \frac{1}{2}at^2 \Rightarrow t = \sqrt{\frac{2s}{a}}$$

As the car A starts with acceleration a_1 and car B with speed u.

$\therefore$ Car A will reach at time $(T - t)$ and with speed $(u + v)$.

For car A, $\quad v + u = \sqrt{2a_1 s}$,

$$T - t = \sqrt{\frac{2s}{a_1}}$$

For car B, $\quad v = \sqrt{2a_2 s},\; T = \sqrt{\frac{2s}{a_2}}$

$$v = \sqrt{2a_1 s} - \sqrt{2a_2 s}$$

$$t = \sqrt{\frac{2s}{a_2}} - \sqrt{\frac{2s}{a_1}}$$

Dividing these equations, we get

$$v = \sqrt{a_1 a_2}$$

151. At 3 s,

$$s = ut + \frac{1}{2}at^2 = 10 \times 3 - \frac{1}{2} \times 5 \times (3)^2$$

$$= 7.5\text{ m}$$

$$v = u + at$$

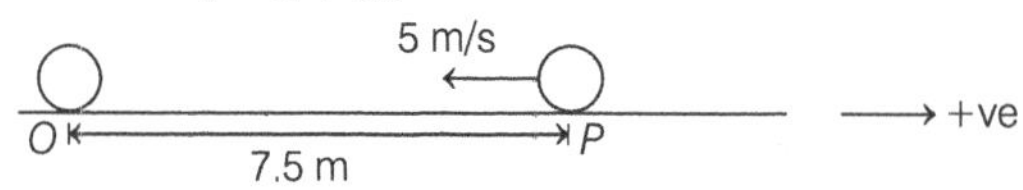

$$t_{PO} = \frac{7.5}{5} = 1.5\text{ s}$$

Therefore, total time = (3 + 1.5) s = 4.5 s

152. Speed $= \sqrt{v_x^2 + v_y^2}$

$$\text{Rate of change of speed} = \frac{2v_x \frac{dv_x}{dt} + 2v_y \frac{dv_y}{dt}}{2\sqrt{v_x^2 + v_y^2}}$$

$$= \frac{v_x a_x + v_y a_y}{\sqrt{v_x^2 + v_y^2}} = \frac{3 \times 2 + 4 \times 1}{\sqrt{3^2 + 4^2}} = 2\text{ m/s}^2$$

153.

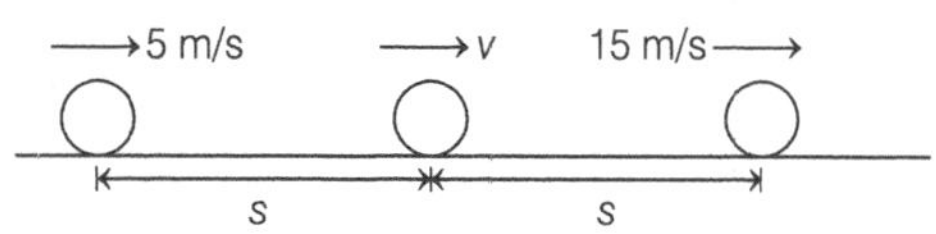

$$(15)^2 = (5)^2 + 2(a)(2s)$$

$$\therefore \quad as = 50\,\frac{\text{m}^2}{\text{s}^2}$$

Now, $v^2 = (5)^2 + 2as = 125$

$$\therefore \quad v = 5\sqrt{5}\text{ m/s}$$

Using $v = u + at$, we can find the time.

154. $h = \frac{1}{2}g\left(\frac{T}{2}\right)^2 = \frac{gT^2}{8}$

$$\frac{3h}{4} = \frac{3gT^2}{32} = \frac{1}{2}gt_{AB}^2$$

$$\therefore \quad t_{AB} = \frac{\sqrt{3}}{4}T$$

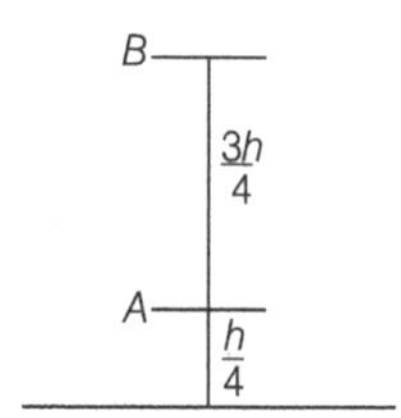

155. The displacement in the vertical direction

$$= \frac{1}{2}gt^2 = \frac{1}{2}(10)(4)^2 = 80 \text{ m}$$

and the displacement in the horizontal direction with

$$a = \frac{4g}{3} = \frac{40}{3} \text{ m/s}^2$$

$$= \frac{1}{2}\frac{40}{3}(4)^2 = 106.6 \text{ m}$$

Therefore, the net displacement

$$= \sqrt{80^2 + 106.6^2} = 133.3 \text{ m}$$

156. $x = \dfrac{1}{(1+t)^{1/2}} = (1+t)^{-1/2}$

$$v = \frac{dx}{dt} = -\frac{1}{2}(1+t)^{-\frac{3}{2}} \quad \text{...(i)}$$

$$a = \frac{dv}{dt} = \frac{3}{4}(1+t)^{-\frac{5}{2}} \quad \text{...(ii)}$$

From Eq. (i), we can see that

$$(1+t) \propto v^{-\frac{2}{3}}$$

Substituting in Eq. (ii), we get

$$a \propto v^{\frac{5}{3}}$$

157. For distance,

$$s = \frac{(v_0 + 0)t_1}{2} \Rightarrow t_1 = \frac{2s}{v_0}$$

Similarly, $t_2 = \dfrac{3s}{v_0}$

Now, $5s = \dfrac{v_0 + 0\,(t_3)}{2} \Rightarrow t_3 = \dfrac{10s}{v_0}$

$$v_{\text{average}} = \frac{s + 3s + 5s}{t_1 + t_2 + t_3} = \frac{9s}{\frac{2s}{v_0} + \frac{3s}{v_0} + \frac{10s}{v_0}} = \frac{3v_0}{5}$$

$$\frac{v_{\text{average}}}{v_0} = \frac{3}{5}$$

158. We have, $60 = u + at_1$ $\quad (\because u = 0)$

$$\Rightarrow \quad a = \frac{60}{t_1}$$

Also, $s_1 = ut_1 + \dfrac{1}{2}\left(\dfrac{60}{t_1}\right)t_1^2 = 30\,t_1$

Now, $v^2 = u^2 + 2as_1$

From $s_1 = \dfrac{1}{2}at_1^2$

$$\Rightarrow \quad a = \frac{2s_1}{t_1^2}, \; v = \frac{2s_1}{t_1}$$

The s_2 distance is travelled,

$$s_2 = vt_2 = \frac{2s_1t_2}{t_1}$$

But the ratio of times is given as

$$\frac{t_2}{t_1} = 8 \quad \text{or} \quad t_2 = 8t_1$$

$$s_2 = 16\,s_1$$

We get, $\quad s_1 = 30t_1$

$\Rightarrow \quad s_2 = 480\,t_1$

and $\quad s_3 = s_1 = 30\,t_1$

Thus, the average speed $= \dfrac{\text{distance travelled}}{\text{time taken}}$

$$= \frac{30t_1 + 480t_1 + 30t_1}{t_1 + 8t_1 + t_1} = 54 \text{ km/h}$$

159. $\dfrac{d^2y}{dt^2} = \beta$ and $\dfrac{d^2x}{dt^2} = 0$

$$y = \alpha x^2$$

$$\frac{dy}{dt} = 2\alpha x \frac{dx}{dt}$$

$$\therefore \quad \frac{d^2y}{dt^2} = 2\alpha\left[x \cdot \frac{d^2x}{dt^2} + \left(\frac{dx}{dt}\right)^2\right]$$

$$\therefore \quad \beta = 2\alpha\,[0 + v_x^2]$$

or $\quad v_x = \sqrt{\dfrac{\beta}{2\alpha}}$

160. $\mathbf{v}_{RM} = \mathbf{v}_R - \mathbf{v}_M$

$$\mathbf{v}_R = \mathbf{v}_{RM} + \mathbf{v}_M$$

$$|\mathbf{v}_R| = \sqrt{v_{RM}^2 + v_M^2 + 2v_{RM}v_M \cos 120^\circ}$$

$$= \sqrt{v^2 + v^2 + 2v \cdot v \cos 120^\circ} \quad \left(\because \cos 120^\circ = \frac{-1}{2}\right)$$

$$= v$$

161. The slope of a displacement-time graph gives the velocity.

As the slope of *P* is increasing, slope is positive. So, acceleration is positive, its average speed increases.

Similarly, in case of the slope of Q is decreasing, hence acceleration is negative, its average speed decreases with time.

∴ Except statement (2), all other statements are correct.

162. Equation of *v versus x* graph is,

$$v = -x + 2$$

differentiating we have,

$$a = \frac{dv}{dt} = -v \quad \text{or} \quad a = x - 2$$

So, option (a) is correct.

Slope is positive and intercept is negative.

163. We know, $a = \dfrac{vdv}{dx}$

$$\int a\,dx = \int v\,dv$$

= Area under the curve from 0 to $x = 1.4$

$$= 0.4 \times 0.4 + \frac{1}{2} \times 0.2 \times 0.4 + 0.4 \times 0.2 + 0.6 \times 0.2$$

$$= \left[\frac{v^2}{2}\right]_u^v$$

$$0.4 = \frac{v^2}{2} - \frac{u^2}{2}$$

Here, $u = 0.8$ m/s

So, $v = 1.2$ m/s

164. $v_x = \frac{dx}{dt} = mt \Rightarrow x = \frac{mt^2}{2} + \alpha$ ...(i)

$v_y = \frac{dy}{dt} = C \Rightarrow y = Ct$ ..(ii)

Eliminating t from Eqs. (i) and (ii), we can find x - y relation, which comes out to be a parabola.

165.

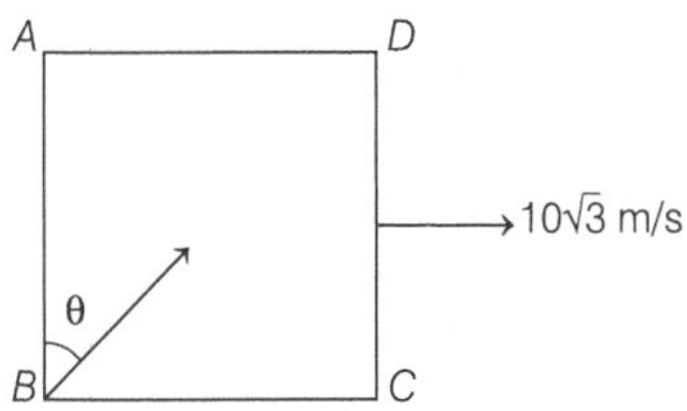

$$u \sin\theta = 10\sqrt{3}$$

$$\Rightarrow \quad \sin\theta = \frac{10\sqrt{3}}{20} = \frac{\sqrt{3}}{2}$$

$$\Rightarrow \quad \theta = 60°$$

166. $s_{23} = ut + \frac{1}{2}at^2 = 20 \times 1 + \frac{1}{2} \times 10 \times 1^2 = 25$ m

$$s_{78} = ut + \frac{1}{2}at^2$$

$$= -10 \times 1 - \frac{1}{2} \times 10 \times 1^2$$

$$= -15 \text{ m}$$

Note

At 2 s, $v = +20$ m/s

At 6 s, $v = 0$

At 7 s, $v = -10$ m/s

167. Let us take vertically downward direction as positive x - axis and east direction as positive y - direction. Then,

$$\mathbf{v}_R = 20\,\hat{\mathbf{i}}$$

$$\mathbf{v}_M = 5\,\hat{\mathbf{j}}$$

$$\mathbf{v}_W = 15\,\hat{\mathbf{j}}$$

Net velocity of rain $\mathbf{v}_N = \mathbf{v}_R + \mathbf{v}_W$

Velocity of rain with respect to man

$$= \mathbf{v}_R - \mathbf{v}_M = (20\,\hat{\mathbf{i}} + 10\,\hat{\mathbf{j}})$$

Its magnitude is $10\sqrt{5}$ m/s and it is at an angle of $\tan^{-1}\left(\frac{1}{2}\right)$ from $\hat{\mathbf{i}}$-direction, or vertical.

168. $v_S = v_{SR} + v_R$

The displacement along x - axis

$$= (v_R - v_S \sin 37°)\ 0.05$$

$$= 0.05 \text{ km} = 50 \text{ m}$$

Time taken during swimming $= \frac{200 \times 10^{-3}}{v_R \cos 37°} = 0.05$ h

So, total time $= \frac{50 \times 10^{-3}}{3 \text{ km/h}} + 0.05 \text{ h} = 4$ min

169. We have given, width of river = 800 m

Time taken = 200 s

Therefore, the speed of boat with respect to ground, v_b

$$= \frac{800}{200} = 4 \text{ m/s}$$

Speed of raft with respect to ground = 3 m/s

$$\mathbf{v}_{br} = v_b - v_r = 4\hat{\mathbf{j}} - 3\hat{\mathbf{i}}$$

$$|v_{br}| = \sqrt{(4)^2 + (3)^2} = 5 \text{ m/s}$$

Now, the distance = 5 × 200 = 1000 m

170. Let u be the initial velocity of swimmer and v be the velocity of river.

$$v^2 + v_r^2 = u^2 \Rightarrow v_r = \sqrt{u^2 - v^2}$$

So, time taken in going up and down,

$$t_1 = 2\left(\frac{b}{\sqrt{u^2 - v^2}}\right) \quad \text{...(i)}$$

Similarly, the time t_2,

$$\frac{b}{u+v} + \frac{b}{u-v} = \frac{2ub}{u^2 - v^2} \quad \text{...(ii)}$$

So, time t_3 taken in crossing a river of width $2b$,

$$t_3 = \frac{2b}{u} \quad \text{...(iii)}$$

From Eqs. (i), (ii) and (iii), we get

$$T_1^2 = T_2 T_3$$

LEVEL 03 Previous Years Questions

171. Given, the initial velocity of a car, $u = 0$

The acceleration of a car, $a = 5$ m/s^2

At $t = 4$ s, $v = u + at$

$\Rightarrow$ $v = 0 + (5)4 \Rightarrow v = 20$ m/s

Thus, the final velocity of car at $t = 4$ s is 20 m/s.

At $t = 4$ s, the ball is dropped out of a window by a person sitting in the car.

The velocity of the ball in the x-direction,

$$v_x = 20 \text{ m/s} \qquad \text{(due to the car)}$$

Therefore, in the y-direction, the acceleration is equal to the acceleration due to gravity,

$$a_y = g = 10 \text{ m/s}^2$$

The velocity of the ball in the y-direction,

$$v_y = u + a_y t$$

$$\Rightarrow \quad v_y = 0 + 10 \times 2$$

$$\Rightarrow \quad v_y = 20 \text{ m/s}$$

Thus, the velocity of the ball in y-direction is 20 m/s.

The net velocity at $t = 6$ s,

$$v = \sqrt{v_x^2 + v_y^2}$$

$$\Rightarrow \quad v = \sqrt{(20)^2 + (20)^2}$$

$$\Rightarrow \quad v = 20\sqrt{2} \text{ m/s}$$

Thus, the velocity of the ball at $t = 6$ s is $20\sqrt{2}$ m/s.

and there is no acceleration in the x-direction,

$$a_x = 0 \text{ ms}^{-2}$$

In y-direction, $a_y = 10 \text{ ms}^{-2}$

Now, we shall determine the net acceleration

at $\quad t = 6 \text{ s}, a = \sqrt{a_x^2 + a_y^2}$

$$\Rightarrow \quad a = \sqrt{(0) + (10)^2}$$

$$\Rightarrow \quad a = 10 \text{ ms}^{-2}$$

172. Distance covered nth seconds is s_n.

Distance covers in $(n+1)$ th seconds is s_{n+1}.

Initial velocity of small block, $u = 0$

Distance cover in nth seconds,

$$s_n = u + \frac{a}{2}(2n - 1)$$

$$\Rightarrow \quad s_n = 0 + \frac{a}{2}(2n - 1)$$

$$\Rightarrow \quad s_n = \frac{a}{2}(2n - 1) \qquad \text{...(i)}$$

Distance cover in $(n+1)$th seconds,

$$s_{n+1} = u + \frac{a}{2}[2(n+1) - 1]$$

$$\Rightarrow \quad s_{n+1} = 0 + \frac{a}{2}(2n + 2 - 1)$$

$$\Rightarrow \quad s_{n+1} = \frac{a}{2}(2n + 1) \qquad \text{...(ii)}$$

On dividing Eq. (i) by Eq. (ii), we get

$$\frac{s_n}{s_{n+1}} = \frac{\frac{a}{2}(2n-1)}{\frac{a}{2}(2n+1)}$$

$$\Rightarrow \quad \frac{s_n}{s_{n+1}} = \frac{(2n-1)}{(2n+1)}$$

173. According to question, time taken by the ball to cross the window,

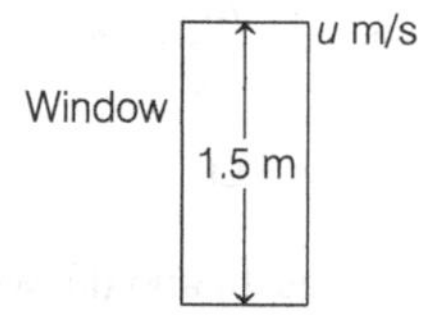

$$t = 0.1 \text{ s}$$

$$h = 1.5 \text{ m}$$

If u be the velocity at the top most point of the window, then from equation of motion,

$$h = ut + \frac{1}{2}gt^2$$

$$\Rightarrow \quad 1.5 = u \times 0.1 + \frac{1}{2} \times 10 \times (0.1)^2$$

$$\Rightarrow \quad 1.5 = 0.1u + 0.05$$

$$\Rightarrow \quad u = \frac{1.5 - 0.05}{0.1} = \frac{1.45}{0.1}$$

$$= 14.5 \text{ m/s}$$

174. Given, $u = 20$ m/s, $v = 80$ m/s and $h = ?$

From kinematic equation of motion,

$$v^2 = u^2 + 2gh$$

$$\Rightarrow \quad h = \frac{v^2 - u^2}{2g}$$

$$= \frac{(80)^2 - (20)^2}{2 \times 10} \qquad (\because \text{ given, } g = 10 \text{ m/s}^2)$$

$$= 300 \text{ m}$$

175. For distance x, the person moves with constant velocity v_1 and for another x distance, he moves with constant velocity of v_2, then

Total distance travelled, $D = x + x = 2x$

Total time-taken, $T = t_1 + t_2$

$$= \frac{x}{v_1} + \frac{x}{v_2} \qquad \left[\because t = \frac{\text{Distance}}{\text{Velocity}}\right]$$

The average velocity,

$$V_{av} = \frac{\text{total distance}}{\text{total time}} = \frac{D}{T}$$

$$v = \frac{2x}{\frac{x}{v_1} + \frac{x}{v_2}} = \frac{2}{\frac{1}{v_1} + \frac{1}{v_2}} \quad [\because v_{av} = v]$$

$$\Rightarrow \frac{1}{v_1} + \frac{1}{v_2} = \frac{2}{v}$$

176. Let h be the height through which the coin is dropped. Then, according to the equation of motion, it is given as

$$h = ut + \frac{1}{2}gt^2$$

$$\Rightarrow \quad t = \sqrt{\frac{2h}{g}} \quad [\because u = 0]$$

$$\Rightarrow \quad t \propto \frac{1}{\sqrt{g}}$$

As the elevator is moving uniformly i.e. its velocity is constant, so the acceleration is zero.

$\therefore$ Relative acceleration of the lift when it is either moving upward or downward is given as, $g' = g \pm a = g \pm 0 = g$

Hence, the time for the coin to reach the floor will remain same i.e. $t_1 = t_2$.

177. According to the question,

For the time duration $0 < t < 1$s, the velocity increase from 0 to 6 ms^{-1}

As the direction of field has been reversed for $1 < t < 2$s : the velocity firstly decreases from 6 ms^{-1} to 0.

Then, for $2 < t < 3$s; as the field strength is same; the magnitude of acceleration would be same, but velocity increases from 0 to – 6 ms^{-1}.

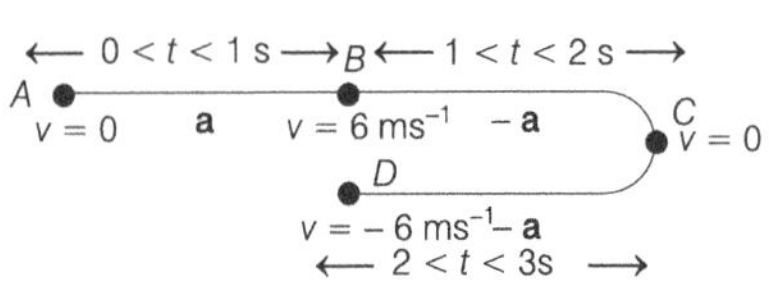

Acceleration of the car,

$$|a| = \left|\frac{v-u}{t}\right| = \frac{6-0}{1} = 6 ms^{-2}$$

The displacement of the particle is given as

$$s = ut + \frac{1}{2}at^2$$

For $t = 0$ to $t = 1$ s,

$$u = 0, a = +6 \text{ m/s}^2$$

$$\Rightarrow \quad s_1 = 0 + \frac{1}{2} \times 6 \times (1)^2 = 3 \text{ m}$$

For $t = 1$ s to $t = 2$ s,

$$u = 6 ms^{-1}, a = -6 ms^{-2}$$

$$\Rightarrow \quad s_2 = 6 \times 1 - \frac{1}{2} \times 6 \times (1)^2$$

$$= 6 - 3 = 3\text{m}$$

For $t = 2$ s to $t = 3$ s,

$$u = 0, a = -6 ms^{-1}$$

$$\Rightarrow \quad s_3 = 0 - \frac{1}{2} \times 6 \times (1)^2 = -3\text{m}$$

$\therefore$ Net displacement, $s = s_1 + s_2 + s_3$

$$= 3 \text{ m} + 3 \text{ m} - 3 \text{ m} = 3 \text{ m}$$

Hence, average velocity

$$= \frac{\text{Net displacement}}{\text{Total time}}$$

$$= \frac{3}{3} = 1 \text{ ms}^{-1}$$

Total distance travelled, $d = 9$m

$$\text{Hence, average speed} = \frac{\text{Total distance}}{\text{Total time}}$$

$$= \frac{9}{3} = 3 \text{ ms}^{-1}$$

178. Speed of walking $= \frac{h}{t_1} = v_1$

Speed of escalator $= \frac{h}{t_2} = v_2$

Time taken when she walks over running escalator

$$\Rightarrow \quad t = \frac{h}{v_1 + v_2}$$

$$\Rightarrow \quad \frac{1}{t} = \frac{v_1}{h} + \frac{v_2}{h} = \frac{1}{t_1} + \frac{1}{t_2}$$

$$\Rightarrow \quad t = \frac{t_1 t_2}{t_1 + t_2}$$

179. Velocity of the particle is given as

$$v = At + Bt^2$$

where A and B are constants.

$$\Rightarrow \quad \frac{dx}{dt} = At + Bt^2 \quad \left[\because v = \frac{dx}{dt}\right]$$

$$\Rightarrow \quad dx = (At + Bt^2)\, dt$$

Integrating both sides, we get

$$\int_{x_1}^{x_2} dx = \int_1^2 (At + Bt^2)\, dt$$

$$\Rightarrow \quad \Delta x = x_2 - x_1 = A\int_1^2 t\, dt + B\int_1^2 t^2 dt$$

$$= A\left[\frac{t^2}{2}\right]_1^2 + B\left[\frac{t^3}{3}\right]_1^2$$

$$= \frac{A}{2}(2^2 - 1^2) + \frac{B}{3}(2^3 - 1^3)$$

$\therefore$ Distance travelled between 1*s* and 2*s* is

$$\Delta x = \frac{A}{2} \times (3) + \frac{B}{3}(7) = \frac{3A}{2} + \frac{7B}{3}$$

180. Velocity of each car is given by

$$v_P = \frac{dx_p(t)}{dt} = a + 2bt$$

and $$v_Q = \frac{dx_Q(t)}{dt} = f - 2t$$

It is given that $v_P = v_Q$

$$\Rightarrow \quad a + 2bt = f - 2t$$

$$t = \frac{f - a}{2(b+1)}$$

181. Given, $v = \beta x^{-2n}$

$$a = \frac{dv}{dt} = \frac{dx}{dt} \cdot \frac{dv}{dx}$$

$$\Rightarrow \quad a = v\frac{dv}{dx} = (\beta x^{-2n})(-2n\beta x^{-2n-1})$$

$$\Rightarrow \quad a = -2n\beta^2 x^{-4n-1}$$

182. Given, $x = 8 + 12t - t^3$

We know $v = \frac{dx}{dt}$

and acceleration, $a = \frac{dv}{dt}$

So, $v = 12 - 3t^2$ and $a = -6t$

At $t = 2$ s

$v = 0$ and $a = -6 \times 2$

$$a = -12 \text{ m/s}^2$$

So, retardation of the particle = 12 m/s^2.

183. Average acceleration

$$a = \frac{\text{Change in velocity}}{\text{Total time}}$$

$$a = \frac{|\mathbf{v}_f - \mathbf{v}_i|}{\Delta t} = \frac{\sqrt{30^2 + 40^2}}{10}$$

$$= \frac{\sqrt{900 + 1600}}{10} = 5 \text{ ms}^{-2}$$

184. Given, $g = 10$ m/s^2 and $h = 20$ m

We have $v = \sqrt{2gh} = \sqrt{2 \times 10 \times 20}$

$$= \sqrt{400} = 20 \text{ m/s}$$

185. For first ball, $u = 0$

$$\therefore \quad s_1 = \frac{1}{2} g t_1^2 = \frac{1}{2} \times g\,(18)^2$$

For second ball, initial velocity = v

$$\therefore \quad s_2 = vt_2 + \frac{1}{2} g t^2$$

$$t_2 = 18 - 6 = 12 \text{ s}$$

$$\Rightarrow \quad s_2 = v \times 12 + \frac{1}{2} g\,(12)^2$$

Here, $s_1 = s_2$

$$\frac{1}{2} g(18)^2 = 12v + \frac{1}{2} g\,(12)^2$$

$$\Rightarrow \quad v = 74 \text{ ms}^{-1}$$

Chapter 05

PROJECTILE MOTION

LEVEL 01 Objective Questions

1 Four bodies P, Q, R and S are projected with same speed having angles of projection 15°, 30°, 45° and 60° with the horizontal, respectively. The body having shortest range is

(a) P (b) Q (c) R (d) S

2 Four balls are projected from ground as shown in the following figures. Which will have maximum range?

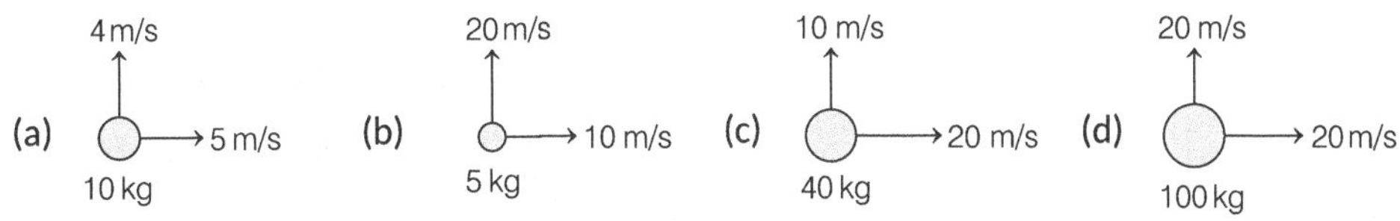

3 The range of a projectile when launched at an angle 15° with the horizontal is 1.5 km. The maximum range of that projectile when launched with the same velocity, will be

(a) 0.75 km (b) 1.5 km
(c) 3.0 km (d) 6.0 km

4 The horizontal and vertical components of the initial velocity of a projectile are 10 m/s and 20 m/s, respectively. The horizontal range of the projectile will be (Take, $g = 10\ \text{m/s}^2$)

(a) 5 m (b) 10 m (c) 20 m (d) 40 m

5 The height y and the distance x along the horizontal plane of a projectile on a certain planet (with no surrounding atmosphere) are given by $y = (8t - 5t^2)$ m and $x = 6t$ m, where t is in seconds. The velocity with which the projectile is projected at $t = 0$ is

(a) 8 ms^{-1} (b) 6 ms^{-1} (c) 10 ms^{-1} (d) 20 ms^{-1}

6 A particle is projected with velocity v at angle θ with horizontal at $t = 0$. What is the magnitude of change in the velocity of the particle when it is at maximum height?

(a) $\frac{v\cos\theta}{2}$ (b) $v\cos\theta$ (c) $v\sin\theta$ (d) $v\tan\theta$

7 A particle is thrown with speed u at an angle α with the horizontal. When the particle makes an angle β with the horizontal, its speed will be

(a) $u\cos\alpha$ (b) $u\cos\alpha \sec\beta$ (c) $u\cos\alpha\cos\beta$ (d) $u \sec\alpha \cos\beta$

8 The horizontal range of a projectile is $4\sqrt{3}$ times its maximum height. Then, the angle of projection is

(a) 90° (b) 60° (c) 45° (d) 30°

9 Two balls of same mass are projected with the same speed one vertically upwards and the other at angle 60° with the vertical. The ratio of their potential energies at the highest point is

(a) 3 : 2 (b) 2 : 1 (c) 4 : 1 (d) 4 : 3

10 Horizontal range and maximum height for a projectile are respectively, 12 m and 4 m. Velocity of projection is

(a) $3\sqrt{\frac{g}{2}}$ (b) $\frac{1}{3}\sqrt{\frac{g}{2}}$ (c) $5\sqrt{\frac{g}{2}}$ (d) $\frac{1}{5}\sqrt{\frac{g}{2}}$

11 A box containing food supplies is released from an aeroplane moving horizontally at a height of 490 m with a velocity of 180 km/h. At what horizontal distance will the box strike on ground with respect to the point on ground just below the releasing point?

(a) 180 m (b) 98 m (c) 500 m (d) 750 m

12 A student stands at the edge of a cliff 50 m above a beach and throws a stone horizontally over the edge with a speed of 18 m/s. How long after being released does the stone strike the beach below the cliff? (Take, $g = 10\text{ m/s}^2$)

(a) 10 s (b) 5 s (c) $\sqrt{5}$ s (d) $\sqrt{10}$ s

13 In Q.no. 12, with what speed does the stone land?

(a) 36.4 m/s (b) 18.0 m/s (c) 9.0 m/s (d) $\sqrt{10}$ m/s

14 In Q.no. 12, at what angle from horizontal does the stone land?

(a) $\tan^{-1}\left(\frac{10}{18}\right)$ (b) $\tan^{-1}\left(\frac{18}{10}\right)$ (c) $\tan^{-1}\left(\frac{10\sqrt{10}}{18}\right)$ (d) $\tan^{-1}\left(\frac{18}{10\sqrt{10}}\right)$

15 A particle is projected horizontally with a speed of $\frac{20}{\sqrt{3}}$ m/s, from some height at $t = 0$. At what time will its velocity make 60° angle with the initial velocity?

(a) 1 s (b) 2 s (c) 1.5 s (d) 2.5 s

16 A ball is thrown from the top of 36 m high tower with velocity 5 m/s at an angle 37° above the horizontal as shown in the following figure. Its horizontal range on the ground is closest to (Take, $g = 10\text{ m/s}^2$)

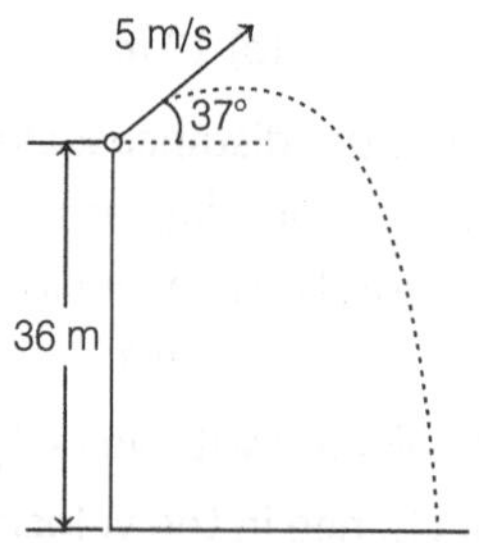

(a) 12 m (b) 18 m (c) 24 m (d) 30 m

17 A projectile is fired at an angle of 45° with the horizontal. Elevation angle of the projectile at its highest point as seen from the point of projection, is

(a) $\tan^{-1}\left(\frac{\sqrt{3}}{2}\right)$ (b) 45° (c) 60° (d) $\tan^{-1}\left(\frac{1}{4}\right)$

LEVEL 02 Objective Questions

18 Stone is projected from ground level. At a height of 0.4 m from the ground, the velocity of the stone in vector form is $\mathbf{v} = (6\hat{\mathbf{i}} + 2\hat{\mathbf{j}})$ m/s (the X-axis is horizontal and Y-axis is vertically upwards). The angle of projection with horizontal is (Take, $g = 10 \text{ m/s}^2$)

(a) 45° (b) 60°
(c) 30° (d) $\tan^{-1}(3/4)$

19 A shooter fires his gun by aiming on a target but the bullet hits the wall vertically 20 m below the target. The bullet is fired from a point that is 20 m away from the wall. The horizontal component of velocity of the bullet velocity is

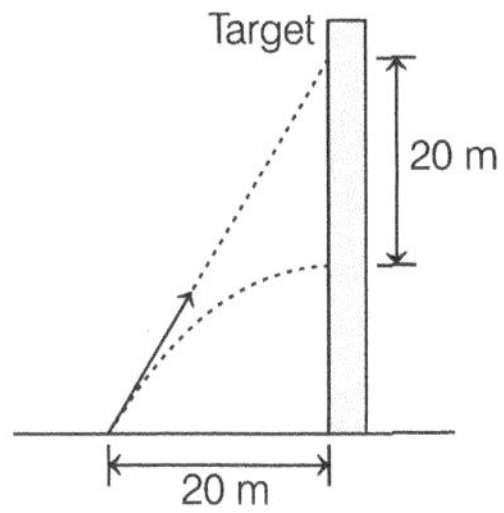

(a) 10 m/s (b) 15 m/s (c) 20 m/s (d) 25 m/s

20 What is the average velocity of a projectile between the instants it crosses half the maximum height? It is projected with a speed u at an angle θ with the horizontal.
(a) $u\sin\theta$ (b) $u\cos\theta$ (c) $u\tan\theta$ (d) u

21 A ball P is projected vertically up. Another similar ball Q is projected at an angle 45°. Both reach the same height during their motion. Then, at the starting point, ratio of kinetic energy of P and Q is

(a) $\frac{1}{2}$ (b) $\frac{1}{4}$ (c) 2 (d) 4

22 A particle A is projected with speed v_A from a point making an angle 60° with the horizontal. At the same instant, a second particle B is thrown vertically upward from a point directly below the maximum height point of parabolic path of A with velocity v_B. If the two particles collide, then the ratio of v_A/v_B should be

(a) 1 (b) $\frac{2}{\sqrt{3}}$ (c) $\frac{\sqrt{3}}{2}$ (d) $\sqrt{3}$

23 Two stones are projected with the same speed but making different angles with the horizontal. Their horizontal ranges are equal. The angle of projection of one is $\frac{\pi}{3}$ and the maximum height reached by it is 102 m. The maximum height reached by the other (in m) is
(a) 336 (b) 224 (c) 56 (d) 34

24 A ball is projected from the ground at an angle θ with the horizontal. After 1 s, it is moving at angle 45° with the horizontal and after 2 s, it is moving horizontally. The speed of projection of the ball is
(a) $10\sqrt{3}\text{ ms}^{-1}$ (b) $20\sqrt{3}\text{ ms}^{-1}$ (c) $10\sqrt{5}\text{ ms}^{-1}$ (d) $20\sqrt{2}\text{ ms}^{-1}$

25 Two balls are thrown from top of a tower at a speed 40 m/s. One is thrown 30° above horizontal and one 30° below horizontal. Find the time interval between the instants, the balls strike the ground in seconds. (Take, $g = 10 \text{ m/s}^2$)

(a) 1 s (b) 2 s (c) 3 s (d) 4 s

26 Two stones are projected simultaneously from a tower at different angles of projection with same speed u. The distance between two stones is increasing at constant rate u. Then, the angle between the initial velocity vectors of the two stones is

(a) 30° (b) 60°
(c) 45° (d) 90°

27 A ball is projected from top of a tower with a velocity of 5 m/s at an angle of 53° to horizontal. Its speed when it is at a height of 0.45 m from the point of projection is

(a) 2 m/s (b) 3 m/s
(c) 4 m/s (d) 2.5 m/s

28 A projectile is thrown with an initial velocity $(x\hat{i} + y\hat{j})$ m/s. If the range of projectile is double the maximum height reached by it, then

(a) $x = 2y$ (b) $y = 2x$
(c) $x = y$ (d) $y = 4x$

29 A projectile is thrown with a velocity of $10\sqrt{2}\text{ ms}^{-1}$ at an angle of 45° with vertical. The interval between the moments when speed is $5\sqrt{5}\text{ ms}^{-1}$, is (Take, $g = 10\text{ ms}^{-2}$)

(a) 1.5 s (b) 0.5 s
(c) 1.0 s (d) 2.0 s

30 A projectile has initially the same horizontal velocity as it would acquire, if it had moved from rest with uniform acceleration of 3 ms^{-2} for 0.5 min. If the maximum height reached by it is 80 m, then the angle of projection is (Take, $g = 10\text{ ms}^{-2}$)

(a) $\tan^{-1}(3)$ (b) $\tan^{-1}(3/2)$
(c) $\tan^{-1}(4/9)$ (d) $\sin^{-1}(4/9)$

31 A car is moving horizontally along a straight line with a uniform velocity of 25 ms^{-1}. A projectile is to be fired from this car in such a way that it will return to it after it has moved 100 m. The speed of the projection must be

(a) 10 ms^{-1} (b) 20 ms^{-1}
(c) 15 ms^{-1} (d) 25 ms^{-1}

32 Two tall buildings are 30 m apart. The speed with which a ball must be thrown horizontally from a window 150 m above the ground in one building, so that it enters a window 27.5 m from the ground in the other building is

(a) 2 ms^{-1} (b) 6 ms^{-1}
(c) 4 ms^{-1} (d) 8 ms^{-1}

33 A particle A is projected from the ground with an initial velocity of 10 m/s at an angle of 60° with horizontal. From what height h should an another particle B be projected horizontally with 5 m/s, so that both the particles collide on ground at point C, if both are projected simultaneously?

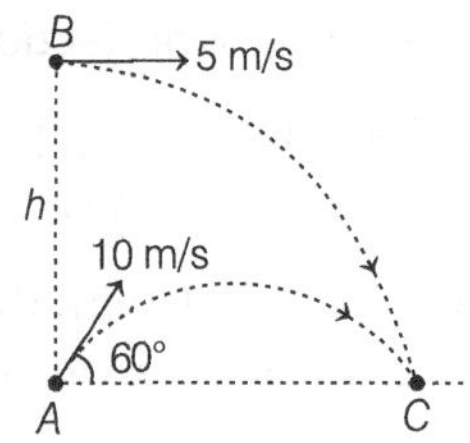

(a) 10 m (b) 30 m (c) 15 m (d) 25 m

34 A child playing on the roof of a 10 m high building throws a ball with a speed of 10 ms^{-1} at an angle of 30° above the horizontal. How far from the throwing point will the ball be at the height of 10 m from the ground? (Take, $g = 10\text{ ms}^{-2}$)

(a) 5.20 m (b) 4.33 m (c) 2.60 m (d) 8.66 m

35 A projectile moves from the ground such that its horizontal displacement is $x = k_1 t$ and vertical displacement is $y = k_1 t(1 - k_2 t)$, where k_1 and k_2 are constants and t is time. The total time of flight (T) and maximum height attained (y_{max}) respectively are

(a) $T = k_1, y_{max} = \frac{k_1}{2k_2}$
(b) $T = \frac{1}{k_2}, y_{max} = \frac{2k_1}{k_2}$
(c) $T = \frac{1}{k_1}, y_{max} = \frac{k_1}{6k_2}$
(d) $T = \frac{1}{k_2}, y_{max} = \frac{k_1}{4k_2}$

36 A projectile is thrown from the origin along XY-plane to have maximum possible horizontal range of 100 m. Then, the position vector of the projectile when its speed is minimum is (Take, y on the vertical axis)

(a) $50\hat{i} + 50\hat{j}$
(b) $50\hat{i} + 100\hat{j}$
(c) $50\hat{i} + 25\hat{j}$
(d) $100\hat{i} + 50\hat{j}$

37 A body is projected from ground at an angle θ_1 with the horizontal with kinetic energy K. What is the potential energy of the body when the velocity makes an angle θ_2 with the horizontal?

(a) $K\frac{\sin^2\theta_1}{\sin^2\theta_2}$
(b) $K\left(1 - \frac{\cos^2\theta_1}{\cos^2\theta_2}\right)$
(c) $K\left(1 - \frac{\sin^2\theta_1}{\sin^2\theta_2}\right)$
(d) $K\frac{\cos^2\theta_1}{\cos^2\theta_2}$

38 A body is thrown with velocity 20 m/s at an angle of 60° with the horizontal. Find the time gap between the two positions of body, where velocity of body makes an angle 30° with horizontal.

(a) 1.15 s
(b) 0.95 s
(c) 1 s
(d) 1.5 s

39 Time taken by the projectile to reach from A to B is t. Then, the distance AB is equal to

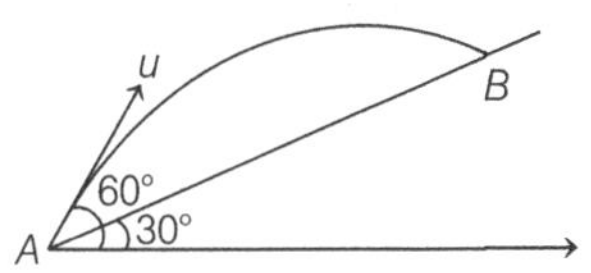

(a) $\frac{ut}{\sqrt{3}}$
(b) $\frac{\sqrt{3}ut}{2}$
(c) $\sqrt{3}\,ut$
(d) $2\,ut$

40 Two particles are projected from the same point on the ground with same speed simultaneously at angles θ_1 and θ_2 with the horizontal. They strike the ground at same point after time t_1 and t_2 respectively, then

(a) $\frac{t_1}{t_2} = \frac{\sin\theta_1}{\sin\theta_2}$
(b) $\frac{t_1}{t_2} = \frac{\cos\theta_2}{\cos\theta_1}$
(c) $\frac{t_1}{t_2} = 1$
(d) All of these

41 Two particles A and B are placed at distance d as shown in the figure alongside. Particle A, on the top of tower of height H is projected horizontally with a velocity u and B is projected along the surface towards the tower simultaneously. If particles meet each other, then the speed of projection of particle B is [Ignore friction]

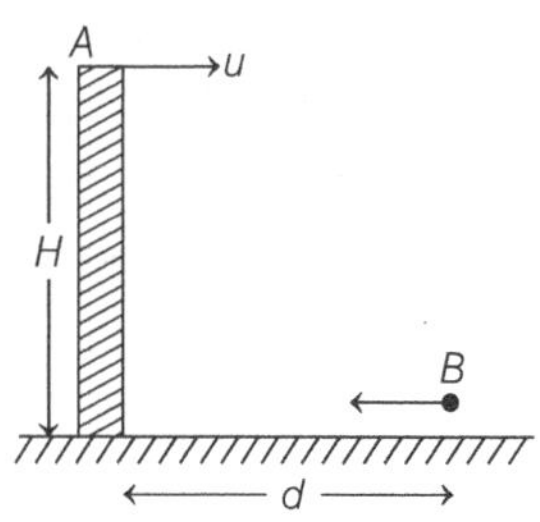

(a) $d\sqrt{\frac{g}{2H}} - u$
(b) $d\sqrt{\frac{g}{2H}}$
(c) $d\sqrt{\frac{g}{2H}} + u$
(d) u

42 The height of a projectile as a function of time is $y(t) = -\frac{1}{2}gt^2 + v_0 t + y_0$. At what time does the projectile reach its maximum height?

(a) $t = \frac{v_0}{2g}$
(b) $t = \frac{v_0}{g}$
(c) $t = \frac{g}{2(y_0 - v_0)}$
(d) $t = \frac{2v_0}{g}$

43 Two projectiles are fired with same speed but at different angles. Their trajectories are shown in the following figure. Find the ratio of their horizontal velocity.

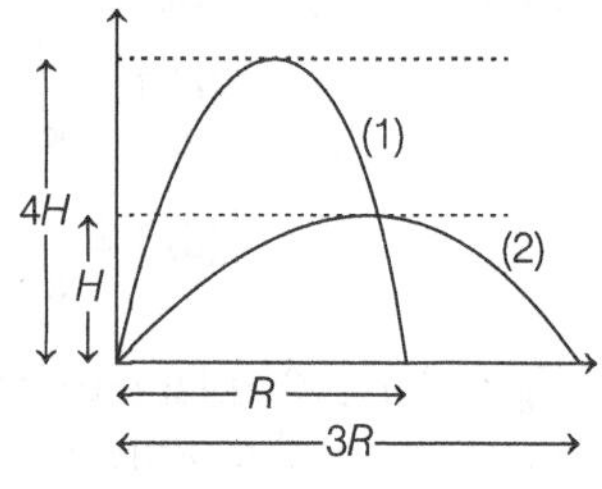

(a) $\frac{u_{x_1}}{u_{x_2}} = \frac{1}{3}$ (b) $\frac{u_{x_1}}{u_{x_2}} = \frac{3}{1}$ (c) $\frac{u_{x_1}}{u_{x_2}} = \frac{1}{6}$ (d) $\frac{u_{x_1}}{u_{x_2}} = \frac{3}{2}$

44 A stone thrown at an angle θ to the horizontal reaches a maximum height h. The time of flight of the stone is

(a) $\frac{\sqrt{2h \sin \theta}}{g}$ (b) $\frac{2\sqrt{2h \sin \theta}}{g}$ (c) $2\sqrt{\frac{2h}{g}}$ (d) $\sqrt{\frac{2h}{g}}$

45 If R is the range of a projectile on a horizontal plane and h its maximum height, the maximum horizontal range with the same velocity of projection is

(a) $2h$ (b) $R^2/8h$ (c) $2R + \frac{h^2}{8R}$ (d) $2h + \frac{R^2}{8h}$

46 An aeroplane is travelling horizontally at a height of 4000 m from the ground. The aeroplane when at a point P, drops a bomb to hit a stationary target Q on the ground. In order that the bomb hits the target, what angle θ must the line PQ make with the vertical? (Take, $g = 10\ \text{m/s}^2$)

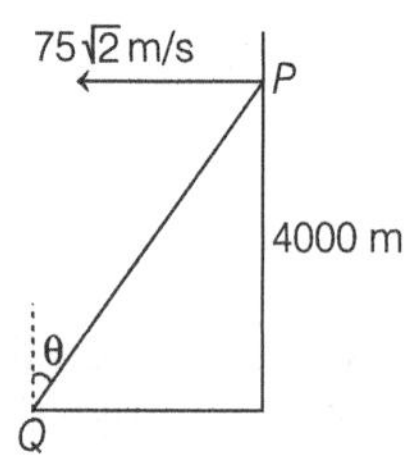

(a) 37° (b) 53° (c) 45° (d) 30°

47 From the top of a tower of height 40 m, a ball is projected upwards with a speed of 20 m/s at an angle of elevation of 30°, then the ratio of the total time taken by the ball to hit the ground to its time of flight upto same horizontal level is (Take, $g = 10\ \text{m/s}^2$)

(a) 2 : 1 (b) 3 : 1 (c) 3 : 2 (d) 4 : 1

48 A ball is hit by a batsman at an angle of 37° as shown in the following figure. The man standing at P should run at what minimum velocity, so that he catches the ball before it strikes the ground. Assume that height of man is negligible in comparison to maximum height of projectile.

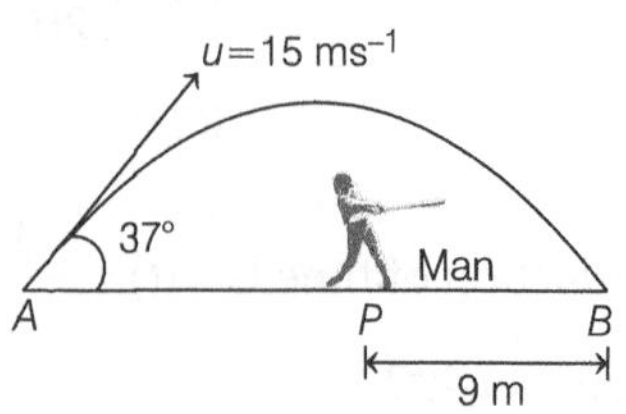

(a) $3\ \text{ms}^{-1}$ (b) $5\ \text{ms}^{-1}$
(c) $9\ \text{ms}^{-1}$ (d) $2.5\ \text{ms}^{-1}$

49 An object of mass m is projected with a momentum p at such an angle that its maximum height is $\frac{1}{4}$ th of its horizontal range. Its minimum kinetic energy in its path will be

(a) $\frac{p^2}{8m}$ (b) $\frac{p^2}{4m}$
(c) $\frac{3p^2}{4m}$ (d) $\frac{p^2}{m}$

50 Anoop is standing on the top of a building of height 25 m. He wants to throw his gun to Ashish who stands on top of another building of height 20 m at distance 15 m from first building. For which horizontal speed of projection, it is possible?

(a) 5 ms^{-1} (b) 10 ms^{-1}
(c) 15 ms^{-1} (d) 20 ms^{-1}

51 A light body is projected with a velocity $(10\hat{\mathbf{i}} + 20\hat{\mathbf{j}} + 20\hat{\mathbf{k}})$ ms^{-1}. Wind blows along X-axis causing an acceleration of 2.5 ms^{-2}. If Y-axis is vertical, then the speed of particle after 2 s will be (Take, $g = 10$ ms^{-2})

(a) 25 ms^{-1} (b) $10\sqrt{5}$ ms^{-1}
(c) 30 ms^{-1} (d) 20 ms^{-1}

52 An object is projected, so that it just clear two walls of height 7.5 m and with separation 50 m from each other. If the time of passing between the walls is 2.5 s, the range of the projectile will be (Take, $g = 10$ m/s^2)

(a) 35 m (b) 70 m
(c) 140 m (d) 57.5 m

53 At high altitude, a body explodes at rest into two equal fragments with one fragment receiving horizontal velocity of 10 m/s. Time taken by the velocity vectors of the fragments to make 90° is (Take, $g = 10$ m/s^2)

(a) 5 s (b) 4 s
(c) 2 s (d) 1 s

54 An object is projected with a velocity of 20 m/s making an angle of 45° with horizontal. The equation for the trajectory is $h = Ax - Bx^2$, where h is height, x is horizontal distance, A and B are constants. The ratio $A : B$ is (Take, $g = 10$ ms^{-2})

(a) 1 : 5 (b) 5 : 1
(c) 1 : 40 (d) 40 : 1

55 Two particles are projected simultaneously from two points, O and O' such that d is the horizontal distance and h is the vertical distance between them. They are projected at the same inclination α to the horizontal with the same speed v. The time after which their separation becomes minimum is

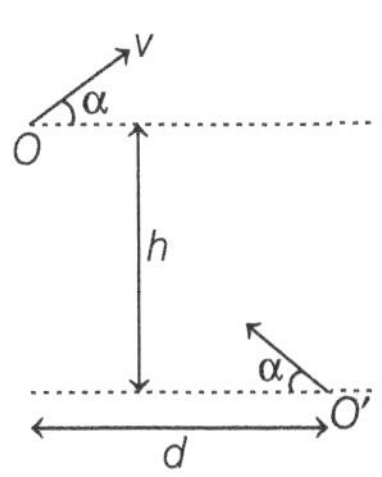

(a) $dv\cos\alpha$ (b) $2d/v\cos\alpha$
(c) $d/2v\cos\alpha$ (d) d/v

56 A projectile passes two points A and B at same height after 3 s and 6 s of its projection. Horizontal separation between the points A and B is 150 m. The horizontal range is closest to

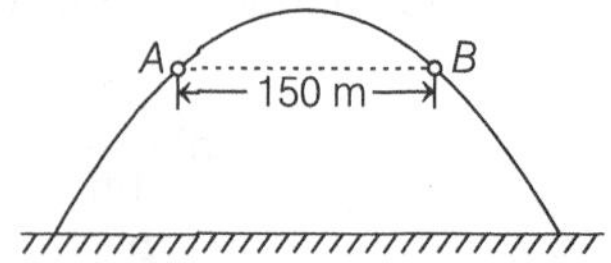

(a) 180 m (b) 240 m
(c) 350 m (d) 450 m

57 A staircase contains three steps each 10 cm high and 20 cm wide as shown in the following figure. The minimum horizontal velocity of a ball rolling off the uppermost plane so as to hit directly the lowest plane is

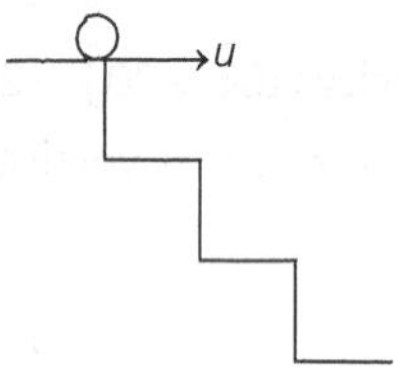

(a) 1 m/s (b) 2 m/s
(c) 3 m/s (d) 4 m/s

58 Two balls are thrown with the same speed from a point O at the same time, so that their horizontal ranges are same. If the difference of the maximum height attained by them is equal to half of the sum of the maximum heights, then the angles of projection for the balls are

(a) 15°, 75° (b) 30°, 60°
(c) 30°, 45° (d) 37°, 53°

59 A ball is thrown from the ground to clear a wall 3 m high at a distance of 6 m and falls 18 m away from the wall. Find the angle of projection of ball.

(a) 30° (b) $\tan^{-1}(2/3)$
(c) $\tan^{-1}(4/3)$ (d) $\tan^{-1}(1/3)$

60 In the figure given below, the maximum height of the projectile will be

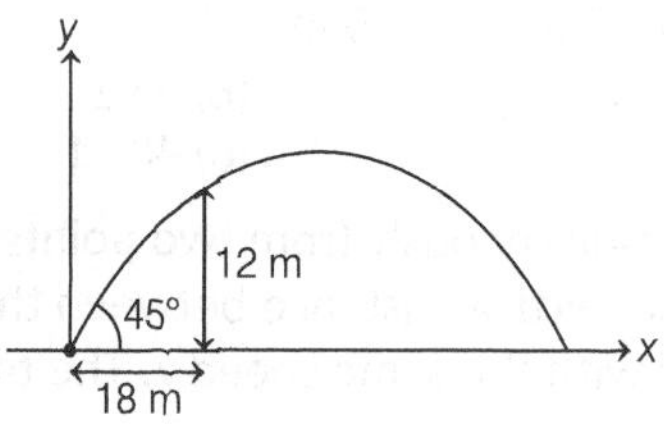

(a) 13.5 m (b) 15.5 m
(c) 18 m (d) 9 m

61 Two particles are simultaneously projected in the same vertical plane from the same point with velocities u and v at angles α and β with horizontal. Find the time that elapses when their velocities are parallel.

(a) $t = \dfrac{uv(\sin\alpha - \sin\beta)}{g(v\cos\beta - u\cos\alpha)}$ (b) $t = \dfrac{uv\sin(\alpha - \beta)}{g(v\cos\beta - u\cos\alpha)}$

(c) $t = \dfrac{uv\sin(\alpha - \beta)}{g[v\cos(\beta - \alpha)]}$ (d) $t = \dfrac{uv\cos(\alpha - \beta)}{g(v\cos\beta - u\sin\alpha)}$

Answer Q.62 to Q.64 by appropriately matching the information given in the three columns of the following table :

A particle of mass m is projected from a point A on the top of the tower as shown in the following figure. H be height of tower from ground surface (in metre), u be velocity of projection (in m/s), θ be angle of projection from horizontal are given in column-1, 2 and 3, respectively.

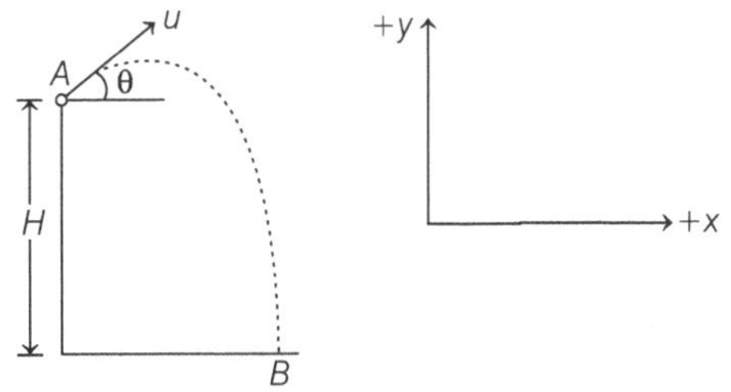

	Column-1 H (m)		Column-2 u (m/s)		Column-3 θ
I	0 m	(i)	15 m/s	(P)	0°
II	20 m	(ii)	20 m/s	(Q)	37°
III	45 m	(iii)	25 m/s	(R)	53°
IV	60 m	(iv)	50 m/s	(S)	90°

62 In which of the following situation, time of flight of particle is maximum?
(a) (II) (iii) (Q) (b) (IV) (ii) S) (c) (III) (iv) (P) (d) (I) (iv) (R)

63 In which of the following situation, the horizontal displacement of particle just before striking the surface is maximum?
(a) (II) (iii) (Q) (b) (I) (iii) (R) (c) (III) (iv) (P) (d) (IV) (iv) (S)

64 In which of the following situation, the vertical component of velocity before striking the surface is maximum?
(a) (II) (iii) (Q) (b) (IV) (ii) (S) (c) (III) (iv) (P) (d) (I) (ii) (R)

65 The horizontal range of a projectile is R and the maximum height attained by it is H. A strong wind now begins to blow in the direction of the motion of the projectile, giving it a constant horizontal acceleration $= \frac{g}{2}$. Under the same conditions of projection, the horizontal range of the projectile will now be
(a) $R + \frac{H}{2}$ (b) $R + H$ (c) $R + \frac{3H}{2}$ (d) $R + 2H$

66 Two particles A and B are thrown simultaneously from different floors of a tower having distance d between them. Velocity of A is u_A at an angle α with horizontal and velocity of B is u_B at an angle β with horizontal. If the two particles collide in mid air, then $\frac{u_A}{u_B}$ is equal to :

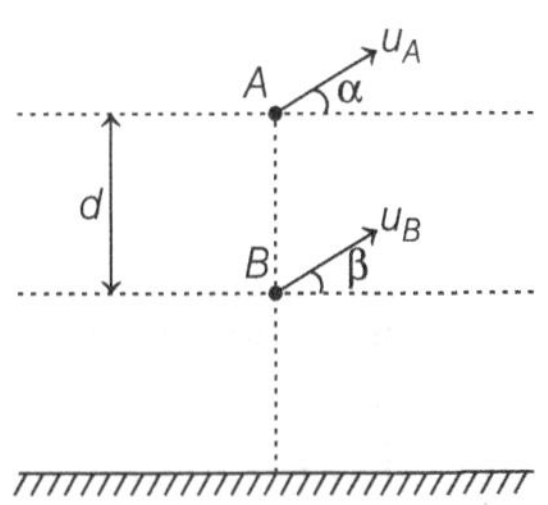

(a) $\frac{\cos\beta}{\cos\alpha}$ (b) $\frac{\sin\alpha}{\sin\beta}$ (c) $\frac{\tan\alpha}{\tan\beta}$ (d) $\frac{\cot\alpha}{\cot\beta}$

67 A boy wants to throw a ball from a point *A* so as to just reach point *B* as shown in the following figure. The minimum velocity with which the boy should throw the ball is (Take, $g = 10 \text{ ms}^{-2}$)

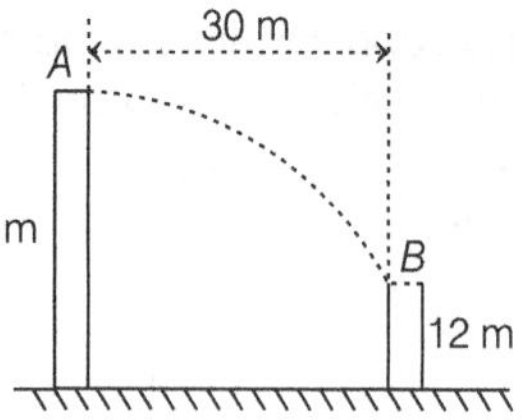

(a) 2.4 ms^{-1}
(b) 23.8 ms^{-1}
(c) 238 ms^{-1}
(d) 47.4 ms^{-1}

68 Two identical balls are set into motion simultaneously from equal heights *h*. While the ball *A* is thrown horizontally with velocity *v*, the ball *B* is just released to fall by itself. Choose the alternative that best represents the motion of *A* and *B* with respect to an observer who moves with velocity $v/2$ with respect to the ground as shown in the figure alongside.

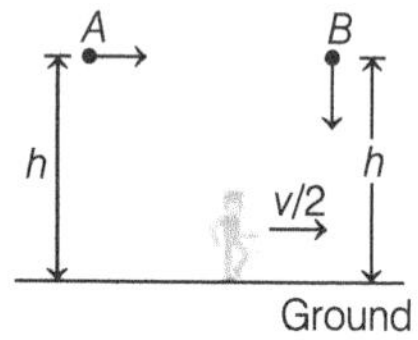

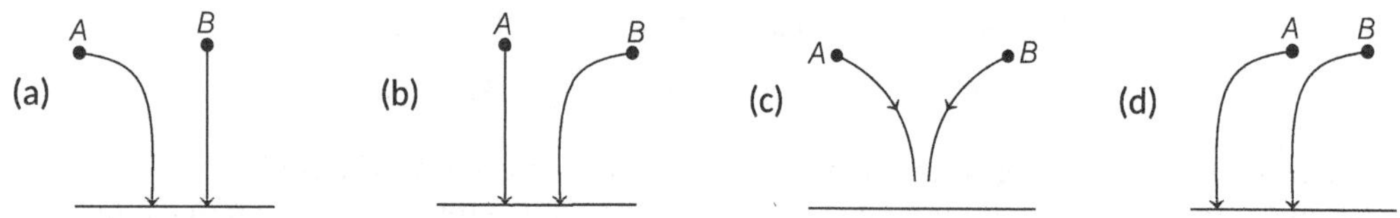

Paragraph (Q69-Q70)

Three balls A, B and C are projected from ground with the same speed at the same angle with the horizontal. The balls A, B and C collides perpendicularly with the wall during their flight as shown in the following figures.

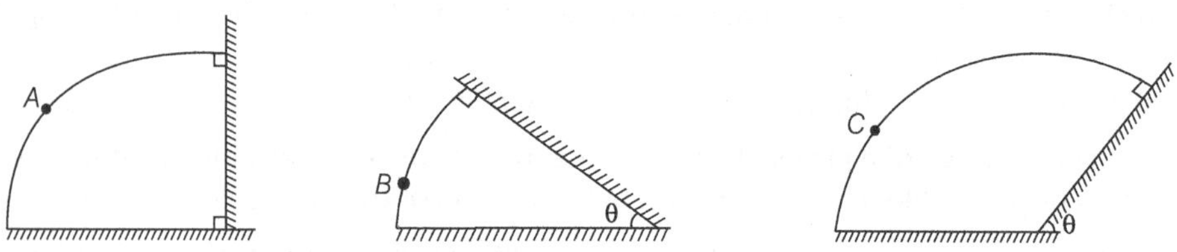

69 If the time taken by the ball *A* to strike the wall is 2 s and that by ball *B* is 1 s, then find the time taken by the ball *C* to strike the inclined plane after projection.

(a) 4 s
(b) 3 s
(c) 6 s
(d) 5 s

70 Find the maximum height attained by ball *B* from the ground during their flight.

(a) 10 m
(b) 15 m
(c) 12 m
(d) 18 m

LEVEL 03 Previous Years Questions

71 Two bullets are fired horizontally and simultaneously towards each other from roof tops of two buildings 100 m apart and of same height of 200 m with the same velocity of 25 m/s. When and where will the two bullets collides? (Take, $g = 10 \text{ m/s}^2$) **[NEET 2019]**

(a) After 2s at a height 180 m
(b) After 2s at a height of 20 m
(c) After 4s at a height of 120 m
(d) They will not collide

72 When an object is shot from the bottom of a long smooth inclined plane kept at an angle 60° with horizontal, it can travel a distance x_1 along the plane. But when the inclination is decreased to 30° and the same object is shot with the same velocity, it can travel x_2 distance. Then $x_1 : x_2$ will be **[NEET 2019]**

(a) $\sqrt{2}:1$ (b) $1:\sqrt{3}$
(c) $1:2\sqrt{3}$ (d) $1:\sqrt{2}$

73 The x and y coordinates of the particle at any time are $x = 5t - 2t^2$ and $y = 10t$ respectively, where x and y are in metres and t in seconds. The acceleration of the particle at $t = 2$ s is **[NEET 2017]**

(a) 0 (b) 5 m/s^2
(c) -4 m/s^2 (d) -8 m/s^2

74 A ship A is moving Westwards with a speed of 10 km h^{-1} and a ship B 100 km South of A, is moving Northwards with a speed of 10 kmh^{-1}. The time after which the distance between them becomes shortest is **[CBSE AIPMT 2015]**

(a) 0 h (b) 5 h
(c) $5\sqrt{2}$ h (d) $10\sqrt{2}$ h

75 The position vector of a particle **R** as a function of time is given by

$$\mathbf{R} = 4\sin(2\pi t)\,\hat{\mathbf{i}} + 4\cos(2\pi t)\,\hat{\mathbf{j}}$$

where R is in metre, t is in seconds and $\hat{\mathbf{i}}$ and $\hat{\mathbf{j}}$ denote unit vectors along x and y-directions, respectively. Which one of the following statements is wrong for the motion of particle? **[CBSE AIPMT 2015]**

(a) Acceleration is along $-\mathbf{R}$

(b) Magnitude of acceleration vector is $\frac{v^2}{R}$, where v is the velocity of particle

(c) Magnitude of the velocity of particle is 8π m/s

(d) Path of the particle is a circle of radius 4 m

76 A projectile is fired from the surface of the earth with a velocity of 5 ms^{-1} at angle θ with the horizontal. Another projectile fired from another planet with a velocity of 3 ms^{-1} at the same angle follows a trajectory which is identical with the trajectory of the projectile fired from the earth. The value of the acceleration due to gravity on the planet is (in ms^{-2})
(Take, $g = 9.8$ ms^{-2}) **[CBSE AIPMT 2014]**

(a) 3.5 (b) 5.9 (c) 16.3 (d) 110.8

77 A particle is moving such that its position co-ordinates (x, y) are (2 m, 3 m) at time $t = 0$, (6 m, 7 m) at time $t = 2$ s and (13 m, 14 m) at time $t = 5$ s. Average velocity vector (v_{av}) from $t = 0$ to $t = 5$ s is **[CBSE AIPMT 2014]**

(a) $\frac{1}{5}(13\hat{\mathbf{i}} + 14\hat{\mathbf{j}})$ (b) $\frac{7}{3}(\hat{\mathbf{i}} + \hat{\mathbf{j}})$

(c) $2(\hat{\mathbf{i}} + \hat{\mathbf{j}})$ (d) $\frac{11}{5}(\hat{\mathbf{i}} + \hat{\mathbf{j}})$

78 The velocity of a projectile at the initial point A is $(2\hat{\mathbf{i}} + 3\hat{\mathbf{j}})$ m/s. Its velocity (in m/s) at point B is **[NEET 2013]**

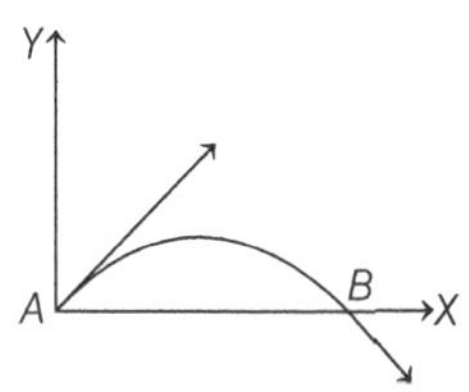

(a) $-2\hat{\mathbf{i}} - 3\hat{\mathbf{j}}$ (b) $-2\hat{\mathbf{i}} + 3\hat{\mathbf{j}}$ (c) $2\hat{\mathbf{i}} - 3\hat{\mathbf{j}}$ (d) $2\hat{\mathbf{i}} + 3\hat{\mathbf{j}}$

79 The horizontal range and the maximum height of a projectile are equal. The angle of projection of the projectile is **[CBSE AIPMT 2012]**

(a) $\theta = \tan^{-1}\left(\frac{1}{4}\right)$ (b) $\theta = \tan^{-1}(4)$

(c) $\theta = \tan^{-1}(2)$ (d) $\theta = 45°$

80 A missile is fired for maximum range with an initial velocity of 20 m/s. If $g = 10\ \text{m/s}^2$, the range of the missile is **[CBSE AIPMT 2011]**

(a) 50 m (b) 60 m
(c) 20 m (d) 40 m

81 A particle has initial velocity $(3\hat{\mathbf{i}} + 4\hat{\mathbf{j}})$ and has acceleration $(0.4\hat{\mathbf{i}} + 0.3\hat{\mathbf{j}})$. Its speed after 10 s is **[CBSE AIPMT 2010]**

(a) 7 unit (b) $7\sqrt{2}$ unit
(c) 8.5 unit (d) 10 unit

ANSWER KEY

Level 1 Objective Questions

1 (a)	2 (d)	3 (c)	4 (d)	5 (c)	6 (c)	7 (b)	8 (d)	9 (c)	10 (c)
11 (c)	12 (d)	13 (a)	14 (c)	15 (b)	16 (a)	17 (d)			

Level 2 Objective Questions

18 (c)	19 (a)	20 (b)	21 (a)	22 (b)	23 (d)	24 (c)	25 (d)	26 (b)	27 (c)
28 (b)	29 (c)	30 (c)	31 (b)	32 (b)	33 (c)	34 (d)	35 (d)	36 (c)	37 (b)
38 (a)	39 (a)	40 (b)	41 (a)	42 (b)	43 (c)	44 (c)	45 (d)	46 (a)	47 (a)
48 (b)	49 (b)	50 (c)	51 (a)	52 (b)	53 (d)	54 (d)	55 (c)	56 (d)	57 (b)
58 (b)	59 (b)	60 (a)	61 (b)	62 (d)	63 (c)	64 (b)	65 (d)	66 (a)	67 (b)
68 (c)	69 (b)	70 (b)							

Level 3 Previous Years Questions

71 (a)	72 (b)	73 (c)	74 (b)	75 (c)	76 (a)	77 (d)	78 (c)	79 (b)	80 (d)
81 (b)									

Solutions

LEVEL 01 Objective Questions

1. Range $= \dfrac{u^2 \sin 2\theta}{g}$

At $\theta = 15°$, the range is minimum.

2. Range $= \dfrac{2u_x \cdot u_y}{g}$

Product $u_x \cdot u_y$ is maximum in case (d).

3. In first case, $R_1 = 1.5$ km, $\theta = 15°$

$$R_1 = 1.5 = \frac{u^2 \sin(2 \times 15°)}{g}$$

$$1.5 = \frac{u^2}{g}\left(\frac{1}{2}\right) \qquad \left(\because \sin 30° = \frac{1}{2}\right)$$

$$\frac{u^2}{2g} = 1.5 \qquad ...(i)$$

Now,

$$R_{max} = \frac{u^2 \sin(2 \times 45°)}{g} = 3 \text{ km} \qquad \text{[using Eq. (i)]}$$

4. Time of flight $= \dfrac{2u}{g} = \dfrac{2(20)}{10} = 4$ s

So, the horizontal range of projectile

$$= vt = 10\,(4) = 40 \text{ m}$$

5. $x = u_x\, t = 6t$

$\therefore \quad u_x = 6$ m/s

$$y = u_y\, t - \frac{1}{2}gt^2 = 8t - 5t^2$$

$\therefore \quad u_y = 8$ m/s

$\therefore \quad u = \sqrt{u_x^2 + u_y^2}$

$= 10$ m/s

6. $\mathbf{v}_f - \mathbf{v}_i = (v\cos\theta)\,\hat{\mathbf{i}} - [(v\cos\theta)\,\hat{\mathbf{i}} + (v\sin\theta)\hat{\mathbf{j}})]$

$= -(v\sin\theta)\,\hat{\mathbf{j}}$

$\therefore$ It's magnitude will be $v\sin\theta$.

7. We know in a projectile, the horizontal speed remains constant,

$$u\cos\alpha = v\cos\beta$$

$$\Rightarrow \quad v = u\cos\alpha\sec\beta$$

8. We know, $R = 4H\cot\theta$

$$4\sqrt{3}H = 4H\cot\theta$$

So, $\theta = 30°$

9. The potential energy of ball must be proportional to the height reached by ball.

In first case, $H_1 = \dfrac{u^2}{2g}$

In second case, $H_2 = \dfrac{u^2\cos^2\theta}{2g} = \dfrac{u^2\cos^2 60°}{2g}$

So, $H_1 : H_2$ or $PE_1 : PE_2 = 4 : 1$

10. Given, horizontal range = 12 m

Maximum height = 4 m

$$\text{Range} = 12 = \frac{u^2 \sin 2\theta}{g} \qquad ...(i)$$

$$\text{Maximum height} = 4 = \frac{u^2\sin^2\theta}{2g} \qquad ...(ii)$$

On solving Eqs. (i) and (ii), we get

$$u = 5\sqrt{\frac{g}{2}}$$

11. Using equation, $s = ut + \dfrac{1}{2}at^2$

$$490 \text{ m} = 5t^2 \qquad (\because u = 0)$$

$$t = 10 \text{ s}$$

So, the box will travel horizontally

$$= 10 \times 180 \times \frac{5}{18} = 500 \text{ m}$$

12. Given, $s_y = 50$ m, $g = 10$ m/s^2, $u = 0$

Using the equation, $s_y = ut + \dfrac{1}{2}gt^2$

$$50 = \frac{10\,t^2}{2}$$

$$t = \sqrt{10} \text{ s}$$

13. $v = \sqrt{v_x^2 + v_y^2} = \sqrt{v_x^2 + (gt)^2}$

$= \sqrt{(18)^2 + (10\sqrt{10})^2} = \sqrt{1324} \approx 36.4$ m/s

14. $\tan\theta = 10\sqrt{\dfrac{10}{18}}$

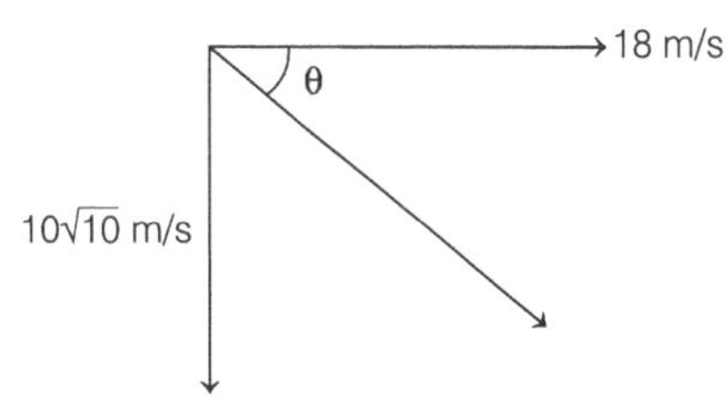

$$\Rightarrow \qquad \theta = \tan^{-1}\left(\frac{10\sqrt{10}}{18}\right)$$

15. In the horizontal direction, the velocity component remains constant,

$$v\cos 60^\circ = \frac{20}{\sqrt{3}}$$

$$v = \frac{40}{\sqrt{3}}\ \text{m/s}$$

Now, in the vertical direction,

$$v^2 = 2gh$$

$$v^2 \sin^2 60^\circ = 2 \times 10\, h$$

$$\frac{40\,(40)}{3} \times \frac{3}{4} = 20\, h \qquad \left(\because \sin 60^\circ = \frac{\sqrt{3}}{2}\right)$$

$$h = 20\ \text{m}$$

Displacement in the vertical direction,

$$20 = \frac{1}{2}(10)\, t^2 \qquad (\because u = 0)$$

$$t = 2\ \text{s}$$

16. The horizontal and vertical components of velocity are

$v_x = 5\cos 37^\circ = 3$ m/s, $v_y = 5\sin 37^\circ = 4$ m/s

Now using $s = ut + \frac{1}{2}at^2$,

$$-36 = 3t - \frac{10t^2}{2}$$

$\Rightarrow \qquad t = 3$ s

Therefore, horizontal range $= 4 \times 3 = 12$ m

17. The height reached by projectile

$$= \frac{u^2\sin^2\theta}{2g} = \frac{u^2\sin^2 45^\circ}{2g} = \frac{u^2}{4g}$$

Range of projectile, $R = \frac{u^2 \sin 2\theta}{g} = \frac{u^2}{g}$

Now, $\quad \tan\alpha = \frac{u^2/4g}{u^2/g} = \frac{1}{4}$

$\Rightarrow \qquad \alpha = \tan^{-1}\left(\frac{1}{4}\right)$

LEVEL 02 Objective Questions

18. $\qquad (2)^2 = v_y^2 - 2(10)(0.4)$

$$v_y = 2\sqrt{3}\ \text{m/s}$$

So, $\tan\theta = \frac{v_y}{v_x} = \frac{2\sqrt{3}}{6} = \frac{1}{\sqrt{3}}$

$\Rightarrow \qquad \theta = 30^\circ$

19. $\frac{1}{2}gt^2 = 20$

$\therefore \qquad \frac{1}{2} \times 10 \times t^2 = 20$ or $t = 2$ s

Now, $\quad u_x t = s_x = 20$ m

$\therefore \qquad u_x = 10$ m/s

20. Average velocity $= \dfrac{\text{Displacement}}{\text{Total time taken}}$

In vertical direction, their is no effect of component of velocity during the displacement.

But in the horizontal direction, the body moves with a constant speed $u\cos\theta$.

Therefore, average velocity will be $u\cos\theta$.

21. As their height is same,

$$\frac{u_P^2}{2g} = \frac{u_Q^2\sin^2 45^\circ}{2g} \Rightarrow \frac{u_P^2}{u_Q^2} = \frac{1}{2}$$

Ratio of kinetic energy,

$$\frac{KE_P}{KE_Q} = \frac{\frac{1}{2}mu_P^2}{\frac{1}{2}mu_Q^2} = \frac{1}{2}$$

22. Maximum height travel by particle A, $h = \dfrac{v_A^2\sin^2 60^\circ}{2g}$

$$h = \frac{3v_A^2}{8g} \qquad \left(\because \sin 60^\circ = \frac{\sqrt{3}}{2}\right)$$

Time taken to reach maximum height,

$$t = \frac{v_A \sin 60^\circ}{g} = \frac{v_A\sqrt{3}}{2g}$$

As the distance travelled = maximum height

$$v_B t - \frac{1}{2}gt^2 = h$$

$$v_B\left(\frac{v_A\sqrt{3}}{2g}\right) - \frac{1}{2}g\left(\frac{v_A\sqrt{3}}{2g}\right)^2 = \frac{3v_A^2}{8g}$$

$$\therefore \qquad \frac{v_A}{v_B} = \frac{2}{\sqrt{3}}$$

23. Let us assume that the angle of projection

$$= \frac{\pi}{3}\ \text{rad}$$

The angle of projection of other object

$$= \frac{\pi}{2} - \frac{\pi}{3} = \frac{\pi}{6}$$

Let the maximum heights be h_1 and h_2.

$$\frac{h_1}{h_2} = \frac{\sin^2(\pi/3)}{\sin^2(\pi/6)}$$

$$\frac{h_1}{h_2} = \frac{3}{1} \Rightarrow h_2 = \frac{h_1}{3} = \frac{102}{3} = 34\ \text{m}$$

24. The horizontal and the vertical components of projectile are as follows

$$x = u\cos\theta t,\ v_x = u\cos\theta$$

$$y = u\sin\theta t - \frac{1}{2}gt^2,\ v_y = u\sin\theta - gt$$

The speed v_x in the horizontal direction is always constant.

The speed is minimum when the vertical component is zero.

$$0 = u\sin\theta - g(2) \Rightarrow u\sin\theta = 2g$$

Now, the ball is moving at an angle 45° at $t = 1$ s,

$$\tan 45° = 1 = \frac{v_y}{v_x} = \frac{u\sin\theta - gt}{u\cos\theta}$$

$$u\cos\theta = u\sin\theta - gt$$
$$= 2g - gt \quad (\because u\sin\theta = 2g)$$
$$= 2g - g(1) \quad (\because t = 1)$$

$\therefore \quad u\cos\theta = g$

$\Rightarrow \quad \tan\theta = 2$

$$\sin\theta = \frac{2}{\sqrt{5}} \text{ and } \cos\theta = \frac{1}{\sqrt{5}}$$

As, $u\sin\theta = 2g$

$$u\left(\frac{2}{\sqrt{5}}\right) = 2g \Rightarrow u = \sqrt{5}\,g = 10\sqrt{5} \text{ m/s}$$

25. Let v_A be the vertical velocity of ball thrown above horizontal and v_B be the velocity of ball thrown below the horizontal.

$$v_A = 40\sin 30° = 20 \text{ m/s}$$
$$v_B = 40\sin 30° = 20 \text{ m/s}$$

At maximum height, $v = 0$

$$0 = 20 - gt \quad (\because v = u + at)$$
$$t = 2 \text{ s}$$

As we know in a projectile, the magnitude of velocity remains same when it goes highest point and when it comes down.

So, the both ball takes same time.

$\therefore$ Total time of journey = 2t = 4 s

26. Let θ be the angle between the initial velocity vectors of two stones and L be the distance between them.

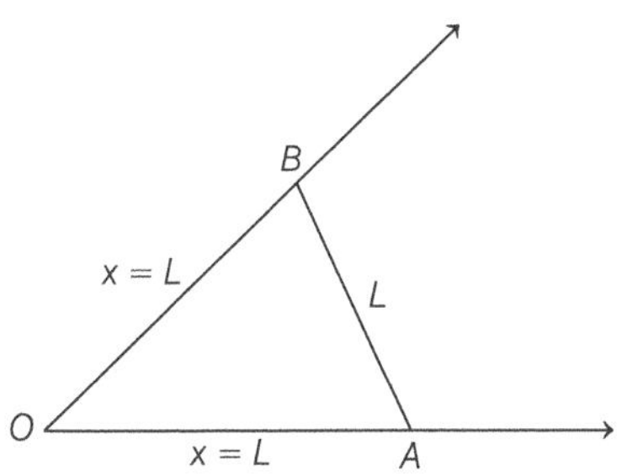

As we know in the horizontal direction, the velocity is constant.

So, $\dfrac{dx}{dt} = u$

Since, their separation is increasing at constant rate,

$$\frac{dL}{dt} = u$$

When $x = L$, OAB will be an equilateral triangle.

Hence, the angle between the velocity vector is 60°.

27. The horizontal and vertical components of velocity are

$$v_y = 5\sin 53° = 4 \text{ m/s}$$

and $\quad v_x = 5\cos 53° = 3$ m/s

Now, the final velocity of ball at certain height,

$$v = \sqrt{16 - 2(10)(0.45)} \quad (\because v^2 = u^2 - 2gh)$$
$$= \sqrt{7} \text{ m/s}$$

The horizontal velocity remains constant.

So, speed $= \sqrt{3^2 + (\sqrt{7})^2} = 4$ m/s

28. Given, initial velocity $= (x\hat{\mathbf{i}} + y\hat{\mathbf{j}})$ m/s

$$R = 2H$$
$$4H\cot\theta = 2H$$
$$\cot\theta = \frac{1}{2} \text{ or } \tan\theta = 2 \quad \text{...(i)}$$

Now, angle of projection, $\theta = \tan^{-1}\left(\dfrac{v_y}{v_x}\right) = \tan^{-1}\left(\dfrac{y}{x}\right)$

$$\therefore \quad \tan\theta = \frac{y}{x} \quad \text{[using Eq. (i)]}$$
$$y = 2x$$

29. The horizontal and vertical components of velocity are

$$v_x = 10\sqrt{2}\cos 45° = 10 \text{ m/s}$$
$$v_y = 10\sqrt{2}\sin 45° = 10 \text{ m/s}$$

Now, speed, $v = 5\sqrt{5}$ m/s

Let v_x and v_y be the velocities in the horizontal and vertical direction.

$$v_x = u_x = 10 \text{ m/s}$$
$$v^2 = v_x^2 + v_y^2$$
$$(5\sqrt{5})^2 = (10)^2 + v_y^2$$
$$v_y = 5 \text{ m/s}$$

In y-direction,

$$5 = 10 - 10\,t_1 \quad (\because v = u + at)$$
$$t_1 = 0.5 \text{ s}$$

Now, in downward motion, $-5 = 10 - 10\,t_2$

$\Rightarrow \quad t_2 = 1.5$ s

So, the time interval = 1.5 − 0.5 = 1 s

30. $u_x = at = 3 \times 30 = 90$ m/s

$$80 = \frac{u_y^2}{2g}$$

$\Rightarrow \quad u_y = 40$ m/s

$$\therefore \quad \tan\theta = \frac{u_y}{u_x} = \frac{4}{9}$$

$$\Rightarrow \quad \theta = \tan^{-1}\left(\frac{4}{9}\right)$$

31. $t = \dfrac{100}{25} = \dfrac{2u_y}{g}$

$\therefore \quad u_y = 20$ m/s

32. $t = \frac{s_x}{u_x} = \sqrt{\frac{2h}{g}}$

$$\frac{30}{u_x} = \sqrt{\frac{2 \times (150 - 27.5)}{9.8}}$$

Solving equation, we get

$$u_x = 6 \text{ m/s}$$

33. For the first projectile,

$$\text{Time of flight} = \frac{2u \sin 60^\circ}{g} = \sqrt{3} \text{ s}$$

Now, the velocity of second projectile in vertical direction, $u_y = 0$

Using equation, $s = ut + \frac{1}{2}at^2 = \frac{1}{2}(10)(\sqrt{3})^2 = 15$ m

34. Range $= \frac{u^2 \sin 2\theta}{g} = \frac{(10)^2 \sin 60^\circ}{10} = 8.66$ m

35. The horizontal and vertical displacements are given as

$$x = k_1 t \text{ and } y = k_1 t\,(1 - k_2 t)$$

As we know, the vertical displacement is zero, so

$$k_1 t(1 - k_2 t) = 0$$

$$T = \frac{1}{k_2}$$

Now, for maximum height, $\frac{dy}{dt} = 0$

$$\frac{dy}{dt} = k_1 - 2k_1 k_2 t = 0$$

$$t = \frac{1}{2k_2}$$

At $t = \frac{1}{2k_2}$, maximum height is attained.

On substituting the value of t in y, we get

$$y_{max} = \frac{k_1}{4k_2}$$

36. We have, maximum range = 100 m at 45°

$$\frac{u^2}{g} = 100 \text{ m}$$

$$H_{max} = \frac{u^2 \sin^2 \theta}{2g} = 25\,\hat{\mathbf{j}}$$

At the highest point, the velocity is minimum, range is $\frac{R}{2}$.

So, the position vector $= 50\,\hat{\mathbf{i}} + 25\,\hat{\mathbf{j}}$

37. Let a body of mass m be projected from ground with speed u m/s.

So, its initial kinetic energy $= \frac{1}{2}mu^2$ ...(i)

As we know in the projectile, the horizontal component of speed remains constant.

$\therefore$ $u \cos \theta_1 = v \cos \theta_2$

$\Rightarrow$ $v = \frac{u \cos \theta_1}{\cos \theta_2}$

Kinetic energy at that point $= \frac{1}{2}mv^2 = \frac{1}{2}\frac{mu^2 \cos^2 \theta_1}{\cos^2 \theta_2}$

From energy conservation,

$$KE_1 = KE_2 + PE$$

$$PE = \frac{1}{2}mu^2 - \frac{1}{2}\frac{mu^2 \cos^2 \theta_1}{\cos^2 \theta_2} = K\left[1 - \frac{\cos^2 \theta_1}{\cos^2 \theta_2}\right]$$

38. As the horizontal velocity remains same,

$$v_i \cos 60^\circ = v_f \cos 30^\circ$$

$$\frac{20}{2} = v_f \times \frac{\sqrt{3}}{2}$$

$\Rightarrow$ $v_f = \frac{20}{\sqrt{3}}$

At this time, the vertical component,

$$v_f \sin 30^\circ = \frac{10}{\sqrt{3}} \text{ m/s}$$

So, time gap = time when vertical velocity is $\pm \frac{10}{\sqrt{3}} \frac{\text{m}}{\text{s}}$

$$-\frac{10}{\sqrt{3}} = \frac{10}{\sqrt{3}} - 10t \qquad (\because v = u + at)$$

$$t = 1.15 \text{ s}$$

39. The horizontal component of velocity,

$$u_x = u \cos 60^\circ = \frac{u}{2}$$

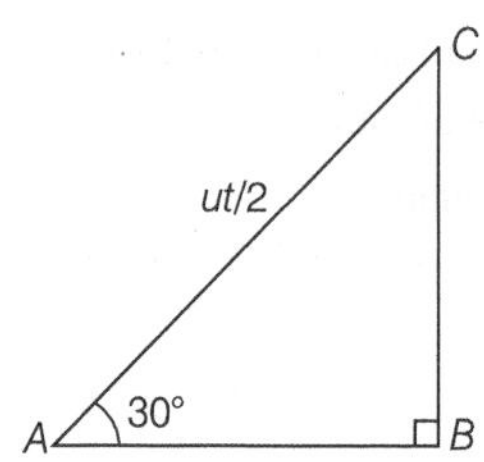

So, $AC = \frac{ut}{2}$

$$AB = AC \sec 30^\circ = \frac{ut}{2} \times \frac{2}{\sqrt{3}} = \frac{ut}{\sqrt{3}} \quad \left(\because \sec 30^\circ = \frac{2}{\sqrt{3}}\right)$$

40. $u_{x_1} t_1 = u_{x_2} t_2$

$\therefore$ $(u \cos \theta_1)\, t_1 = (u \cos \theta_2)\, t_2$

$\Rightarrow$ $\frac{t_1}{t_2} = \frac{\cos \theta_2}{\cos \theta_1}$

41. As we know in a projectile, the horizontal velocity remains constant.

$$t = \frac{d}{v + u}$$

$$\text{Time} = \sqrt{\frac{2H}{g}}$$

So, $\dfrac{d^2}{(v+u)^2} = \dfrac{2H}{g}$

$$v = d\sqrt{\frac{g}{2H}} - u$$

42. Given, $y(t) = -\dfrac{1}{2}gt^2 + v_0 t + y_0$

At maximum height, $\dfrac{d(y)}{dt} = 0$

$\Rightarrow \quad -gt + v_0 = 0 \Rightarrow t = \dfrac{v_0}{g}$

43. $h_1 = 4h_2$ (h = maximum height)

$\Rightarrow \quad \dfrac{u_{y_1}^2}{2g} = \dfrac{4u_{y_2}^2}{2g}$

$\Rightarrow \quad u_{y_1} = 2u_{y_2}$

$\therefore \quad T_1 = 2T_2$

$R_2 = 3R_1$

$\therefore \quad u_{x_2}T_2 = 3[u_{x_1}T_1]$

or $\quad \dfrac{u_{x_1}}{u_{x_2}} = \dfrac{1}{3}\left(\dfrac{T_2}{T_1}\right) = \dfrac{1}{6}$

44. $s = ut - \dfrac{1}{2}at^2$

$$h = \frac{1}{2}(-g)\left(\frac{T}{2}\right)^2$$

$$T = 2\sqrt{\frac{2h}{g}}$$

45. Let u be the velocity of projection and θ be the angle of projection.

Range, $R = \dfrac{u^2}{g}\sin 2\theta$ ...(i)

Maximum height, $h = \dfrac{u^2 \sin^2\theta}{2g}$ ...(ii)

From Eqs. (i) and (ii), we get

$\dfrac{h}{R} = \dfrac{\tan\theta}{4} \Rightarrow \tan\theta = \dfrac{4h}{R}$

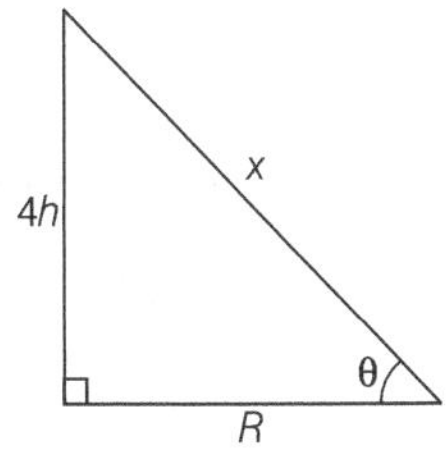

Here, $x = \sqrt{R^2 + (4h)^2}$

$= \sqrt{R^2 + 16h^2}$

$\sin\theta = \dfrac{4h}{\sqrt{R^2 + 16h^2}}$ ...(iii)

Now, maximum range, $R_{max} = \dfrac{u^2}{g}$

From Eq. (ii), we get

$\dfrac{u^2}{g} = \dfrac{2h}{\sin^2\theta}$ ...(iv)

Now, from Eqs. (iii) and (iv), we get

$$R_{max} = \frac{R^2}{8h} + 2h$$

46. $\tan\theta = \dfrac{\text{Horizontal displacement}}{\text{Vertical displacement}} = \dfrac{(75\sqrt{2})\,t}{h}$

$$= \frac{(75\sqrt{2})\sqrt{\frac{2h}{g}}}{h} = \frac{150}{\sqrt{gh}} = \frac{150}{200} = \frac{3}{4}$$

$\therefore \quad \theta = 37°$

47. Time of flight $= \dfrac{2v\sin\theta}{g} = \dfrac{2 \times 20 \sin 30°}{10} = 2\text{ s}$

Now, the total time taken by ball to hit the ground,

$$s = ut + \frac{1}{2}at^2$$

$$40 = -20\sin 30°\, t + \frac{1}{2}10t^2$$

$$t = 4\text{ s}$$

So, $\dfrac{\text{total time taken by ball}}{\text{time of flight}} = 2:1$

48. Time of flight $= \dfrac{2u\sin 37°}{g} = 1.8\text{ s}$

So, the speed of man $= \dfrac{9}{1.8} = 5\text{ m/s}$

49. Given that,

$$H = \frac{R}{4}$$

$$\frac{v^2\sin^2\theta}{2g} = \frac{v^2\sin 2\theta}{4g}$$

On solving, we get

$$\theta = 45°$$

Thus, the x and y-components of initial velocity are $\dfrac{v}{\sqrt{2}}$ each.

At the maximum height, the velocity of y-component is zero, thus the point of minimum kinetic energy

$$= \frac{1}{2}m\left(\frac{v_0}{\sqrt{2}}\right)^2 = \frac{p^2}{4m}$$

50. In vertical direction,

$$s = \frac{1}{2}gt^2$$

$$(25 - 20) = \frac{1}{2} \times 10 \times t^2$$

$\therefore \quad t = 1\text{ s}$

In horizontal direction,

$v \times t = 15\text{ m}$ $\quad (\because t = 1\text{ s})$

$\therefore \quad v = 15\text{ m/s}$

51. Let v_x and v_y be the velocity of particles.

As the wind blows, $a_x = 2.5\text{ m/s}^2$

So, $v_x = u_x + a_x t$

$= 10 + 2.5(2) = 15\text{ m/s}$

Similarly, the velocity v_y in y-direction,

$v_y = u_y - gt$
$v_y = 20 - 10\,(2) = 0$

So, velocity $= 15\,\hat{\mathbf{i}} + 20\,\hat{\mathbf{k}} = \sqrt{15^2 + 20^2} = 25$ m/s

52. The horizontal component of projectile motion

$$= \frac{50}{2.5} = 20 \text{ m/s}$$

To have a range of 50 m, maximum height should be reached in $\frac{2.5}{2} = 1.25$ s

Vertical component at that height

$$= gt = 10 \times 1.25 = 12.5 \text{ m/s}$$

Height attained above the walls

$$= \frac{v^2}{2g} = \frac{(12.5)^2}{2g} = 7.8125 \text{ m}$$

So, total height $= (7.812 + 7.5) = 15.31$ m

Time to reach maximum height from ground

$$= \sqrt{\frac{2h}{g}} = \sqrt{\frac{2 \times 15.31}{10}} = 1.75$$

Total time in air $= 1.75 \times 2 = 3.5$ s

Range $= 3.5 \times 20 = 70$ m

53. Initial momentum of system is zero.

After the blast, the momentum remains conserved.

$m_1v_1 = m_2v_2$ (as mass is same)

$v_1 = v_2$

Also, v_1 and v_2 are in opposite directions.

When the velocity vectors are perpendicular, their dot product will be zero.

$$(-10\hat{\mathbf{i}} - gt\hat{\mathbf{j}}) \cdot (10\hat{\mathbf{i}} - gt\hat{\mathbf{j}}) = 0 \;\Rightarrow\; t = \sqrt{\frac{100}{100}} = 1 \text{ s}$$

54. The horizontal and vertical velocity,

$v_x = 20 \cos 45° = 10\sqrt{2}$ m/s, $v_y = 20 \sin 45° = 10\sqrt{2}$ m/s

Now, using $s = ut + \frac{1}{2}at^2$

$$s = 10\sqrt{2}t - \frac{1}{2} \times 10 \times t^2 \qquad \text{...(i)}$$

Comparing with the equation of trajectory, we get

$$x = 10\sqrt{2}t \;\Rightarrow\; t = \frac{x}{10\sqrt{2}}$$

Substituting t in Eq. (i), we get

$$h = x - \frac{x^2}{40}$$

On comparing with equation of projectile,

$$A = 1,\; B = \frac{1}{40}$$

Hence, $A : B = 1 : \frac{1}{40} = 40 : 1$

55. In the vertical direction, the component of velocity of both particles is same,

$$v_y O/O' = 0$$

Now, in the horizontal direction,

$$v_x\, O/O' = 2v \cos \alpha$$

Since, we know perpendicular distance is the minimum distance.

So, time $= \dfrac{d}{2v \cos \alpha}$

56. Time of flight $= 3$ s $+ 6$ s $= 9$ s

For horizontal range, $t = 6 - 3 = 3$ s

So, horizontal velocity, $u_x = \frac{150}{3} = 50$ m/s

Now, horizontal range $= u_x t = 50 \times 9 = 450$ m

57. At the minimum velocity, the ball will move just touching point E.

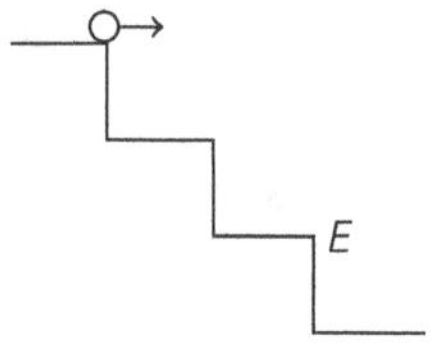

$\therefore\; x = 40, y = -20, \theta = 0°$

So, $y = x \tan\theta - \dfrac{gx^2}{2u^2 \cos^2\theta}$

On substituting the values, we get

$$u = 2 \text{ m/s}$$

58. Let H_1 and H_2 be the height travelled by two balls.

For same range, projection angles are complementary

$$H_1 = \frac{u^2 \sin^2\theta}{2g}$$

$$H_2 = \frac{u^2 \sin^2(90° - \theta)}{2g} = \frac{u^2 \cos^2\theta}{2g}$$

Now, we have given,

$$H_1 - H_2 = \frac{1}{2}(H_1 + H_2)$$

$$2H_1 - 2H_2 = H_1 + H_2$$

$$\therefore \quad H_1 = 3H_2$$

$$\frac{u^2 \sin^2\theta}{2g} = \frac{3 \cdot u^2 \cos^2\theta}{2g}$$

$$\tan\theta = \sqrt{3} \;\Rightarrow\; \theta = 60°$$

Another angle $= 90° - \theta = 30°$

59. The horizontal range, $R = 6 + 18 = 24$ m

$$24 = \frac{u^2 \sin 2\theta}{g}$$

$$u^2 = \frac{24 \times 10}{2 \sin\theta \cos\theta} = \frac{120}{\sin\theta \cos\theta}$$

Now, we know,

$$y = x \tan\theta - \frac{gx^2}{2u^2 \cos^2\theta}$$

$$3 = 6 \tan\theta - \frac{360}{2u^2 \cos^2\theta}$$

On eliminating u^2, we get

$$\theta = \tan^{-1}\left(\frac{2}{3}\right)$$

60. The equation of projectile,

$$y = x\tan\theta - \frac{gx^2}{2u^2\cos^2\theta}$$

Substituting the given values,

$$12 = 18\tan 45^\circ - \frac{10\,(18)^2}{2u^2\cos^2 45^\circ}$$

$$u = 23.2 \text{ m/s}$$

Now, maximum height $= \dfrac{u^2\sin^2\theta}{2g}$

$$= \frac{(23.2)^2\sin^2 45^\circ}{20} = 13.5 \text{ m}$$

61. Let u and v be the velocities of both the particles and α and β be the angle of projection.

So, $v_1 = (u\cos\alpha)\,\hat{i} + (u\sin\alpha - gt)\,\hat{j}$

$v_2 = (v\cos\beta)\,\hat{i} + (v\sin\beta - gt)\,\hat{j}$

The two velocity vectors must be parallel when the coefficients of $\hat{i}$ and $\hat{j}$ are equal.

$$\frac{u\cos\alpha}{v\cos\beta} = \frac{u\sin\alpha - gt}{v\sin\beta - gt}$$

On solving this equation, we get

$$t = \frac{uv\sin(\alpha - \beta)}{g\,(v\cos\beta - u\cos\alpha)}$$

62. to 64. For time :

Use $s = ut + \frac{1}{2}at^2$,

in vertical direction with proper sign convention.

For horizontal displacement :

Use $s_x = u_x t$

For vertical component of velocity :

Use $v_y = \sqrt{u_y^2 + 2gh}$

65. Let the speed of projectile be u and angle of projection be θ.

Initial horizontal component, $u_x = u\cos\theta$

Now, the acceleration in horizontal direction, $a = \dfrac{g}{2}$

So, the horizontal range of projectile $= u_x T + \dfrac{1}{2}aT^2$ ...(i)

Here, $T = \dfrac{2u\sin\theta}{g}$

On substituting T in Eq. (i), we get

$$R' = \frac{u^2\sin^2\theta}{g} + \frac{2u^2\sin^2\theta}{2g}$$

$$\left[\because R = \frac{u^2\sin 2\theta}{g} \text{ and } H = \frac{u^2\sin^2\theta}{2g}\right]$$

$$R' = R + 2H$$

66. The distance travelled by both the particles in the horizontal direction must be same for the condition of collision.

$$\therefore \quad u_A\cos\alpha t = u_B\cos\beta t$$

$$\frac{u_A}{u_B} = \frac{\cos\beta}{\cos\alpha}$$

67. Using the equation $s = ut + \frac{1}{2}at^2$ in vertical direction,

$$8 = \frac{1}{2}(10)\,t^2 \qquad (\because u = 0)$$

$$t = \sqrt{\frac{8}{5}} \text{ s}$$

Now, horizontal distance, $u_x = \dfrac{30 \times \sqrt{5}}{\sqrt{8}}$

$$= 23.71 \text{ m/s}$$

68. Relative velocity of A, $v_A = v - \dfrac{v}{2} = \dfrac{v}{2}$

Similarly, the relative velocity of B, $v_B = 0 - \dfrac{v}{2} = -\dfrac{v}{2}$

$$a_{B\,\text{relative}} = g$$

So, option (c) describes the best motion of A and B with respect to each other.

69. In ball A,

$$1 = \frac{u\sin\theta_1}{g} \Rightarrow u\sin\theta_1 = 10 \qquad \text{...(i)}$$

Now, the velocity in the horizontal direction remains constant

$$= u\cos\theta_1$$

Velocity in vertical direction,

$$v = u\sin\theta_1 - gt$$

Now, $\tan(90^\circ - \theta) = \dfrac{u\sin\theta_1 - 10}{u\cos\theta_1}$

$$\cot\theta = \frac{u\sin\theta_1 - 10}{u\cos\theta_1} = \frac{10}{u\cos\theta_1}$$

$$u\cos\theta_1 = 10\tan\theta_1 \qquad \text{...(ii)}$$

For the inclined plane,

$$\cot\theta = \frac{gt - u\sin\alpha}{u\cos\alpha}$$

$$\frac{1}{\tan\theta} = \frac{10\,t - 20}{10\tan\theta}$$

$$t = 3 \text{ s}$$

70. The maximum height attained by the ball B with $u\sin\theta_1 = 20$ and $t = 1$ s.

$$s = ut + \frac{1}{2}at^2$$

$$s = 20\,(1) - 5\,(1)^2$$

$$= 15 \text{ m}$$

LEVEL 03 Previous Years Questions

71. Given, distance between the two buildings

$$d = 100 \text{ m}$$

height of each tower, $h = 200$ m

speed of each bullet, $v = 25 \text{ ms}^{-1}$

The situation can be shown as below

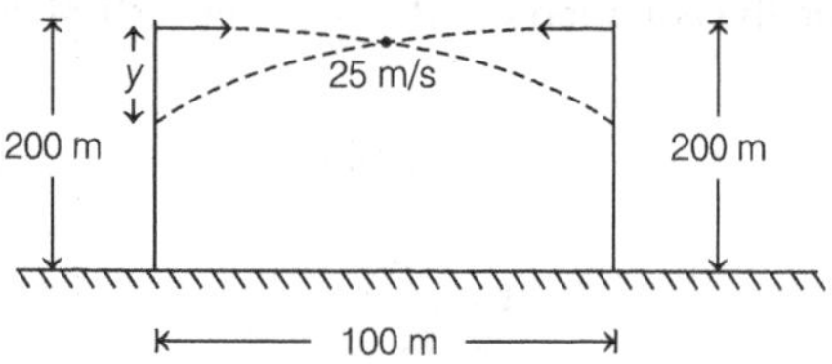

where, x is the vertical distance travelled from the top of the building and t is the time at which they collide.

As two bullets are fired towards each other, so their relative velocity will be

$$v_{rel} = 25 - (-25) = 50 \text{ ms}^{-1}$$

Then time, $t = \frac{d}{v_{rel}} = \frac{100}{50} = 2 \text{ s}$

The distance or height at which they collide is calculated from equation of motion,

$$y = ut + \frac{1}{2}at^2$$

$$y = -\frac{1}{2}gt^2$$

$$y = -\frac{1}{2} \times 10(2)^2 = -20 \text{ m}$$

The negative sign shows that the bullets will collide 20 m below the top of tower i.e. at a height of $(200 - 20) = 180$ m from the ground after 2 s.

72. The motion of object shot in two cases can be depicted as shown below

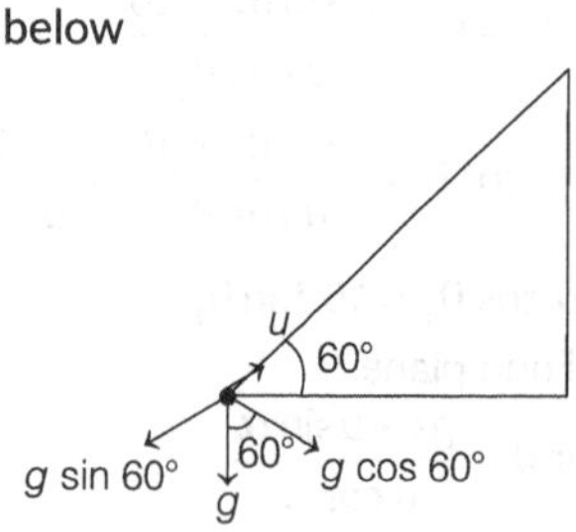

Case I

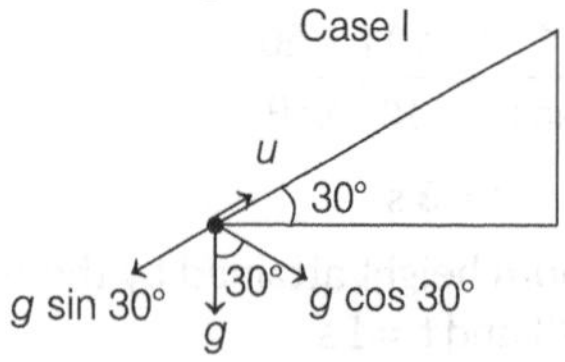

Case II

Using third equation of motion,

$$v^2 = u^2 - 2gh \quad \text{... (i)}$$

As the object stops finally, so

$$v = 0$$

For inclined motion,

$$g = g\sin\theta \text{ and } h = x$$

Substituting these values in Eq. (i), we get

$$\Rightarrow u^2 = 2g\sin\theta\, x \Rightarrow x = \frac{u^2}{2g\sin\theta}$$

For case (I), $x_1 = \frac{u^2}{2g\sin 60°}$

For case (II), $x_2 = \frac{u^2}{2g\sin 30°}$

$$\Rightarrow \quad \frac{x_1}{x_2} = \frac{u^2}{2g\sin 60°} \times \frac{2g\sin 30°}{u^2}$$

$$= \frac{1}{2} \times \frac{2}{\sqrt{3}} = \frac{1}{\sqrt{3}} \text{ or } 1 : \sqrt{3}$$

73. Given, $x = 5t - 2t^2$

Velocity of the particle,

$$v_x = \frac{dx}{dt} = \frac{d}{dt}(5t - 2t^2)$$

$$= 5 - 4t$$

Acceleration, $a_x = \frac{d}{dt}v_x = -4 \text{ ms}^{-2}$

Also, $y = 10t$

Velocity, $v_y = \frac{dy}{dt} = 10$

$\therefore$ Acceleration, $a_y = \frac{dv_y}{dt} = 0$

$\therefore$ Net acceleration of the particle,

$$\mathbf{a}_{net} = a_x\hat{\mathbf{i}} + a_y\hat{\mathbf{j}} = (-4 \text{ ms}^{-2})\,\hat{\mathbf{i}}$$

or $\quad \mathbf{a}_{net} = -4\,\hat{\mathbf{i}} \text{ ms}^{-2}$

74. After t hours, ships A and B are as shown below

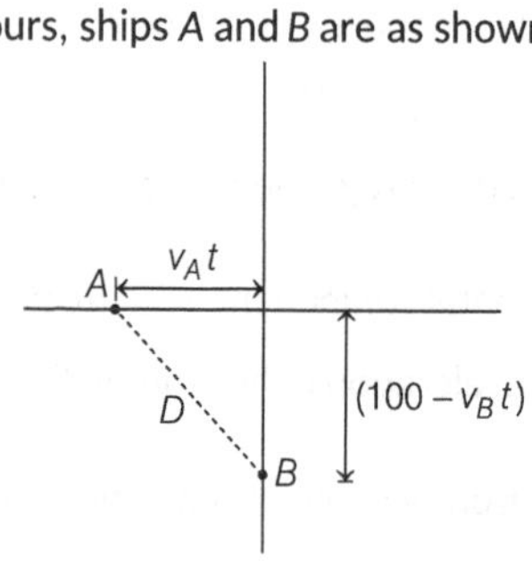

Distance between ships,

$$D = \sqrt{[(v_A t^2) + (100 - v_B t)^2]}$$

$$= \sqrt{(200t^2 - 2000t + 100^2)}$$

Distance is minimum, when $\frac{d}{dt}D = 0$

$$\Rightarrow \quad \frac{1}{2D} \cdot (400t - 2000) = 0$$

or $\quad 400t = 2000$

$\Rightarrow \quad t = 5$ h

75. (i) The position vector of a particle **R** as a function of time is given by

$$\mathbf{R} = 4\sin(2\pi t)\,\hat{\mathbf{i}} + 4\cos(2\pi t)\hat{\mathbf{j}}$$

X-axis component, $x = 4\sin 2\pi t$ ...(i)

Y-axis component, $y = 4\cos 2\pi t$ ...(ii)

Squaring and adding both equations, we get

$$x^2 + y^2 = 4^2[\sin^2(2\pi t) + \cos^2(2\pi t)]$$

i.e. $x^2 + y^2 = 4^2$ i.e. equation of circle and radius is 4 m.

(ii) Acceleration vector, $\mathbf{a} = \frac{v^2}{R}(-\mathbf{R})$, while v is velocity of a particle.

(iii) Magnitude of acceleration vector,

$$a = \frac{v^2}{R}$$

(iv) As, we have $v_x = +4(\cos 2\pi t)\,2\pi$ and

$$v_y = -4(\sin 2\pi t)\,2\pi$$

Net resultant velocity,

$$v = \sqrt{v_x^2 + v_y^2}$$

$$= \sqrt{(8\pi)^2(\cos^2 2\pi t + \sin^2 2\pi t)}$$

$$v = 8\pi \qquad [\because \cos^2 2\pi t + \sin^2 2\pi t = 1]$$

So, option (c) is incorrect.

76. Same or identical trajectory means same maximum heights,

$$\Rightarrow \quad \frac{u_1^2\sin^2\theta}{2g_1} = \frac{u_2^2\sin^2\theta}{2g_2}$$

So, $$\frac{5^2}{9.8} = \frac{3^2}{g_2}$$

or $$g_2 = 3.5\ \text{ms}^{-2}$$

77. Given, position vector of the particle at $t = 0$ is $(2\hat{\mathbf{i}} + 3\hat{\mathbf{j}})$ and $t = 5$ s is $(13\hat{\mathbf{i}} + 14\hat{\mathbf{j}})$

Average velocity vector,

$$\mathbf{v}_{av} = \frac{\text{Net displacement}}{\text{Time taken}}$$

$$= \frac{(13-2)\hat{\mathbf{i}} + (14-3)\hat{\mathbf{j}}}{5}$$

$$= \frac{11\hat{\mathbf{i}} + 11\hat{\mathbf{j}}}{5} = \frac{11}{5}(\hat{\mathbf{i}} + \hat{\mathbf{j}})$$

78. **Concept** As we know that in projectile motion, only velocity of y component change, whereas velocity of x component remains constant.

From the figure, the x-component remains unchanged, while the y-component is reversed. Thus, the velocity at point B is $(2\hat{\mathbf{i}} - 3\hat{\mathbf{j}})$ ms^{-1}.

79. Given, horizontal range

$$R = \text{vertical maximum height } H$$

$$\text{Range, } R = \frac{u^2(2\sin\theta\cos\theta)}{g}$$

$$\text{Height, } H = \frac{u^2\sin^2\theta}{2g}$$

$$\text{Hence, } \frac{u^2(2\sin\theta\cos\theta)}{g} = \frac{u^2\sin^2\theta}{2g}$$

$$2\cos\theta = \frac{\sin\theta}{2}$$

$$\tan\theta = 4$$

$$\Rightarrow \quad \theta = \tan^{-1}(4)$$

80. For maximum range of projectile, θ will be 45° by the law of projectile motion.

So, maximum range, $R_{max} = \frac{u^2}{g}$

Given, $u = 20\ \text{ms}^{-1}$ and $g = 10\ \text{ms}^{-2}$

$$R_{max} = \frac{(20)^2}{10} = \frac{400}{10}$$

$$R_{max} = 40\ \text{m}$$

81. Given, initial velocity $(u) = 3\hat{\mathbf{i}} + 4\hat{\mathbf{j}}$

Final velocity $(v) = ?$

Acceleration $(a) = (0.4\,\hat{\mathbf{i}} + 0.3\hat{\mathbf{j}})$

Time $(t) = 10$ s

From first equation of motion, $v = u + at$

$$v = 3\hat{\mathbf{i}} + 4\hat{\mathbf{j}} + (0.4\hat{\mathbf{i}} + 0.3\hat{\mathbf{j}})\,10$$

$$v = 7\hat{\mathbf{i}} + 7\hat{\mathbf{j}} \Rightarrow |v| = 7\sqrt{2}$$

Chapter 06

LAWS OF MOTION

LEVEL 01 Objective Questions

1 In the figure shown, a balloon is pressed against a wall. It is in equilibrium and in maximum compressed state.

$\mathbf{F}_1$ = force of balloon on hand of man,

$\mathbf{F}_2$ = force of balloon on wall,

$\mathbf{F}_3$ = friction force,

$\mathbf{F}_4$ = weight of balloon.

Choose the correct statement(s).

(A) $\mathbf{F}_1$ and $\mathbf{F}_2$ are action-reaction pairs.

(B) $\mathbf{F}_3$ and $\mathbf{F}_4$ are action-reaction pairs.

(C) $\mathbf{F}_2$ and $\mathbf{F}_3$ are action-reaction pairs.

(D) $\mathbf{F}_4$ and gravitational force exerted by balloon on earth are action-reaction pairs.

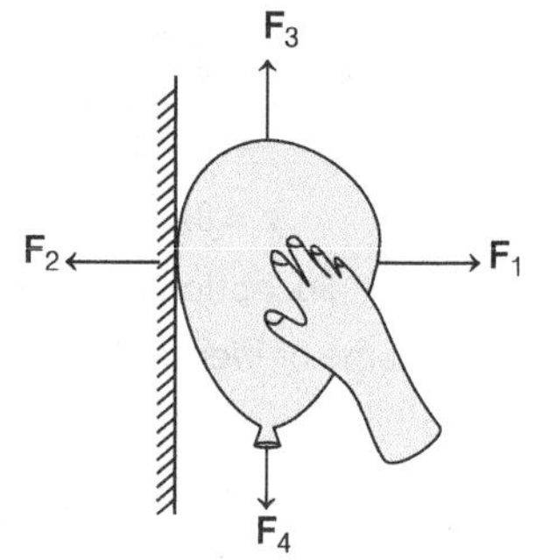

(a) Only A and D
(b) Only A, B and D
(c) Only D
(d) Only C and D

2 A block is in the state of rest on an inclined plane as shown. Find which direction best describes the net force applied by inclined plane on the block.

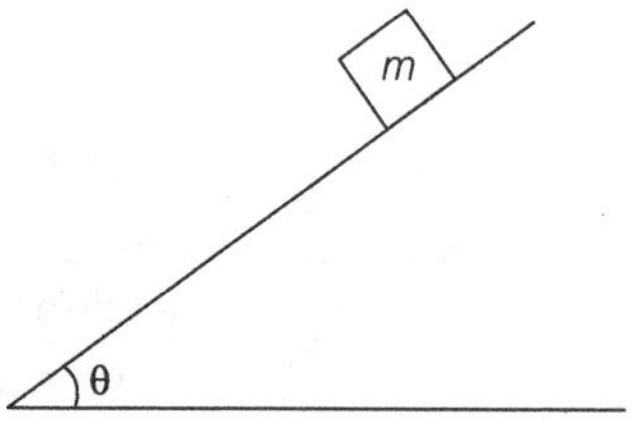

(a) ⟵ (horizontally)

(b) ↙ (along inclined)

(c) ↖ (normal to inclined)

(d) ↑ (vertically)

3 A force F_1 accelerates a particle from rest to a velocity v. Another force F_2 decelerates the same particle from v to rest, then

(a) F_1 is always equal to F_2

(b) F_2 is greater than F_1

(c) F_2 may be smaller than, greater than or equal to F_1

(d) F_2 cannot be equal to F_1

4 A block of mass M is pulled along a horizontal frictionless surface by a rope of mass m. If force F is applied at one end of the rope, the force which the rope exerts on the block is

(a) $\frac{F}{M+m}$ (b) F

(c) $\frac{FM}{M+m}$ (d) zero

5 A body of mass 4 kg is at rest on a horizontal table. The coefficient of friction between the body and the table is 0.2. A horizontal force of 6 N is applied on the body. The acceleration of the body is

(a) $0\ \text{m/s}^2$ (b) $0.5\ \text{m/s}^2$

(c) $1.5\ \text{m/s}^2$ (d) $2\ \text{m/s}^2$

6 A block of mass m is placed at rest on a rough horizontal surface with the coefficient of friction μ.

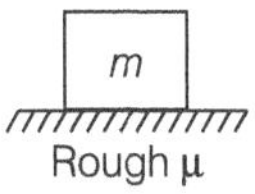

The following figure shows four different ways in which a constant force F may be applied on the block. Identify the situation in which the block can be moved with the least effort.

(a) 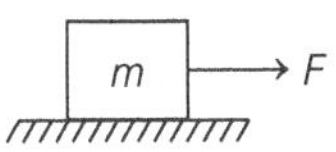

(b) 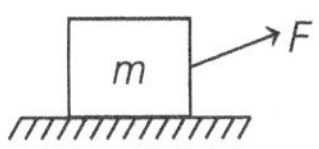

(c) 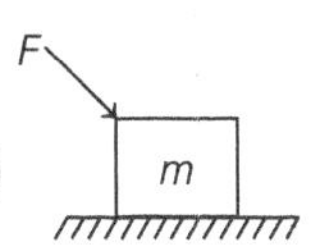

(d) 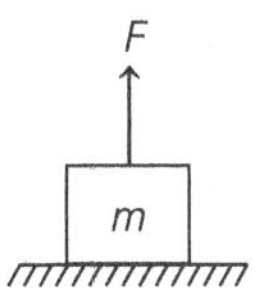

7 A horizontal force of 10.0 N is acting on a 10 kg box that is sliding to the right along the floor with velocity v. The coefficient of kinetic friction between the box and the floor is 0.20. The box is having

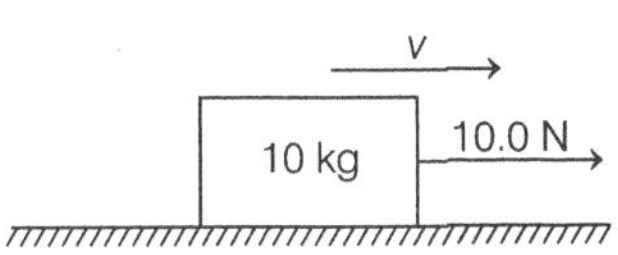

(a) acceleration towards left (b) acceleration towards right

(c) constant speed and constant velocity (d) constant speed but not constant velocity

8 In the arrangement shown in figure, pulley is smooth and massless and all the strings are light let F_1 be the force exerted on the pulley in case (i) and F_2 the force in case (ii). Then,

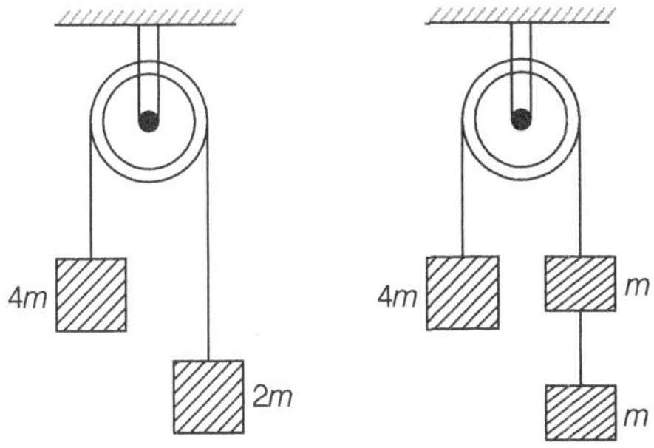

(a) $F_1 > F_2$ (b) $F_1 < F_2$

(c) $F_1 = F_2$ (d) $F_1 = 2F_2$

9 A monkey is descending from the branch of a tree with constant acceleration. If the breaking strength is 75% of the weight of the monkey, the minimum acceleration with which the monkey can slide down without breaking the branch is

(a) g (b) $3g/4$

(c) $g/4$ (d) $g/2$

10 Two inclined frictionless tracks, one gradual and the other steep meet at A from where two stones are allowed to slide down from rest as shown in figure. Which of the following statement is correct?

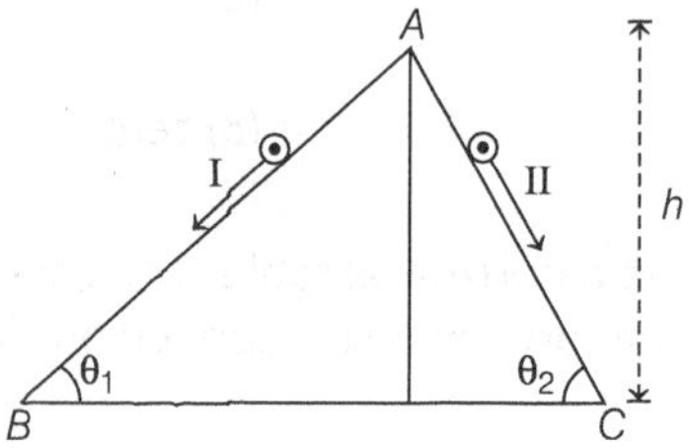

(a) Both the stones reach the bottom at the same time but not with the same speed.
(b) Both the stones reach the bottom with the same speed and stone I reaches the bottom earlier than stone II.
(c) Both the stones reach the bottom with the same speed and stone II reaches the bottom earlier than stone I.
(d) Both the stones reach the bottom at different times and with different speeds.

11 A 10 N force is applied on B. All surfaces are smooth. Mass of A and B are 5 kg each. What is the acceleration of A ?

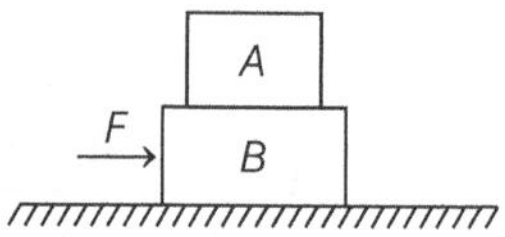

(a) 0 m/s^2 (b) 1 m/s^2
(c) 2 m/s^2 (d) 3 m/s^2

12 What will be the maximum acceleration of the truck, so that the block of mass 5 kg placed inside the truck does not slip, if the coefficient of friction $\mu = 0.3$?

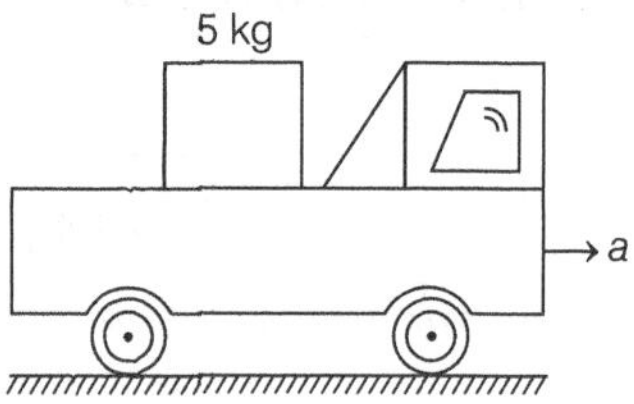

(a) 3 m/s^2 (b) 2 m/s^2 (c) 1 m/s^2 (d) Zero

13 Mass of a monkey is 50 kg. It climbs up a rope having breaking strength 800 N. In which of the following cases, the rope will break? (Take, $g = 10 \text{ m/s}^2$)
(a) climbs down with an acceleration of 6 ms^{-2}
(b) climbs up with an acceleration of 5 ms^{-2}
(c) climbs up with an acceleration of 6.5 ms^{-2}
(d) All of the above

14 In the arrangement shown in figure. The frictional force on 2 kg block is

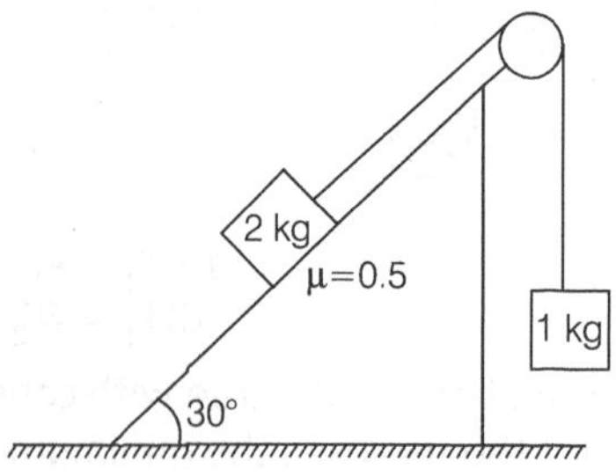

(a) 10 N (b) $5\sqrt{3}$ N
(c) zero (d) 5 N

15 Figure shows two identical blocks of mass m kept on a smooth floor. Block A is connected to front wall with a just straight string and block B is connected to rear wall with a relaxed spring. Assume that the floor of the train car is smooth and exerts no horizontal forces on the block. Mark the correct statement(s). Initially train is moving with constant speed.

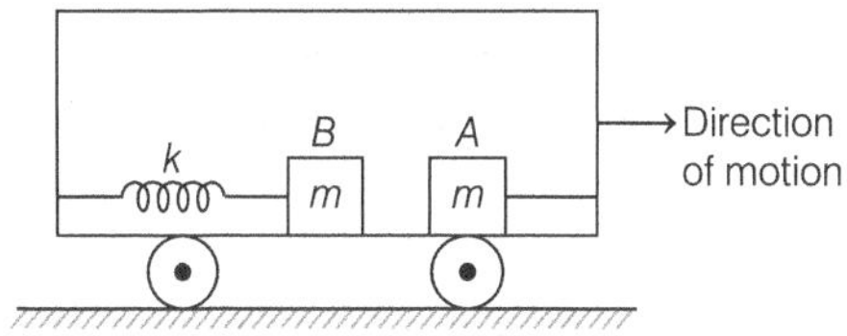

(a) When train moves with constant velocity, then force on block B is rightwards.
(b) When train speeds up, string gets taut and spring is compressed such that force exerted by each on respective block is same.
(c) When train slows down, block A moves towards front wall.
(d) When train slows down, resultant force on block A is non-zero in ground frame.

16 A block of mass 2 kg slides down an inclined plane of inclination 30°. The coefficient of friction between block and plane is 0.5. The contact force between block and plane is (Take, $g = 10\ \text{m/s}^2$)

(a) 20 N (b) $10\sqrt{3}$ N
(c) $5\sqrt{7}$ N (d) $5\sqrt{15}$ N

17 If coefficient of friction between block and the surface is $\frac{1}{\sqrt{3}}$ and m is such a mass that the 4 kg block is moving up the plane with constant speed, then m is

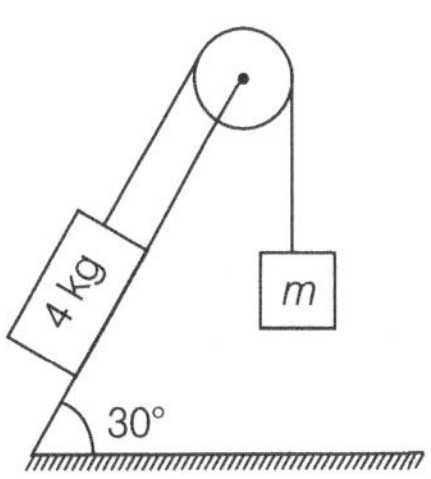

(a) 2 kg (b) 2.8 kg
(c) 3.4 kg (d) 4 kg

18 Two blocks, each having a mass M, rest on frictionless surface as shown in the figure. If the pulleys are light and frictionless and M on the incline is allowed to move down, then the tension in the string will be

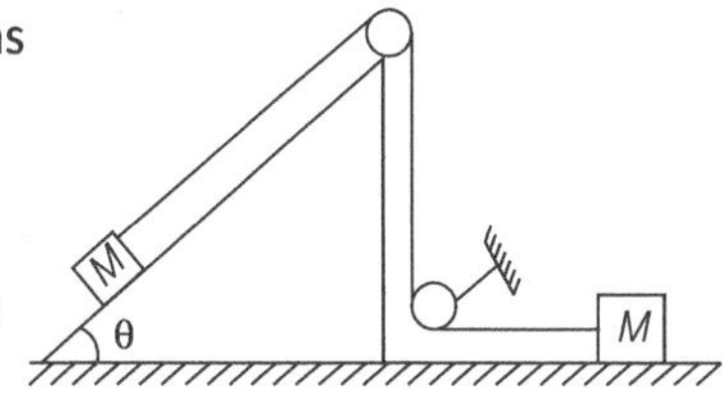

(a) $2.3\ Mg \sin\theta$ (b) $1.3\ Mg \sin\theta$
(c) $\frac{Mg}{2} \sin\theta$ (d) $2\ Mg \sin\theta$

19 In the arrangements shown in figure, the mass of block A, so that the system remains at rest is (Assume friction is negligible)

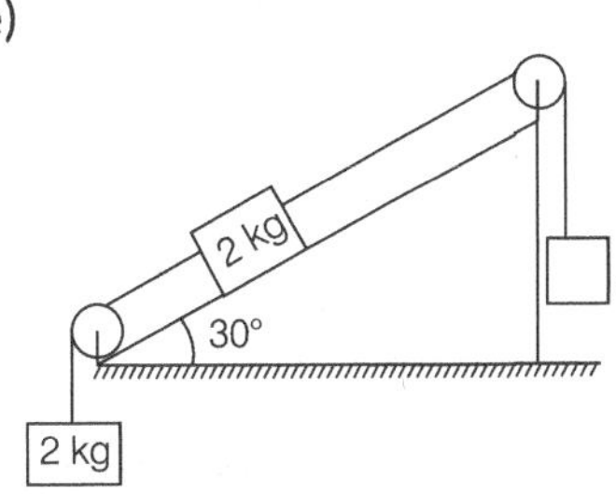

(a) $4\sqrt{3}$ kg (b) 4 kg
(c) $3\sqrt{3}$ kg (d) 3 kg

20 In the following figure, what is the minimum coefficient of friction needed between the block and fixed inclined plane, so that the system does not move?

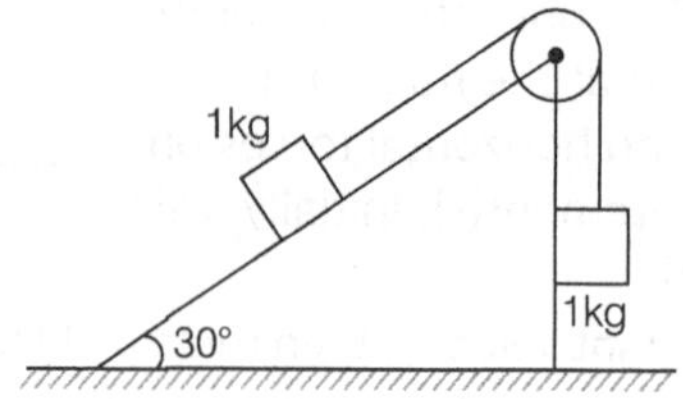

(a) $\sqrt{3}$ (b) $1/\sqrt{3}$ (c) $1/\sqrt{2}$ (d) $\sqrt{2}$

21 Two masses M_1 and M_2 are attached to the ends of a light string which passes over a massless pulley attached to the top of a double inclined smooth plane of angles of inclination α and β. The tension in the string is ($M_1 > M_2$ and $\alpha > \beta$)

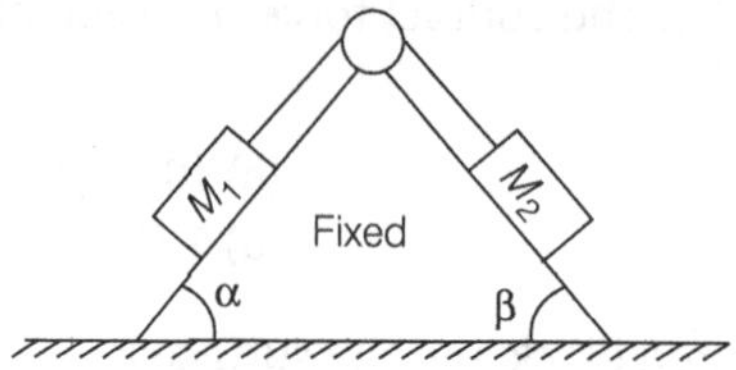

(a) $\dfrac{M_2 (\sin\beta) g}{M_1 + M_2}$ (b) $\dfrac{M_1 (\sin\alpha) g}{M_1 + M_2}$

(c) $\dfrac{M_1 M_2 (\sin\beta + \sin\alpha) g}{M_1 + M_2}$ (d) zero

22 A particle starts from rest point P and follow a path PQ on the three surfaces as shown in figures. If time taken by the particles in three cases are t_1, t_2 and t_3, then select the correct statement.

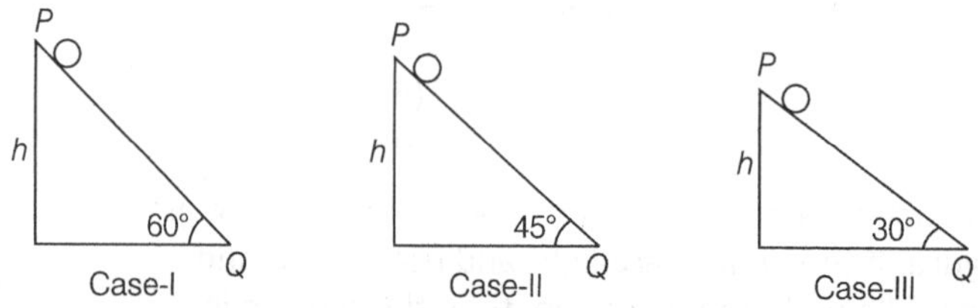

(a) $t_3 > t_2 > t_1$ (b) $t_1 > t_2 > t_3$ (c) $t_1 = t_2 = t_3$ (d) $t_2 > t_1 > t_3$

23 The coefficient of friction between the block m_2 and the inclined plane is μ. If $\dfrac{m_1}{m_2} = \sin\theta$ (shown in figure), then the mass m_1 moves downwards with acceleration

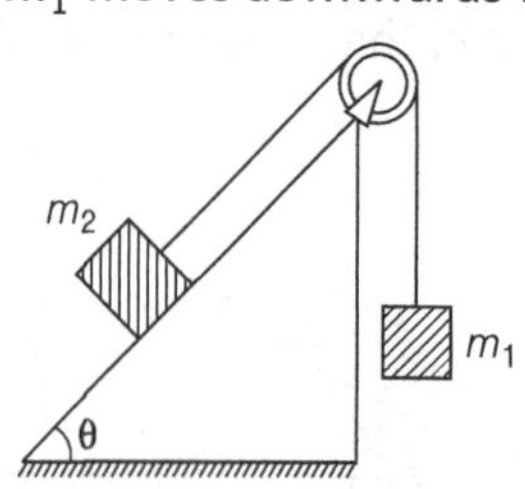

(a) $\mu g \cos\theta$ (b) $\dfrac{\mu m_1}{m_1 + m_2} g \cos\theta$

(c) $\dfrac{\mu m_2}{m_1 + m_2} g \cos\theta$ (d) zero

24 In the arrangement shown in figure, acceleration of the block is

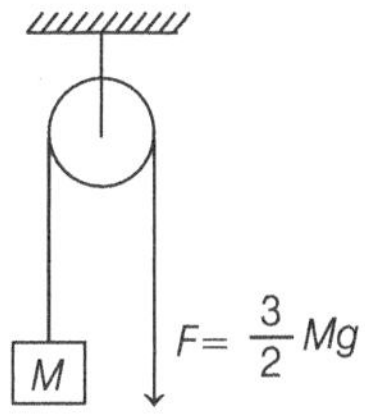

(a) zero (b) $g/2$ (c) $3g/2$ (d) g

25 In which of the following cases, the magnitude of acceleration of the block A will be maximum? (Neglect friction, mass of pulley and string)

(a)

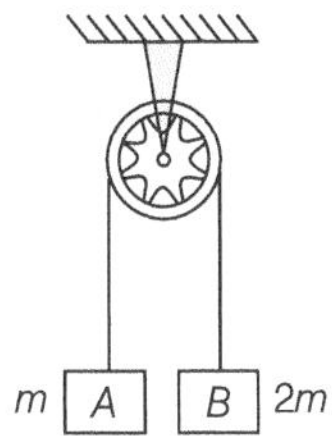

(b)

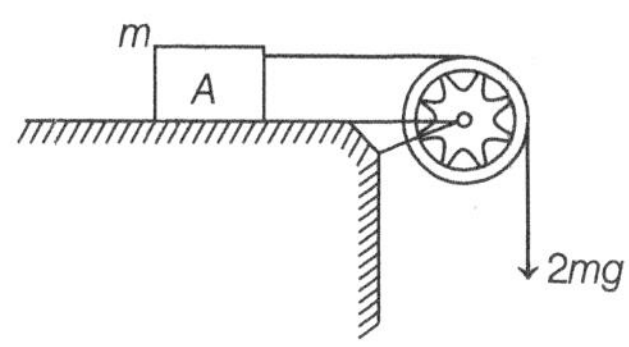

(c)

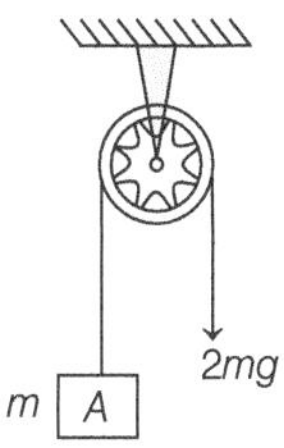

(d)

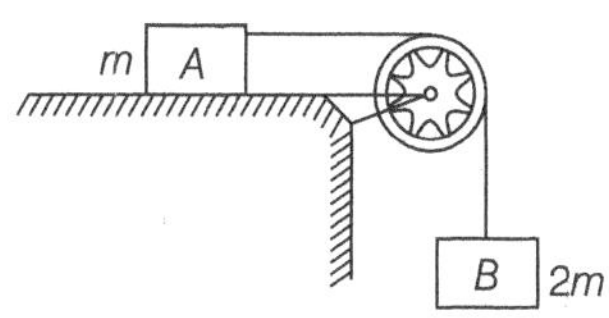

26 A monkey climbs up and another monkey climbs down a rope passing over pulley with same uniform acceleration separately. If the respective masses of monkeys are in the ratio 2 : 3, the common acceleration must be

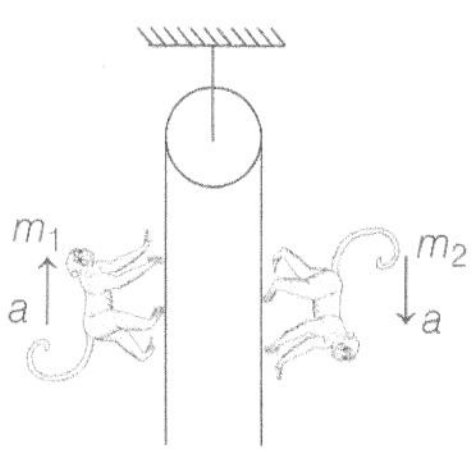

(a) $\frac{g}{5}$ (b) $6g$ (c) $\frac{g}{2}$ (d) g

27 Blocks A and B of masses 2 kg and 4 kg are suspended through a string using a pulley, inside an elevator moving downward with constant acceleration 2 m/s^2. The tension in the string which is joining the two blocks is :

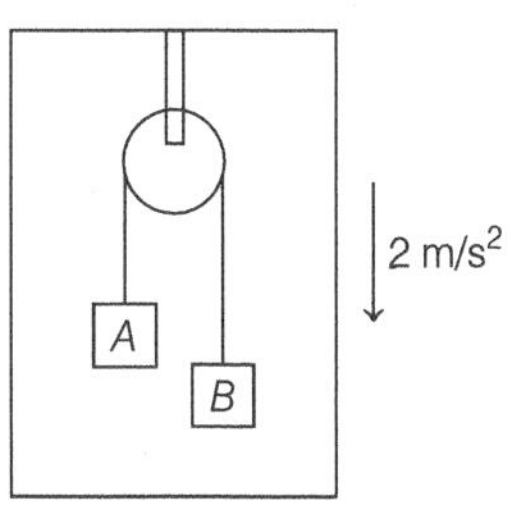

(a) $\frac{64}{3}$ N (b) $\frac{32}{3}$ N (c) $\frac{8}{3}$ N (d) $\frac{16}{3}$ N

28 In the diagram shown, block A of mass 2 kg is hanging from the string passing over a smooth pulley and block B is placed on the top of a table. If the reaction of the table is 10 N, mass of block B is (Take, $g = 10 \text{ m/s}^2$)

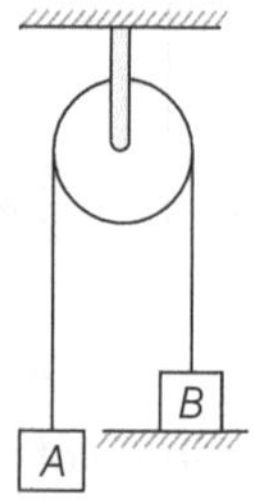

(a) 1 kg (b) 2 kg (c) 3 kg (d) 4 kg

29 The velocity of A, at an instant is 4 m/s rightwards. Then, the velocity of block B is

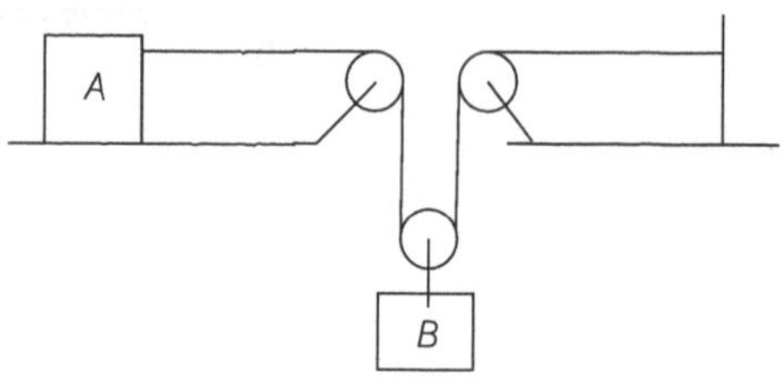

(a) 4 m/s downwards (b) 2 m/s upwards (c) 2 m/s downwards (d) 1 m/s upwards

30 In the figure, a man of true mass M is standing on a weighing machine placed in a cabin. The cabin is joined by a string with a body of mass m. Assuming no friction and negligible mass of cabin and weighing machine, the measured weight of man is

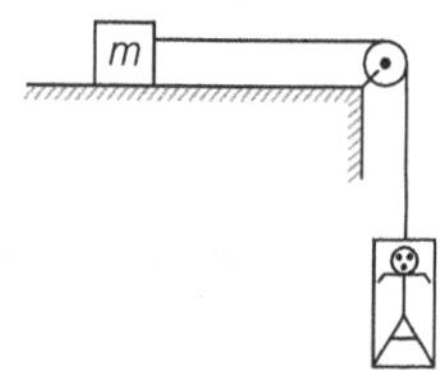

(a) $\frac{Mmg}{M+m}$ (b) $\frac{Mmg}{M-m}$

(c) Mg (d) mg

31 A force F is applied on block A as shown. The contact force between the blocks A and B and between the blocks B and C respectively are (Assume frictionless surface)

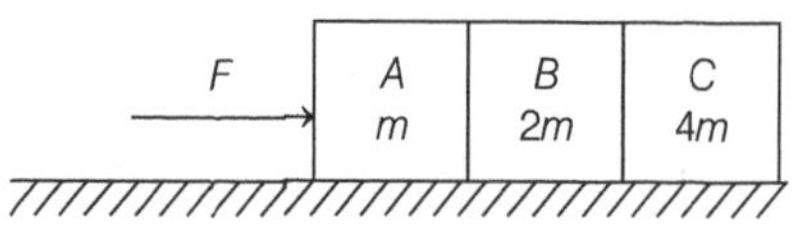

(a) $\frac{F}{7}, \frac{2F}{7}$ (b) $\frac{6F}{7}, \frac{4F}{7}$

(c) $F, \frac{F}{2}$ (d) $\frac{4F}{7}, \frac{6F}{7}$

32 If the block is in equilibrium, then values of T_1 and T_2 are

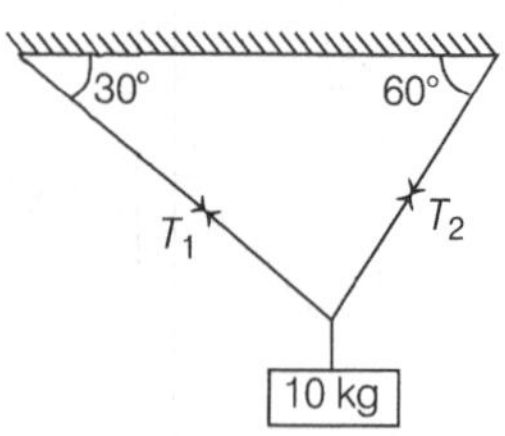

(a) 50 N, $50\sqrt{3}$ N (b) 80 N, 60 N
(c) 30 N, 40 N (d) 100 N, 0 N

33 If $T_3 = 36$ N, then value of T_2 is

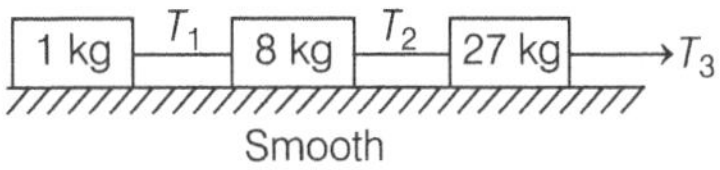

(a) 18 N (b) 9 N
(c) 3.375 N (d) 1.75 N

34 A block of mass 10 kg is moving with an acceleration of 2 m/s^2 on a rough surface as shown. The value of coefficient of kinetic friction is

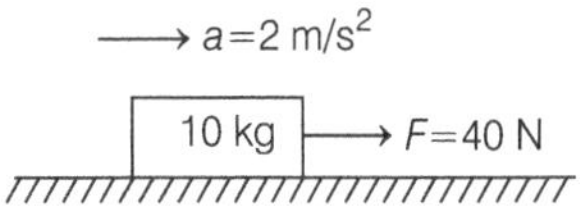

(a) 0.2 (b) 0.4
(c) 0.5 (d) 0.1

35 A mass M is hung with a light inextensible string. Tension in horizontal string is

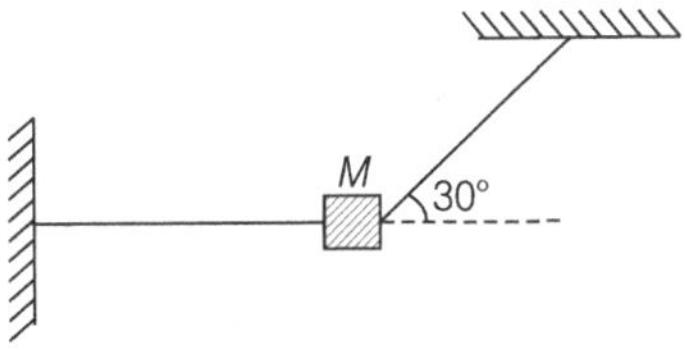

(a) $\sqrt{3}\,Mg$ (b) $\sqrt{2}\,Mg$ (c) $\frac{Mg}{\sqrt{3}}$ (d) $\frac{Mg}{2}$

36 A body of weight 50 N is pulled with a force of $20\sqrt{2}$ N. If the body remains stationary, then the frictional force acting on the body and the normal reaction force will be

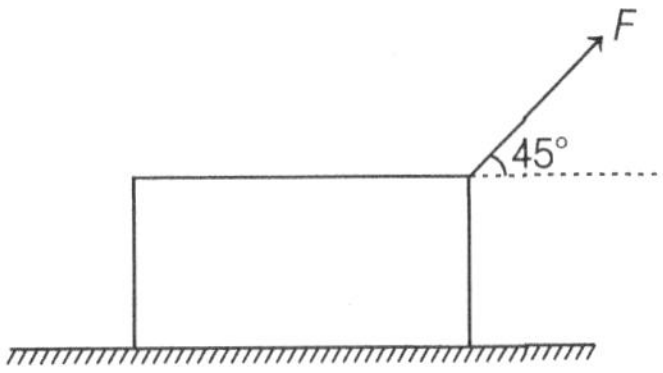

(a) 10 N, 15 N (b) 5 N, 6N (c) 2 N, 3 N (d) 20 N, 30 N

37 An arrangement shown in figure has four identical blocks connected with horizontal strings being pulled by a force F on one side. Tension in the various strings are T_1, T_2 and T_3 and surface is smooth. Choose the correct option.

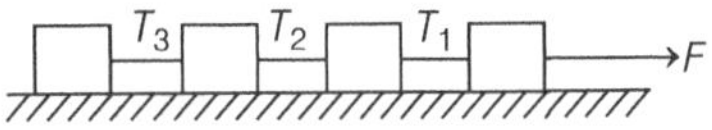

(a) $T_1 + T_2 + T_3 = F$ (b) $T_1 + T_2 + T_3 = 2F$ (c) $2(T_1 + T_2 + T_3) = F$ (d) $2(T_1 + T_2 + T_3) = 3F$

38 A block of mass 10 kg is kept over a rough surface and a force $F = 4t$ is applied on it. Then, the minimum value of t for which the block will start moving is (Take, $g = 10$ m/s^2)

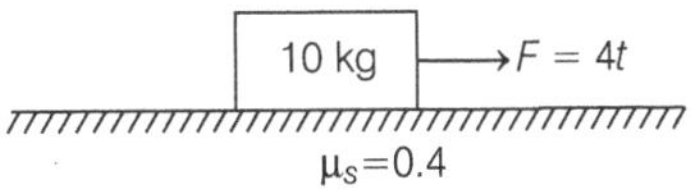

(a) 4 s (b) 10 s (c) 8 s (d) 5 s

39 A horizontal force of 10 N is applied on three blocks A, B and C of masses 2 kg, 3 kg and 5 kg kept on a horizontal smooth surface as shown in figure. Find the net force acting on 2 kg block.

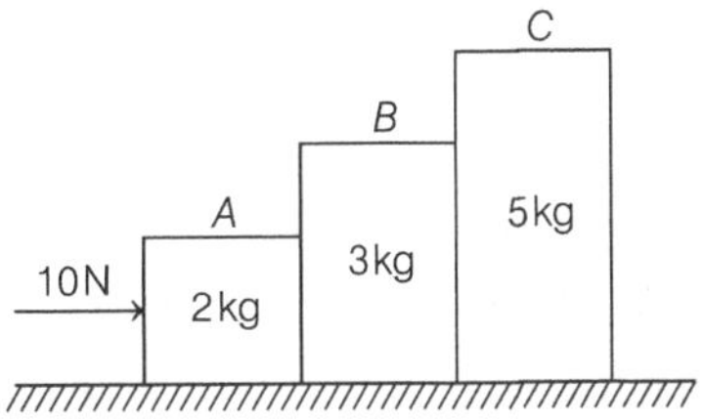

(a) 8 N (b) 10 N (c) 2 N (d) 5 N

40 A force $\mathbf{F} = \hat{\mathbf{i}} + 4\hat{\mathbf{j}}$ acts on a block as shown. The force of friction acting on the block is (Take, $g = 10 \text{ m/s}^2$)

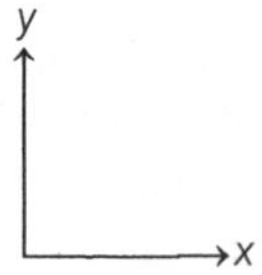

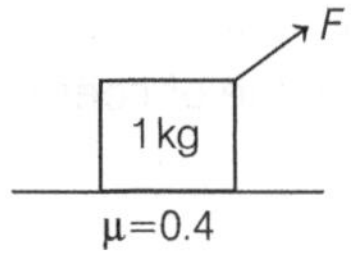

(a) $-\hat{\mathbf{i}}$ (b) $-1.8\hat{\mathbf{i}}$ (c) $-2.4\hat{\mathbf{i}}$ (d) $-3\hat{\mathbf{i}}$

41 A force produces an acceleration of 4 m/s^2 in a body and the same force produces an acceleration of 6 m/s^2 in another body. If the two bodies are combined and the same force is applied to the combination, then acceleration produced in the combined body is

(a) 10 m/s^2 (b) 2 m/s^2
(c) 2.4 m/s^2 (d) 5.4 m/s^2

42 If force $F = 24$ N is applied on the lower block, then force of friction between the two blocks is

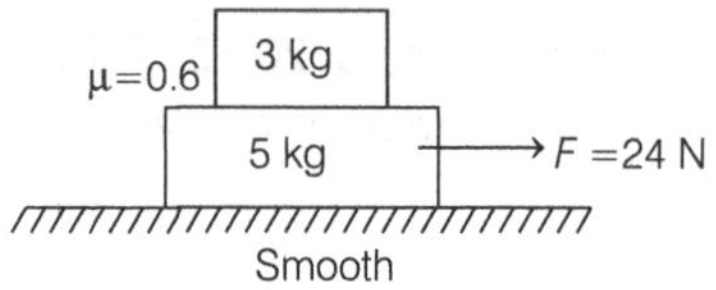

(a) 18 N (b) 9 N (c) 15 N (d) 6 N

43 In the system shown in figure, force of friction between the two blocks is

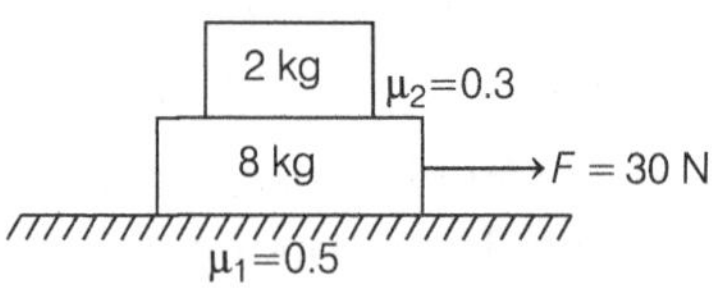

(a) 6 N (b) 10 N (c) 2 N (d) zero

44 A block of mass 4 kg is kept on ground. The coefficient of friction between the block and the ground is 0.8. The external force of magnitude 30 N is applied parallel to the ground. The resultant force exerted by the ground on the block is (Take, $g = 10 \text{ m/s}^2$)

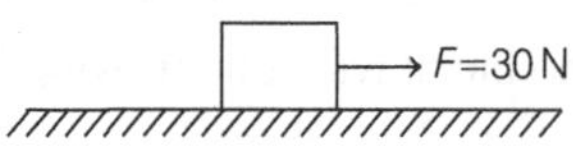

(a) 40 N (b) 30 N (c) zero (d) 50 N

45 Consider the situation as shown in the figure.

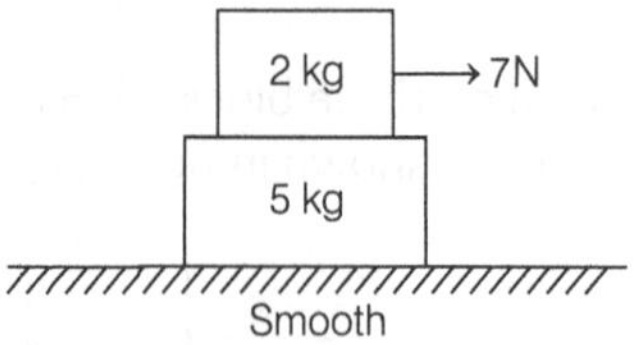

The coefficient of friction between the blocks is 0.5. The acceleration of the lower block is

(a) 1 m/s^2 (b) 1.5 m/s^2 (c) 2 m/s^2 (d) 2.5 m/s^2

46 A body moves along a rough horizontal surface with a velocity of 6 m/s. If the body comes to rest after travelling 9 m, then coefficient of kinetic friction between body and the surface is

(a) 0.5 (b) 0.4 (c) 0.6 (d) 0.2

47 If coefficient of friction between block and the surface is μ and $F = \mu\, mg/2$, then tension in the string is

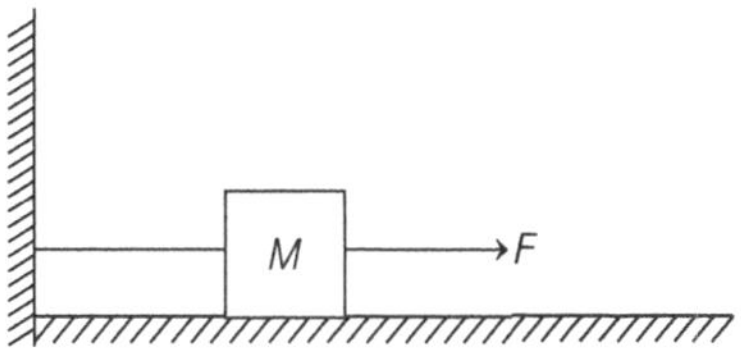

(a) zero (b) $\frac{3\,\mu mg}{2}$ (c) $\frac{\mu mg}{2}$ (d) μmg

48 A particle is projected up along a rough inclined plane having angle of inclination 45°. If the coefficient of friction is $\frac{1}{2}$, then retardation of particle is

(a) $\frac{g}{2\sqrt{2}}$ (b) $\frac{g}{\sqrt{2}}$ (c) $\frac{g}{\sqrt{2}}\left(1-\frac{1}{2}\right)$ (d) $\frac{g}{\sqrt{2}}\left(1+\frac{1}{2}\right)$

49 A uniform disk of radius R and mass m is connected to a wall by string of length $2R$. The normal reaction of the wall is

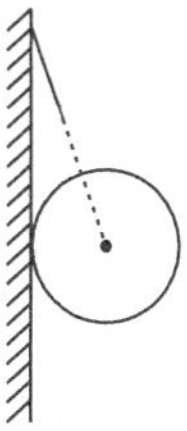

(a) mg
(b) $\frac{mg}{2}$
(c) $\frac{mg}{\sqrt{3}}$
(d) $2\,mg$

50 A body is sliding down an inclined plane (angle of inclination 45°). If the coefficient of friction is 0.5 and $g = 9.8\ \text{m/s}^2$, then the downward acceleration of the body (in m/s^2) is

(a) $\frac{4.9}{\sqrt{2}}$ (b) $4.9\sqrt{2}$ (c) $19.6\sqrt{2}$ (d) 4.9

51 A uniform chain of length $2L$ is hanging in equilibrium position. If end B is given a slightly downward displacement, the imbalance causes an acceleration. Here, pulley is small and smooth and string is inextensible. The acceleration of end B when it has been displaced by distance x, is

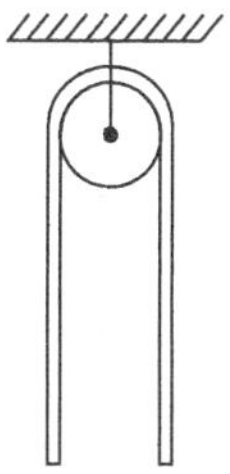

(a) $\frac{x}{L}g$ (b) $\frac{2x}{L}g$
(c) $\frac{x}{2}g$ (d) g

52 A spring balance fastened to the roof of a lift accelerating upward indicates 120 N as the weight of a 80 N body. The acceleration of the lift is (Take, $g = 10\ \text{m/s}^2$)

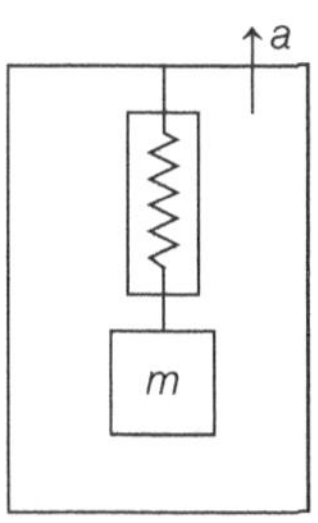

(a) $5\ \text{m/s}^2$ (b) $\frac{20}{3}\ \text{m/s}^2$ (c) $\frac{10}{3}\ \text{m/s}^2$ (d) $4\ \text{m/s}^2$

53 An elevator and its load have a total mass of 300 kg. If the elevator originally moving downward at 10 m/s is brought to rest with constant deceleration in a distance of 25 m, the tension in the supporting cable will be (Take, $g = 10\ m/s^2$)

(a) 8000 N (b) 6400 N (c) 11200 N (d) 3600 N

54 A monkey of mass 20 kg is holding a vertical rope. The rope will break, if the mass suspended from it exceed 25 kg. What is the maximum acceleration with which the monkey can climb up along the rope? (Take, $g = 10\ m/s^2$)

(a) $10\ m/s^2$ (b) $25\ m/s^2$ (c) $2.5\ m/s^2$ (d) $5\ m/s^2$

55 The minimum acceleration that must be imparted to the cart in the figure, so that the block *A* will not fall (μ is the coefficient of friction between the surfaces of block and cart) is given by

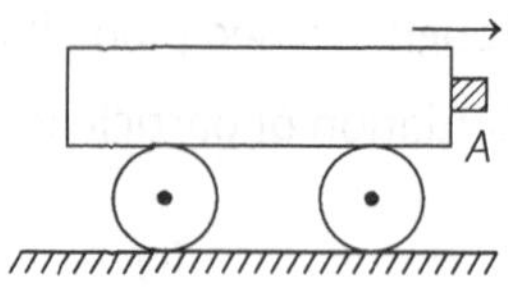

(a) μg (b) g/μ (c) $g/\sqrt{\mu}$ (d) μ/g

56 A horizontal force of 10 N is necessary to just hold a block stationary against a wall. The coefficient of friction between the block and the wall is 0.2. The weight of the block is

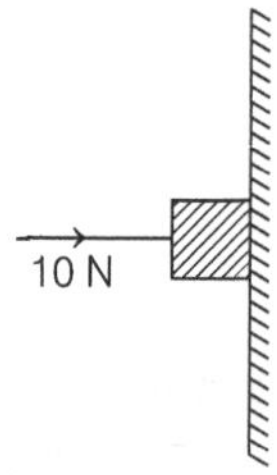

(a) 20 N (b) 50 N (c) 100 N (d) 2 N

57 Find acceleration of the block in given figure. Neglect friction.

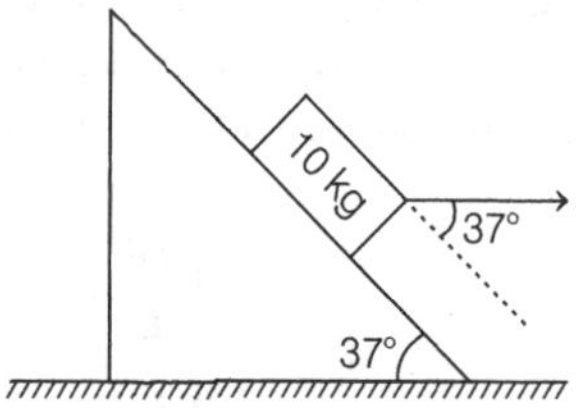

(a) $6\ m/s^2$ (b) $11\ m/s^2$ (c) $1\ m/s^2$ (d) $10\ m/s^2$

58 Two blocks *A* and *B* are as shown in figure. The minimum horizontal force *F* applied to the block *B* for which slipping just begins between *B* and ground is

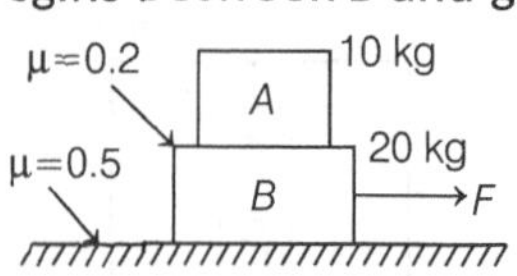

(a) 20 N (b) 120 N (c) 150 N (d) 170 N

59 A metre stick is balanced on a knife edge at its centre. When two coins, each of mass 5 g are put one on top of the other at the 12 cm mark, the stick is found to be balanced at 45 cm. The mass of the metre stick is

(a) 56 g (b) 66 g (c) 76 g (d) 86 g

60 An elevator is descending with an acceleration $3\ m/s^2$. The mass of block A is 2 kg and B is 3 kg. The normal force exerted by block A on block B is

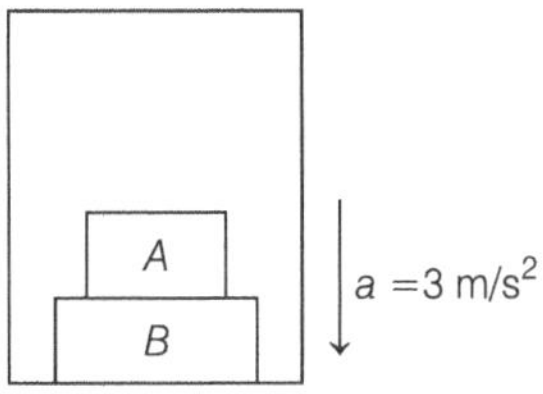

(a) 20 N
(b) 15 N
(c) 50 N
(d) 14 N

61 A light string passing over a smooth light pulley connects two blocks of masses m_1 and m_2 (vertically). If the acceleration of the blocks is $g/8$, then the ratio of the masses is

(a) 8 : 1 (b) 9 : 7
(c) 4 : 3 (d) 5 : 3

62 A ball of mass 1 kg hangs in equilibrium from two strings OA and OB as shown in figure. The ratio of tensions in strings OA and OB is (Take, $g = 10\ ms^{-2}$)

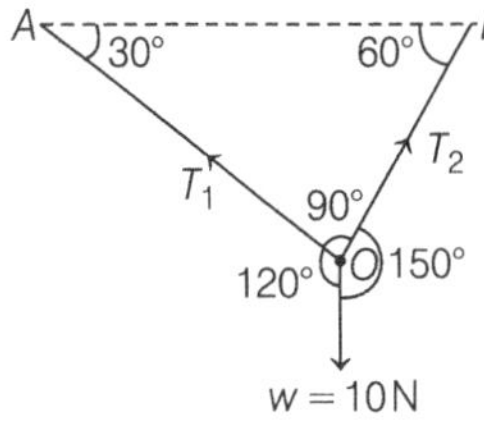

(a) $\sqrt{3}$ (b) $\frac{\sqrt{3}}{2}$
(c) $\frac{1}{\sqrt{3}}$ (d) $\frac{1}{2\sqrt{3}}$

63 In the system shown in figure, the acceleration of the 1 kg mass is

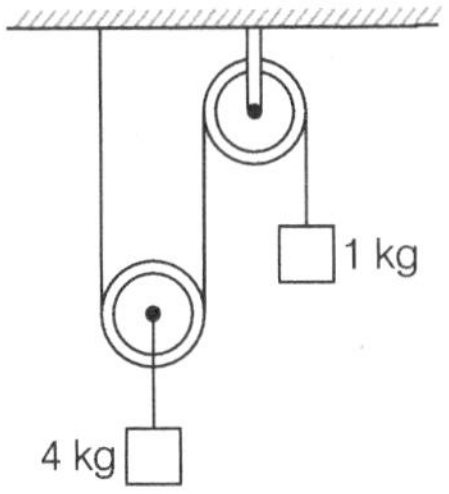

(a) $\frac{g}{4}$ downward (b) $\frac{g}{4}$ upward (c) $\frac{g}{2}$ downward (d) $\frac{g}{2}$ upward

64 A rope of length 10 m and linear mass density 4 kg/m is lying lengthwise on a horizontal smooth table. One end of the rope is pulled horizontally by a force of 40 N. The tension in the rope at a point 4 m from point of application of force will be

(a) 40 N (b) 24 N
(c) 49 N (d) 15 N

65 A block of mass 4 kg rests on an inclined plane. The inclination of the plane is gradually increased, it is found that when the inclination is 3 in 5, $\left(\sin\theta = \frac{3}{5}\right)$, the block just begins to slide down the plane. The coefficient of friction between the block and the plane is

(a) 0.4 (b) 0.6
(c) 0.8 (d) 0.75

66 Consider the shown arrangement, where the blocks A and B connected by means of a uniform string is being moved vertically up by the force F. Each block weighs 2 kg while the mass of string is 1000 g. The tension at bottom of the string equals

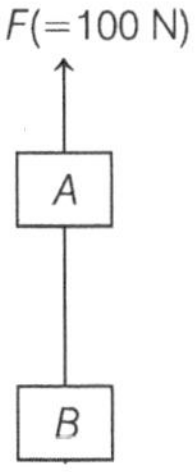

(a) 20 N
(b) 40 N
(c) 10 N
(d) 270 N

67 Calculate the tensions T_1, T_2 and T_3 in the three threads shown in the following figure. (All threads are massless) (Take, $g = 10 \text{ m/s}^2$)

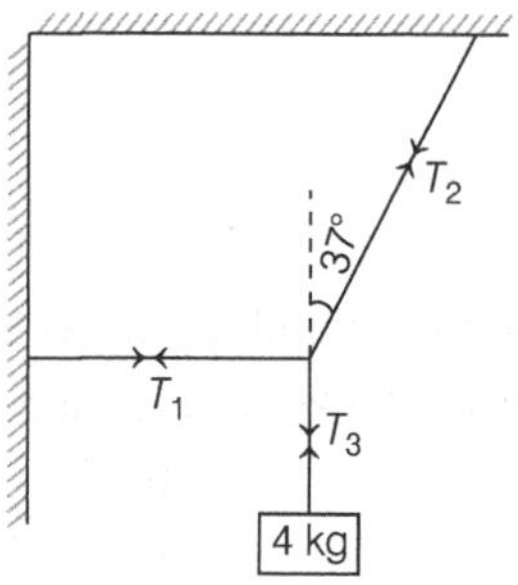

(a) 30 N, 40 N, 50 N
(b) 50 N, 30 N, 40 N
(c) 35 N, 45 N, 40 N
(d) 30 N, 50 N, 40 N

68 A block slides down on an inclined surface of inclination 30° with the horizontal. Starting from rest, it covers 8 m in the first two seconds. The coefficient of friction is (Take, $g = 10 \text{ ms}^{-2}$)

(a) $\frac{1}{5\sqrt{3}}$
(b) $\frac{1}{2\sqrt{3}}$
(c) $\frac{1}{\sqrt{3}}$
(d) $\frac{\sqrt{3}}{5}$

69 Two blocks are in contact on a frictionless table. One has mass m and the other $2m$. Same force F is applied from left and right on m and $2m$. The ratio of contact force between the blocks in the two cases will be

(a) 1 : 4
(b) 1 : 2
(c) 1 : 1
(d) 2 : 3

70 A block rests on a rough inclined plane making an angle of 30° with the horizontal. The coefficient of static friction between the block and the plane is 0.8. If the frictional force on the block is 10 N, the mass of the block (in kg) is (Take, $g = 10 \text{ m/s}^2$)

(a) 2.0
(b) 4.0
(c) 1.6
(d) 2.5

LEVEL 02 Objective Questions

71 A trolley is being pulled up on a smooth inclined plane by a man sitting on it (as shown in figure). He applies a force of 250 N. If the combined mass of the man and trolley is 100 kg, the acceleration of the trolley will be ($\sin 15° = 0.26$) (Take, $g = 10 \text{ m/s}^2$)

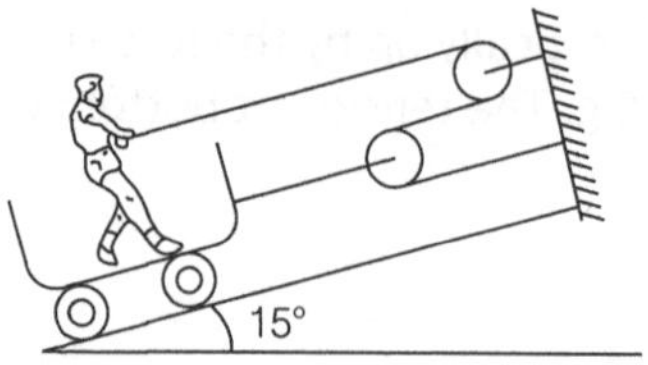

(a) 2.4 m/s^2
(b) 9.4 m/s^2
(c) 6.9 m/s^2
(d) 4.9 m/s^2

72 A lift is accelerating up with acceleration $g/2$ as shown in the figure. Find the reading of the spring balance present in the lift. (Take, $g = 10\text{ m/s}^2$)

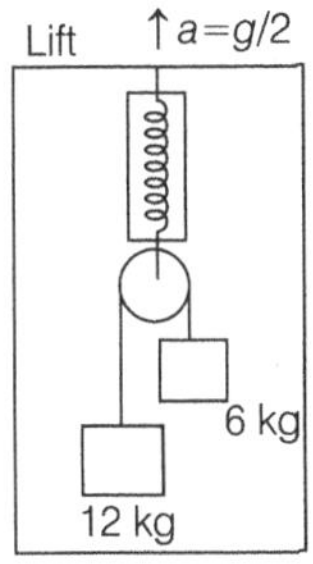

(a) 18 kg
(b) 16 kg
(c) 24 kg
(d) 36 kg

73 In figure, acceleration of mass m_1 is (all surfaces are smooth)

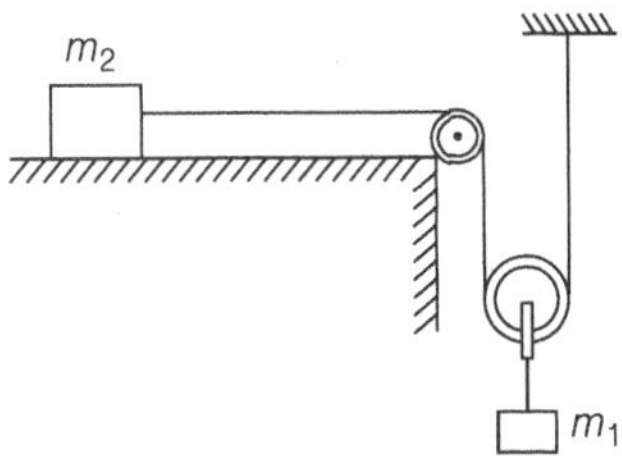

(a) $\dfrac{m_1 g}{m_1 + m_2}$
(b) $\dfrac{m_2 g}{m_1 + 4m_2}$
(c) $\dfrac{m_1 g}{m_1 + 4m_2}$
(d) $\dfrac{m_1 g}{4m_1 + m_2}$

74 A block of mass M slides down a rough inclined surface of inclination θ and reaches the bottom at speed v. However, if it slides down a smooth inclined surface of same length and same inclination, it reaches the bottom with speed kv. Coefficient of friction between the block and the rough incline is

(a) $\sin\theta\ (k^2 - 1)$
(b) $\left(\dfrac{k^2 - 1}{k^2}\right)\tan\theta$
(c) $\left(\dfrac{k^2 + 1}{k^2}\right)\cos\theta$
(d) $\dfrac{k^2 - 1}{k^2}$

75 In the figure shown, if mass of the rope is 2 kg, then tension at the mid-point of the rope is

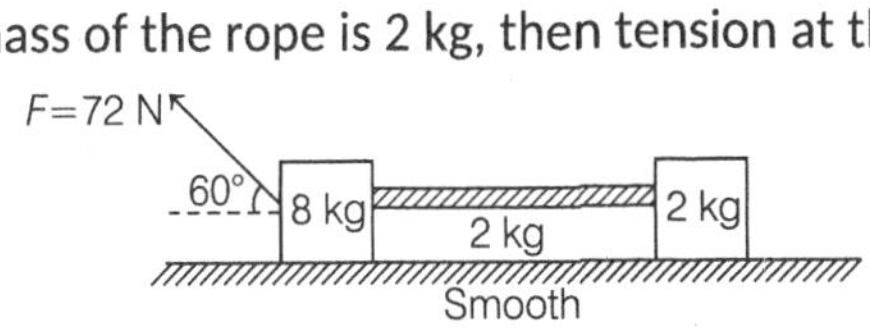

(a) 12 N
(b) 9 N
(c) 6 N
(d) 18 N

76 In the given diagram with what force must the man pull the rope to hold the plank in position? Mass of the man is 80 kg, neglect the weight of rope, plank and pulley. (Take, $g = 10\text{ m/s}^2$)

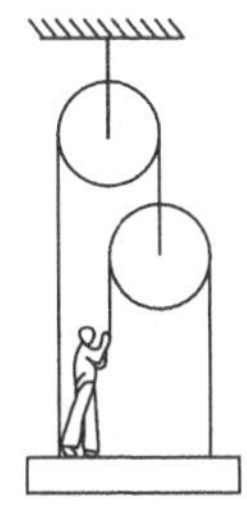

(a) 200 N
(b) 300 N
(c) 600 N
(d) 266.66 N

77 A train having 60 compartments is pulled by engine with a force 6×10^4 N. If mass of each compartment is 4000 kg, the tension in the coupling between 40th and 41st compartment is
(a) 4×10^4 N (b) 3×10^4 N (c) 2.5×10^4 N (d) 2×10^4 N

78 A block is placed on the top of a smooth inclined plane of inclination θ kept on the floor of a lift. When the lift is descending with a retardation a, the block is released. The acceleration of the block relative to the incline is
(a) $g\sin\theta$ (b) $a\sin\theta$ (c) $(g-a)\sin\theta$ (d) $(g+a)\sin\theta$

79 In the figure shown, 100 kg block is moving up with constant velocity, then tension at point P is (Take, $g = 9.8\text{ m/s}^2$)

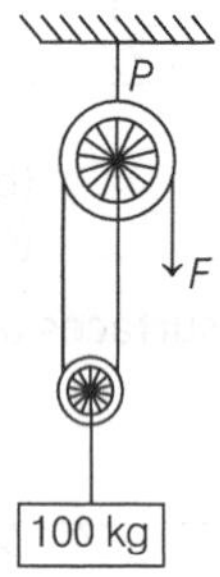

(a) 1330 N (b) 490 N (c) 1470 N (d) 980 N

80 If a ladder weighing 250 N is placed against a smooth vertical wall. The coefficient of friction between ladder and the floor is 0.3, then what is the maximum force of friction available at the point of contact between the ladder and the floor?
(a) 75 N (b) 50 N (c) 35 N (d) 25 N

81 A uniform horizontal meter scale of mass M is suspended by two vertical strings attached to its two ends. A block of mass 3m is placed on the 60 cm mark. The tensions in the two strings are in the ratio
(a) 17 : 19 (b) 19 : 21 (c) 17 : 23 (d) 19 : 23

82 A uniform rod of length l and mass M is hung from two strings of equal length from a ceiling as shown. Tension in the two strings is

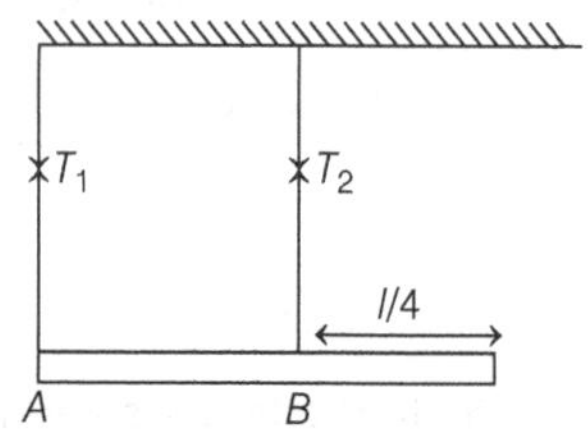

(a) $T_1 = \frac{Mg}{3}, T_2 = \frac{2Mg}{3}$ (b) $T_1 = T_2 = \frac{Mg}{2}$
(c) $T_1 = \frac{2Mg}{3}, T_2 = \frac{Mg}{3}$ (d) $T_1 = Mg, T_2 = 0$

83 An empty plastic box of mass m is found to accelerate up at the rate of $g/6$ when placed deep inside water. How much sand should be put inside the box, so that it may accelerate down at the rate of $g/6$?
(a) $\frac{2m}{3}$ (b) $\frac{2m}{5}$ (c) $\frac{3m}{4}$ (d) $\frac{3m}{5}$

84 A block slides down an inclined plane of inclination θ with constant velocity. It is then projected up the inclined plane with an initial speed u. The distance up the plane, it will move before coming to rest is
(a) $\frac{u^2}{g}$ (b) $\frac{u^2}{g\sin\theta}$ (c) $\frac{u^2}{2g\sin\theta}$ (d) $\frac{u^2}{4g\sin\theta}$

85 A 10 kg stone is suspended with a rope of breaking strength 30 kg-wt. The minimum time in which the stone can be raised through a height 10 m starting from rest is (Take, $g = 10$ N/kg)
(a) 0.5 s (b) 1.0 s (c) $\sqrt{\frac{2}{3}}$ s (d) 2.0 s

86 Find the acceleration of block B.

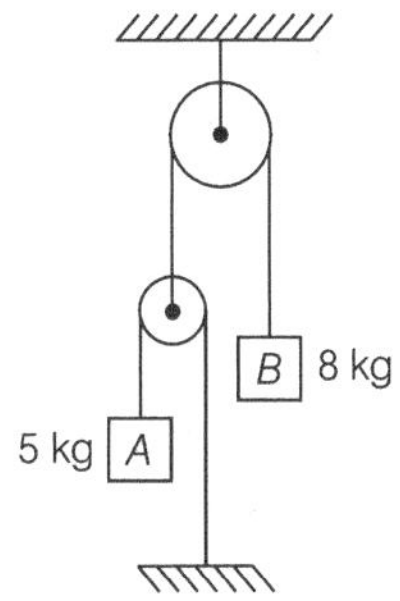

(a) 0 (b) $\frac{5}{2}$ m/s^2 (c) $\frac{5}{7}$ m/s^2 (d) $\frac{5}{14}$ m/s^2

87 In the system shown in figure, all surfaces are smooth, pulley and string are massless. The string between two pulley and between pulley and block of mass $5m$ is parallel to inclined surface. If system is released from rest, then acceleration of the block is

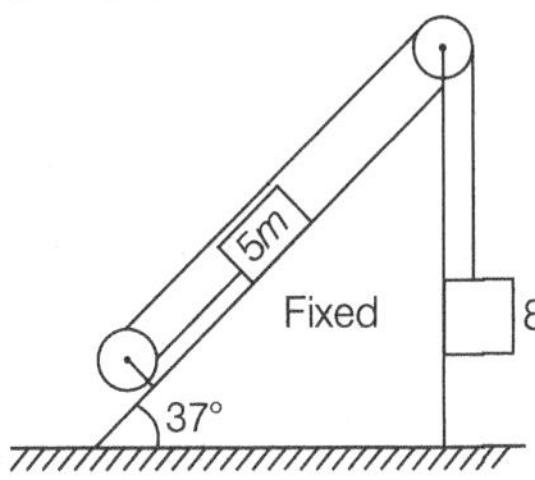

(a) $\frac{11g}{13}$ (b) $\frac{8g}{13}$ (c) $\frac{5g}{13}$ (d) None of these

88 A block is placed on a rough plank moving with velocity $3\hat{j}$ and velocity of block relative to plank is $3\hat{i}+4\hat{j}$, find direction of kinetic friction force acting on the block.

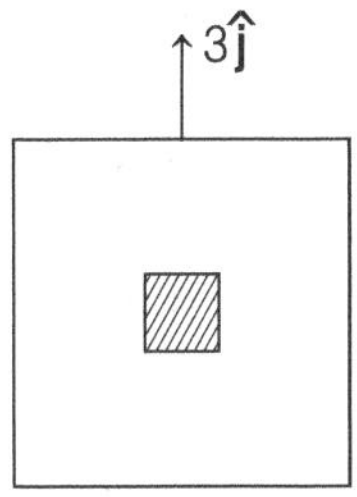

(a) $\frac{-(3\hat{i}+\hat{j})}{\sqrt{10}}$ (b) $\frac{-(3\hat{i}+4\hat{j})}{5}$ (c) $\frac{(3\hat{i}+4\hat{j})}{5}$ (d) $\frac{3\hat{i}+\hat{j}}{\sqrt{10}}$

89 If all the pulleys are massless and strings are ideal, then tension in the string connected with 2 kg block is (All surfaces are smooth)

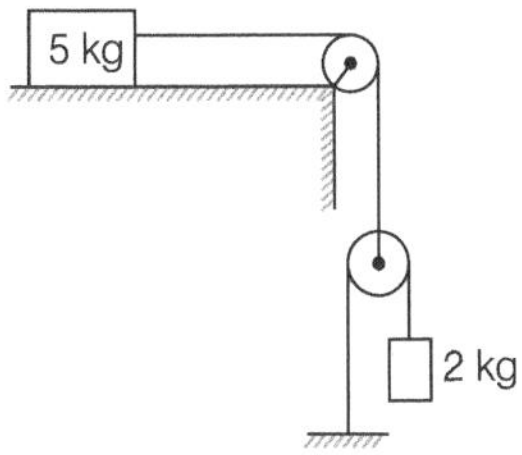

(a) $\frac{50}{8}$ N (b) $\frac{100}{13}$ N (c) $\frac{60}{3}$ N (d) $\frac{120}{7}$ N

90 A 14 kg block is hanged using a system of pulleys as shown in figure. Tension in string connecting ceiling and topmost pulley is

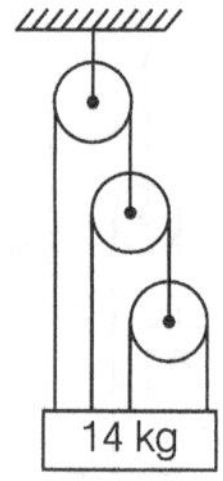

(a) 17.5 N (b) 70 N
(c) 140 N (d) 280 N

91 Calculate the acceleration of the blocks A and B.

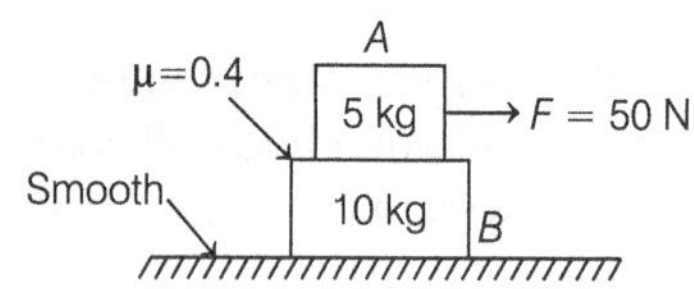

(a) $a_A = 2\ \text{m/s}^2$, $a_B = 6\ \text{m/s}^2$
(b) $a_A = 6\ \text{m/s}^2$, $a_B = 2\ \text{m/s}^2$
(c) $a_A = a_B = 3\ \text{m/s}^2$
(d) $a_A = a_B = 1\ \text{m/s}^2$

92 A block is kept on a rough inclined plane whose angle of inclination is greater than the angle of repose, then find the minimum force F required, which is applied normal to the incline that will keep it in equilibrium.

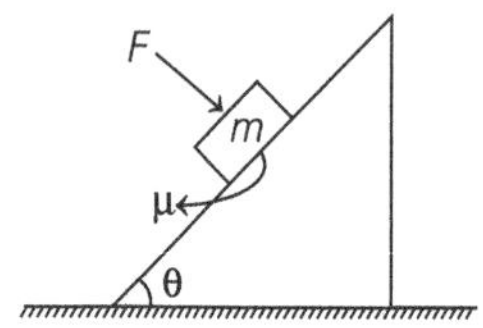

(a) $mg(\sin\theta - \mu\cos\theta)$ (b) $mg(\sin\theta + \mu\cos\theta)$
(c) $\frac{mg}{\mu}(\sin\theta - \mu\cos\theta)$ (d) $mg\frac{(\sin\theta - \mu\cos\theta)}{(\cos\theta + \mu\sin\theta)}$

93 The 50 kg homogeneous smooth sphere rests on the 30° incline A and bears against the smooth vertical wall B. Calculate the contact forces at A and B.

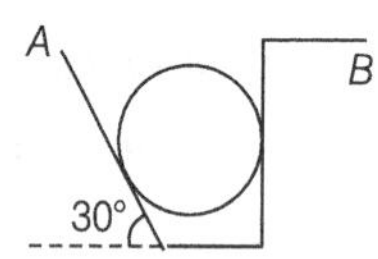

(a) $N_B = \frac{1000}{\sqrt{3}}$ N, $N_A = \frac{500}{\sqrt{3}}$ N (b) $N_A = \frac{1000}{\sqrt{3}}$ N, $N_B = \frac{500}{\sqrt{3}}$ N
(c) $N_A = \frac{100}{\sqrt{3}}$ N, $N_B = \frac{500}{\sqrt{3}}$ N (d) $N_A = \frac{1000}{\sqrt{3}}$ N, $N_B = \frac{50}{\sqrt{3}}$ N

94 In the given diagram, with what force must the man pull the rope to hold the plank in position? Weight of the man is 60 kg. Neglect the weight of plank, rope and pulley.

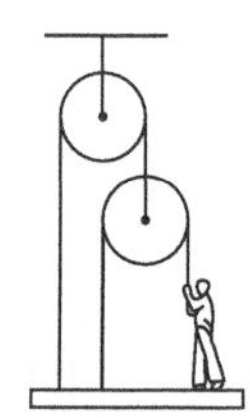

(a) 150 N (b) 300 N
(c) 600 N (d) 120 N

95 A block of mass 1 kg is placed on a rough inclined plane at angle $\theta = 30°$ with horizontal. The block is connected with a string as shown in figure. If $\mu_s = 3/4$, find the tension in string.

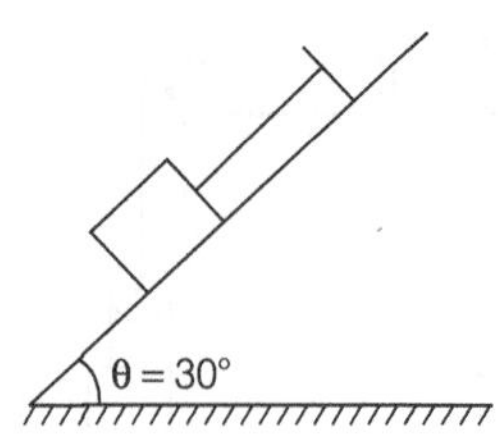

(a) 5 N (b) 0 N
(c) 10 N (d) 2 N

96 A smooth ring A can slide on a fixed horizontal rod as shown. The pulley is fixed. If some instant velocity of ring is v, find the velocity of block at that instant.

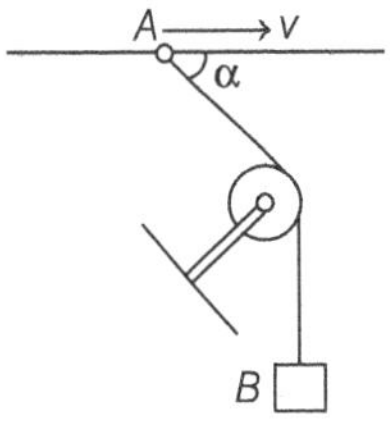

(a) $\frac{v}{\cos\alpha}$ (b) $v \cos \alpha$

(c) $v \sin\alpha$ (d) $\frac{v}{\sin\alpha}$

97 In the set up shown, find acceleration of the block C. Given, $a_A = 2 \text{m/s}^2$ and a_B with respect to pulley is 1m/s^2 downwards.

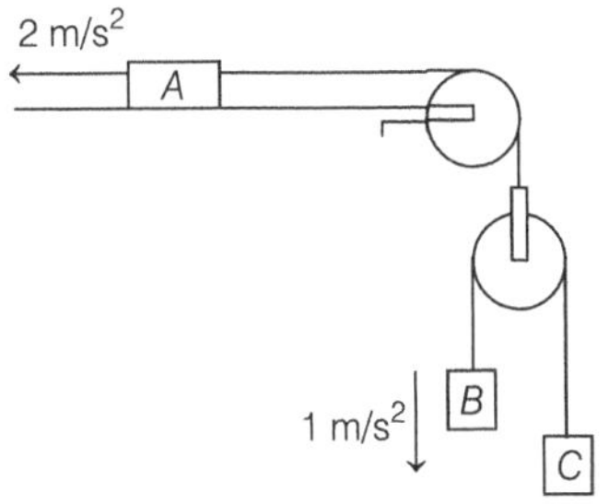

(a) $3 \text{m/s}^2 \uparrow$ (b) $3 \text{m/s}^2 \downarrow$ (c) $5 \text{m/s}^2 \uparrow$ (d) $5 \text{m/s}^2 \downarrow$

98 Figure shows four blocks that are being pulled along a smooth horizontal surface. The masses of the blocks and tension in one cord are given. The pulling force F is

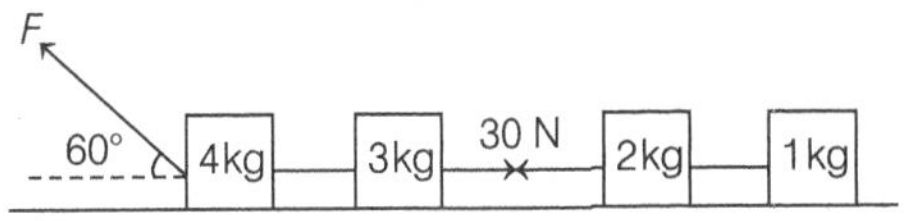

(a) 50 N (b) 100 N (c) 125 N (d) 200 N

99 In each case $m_1 = 4$ kg and $m_2 = 3$ kg. If a_1 and a_2 are the respective accelerations of the block m_1 in given situations, then (neglect friction)

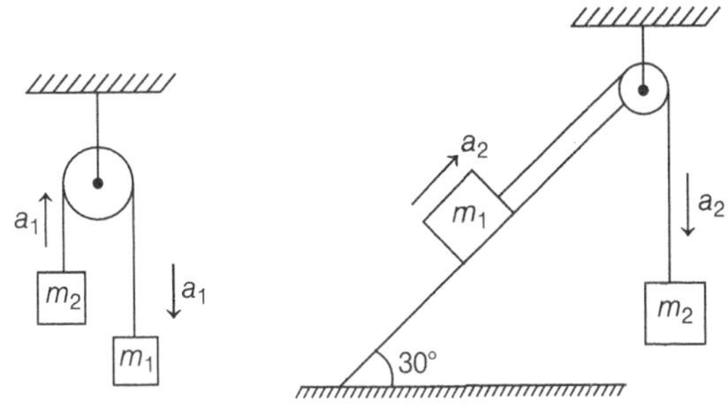

(a) $a_1 < a_2$ (b) $a_1 > a_2$ (c) $a_1 = a_2$ (d) $a_1 = a_2 = 0$

100 If the system shown in figure is in equilibrium, then value of T_1 is (Take, $g = 10 \text{ ms}^{-2}$)

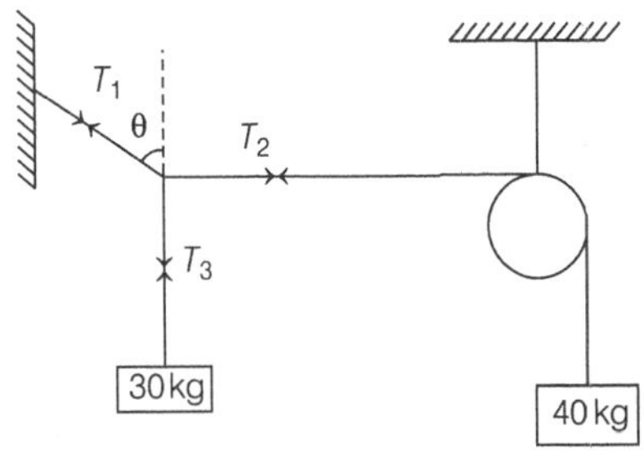

(a) 300 N (b) 400 N (c) 500 N (d) 700 N

101 In figure, blocks A and B move with velocities v_1 and v_2 along horizontal direction. Find the ratio of v_1/v_2.

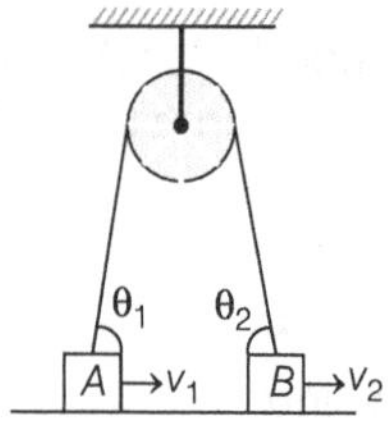

(a) $\frac{\sin\theta_1}{\sin\theta_2}$ (b) $\frac{\sin\theta_2}{\sin\theta_1}$
(c) $\frac{\cos\theta_2}{\cos\theta_1}$ (d) $\frac{\cos\theta_1}{\cos\theta_2}$

102 Block B has mass m and is released from rest, when it is on top of wedge A, which has a mass 3m. Determine the tension in cord CD while B is sliding down A. Neglect friction.

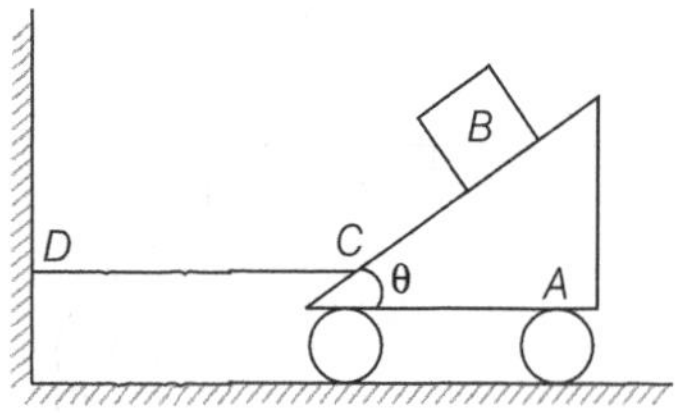

(a) $2\,mg\cos\theta$ (b) $\frac{mg}{2}\cos\theta$ (c) $\frac{mg}{2}\sin 2\theta$ (d) $mg\sin 2\theta$

103 In figure, a person want to rise a block lying on the ground to a height h. In both the cases, if the time required is same, then in which case, he has to exert more force? Assume pulleys and strings light.

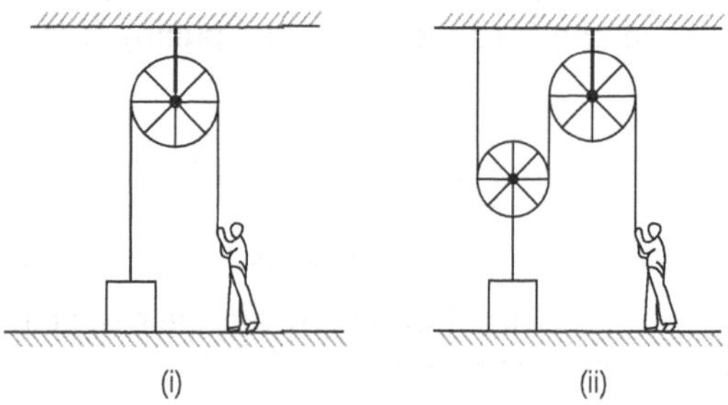

(a) (i) (b) (ii)
(c) Same in both (d) Cannot be determined

104 A body slides on an inclined plane. If height of inclined plane is h and length is l and angle of inclination is θ, then time is taken for travelling from upper point to lower point (Neglect friction)

(a) $\sqrt{\frac{2h}{g}}$ (b) $\sqrt{\frac{2l}{g}}$ (c) $\frac{1}{\sin\theta}\sqrt{\frac{2h}{g}}$ (d) $\sin\theta\sqrt{\frac{2h}{g}}$

105 In the figure given below, the correct relation between acceleration a_1, a_2 and a_3 is

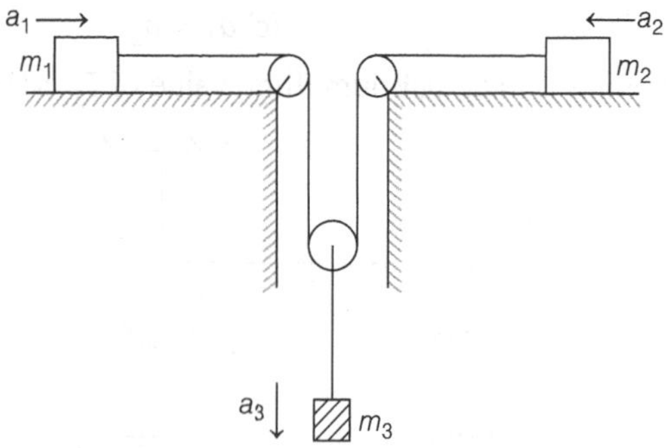

(a) $2a_1 + 2a_2 = a_3$ (b) $a_1 + a_2 = 2a_3$ (c) $a_1 + a_2 = a_3$ (d) $a_1 = a_2 = a_3$

106 A block of mass m is lying on an inclined plane. The coefficient of friction between the plane and the block is μ. The minimum force (F) required to move the block up the inclined plane will be

(a) $mg\sin\theta + \mu mg\cos\theta$ (b) $mg\cos\theta - \mu mg\sin\theta$

(c) $mg\sin\theta - \mu mg\cos\theta$ (d) $mg\cos\theta + \mu mg\sin\theta$

107 In a given figure, two masses m_1 and m_2 ($m_2 > m_1$) are at rest in equilibrium position. Find the tension in string AB.

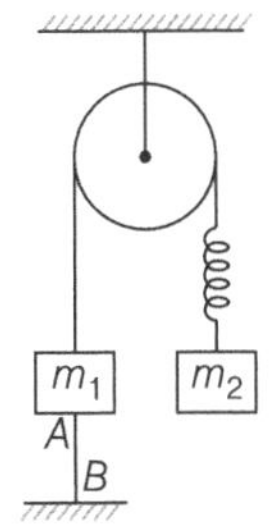

(a) $m_1 g$ (b) $m_2 g$ (c) $(m_1 + m_2)g$ (d) $(m_2 - m_1)g$

108 Two blocks of masses 2 kg and 3 kg are placed on the floor of a lift as shown in the figure. The normal reaction between 3 kg block and the floor of the lift is 40 N. Which of the following statement(s) is/are correct? (Take, $g = 10 \text{ m/s}^2$)

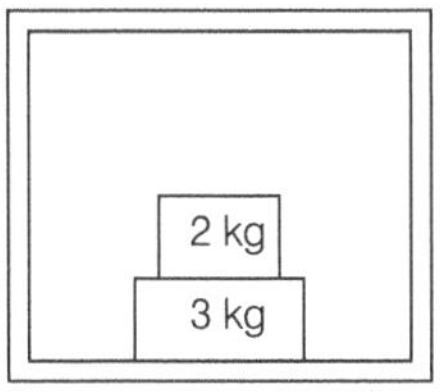

(a) The lift is moving with an upward acceleration.

(b) The lift is moving with a downward acceleration.

(c) The lift is either at rest of moving with uniform speed.

(d) Force of normal reaction applied by the 2 kg block on the 3 kg block is 18 N.

109 Reading of the spring balance as shown in figure, is (assume strings and spring are ideal and neglect friction, $g = 10 \text{ m/s}^2$)

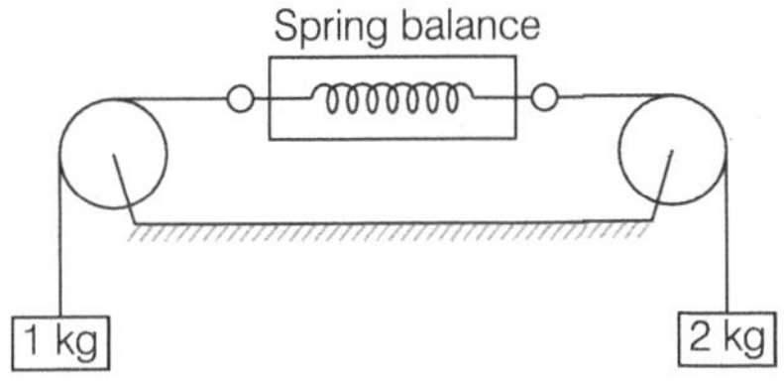

(a) 20 N (b) 10 N (c) $\frac{40}{3}$ N (d) zero

110 If acceleration of block B is 4 m/s^2 upward and that of C is 6 m/s^2 downward. Find acceleration of A.

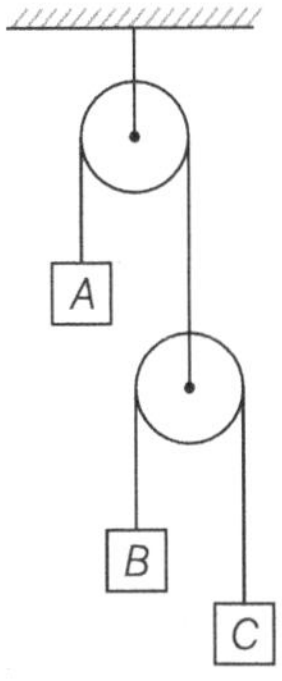

(a) 2 m/s^2 upward (b) 1 m/s^2 upward

(c) 1 m/s^2 downward (d) 2 m/s^2 downward

111 A mass of 6 kg is suspended by a rope of length 2 m from a ceiling. A force of 50 N is applied in horizontal direction at the mid-point of the rope. What is the angle between the rope and the vertical in equilibrium?

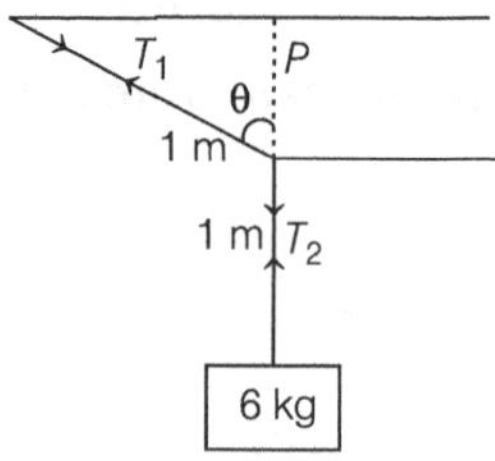

(a) $\tan^{-1}\left(\frac{4}{5}\right)$ (b) $\tan^{-1}\left(\frac{5}{4}\right)$ (c) $\tan^{-1}\left(\frac{5}{6}\right)$ (d) None of these

112 In the arrangement shown, block *A* of mass 15 kg is supported in equilibrium by the block *B*. Mass of the block *B* is [Assume pulleys are massless and smooth]

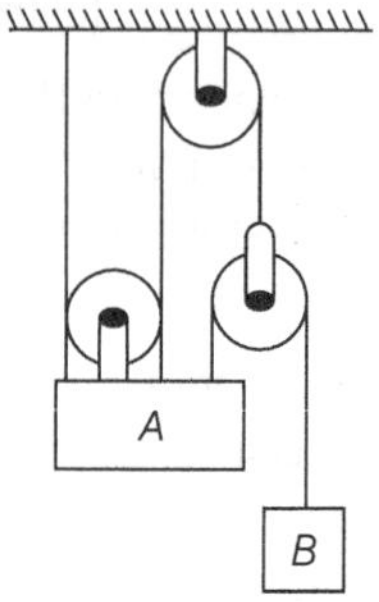

(a) 2 kg (b) 3 kg (c) 4 kg (d) 5 kg

113 Two identical blocks *A* and *B* each of mass 10 kg are joined together with spring of unknown stiffness as shown in figure. The coefficient of friction between each block and surface is 0.1. The block *B* is pulled with a force of 40 N and the block *A* accelerates with 1.2 ms^{-2}. The acceleration of block *B* is

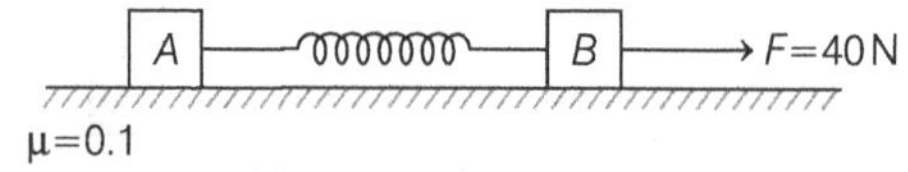

(a) 1.2 ms^{-2} (b) 1.8 ms^{-2}
(c) 0.8 ms^{-2} (d) 1.5 ms^{-2}

114 A block of mass 5 kg is placed on a wedge having inclination of 37° with the horizontal. Coefficient of friction between block and wedge is 0.8. Then, select the correct statement.

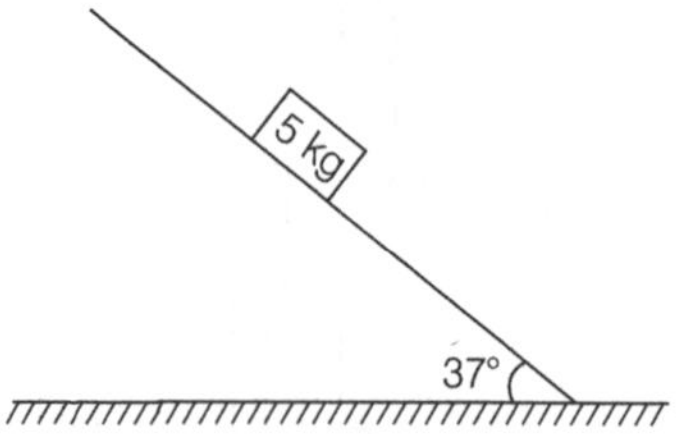

(a) Friction force between block and the wedge is 32 N.
(b) Friction force between block and the wedge is 30 N.
(c) Contact force between block and wedge is 80 N.
(d) Contact force between block and wedge is 40 N.

115 A block of mass 4 kg is placed on a rough horizontal plane. A time dependent force $F = kt^2$ acts on the block, where $k = 2\ \text{N/s}^2$. Coefficient of friction $\mu = 0.8$. Force of friction between block and the plane at $t = 2\text{s}$ is

(a) 8 N (b) 4 N
(c) 2 N (d) 32 N

116 In the balance machine shown in the figure, which arm will move downward after the system is released from rest? (Assuming string, arm of balance machine and the pulley to be light)

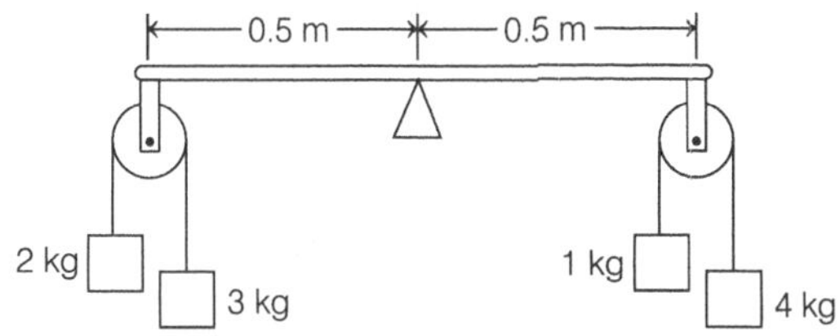

(a) Left (b) Right (c) Remains stationary (d) Data insufficient

117 Figure shows a boy on a horizontal platform A on a smooth horizontal surface, holding a rope attached to a box B. Boy pulls the rope with a constant force of 50 N. The combined mass of platform A and boy is 250 kg and that of box B is 500 kg. The velocity of A relative to the box B, 5 s after the boy on A begins to pull the rope, will be

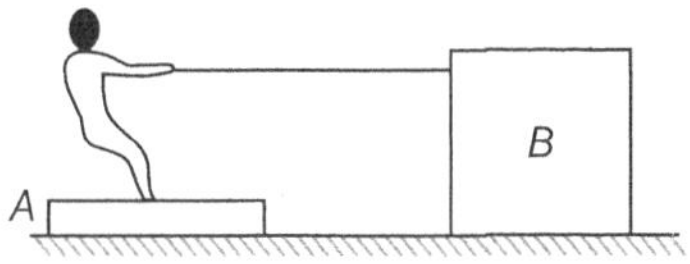

(a) 1 m/s (b) 1.5 m/s (c) 2 m/s (d) 0.5 m/s

118 Friction is absent everywhere. Find acceleration of blocks A, B and C as shown instant.

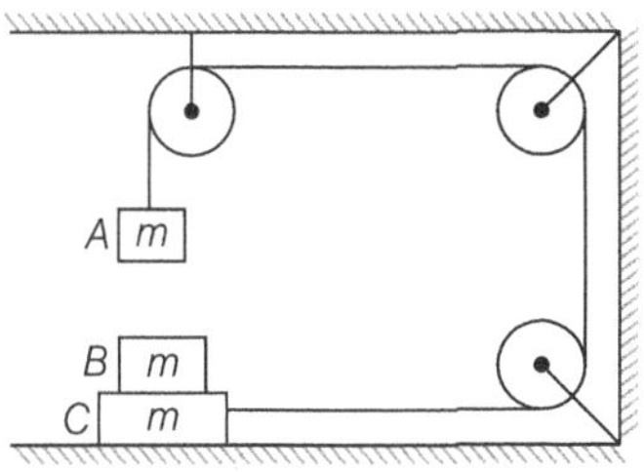

(a) $a_A = \frac{g}{2};\ a_B = 0;\ a_C = \frac{g}{2}$ (b) $a_A = g;\ a_B = 0;\ a_C = g$
(c) $a_A = \frac{g}{3};\ a_B = \frac{g}{3};\ a_C = \frac{g}{3}$ (d) $a_A = \frac{g}{3};\ a_B = 0;\ a_C = \frac{g}{3}$

119 A block of mass m is suspended through a vertical spring of spring constant k as shown in the figure. If the block is pulled down by a distance $\frac{mg}{4k}$ from equilibrium position and released, then the initial acceleration of the block will be

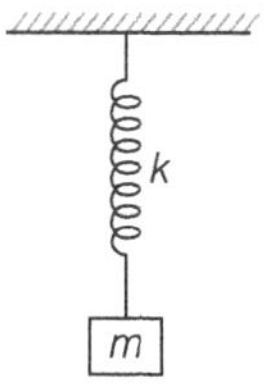

(a) $g/3$ (b) $g/4$
(c) $g/5$ (d) g

120 A man is hanging from the free end of a massless rope of length 40 m. The mass of the man is 20 kg and the maximum tension that the rope can bear is 300 N. Then, starting from rest, the minimum time in which man can reach the other end of the rope is (Take, $g = 10\ \text{m/s}^2$)

(a) 2 s (b) 4 s
(c) 6 s (d) 8 s

121 A block of mass 4 kg is kept on a smooth inclined plane as shown and a horizontal force is applied on it. Which of the following forces can push the block up the inclined plane?

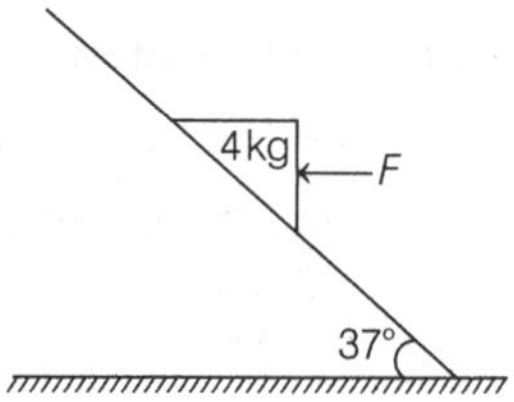

(a) 35 N
(b) 20 N
(c) 15 N
(d) 24 N

122 A block of mass 8 kg is at rest on level ground for which $\mu = 0.6$. A force is applied on it in horizontal direction. The force varies with time as shown in the graph. After what time does the block start moving?

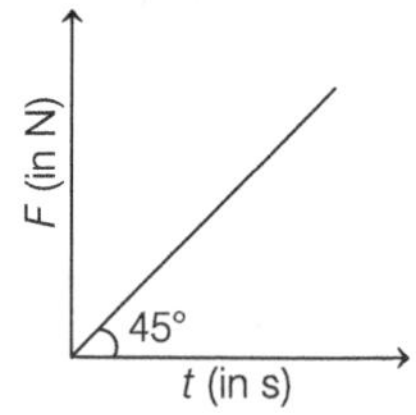

(a) $t = 30$ s (b) $t = 24$ s
(c) $t = \frac{48}{\sqrt{2}}$ s (d) $t = 48$ s

123 A smooth cylinder of mass m and radius R is resting on two corner edges A and B as shown in figure. The relation between normal reaction at the edges A and B is

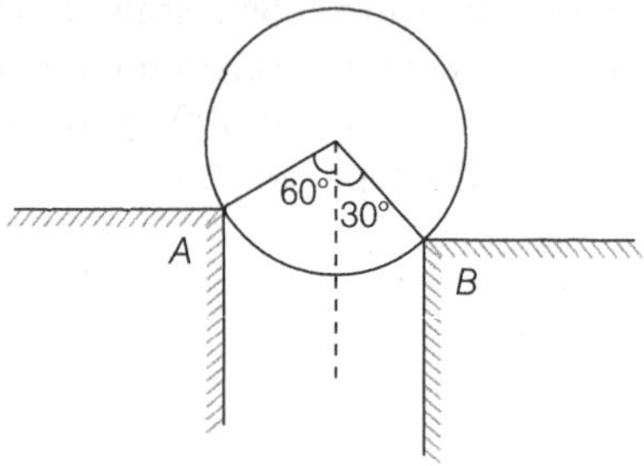

(a) $N_A = \sqrt{2}\, N_B$ (b) $N_B = \frac{2\sqrt{3}\, N_A}{5}$ (c) $N_A = \frac{N_B}{2}$ (d) $N_B = \sqrt{3}\, N_A$

124 In the situation shown in figure, surfaces are frictionless. Reading of the spring balance will be

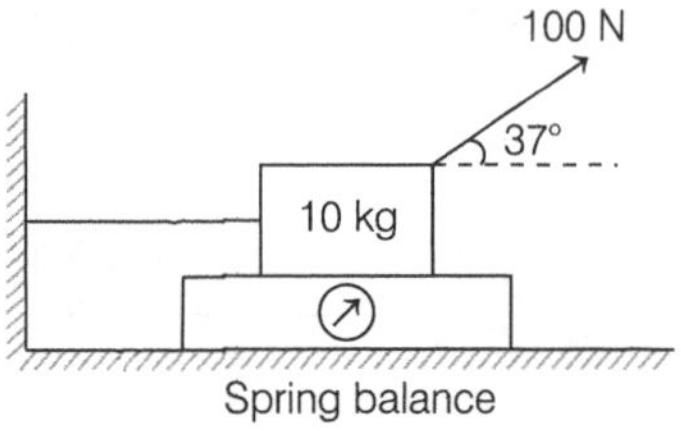

(a) 100 N (b) 40 N (c) 180 N (d) 120 N

125 Two blocks A and B are connected by a light inextensible string passing over a fixed smooth pulley as shown in figure. The coefficient of friction between the block A and the horizontal table is $\mu = 0.5$. If the block A is just to slip, find the ratio of the masses of the blocks.

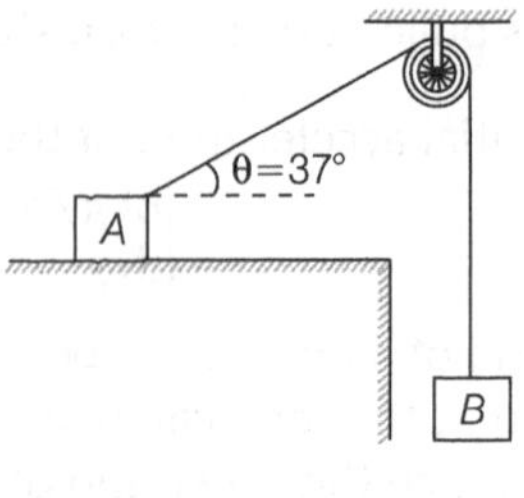

(a) $\frac{11}{5}$ (b) 2 (c) $\frac{17}{5}$ (d) 3

126 Figure shows two blocks in contact sliding down an inclined surface of inclination 30°. The friction coefficient between the block of mass 2 kg and the incline is $\mu_1 = 0.20$ and that between the block of mass 4 kg and the incline is $\mu_2 = 0.30$. Find the acceleration of 2.0 kg block. (Take, $g = 10\ \text{ms}^{-2}$)

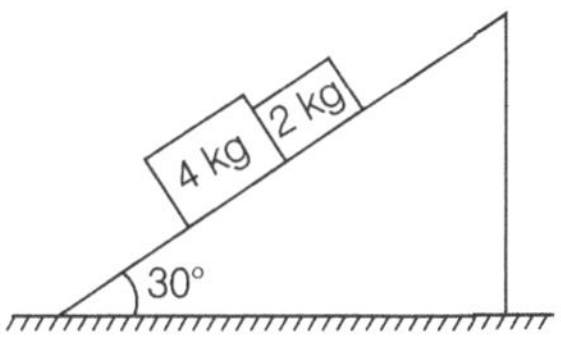

(a) $5.4\ \text{m/s}^2$ (b) $4.5\ \text{m/s}^2$ (c) $2.7\ \text{m/s}^2$ (d) 0

127 A block pressed against the vertical wall is in equilibrium. The minimum coefficient of friction is

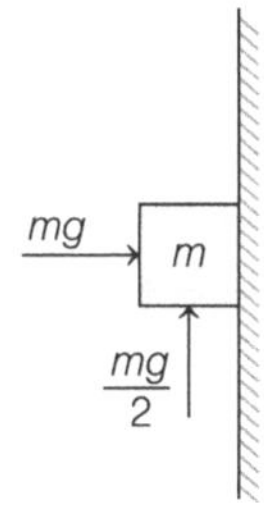

(a) 0.4 (b) 0.2 (c) 0.5 (d) None of these

128 Calculate the acceleration of the mass 12 kg shown in the figure. Also, calculate the tension in the string connecting the 12 kg mass. The strings are weightless and inextensible, the pulleys are weightless and frictionless. All surface are smooth.

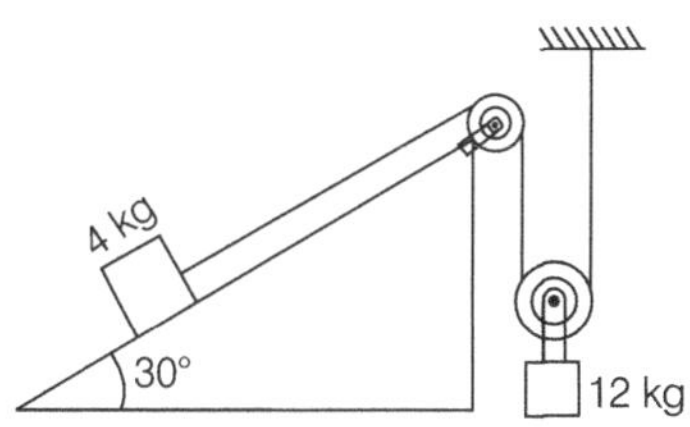

(a) $\frac{g}{7}\ \text{m/s}^2, \frac{56g}{5}\ \text{N}$ (b) $\frac{2g}{7}\ \text{m/s}^2, \frac{60g}{7}\ \text{N}$ (c) $\frac{g}{3}\ \text{m/s}^2, \frac{5}{56g}\ \text{N}$ (d) $\frac{g}{14}\ \text{m/s}^2, \frac{5}{56g}\ \text{N}$

129 A conveyor belt is moving at a constant speed of 2 m/s. A box is gently dropped on it. The coeffcient of friction between them is $\mu = 0.5$. The distance that the box will move relative to belt before coming to rest on it, (Take, $g = 10\ \text{ms}^{-2}$) is

(a) 0.4 m (b) 1.2 m (c) 0.6 m (d) zero

130 A bar of mass m_1 is placed on a plank of mass m_2 which rests on a smooth horizontal plane. Coefficient of friction between two blocks is μ. The plank is subjected to a horizontal force F depending on time t as $F = at$, where a is constant. The moment of time t_0 at which the plank starts sliding is

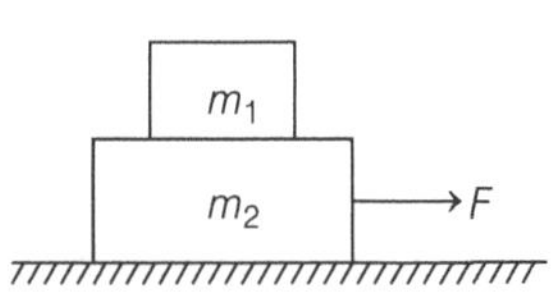

(a) $\frac{a\mu g}{m_1 + m_2}$ (b) $\frac{(m_1 + m_2)\mu g}{a}$ (c) $\frac{(m_1 + m_2)g}{\mu a}$ (d) $\frac{\mu a}{(m_1 + m_2)g}$

131 Three boys are pushing horizontally a box placed on a horizontal table. One is pushing towards north with $5\sqrt{3}$ N force. The second is pushing towards east and the third pushes with a force of 10 N, such that the box is in equilibrium. Find the magnitude of the force, second boy is applying on the box is

(a) 5 N (b) $5\sqrt{3}$ N
(c) 10 N (d) $10\sqrt{3}$ N

132 Two identical heavy spheres of equal mass are placed on a smooth cup of radius $3r$, where r is radius of each sphere. Then, the ratio of reaction force between cup and any sphere to reaction force between two sphere is

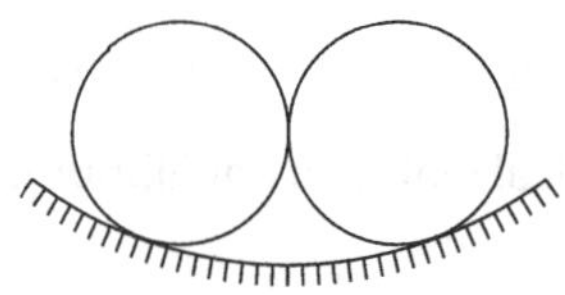

(a) 1 (b) 2
(c) 3 (d) $\sqrt{3}$

133 In which of the following cases, the contact force between A and B is maximum ? (Take, $m_A = m_B = 1$ kg, $g = 10$ ms^{-2})

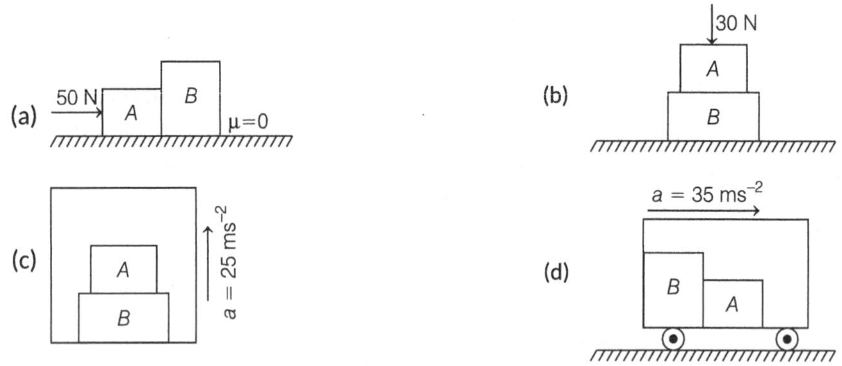

134 A block of mass 2 kg is pushed normally against a rough vertical wall with a force of 40 N, coefficient of static friction being 0.5. Another horizontal force of 15 N is applied on the block in a direction parallel to the wall. Find the frictional force exerted by the wall on the block and acceleration of the block.

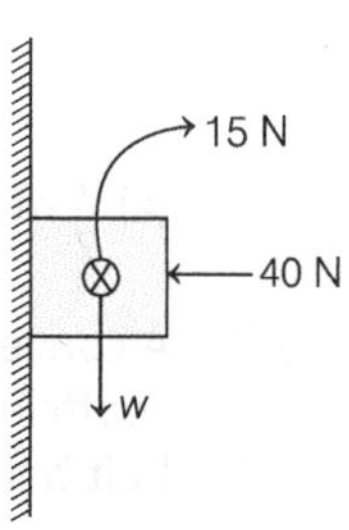

(a) 20 N, 2 m/s^2
(b) 25 N, 2.5 m/s^2
(c) 20 N, 2.5 m/s^2
(d) 25 N, 2 m/s^2

135 In figure system is at rest. The tension in the rope (rope is light) is

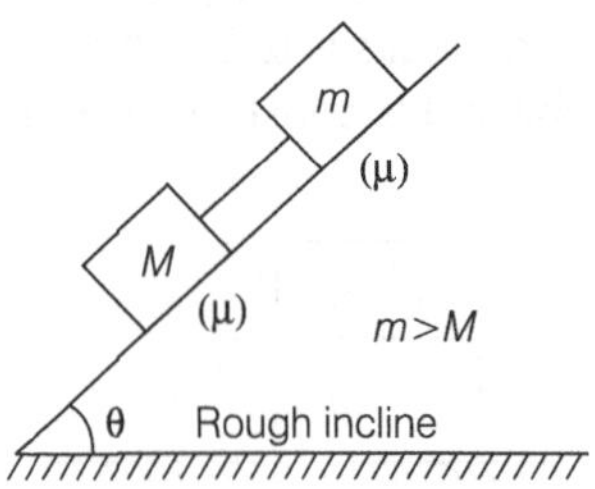

(a) $(M+m)\,g\sin\theta$ (b) $(M+m)\,g\sin\theta - \mu\,mg\cos\theta$
(c) zero (d) $(M+m)\,g\cos\theta$

136 A bead slides on a frictionless straight rod fixed between points A and B of a vertical circular loop of radius r as shown in the figure. Line BC is a vertical diameter.

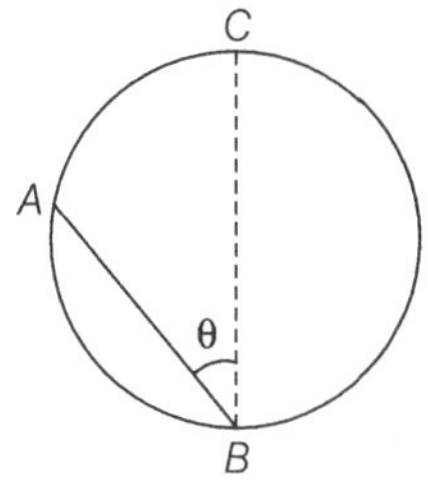

Denoting acceleration due to gravity by g, which of the following is correct expression for the time taken by the bead to slide from A to B.

(a) $\sqrt{\frac{4r}{g}}$ (b) $\sqrt{\frac{4r}{g\cos\theta}}$

(c) $\sqrt{\frac{4r\tan\theta}{g}}$ (d) $\sqrt{\frac{4r\cot\theta}{g}}$

137 A conveyor belt is moving horizontally at a speed of 4 ms^{-1}. A box of mass 20 kg is gently laid on it. It takes 0.1 s for the box to come to rest. If the belt continues to move uniformly, then the distance moved by the box on the conveyor belt in above 0.1 s is

(a) zero (b) 0.2 m (c) 0.4 m (d) 0.8 m

138 In the set up shown, a 200 N block is supported in equilibrium with the help of strings and a spring. At point O, extension in the spring is 4 cm. Force constant of the spring is closest to (Take, $g = 10$ m/s^2)

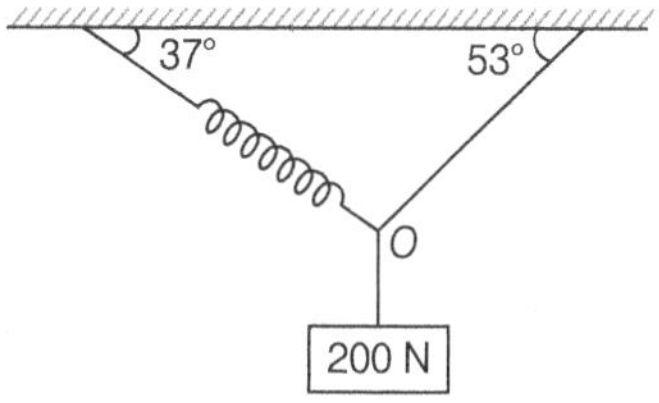

(a) 30 N/m (b) 2500 N/m
(c) 3000 N/m (d) 4000 N/m

139 A block of mass $2m$ hangs by an inextensible light string that passes over a smooth pulley. Another mass m hangs from the other end of the string and is connected to a vertical massless spring. The system is stationary. Just after the string is cut, the accelerations of m and $2m$ respectively are

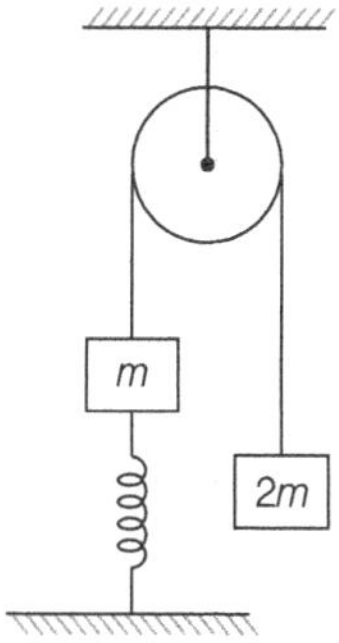

(a) 0, 0
(b) 0, $g\downarrow$
(c) $\frac{g}{2}\downarrow, g\downarrow$
(d) $2g\downarrow, g\downarrow$

140 In the shown arrangement, if f_A, f_B and T be the frictional forces on block A, block B and tension in the string respectively, then their values are

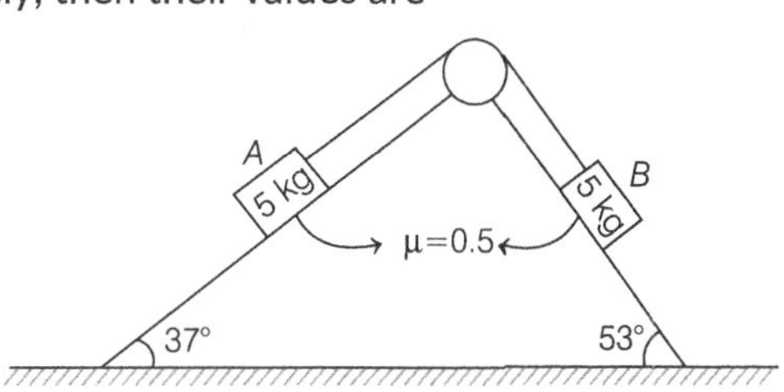

(a) $f_A = 20$ N, $f_B = 15$ N, $T = 25$ N (b) $f_A = 20$ N, $f_B = 10$ N, $T = 10$ N
(c) $f_A = 20$ N, $f_B = 15$ N, $T = 5$ N (d) $f_A = 15$ N, $f_B = 5$ N, $T = 20$ N

141 In shown situation, elevator is moving upward with acceleration of 5 m/s^2.

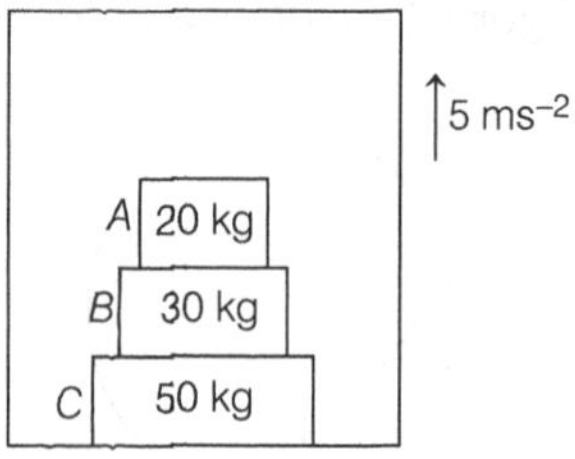

	Column-I		Column-II
(a)	Net force acting on *B*	(p)	150 N
(b)	Normal reaction between *A* and *B*	(q)	300 N
(c)	Normal reaction between *B* and *C*	(r)	450 N
(d)	Normal reaction between C and elevator	(s)	750 N
		(t)	1500 N

Codes

(a) (a)-p, (b)-q, (c)-s, (d)-t (b) (a)-q, (b)-q, (c)-r, (d)-t
(c) (a)-p, (b)-r, (c)-s, (d)-s (d) (a)-q, (b)-r, (c)-r, (d)-t

142 The ratio of tension in the string connected to the block of mass m_2 in figure (i) and figure (ii) respectively, is (friction is absent everywhere) [Take, $m_1 = 2$ kg, $m_2 = 8$ kg and $F = 50$ N]

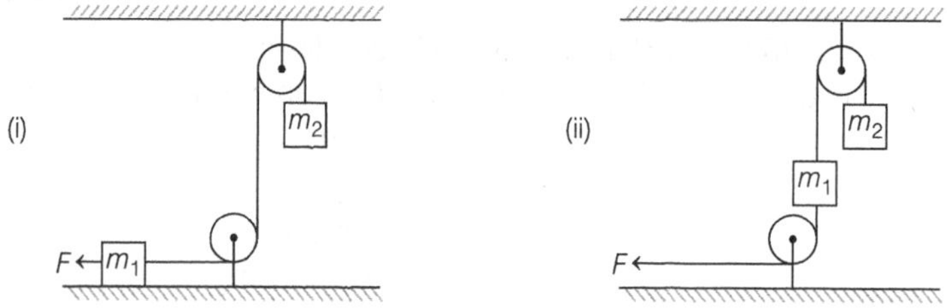

(a) 7 : 9 (b) 9 : 7 (c) 2 : 7 (d) 7 : 2

143 A block of mass 10 kg is held on a vertical wall of lift which is going upward with acceleration of 2 m/s^2. Find the minimum value of *F*, so that we can hold the block stationary with respect to lift. Friction coefficient between block and wall of lift is 0.5.

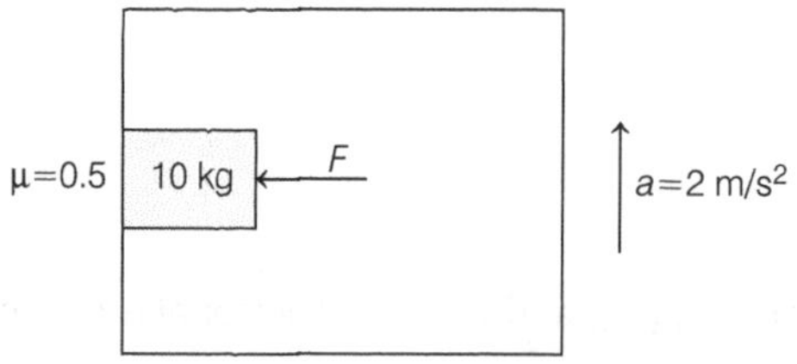

(a) 40 N (b) 50 N (c) 240 N (d) 60 N

144 A cubical block *A* of mass 10 kg and a block *B* of mass 2 kg are connected *via* string-pulley system as shown in the figure. If one end of the string is pulled with a force 20 N as shown, then the acceleration of block *A* will be (All surfaces are smooth. Pulley and strings are ideal)

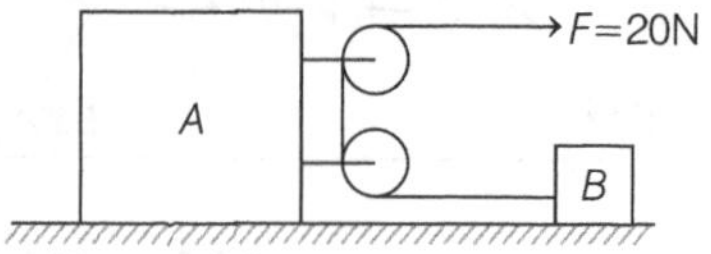

(a) 4 m/s^2 (b) 2 m/s^2 (c) 1 m/s^2 (d) 8 m/s^2

145 What is the minimum value of applied force F, so that both the block lifts up? Consider string to be ideal and pulley to be massless and frictionless.

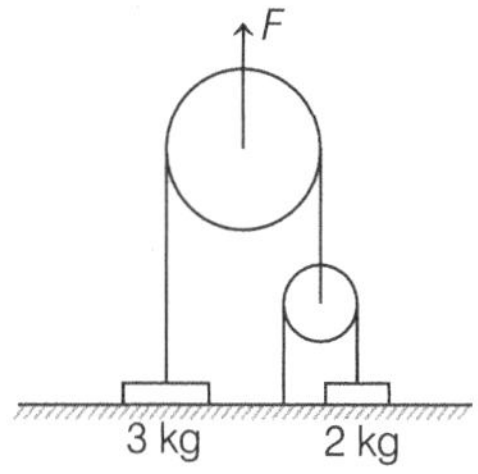

(a) $4g$
(b) $5g$
(c) $7g$
(d) $8g$

146 In figure shown, both blocks are released from rest. Length of 4 kg block is 2 m and of 1 kg is 4 m. Find the time (in seconds) they take to cross each other. Assume pulley to be light and string to be light and inextensible.

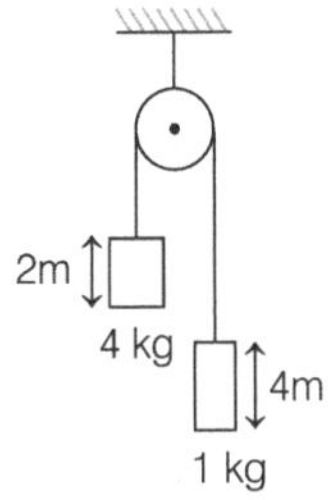

(a) 1 s (b) $\frac{2}{\sqrt{6}}$ s (c) 2 s (d) 3 s

147 Two steel balls A and B are placed inside a right circular cylinder of diameter 54 cm making contacts at points P, Q and R as shown. The radius $r_A = 12$ cm and $r_B = 18$ cm. The masses are $m_A = 15$ kg and $m_B = 60$ kg. The forces exerted by the floor at the point Q and the wall at R are, respectively. (Take, $g = 10$ m/s^2)

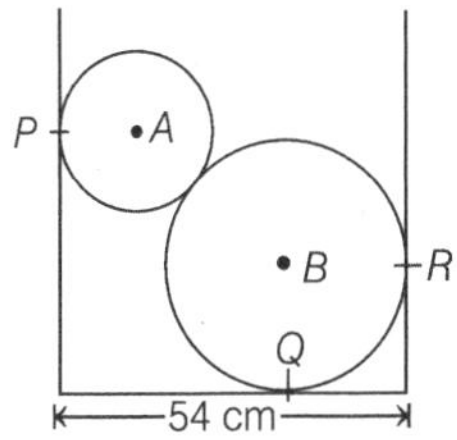

(a) 600 N, 150 N (b) 750 N, 150 N (c) 600 N, 200 N (d) 750 N, 200 N

148 A rock-climber (49 kg) is managing to be at rest between two vertical rocks pressing on rock A by his feet and the other B by his back. If the coefficient of friction between his shoes and wall A is 1.2 while between his back and wall B is 0.8 and limiting friction acts at all the contacts, then what per cent of the weight is supported by the frictional force on legs?

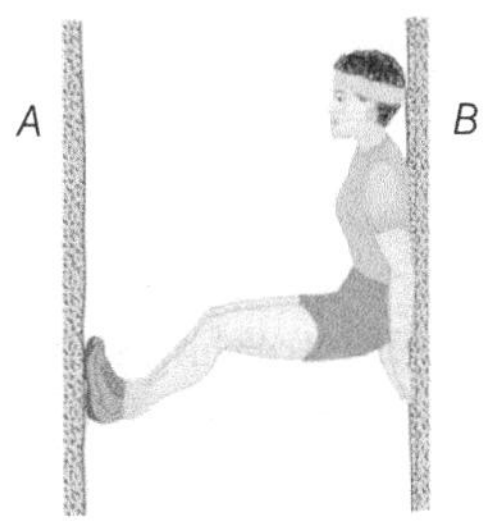

(a) 60% (b) 40% (c) 50% (d) None of these

149 Five forces $\mathbf{F}_1$, $\mathbf{F}_2$, $\mathbf{F}_3$, $\mathbf{F}_4$ and $\mathbf{F}_5$ are acting on a particle of mass 2 kg, so that it is moving with 4 m/s^2 in east direction. If $\mathbf{F}_1$ is removed, then the acceleration becomes 7 m/s^2 in north, then the acceleration of the block, if only $\mathbf{F}_1$ is acting will be

(a) 16 m/s^2 (b) $\sqrt{65}$ m/s^2 (c) $\sqrt{260}$ m/s^2 (d) $\sqrt{33}$ m/s^2

150 Figure shows a man of mass 50 kg standing on a light weighing machine kept in a box of mass 30 kg. The box is hanging from a pulley fixed to the ceiling through a light rope, the other end of which is held by the man himself. If the man manages to keep the box at rest, the weight shown by the machine is

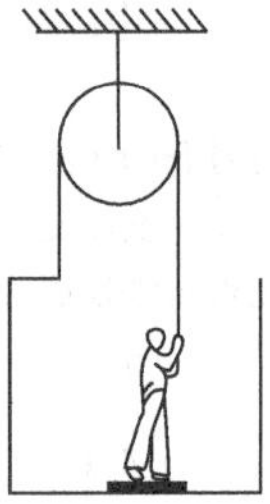

(a) 500 N (b) 400 N (c) 300 N (d) 100 N

151 Two identical bodies M_2 and M_3 each of 4.0 kg are tied to a massless inextensible string which is made to pass around pulleys P_1 and P_2 as shown in figure. Angle $ABC = 37°$. The coefficient of kinetic friction between the bodies and the surface on which they slide is 0.25. If the body M_1 moves down with uniform speed, neglecting the masses and friction of pulleys, $M_1 =$

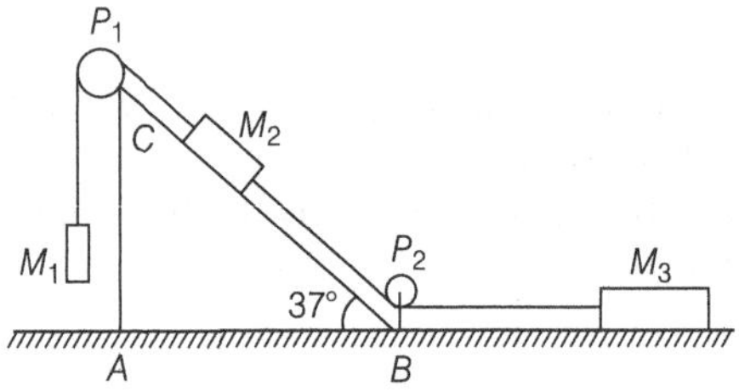

(a) 36.8 kg (b) 4.2 kg (c) 9.8 kg (d) 2.1 kg

152 Two blocks A and B are separated by some distance and tied by a string as shown in the figure. The force of friction on both the blocks at $t = 2$ s is

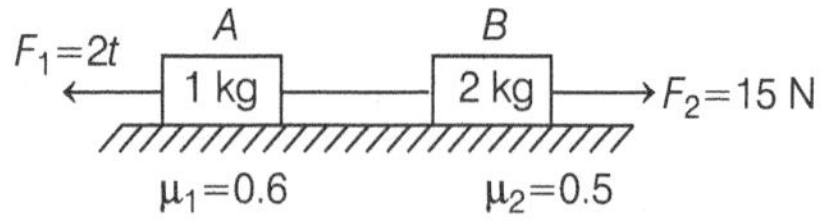

(a) 4 N (→), 5 N (←) (b) 2 N (→), 5 N (←)
(c) 0 N (→), 10 N (←) (d) 1 N (←), 10 N (←)

153 In an arrangement shown in figure, coefficient of friction between the two blocks is $\mu = 1/2$. The force of friction acting between the two blocks is

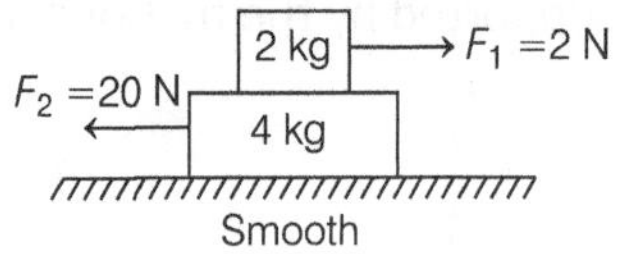

(a) 8 N (b) 10 N (c) 6 N (d) 4 N

154 A lift is moving in upward direction. The total mass of the lift and the passengers is 1600 kg. The variation of the velocity of the lift is as shown in the figure. The tension in the rope at $t =$ 8th second will be

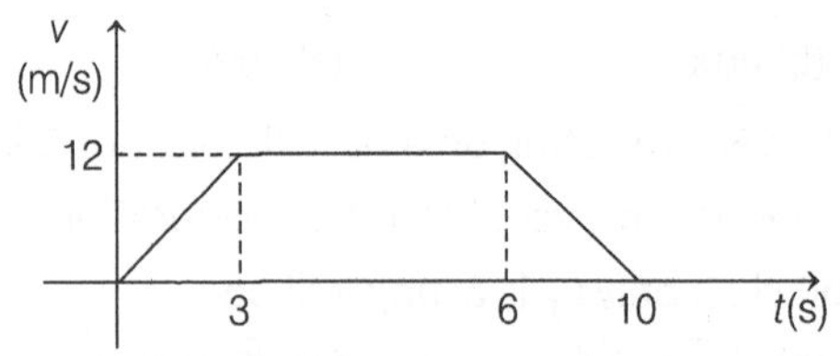

(a) 11200 N (b) 16000 N (c) 4800 N (d) 12000 N

155 Consider the situation as shown in the figure.

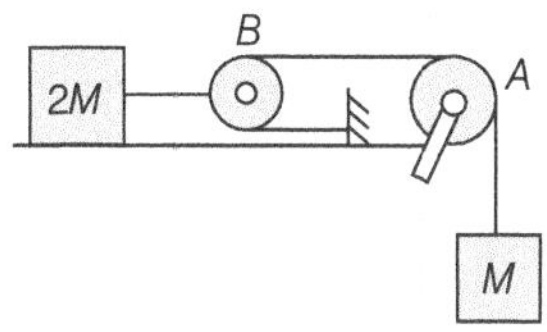

(a) The acceleration of M is $2g/3$
(b) The tension in the string connecting A and B is $Mg/3$
(c) The reaction on pulley A is $\dfrac{\sqrt{2}\,Mg}{3}$ at an angle 45° with the horizontal
(d) All the above options are correct

156 The collar A is free to slide along the smooth rod B mounted in the frame. The plane of the frame is vertical and fixed inside a vertical surface as shown. The horizontal acceleration a of the frame necessary to maintain the collar in a fixed position on the shaft is

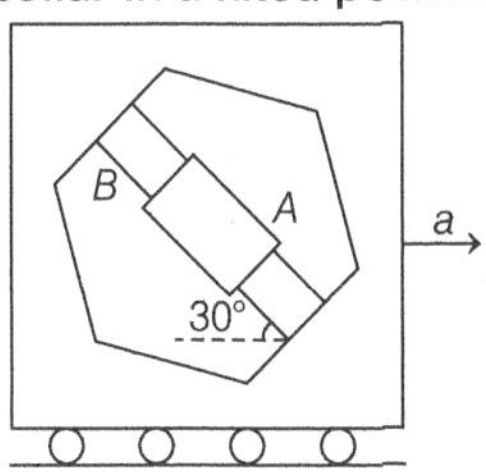

(a) $\dfrac{g}{\sqrt{3}}$ (b) $\dfrac{3g}{\sqrt{2}}$ (c) $\dfrac{g}{2\sqrt{3}}$ (d) $\dfrac{g}{\sqrt{5}}$

157 In the figure, $m_A = m_B = m_C = 60$ kg. The coefficient of friction between C and ground is 0.5, B and ground is 0.3, A and B is 0.4. C is pulling with the maximum possible force without moving. Then, the tension in the string connected to A will be

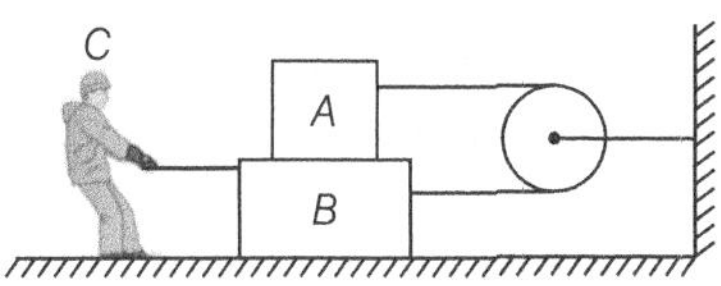

(a) 120 N (b) 60 N (c) 100 N (d) zero

158 A block slides down a frictionless plane inclined at an angle θ. For what value of θ, the horizontal component of acceleration of the block is maximum?
(a) $\theta = 45°$ (b) $\theta = 90°$ (c) $\theta = 53°$ (d) $\theta = 75°$

159 Two particles of mass 1.5 kg each are tied at the ends of a light string of length 10 m. The whole system is kept on a frictionless horizontal surface with the string held tight, so that each mass is at a distance 5 m from the centre P (as shown in figure). Now, the mid-point of the string is pulled vertical upwards with a constant force of 10 N. As a result, the particles move towards each other on the surface. The magnitude of relative acceleration, when the separation between them becomes 6 m, is

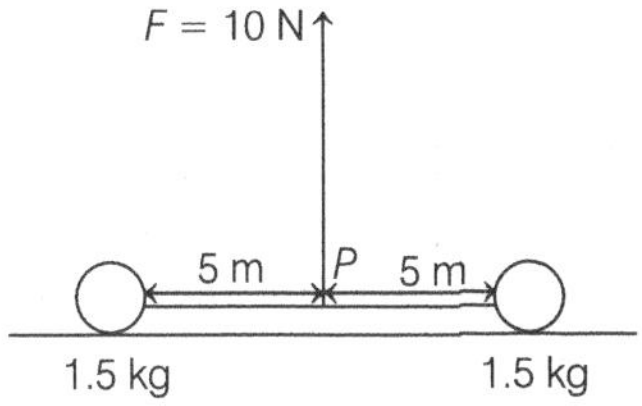

(a) 2.5 m/s^2 (b) 5 m/s^2 (c) 10 m/s^2 (d) 20 m/s^2

160 A perfectly straight portion of a uniform rope has mass M and length L. At end A of the segment, the tension in the rope is T_A; at end B, it is $T_B (> T_A)$. The tension in the rope at a distance $L/5$ from end A is
(a) $T_B - T_A$ (b) $(T_A + T_B)/5$ (c) $(4T_A + T_B)/5$ (d) $(T_B - T_A)/5$

161 Figure shows N identical blocks connected with identical strings on a smooth horizontal surface. A constant force is pulling the blocks horizontally. If tension in string connecting 4th and 5th block is two times that between 8th and 9th block, what is total number of blocks ?

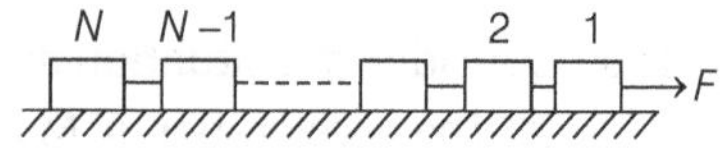

(a) 6 (b) 8 (c) 10 (d) 12

162 In given figure, extension in the spring will be (Take, $g = 10 \text{ m/s}^2$)

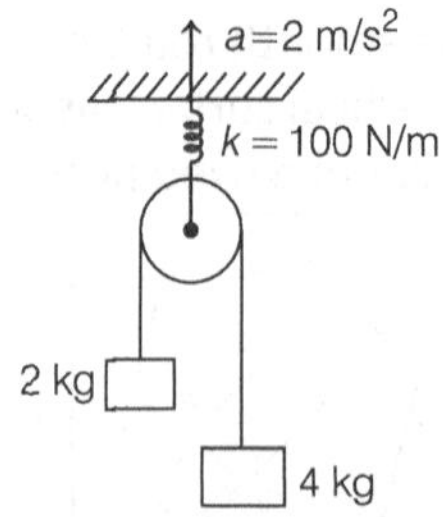

(a) 64 cm (b) 16 cm (c) 32 cm (d) None of these

163 The system is pushed by a force F as shown in figure. All surfaces are smooth except between B and C. Friction coefficient between B and C is μ. Minimum value of F to prevent block B from downward slipping is

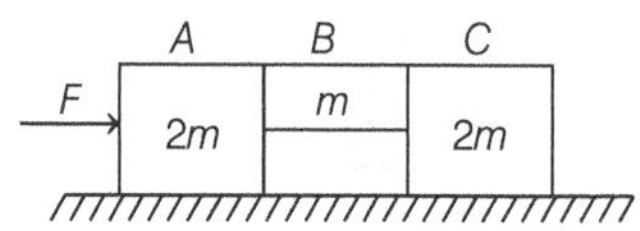

(a) $\left(\frac{3}{2\mu}\right) mg$ (b) $\left(\frac{5}{2\mu}\right) mg$ (c) $\left(\frac{5}{2}\right) \mu mg$ (d) $\left(\frac{3}{2}\right) \mu mg$

164 A body of mass m rests on a horizontal floor with which it has coefficient of static friction μ. It is desired to make the body move by applying the minimum possible force F. The magnitude of F is

(a) $\mu\, mg$ (b) $\frac{\sqrt{1+\mu^2}}{\mu} mg$ (c) $\mu\sqrt{1+\mu^2}\, mg$ (d) $\frac{\mu\, mg}{\sqrt{1+\mu^2}}$

165 What should be the value of F, so that block of mass 1 kg remains in equilibrium? The coefficient of friction between 1 kg and 4 kg block is 0.5 and except it all surfaces are smooth.

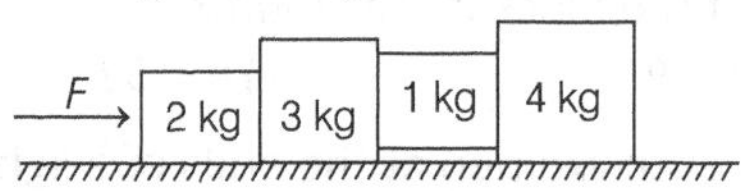

(a) 220 N (b) 170 N (c) 135 N (d) 50 N

166 Contact force between blocks B and C is

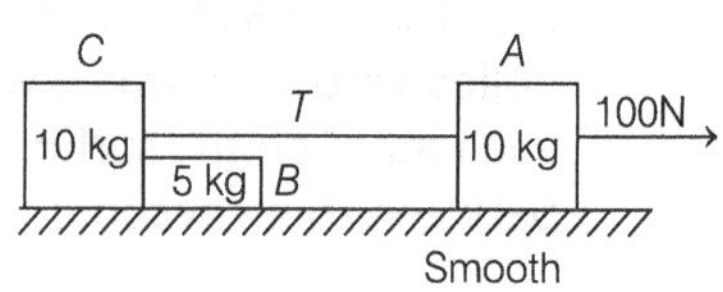

(a) 10 N
(b) 20 N
(c) 40 N
(d) 50 N

167 Two blocks are suspended in a lift that has an upward acceleration of 2 m/s^2. If tension forces in strings A and B bear relation $T_A - T_B = 6$ N, what can be concluded regarding value of masses m_1 and m_2?

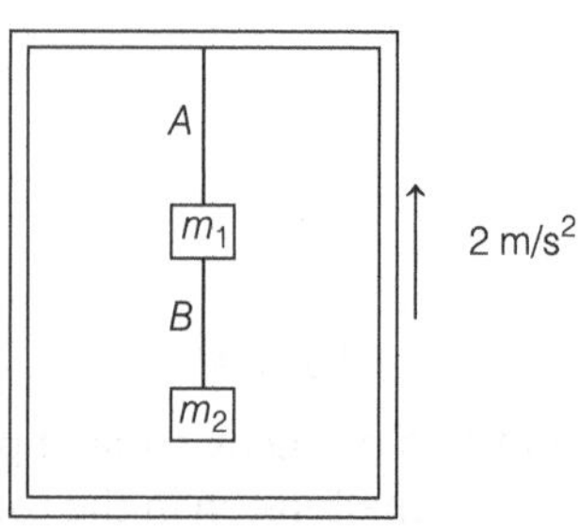

(a) $m_1 - m_2 = 0.5$ kg.
(b) $m_2 = 0.5$ kg and m_1 cannot be determined from the given information.
(c) $m_1 = 0.5$ kg and m_2 cannot be determined from the given information.
(d) Both m_1 and m_2 cannot be determined from the given information.

168 On a rough table, three blocks (including the first block) are placed as shown in the figure. Mass of each block is m and coefficient of friction for each block is μ. A force F is applied on the first block so as to move the system. The minimum value of F should be

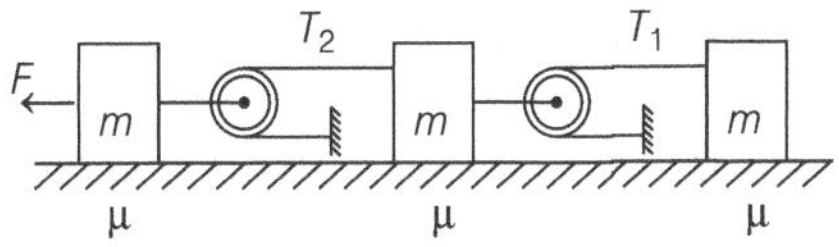

(a) $8\mu mg$ (b) $9\mu mg$ (c) $7\mu mg$ (d) $5\mu mg$

169 A man is raising himself and the crate on which he stands with an acceleration of 5 ms^{-2} by a massless rope and pulley arrangement. Mass of the man is 100 kg and that of the crate is 50 kg. If $g = 10\ ms^{-2}$, the tension in the rope is

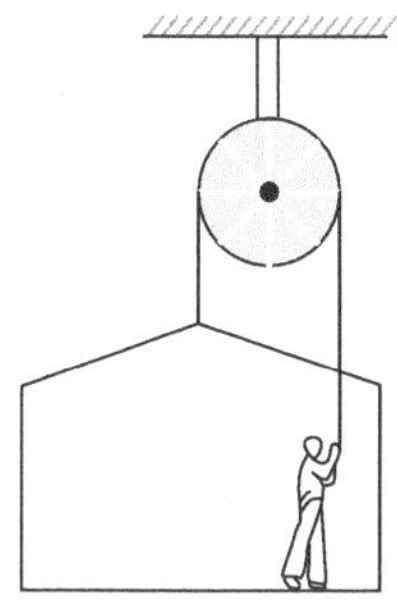

(a) 2250 N
(b) 1125 N
(c) 750 N
(d) 375 N

170 The force acting on the block pushes it. Then, pushing of the block will be possible along the surface, if

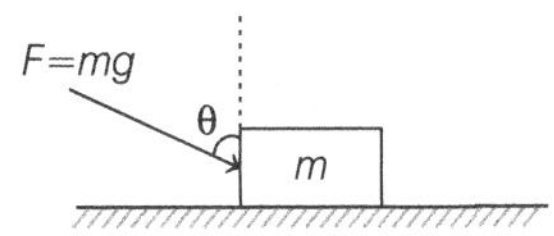

(a) $\tan\theta \geq \mu$ (b) $\cot\theta \geq \mu$ (c) $\tan\frac{\theta}{2} \geq \mu$ (d) $\cot\frac{\theta}{2} \geq \mu$

171 A block of mass m is placed on a surface with a vertical cross-section given by $y = \frac{x^3}{6}$. If the coefficient of friction is 0.5, the maximum height above the ground at which the block can be placed without slipping is

(a) $\frac{1}{6}$ m (b) $\frac{2}{3}$ m (c) $\frac{1}{3}$ m (d) $\frac{1}{2}$ m

172 In the figure shown, blocks are initially at rest on the floor. Mark the correct statement.

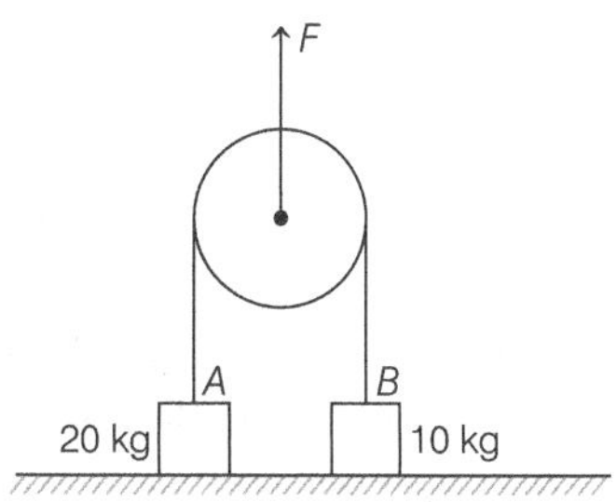

(a) When applied force on pulley is 200 N, the normal reaction on block A is zero.
(b) When applied force is 400 N, acceleration of B is 10 m/s^2.
(c) When applied force is 240 N, acceleration of pulley is 2 m/s^2.
(d) For a force F = 100 N, normal reaction on B is zero.

173 An arrangement of two blocks of same mass m kept inside a lift which is moving with acceleration a in vertical downward direction. Find acceleration of block A with respect to ground.

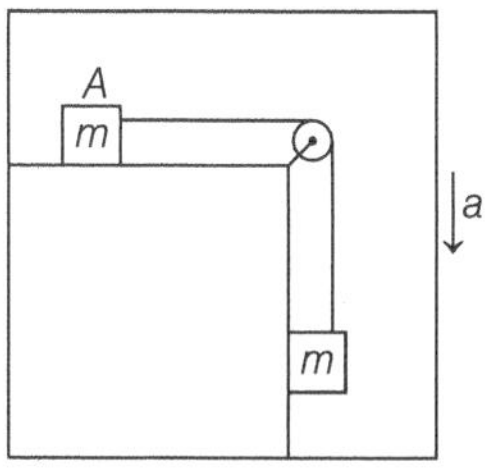

(a) $\frac{1}{2}\sqrt{5a^2 + 2ga + g^2}$
(b) $\frac{1}{2}\sqrt{5a^2 - 2ga + g^2}$
(c) $\frac{1}{2}\sqrt{5a^2 - 2ga - g^2}$
(d) None of the above

174 Two smooth cylindrical bars weighing w each lie next to each other in contact. A similar third bar is placed over the two bars as shown in figure. Neglecting friction, the minimum horizontal force on each lower bar necessary to keep them together is

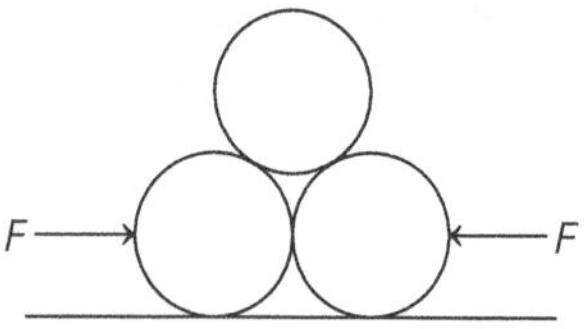

(a) $w/2$ (b) w
(c) $w/\sqrt{3}$ (d) $w/(2\sqrt{3})$

175 A man of mass 60 kg starts climbing up a massless rope by exerting 900 N on the string. After 2 s, he is exhausted. He stops climbing and holds on to the rope, choose incorrect option.

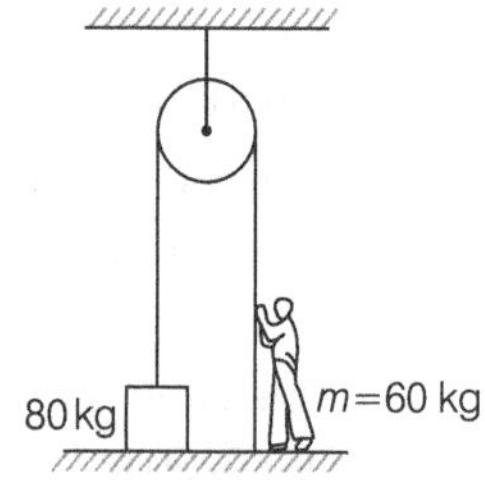

(a) String will become slack for some time.
(b) The speed of man at end of 2 s is 10 m/s.
(c) Just after $t = 2$ s, the acceleration of man and block will be same in magnitude.
(d) The acceleration of man just after 2 s is $\frac{2g}{7}$.

176 In figure, a block of mass m is released from rest when spring was in its natural length. The pulley also has mass m but it is frictionless. Suppose the value of m is such that finally, it is just able to lift the block M up after releasing it. Choose incorrect option.

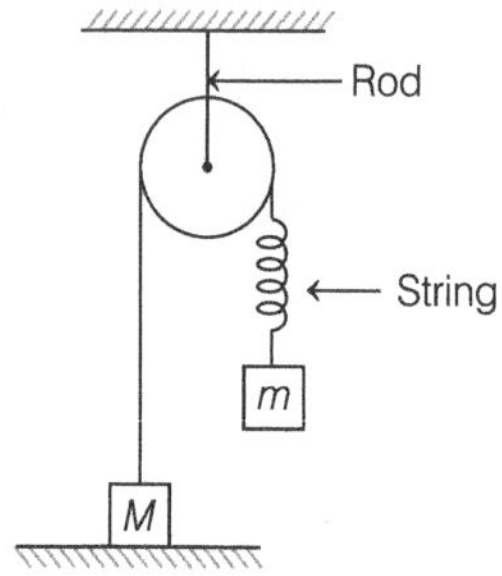

(a) The weight of m required to just lift M is $\frac{M}{2}g$.
(b) The tension in the rod, when m has zero acceleration is $\frac{M}{2}g$.
(c) The normal force acting on M when m has zero acceleration is $\frac{M}{2}g$.
(d) The tension in the string, when displacement of m is maximum possible is Mg.

177 All surfaces are frictionless. The masses of A, B and C are 1 kg, 2 kg and 3 kg, respectively. If the acceleration of block C is a_1, a_2 and a_3 respectively, in three cases, then (blocks are not attached to each other)

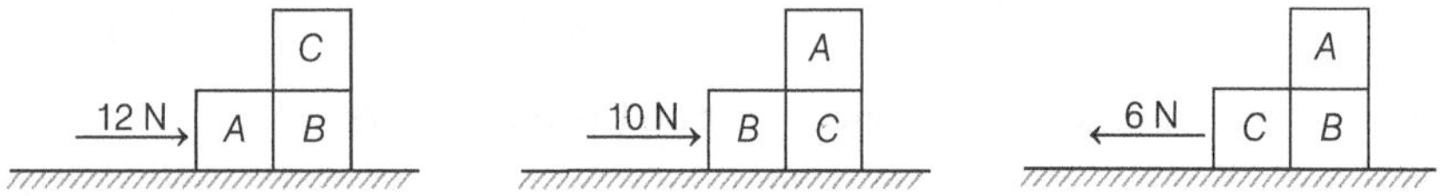

(a) $a_1 + a_2 + a_3 = 4\ m/s^2$ (b) $a_2 = a_3$
(c) $a_2 = a_3 > a_1$ (d) All of these

178 Two smooth spheres each of the radius 5 cm and weight w rest one on the other inside a fixed smooth cylinder of radius 8 cm. The reactions between the spheres and the vertical side of the cylinder are

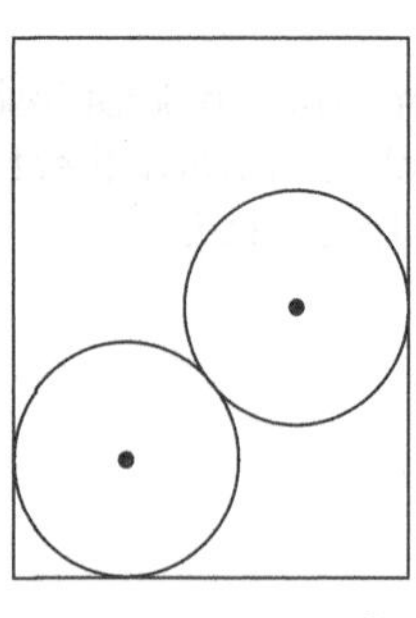

(a) $\frac{w}{4}$ and $\frac{3w}{4}$ (b) $\frac{w}{4}$ and $\frac{w}{4}$ (c) $\frac{3w}{4}$ and $\frac{3w}{4}$ (d) w and w

LEVEL 03 Previous Years Questions

179 A ball of mass 0.15 kg is dropped from a height 10 m, strikes the ground and rebounds to the same height. The magnitude of impulse imparted to the ball is nearly (Take, $g = 10\ \text{m/s}^2$) [NEET 2021]

(a) 0 (b) 4.2 kg-m/s (c) 2.1 kg-m/s (d) 1.4 kg-m/s

180 Calculate the acceleration of the block and trolly system shown in the figure. The coefficient of kinetic friction between the trolly and the surface is 0.05. (Take, $g = 10\ \text{m/s}^2$, mass of the string is negligible and no other friction exists). [NEET 2020]

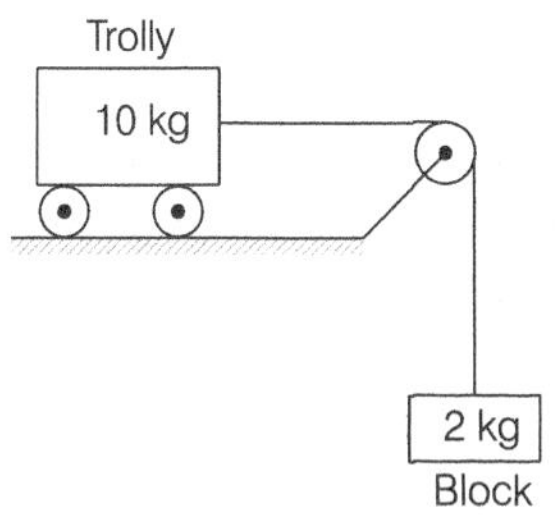

(a) $1.25\ \text{m/s}^2$ (b) $1.50\ \text{m/s}^2$ (c) $1.66\ \text{m/s}^2$ (d) $1.00\ \text{m/s}^2$

181 Two bodies of mass 4 kg and 6 kg are tied to the ends of a massless string. The string passes over a pulley which is frictionless (see figure). The acceleration of the system in terms of acceleration due to gravity g is [NEET 2020]

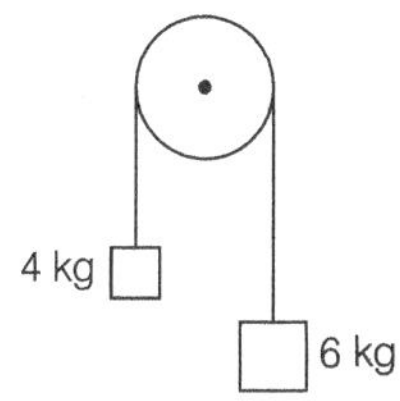

(a) $g/2$ (b) $g/5$ (c) $g/10$ (d) g

182 A truck is stationary and has a bob suspended by a light string, in a frame attached to the truck. The truck, suddenly moves to the right with an acceleration of a. The pendulum will tilt [NEET 2020]

(a) to the left and the angle of inclination of the pendulum with the vertical is $\sin^{-1}\left(\frac{g}{a}\right)$

(b) to the left and angle of inclination of the pendulum with the vertical is $\tan^{-1}\left(\frac{a}{g}\right)$

(c) to the left and angle of inclination of the pendulum with the vertical is $\sin^{1-}\left(\frac{a}{g}\right)$

(d) to the left and angle of inclination of the pendulum with the vertical is $\tan^{-1}\left(\frac{g}{a}\right)$

183 A particle moving with velocity **v** is acted by three forces shown by the vector triangle PQR. The velocity of the particle will [NEET 2019]

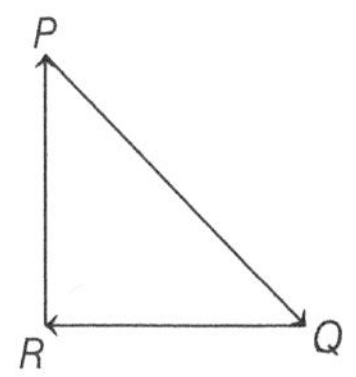

(a) decrease (b) remain constant
(c) change according to the smallest force QR (d) increase

184 Two particles *A* and *B* are moving in uniform circular motion in concentric circles of radii r_A and r_B with speed v_A and v_B, respectively. Their time period of rotation is the same. The ratio of angular speed of *A* to that of *B* will be **[NEET 2019]**

(a) $v_A : v_B$
(b) $r_B : r_A$
(c) $1 : 1$
(d) $r_A : r_B$

185 A body of mass *m* is kept on a rough horizontal surface (coefficient of friction = μ). Horizontal force is applied on the body, but it does not move. The resultant of normal reaction and the frictional force acting on the object is given *F*, where *F* is **[NEET 2019]**

(a) $|\mathbf{F}| = mg + \mu mg$
(b) $|\mathbf{F}| = \mu mg$
(c) $|\mathbf{F}| \le mg\sqrt{1+\mu^2}$
(d) $|\mathbf{F}| = mg$

186 A block of mass *m* is placed on a smooth inclined wedge *ABC* of inclination θ as shown in the figure. The wedge is given an acceleration *a* towards the right. The relation between *a* and θ for the block to remain stationary on the wedge is **[NEET 2018]**

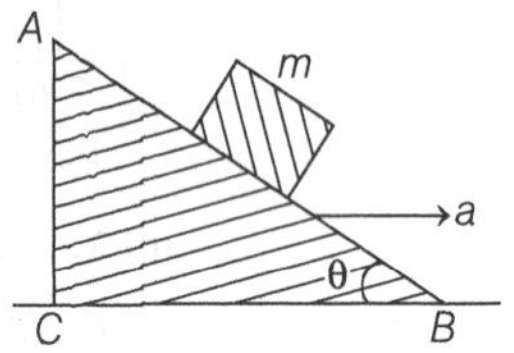

(a) $a = g\cos\theta$
(b) $a = \dfrac{g}{\sin\theta}$
(c) $a = \dfrac{g}{\operatorname{cosec}\theta}$
(d) $a = g\tan\theta$

187 Which one of the following statements is **incorrect**? **[NEET 2018]**

(a) Frictional force opposes the relative motion.
(b) Limiting value of static friction is directly proportional to normal reaction.
(c) Rolling friction is smaller than sliding friction.
(d) Coefficient of sliding friction has dimensions of length.

188 Two blocks *A* and *B* of masses 3*m* and *m* respectively, are connected by a massless and inextensible string. The whole system is suspended by a massless spring as shown in figure. The magnitudes of acceleration of *A* and *B* immediately after the string is cut, are respectively **[NEET 2017]**

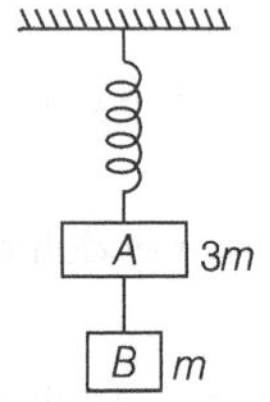

(a) $g, \dfrac{g}{3}$
(b) $\dfrac{g}{3}, g$
(c) g, g
(d) $\dfrac{g}{3}, \dfrac{g}{3}$

189 A rigid ball of mass *m* strikes a rigid wall at 60° and gets reflected without loss of speed as shown in the figure. The value of impulse imparted by the wall on the ball will be **[NEET 2016]**

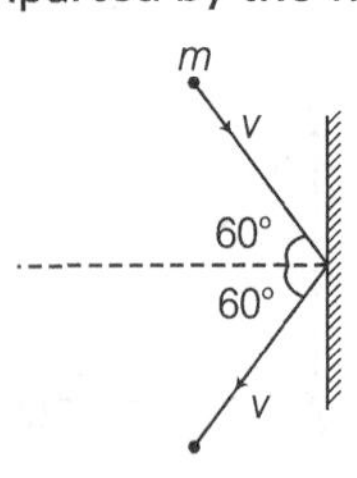

(a) mv
(b) $2mv$
(c) $mv/2$
(d) $mv/3$

190 A bullet of mass 10 g moving horizontal with a velocity of 400 m/s strikes a wood block of mass 2 kg which is suspended by light inextensible string of length 5 m. As result, the centre of gravity of the block found to rise a vertical distance of 10 cm. The speed of the bullet after it emerges of horizontally from the block will be **[NEET 2016]**

(a) 100 m/s (b) 80 m/s (c) 120 m/s (d) 160 m/s

191 Three blocks *A*, *B* and *C* of masses 4 kg, 2 kg and 1 kg respectively, are in contact on a frictionless surface, as shown. If a force of 14 N is applied on the 4 kg block, then the contact force between *A* and *B* is **[CBSE AIPMT 2015]**

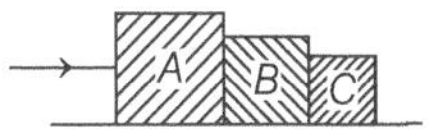

(a) 2 N (b) 6 N (c) 8 N (d) 18 N

192 A block *A* of mass m_1 rests on a horizontal table. A light string connected to it passes over a frictionless pulley at the edge of table and from its other end another block *B* of mass m_2 is suspended. The coefficient of kinetic friction between the block and the table is μ_k. When the block *A* is sliding on the table, the tension in the string is **[CBSE AIPMT 2015]**

(a) $\dfrac{(m_2 + \mu_k m_1)g}{(m_1 + m_2)}$ (b) $\dfrac{(m_2 - \mu_k m_1)g}{(m_1 + m_2)}$

(c) $\dfrac{m_1 m_2 (1 + \mu_k)g}{(m_1 + m_2)}$ (d) $\dfrac{m_1 m_2 (1 - \mu_k)g}{(m_1 + m_2)}$

193 A plank with a box on it at one end is gradually raised about the other end. As the angle of inclination with the horizontal reaches 30°, the box starts to slip and slides 4.0 m down the plank in 4.0 s. The coefficients of static and kinetic friction between the box and the plank will be, respectively **[CBSE AIPMT 2015]**

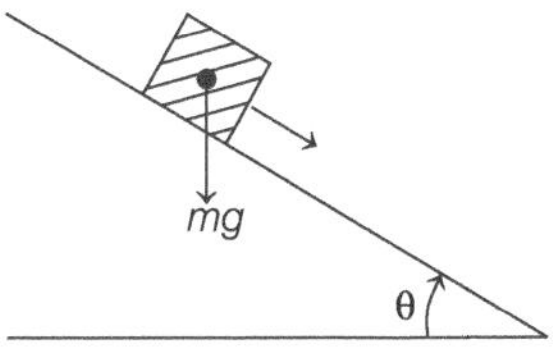

(a) 0.6 and 0.6 (b) 0.6 and 0.5 (c) 0.5 and 0.6 (d) 0.4 and 0.3

194 The force *F* acting on a particle of mass *m* is indicated by the force-time graph shown below. The change in momentum of the particle over the time interval from 0 to 8 s is **[CBSE AIPMT 2014]**

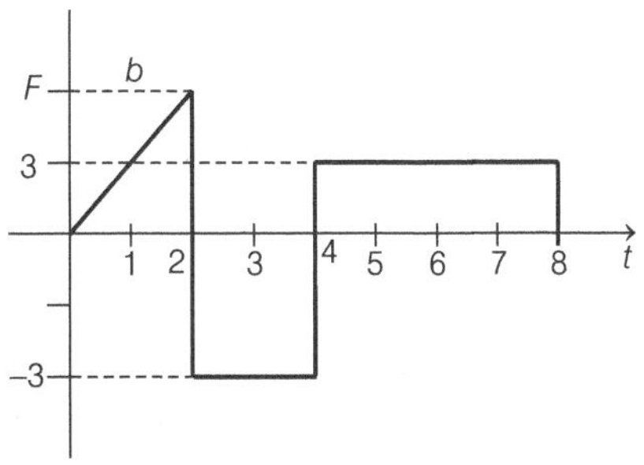

(a) 24 N-s (b) 20 N-s (c) 12 N-s (d) 6 N-s

195 A balloon with mass *m* is descending down with an acceleration *a* (where, $a < g$). How much mass should be removed from it, so that it starts moving up with an acceleration *a*? **[CBSE AIPMT 2014]**

(a) $\dfrac{2ma}{g + a}$ (b) $\dfrac{2ma}{g - a}$ (c) $\dfrac{ma}{g + a}$ (d) $\dfrac{ma}{g - a}$

196 A system consists of three masses m_1, m_2 and m_3 connected by a string passing over a pulley P. The mass m_1 hangs freely and m_2 and m_3 are on a rough horizontal table (the coefficient of friction $= \mu$). The pulley is frictionless and of negligible mass. The downward acceleration of mass m_1 is (Assume, $m_1 = m_2 = m_3 = m$) **[CBSE AIPMT 2014]**

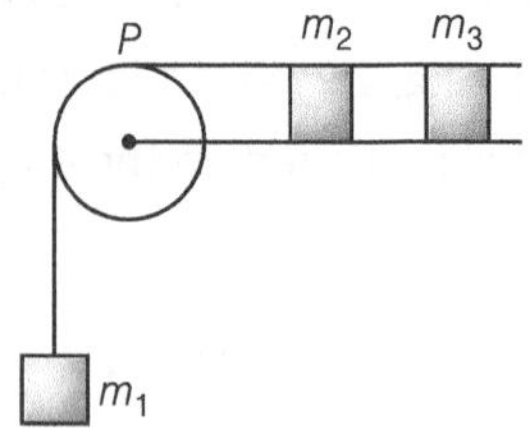

(a) $\frac{g(1-g\mu)}{9}$ (b) $\frac{2g\mu}{3}$ (c) $\frac{g(1-2\mu)}{3}$ (d) $\frac{g(1-2\mu)}{2}$

197 An explosion breaks a rock into three parts in a horizontal plane. Two of them go off at right angles to each other. The first part of mass 1 kg moves with a speed of 12 ms^{-1} and the second part of mass 2 kg moves with 8 ms^{-1} speed. If the third part flies off with 4 ms^{-1} speed, then its mass is **[NEET 2013]**

(a) 3 kg (b) 5 kg
(c) 7 kg (d) 17 kg

198 Three blocks with masses m, $2m$ and $3m$ are connected by strings, as shown in the figure. After an upward force F is applied on block m, the masses move upward at constant speed v. What is the net force on the block of mass $2m$? (g is the acceleration due to gravity). **[NEET 2013]**

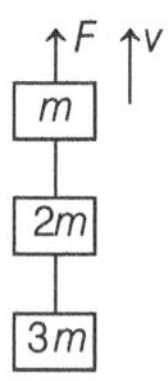

(a) Zero (b) 2 mg
(c) 3 mg (d) 6 mg

199 The upper half of an inclined plane of inclination θ is perfectly smooth while lower half is rough. A block starting from rest at the top of the plane will again come to rest at the bottom, if the coefficient of friction between the block and lower half of the plane is given by **[NEET 2013]**

(a) $\mu = \frac{1}{\tan\theta}$ (b) $\mu = \frac{2}{\tan\theta}$
(c) $\mu = 2\tan\theta$ (d) $\mu = \tan\theta$

200 Two spheres A and B of masses m_1 and m_2, respectively collide. A is at rest initially and B is moving with velocity v along x-axis. After collision, B has a velocity $\frac{v}{2}$ in a direction perpendicular to the original direction. The mass A moves after collision in the direction **[CBSE AIPMT 2012]**

(a) same as that of B (b) opposite to that of B
(c) $\theta = \tan^{-1}\left(\frac{1}{2}\right)$ to the x-axis (d) $\theta = \tan^{-1}\left(\frac{-1}{2}\right)$ to the x-axis

201 A person of mass 60 kg is inside a lift of mass 940 kg and presses the button on control panel. The lift starts moving upwards with an acceleration 1.0 m/s^2. If $g = 10$ m/s^2, the tension in the supporting cable is **[CBSE AIPMT 2011]**

(a) 9680 N (b) 11000 N
(c) 1200 N (d) 8600 N

202 A man of 50 kg mass is standing in a gravity free space at a height of 10 m above the floor. He throws a stone of 0.5 kg mass downwards with a speed 2 ms^{-1}. When the stone reaches the floor, the distance of the man above the floor will be [CBSE AIPMT 2010]

(a) 9.9 m (b) 10.1 m
(c) 10 m (d) 20 m

203 A block of mass m is in contact with the cart C as shown in the figure.

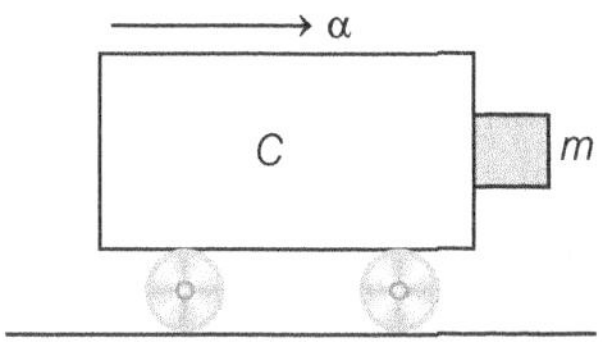

The coefficient of static friction between the block and the cart is μ. The acceleration α of the cart that will prevent the block from falling satisfies [CBSE AIPMT 2010]

(a) $\alpha > \frac{mg}{\mu}$ (b) $\alpha > \frac{g}{\mu m}$
(c) $\alpha \geq \frac{g}{\mu}$ (d) $\alpha < \frac{g}{\mu}$

ANSWER KEY

Level 1 Objective Questions

1 (a)	2 (d)	3 (c)	4 (c)	5 (a)	6 (b)	7 (a)	8 (c)	9 (c)	10 (c)
11 (a)	12 (a)	13 (c)	14 (c)	15 (c)	16 (d)	17 (d)	18 (c)	19 (d)	20 (b)
21 (c)	22 (a)	23 (d)	24 (b)	25 (b)	26 (a)	27 (a)	28 (c)	29 (c)	30 (a)
31 (b)	32 (a)	33 (b)	34 (a)	35 (a)	36 (d)	37 (d)	38 (b)	39 (c)	40 (a)
41 (c)	42 (b)	43 (d)	44 (d)	45 (a)	46 (d)	47 (a)	48 (d)	49 (c)	50 (a)
51 (a)	52 (a)	53 (d)	54 (c)	55 (b)	56 (d)	57 (d)	58 (c)	59 (b)	60 (d)
61 (b)	62 (c)	63 (d)	64 (b)	65 (d)	66 (b)	67 (d)	68 (a)	69 (b)	70 (a)

Level 2 Objective Questions

71 (d)	72 (c)	73 (c)	74 (b)	75 (b)	76 (a)	77 (d)	78 (d)	79 (c)	80 (a)
81 (c)	82 (a)	83 (b)	84 (d)	85 (b)	86 (c)	87 (a)	88 (b)	89 (b)	90 (c)
91 (b)	92 (c)	93 (b)	94 (a)	95 (b)	96 (b)	97 (a)	98 (d)	99 (c)	100 (c)
101 (c)	102 (c)	103 (a)	104 (c)	105 (b)	106 (a)	107 (d)	108 (b)	109 (c)	110 (b)
111 (c)	112 (b)	113 (c)	114 (b)	115 (a)	116 (a)	117 (b)	118 (a)	119 (b)	120 (b)
121 (a)	122 (d)	123 (d)	124 (b)	125 (a)	126 (c)	127 (c)	128 (b)	129 (a)	130 (b)
131 (a)	132 (b)	133 (b)	134 (c)	135 (c)	136 (a)	137 (b)	138 (c)	139 (d)	140 (a)
141 (a)	142 (a)	143 (c)	144 (a)	145 (d)	146 (a)	147 (d)	148 (a)	149 (b)	150 (d)
151 (b)	152 (d)	153 (a)	154 (a)	155 (d)	156 (a)	157 (d)	158 (a)	159 (b)	160 (c)
161 (d)	162 (a)	163 (b)	164 (d)	165 (d)	166 (b)	167 (c)	168 (c)	169 (b)	170 (c)
171 (a)	172 (b)	173 (b)	174 (d)	175 (d)	176 (b)	177 (d)	178 (c)		

Level 3 Previous Years Questions

179 (b)	180 (a)	181 (b)	182 (b)	183 (b)	184 (c)	185 (c)	186 (d)	187 (d)	188 (b)
189 (a)	190 (c)	191 (b)	192 (c)	193 (b)	194 (c)	195 (a)	196 (c)	197 (b)	198 (a)
199 (c)	200 (c)	201 (b)	202 (b)	203 (c)					

Solutions

LEVEL 01 Objective Questions

1. Action-reaction force act on two different bodies.
2. Block is at rest. Therefore, net force (resultant of normal reaction and friction) is in upward direction.
3. To decelerate the particle, a force must be applied in the opposite direction, irrespective of its magnitude. Hence, F_2 may be greater than, smaller than or equal to F_1.
4. From Newton's second law,
$$F = (m+M)\,a$$
$$\Rightarrow \quad a = \frac{F}{m+M}$$
So, the force exerted by the rope,
$$F = Ma = \frac{FM}{m+M}$$
5. Frictional force on a body, $f = \mu mg$
$$= 0.2\,(4)\,(10) = 8\text{ N}$$
Since, the frictional force is more than the applied force.
So, its acceleration is zero.
6. In figure (b), normal reaction and therefore maximum value of friction decreases.
7. Frictional force acting against the box motion,
$$f = \mu mg = 0.20\,(10)(10)$$
$$= 20\text{ N}$$
So, the frictional force is more than the applied force, the box will accelerate to left and its speed decreases due to friction.
8. The net force on the system is same in both the cases, so the same acceleration will result similar tension in both string.
$$\therefore \quad F_1 = F_2$$
9. The tension sustained by branch $= 75\% = \dfrac{3Mg}{4}$
$$M(g-a) = \frac{3Mg}{4}$$
$$a = \frac{g}{4}$$
10. $a_1 = g\sin\theta_1$
and $\quad a_2 = g\sin\theta_2$
Since, $\quad \theta_2 > \theta_1$
$\because \quad a_2 > a_1$
$\therefore \quad t_2 < t_1$
From the energy conservation,
speed of the stone will be same in both the cases.
11. Block *A* will move by friction, when force is applied on *B*. Here, friction is zero. Hence,
$$a_A = 0$$
12. $F_r = ma$
$$\mu N = ma$$
$$0.3\,(50) = 5a$$
$$\Rightarrow \quad a = 3\text{ m/s}^2$$
13. Mass of a monkey = 50 kg
Maximum tension in string, $T_{max} = 800$ N
If monkey climbs with an acceleration of 6.5 m/s^2,
then from Newton's second law,
$$T - mg = ma$$
$$T = m(g+a)$$
$$= 50\,(10+6.5)$$
$$= 825\text{ N}$$
So, $T > T_{max}$. Hence, rope will break.
14. Acceleration of system $= \dfrac{mg\sin\theta - mg}{m}$
$$= \frac{10-10}{m+2} = 0$$
So, frictional force on the block will be zero.
15. Train slows down means acceleration of train car is backwards. So, pseudo force is in forward direction.
16. Contact force $= \sqrt{N^2 + f_r^2}$
$$= \sqrt{(mg\cos\theta)^2 + (\mu mg\cos\theta)^2}$$
$$= \sqrt{(2\times 10\cos 30^\circ)^2 + [0.5\,(2)\,(10)\cos 30^\circ]^2}$$
$$= 5\sqrt{15}\text{ N}$$
17. $T = mg$
$$T = mg\sin\theta + f_r$$
$$T = mg\sin\theta + \mu mg\cos\theta$$
$$= 4\,(10)\sin 30^\circ + \frac{(4)(10)}{\sqrt{3}}\cos 30^\circ$$
$$mg = 40\text{ N}$$
$$m = 4\text{ kg}$$
18. $Mg\sin\theta - T = Ma$
$\because \quad T = Ma$
$$Mg\sin\theta - T = T$$
$$Mg\sin\theta = 2T$$
So, we get
$$T = \frac{Mg\sin\theta}{2}$$

19. Let T_1 be the tension connected with 2 kg block and T_2 with the block A.

$$T_2 = T_1 + mg \sin 30°$$

$$T_2 = 20 + 20\left(\frac{1}{2}\right) = 30 \text{ N}$$

Now, $T_2 = mg$

$$30 = m(10) \Rightarrow m = 3 \text{ kg}$$

20. For system to be at rest,

$$mg \sin 30° = \mu mg \cos 30°$$

$$1(10)\left(\frac{1}{2}\right) = \mu(1)(10)\left(\frac{\sqrt{3}}{2}\right)$$

$$\mu = \frac{1}{\sqrt{3}}$$

21. Since, $M_1 > M_2$

$$M_1 g \sin\alpha - T = M_1 a \quad \text{...(i)}$$

$$T - M_2 g \sin\beta = M_2 a \quad \text{...(ii)}$$

On solving Eqs. (i) and (ii), we get

$$a = \frac{(M_1 \sin\alpha - M_2 \sin\beta)\cdot g}{M_1 + M_2}$$

Substituting a in Eq. (i), we get

$$T = \frac{M_1 M_2 (\sin\beta + \sin\alpha)}{(M_1 + M_2)} g$$

22. $t = \sqrt{\frac{2s}{a}} = \sqrt{\frac{2(h/\sin\theta)}{g \sin\theta}}$

or $\quad t \propto \frac{1}{\sin\theta}$

$\sin 60°$ is maximum. Therefore, t_1 is minimum.

23. Net pulling force $= m_1 g - m_2 g \sin\theta$

$$= 0 \qquad \left(\text{because } \frac{m_1}{m_2} = \sin\theta\right)$$

$\therefore$ System is at rest.

24. $a = \frac{\text{Net force}}{\text{Total mass}} = \frac{\frac{3}{2}Mg - Mg}{M} = \frac{g}{2}$

25. Case-I

$$2mg - T = 2ma$$

$$T - mg = ma$$

$$\therefore \quad a = \frac{g}{3}$$

Case - II

$$2mg = ma$$

$$a = 2g$$

Case - III

$$a = \frac{2mg - mg}{3m} = \frac{g}{3}$$

Case - IV

$$2mg - T = 2ma$$

$$T = ma$$

$$a = \frac{2}{3}g$$

So, in case II, the magnitude of acceleration is maximum.

26. Let tension T be in the rope.

When monkey climbs up with an acceleration a,

$$T - m_1 g = m_1 a \quad \text{...(i)}$$

When the other monkey climbs down with the acceleration a,

$$m_2 g - T = m_2 a \quad \text{...(ii)}$$

On solving Eqs. (i) and (ii), we get

$$m_2(g - a) = m_1(g + a)$$

Since, the ratio of masses of monkey is $m_1 : m_2 = 2:3$, we get

$$5a = g \Rightarrow a = \frac{g}{5}$$

27. $g_{\text{effective}} = g' = g - a$

$$= 10 - 2 = 8 \text{ ms}^{-2}$$

As, $\quad T - m_A g' = m_A a \quad$ ...(i)

and $\quad m_B g' - T = m_B a \quad$ ...(ii)

$$\Rightarrow \quad a = \left(\frac{m_B - m_A}{m_B + m_A}\right)\cdot g'$$

$$\Rightarrow \quad a = \frac{2}{6} \times 8 = \frac{8}{3} \text{ ms}^{-2}$$

From Eq. (i),

$$T = m_A(g' + a)$$

$$= 2\left(8 + \frac{8}{3}\right) = \frac{2}{3} \times 32 = \frac{64}{3} \text{ N}$$

28. Net force acting on the block = 20 + 10 = 30 N

So, mass of block $B = \frac{30}{10} = 3$ kg

29. Let x be the distance descended by block A and y be the distance descended be block B.

$$y = \frac{x + 0}{2} = \frac{x}{2}$$

Now, $\quad \frac{dy}{dt} = \frac{1}{2}\frac{dx}{dt} = \frac{1}{2}(4) = 2 \text{ m/s} \quad \text{(downwards)}$

30. Weight measured by the weighing machine

$$= M(g - a) \quad \text{...(i)}$$

From FBD of lift,

$$T = M(g - a) \quad \text{...(ii)}$$

From pulley, $T = ma$

$$ma = m(g - a) \quad \text{...(iii)}$$

So, $a = \frac{Mg}{M + m}$

Substituting in Eq. (i), we get

$$= \frac{Mmg}{M + m}$$

31. $a = \frac{F}{m + 2m + 4m} = \frac{F}{7m}$

Let the normal force between A and B is N_1, and between B and C is N_2.

$$N_1 = (2m + 4m)\, a = \frac{6F}{7}$$

$$N_2 = 4ma = \frac{4F}{7}$$

32. Along x-direction,

$$T_1 \cos 30° - T_2 \cos 60° = 0 \quad \text{...(i)}$$

Along y-direction,

$$T_1 \sin 30° + T_2 \sin 60° = 100 \quad \text{...(ii)}$$

On solving Eqs. (i) and (ii), we get

$$T_1 = 50 \text{ N}, T_2 = 50\sqrt{3} \text{ N}$$

33. The acceleration of the system $= \frac{F_{net}}{m} = \frac{36}{1+8+27}$

$= 1 \text{ m/s}^2$

Now, the tension T_2,

$$T_2 = (8 + 1) \times a$$

$$\Rightarrow \quad T_2 = 9 \times 1 = 9 \text{ N}$$

34. $F_{net} = ma$

$$\Rightarrow \quad 40 - \mu \times 10 \times 10 = 10 \times 2$$

$$\Rightarrow \quad \mu = \frac{20}{100} = 0.2$$

35. Let T_1 be the tension in the horizontal part of string and T_2 in other string.

$$T_2 \sin 30° = Mg$$

$$T_2 = 2\,Mg$$

$$T_1 = T_2 \cos 30°$$

$$= 2\,Mg \times \frac{\sqrt{3}}{2} = \sqrt{3}\,Mg$$

36. Friction force,

$$f_r = 20\sqrt{2} \cos 45° = 20 \text{ N}$$

For normal force,

$$N = mg - f \sin 45° = 50 - 20$$

$$N = 30 \text{ N}$$

37. Acceleration of the system, $a = \frac{F}{4m}$ ...(i)

$$T_3 = ma = \frac{F}{4} \quad \text{...(ii)}$$

Now for T_2,

$$T_2 - T_3 = ma$$

$$T_2 - \frac{F}{4} = m\left(\frac{F}{4m}\right) \quad \text{[using Eqs.(i) and (ii)]}$$

$$T_2 = \frac{F}{2}$$

For T_1, $\quad T_1 - T_2 = m\left(\frac{F}{4m}\right)$

$$T_1 - \frac{F}{2} = m\left(\frac{F}{4m}\right) \quad \left[\because T_2 = \frac{F}{2}\right]$$

$$T_1 = \frac{3F}{4}$$

So, $T_1 + T_2 + T_3 = \frac{3F}{4} + \frac{F}{2} + \frac{F}{4}$

$$2\,(T_1 + T_2 + T_3) = 3F$$

38. Applied force $4t$ must be greater than the limiting frictional force for the block to move,

$$4t = \mu_s mg$$

$$4t = 0.4(10)(10)$$

$$t = 10 \text{ s}$$

39. Acceleration of a system,

$$a = \frac{\text{Net force}}{\text{Total mass}}$$

$$= \frac{10}{2+3+5} = 1 \text{ m/s}^2$$

F_{net} (on 2 kg block)

$$= m \times a = 2 \times 1$$

$$= 2 \text{ N}$$

40. The maximum value of friction $= \mu N$

Here, $\quad N + 4 = 10 \Rightarrow N = 6$

$$F_{max} = 0.4 \times 6 = 2.4 \text{ N}$$

The applied force in the horizontal direction is 1N.

$$\therefore \quad F = f_{max}$$

$$F = 1 \text{ N or } \mathbf{f} = -\hat{\mathbf{i}}$$

41. We have, acceleration of 4 m/s^2 on body 1 and 6 m/s^2 on body 2.

$$F = 4m_1, F = 6m_2$$

$$m_2 = \frac{2}{3}m_1$$

Now, for the combined body,

$$a = \frac{F}{m_1 + m_2} = \frac{F}{m_1 + \frac{2}{3}m_1}$$

$$= \frac{3 \times 4}{5} = 2.4 \text{ m/s}^2$$

42. Let a = acceleration of system of blocks,

$$F = (5 + 3)a$$

$$\Rightarrow \quad \frac{24}{8} = a$$

$$\Rightarrow \quad a = 3 \text{ m/s}^2$$

Now, the pseudo force on block of mass 3 kg,

$$f_{pseudo} = 3\,(3) = 9 \text{ N}$$

Now, we have, $f_{limiting} = \mu mg$

$$= 0.6\,(3)\,(10) = 18 \text{ N}$$

So, $f_{required} < f_{min}$. So, blocks move together.

Hence, the force of friction between the blocks is 9 N.

43. Maximum friction at bottommost point > 30 N.

So, blocks will not move.

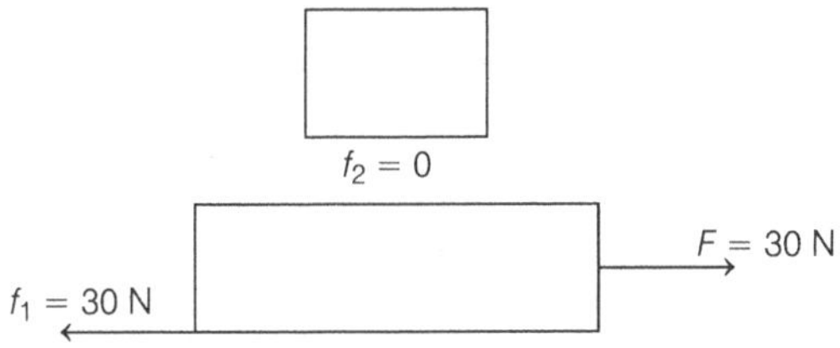

44. Normal force on block = 40 N

Frictional force acting on block = μN

$$= 0.8\,(40) = 32 \text{ N}$$

Since, the applied force is less than the frictional force, so friction on block is 30N.

So, the net force exerted by ground

$$= \sqrt{30^2 + 40^2} = 50 \text{ N}$$

45. $F = \mu mg = 0.5 \times 2 \times 10 = 10$ N

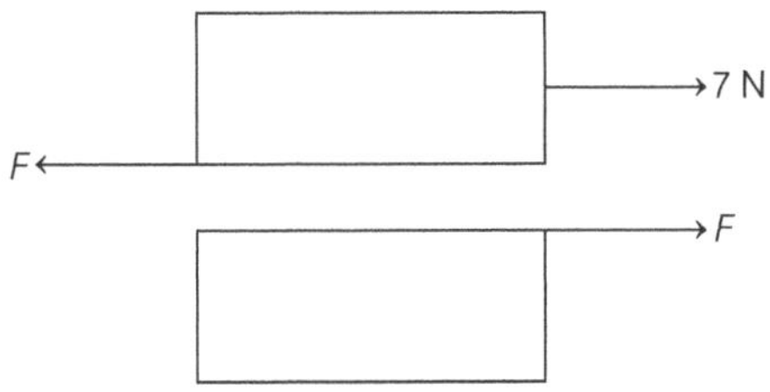

Since 7N < F, there is no relative motion between the blocks.

So, the acceleration of lower block will be the acceleration of system,

$$a = \frac{7}{2+5} = \frac{7}{7} = 1 \text{ m/s}^2$$

46. We have, $a = \mu_k g$

Now, $v^2 = u^2 + 2as$

$$0 = (6)^2 + 2\,(\mu_k)\,(-g)\,(9)$$

$\Rightarrow \quad \mu_k = 0.2$

47. Since, $F = \dfrac{\mu mg}{2} < \mu mg$

$$F = T + f$$

$$\frac{\mu mg}{2} = T + \frac{\mu mg}{2}$$

$$T = 0$$

48. The normal force, $N = \dfrac{mg}{\sqrt{2}}$

$\therefore$ Friction, $f = \mu N = \dfrac{mg}{2\sqrt{2}}$ $\quad \left(\because \mu = \dfrac{1}{2}\right)$

So, the net force opposing the motion = ma

$$\therefore \quad ma = \frac{mg}{\sqrt{2}} + \frac{mg}{2\sqrt{2}}$$

$$a = \frac{g}{\sqrt{2}}\left(1 + \frac{1}{2}\right)$$

49. $mg = T\cos 30°$

$$N = T\sin 30°$$

$$\Rightarrow \quad N = \frac{mg}{\sqrt{3}}$$

50. $F = mg\sin\theta - \mu mg\cos\theta = ma$

$$\Rightarrow \quad a = \frac{g}{\sqrt{2}} - 0.5\,g\frac{1}{\sqrt{2}} = \frac{4.9}{\sqrt{2}}$$

51. Force causing acceleration

= weight of chain of length $2x$

= (mass per unit length) $\times 2x \times g$

$= \lambda \cdot 2x \cdot g$

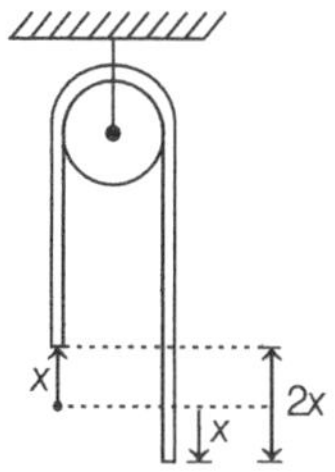

If a = Initial acceleration of chain

Then, $\quad F_{net} = m \cdot a$

$$\Rightarrow \quad \lambda \cdot 2x \cdot g = \lambda \cdot 2L \cdot a$$

$$\Rightarrow \quad a = \frac{x}{L} \cdot g$$

52. $m\,(g + a) = 120$

Here, $mg = 80$

$$80 + ma = 120$$

$$\Rightarrow \quad a = \frac{40}{8} = 5 \text{ m/s}^2 \quad \text{[upwards]}$$

53. $0 = (10)^2 - (2)(a)(25)$

$$\therefore \quad a = 2 \text{ m/s}^2 \quad \text{(upwards)}$$

$$T - mg = ma$$

$$\therefore \quad T = m\,(g + a)$$

$$= 300\,(10 + 2)$$

$$= 3600 \text{ N}$$

54. Let the acceleration of monkey be a.

Now, permissible tension,

$$T = 25 \times 10 = 250 \text{ N}$$

For monkey to climb upwards,

$$T - mg = ma$$

$$\Rightarrow \quad T = m\,(g + a)$$

$$250 = 20\,(10 + a)$$

$$a = 2.5 \text{ m/s}^2$$

55. We have, $\quad N = ma \quad$ ...(i)

$$f = \mu N = mg \quad \text{...(ii)}$$

From Eqs. (i) and (ii), we get

$$mg = \mu\,(ma)$$

$$a = \frac{g}{\mu}$$

56. For the block to be stationary,

$$f_r = mg = w$$

$$\mu N = mg = w$$

$$0.2\,(10) = w$$

$$\Rightarrow \quad w = 2 \text{ N}$$

57.

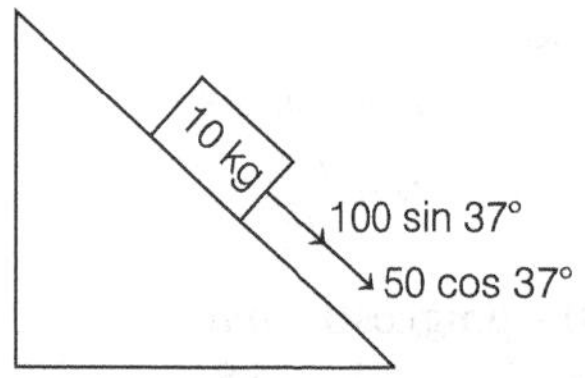

$$\text{Acceleration of the block} = \frac{\text{Net force}}{\text{Total mass}}$$
$$= \frac{100 \sin 37° + 50 \cos 37°}{10}$$
$$= \frac{100}{10} = 10 \text{ m/s}^2$$

58. F = Maximum force of friction between B and ground
$$= \mu (m_A + m_B) g$$
$$= 0.5 (10 + 20)(10)$$
$$= 150 \text{ N}$$

59. For the metre stick to be in equilibrium, taking mass moments,

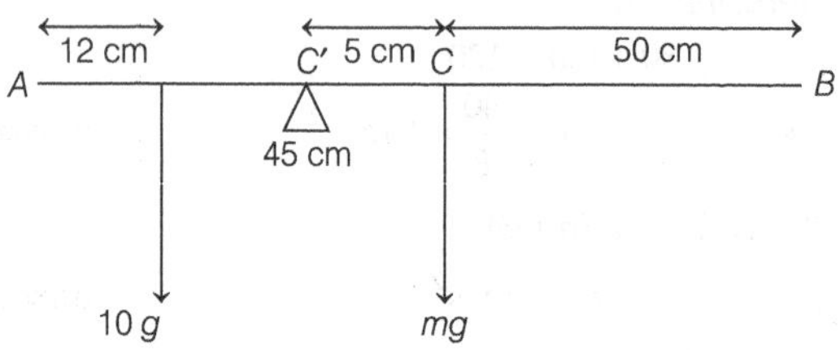

$$10\,g(45 - 12) = mg(50 - 45)$$
$$m = 66 \text{ g}$$

60. As the elevator is descending, a pseudo force will act in the upward direction.
$$mg - N = ma$$
$$N = m(g - a)$$
$$= 2(10 - 3) = 14 \text{ N}$$

61. From Newton's second law,
$$T - m_1 g = m_1 a \quad \text{...(i)}$$
$$m_2 g - T = m_2 a \quad \text{...(ii)}$$
On solving Eqs. (i) and (ii), we get
$$a = \frac{m_2 - m_1}{m_1 + m_2} g = \frac{g}{8}$$
So, we get $\dfrac{m_2}{m_1} = \dfrac{9}{7}$

62. Using Lami's theorem,
$$\frac{T_1}{\sin 150°} = \frac{T_2}{\sin 120°} = \frac{10}{\sin 90°}$$
We get, $T_2 = 5\sqrt{3}$ N, $T_1 = 5$ N

So, $\dfrac{T_1}{T_2} = \dfrac{1}{\sqrt{3}}$

63. If a is the downward acceleration of 4 kg block, the upward acceleration of 1 kg block must be $2a$.
$$4g - 2T = 4a \quad \text{...(i)}$$
$$T - g = 2a \quad \text{...(ii)}$$
From Eqs. (i) and (ii), we get
$$a = \frac{g}{4}$$
So, the acceleration of 1 kg block $= 2a = \dfrac{g}{2}$ upwards.

64. Acceleration of rope
$$= \frac{F}{M} = \frac{40}{10 \times 4} = 1 \text{ m/s}^2$$

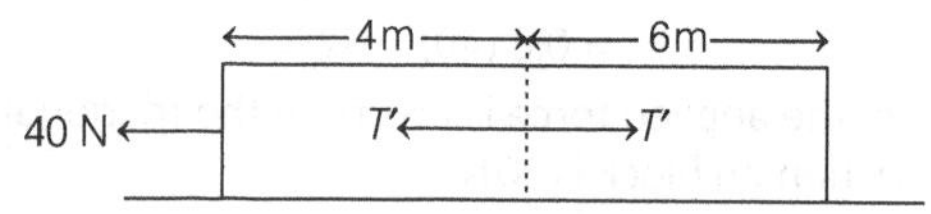

Tension T' at a distance 4 m from end of rope is pulling 6 m length of rope.

$\therefore \quad T' = ma$

$\Rightarrow \quad T' = 6 \times 4 \times 1$

$= 24$ N

65. We know, coefficient of friction, when block is at point of sliding,

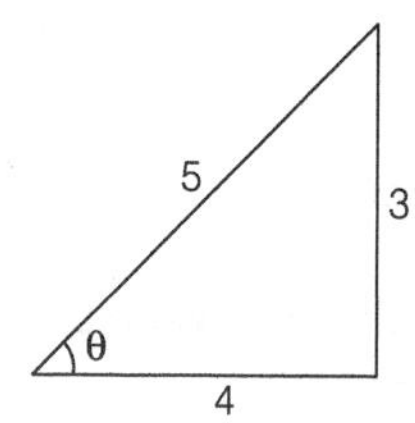

$$\mu = \tan\theta = \frac{3}{4} = 0.75$$

66.

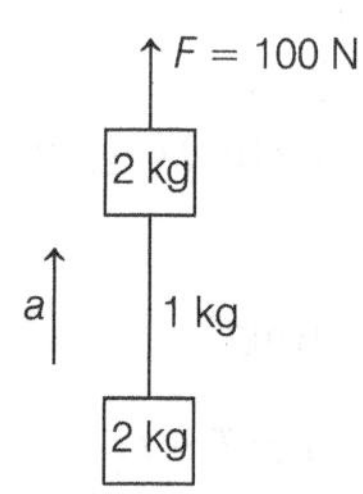

$$a = \frac{100 - (2 + 1 + 2)g}{2 + 1 + 2} = 10 \text{ m/s}^2$$

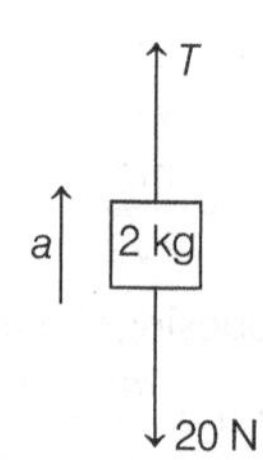

$$T - 20 = 2 \times a = 20$$
$\therefore \quad T = 40$ N

67. For system to be in equilibrium,

$$T_3 = 4g = 40 \text{ N}$$
$$T_2 \cos 37° = T_3 = 40$$
$$T_2 = 50 \text{ N}$$
$$T_1 = T_2 \sin 37° = 50 \times \frac{3}{5} = 30 \text{ N}$$
$$\therefore \quad T_1 = 30 \text{ N}, T_2 = 50 \text{ N}, T_3 = 40 \text{ N}$$

68. We have, $f = \mu mg\cos\theta$

Acceleration of block, $a = \dfrac{mg\sin\theta - \mu mg\cos\theta}{m}$

$$= g\sin\theta - \mu g\cos\theta$$
$$= 10 \sin 30° - \mu \times 10 \cos 30°$$
$$= 5 - 5\sqrt{3}\,\mu \quad \text{...(i)}$$

Now, using $s = ut + \frac{1}{2}at^2$

$$8 = \frac{1}{2}a(2)^2$$
$$a = 4 \text{ m/s}^2$$

Substituting in equation (i), we get

$$4 = 5 - 5\sqrt{3}\,\mu \Rightarrow \mu = \frac{1}{5\sqrt{3}}$$

69. Acceleration of system in both cases; $a = \dfrac{\text{Net force}}{\text{Total mass}}$

$$= \frac{F}{2m+m} = \frac{F}{3m}$$

Let N_1 be the normal force in case (I) and N_2 in case (II).

$$N_1 = 2m \cdot a$$
$$N_2 = m \cdot a$$
$$\therefore \quad \text{Ratio } N_2 / N_1 = \frac{ma}{2ma} = \frac{1}{2}$$

70. As the block rests on a rough inclined plane,

Force of friction $= mg\sin\theta$

$$10 = m \times 10 \sin 30°$$
$$m = 2 \text{ kg}$$

LEVEL 02 Objective Questions

71. $3T - mg \sin\theta = ma$

$$3\,(250) - 100 \times 10 \times \sin 15° = 100\,a$$
$$a = 4.9 \text{ m/s}^2$$

72. Let the tension be T in the string.

Now, for block having mass 12 kg,

$$12g + \frac{12g}{2} - T - 12a = 0 \quad \text{...(i)}$$

For the block having mass 6 kg,

$$T - 6g - \frac{6g}{2} - 6a = 0 \quad \text{...(ii)}$$

On solving Eqs. (i) and (ii), we get

$$T = 120 \text{ N}$$

Since, the string is attached from both sides,

$$2T = 240$$

The reading in the spring $= \dfrac{240}{g} = 24$ kg

73. If a_{m_1} is a, then a_{m_2} will be $2a$.

For m_1 :

$$m_1 g - 2T = m_1 a \quad \text{...(i)}$$

For m_2 :

$$T = m_2\,(2a) \quad \text{...(ii)}$$

Solving Eqs. (i) and (ii), we get

$$a = \frac{m_1 g}{m_1 + 4m_2}$$

74. $Mg\sin\theta - f = Ma$

$Mg\sin\theta - \mu Mg\cos\theta = Ma \qquad (\because f = \mu N = \mu\, Mg \cos\theta)$

$$a_r = g(\sin\theta - \mu\cos\theta)$$

For smooth surface, $\mu = 0$

$$a_s = g\sin\theta$$

Now, using $\quad v = \sqrt{2as} \propto \sqrt{a}$

$$\frac{v_R}{v_S} = \frac{\sqrt{a_R}}{\sqrt{a_S}}$$
$$\Rightarrow \quad \frac{v}{kv} = \sqrt{\frac{g(\sin\theta - \mu\cos\theta)}{g\sin\theta}}$$
$$\Rightarrow \quad \mu = \tan\theta\left(\frac{k^2-1}{k^2}\right)$$

75. Acceleration of the system,

$$a = \frac{\text{Net force}}{\text{Total mass}}$$
$$= \frac{72 \cos 60°}{8+2+2} = \frac{36}{12} = 3 \text{ m/s}^2$$

Now, FBD of left part

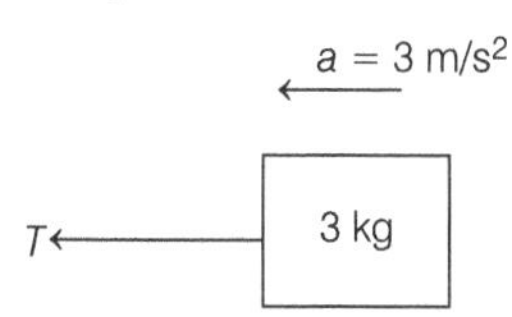

So, tension at the mid-point, $T = ma$

$$= 3 \times 3 = 9 \text{ N}$$

76. For the plank to be in position,

the net force = weight of man

$$F + F + 2F = 80 \times 10$$
$$F = 200 \text{ N}$$

77. A train is having 60 compartments, the mass of each compartment = 4000 kg

So, total mass = $4000 \times 60 = 240000$ kg

$$\therefore \quad F = ma$$
$$\Rightarrow \quad a = \frac{6 \times 10^4}{240000} = \frac{1}{4} \text{ m/s}^2$$

Now, let tension T in the coupling between 40th and 41st compartment,

$$T = 4000 \times 20 \times \frac{1}{4} = 2 \times 10^4 \text{ N}$$

78. $g_{\text{effective}} = g - (-a) = g + a$

So, the acceleration of block relative to inclined is,

$$a = (g + a) \sin\theta$$

79. $2T = 100$

$$T = 50g$$

At point P, tension $= 3T$

$$= 3\,(50)\,(9.8) = 1470 \text{ N}$$

80. Normal force in the vertical direction,

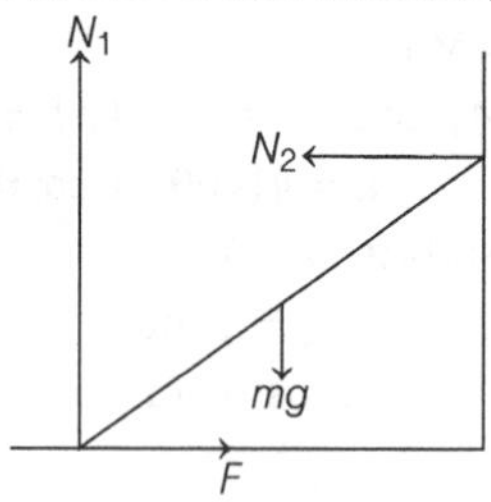

$$N_1 = mg = 250 \text{ N}$$

So, the frictional force at contact point on the floor

$$= 0.3 \times 250 = 75 \text{ N}$$

81. $T_2 + T_1 = 4mg$ ...(i)

Now, $\sum \tau = 0$

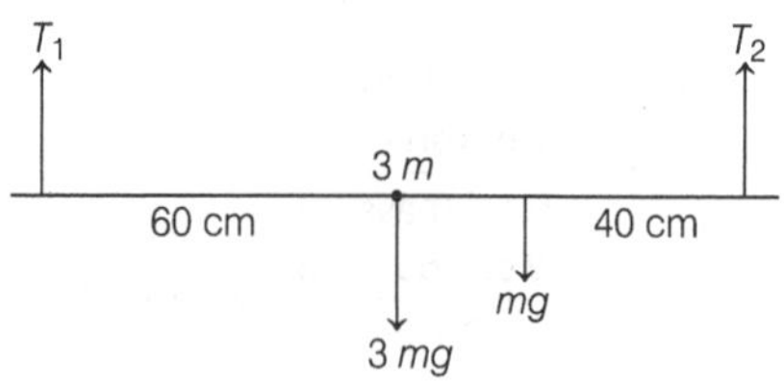

$$= \frac{T_2}{2} - T_1\left(\frac{1}{2}\right) - 3mg\left(\frac{10}{100}\right)$$

$$T_2 - T_1 = \frac{3mg}{5} \quad \text{...(ii)}$$

From Eqs. (i) and (ii), we get

$$T_1 : T_2 = 17 : 23$$

82. Let T_1 and T_2 be the tension in the left and the right string. For rod to be in equilibrium, then $\mathbf{F}_{\text{net}} = 0$, $\tau_{\text{net}} = 0$

From $\mathbf{F}_{\text{net}} = 0$,

$$Mg = T_1 + T_2 \quad \text{...(i)}$$

From $\tau_{\text{net}} = 0$,

$$\frac{MgL}{2} - \frac{3L}{4} T_2 = 0$$

$$T_2 = \frac{2Mg}{3}$$

From Eq. (i), we get

$$Mg = T_1 + \frac{2Mg}{3} \Rightarrow T_1 = \frac{Mg}{3}$$

83. Let F be the buoyant force.

$$F - Mg = \frac{Mg}{6} \Rightarrow F = \frac{7Mg}{6} \quad \text{...(i)}$$

Now, the mass M of sand is put to accelerate at rate of $\frac{g}{6}$ down.

$$(m + M)g - F = (m + M)\frac{g}{6}$$

Substituting the value of F from Eq. (i), we get

$$M = \frac{2m}{5}$$

84. $N = mg\cos\theta$

$$f_r = mg\sin\theta$$

$$\tan\theta = \frac{F}{N} = \mu$$

Now, $v^2 = u^2 + 2as$

$$u^2 = 2s(g\sin\theta + \mu g\cos\theta)$$

$$u^2 = 2s\,(\sin\theta + \tan\theta \times \cos\theta)\,g$$

$$u^2 = 4\,gs\sin\theta$$

$$\Rightarrow \quad s = \frac{u^2}{4g\sin\theta}$$

85. Maximum permissible acceleration $= \frac{F_{\max}}{m} = \frac{30 \times 10}{10}$

$$= 30 \text{ m/s}^2$$

Now, the acceleration (a) upto which the stone can be lifted is

$$= 30 - 10 = 20 \text{ m/s}^2$$

$$\therefore \quad t = \sqrt{\frac{2s}{a}} = \sqrt{\frac{2 \times 10}{20}} = 1 \text{ s}$$

86.

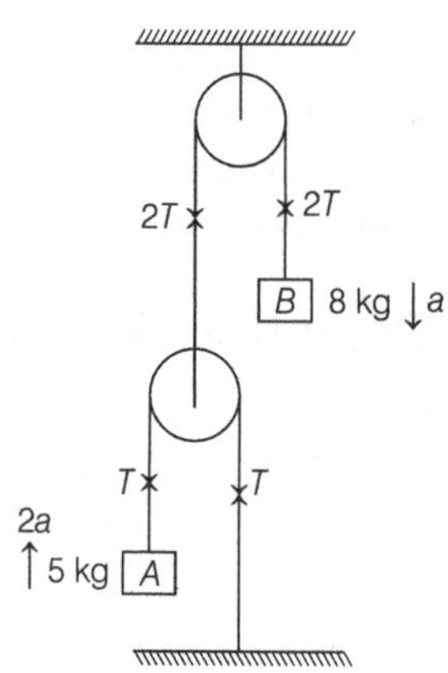

$$80 - 2T = 8a \quad \text{...(i)}$$

$$T - 50 = 5(2a) \quad \text{...(ii)}$$

Solving Eqs. (i) and (ii), we get

$$a = -\frac{5}{7} \text{ m/s}^2$$

$\therefore$ Acceleration of B is $\frac{5}{7}$ m/s^2.

87. Acceleration and tension will be same for both the blocks.

$$8\,mg - T = 8\,ma \quad \text{...(i)}$$

$$5\,mg \sin 37^\circ + T = 5\,ma$$

or
$$3\,mg + T = 5\,ma \quad \text{...(ii)}$$

Solving Eqs. (i) and (ii), we get

$$a = \frac{11g}{13}$$

88. The velocity of block with respect to plank,

$$\mathbf{v}_{BP} = 3\hat{\mathbf{i}} + 4\hat{\mathbf{j}}$$

Now, the direction of friction is opposite to that of motion or along the unit vector in the opposite direction of motion.

$$\hat{\mathbf{n}} = -\frac{(3\hat{\mathbf{i}} + 4\hat{\mathbf{j}})}{5}$$

89.

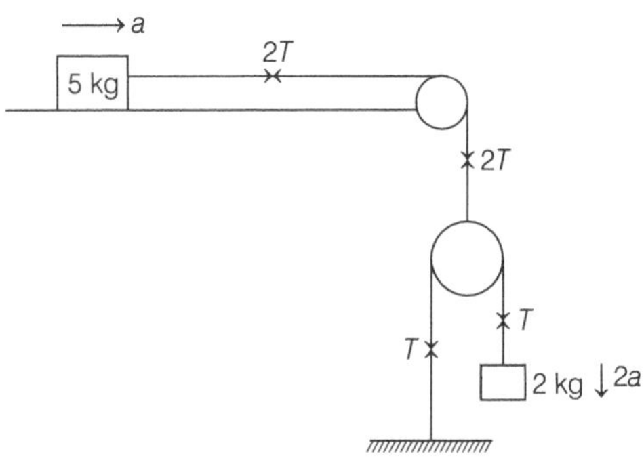

$$20 - T = 2\,(2a) \quad \text{...(i)}$$

$$2T = 5\,(a) \quad \text{...(ii)}$$

Solving Eqs. (i) and (ii), we get

$$T = \frac{100}{13}\text{ N}$$

90.

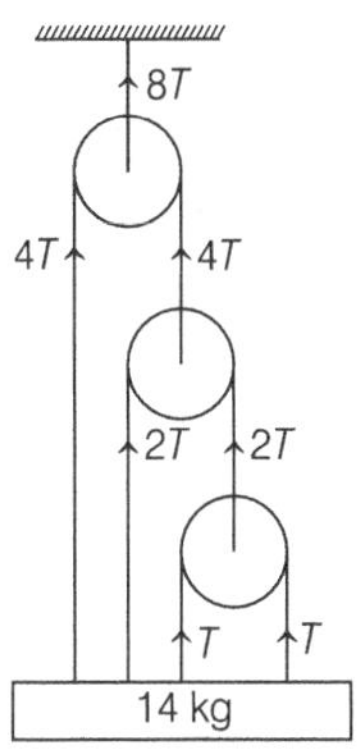

$$8T = 140\text{ N}$$

91. Maximum value of friction between A and B

$$= 0.4 \times 5 \times 10$$

$$= 20\text{ N}$$

Block B moves with this friction.

$$\therefore \quad (a_B)_{max} = \frac{20}{10}$$

$$= 2\text{ m/s}^2$$

If there is no slip between two blocks and they move with a common acceleration, then that common acceleration will be,

$$a = \frac{50}{15}\text{ m/s}^2$$

This, $a > 2\text{ m/s}^2$

So, there will be sliding between two blocks and FBD of two blocks is as shown below

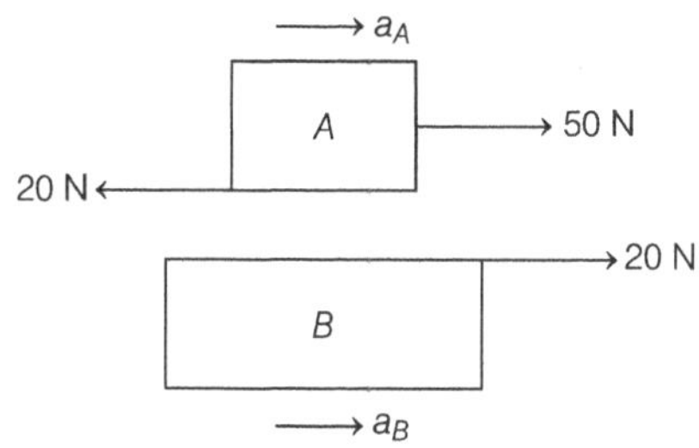

$$a_A = \frac{50 - 20}{5} = 6\text{ m/s}^2$$

$$a_B = \frac{20}{10} = 2\text{ m/s}^2$$

92.

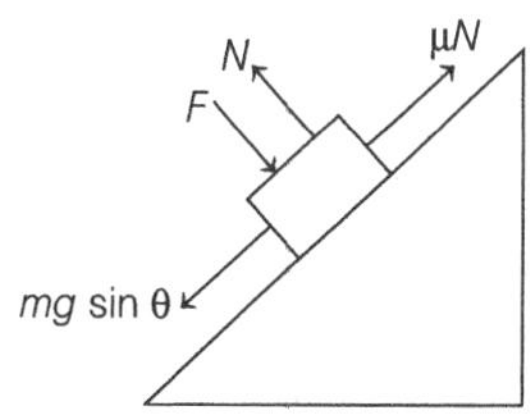

Normal force, $N - F = mg \cos \theta$

$$N = mg \cos \theta + F$$

Friction force, $f_r = \mu N = \mu\,(mg \cos \theta + F)$

For the block to be in equilibrium,

$$mg \sin \theta = \mu\,(mg \cos \theta + F)$$

$$\therefore \quad F = \frac{mg}{\mu}(\sin \theta - \mu \cos \theta)$$

93.

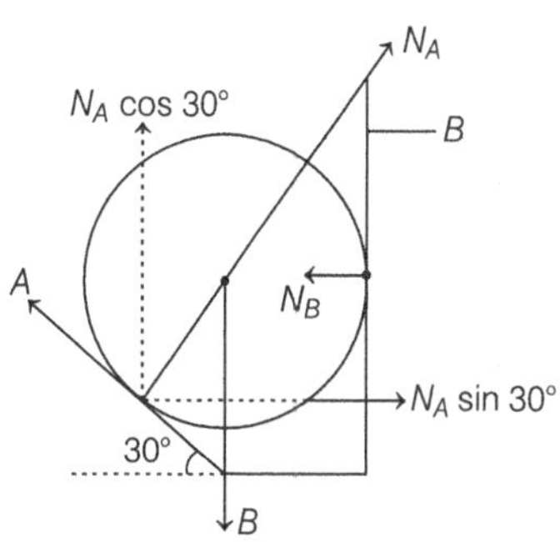

$$N_A \cos 30^\circ = 50\,g$$

$\Rightarrow \quad N_A = \frac{1000}{\sqrt{3}}$ N

and $\quad N_B = N_A \sin 30° = \frac{500}{\sqrt{3}}$ N

94. Force with which man should have pulled rope,

$$4F = 600 \Rightarrow F = 150 \text{ N}$$

95. As we know the angle of repose

$$= \tan^{-1}(\mu_s) = \tan^{-1}\left(\frac{3}{4}\right) = 37°$$

Here, the angle of inclination is less than the angle of repose, so block will not slide. Hence, the tension in string is zero.

96. For the length of string to be constant,

$$v_{block} = v \cos \alpha$$

97. Let a be the acceleration of blocks B and C. For block B, net acceleration $= 1 \text{ m/s}^2$

$$2 - a = 1$$

$\Rightarrow \quad a = 1 \text{ m/s}^2$

Now, acceleration of block C,

$$a_c = 2 + a = 3 \text{ m/s}^2 \uparrow$$

98. For block of mass 4 kg,

$$F \cos 60° - T_1 = 4a \quad \text{...(i)}$$

Now, for block of mass 3 kg,

$$T_1 - 30 = 3a$$

$\Rightarrow \quad T_1 = 30 + 3a$

Similarly, for block of mass 1 kg,

$$30 - T_2 = 2a$$

$$30 - a = 2a \quad (\because T_2 = a)$$

$\Rightarrow \quad a = 10 \text{ m/s}^2$

Now, substituting the value of T_1 and a in Eq. (i), we get

$$F = 200 \text{ N}$$

99. $a = \frac{\text{Net pulling force}}{\text{Total mass}} = \frac{F}{m_T}$ (say)

Total mass in both cases is same.

$$F_1 = (m_1 - m_2)g = 10 \text{ N}$$

$$F_2 = (m_2 - m_1 \sin 30°)\, g = 10 \text{ N}$$

$\therefore \quad a_1 = a_2$

100. For system to be in equilibrium,

$$T_1 \sin\theta = T_2 = 400 \text{ N}$$

$$T_1 \cos\theta = T_3 = 300 \text{ N}$$

So, $\tan \theta = \frac{4}{3}$

$\therefore \quad T_1 \sin\theta = 400$

$$T_1\left(\frac{4}{5}\right) = 400$$

$\Rightarrow \quad T_1 = 500 \text{ N}$

101. For the length of a string to be constant, the horizontal components must be equal.

$$v_1 \cos\theta_1 = v_2 \cos\theta_2$$

$\Rightarrow \quad \frac{v_1}{v_2} = \frac{\cos\theta_2}{\cos\theta_1}$

102. Normal reaction between A and B,

$$N = mg \cos \theta$$

Now, horizontal component of this N acting on A is balanced by the tension in the string CD. Thus,

$$T = N \sin \theta$$

$$= mg \sin \theta \cdot \cos \theta$$

$$= \frac{mg}{2} \sin 2\theta$$

103. Let F_1 and F_2 be the force applied by the man.

In first case, $F_1 - mg = ma$

$$F_1 = mg + ma \quad \text{...(i)}$$

In second case, $2F_2 - mg = ma$

$$F_2 = \frac{mg + ma}{2} \quad \text{...(ii)}$$

From Eqs. (i) and (ii), we get

$$F_1 > F_2$$

104. $s = ut + \frac{1}{2}at^2 \Rightarrow l = \frac{1}{2}(g\sin\theta)t^2$

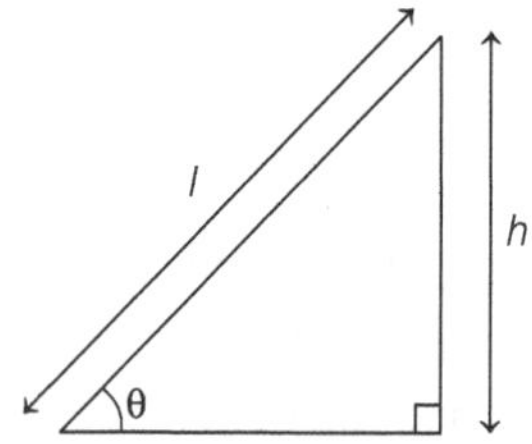

From right angle triangle,

$$l = \frac{h}{\sin\theta}$$

$$\frac{h}{\sin\theta} = \frac{1}{2} g \sin\theta\, t^2$$

$\Rightarrow \quad t = \frac{1}{\sin\theta}\sqrt{\frac{2h}{g}}$

105. Let the displacement of m_1 be x_1 and m_2 be x_2 and the displacement of m_3 be y.

$$y = \frac{x_1 + x_2}{2}$$

$$2y = x_1 + x_2$$

Differentiating this, we get

$$2\frac{dy}{dt} = \frac{dx_1}{dt} + \frac{dx_2}{dt}$$

Differentiating again,

$$2\frac{d^2y}{dt^2} = \frac{dx_1^2}{dt} + \frac{dx_2^2}{dt}$$

$$2a_3 = a_1 + a_2$$

106. We have, $N = mg\cos\theta$

So, the minimum force F required to move the block,

$$F = mg\sin\theta + f_r$$

$$F = mg\sin\theta + \mu mg\cos\theta \quad (\because f_r = \mu N)$$

107. Let T_1 be the tension in the string connected with m_2 and T_2 with m_1.

$$T_1 = m_2 g$$
$$T_2 + m_1 g = T_1$$
$$T_2 + m_1 g = m_2 g$$
$$T_2 = (m_2 - m_1)g$$

108. From ground frame of reference

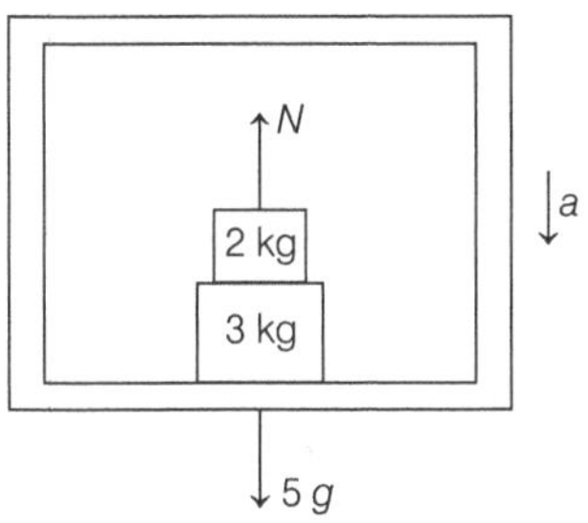

If m is the combined mass of 3 kg and 2 kg block,

$$mg - N = ma$$
$$5\,(10) - 40 = 5a$$
$$a = 2\text{ m/s}^2$$

Therefore, lift is moving downwards with acceleration,

$$a = 2\text{ m/s}^2$$

109.
$$2g - T = 2a \quad \text{...(i)}$$
$$T - g = a \quad \text{...(ii)}$$

From Eqs. (i) and (ii), we get

$$T = \frac{4g}{3}$$

So, the reading in the spring balance

$$= T = \frac{40}{3}\text{ N}$$

110.

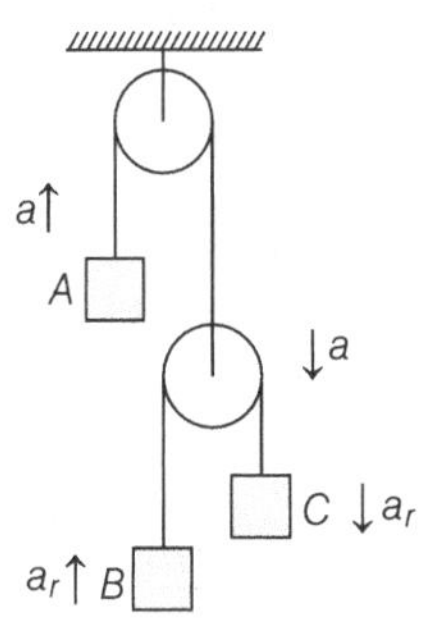

$$a_r - a = +4\text{ m/s}^2 \quad \text{...(i)}$$
$$a_r + a = +6\text{ m/s}^2 \quad \text{...(ii)}$$

Solving Eqs. (i) and (ii), we get

$$a = 1\text{ m/s}^2 \quad \text{(upward)}$$

111. $T_1 \sin\theta = 50$ ...(i)

$$T_1 \cos\theta = T_2 = 60 \quad \text{...(ii)}$$

On dividing Eq. (i) by Eq. (ii), we get

$$\tan\theta = \frac{5}{6}$$

or $$\theta = \tan^{-1}\left(\frac{5}{6}\right)$$

112. $T = mg$

$$2T + 2T + T = 15g$$
$$5T = 15g$$
$$m = 3\text{ kg}$$

113. Kinetic friction on both blocks,

$$f = 0.1 \times 10 \times 10 = 10\text{ N}$$

F_s = spring force

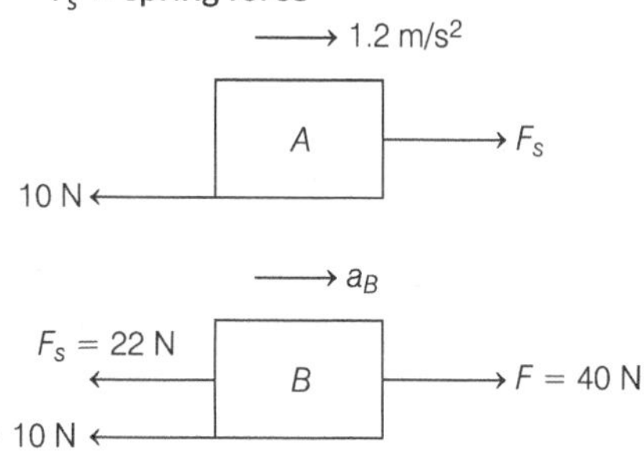

$$F_s - 10 = 10 \times 1.2$$

$$\therefore \quad F_s = 22\text{ N}$$

$$a_B = \frac{40 - 22 - 10}{10}$$
$$= 0.8\text{ m/s}^2$$

114.

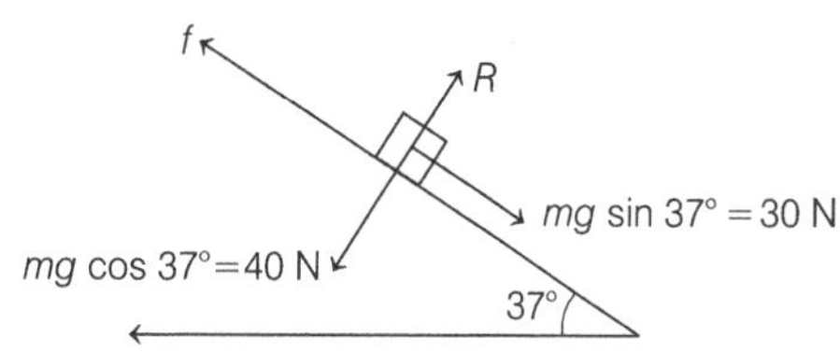

Normal reaction, $R = 40$ N

Maximum value of friction, $f_{max} = \mu R = 32$ N

Since, $f_{max} > 30$ N

$\therefore$ Block is at rest and $f = 30$ N.

Net contact force = Weight or 50 N in upward direction.

115. $F = kt^2$

$$= 2t^2 \quad (\because k = 2)$$
$$F_{max} = \mu mg$$
$$= 0.8\,(4)(10)$$
$$= 32\text{ N}$$

Applied force, $F = 2t^2 = 8$ N

At $t = 2$ s,

So, $F < F_{max}$, hence the block will not move. Frictional force between block and plane is 8 N.

116. Let T_1 be tension in left part of system and T_2 in other part.

We know, $T = \frac{2m_1m_2g}{m_1+m_2}$

So, we get $T_1 = 24$ N, $T_2 = 16$ N

$2T_1 = 48$ N, $2T_2 = 32$ N

$$\Sigma\tau_0 = 0$$
$$= 2T_1 \times 10.5 - 2T_2(10.5)$$
$$= (48)\times 10.5 - 32(10.5)$$
$$= 168$$

Since, the value is positive, so the left part will move downward.

117. From Newton's second law,

$$F = ma_1$$
$$\Rightarrow \quad a_1 = \frac{50}{250} = 0.2 \text{ m/s}^2$$

Similarly, $a_2 = \frac{50}{500} = 0.1 \text{ m/s}^2$

Now, relative acceleration $= 0.2 + 0.1 = 0.3 \text{ m/s}^2$

$$\therefore \quad v = at$$
$$v = 0.3\ (5) = 1.5 \text{ m/s}$$

118. B will move due to friction and friction is absent. So,

$$a_B = 0$$
$$a_A = a_C = \frac{\text{Net pulling force}}{\text{Total mass}}$$
$$= \frac{m_A g}{m_A + m_C} = \frac{g}{2}$$

119. At equilibrium position,

$$kx = mg$$
$$\Rightarrow \quad x = \frac{mg}{k}$$

Now, x' be the new extension in the spring,

$$x' = \frac{mg}{k} + \frac{mg}{4k}$$
$$= \frac{5mg}{4k} \quad \text{(i)}$$

Now, $kx' - mg = ma$

$$k \times \frac{5mg}{4k} - mg = ma \quad [\because \text{using Eq. (i)}]$$
$$\Rightarrow \quad a = \frac{g}{4}$$

120. $T_{max} = w + F$

$$300 = 200 + F$$
$$F = 100 \text{ N}$$
$$\therefore \quad a = \frac{F}{m} = \frac{100}{20} = 5 \text{ m/s}^2$$

Now, using $s = ut + \frac{1}{2}at^2$

$$40 = \frac{1}{2}(5)t^2 \quad (\because u = 0)$$
$$\Rightarrow \quad t = 4 \text{ s}$$

121. $F\cos\theta \geq mg\sin\theta$

$$F \geq mg \tan\theta$$
$$F \geq 4\ (10)\left(\frac{3}{4}\right)$$

$\therefore F = 35$ N can push the block up the inclined plane.

122. $f_{max} = 0.6 \times 8 \times 10 = 48$ N

Block will start moving, when $F = f_{max}$

Angle 45° means :

$$t = F = f_{max} = 48$$
$$\therefore \quad t = 48 \text{ s}$$

123. For the system to be in equilibrium,

$$N_B \sin 30° = N_A \sin 60°$$
$$N_B = \sqrt{3}\, N_A$$

124. The reading of the spring balance will be

$$N + 100 \sin 37° = 100$$
$$N = 100 - 100\left(\frac{3}{5}\right)$$
$$= 40 \text{ N}$$

125. From FBD of A,

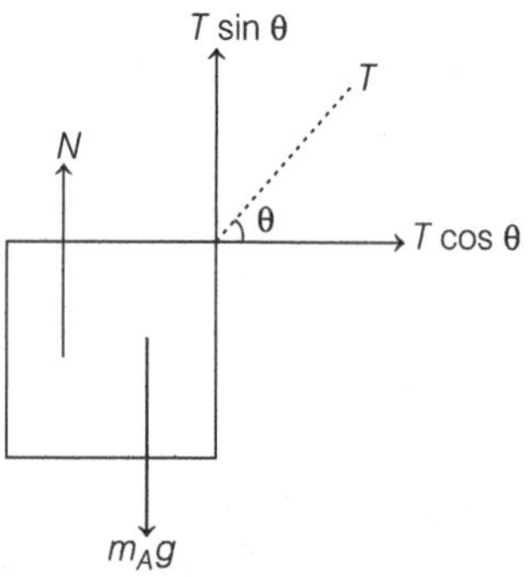

$$N + T\sin\theta = m_A g \quad \text{...(i)}$$
$$T\cos\theta = F_{max} = \mu N \quad \text{...(ii)}$$

From Eq. (i), we get

$$N = m_A g - T\sin\theta \quad \text{...(iii)}$$

From Eqs. (ii) and (iii), we get

$$T\cos\theta = \mu m_A g - \mu T\sin\theta \quad \text{...(iv)}$$
$$T(\cos\theta + \mu\sin\theta) = \mu m_A g \quad \text{...(v)}$$

On taking ratio of Eqs. (iv) and (v), we get

$$\frac{m_A}{m_B} = \frac{11}{5}$$

126. Considering the blocks as a system, net force down the plane $= (4+2)\, g\sin 30° - (\mu_1(2)g\cos 30° + \mu_2(4)g\cos 30°)$

$$= 30 - 13.76 = 16.24 \text{ N}$$

So, the acceleration of both the block will be

$$a = \frac{\text{net force}}{\text{total mass}} = \frac{16.24}{6} = 2.7 \text{ m/s}^2$$

127. Since, the block is in equilibrium,

$$f + \frac{mg}{2} = mg$$

$$\mu N = mg - \frac{mg}{2}$$

$$\mu mg = \frac{mg}{2}$$

$$\mu = 0.5$$

128. $$T - 2g = 4\,(2a) \quad \text{...(i)}$$

$$12g - 2T = 12a \quad \text{...(ii)}$$

From Eqs. (i) and (ii), we get

$$a = \frac{2g}{7} \text{ m/s}^2$$

Substituting the value of a in Eq. (i), we get

$$T = \frac{60}{7} g \text{ N}$$

129. We have, $a = \mu g = 0.5\,(10) = 5 \text{ m/s}^2$

So, $v^2 = u^2 + 2as$

$$0 = 2^2 + 2\,(5)\,s$$

$s = 0.4$ m with respect to belt.

130. We have, $a_{max} = \mu g$

So, the force at an instant when the slipping starts is,

$$F = (m_1 + m_2)(\mu g) \Rightarrow t_0 = \frac{(m_1 + m_2)}{a} \mu g$$

131. Let F be the force applied by the second boy on the box.

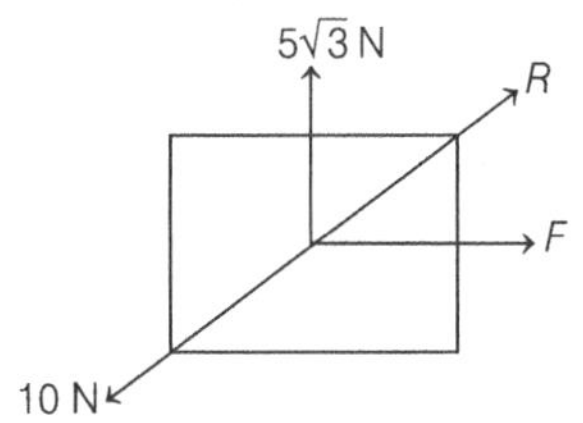

For box to be in equilibrium,

$$10 = \sqrt{(5\sqrt{3})^2 + F^2}$$

$$F = 5 \text{ N}$$

132. $N_2 = N_1 \sin 30°$

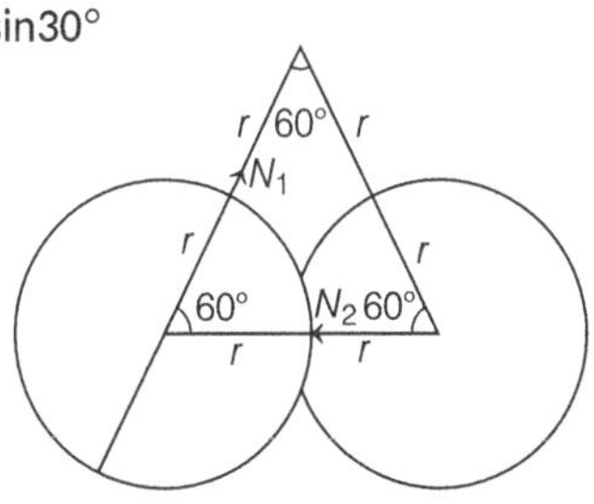

$$\frac{N_1}{N_2} = 2$$

133. Case-I $a = \frac{50}{1+1} = 25 \text{ m/s}^2$

Contact force between a and b is,

$$F = 1 \times 25 = 25 \text{ N}$$

Case-II Contact force between A and B is,

$$50 - 10 = 40 \text{ N}$$

Case-III $$N_{AB} - 10 = 25$$

$$N_{AB} = 35 \text{ N}$$

Case-IV $N_{AB} = 35\,(1) = 35$ N

So, the contact force between A and B is maximum in the second case.

134. $N = 40$ N

$$f_{max} = \mu N = 20 \text{ N}$$

Force tangential to the wall,

$$F = \sqrt{(15)^2 + (20)^2} \quad (\because w = 20 \text{ N})$$

$$= 25 \text{ N}$$

Note : 15 N and $w = 20$ N are perpendicular to each other and along the wall.

Since, $F > f_{max}$

$\therefore$ Force of friction that will act on the block,

$$f = f_{max} = 20 \text{ N}$$

and $$a = \frac{F - f}{m} = \frac{25 - 20}{2}$$

$$= 2.5 \text{ m/s}^2$$

Note : Direction of a is in the direction of F.

135. $$T - f_1 + mg\sin\theta = ma \quad \text{...(i)}$$

$$mg\sin\theta - T - f_2 = Ma \quad \text{...(ii)}$$

On solving Eqs. (i) and (ii), we get

$$T = 0$$

136. From figure, we have

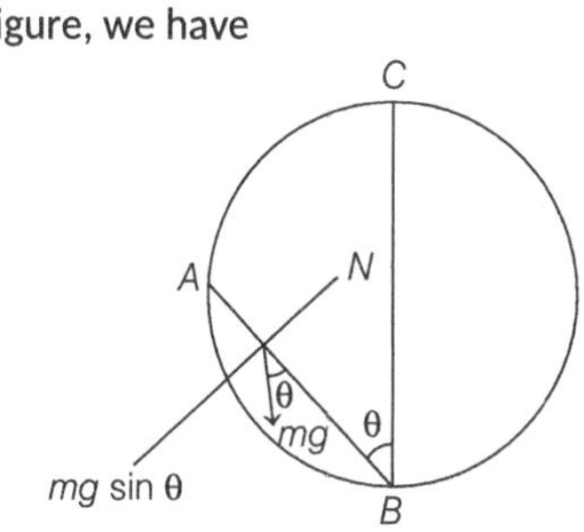

$$BC = 2r,\ AB = BC\cos\theta$$

$$AB = 2r\cos\theta$$

Now, $mg\cos\theta = ma$

$$\Rightarrow \quad a = g\cos\theta$$

Using $s = ut + \frac{1}{2}at^2$

$$s = 0 + \frac{1}{2} g\cos\theta\, t^2 \quad (\because u = 0)$$

$$2r\cos\theta = \frac{1}{2}g\cos\theta\, t^2$$

$$\Rightarrow \qquad t = \sqrt{\frac{4r}{g}}$$

137. We have, $u = 4$ m/s, $t = 0.1$ s

$\therefore \qquad v = u + at$

$a = -40$ m/s $\qquad (\because v = 0)$

$$s = 4(0.1) + \frac{1}{2}(-40)(0.1)^2 = 0.2 \text{ m}$$

138. The tension (T_1) in the spring $= kx$

$$T_1 = k(0.4)$$

In horizontal direction,

$$T_1 \cos 37° = T_2 \cos 53°$$

$$0.4\,k\,(4) = 3T_2 \qquad \text{...(i)}$$

In vertical direction,

$$T_1 \sin 37° + T_2 \sin 53° = 200$$

$$\frac{3T_1}{5} + \frac{4T_2}{5} = 200 \qquad \text{...(ii)}$$

On solving Eqs. (i) and (ii), we get

$$k = 3000 \text{ N/m}$$

139. For block of mass $2m$,

$$T = 2mg$$

For block of mass m, $T = kx + mg$

$$kx = 2mg - mg = mg$$

If the spring is cut, then tension in string will become zero.

$\therefore$ Acceleration of block having mass, $2m = \dfrac{\text{Net force}}{\text{Total mass}}$

$$= \frac{2mg}{2m} = g\downarrow$$

Similarly, acceleration of block having mass,

$$m = \frac{\text{net force}}{\text{total mass}} = \frac{kx + mg}{m} = \frac{mg + mg}{m} = 2g\downarrow$$

140. Net pulling force,

$$F = 5 \times 10 \times \sin 53° - 5 \times 10 \times \sin 37°$$

$$= 10 \text{ N}$$

$$(f_A)_{max} = 0.5 \times 5 \times 10 \times \cos 37° = 20 \text{ N}$$

$$(f_B)_{max} = 0.5 \times 5 \times 10 \times \cos 53° = 15 \text{ N}$$

Since, $(f_A)_{max} + (f_B)_{max} > F$

$\therefore$ System will not move and free body diagram is as shown below.

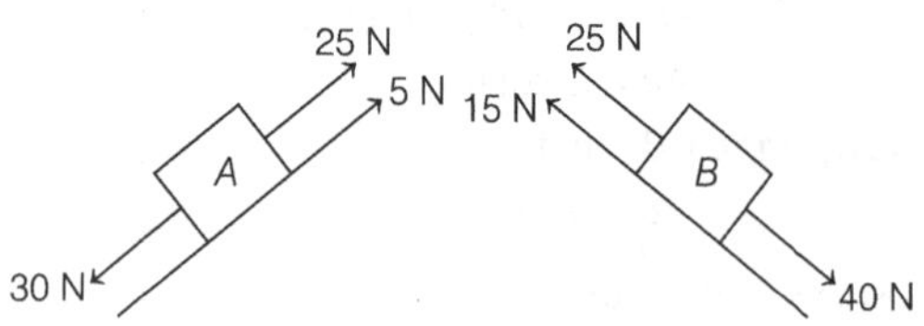

$$T = 25 \text{ N}$$

$$(f_B) = (f_B)_{max} = 15 \text{ N}$$

141. Normal reaction between A and B,

$$N - 20g = 20\,(5)$$

$$N = 300 \text{ N}$$

Normal reaction between B and C,

$$N - (300 + 30g) = 30(5)$$

$$N = 750 \text{ N}$$

Normal reaction between C and elevator,

$$N - 750 - 50g = 50(5)$$

$$N = 1500 \text{ N}$$

Net force action on B is $= ma_{net}$

$$= 30\,(5) = 150 \text{ N}$$

142. Let T be the tension in the figure (i).

$$F - T_1 = m_1 a$$

$$50 - T_1 = 2a \qquad \text{...(i)}$$

$$T_1 - m_2 = m_2 a$$

$$T_1 - 80 = 8a \qquad \text{...(ii)}$$

On solving Eqs. (i) and (ii), we get

$$T_1 = 56 \text{ N}$$

For figure (ii),

$$F + 20 - T_2 = 2a \qquad \text{...(iii)}$$

$$T_2 - 80 = 8a \qquad \text{...(iv)}$$

On solving Eqs. (iii) and (iv), we get

$$T_2 = 72$$

$$\therefore \qquad \frac{T_1}{T_2} = \frac{7}{9}$$

143. Frictional force, $f_r = \mu N$

Here, $N = F$

$\therefore \qquad f_r = \mu F$

Now, $f_r = M\,(g + a)$

$$\mu F = M(g + a)$$

$$0.5\,F = 10\,(10 + 2)$$

$$F = 240 \text{ N}$$

144.

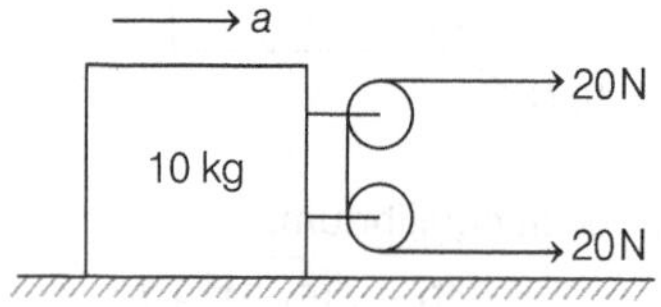

$$a = \frac{40}{10} = 4 \text{ m/s}^2$$

145. Let T_1 be the tension connected with mass 3 kg and T_2 with mass 2 kg.

$$T_1 = 2T_2$$

$$T_1 = 2 \times 2g = 4g$$

$$T_2 = 2g$$

So, $T_2 = 4g$(min)

So, minimum force, $F = 4g \times 2 = 8g$

146. Acceleration, $a = \dfrac{4g - 1g}{4+1} = 6\ \text{m/s}^2$

Now, the relative acceleration of A with respect to block B,

$$a = 12\ \text{m/s}^2$$

$$s = ut + \frac{1}{2}at^2$$

$$s = \frac{1}{2} \times 12t^2 \qquad [\because u = 0]$$

$$6 = 6t^2$$

[$\because s$ = length of block A + length of block B

$s = (2+4)\ \text{m} = 6\text{m}$]

$$t = 1\ \text{s}$$

147. We have, $AB = 30$ cm

$$BC = 54 - 30 = 24\ \text{cm}$$

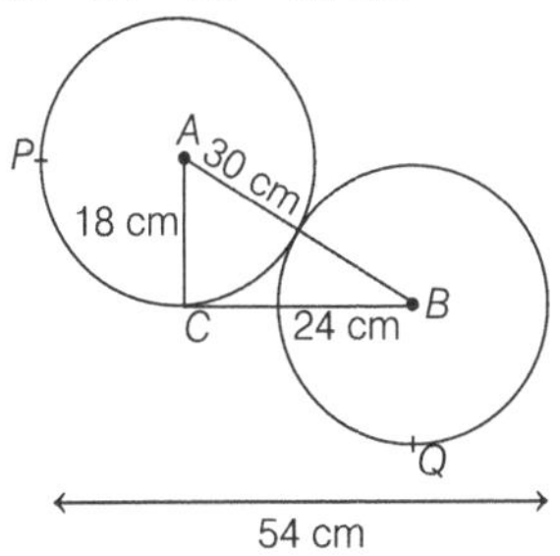

From ΔABC,

$$AC = \sqrt{(30)^2 - (24)^2} = 18\ \text{cm}$$

Now, the FBD of A,

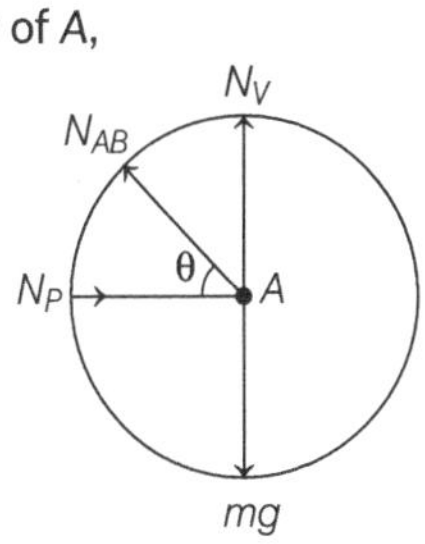

$$N_V = m_A g = 150\ \text{N}$$

FBD of B,

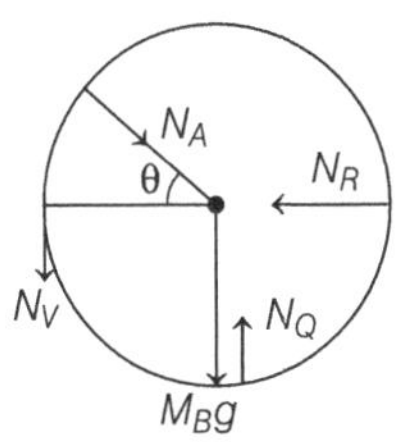

$$N_Q = 150 + 60 \times 10 = 750\ \text{N}$$

Now, $\Sigma F_x = 0$

$$N_R = N_{AB} \cos\theta$$

$$N_R = \frac{250 \times 24}{30} = 200\ \text{N}$$

148. Let f_A and f_B be the friction between the shoes and wall, and between his back and wall respectively.

As the climber is at rest, $N_A = N_B = N$

$$f_A + f_B = mg$$

$$(\mu_A + \mu_B)N = mg$$

$$N = \frac{mg}{\mu_A + \mu_B} = \frac{49 \times 10}{1.2 + 0.8} = 245\ \text{N}$$

Now, frictional force f_A between shoes and wall is,

$$f_A = 1.2 \times 245 = 294\ \text{N}$$

$\therefore$ Percentage of weight supported by frictional force on legs

$$= \frac{294}{490} \times 100\% = 60\%$$

149. We have,

$$\mathbf{F}_1 + \mathbf{F}_2 + \mathbf{F}_3 + \mathbf{F}_4 + \mathbf{F}_5 = ma$$

$$= 4\,(2) = 8\hat{\mathbf{i}} \qquad \text{...(i)}$$

Now, the force $\mathbf{F}_1$ is removed,

$$\mathbf{F}_2 + \mathbf{F}_3 + \mathbf{F}_4 + \mathbf{F}_5 = 2\,(7) = 14\hat{\mathbf{j}} \qquad \text{...(ii)}$$

From Eqs. (i) and (ii), we get

$$\mathbf{F}_1 = 8\hat{\mathbf{i}} - 14\hat{\mathbf{j}}$$

$$a = \frac{\mathbf{F}_1}{m} = \frac{8\hat{\mathbf{i}} - 14\hat{\mathbf{j}}}{2}$$

$$= 4\hat{\mathbf{i}} - 7\hat{\mathbf{j}}$$

$$|\mathbf{a}| = \sqrt{65}\ \text{m/s}^2$$

150. FBD of man

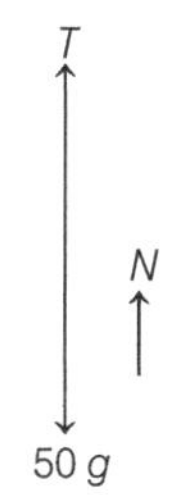

$$T + N = 50g \qquad \text{...(i)}$$

Similarly, the FBD of box,

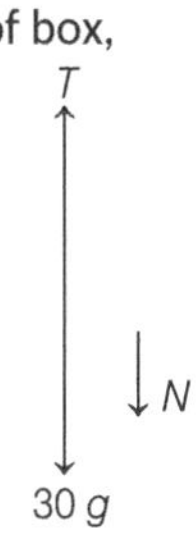

$$T = N + 30g \qquad \text{...(ii)}$$

On solving Eqs. (i) and (ii), we get

$$N = 10g = 100\ \text{N}$$

151. $T_1 = M_1 g$

$$T_3 = \mu M_3 g \qquad [\because N = M_3 g]$$

$$= 0.25 \times 4 \times 10 = 10\ \text{N}$$

$$T_1 = M_2 g \sin 37^\circ + \mu M_2 g \cos 37^\circ + T_3$$

$$M_1 g = 4(10)\left(\frac{3}{5}\right) + 0.25\,(4)\,(10)\left(\frac{4}{5}\right) + 10$$

$$M_1 = 4.2 \text{ kg}$$

152. At time 2 s, $F_1 = 2\,(2) = 4$ N

$$f_1 = \mu_1 m_1 g = 0.6\,(1)(10) = 6 \text{ N}$$

$$f_2 = \mu_2 m_2 g = 0.5\,(2)(10) = 10 \text{ N}$$

As the total frictional force (10 + 6) N is more than that of the net force applied.

So, the system will not move.

For block *B*,

$$15 - 10 - T = 0$$

$$\Rightarrow \quad T = 5 \text{ N}$$

For block *A*, $T - f - 4 = 0$

$$5 - f - 4 = 0$$

$$\Rightarrow \quad f = 1 \text{ N} \leftarrow$$

So, frictional forces on both blocks are (1N ←) and (10N ←).

153. The maximum frictional force acting between the blocks is,

$$f_{max} = \mu mg = \frac{1}{2} \times 2 \times 10$$

$$= 10 \text{ N}$$

So, the acceleration of system

$$= \frac{\text{Net force}}{m_1 + m_2} = \frac{20 - 2}{4 + 2} = 3 \text{ m/s}^2$$

The system has an acceleration towards left side, so the frictional force will act in the rightward direction.

So, the force of friction acting between the blocks

$$= 2 \times 3 + 2 = 8 \text{ N}$$

154. The slope of v - t graph gives acceleration,

$$a = \frac{12 - 0}{10 - 6} = 3 \text{ m/s}^2$$

$$mg - T = ma$$

$$T = m\,(g - a)$$

$$= 1600\,(10 - 3) = 11200 \text{ N}$$

155.

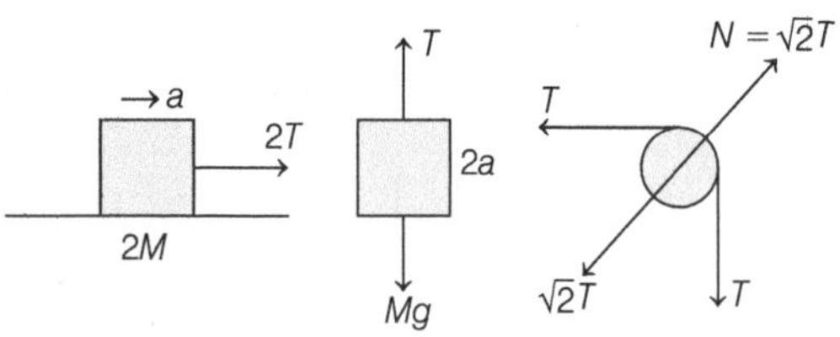

$$2T = 2Ma \qquad \text{...(i)}$$

$$Mg - T = M(2a) \qquad \text{...(ii)}$$

Solving Eqs. (i) and (ii), we get

$$a = \frac{g}{3},\; 2a = \frac{2g}{3},\; T = \frac{Mg}{3}$$

$$2\sqrt{2}\,T = \frac{\sqrt{2}Mg}{3}$$

156. Pseudo force = *ma*

For the collar to be in fixed position,

$$N\cos 30° = mg \qquad \text{...(i)}$$

$$N\sin 30° = ma \qquad \text{...(ii)}$$

On dividing Eq. (i) by Eq. (ii), we get

$$\therefore \quad a = \frac{g}{\sqrt{3}} \text{ m/s}^2$$

157. Maximum values of friction

(i) between *C* and ground,

$$(f_1)_{max} = 0.5 \times 60 \times 10 = 300 \text{ N}$$

(ii) between *B* and ground,

$$(f_2)_{max} = 0.3 \times 120 \times 10 = 360 \text{ N}$$

(iii) between *A* and *B*,

$$(f_3)_{max} = 0.4 \times 60 \times 10 = 240 \text{ N}$$

Now, free body diagrams are as shown below

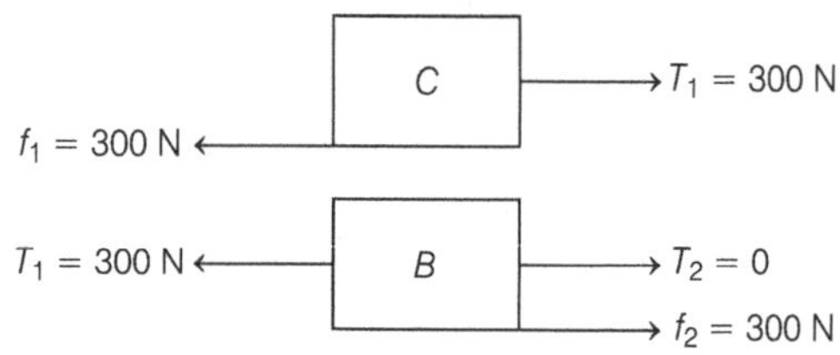

So, tension T_2 connected to *A* is zero.

158. $a = g\sin\theta$

Now, its horizontal component,

$$a_x = g\sin\theta\cos\theta = \frac{1}{2} g \sin 2\theta$$

$$(a_x)_{max} = \frac{g}{2}$$

So, $(a_x)_{max} = \frac{g}{2}$, when value of $\theta = 45°$

$$\therefore \quad \theta = 45°$$

159.

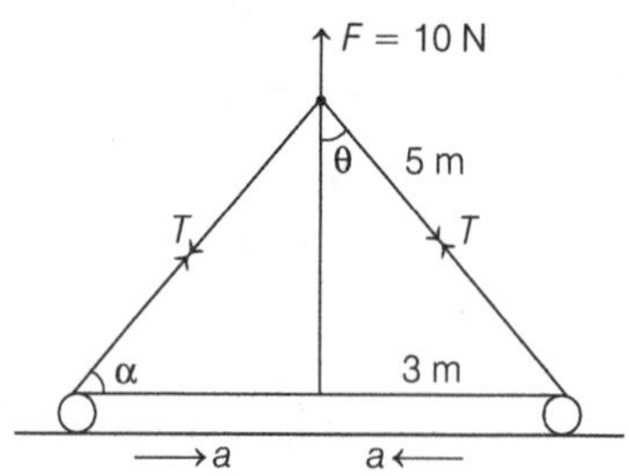

$$\sin\theta = \frac{3}{5}$$

$$\therefore \quad \theta = 37°$$

$$\text{or} \quad \alpha = 53°$$

$$10 = 2T\cos\theta = (2T)\left(\frac{4}{5}\right)$$

$\therefore \quad T = \frac{50}{8}$ N

Relative acceleration = $2a$

$$= \frac{2T\cos\alpha}{m} = \frac{2 \times \frac{50}{8} \times \frac{3}{5}}{1.5}$$

$$= 5 \text{ m/s}^2$$

160. For the complete system,

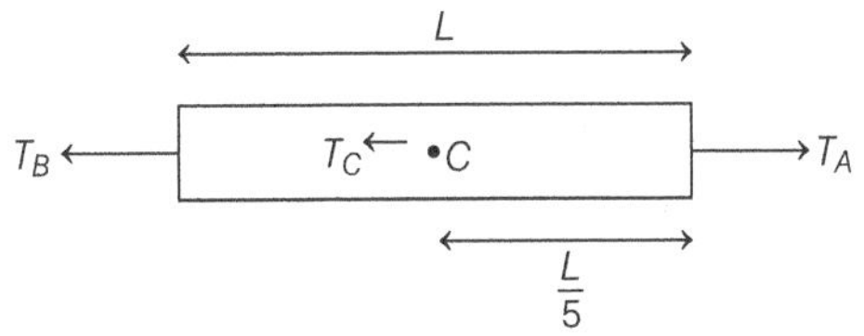

$$T_B - T_A = Ma$$

$$T_A = T_B - Ma \quad \text{...(i)}$$

Now, considering right part as a system,

$$T_C - T_A = \frac{M}{5}a \quad \text{...(ii)}$$

From Eqs. (i) and (ii), we get

$$T_C = \frac{(4T_A + T_B)}{5}$$

161. We have given,

$$T_4 = 2T_8$$

Let NM be the mass of total carts.

So, its acceleration, $a = \frac{F}{NM}$

Now, $T_4 = (N-4)M\frac{F}{MN}$

$$T_4 = (N-4)\frac{F}{N}$$

Similarly, $T_8 = (N-8)\frac{F}{N}$

So, $T_4 = 2T_8$

$$\therefore \quad (N-4)\frac{F}{N} = 2(N-8)\frac{F}{N}$$

$$N = 12$$

So, total number of blocks is 12.

162. From Newton's second law,

$$mg - T = ma$$

$$4 \times 10 - T = 4 \times 2$$

$$T = 32 \text{ N}$$

Now, tension in the spring,

$$T' = 2T = kx$$

$$64 = 100\,x$$

$$x = 64 \text{ cm}$$

163. Common acceleration of system $= \frac{F}{5m}$

Now, let N_2 be the normal force between B and C.

$$N_2 = 2ma = 2m\left(\frac{F}{5m}\right) = \frac{2F}{5}$$

Maximum frictional force, $f_{max} = \mu N_2$

$$= \frac{2}{5}\mu F$$

$$\therefore \quad \frac{2}{5}\mu F \geq mg \Rightarrow F = \left(\frac{5}{2\mu}\right)mg$$

164. Let F be the force applied at an angle θ.

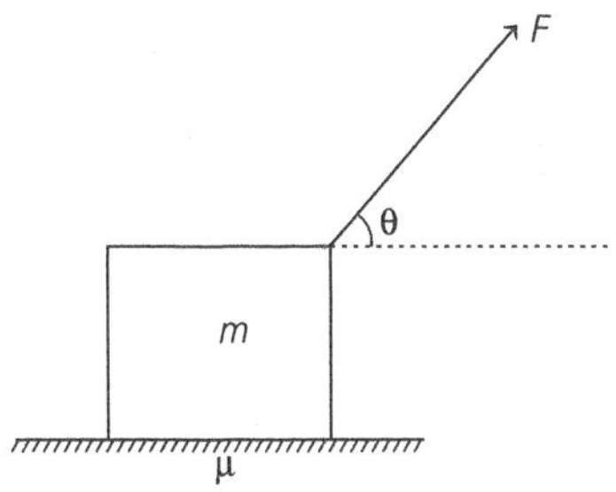

Normal reaction, $N = mg - F\sin\theta$

For the block to move,

$$F\cos\theta = f_r = \mu N$$

$$F\cos\theta = \mu(mg - F\sin\theta)$$

$$F = \frac{\mu mg}{\cos\theta + \mu\sin\theta} \quad \text{...(i)}$$

For F to be minimum, denominator should be maximum.

$$\frac{d}{d\theta}(\cos\theta + \mu\sin\theta) = 0$$

$$-\sin\theta + \mu\cos\theta = 0$$

$$\theta = \tan^{-1}(\mu)$$

Substituting the value of θ in Eq. (i), we get

$$F_{min} = mg\sin\theta$$

$$F_{min} = \frac{\mu mg}{\sqrt{1+\mu^2}} \quad \left(\because \sin\theta = \frac{\mu}{\sqrt{1+\mu^2}}\right)$$

165. Acceleration of the system, $a = \frac{\text{Net force}}{\text{Total mass}}$

$$= \frac{F}{10} \text{ m/s}^2$$

$$N = 4a = \frac{4F}{10}$$

For the 1 kg block to be in equilibrium,

$$mg = \mu N$$

$$10 = 0.5 \times \frac{4F}{10}$$

$$F = 50 \text{ N}$$

166. Acceleration of the system, $a = \frac{\text{Net force}}{\text{Total mass}}$

$$= \frac{100}{10+5+10} = 4 \text{ m/s}^2$$

Now, FBD of block A,

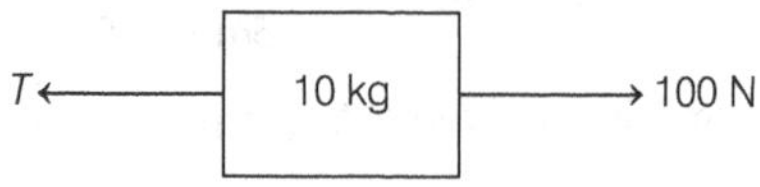

$$100 - T = 10a$$
$$100 - T = 10 \times 4$$
$$\therefore \quad T = 60 \text{ N}$$

Now, FBD of block C,

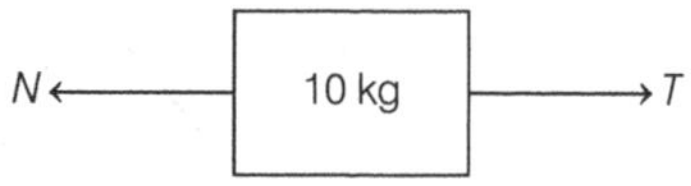

$$T - N = 10\,a$$
$$60 - N = 10 \times 4$$
$$\therefore \quad N = 20 \text{ N}$$

So, contact force between blocks B and C is 20 N.

167. $T_B = m_2(g + a)$

$$= m_2(10 + 2) = 12m_2$$
$$T_A = T_B + m_1 g + m_1 a$$
$$= 12m_2 + 10m_1 + 2m_1$$
$$= 12m_2 + 12m_1$$

Now, we have

$$T_A - T_B = 6$$
$$6 = 12\,m_1$$
$$m_1 = 0.5 \text{ kg}$$

Hence $m_1 = 0.5$ kg, m_2 cannot be determined from the given information.

168. For block having tension T_1,

$$T_1 \geq \mu mg$$

For block having tension T_2,

$$T_2 \geq 2T_1 + \mu mg$$
$$T_2 \geq 3\mu mg$$

Now, for block first on which force is applied,

$$F \geq 2T_2 + \mu mg$$
$$\geq 2\,(3\mu mg) + \mu mg$$
$$\geq 7\,\mu mg$$

169.

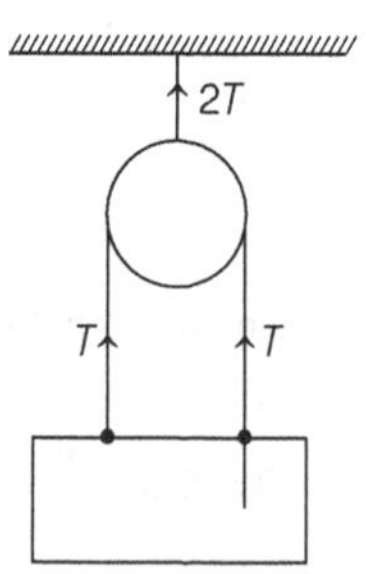

Tension in the rope, $2T = (100 + 50)\,(10 + 5)$

$$T = \frac{2250}{2} = 1125 \text{ N}$$

170. $N = mg(1 + \cos\theta)$

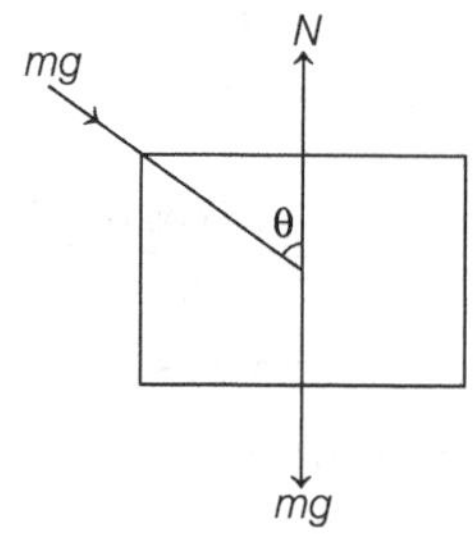

$$F = \mu mg = \mu mg(1 + \cos\theta)$$

Now, $\quad mg\sin\theta \geq F$

$$\sin\theta \geq \mu(1 + \cos\theta)$$
$$2\sin\frac{\theta}{2}\cos\frac{\theta}{2} \geq \mu\left(2\cos^2\frac{\theta}{2}\right)$$
$$\tan\frac{\theta}{2} \geq \mu$$

171. Since, the block is under limiting friction,

$$\mu = \tan\theta = 0.5$$

Now, we have

$$y = \frac{x^3}{6}$$
$$\frac{dy}{dx} = \frac{x^2}{2} = \text{Slope}$$
$$\tan\theta = 0.5 = \frac{x^2}{2}$$
$$x = 1 \text{ m}$$

and $\quad y = \frac{1}{6}$ m

172. $T = \frac{F}{2}$ ...(i)

$$a_B = \frac{T - m_B g}{m_B} = \frac{F}{2m_B} - g \quad \text{[using Eq. (i)]}$$
$$= \frac{400}{2 \times 10} - 10$$
$$= 10 \text{ m/s}^2$$

173. Assume lift to be at rest.

Now, we will have to apply pseudo forces ($F = ma$) on both blocks. Only that pseudo force is shown which is of our use.

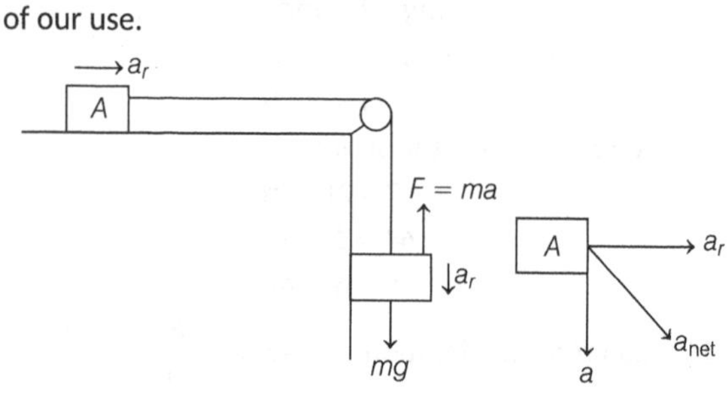

$$a_r = \frac{\text{Net pulling force}}{\text{Total mass}}$$
$$= \frac{mg - ma}{2m} = \frac{1}{2}(g - a)$$

Now, a and a_r are mutually perpendicular and net acceleration is vector sum of these two. Hence,

$$a_{net} = \sqrt{a^2 + a_r^2}$$

174. In vertical direction,

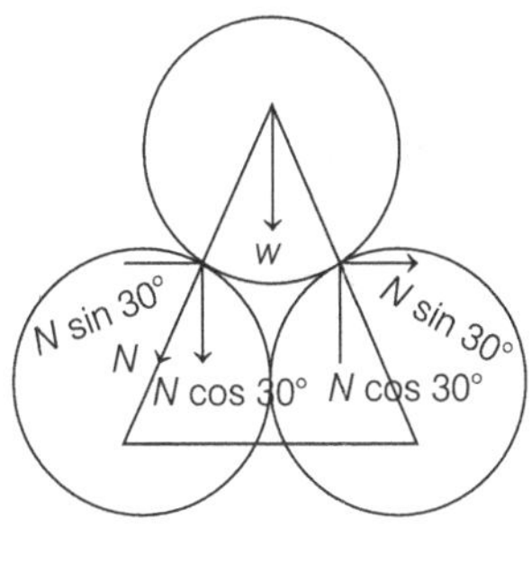

$$w = 2N\cos 30°$$
$$N = \frac{w}{\sqrt{3}} \quad \text{...(i)}$$

Now, $F = N\sin 30°$

$$= \frac{w}{2\sqrt{3}} \quad \text{[using Eq. (i)]}$$

175. $a_{man} = a_m = \dfrac{900 - 600}{60} = 5 \text{ m/s}^2$

$$a_{block} = a_B = \frac{900 - 800}{80} = 1.25 \text{ m/s}^2$$

After 2.0 s, velocity of man and block are

$v_m = a_m t = 10$ m/s, $v_B = a_B t = 2.5$ m/s

Just after 2 s

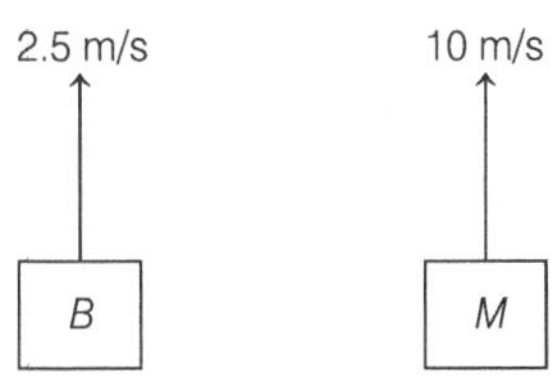

So, the string will slack and $a_B = a_m = g$

176. Let the maximum downward displacement of m be x_0.

So, $\frac{1}{2}kx_0^2 = mgx_0$

$$x_0 = \frac{2mg}{k} \quad \text{...(i)}$$

To lift the block, $kx_0 = Mg$ [using Eq. (i)]

$$2mg = Mg$$
$$mg = \frac{Mg}{2}$$

Now, the normal force,

$$N + kx = Mg$$
$$N = Mg - kx = Mg - \frac{Mg}{2} = \frac{M}{2}g$$

Tension, $T = kx_0 = Mg = 2mg$

Tension T when m is in equilibrium,

$$kx = mg$$
$$T = 2kx + mg = 3mg = \frac{3M}{2}g$$

Hence, the option (b) is wrong.

177. $a_1 = 0$, $a_2 = \dfrac{10}{2+3} = 2 \text{ m/s}^2$, $a_3 = \dfrac{6}{3} = 2 \text{ m/s}^2$

178. In horizontal direction,

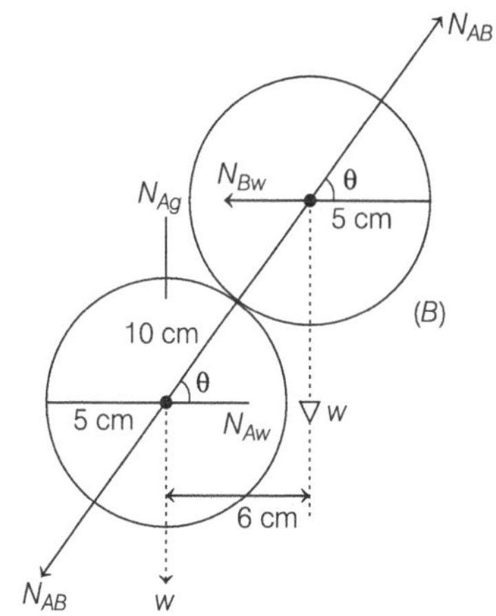

(A)

$$N_{Aw} = N_{Bw}$$
$$\cos\theta = \frac{6}{10} = \frac{3}{5}$$
$$\Rightarrow \quad \theta = 53°$$

From sphere B,

$$w = N_{AB}\sin\theta \quad (\because \theta = 53°)$$
$$N_{Bw} = N_{AB}\cos\theta$$
$$= \frac{w}{\frac{4}{5}} \times \frac{3}{5} = \frac{3w}{4}$$
$$N_{Aw} = N_{Bw} = \frac{3w}{4}$$

LEVEL 03 Previous Years Questions

179. Given, the mass of the ball dropped from the height, $m = 0.15$ kg

The height from the ball dropped,

$$h = 10 \text{ m}$$

Now, the velocity of the ball just before striking to the ground is

$$v_1 = -\sqrt{2gh}$$
$$\Rightarrow \quad v_1 = -\sqrt{2(10)(10)}$$
$$\Rightarrow \quad v_1 = -10\sqrt{2} \text{ m/s}$$

For case (2), ball rebounds to the same height.

The velocity with which the ball just reaches to the same height,

$$v_2 = \sqrt{2gh}$$

$$\Rightarrow \quad v_2 = \sqrt{2(10)(10)}$$

$$\Rightarrow \quad v_2 = 10\sqrt{2} \text{ m/s}$$

Now, magnitude of the impulse imparted to the ball,

$$|\text{Impulse}| = m\,|\Delta \bar{v}| = m\,|(v_2 - v_1)|$$

$$= 0.15\,|10\sqrt{2} - (-10\sqrt{2})|$$

$$= 0.15\,|2 \times 10\sqrt{2}|$$

$$= 4.2 \text{ kg-m/s}$$

180. The given situation is shown in the following diagram.

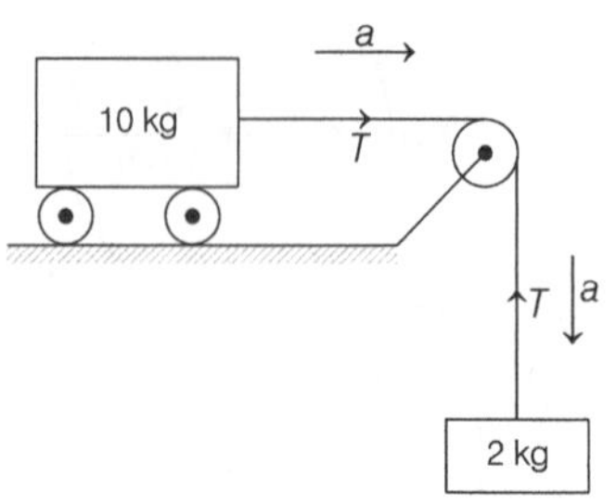

If 'a' be the acceleration of the system, then equation of motion of 10 kg trolly,

$$T - \mu R = 10a$$

$$\Rightarrow \quad T - 0.05 \times 10g = 10a \qquad [\because \mu = 0.05, R = 10g]$$

$$\Rightarrow \quad T - 0.05 \times 10 \times 10 = 10a$$

$$\Rightarrow \quad T - 5 = 10a \qquad \dots \text{(i)}$$

Equation of motion of 2 kg block,

$$2g - T = 2a$$

$$2 \times 10 - T = 2a$$

$$20 - T = 2a \qquad \dots \text{(ii)}$$

Adding Eqs. (i) and (ii), we have

$$20 - 5 = 12a$$

$$\Rightarrow \quad 15 = 12a$$

$$\Rightarrow \quad a = \frac{15}{12} = \frac{5}{4} = 1.25 \text{ ms}^{-2}$$

181. Given, $m_1 = 4$ kg, $m_2 = 6$ kg and $a = ?$

From the given free body diagram, the relation for acceleration of the given system can be given as

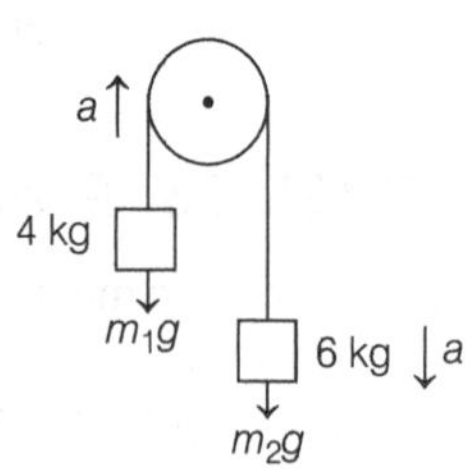

$$a = \left(\frac{m_2 - m_1}{m_1 + m_2}\right) \cdot g$$

$$= \left(\frac{6-4}{4+6}\right) \times g = \frac{g}{5}$$

Hence, correct option is (b).

182. As the truck move to the right, so the bob will move to the left due to inertia of rest with acceleration a.

Thus, the given situation can be drawn

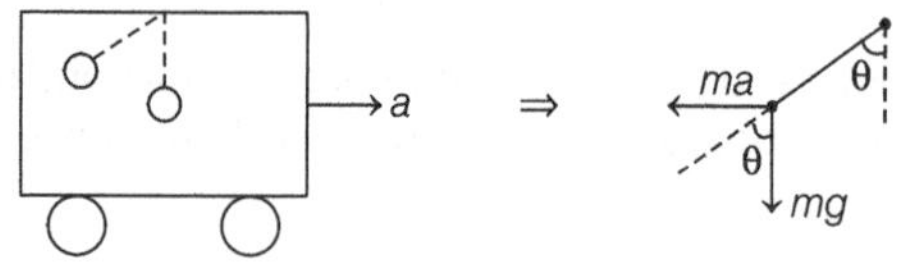

(a) (b)

From the above diagram (b) as the string moves by an angle of θ with the vertical, then the tangent angle is

$$\tan\theta = \frac{ma}{mg} \Rightarrow \theta = \tan^{-1}\left(\frac{a}{g}\right)$$

183. As the three forces are represented by three sides of a triangle taken in order, then their resultant will be zero.

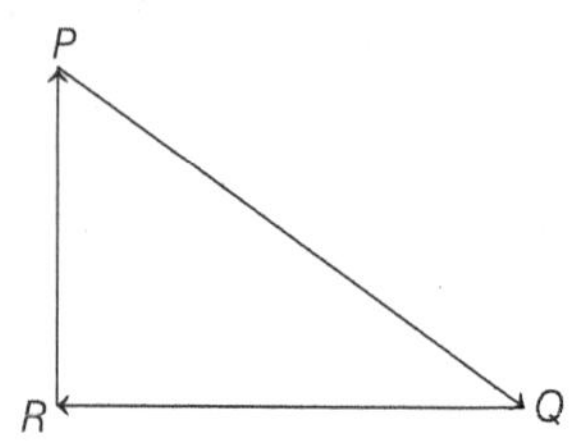

$$\Rightarrow \quad \mathbf{F}_{\text{net}} = \mathbf{F}_{PQ} + \mathbf{F}_{QR} + \mathbf{F}_{RP} = 0$$

$$\mathbf{F}_{\text{net}} = m \times \mathbf{a} = m\frac{d\mathbf{v}}{dt} = 0 \Rightarrow \frac{d\mathbf{v}}{dt} = 0$$

or $\quad \mathbf{v} = \text{constant}$

So, the velocity of particle remain constant.

184. The angular speed of a particle in a uniform circular motion is given by

$\omega = \dfrac{2\pi}{T}$, where T is the time period of rotation

For particle A, $\omega_A = \dfrac{2\pi}{T_A}$

For particle B, $\omega_B = \dfrac{2\pi}{T_B}$

$$\therefore \quad \frac{\omega_A}{\omega_B} = \frac{2\pi}{T_A} \times \frac{T_B}{2\pi} = \frac{T_B}{T_A}$$

$$= \frac{1}{1} \text{ or } 1:1 \qquad [\because T_A = T_B \text{ (given)}]$$

185. The situation can be drawn as

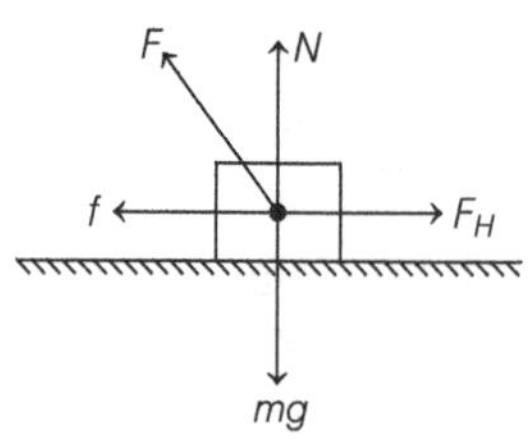

The frictional force, $f = \mu N = \mu mg \qquad [\because N = mg]$

From Free body diagram (FBD), the resultant force is

$$|\mathbf{F}| = \sqrt{N^2 + f^2}$$
$$= \sqrt{(mg)^2 + (\mu mg)^2}$$
$$= mg\sqrt{1 + \mu^2}$$

This is the minimum force required to move the object. But as the body is not moving

$\therefore \quad |\mathbf{F}| \le mg\sqrt{1 + \mu^2}$

186. According to the question, the FBD of the given condition will be

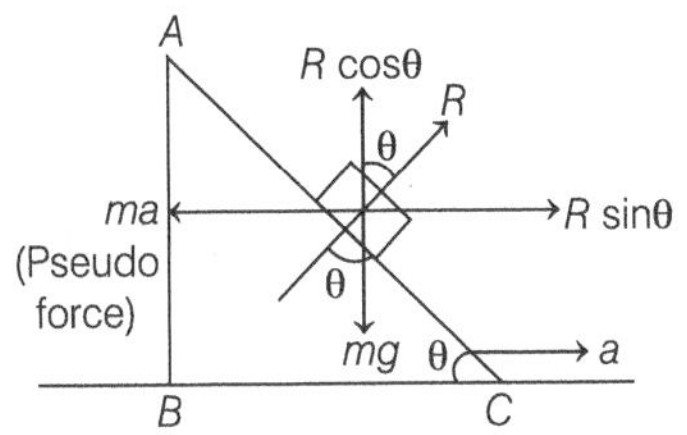

Since, the wedge is accelerating towards right with a, thus a pseudo force ma acts in the left direction. As, the system is in equilibrium,

$\therefore \quad \Sigma F_x = 0$

or $\quad \Sigma F_y = 0$

$\Rightarrow \quad R \sin\theta = ma$

or $\quad mg \sin\theta = ma \quad$ ...(i)

Similarly, $\quad R\cos\theta = mg$

or $\quad mg\cos\theta = mg \quad$ ...(ii)

Dividing Eq. (i) by Eq (ii), we get

$$\frac{mg\sin\theta}{mg\cos\theta} = \frac{ma}{mg}$$

$\Rightarrow \quad \tan\theta = \frac{a}{g}$

or $\quad a = g\tan\theta$

$\therefore$ The relation between a and g for the block to remain stationary on the wedge is $a = g\tan\theta$.

187. The opposing force that comes into play when one body is actually sliding over the surface of the other body is called sliding friction.

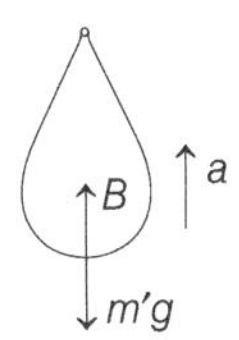

The coefficient of sliding is given as

$$\mu_S = \frac{N}{F_{\text{sliding}}}$$

where, N is the normal reaction and F_{sliding} is the sliding force.

As, the dimensions of N and F_{sliding} are same. Thus, μ_S is a dimensionless quantity.

Hence, statement(d) is incorrect.

188. Initially system is in equilibrium with a total weight of $4mg$ over spring.

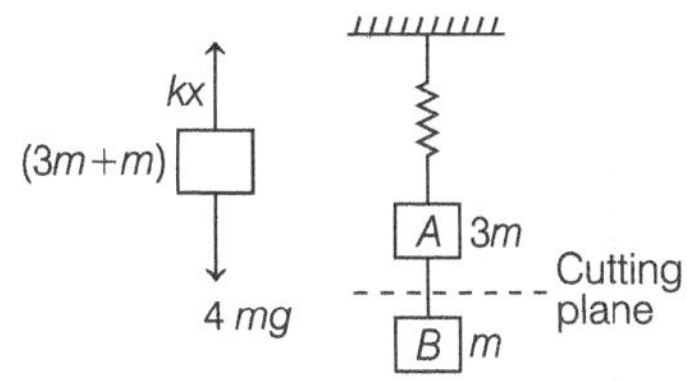

$\therefore \quad kx = 4mg$

When string is cut at the location as shown above.

Free body diagram for m is

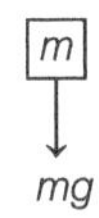

So, force on mass $m = mg$

$\therefore$ Acceleration of mass, $m = g$

For mass $3m$; free body diagram is

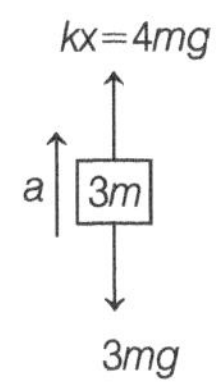

If a = acceleration of block of mass $3m$, then

$$F_{\text{net}} = 4mg - 3mg$$

$\Rightarrow \quad 3m \cdot a_A = mg$ or $a_A = \frac{g}{3}$

So, accelerations for blocks A and B are

$$a_A = \frac{g}{3} \text{ and } a_B = g$$

189. As there is no change in y-component of velocity, impulse is imparted due to change in x-component of momentum of ball.

$$J = \Delta p = mv_f - mv_i$$
$$= mv\cos 60° - (-mv\cos 60°)$$
$$= 2mv\cos 60° = 2mv \times \frac{1}{2} = mv$$

190. According to the law of conservation of momentum,

$p_i = p_f$

$\Rightarrow \quad (0.01) \times 400 + 0 = 2v + (0.01)v' \quad$ ...(i)

Also velocity v of the block just after the collision is

$$v = \sqrt{2gh}$$
$$= \sqrt{2 \times 10 \times 0.1} = \sqrt{2} \quad \text{...(ii)}$$

$\Rightarrow$ From Eqs. (i) and (ii), we have

$$4 = 2\sqrt{2} + (0.01)v'$$
$$1.2 = (0.01)v'$$
$$v' \approx 120 \text{ m/s}$$

191. Given, $m_A = 4$ kg

$$m_B = 2\text{ kg} \Rightarrow m_C = 1\text{ kg}$$

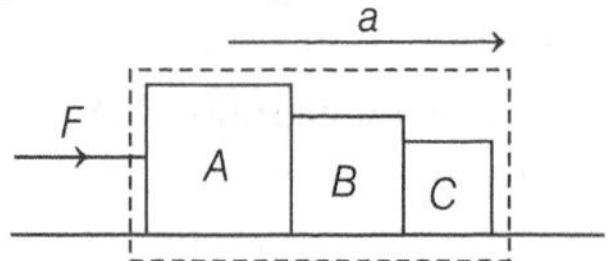

So, total mass $(M) = 4 + 2 + 1 = 7$ kg

Now, $F = Ma \Rightarrow 14 = 7a$

$\Rightarrow$ $a = 2\text{ m/s}^2$

FBD of block A,

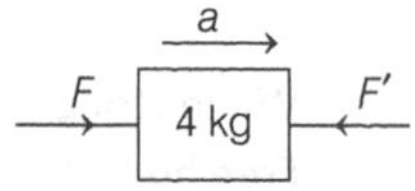

$$F - F' = 4a$$

$$\Rightarrow F' = 14 - 4 \times 2$$

$$\Rightarrow F' = 6\text{ N}$$

192. FBD of block A,

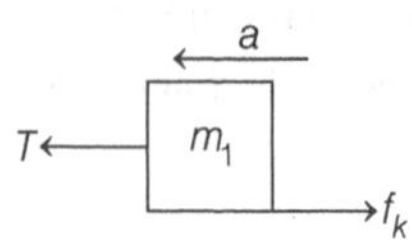

$$T - m_1 a = f_k \quad \text{...(i)}$$

FBD of block B,

$$m_2 g - T = m_2 a \quad \text{...(ii)}$$

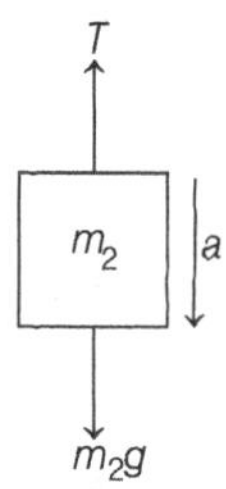

Adding Eqs. (i) and (ii), we get

$$m_2 g - m_1 a = m_2 a + f_k$$

$$\Rightarrow m_2 g - m_1 a = m_2 a + \mu_k m_1 g$$

$$\Rightarrow a = \frac{(m_2 - \mu_k m_1)g}{m_1 + m_2}$$

From Eq. (ii), $T = m_2(g - a)$

$$= m_2\left[1 - \frac{(m_2 - \mu_k m_1)}{m_1 + m_2}\right]g$$

$$T = \frac{m_1 m_2 (1 + \mu_k)}{m_1 + m_2} g$$

193. Given a plank with a box on its one end is gradually raised about the end having angle of inclination of 30°, the box starts to slip and slides down 4 m the plank in 4 s as shown in adjacent figure.

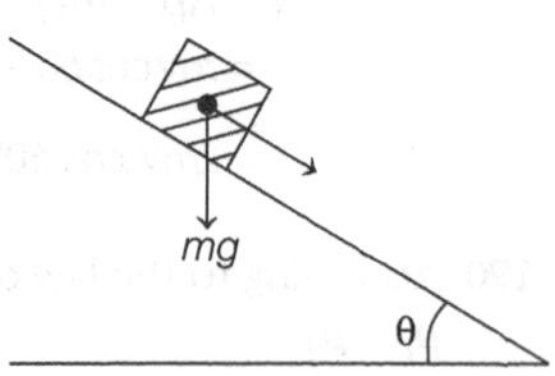

The coefficient of static friction,

$$\mu_s = \tan 30° = \frac{1}{\sqrt{3}} = 0.6$$

So, distance covered by a plank,

$$s = ut + \frac{1}{2}at^2$$

Here, $u = 0$ and $a = g(\sin\theta - \mu\cos\theta)$

$\therefore$ $4 = \frac{1}{2} g(\sin 30° - \mu_k \cos 30°)(4)^2$

$$\Rightarrow 0.5 = 10 \times \frac{1}{2} - \mu_K \times 10 \times \frac{\sqrt{3}}{2}$$

$$\Rightarrow 5\sqrt{3}\,\mu_K = 4.5 \Rightarrow \mu_K = 0.51$$

Thus, coefficient of kinetic friction between the box and the plank is 0.51.

194. The area under F-t graph gives change in momentum.

For 0 to 2 s, $\Delta p_1 = \frac{1}{2} \times 2 \times 6 = 6$ kg-m/s

For 2 to 4 s, $\Delta p_2 = 2 \times -3 = -6$ kg-m/s

For 4 to 8 s, $\Delta p_3 = 4 \times 3 = 12$ kg-m/s

So, total change in momentum for 0 to 8 s

$$\Delta p_{net} = \Delta p_1 + \Delta p_2 + \Delta p_3$$
$$= (+6 - 6 + 12)$$
$$= 12\text{ kg-m/s} = 12\text{ N-s}$$

195. When, the balloon is descending down with acceleration a

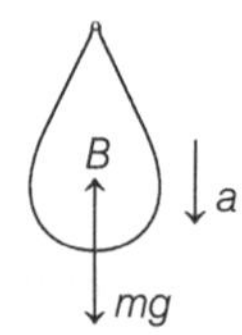

So, from free body diagram

$$mg - B = m \times a \quad \text{...(i)}$$

[$B \to$ Buoyant force]

Here, we should assume that while removing some mass, the volume of balloon and hence,buoyant force will not change.

Let the new mass of the balloon be m'.

So, mass removed $(m - m')$

So, $B - m'g = m' \times a$ ...(ii)

Solving Eqs. (i) and (ii), we get

$$mg - m'g = ma + m'a$$

$$(mg - ma) = m'(g + a)$$

$$\Rightarrow m(g - a) = m'(g + a)$$

$$m' = \frac{m(g - a)}{g + a}$$

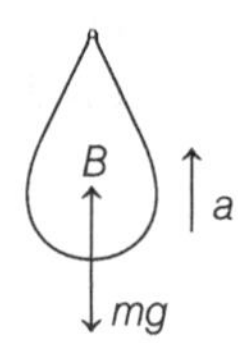

So, mass removed $= \Delta m = m - m'$

$$= m\left[1 - \frac{(g - a)}{(g + a)}\right]$$

$$= m\left[\frac{(g + a) - (g - a)}{(g + a)}\right]$$

$$= m\left[\frac{g + a - g + a}{g + a}\right] \Rightarrow \Delta m = \frac{2ma}{g + a}$$

196. First of all consider the forces on the blocks

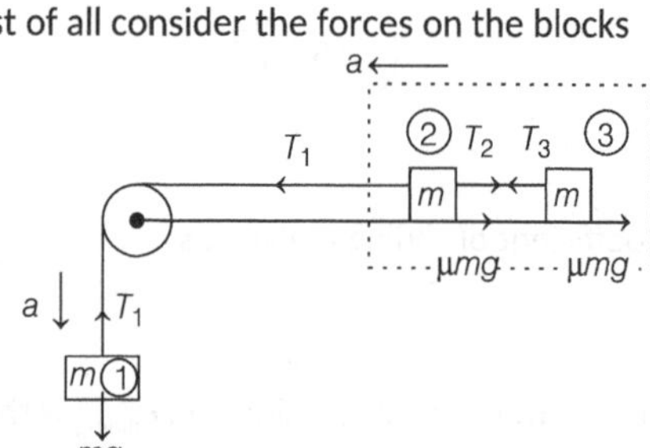

For the 1st block, $[\because m_1 = m_2 = m_3]$

$$mg - T_1 = m \times a \quad ...(i)$$

Let us consider 2nd and 3rd block as a system.

So, $$T_1 - 2\mu mg = 2m \times a \quad ...(ii)$$

Addiing Eqs. (i) and (ii)

$$mg\,(1 - 2\mu) = 3m \times a$$

$$\Rightarrow \quad a = \frac{g}{3}(1 - 2\mu)$$

197. Concept Momentum is conserved before and after collision.

We have, $$\mathbf{p}_1 + \mathbf{p}_2 + \mathbf{p}_3 = 0 \quad [\because p = mv]$$

$$\therefore \quad 1 \times 12\hat{\mathbf{i}} + 2 \times 8\hat{\mathbf{j}} + \mathbf{p}_3 = 0$$

$$\Rightarrow \quad 12\hat{\mathbf{i}} + 16\hat{\mathbf{j}} + \mathbf{p}_3 = 0$$

$$\Rightarrow \quad \mathbf{p}_3 = -(12\hat{\mathbf{i}} + 16\hat{\mathbf{j}})$$

$$\therefore \quad \mathbf{p}_3 = \sqrt{(12)^2 + (16)^2} = \sqrt{144 + 256}$$

$$= 20 \text{ kg-m/s}$$

Now, $$\mathbf{p}_3 = m_3 v_3$$

$$\Rightarrow \quad m_3 = \frac{\mathbf{p}_3}{v_3} = \frac{20}{4} = 5 \text{ kg}$$

198. Since, all the blocks are moving with constant velocity and we know that, if velocity is constant, acceleration of the body becomes zero. Hence, the net force on all the blocks will be zero.

199. Net work done by the block in going from top to bottom of the inclined plane, must be equal to the work done against frictional force.

$$mg \sin\theta \cdot L = \mu\, mg \cos\theta\, \frac{L}{2}$$

or $$\mu = \frac{mg \sin\theta \cdot L}{mg \cos\theta \cdot \frac{L}{2}}$$

$$= 2\,\frac{\sin\theta}{\cos\theta} = 2\tan\theta$$

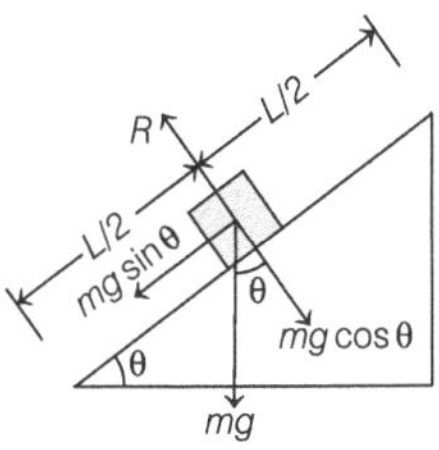

$$\mu = 2\tan\theta$$

200. Here, $\mathbf{p}_i = m_2 v\hat{\mathbf{i}} + m_1 \times 0$

$$\mathbf{p}_f = m_2 \frac{v}{2}\hat{\mathbf{j}} + m_1 \times \mathbf{v}_1$$

Law of conservation of momentum,

$$\mathbf{p}_i = \mathbf{p}_f$$

$$m_2 v\hat{\mathbf{i}} = m_2 \frac{v}{2}\hat{\mathbf{j}} + m_1 \times \mathbf{v}_1$$

$$\mathbf{v}_1 = \frac{m_2}{m_1} v\hat{\mathbf{i}} - \frac{m_2}{m_1}\frac{v}{2}\hat{\mathbf{j}}$$

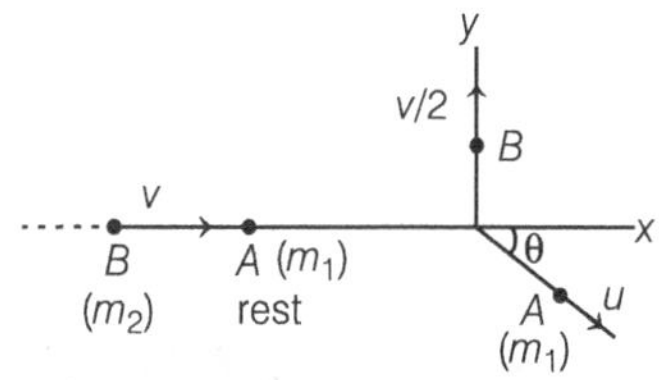

From this equation, we can find

$$\tan\theta = \frac{y}{x} = \frac{1}{2}, \theta = \tan^{-1}\left(\frac{1}{2}\right) \text{ to the } x\text{-axis.}$$

201. Total mass (m)

= Mass of lift + Mass of person

= 940 + 60 = 1000 kg

So, from the free body diagram

$$T - mg = ma$$

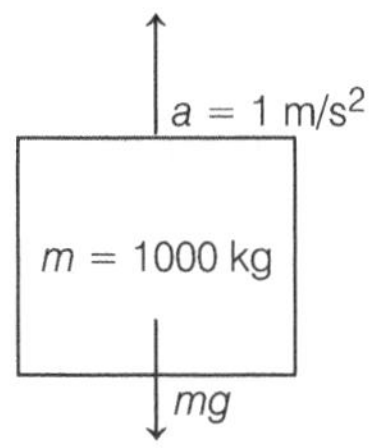

Hence, $T - 1000 \times 10 = 1000 \times 1$

$$T = 11000 \text{ N}$$

202. As, mr = constant

$$m_1 r_1 = m_2 r_2$$

$$\Rightarrow \quad r_2 = \frac{m_1 r_1}{m_2} = \frac{0.5 \times 10}{50} = 0.1$$

So, distance travelled by the man will be

10 + 0.1 = 10.1 m

203. When, a cart moves with some acceleration towards right, then a pseudo force ($m\alpha$) acts on block towards left. This force ($m\alpha$) is action force by a block on cart

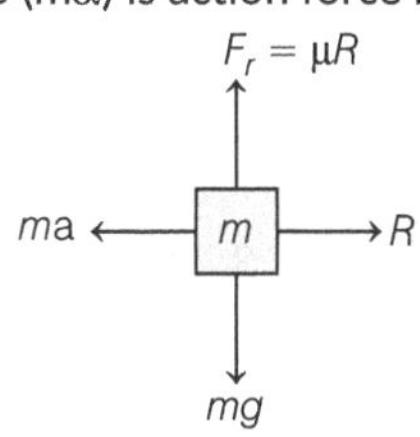

Now, block will remain static with respect to cart, if frictional force $\mu R \geq mg$

$$\Rightarrow \quad \mu\, m\alpha \geq mg \quad [\text{as } R = m\alpha]$$

$$\Rightarrow \quad \alpha \geq \frac{g}{\mu}$$

Chapter 07

WORK, ENERGY AND POWER

LEVEL 01 Objective Questions

1 A force $\mathbf{F} = 6\hat{\mathbf{i}} + 2\hat{\mathbf{j}} - 3\hat{\mathbf{k}}$ acts on a particle while it is displaced perpendicularly to force with displacement $\mathbf{s} = 2\hat{\mathbf{i}} - 3\hat{\mathbf{j}} - x\hat{\mathbf{k}}$, the value of x is

(a) -2 (b) $\frac{1}{2}$ (c) 6 (d) 2

2 A particle of mass 2 kg is moving on x-axis. Its position-time graph is as shown below, find the net work done by the particle in 0 to 4 s.

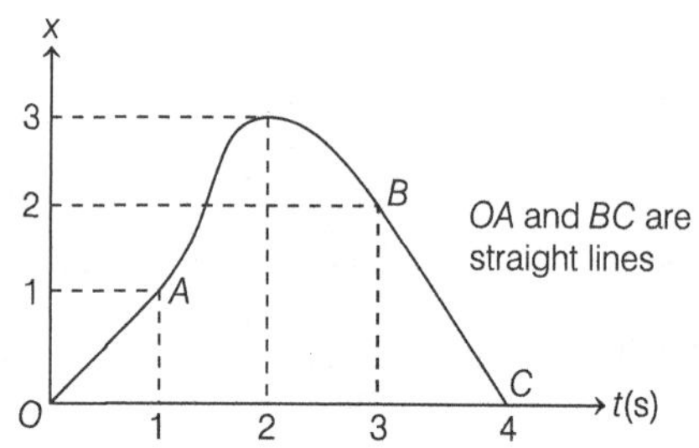

(a) 0 J (b) 2 J (c) 3 J (d) 4 J

3 A block of mass 1 kg slides down a curved track i.e. one quadrant of a circle of radius 1 m. Its speed at the bottom is 2 m/s. The work done by frictional force is

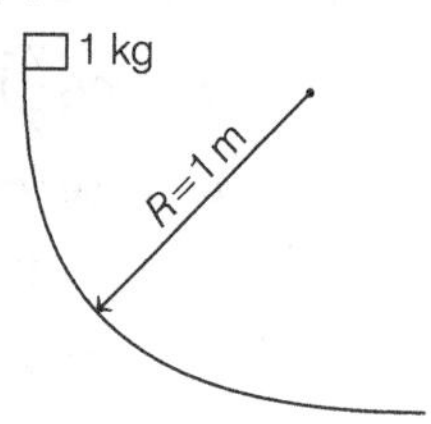

(a) – 8 J
(b) –10 J
(c) 8 J
(d) 10 J

4 When the load on a wire is slowly increased from 20 N to 40 N, the elongation increases from 0.6 mm to 1.0 mm. Work done during the extension is

(a) 4×10^{-3} J (b) 1.4×10^{-2} J (c) 8×10^{-3} J (d) 2.8×10^{-2} J

5 A man displaces a block by 5 m on a rough surface ($\mu = 1/2$) by applying a force of 50 N acting at 37° to the horizontal. The work done by the applied force is

(a) 200 J (b) 100 J (c) 150 J (d) 50 J

6 A block of mass 5 kg is being raised vertically upwards by the help of a string attached to it with an acceleration of 2 ms^{-2}. The work done by the tension in the string, if the block rises by 2.5 m is

(a) 147.5 J (b) 152.7 J (c) 125.5 J (d) 135.5 J

7 A block of mass m is suspended by a light string from an elevator. The elevator starts accelerating downwards from rest with uniform acceleration a. The work done by tension in the string on the block during first t second is ($a < g$)

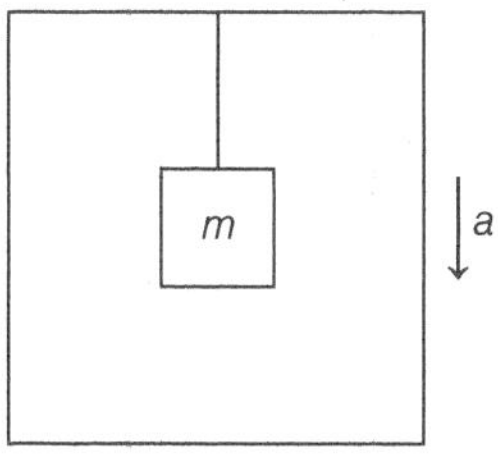

(a) $\frac{m}{2}(a-g)\,at^2$ (b) $m(a-g)\,at^2$

(c) $\frac{m}{2}(g+a)\,at^2$ (d) $m(g+a)\,at^2$

8 An object of mass 5 kg falls from rest through a vertical distance of 20 m and attains a velocity of 10 m/s. Work done by air resistance on the object is (Take, $g = 10\ m/s^2$)

(a) −250 J (b) −750 J (c) 500 J (d) 1000 J

9 A body starts from rest and acquires a velocity v in time T. The work done on the body in time t will be proportional to

(a) $\frac{v}{T}t$ (b) $\frac{v^2t^2}{T}$ (c) $\frac{v^2}{T^2}t$ (d) $\frac{v^2}{T^2}t^2$

10 The energy required to accelerate a car from 10 m/s to 20 m/s is n times the energy required to accelerate the same car from rest to 10 m/s, where n is

(a) 1 (b) 4
(c) 2 (d) 3

11 A block is moved from rest through a distance of 4 m along a straight line path. The mass of the block is 5 kg and the force acting on it is 20 N. If kinetic energy acquired by the block is 40 J, at what angle to the path is the force acting?

(a) 30° (b) 60°
(c) 45° (d) 0°

12 A body moves from rest with a constant acceleration. Which one of the following graphs in figure represents the variation of its kinetic energy K with the distance travelled (x)?

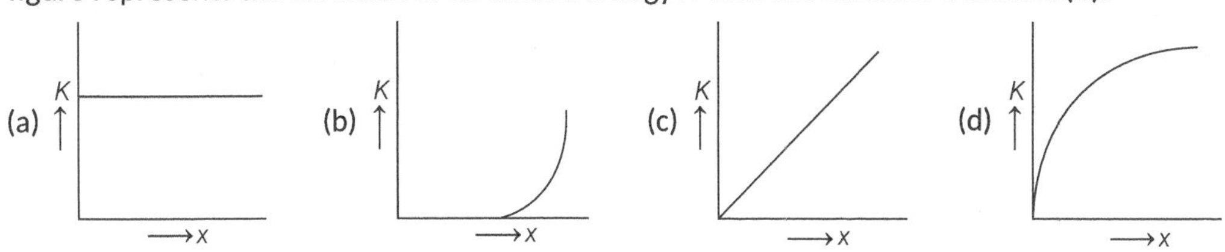

13 A body is projected vertically up with certain velocity. At some point P in its path, the ratio of its potential energy to kinetic energy is 36 : 64. The ratio of velocity of projection to velocity at P is

(a) $\frac{5}{4}$ (b) $\frac{4}{3}$ (c) $\frac{9}{4}$ (d) $\frac{3}{2}$

14 A force $F = -6x^3$ is acting on a block moving along X-axis. Work done by this force is

(a) negative in displacing the block from $x = 3$ to $x = 1$
(b) negative in displacing the block from $x = -3$ to $x = -1$
(c) negative in displacing the block from $x = 0$ to $x = 4$
(d) non-zero in displacing the block from $x = -2$ to $x = 2$

15 A plate of uniform thickness mass m, length b and breadth a is initially lying on a horizontal floor with length parallel to the floor and breadth perpendicular to the floor. The work done to change the position as breadth parallel to the floor and length perpendicular to the floor is

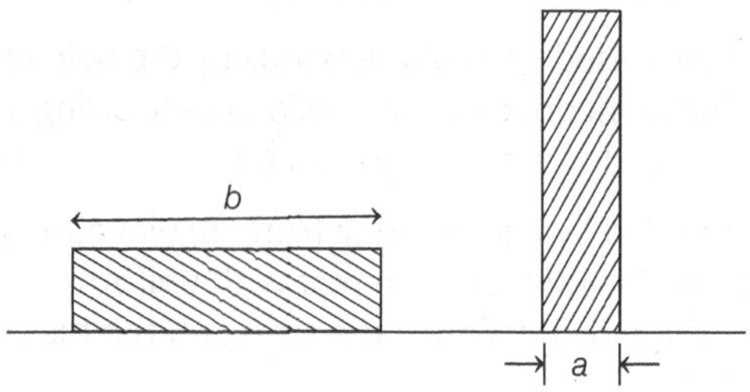

(a) $mg\left[\frac{b}{2}\right]$ (b) $mg\left[a+\frac{b}{2}\right]$ (c) $mg\left[\frac{b-a}{2}\right]$ (d) $mg\left[\frac{b+a}{2}\right]$

16 A body of mass 1 kg is thrown upwards with a velocity 20 m/s. It momentarily comes to rest after attaining a height of 18 m. How much energy is lost due to air friction? (Take, $g = 10 \text{ m/s}^2$)

(a) 20 J (b) 30 J (c) 40 J (d) 10 J

17 A particle with total energy E moves in one-dimensional region, where the potential energy is $U(x)$. The speed of the particle is zero, where

(a) $U(x) = 0$ (b) $U(x) = E$
(c) $\frac{dU(x)}{dx} = 0$ (d) $\frac{d^2U(x)}{dx^2} = 0$

18 As a particle moves along the x-axis, it is acted upon by a conservative force. The potential energy is shown as a function of the co-ordinate x of the particle. Rank the labelled regions according to the magnitude of the force from least to greatest. (Take, $AB = BC = CD$)

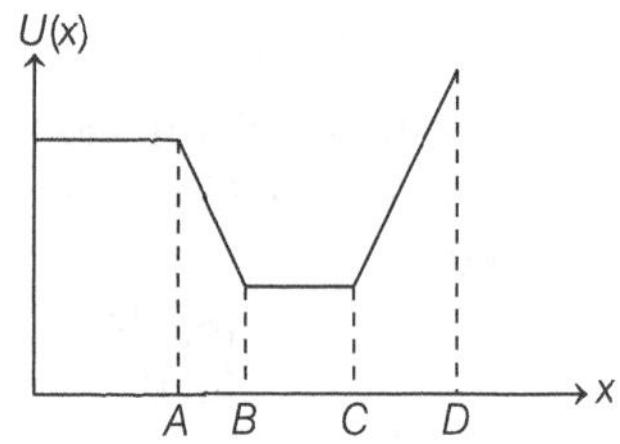

(a) AB, BC, CD (b) AB, CD, BC (c) BC, CD, AB (d) BC, AB, CD

19 Particle of mass 1 kg is moving on the x- axis, such that force on a particle varies with position as shown in the following graph. Select the position of stable equilibrium for the particle.

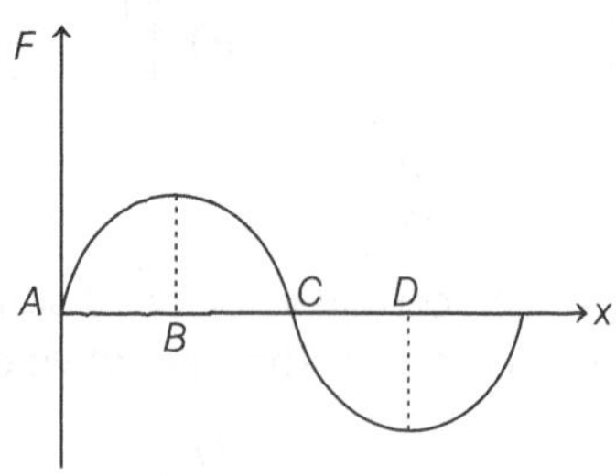

(a) A (b) B (c) C (d) D

20 A particle is released from rest at origin. It moves under the action of conservative force. If potential energy of particle is given as $U = x^2 - 2x$, then its kinetic energy at $x = 2$ is

(a) 2 J (b) 1 J
(c) 1.5 J (d) 0 J

21 The potential energy corresponding to a conservative force is given by $U = \alpha x^2 - \beta x$, where α and β are positive constants. Find the value of x for equilibrium and decide the type of equilibrium.

(a) $\frac{\beta}{2\alpha}$, stable (b) β, unstable (c) $\frac{\beta}{2\alpha}$, unstable (d) $\frac{\alpha}{\beta}$, unstable

22 Track *OABCD* (as shown in the following figure) is smooth and fixed in vertical plane. What minimum speed has to be given to a particle lying at point *A*, so that it can reach point *C*?

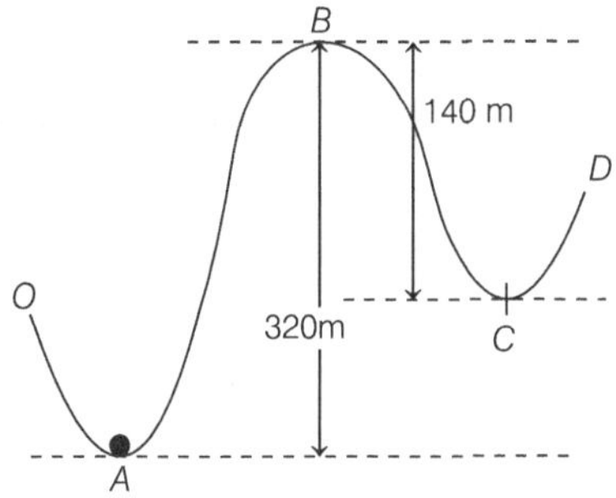

(a) 60 m/s (b) 100 m/s (c) 70 m/s (d) 80 m/s

23 The potential energy function in one-dimension is given as shown in the following figure. If total energy of the particle is 5 J, then in the region from x_1 to x_2, the value of its kinetic energy will be

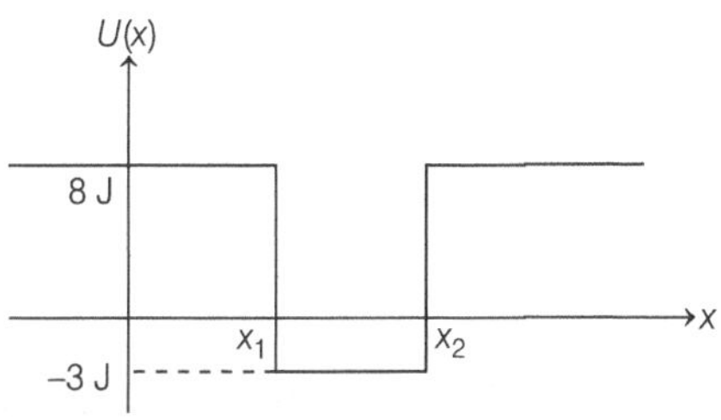

(a) 8 J (b) 11 J (c) 5 J (d) 0 J

24 A spring of spring constant *k* is cut into three parts of lengths in the ratio 1 : 2 : 1. The algebraic addition of spring constants of these springs will be

(a) k (b) $10k$
(c) $8k$ (d) $6k$

25 Block of mass *m* is lowered from natural length of spring slowly by an external agent to equilibrium position. The extension present in the massless spring was calculated by two students writing the equations:

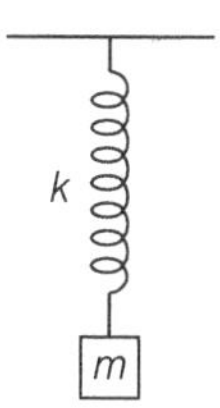

Student A $\frac{1}{2}kx^2 = mgx$

$$\therefore \quad x = \frac{2mg}{k}$$

Student B $mg = kx$

$$\therefore \quad x = \frac{mg}{k}$$

(a) Student A is incorrect, Student B is correct.
(b) Student A is correct, Student B is incorrect.
(c) Both are correct
(d) Both are incorrect

26 Two blocks A and B of masses m and $2m$ placed on smooth horizontal surface are connected with light spring. The two blocks are given velocities as shown below, when spring is at natural length. Find maximum extension.

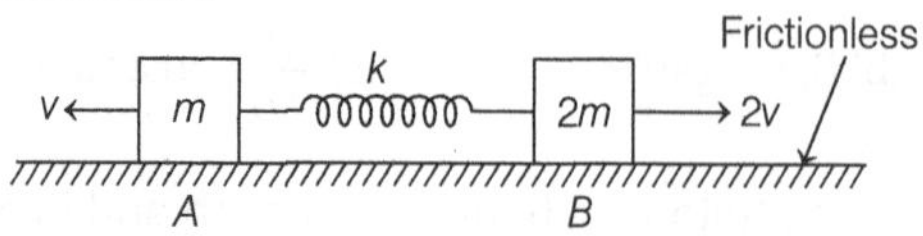

(a) $v\sqrt{\frac{3m}{k}}$ (b) $v\sqrt{\frac{4m}{k}}$ (c) $v\sqrt{\frac{2}{3}\left(\frac{m}{k}\right)}$ (d) $v\sqrt{\frac{2m}{k}}$

27 A ball of mass 1 kg is dropped from a tower. Find power of gravitational force at time $t = 2$ s. (Take, $g = 10\ \text{m/s}^2$)

(a) 100 W (b) 400 W
(c) 200 W (d) 50 W

28 A force of $(2\hat{\mathbf{i}} + 3\hat{\mathbf{j}} + 4\hat{\mathbf{k}})$ N acts on a body for 4 s and produces a displacement of $(3\hat{\mathbf{i}} + 4\hat{\mathbf{j}} + 5\hat{\mathbf{k}})$ m. The average power used is

(a) 9.5 W (b) 7.5 W
(c) 6.5 W (d) 4.5 W

29 A sports car accelerates from zero to a certain speed in t seconds. How long does it take to accelerate to twice that speed starting from rest, assuming the power of the engine to be constant (independent of velocity) and neglecting any resistance to motion ?

(a) $\sqrt{2}\,t$ s (b) $2t$ s
(c) $3t$ s (d) $4t$ s

30 A body of mass m starting from rest from origin moves along x-axis with constant power P. Calculate the relation between velocity and distance.

(a) $x \propto v^{1/2}$ (b) $x \propto v^2$
(c) $x \propto v$ (d) $x \propto v^3$

31 A block of mass 10 kg accelerates uniformly from rest to a speed of 2 m/s in 20 s. The average power developed in time interval of 0 to 20 s is

(a) 10 W (b) 1 W
(c) 20 W (d) 2 W

32 Power applied to a particle varies with time as $P = (4t^3 - 5t + 2)$ W, where t is in seconds. Find the change in its kinetic energy between time $t = 2$ s and $t = 4$ s.

(a) 212 J (b) 213 J
(c) 214 J (d) 215 J

33 In the system shown, coefficient of friction (μ) between 10 kg block and the surface is 0.2. If the system is released from rest, then velocity of blocks when 5 kg block has descended by 4 m is (Take, $g = 10\ \text{m/s}^2$)

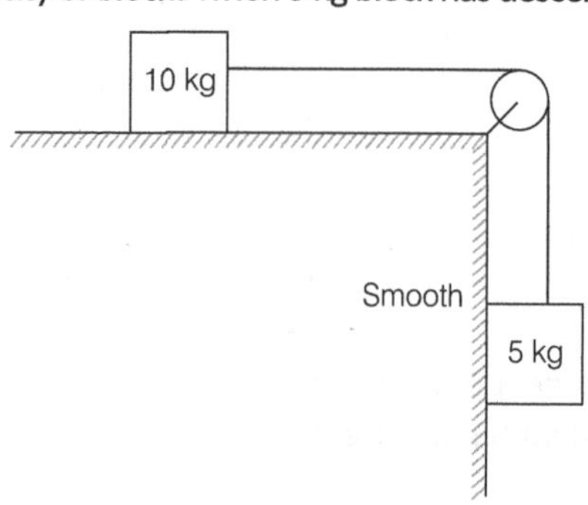

(a) 4 m/s (b) 2 m/s (c) $\sqrt{2}$ m/s (d) $2\sqrt{2}$ m/s

34 In the situation shown in figure, force constant of the spring is k and the surface is smooth. If the system is released from rest with the spring unstretched, then maximum extension in the spring is

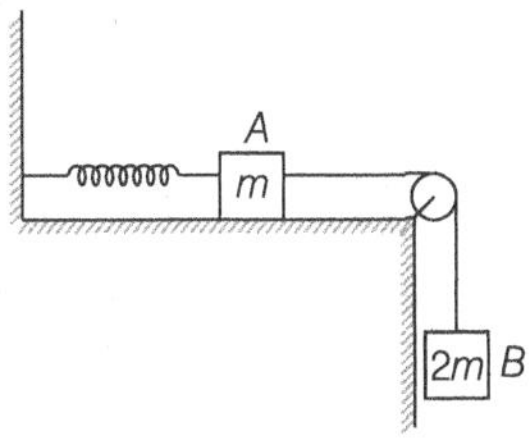

(a) $\frac{4mg}{k}$ (b) $\frac{2mg}{k}$ (c) $\frac{mg}{k}$ (d) $\frac{mg}{2k}$

35 A stone thrown vertically upwards with an initial velocity u from the top of a tower, reaches the ground with a velocity $3u$. The height of the tower is

(a) $\frac{3u^2}{g}$ (b) $\frac{4u^2}{g}$ (c) $\frac{6u^2}{g}$ (d) $\frac{9u^2}{g}$

36 The potential energy of a spring when stretched through a distance x is 10 J. What is the amount of work done on the same spring to stretch it through an additional distance x?

(a) 10 J (b) 20 J (c) 30 J (d) 40 J

37 In the track shown in figure, section AB is a quadrant of a circle of 1 m radius in vertical plane. A block is released at A and slides without friction until it reaches at B. After B, it moves on a rough horizontal floor and comes to rest at D, 3 m from B. The coefficient of friction between floor and the body will be

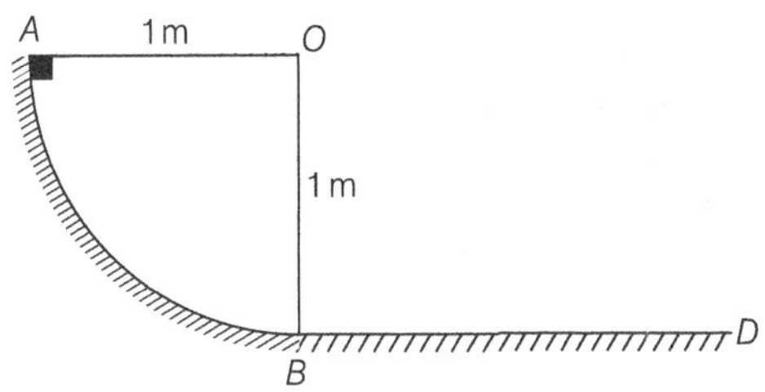

(a) 1/3 (b) 2/3
(c) 1/4 (d) 3/8

38 A body of mass m dropped from a height h reaches the ground with a speed of $\sqrt{2gh}$. The work done by air drag is

(a) $-0.2\ mgh$ (b) zero
(c) $-0.04\ mgh$ (d) mgh

39 The figure shows a smooth curved track terminating in a smooth horizontal part. A spring of spring constant 400 N/m is attached at one end to a wedge fixed rigidly with the horizontal part. A 40 g mass is released from rest at a height of 5 m on the curved track. The maximum compression of the spring will be

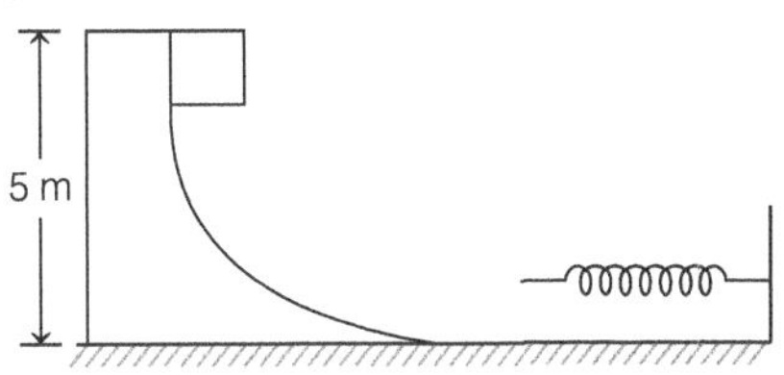

(a) 10 cm (b) 2 cm (c) 3 cm (d) 4 cm

40 A ball of mass m is thrown upward with a velocity v. If air exerts an average resisting force F, the velocity with which the ball returns to the thrower is

(a) $v\sqrt{\frac{mg}{mg+F}}$ (b) $v\sqrt{\frac{F}{mg+F}}$ (c) $v\sqrt{\frac{mg-F}{mg+F}}$ (d) $v\sqrt{\frac{mg+F}{mg-F}}$

41 A particle of mass 0.01 kg travels along a space curve with velocity given by $4\hat{\mathbf{i}}+16\hat{\mathbf{k}}$ m/s. After some time, its velocity becomes $8\hat{\mathbf{i}}+20\hat{\mathbf{j}}$ m/s due to the action of a conservative force. The total work done on the particle during this interval of time is

(a) 0.32 J (b) 6.9 J
(c) 9.6 J (d) 0.96 J

42 A pendulum of length 2 m is left at P. When it reaches Q. It looses 10% of its total energy due to air resistance. The velocity at Q is

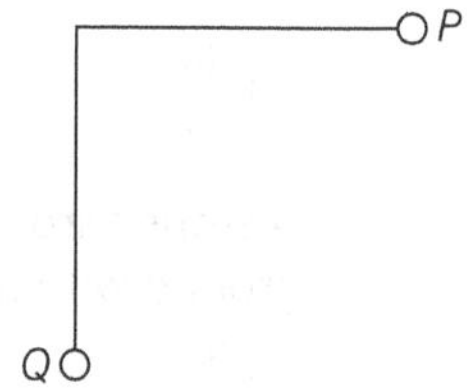

(a) 6 m/s
(b) 1 m/s
(c) 2 m/s
(d) 8 m/s

43 A block is released from rest from the top of a rough curved track as shown in figure. It comes to rest at some point on the horizontal part. If its mass is 200 g, calculate work by friction (in J).

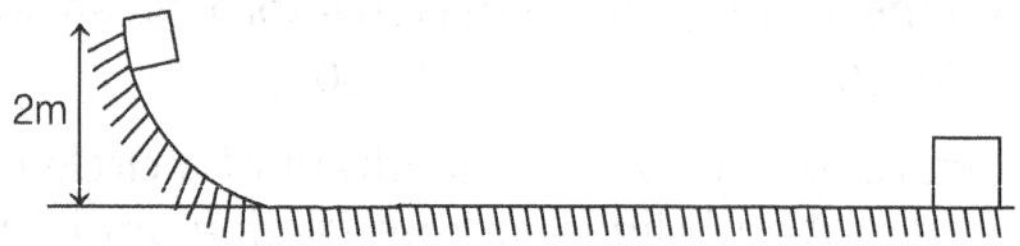

(a) 2 (b) – 2
(c) 4 (d) – 4

44 In the figure, a block slides along a track from one level to a higher level, by moving through an intermediate valley. The track is frictionless until the block reaches the higher level. There is a frictional force which stops the block in a horizontal distance d. The block's initial speed v_0 is 8 m/s, the height difference h is 2 m and the coefficient of kinetic friction μ is 0.3. The value of d is

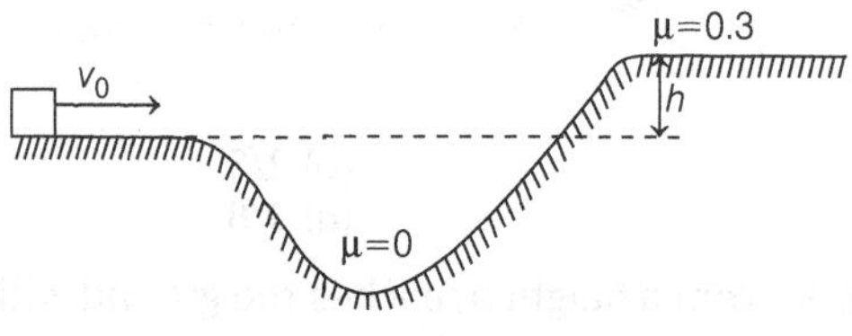

(a) 4 m (b) 3 m (c) 5 m (d) 8 m

45 A ball of mass m is dropped from a height h on a platform fixed at the top of a vertical spring. The platform is displaced by a distance x. The spring constant is

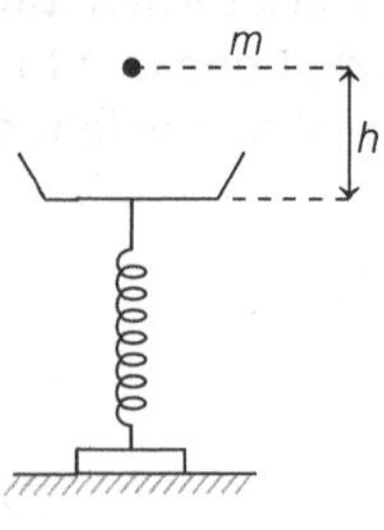

(a) $\frac{2mg}{x}$ (b) $\frac{2mgh}{x^2}$ (c) $\frac{2mg(h+x)}{x^2}$ (d) $\frac{2mg(h+x)}{h^2}$

LEVEL 02 Objective Questions

46 If a person is pushing a box by a force **F** inside a moving train, the work done in the frame of earth will be

$\mathbf{s}_0$ = displacement of the train relative to ground,

$\mathbf{s}$ = displacement of the box with respect to train.

(a) zero (b) $\mathbf{F} \cdot (\mathbf{s} + \mathbf{s}_0)$ (c) $\mathbf{F} \cdot \mathbf{s}$ (d) $\mathbf{F} \cdot \mathbf{s}_0$

47 Force acting on a particle is $(3\hat{\mathbf{i}} + 2\hat{\mathbf{j}})$ N. Work done by force is zero, when a particle is moved on the line $2y + kx = 5$. Here, value of k is

(a) 2 (b) 4 (c) 3 (d) 6

48 A block of mass 1 kg is pushed up a surface inclined to horizontal at an angle of 30° by a force of 10 N parallel to the inclined surface as shown in the figure.

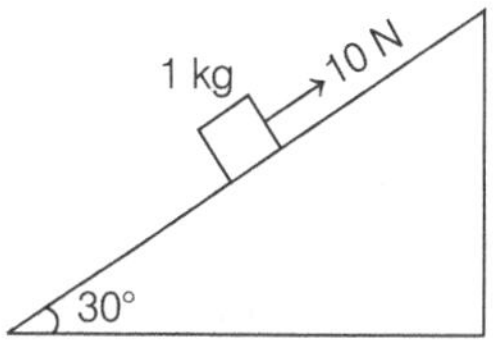

The coefficient of friction between block and the incline is 0.1. If the block is pushed up by 10 m along the inclined, the work done against force of friction is

(a) 8.7 J (b) 10.7 J
(c) 7.8 J (d) 12.7 J

49 A block of mass m is pulled along a horizontal surface by applying a force at an angle θ with the horizontal. If the block travels with a uniform velocity and has a displacement d and the coefficient of friction is μ, then the minimum work done by the applied force is

(a) $\dfrac{\mu mgd}{\cos\theta + \mu\sin\theta}$ (b) $\dfrac{\mu mgd\cos\theta}{\cos\theta + \mu\sin\theta}$ (c) $\dfrac{\mu mgd\sin\theta}{\cos\theta + \mu\sin\theta}$ (d) $\dfrac{\mu mgd\cos\theta}{\cos\theta - \mu\sin\theta}$

50 A small block of mass m is kept on a rough inclined surface of inclination θ fixed in an elevator. The elevator goes up with a uniform velocity v and the block does not slide on the wedge. The work done by the force of friction on the block in time t will be

(a) zero (b) $mgvt\cos^2\theta$ (c) $mgvt\sin^2\theta$ (d) $mgvt\sin 2\theta$

51 A box is thrown on a rough surface such that friction force (f) acting on it is linearly varying with distance travelled (x) by it as $f = ax + b$. If before coming to rest the box travels a distance s, the magnitude of work done by the friction on the box is

(a) $as + b$ (b) $\frac{1}{2}as^2 + bs$
(c) $a^2s + b$ (d) None of these

52 A body is displaced from (1, 1) to (2, 2) along the line $y = x$ under the force $\mathbf{F} = x\hat{\mathbf{j}} + y\hat{\mathbf{i}}$, then find the work done by this force for above displacement. (All units are in SI)

(a) 2 unit (b) 3 unit (c) 4 unit (d) 5 unit

53 In the figure shown, find the accelerations of blocks A, B and C immediately after string connecting block A with ceiling is cut.

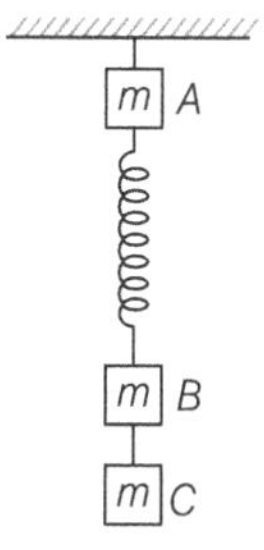

(a) $a_A = 3g\downarrow, a_B = g\downarrow, a_C = g\downarrow$
(b) $a_A = 3g\downarrow, a_B = a_C = 0$
(c) $a_A = 2g\downarrow, a_B = a_C = 0$
(d) $a_A = 3g\downarrow, a_B = 0, a_C = g\downarrow$

54 The force exerted by a special compression device is given as function of compression x, as $F_x(x) = kx(x - l)$ for $0 \le x \le l$, where l is the maximum possible compression and k is a negative constant. The force exerted by the device under compression is maximum when compression is

(a) 0 (b) $l/4$ (c) $l/\sqrt{2}$ (d) $l/2$

55 A point object of mass 2 kg is moved from point A to point B very slowly on a curved path by applying a tangential force on a curved path as shown in figure. Find the work done by external force in moving the body. (Take, $\mu_s = 0.3, \mu_k = 0.1$ and $g = 10 \text{ m/s}^2$)

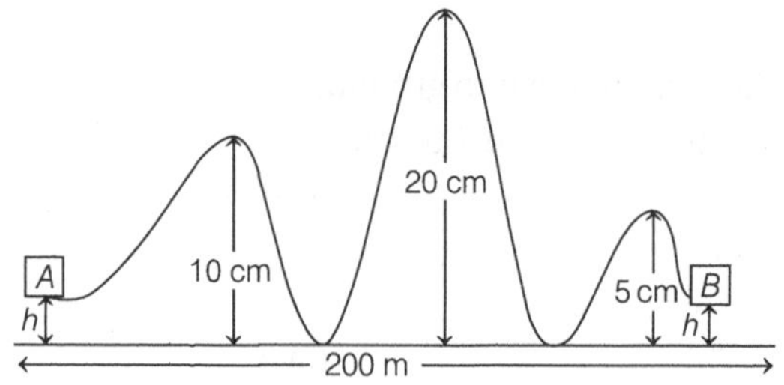

(a) 100 J
(b) 200 J
(c) 300 J
(d) 400 J

56 A street car moves rectilinearly from station A to the next station B with an acceleration varying according to the law $f = a - bx$, where a and b are positive constants and x is its distance from station A. The distance AB is then equal to

(a) $\frac{2a}{b}$ (b) $\frac{a}{b}$ (c) $\frac{a}{2b}$ (d) $2a - b$

57 A variable force given by the two dimensional vector $\mathbf{F} = (3x^2 \hat{\mathbf{i}} + 4\hat{\mathbf{j}})$ acts on a particle. The force is in newtons and x is in metres. What is the change in the kinetic energy of the particle as it moves from the point with coordinates (2, 3) to (3, 0)? (The coordinates are in metres)

(a) -7 J
(b) Zero
(c) 7 J
(d) 19 J

58 If the block in the shown arrangement is acted upon by a constant force F for $t \geq 0$, its maximum speed will be

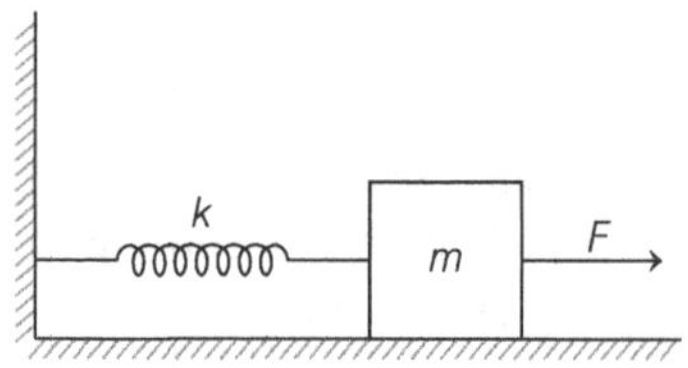

(a) $\frac{F}{\sqrt{mk}}$ (b) $\frac{2F}{\sqrt{mk}}$ (c) $\frac{F}{\sqrt{2mk}}$ (d) $\frac{\sqrt{2}F}{\sqrt{mk}}$

59 A particle moves in a straight line with retardation proportional to its displacement. Its loss of kinetic energy for any displacement x is proportional to

(a) x^2 (b) e^x (c) x (d) $\log_e x$

60 The potential energy U of two atoms of a diatomic molecule as a function of distance r between the atoms is shown in the given figure.

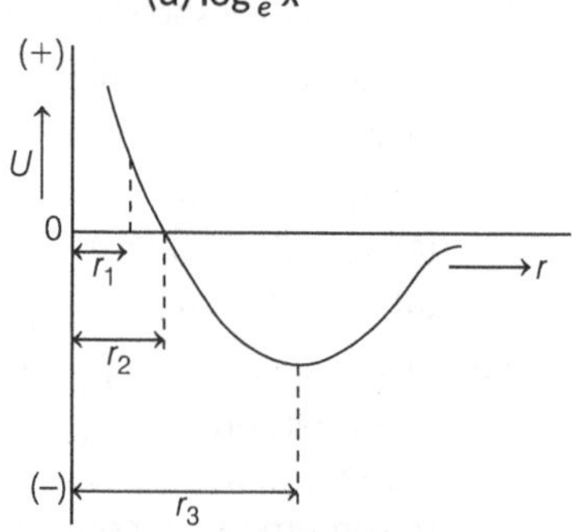

Read the following statements carefully.

1. The equilibrium separation distance between the atoms is equal to r_2.
2. At $r = r_1$, the force between the atom is repulsive.
3. For $r > r_2$, the force between the atoms is attractive.

Which of the above statements is true?

(a) 1 only (b) 2 only (c) 3 only (d) 2 and 3

61 The potential energy $U(x)$ of a particle moving along x-axis is shown in figure. Which one of the following graphs represents the variation of kinetic energy of this particle, if its total mechanical energy is E_0?

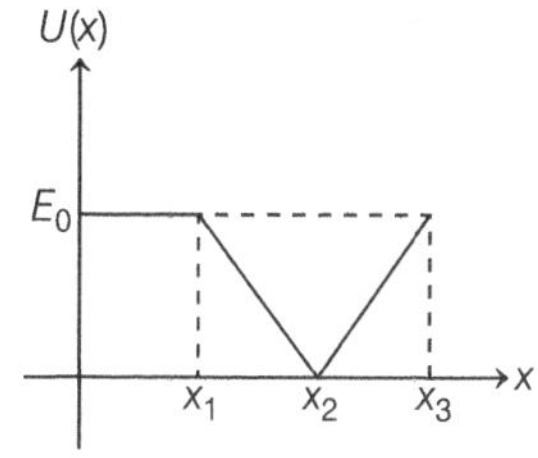

(a) 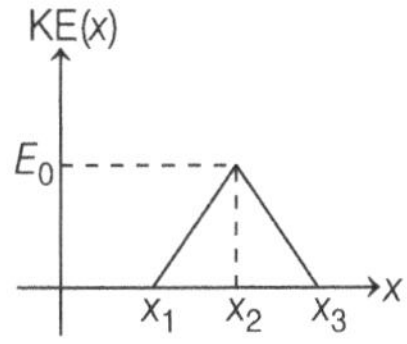

(b) 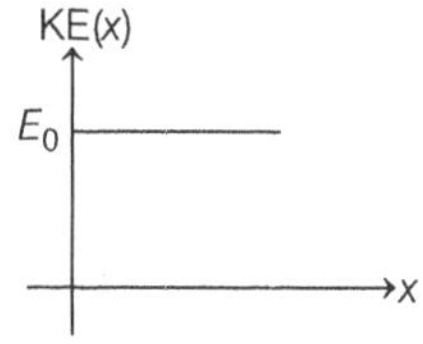

(c) 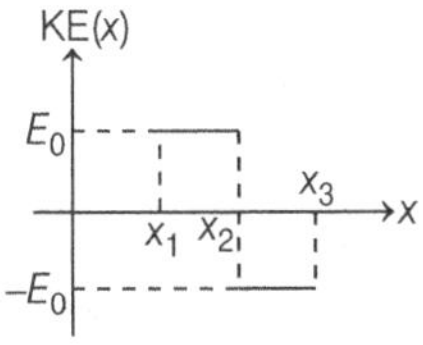

(d) 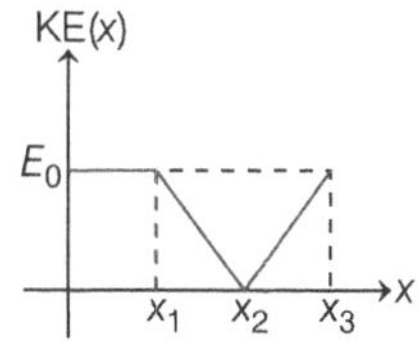

62 The potential energy of a particle oscillating along x-axis is given as $U = 20 + (x - 2)^2$, where U is in joules and x in metres. Total mechanical energy of the particle is 36 J. Maximum kinetic energy of the particle is

(a) 24 J (b) 36 J (c) 16 J (d) 20 J

63 The potential energy of a 1 kg particle free to move along the x-axis is given by

$$U(x) = \left(\frac{x^2}{4} - \frac{x}{2}\right) \text{J}$$

The total mechanical energy of the particle is 2 J. Then, the maximuim speed (in m/s) is

(a) $\frac{1}{\sqrt{2}}$ (b) 2 (c) $\frac{3}{\sqrt{2}}$ (d) $\sqrt{2}$

64 Potential energy of a particle of mass 1 kg moving along x-axis is given by $U = \left(\frac{x^3}{3} - 4x + 12\right)$ joule, where x is in metres. For this force field,

(i) $x = 2$ m and $x = -2$ m are positions of equilibrium

(ii) $x = 2$ m is the position of stable equilibrium

(iii) $x = -2$ m is the position of unstable equilibrium

(iv) potential energy is minimum at $x = 2$ m

(a) (i, ii) (b) (ii, iv)
(c) (i, ii, iii) (d) (i, ii, iii, iv)

65

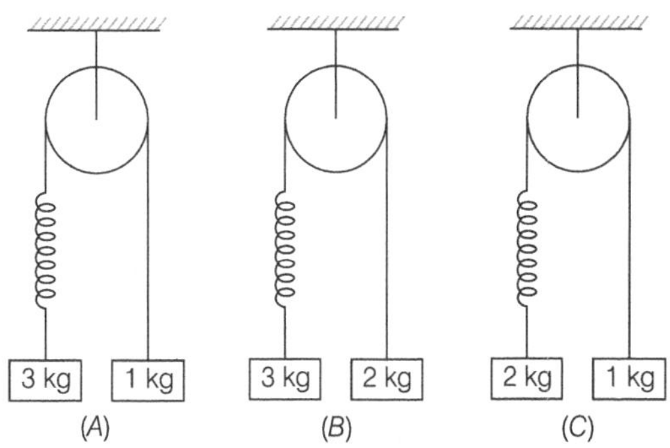

In three cases A, B and C change in the length of spring is x_1, x_2 and x_3 respectively, then

(a) $x_2 > x_1 > x_3$ (b) $x_3 > x_1 > x_2$
(c) $x_1 > x_2 > x_3$ (d) $x_1 = 0, x_3 > x_2$

66 Power due to a force acting on a block varies with time t as shown in figure. Then, angle between force acting on the block and its velocity is

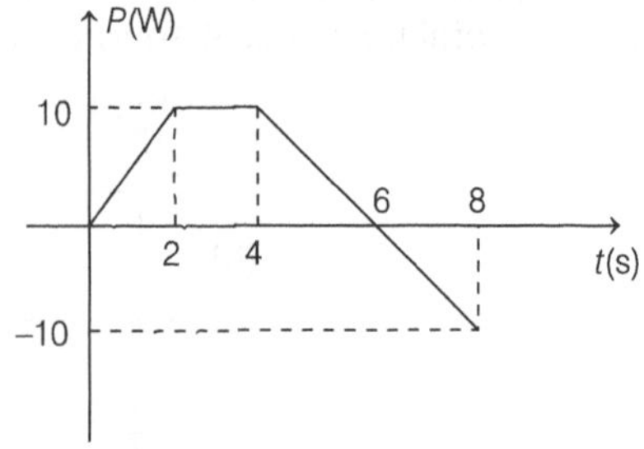

(a) obtuse at $t = 1$ s (b) 90° at $t = 6$ s (c) acute at $t = 7$ s (d) 90° at $t = 3$ s

67 An electric pump is used to fill an overhead tank of capacity 9 m^3, kept at a height of 10 m above the ground. If the pump takes 5 min to fill the tank by consuming 10 kW power, the efficiency of the pump should be (Take, $g = 10\ ms^{-2}$)
(a) 60% (b) 40% (c) 20% (d) 30%

68 A chain of length L is released from rest on a smooth fixed inclined plane with $x = 0$, as shown in the figure. Determine velocity v of the chain when a half of the length has fallen. $\left(\text{Take, } L = \frac{4}{25}\ \text{m}\right)$

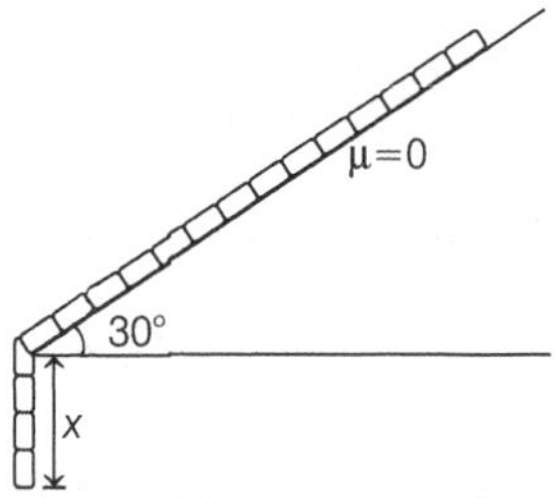

(a) 1 m/s (b) 4 m/s (c) 5 m/s (d) 8 m/s

69 A block is released from rest from a height $h = 5$ m. After travelling through the smooth curved surface, it moves on the rough horizontal surface through a length $l = 8$ m and climbs onto the other smooth curved surface through a height h'. If $\mu = 0.5$, find h'.

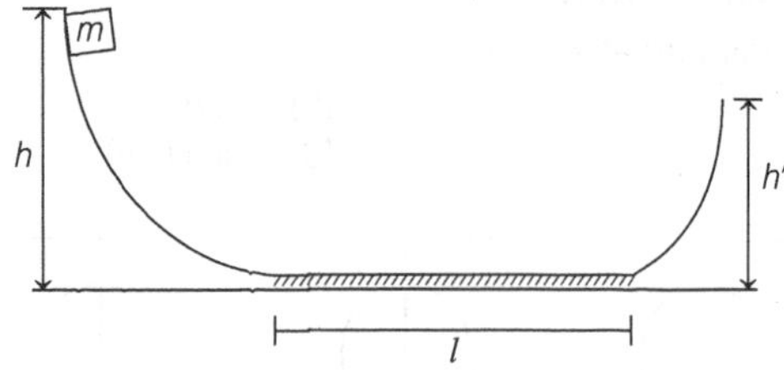

(a) 2 m (b) 3 m (c) 1 m (d) zero

70 A pump motor is used to deliver water at a certain rate from a given pipe. To obtain n times water from the same pipe in the same time, the amount by which the power of the motor should be increased?
(a) $n^{1/2}$ (b) n^2 (c) n^3 (d) n

71 A block of mass 2 kg is pulled up on a smooth inclined plane of angle 30° with the horizontal. If the block moves with an acceleration of 1.0 m/s^2, the power delivered by the pulling force at a time 4 s after the motion starts will be
(a) 12 W (b) 24 W (c) 36 W (d) 48 W

72 System shown in figure is released from rest. Pulley and spring is massless and friction is absent everywhere. The speed of 5 kg block, when 2 kg block leaves the contact with ground is (Take, force constant of spring, $k = 40$ N/m and $g = 10$ m/s^2)

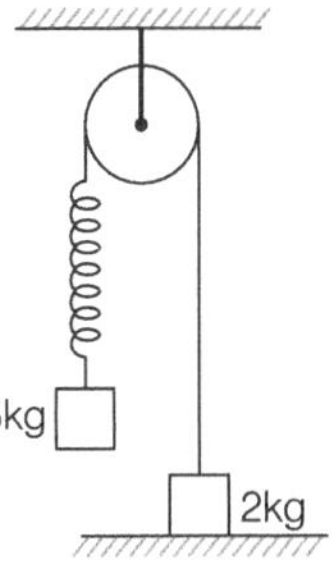

(a) $\sqrt{2}$ m/s (b) $2\sqrt{2}$ m/s (c) 2 m/s (d) $4\sqrt{2}$ m/s

73 This question has Statement-1 and Statement-2. Of the four choices given after the statements, choose the one that best describes the two statements.

If two springs S_1 and S_2 of force constants k_1 and k_2 respectively, are stretched by the same force, it is found that more work is done on spring S_1 than on spring S_2.

Statement-1 If stretched by the same amount, work done on S_1 will be more than that on S_2.

Statement-2 $k_1 < k_2$.

(a) Statement-1 is true, Statement-2 is true and Statement-2 is not the correct explanation of Statement-1.
(b) Statement-1 is false, Statement-2 is true
(c) Statement-1 is true, Statement-2 is false
(d) Statement-1 is true, Statement-2 is true and Statement-2 is the correct explanation of Statement-1.

74 A particle is subjected to a conservative force as shown in the graphs. Which of the following option is **incorrect**?

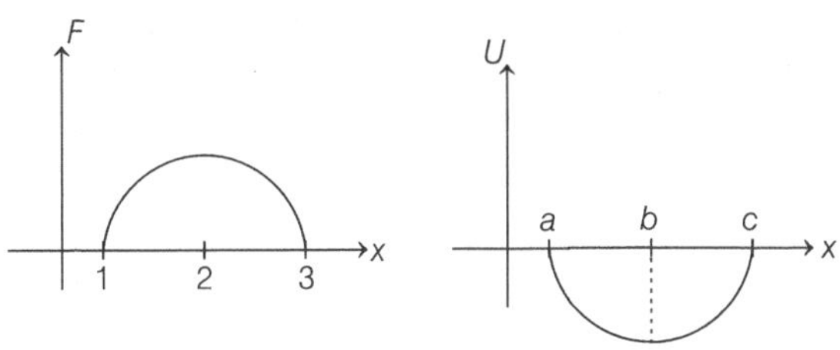

(a) Particle is in stable equilibrium at points 3 and *b*.
(b) Particle is in neutral equilibrium at points *b* and 2.
(c) No power is delivered by the force to the particle at 1, 3 and *b*.
(d) Particle has maximum kinetic energy at position *b*.

75 The kinetic energy (KE) *versus* position (x) for a particle moving in a straight line is shown in figure. Then, the acceleration of particle is

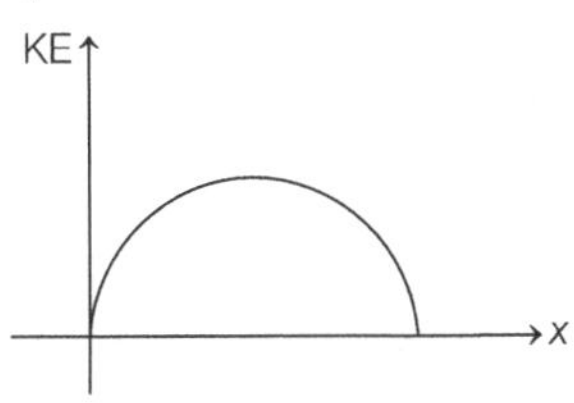

(a) uniform and positive (b) uniform and negative
(c) non-uniform (d) Insufficient data

76 A particle of mass $\frac{10}{7}$ kg is moving in the positive direction of x-axis. At initial position $x = 0$, its velocity is 1 m/s, then its velocity at $x = 10$ m is (use the given graph)

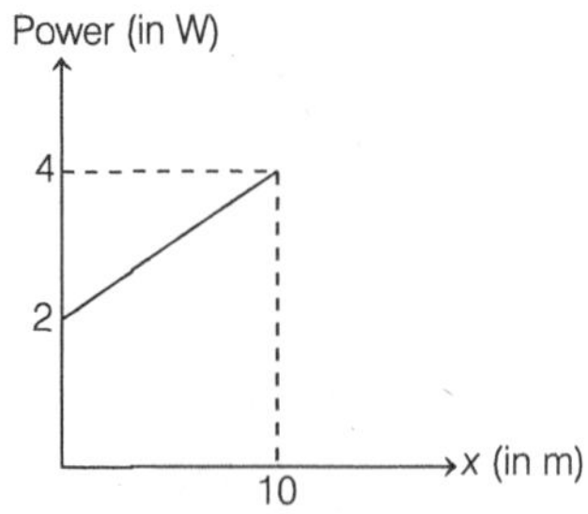

(a) 4 m/s (b) 2 m/s (c) $3\sqrt{2}$ m/s (d) $\frac{100}{3}$ m/s

77 A block of mass $m = 0.1$ kg is released from a height of 4 m on a curved smooth surface. On the horizontal surface, path AB is smooth and BC offers coefficient of friction $\mu = 0.1$. If the impact of block with the vertical wall at C be perfectly elastic, the total distance covered by the block on the horizontal surface before coming to rest will be (Take, $g = 10 \text{ ms}^{-2}$)

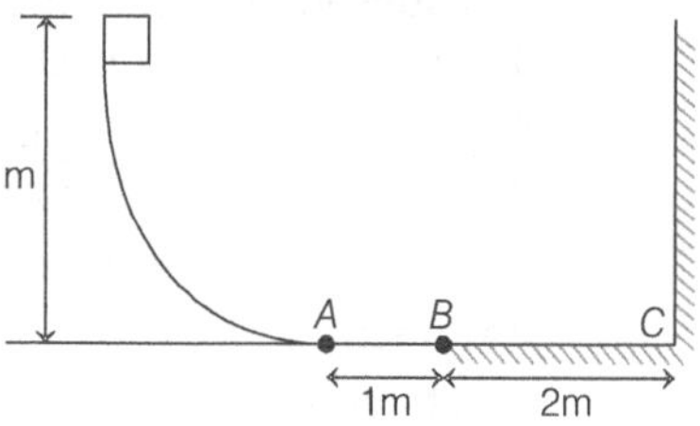

(a) 29 m (b) 49 m (c) 59 m (d) 109 m

78 The kinetic energy of a particle moving along a straight line increases uniformly with respect to the distance travelled by it. The force acting on the particle is (v is the instantaneous speed of particle)

(a) constant (b) proportional to v

(c) proportional to v^2 (d) inversely proportional to v

79 A block of mass m is placed on a rough inclined plane. The plane is moving with a constant velocity v. The block is not sliding on the inclined plane. Work done by force of friction (with respect to ground) in time t is

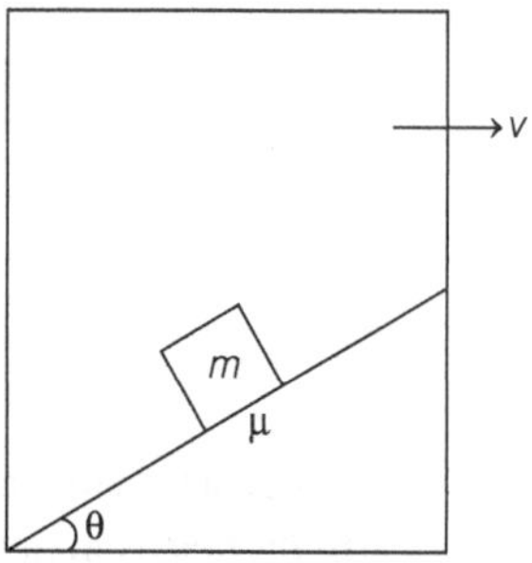

(a) $(\mu mg\cos\theta)vt$

(b) zero

(c) $(mgvt)\sin\theta\cos\theta$

(d) $(mg\sin\theta)vt$

LEVEL 03 Previous Years Questions

80 A particle is released from height S from the surface of the earth. At a certain height, its kinetic energy is three times its potential energy. The height from the surface of earth and the speed of the particle at that instant are respectively **[NEET 2021]**

(a) $\frac{S}{4}, \frac{3gS}{2}$ (b) $\frac{S}{4}, \frac{\sqrt{3gS}}{2}$ (c) $\frac{S}{2}, \frac{\sqrt{3gS}}{2}$ (d) $\frac{S}{4}, \sqrt{\frac{3gS}{2}}$

81 Water falls from a height of 60 m at the rate of 15 kg/s to operate a turbine. The losses due to frictional force are 10% of the input energy. How much power is generated by the turbine? (Take, $g = 10 \text{ m/s}^2$) **[NEET 2021]**

(a) 10.2 kW (b) 8.1 kW (c) 12.3 kW (d) 7.0 kW

82 The energy required to break one bond in DNA is 10^{-20} J. This value (in eV) is nearly [NEET 2020]

(a) 0.6 (b) 0.06 (c) 0.006 (d) 6

83 A point mass m is moved in a vertical circle of radius r with the help of a string. The velocity of the mass is $\sqrt{7gr}$ at the lowest point. The tension in the string at the lowest point is [NEET 2020]

(a) 6 mg (b) 7 mg (c) 8 mg (d) 1 mg

84 A force $F = 20 + 10y$ acts on a particle in y-direction, where F is in newton and y in metre. Work done by this force to move the particle from $y = 0$ to $y = 1$ m is [NEET 2019]

(a) 5 J (b) 25 J (c) 20 J (d) 30 J

85 An object of mass 500 g, initially at rest acted upon by a variable force whose x component varies with x in the manner shown. The velocities of the object a point $x = 8$ m and $x = 12$ m, would be the respective values of (nearly) [NEET (Odisha) 2019]

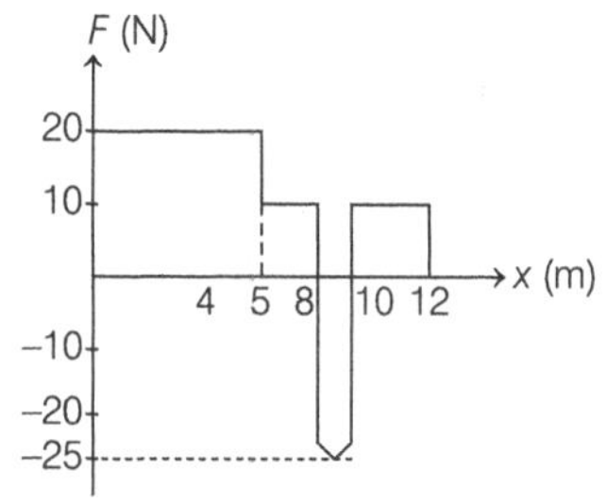

(a) 18 m/s and 24.4 m/s
(b) 23 m/s and 24.4 m/s
(c) 23 m/s and 20.6 m/s
(d) 18 m/s and 20.6 m/s

86 A mass m is attached to a thin wire and whirled in a vertical circle. The wire is most likely to break when [NEET 2019]

(a) the wire is horizontal
(b) the mass is at the lowest point
(c) inclined at an angle of 60° from vertical
(d) the mass is at the highest point

87 A body initially at rest and sliding along a frictionless track from a height h (as shown in the figure) just completes a vertical circle of diameter $AB = D$. The height h is equal to [NEET 2018]

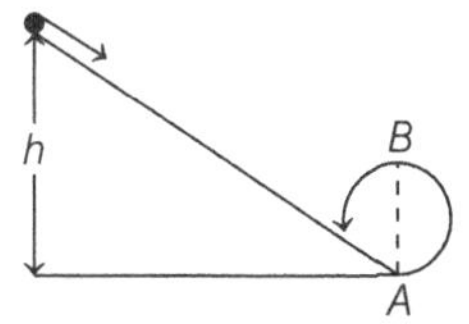

(a) $\frac{7}{5}D$ (b) D (c) $\frac{3}{2}D$ (d) $\frac{5}{4}D$

88 Consider a drop of rain water having mass 1 g falling from a height of 1 km. It hits the ground with a speed of 50 m/s. Take g constant with a value of 10 m/s^2. The work done by the (i) gravitational force and the (ii) resistive force of air is [NEET 2017]

(a) (i) –10 J, (ii) –8.25 J
(b) (i) 1.25 J, (ii) –8.25 J
(c) (i) 100 J, (ii) 8.75 J
(d) (i) 10 J, (ii) – 8.75 J

89 A particle of mass 10 g moves along a circle of radius 6.4 cm with a constant tangential acceleration. What is the magnitude of this acceleration, if the kinetic energy of the particle becomes equal to 8×10^{-4} J by the end of the second revolution after the beginning of the motion? [NEET 2016]

(a) 0.15 m/s^2 (b) 0.18 m/s^2 (c) 0.2 m/s^2 (d) 0.1 m/s^2

90 A particle moves from a point $(-2\hat{i} + 5\hat{j})$ to $(4\hat{j} + 3\hat{k})$ when a force of $(4\hat{i} + 3\hat{j})$ N is applied. How much work has been done by the force? [NEET 2016]

(a) 8 J (b) 11 J (c) 5 J (d) 2 J

91 A body of mass 1 kg begins to move under the action of a time dependent force $\mathbf{F} = (2t\hat{i} + 3t^2\hat{j})$ N, where $\hat{i}$ and $\hat{j}$ are unit vectors along X and Y axes. What power will be developed by the force at the time (t)? [NEET 2016]

(a) $(2t^2 + 4t^4)$ W (b) $(2t^3 + 3t^4)$ W (c) $(2t^3 + 3t^5)$ W (d) $(2t + 3t^3)$ W

92 What is the minimum velocity with which a body of mass m must enter a vertical loop of radius R so that it can complete the loop? **[NEET 2016]**

(a) $\sqrt{2gR}$ (b) $\sqrt{3gR}$ (c) $\sqrt{5gR}$ (d) $\sqrt{gR}$

93 Two similar springs P and Q have spring constants K_P and K_Q, such that $K_P > K_Q$. They are stretched, first by the same amount (case a), then by the same force (case b). The work done by the springs W_P and W_Q are related as, in case (a) and case (b), respectively **[CBSE AIPMT 2015]**

(a) $W_P = W_Q$; $W_P > W_Q$ (b) $W_P = W_Q$; $W_P = W_Q$
(c) $W_P > W_Q$; $W_Q > W_P$ (d) $W_P < W_Q$; $W_Q < W_P$

94 A block of mass 10 kg, moving in x-direction with a constant speed of 10 ms^{-1}, is subjected to a retarding force $F = 0.1\, x$ J/m during its travel from $x = 20$ m to 30 m. Its final KE will be **[CBSE AIPMT 2015]**

(a) 475 J (b) 450 J (c) 275 J (d) 250 J

95 Two particles of masses m_1, m_2 move with initial velocities u_1 and u_2. On collision, one of the particles get excited to higher level, after absorbing energy ε. If final velocities of particles be v_1 and v_2, then we must have **[CBSE AIPMT 2015]**

(a) $m_1^2 u_1 + m_2^2 u_2 - \varepsilon = m_1^2 v_1 + m_2^2 v_2$

(b) $\frac{1}{2} m_1 u_1^2 + \frac{1}{2} m_2 u_2^2 = \frac{1}{2} m_1 v_1^2 + \frac{1}{2} m_2 v_2^2 - \varepsilon$

(c) $\frac{1}{2} m_1 u_1^2 + \frac{1}{2} m_2 u_2^2 - \varepsilon = \frac{1}{2} m_1 v_1^2 + \frac{1}{2} m_2 v_2^2$

(d) $\frac{1}{2} m_1^2 u_1^2 + \frac{1}{2} m_2^2 u_2^2 + \varepsilon = \frac{1}{2} m_1^2 v_1^2 + \frac{1}{2} m_2^2 v_2^2$

96 A mass m moves in a circle on a smooth horizontal plane with velocity v_0 at a radius R_0. The mass is attached to a string which passes through a smooth hole in the plane as shown.

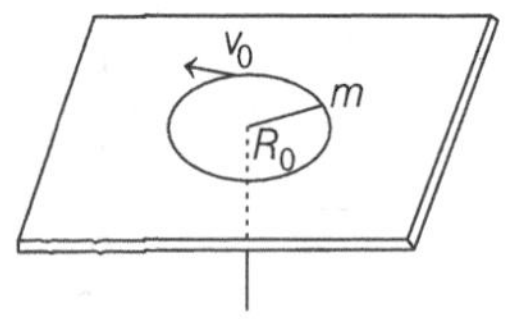

The tension in the string is increased gradually and finally m moves in a circle of radius $\frac{R_0}{2}$. The final value of the kinetic energy is **[CBSE AIPMT 2015]**

(a) mv_0^2 (b) $\frac{1}{4} mv_0^2$ (c) $2\, mv_0^2$ (d) $\frac{1}{2} mv_0^2$

97 A ball is thrown vertically downwards from a height of 20 m with an initial velocity v_0. It collides with the ground, loses 50% of its energy in collision and rebounds to the same height. The initial velocity v_0 is (Take, $g = 10\ ms^{-2}$) **[CBSE AIPMT 2015]**

(a) 14 ms^{-1} (b) 20 ms^{-1} (c) 28 ms^{-1} (d) 10 ms^{-1}

98 A particle of mass m is driven by a machine that delivers a constant power k watts. If the particle starts from rest, the force on the particle at time t is **[CBSE AIPMT 2015]**

(a) $\sqrt{\frac{mk}{2}}\, t^{-1/2}$ (b) $\sqrt{mk}\, t^{-1/2}$ (c) $\sqrt{2mk}\, t^{-1/2}$ (d) $\frac{1}{2}\sqrt{mk}\, t^{-1/2}$

99 The heart of a man pumps 5 L of blood through the arteries per minute at a pressure of 150 mm of mercury. If the density of mercury be 13.6×10^3 kg/m^3 and $g = 10$ m/s^2, then the power of heart (in W) is **[CBSE AIPMT 2015]**

(a) 1.70 (b) 2.35
(c) 3.0 (d) 1.50

100 A uniform force of $(3\hat{\mathbf{i}}+\hat{\mathbf{j}})$ N acts on a particle of mass 2 kg. Hence, the particle is displaced from position $(2\hat{\mathbf{i}}+\hat{\mathbf{k}})$ m to position $(4\hat{\mathbf{i}}+3\hat{\mathbf{j}}-\hat{\mathbf{k}})$ m. The work done by the force on the particle is **[NEET 2013]**

(a) 9 J (b) 6 J (c) 13 J (d) 15 J

101 The potential energy of a particle in a force field is $U=\frac{A}{r^2}-\frac{B}{r}$, where A and B are positive constants and r is the distance of particle from the centre of the field. For stable equilibrium, the distance of the particle is **[CBSE AIPMT 2012]**

(a) $B/2A$ (b) $2A/B$ (c) A/B (d) B/A

102 Force F on a particle moving in a straight line varies with distance d as shown in the figure. The work done on the particle during its displacement of 12 m is **[CBSE AIPMT 2011]**

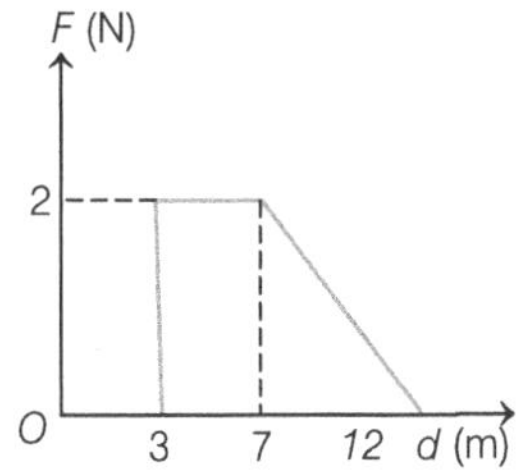

(a) 21 J (b) 26 J (c) 13 J (d) 18 J

103 An engine pumps water through a hose pipe. Water passes through the pipe and leaves it with a velocity of 2 ms^{-1}. The mass per unit length of water in the pipe is 100 kg m^{-1}. What is the power of the engine? **[CBSE AIPMT 2010]**

(a) 400 W (b) 200 W (c) 100 W (d) 800 W

ANSWER KEY

Level 1 Objective Questions

1 (a)	2 (c)	3 (a)	4 (b)	5 (a)	6 (a)	7 (a)	8 (b)	9 (d)	10 (d)
11 (b)	12 (a)	13 (a)	14 (c)	15 (c)	16 (a)	17 (b)	18 (d)	19 (c)	20 (d)
21 (a)	22 (d)	23 (a)	24 (b)	25 (a)	26 (c)	27 (c)	28 (a)	29 (d)	30 (d)
31 (b)	32 (c)	33 (a)	34 (a)	35 (b)	36 (c)	37 (a)	38 (b)	39 (a)	40 (c)
41 (d)	42 (a)	43 (d)	44 (a)	45 (c)					

Level 2 Objective Questions

46 (b)	47 (c)	48 (a)	49 (b)	50 (c)	51 (b)	52 (b)	53 (b)	54 (d)	55 (d)
56 (a)	57 (c)	58 (a)	59 (a)	60 (b)	61 (a)	62 (c)	63 (c)	64 (d)	65 (a)
66 (b)	67 (d)	68 (a)	69 (c)	70 (c)	71 (d)	72 (b)	73 (d)	74 (b)	75 (c)
76 (a)	77 (c)	78 (a)	79 (c)						

Level 3 Previous Years Questions

80 (d)	81 (b)	82 (b)	83 (c)	84 (b)	85 (c)	86 (b)	87 (d)	88 (d)	89 (d)
90 (c)	91 (c)	92 (c)	93 (c)	94 (a)	95 (c)	96 (c)	97 (b)	98 (a)	99 (a)
100 (a)	101 (b)	102 (c)	103 (d)						

Solutions

LEVEL 01 Objective Questions

1. $\mathbf{W} = \mathbf{F} \cdot d\mathbf{s}$

$\Rightarrow \quad 0 = (6\hat{\mathbf{i}} + 2\hat{\mathbf{j}} - 3\hat{\mathbf{k}}) \cdot (2\hat{\mathbf{i}} - 3\hat{\mathbf{j}} - x\hat{\mathbf{k}})$

$12 - 6 + 3x = 0$

$x = -2$

2. The slope of a x - t graph gives velocity

at $t = 0$, $v = 1$ m/s

at $t = 4$ s, $v = -2$ m/s

The net work done by the particle $= K_f - K_i$

$$= \frac{1\,(2)}{2}(2^2 - 1^2) = 3 \text{ J}$$

3. $W = E_f - E_i = \frac{1}{2}mv^2 - mgh$

$$= \frac{1}{2} \times 1 \times (2)^2 - (1)(10)(1)$$

$$= -8 \text{ J}$$

4. Work done during extension = Change in potential energy

$$= \frac{1}{2} \times \text{Stress} \times \text{Strain} \times \text{Volume}$$

$$= \frac{1}{2} \times \frac{40}{A} \times \frac{10^{-3}}{L} \times AL - \frac{1}{2} \times \frac{20}{A} \times \frac{0.6 \times 10^{-3}}{L} \times AL$$

$$\approx 1.4 \times 10^{-2} \text{ J}$$

5. The work done by the applied force

$$= F \cdot d \cos\theta = 50 \cdot 5 \cos 37^\circ = 200 \text{ J}$$

6. Since, the mass m is raised vertically upwards with acceleration a,

$T - mg = ma$

$\Rightarrow \quad T = m\,(a + g) = 5\,(2 + 9.8) = 59$ N

So, work done $= T \cdot s = (59)(2.5) = 147.5$ J

7. Tension, $T = m\,(g - a)$

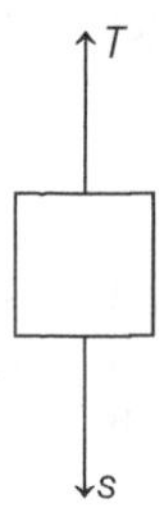

Displacement in time t,

$$s = \frac{1}{2}at^2$$

$\therefore \quad W = T \cdot s \cos 180^\circ$

$$= m(g - a)\frac{1}{2}at^2(-1) = \frac{m}{2}(a - g)\,at^2$$

8. W = Gain in kinetic energy − Loss in potential energy

$$= \frac{1}{2} \times 5 \times (10)^2 - (5)(10)(20)$$

$$= -750 \text{ J}$$

9. The acceleration of body, $a = \frac{v}{T}$ ...(i)

So, work done, $W = F \cdot s$

As the body starts from rest, therefore its initial velocity will be zero.

$\because$ From Newton's 2nd equation of motion,

$$s = ut + \frac{1}{2}at^2$$

$$s = \frac{1}{2}at^2 \qquad (\because u = 0)$$

$\Rightarrow \quad W = ma\left(\frac{1}{2}at^2\right)$ ...(ii)

From Eqs. (i) and (ii), we get

$$= \frac{1}{2}\frac{mv^2}{T^2} \cdot t^2 \;\Rightarrow\; W \propto \frac{v^2t^2}{T^2}$$

10. $\frac{1}{2}m[20^2 - 10^2] = n\frac{1}{2}m(10)^2 \;\Rightarrow\; n = 3$

11. From work-energy theorem,

$$\Delta W = \Delta \text{KE}$$

$$F \cdot s \cos\theta = \text{KE}$$

$$\cos\theta = \frac{\text{KE}}{F \cdot s} \;\Rightarrow\; \cos\theta = \frac{40}{20 \times 4}$$

$\Rightarrow \quad \theta = 60^\circ \;\Rightarrow\; \cos\theta = \frac{1}{2}$

12. $\text{KE} = F \cdot s \;\Rightarrow\; \text{KE} \propto s$

So, the graph will be straight line.

13. Let kinetic energy at point of projection is $K_i = K$. Then, kinetic energy at P will be

$$K_P = \left(\frac{64}{36 + 64}\right)K = 0.64\,K$$

Now, $\quad \frac{K_i}{K_P} = \frac{K}{0.64\,K} = \frac{25}{16}$

$\therefore \quad \frac{v_i}{v_P} = \sqrt{\frac{K_i}{K_P}} = \frac{5}{4}$

14. Given, $F = -6x^3$

$$dW = \int F \cdot dx$$

$$W = \int_0^4 -6x^3 dx = -384 \text{ J}$$

Hence, the work is −ve in displacing the particle from $x = 0$ to 4.

15. The initial height of centre of gravity $= \frac{a}{2}$

Final height of centre of gravity $= \frac{b}{2}$

So, displacement $= \frac{b}{2} - \frac{a}{2}$

Work done $= mg\left[\frac{b}{2}-\frac{a}{2}\right]$ or $mg\left[\frac{b-a}{2}\right]$

16. The energy lost due to air friction $= \frac{1}{2}mv^2 - mgh$

$$= \frac{1}{2}(1)(20)^2 - 1(10)(18) = 20 \text{ J}$$

17. Since, TE = PE + KE

If speed = 0, then KE = 0

So, $U(x) = E$

18. $F = -\frac{dU}{dx}$

So, the magnitude of force from least to greatest is BC, AB, CD.

19. At C, $F = 0$

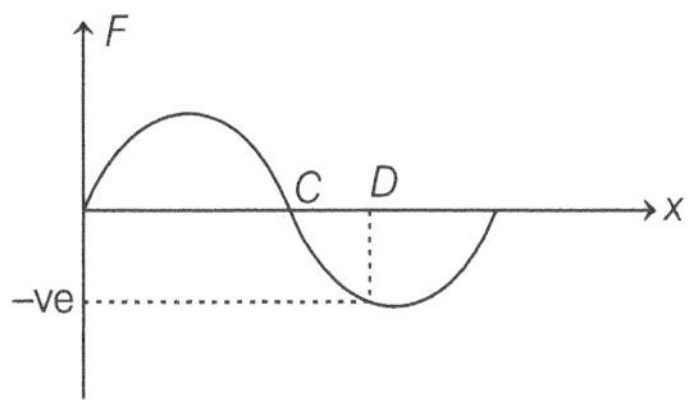

When particle is displaced in positive direction (towards D), force is in negative direction (can be seen from the graph). So, force is restoring in nature. Hence, at C, equilibrium is stable.

20. At $x = 0$, KE = 0, PE = 0

$\therefore$ ME = 0

At $x = 2$, PE = 0

$\therefore$ KE should also be zero, as only conservative forces are acting.

So, ME should remain conserved.

21. Given, $U = \alpha x^2 - \beta x$

$$F = -\frac{dU}{dx} = \beta - 2\alpha x = 0$$

$$x = \frac{\beta}{2\alpha}$$

At $x = \frac{\beta}{2\alpha}$, U is minimum as $\frac{d^2U}{dx^2} < 0$ and hence, stable equilibrium.

22. If the particle reaches B, it will itself go to C.

So, from energy conservation,

$$\frac{1}{2}mv^2 = mgh$$

$\Rightarrow$ $v = \sqrt{2gh} = \sqrt{2(10)(320)}$

$$= 80 \text{ m/s}$$

23. KE = TE − PE = 5 − (−3) = 8 J

Note : Potential energy function means, only conservative forces are acting.

So, mechanical energy is conserved.

24. $k \propto \frac{1}{l}$

$$\frac{k_1}{k} = \frac{k_3}{k} = \frac{l}{l/4} = 4$$

$\therefore$ $k_1 = k_3 = 4k$

Similarly,

$$\frac{k_2}{k} = \frac{l}{(2l/4)} = 2$$

$\therefore$ $k_2 = 2k$

Now, $k_1 + k_2 + k_3 = 10k$

25. Energy balance is correct only, when there is no dissipative force. In given case, an external agent is present. So, student A will be incorrect.

26. $v_{CM} = \frac{2m(2v) + m(-v)}{3m}$

$\Rightarrow$ $v_{CM} = v$,

Reduced or equivalent mass, $m_u = \frac{m_1 \times m_2}{m_1 + m_2} = \frac{2m}{3}$

Now, the conservation of mechanical energy,

$$\frac{1}{2}m_u(v_{CM})^2 = \frac{1}{2}kx^2$$

$$\frac{1}{2}\left(\frac{2m}{3}\right)(v^2) = \frac{1}{2}kx^2$$

$\Rightarrow$ $\frac{2}{3}mv^2 = kx^2$

$\Rightarrow$ $x = \sqrt{\frac{2m}{3k}} \cdot v$

So, the maximum extension in spring is, $\sqrt{\frac{2m}{3k}} \cdot v$.

27. Power $= Fv\cos\theta$

$= Fv$ $\quad (\because \theta = 0°)$

$= 10 \times 10 \times 2$ $\quad (\because v = at)$

$= 200$ W

28. Work done by force $= \mathbf{F} \cdot \mathbf{s}$

$$= (2\hat{i} + 3\hat{j} + 4\hat{k}) \cdot (3\hat{i} + 4\hat{j} + 5\hat{k})$$

$$= 38 \text{ J}$$

$$\text{Power} = \frac{\text{Work done by force}}{\text{Time taken}}$$

$$= \frac{38}{4} = 9.5 \text{ W}$$

29. We have, $P = \frac{\text{Work done}}{\text{Time taken}} = \frac{\frac{1}{2}mv^2}{t}$

$$Pt = \frac{1}{2}mv^2$$

So, $t \propto v^2$

if v becomes twice, then t will be 4 times.

30. $P = Fv = (ma)v$

$\therefore \quad a = \dfrac{vdv}{dx}$

$$P = m\left(v\frac{dv}{dx}\right)v = mv^2\frac{dv}{dx}$$

On integrating both sides, we get

$$\frac{P}{m}\int dx = \int v^2 dv \Rightarrow \frac{P}{m}\cdot x = \frac{v^3}{3}$$

$\therefore \quad x \propto v^3$

31. $a = \dfrac{v}{t} = \dfrac{2-0}{20} = 0.1 \text{ m/s}^2$

Now, displacement of block

$$= \frac{1}{2}at^2 = \frac{1}{2}(0.1)(20)^2 = 20 \text{ m}$$

$\therefore$ Power $= \dfrac{\text{Work done}}{\text{Time taken}}$

$$= \frac{F\cdot ds}{t} = \frac{10(0.1)(20)}{20} \quad (\because F = ma)$$

$= 1$ W

32. Given, $P = 4t^3 - 5t + 2$

$$\int dW = \int Pdt$$

$\Rightarrow \quad W = \displaystyle\int_2^4 (4t^3 - 5t + 2)\, dt$

$\Rightarrow \quad W = \left[t^4 - \dfrac{5}{2}t^2 + 2t\right]_2^4$

$\Rightarrow \quad W = [(4)^4 - (2)^4] - \dfrac{5}{2}[(4)^2 - (2)^2] + 2(4-2)$

$\Rightarrow \quad W = [256 - 16] - \dfrac{5}{2}[16 - 4] + 4$

$\Rightarrow \quad W = 240 - 30 + 4$

$= 214$ J

From work-energy theorem,

$$W = \Delta K$$

$\therefore \quad \Delta K = 214$ J

33. The work done by all the forces $= \Delta K$

$$W_g + W_f = \frac{1}{2}(m_1 + m_2)v^2$$

$$5(10)(4) - 0.2(10)(10)(4) = \frac{10v^2}{2} + \frac{5v^2}{2}$$

$\Rightarrow \quad v = 4$ m/s

34. Applying conservation of mechanical energy, decrease in gravitational energy of block B = increase in spring potential energy

$$2mgx = \frac{1}{2}kx^2$$

$\Rightarrow \quad x = \dfrac{4mg}{k}$

35. Work done by all the forces $= K_f - K_i$

$$mgh = \frac{1}{2}m(v_f^2 - v_i^2)$$

$$gh = \frac{1}{2}[(3u)^2 - u^2]$$

$\Rightarrow \quad h = \dfrac{4u^2}{g}$

36. $F = \dfrac{1}{2}kx^2$

In first case $= \dfrac{1}{2}kx^2 = 10$ J

In second case $= \dfrac{1}{2}k(2x)^2 = 4 \times 10 = 40$ J

So, the amount of work done to stretch same spring from extension x to $2x = 40 - 10 = 30$ J

37. The velocity at point B is v_B.

Applying conservation of energy,

$$\frac{1}{2}mv_B^2 = mgh$$

$\Rightarrow \quad v_B = \sqrt{20}$ m/s

Now, let a be the retardation provided by friction,

$$a = \mu g$$

Using $\quad v^2 = u^2 - 2as$

$\Rightarrow \quad 0 = 20 - 2a(3)$

$\Rightarrow \quad a = \dfrac{10}{3}$ m/s^2

$\therefore \quad a = \mu g = \dfrac{10}{3}$

$\Rightarrow \quad \mu = \dfrac{1}{3}$

38. Work done by air drag $= \dfrac{1}{2}mv^2 - mgh$

$$= \frac{1}{2}m(\sqrt{2gh})^2 - mgh$$

$= 0$

39. From energy conservation,

$$mgh = \frac{1}{2}kx^2$$

$$0.04(10)(5) = \frac{1}{2}(400)x^2$$

$\Rightarrow \quad x = \dfrac{1}{10}$ m

$\Rightarrow \quad x = \dfrac{1}{10} \times 100$ cm

$\Rightarrow \quad x = 10$ cm

40. For the upward motion,

Net force $= mg + F$

Retardation $(a) = \dfrac{mg + F}{m}$

So, distance covered, $s = \dfrac{v^2}{2a} = \dfrac{v^2 m}{2(mg + F)}$

Now, for downward motion, net force $= mg - F$

$$a' = \frac{mg - F}{m}$$

So, distance, $s' = \dfrac{v'^2}{2a'} = \dfrac{v'^2 m}{2\,(mg - F)}$

As, $s = s'$

$$\frac{v^2 m}{2\,(mg + F)} = \frac{v'^2 m}{2\,(mg - F)}$$

$$\therefore \quad v' = v\sqrt{\frac{mg - F}{mg + F}}$$

41. We have, initial velocity, $v_1 = 4\hat{\mathbf{i}} + 16\,\hat{\mathbf{k}}$

$$|v_1| = \sqrt{4^2 + 16^2} = \sqrt{272} \text{ m/s}$$

Final velocity, $v_2 = 8\hat{\mathbf{i}} + 20\,\hat{\mathbf{j}}$

$$|v_2| = \sqrt{8^2 + 20^2} = \sqrt{464} \text{ m/s}$$

So, the work done $= K_f - K_i$

$$= \frac{1}{2}(0.01)\,(464 - 272) = 0.96 \text{ J}$$

42. Energy at point $P = mgh = m(10)(2) = 20\,m$

Now, 10% of energy is lost.

$\therefore$ Remaining energy = KE

$$(0.9)(20)\,m = \frac{1}{2}mv^2$$

$$\Rightarrow \quad v = 6 \text{ m/s}$$

43. Work done by frictional force = – ve of potential energy

$$= -\,0.2 \times 10 \times 2 = -\,4 \text{ J}$$

44. The work done by frictional force $= \dfrac{1}{2}mv^2 - mgh$

$$0.3\,(m)(g)(d) = \frac{1}{2}m\,(8)^2 - m\,(10)\,(2)$$

$$\Rightarrow \quad d = 4\text{m}$$

45. As the particle is dropped from height h, the spring is compressed by x.

So, $mg\,(h + x) = \dfrac{1}{2}kx^2$

$$\Rightarrow \quad k = \frac{2mg\,(h + x)}{x^2}$$

LEVEL 02 Objective Questions

46. We have, displacement of train $= \mathbf{s}_0$

Displacement of box with respect to train $= \mathbf{s}$

The total displacement in ground frame $= \mathbf{s} + \mathbf{s}_0$

$$\text{Work done} = \mathbf{F}\cdot(\mathbf{s} + \mathbf{s}_0)$$

47.

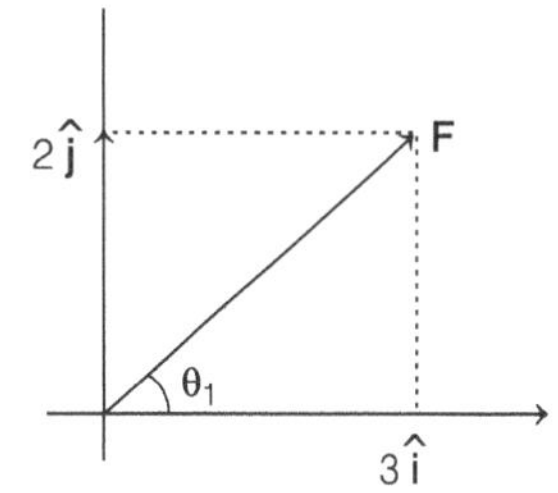

$$\tan\theta_1 = \frac{2}{3}$$

Given line is, $y = -\dfrac{k}{2}x + 2.5$

Slope or $\tan\theta_2 = -\dfrac{k}{2}$

$$W = 0, \quad \text{if } \mathbf{F} \perp \mathbf{s}$$

or product of two slopes = – 1

$$\therefore \quad \left(\frac{2}{3}\right)\left(-\frac{k}{2}\right) = -1$$

$$\therefore \quad k = 3$$

48. Work done against force of friction $= \mu\, mg\cos\theta \cdot d$

$$= 0.1\,(1)\,(10)\left(\frac{\sqrt{3}}{2}\right)(10)$$

$$= 8.7 \text{ J}$$

49. We have, frictional force, $f = \mu N$

$$= \mu\,(mg - F\sin\theta)$$

Now, for the horizontal direction,

$$F\cos\theta \ge f$$

$$F\cos\theta \ge \mu\,(mg - F\sin\theta)$$

$$F \ge \frac{\mu\, mg}{\cos\theta + \mu\sin\theta}$$

$\therefore$ So, minimum work $= W = F\cdot ds = Fd\cos\theta$

$$= \frac{\mu\, mg\, d\cos\theta}{\cos\theta + \mu\sin\theta}$$

50. The displacement of block in time t, $d = vt$

Work done by frictional force $= F\cdot d\cos\theta$

$$= (mg\sin\theta)\,(vt)\cos(90^\circ - \theta)$$

$$= mg\,vt\sin^2\theta$$

51. Given, $f = ax + b$

$$W = f\cdot dx = \int_0^s (ax + b)\,dx$$

$$= \frac{as^2}{2} + bs$$

52. We have, $\mathbf{F} = x\hat{\mathbf{j}} + y\hat{\mathbf{i}}$

$$d\mathbf{r} = dx\hat{\mathbf{i}} + dy\hat{\mathbf{j}}$$

$$W = \mathbf{F}\cdot d\mathbf{r}$$

$$= (x\hat{j} + y\hat{i}) \cdot (dx\hat{i} + dy\hat{j})$$
$$= xdy + ydx$$

As $x = y$,
$$dx = dy$$

So, $$W = \int_1^2 xdx + \int_1^2 ydy = 3 \text{ unit}$$

53. Tension in the topmost string will be,

T = net downwards force = 3 mg.

Immediately after this string is cut, this T acting on A in upward direction vanishes. Hence,
$$a_A = \frac{3mg}{m} = 3g \quad \text{(downwards)}$$

There is no change in any force acting on B and C.

Hence, $a_B = a_C = 0$

54. $F = kx^2 - kxl$

F is maximum, when
$$\frac{dF}{dx} = 0$$
or $$2kx - kl = 0$$
or $$x = \frac{l}{2}$$

55. As point object is moving slowly from A to B.

∴ Kinetic energy is 0 at point A and B

∵ A and B are at same height from ground

∴ Work done by gravity, $W_g = 0$

Work done by friction,

$W_f = -\mu_k\, mgl = -0.1 \times 2 \times 10 \times 200 = -400$ J

Assume that the work done by tangential force be W_T

∴ By applying work-energy theorem,
$$W_g + W_f + W_T = \Delta KE$$
$$\Rightarrow \quad 0 - 400 + W_T = 0 - 0$$
$$\Rightarrow \quad W_T = 400 \text{ J}$$

56. $f = ma = m \cdot \left(v \cdot \frac{dv}{dx}\right) = (a - bx) \cdot m$

Integration gives,
$$\frac{1}{2}mv^2 = \left(ax - \frac{bx^2}{2}\right)m + C$$

At $x = 0$, $v = 0$
$$\Rightarrow \quad C = 0$$
$$\therefore \quad \frac{1}{2}mv^2 = \left(ax - \frac{bx^2}{2}\right) \cdot m$$

v is again zero at $x = \frac{2a}{b}$

∴ Distance $AB = \frac{2a}{b}$

57. The work done = change in kinetic energy
$$W = \int \mathbf{F} \cdot d\mathbf{s}$$

Here, $\mathbf{F} = 3x^2\hat{\mathbf{i}} + 4\hat{\mathbf{j}}$
$$d\mathbf{s} = dx \cdot \hat{\mathbf{i}} + dy \cdot \hat{\mathbf{j}}$$
$$= \int_2^3 3x^2\, dx + \int_3^0 4\, dy$$
$$= 7 \text{ J}$$

58. From work-energy theorem,
$$\text{work done} = \Delta KE$$
$$Fx - \frac{1}{2}kx^2 = \frac{1}{2}mv^2$$
$$\Rightarrow \quad v = \frac{F}{\sqrt{km}}$$

59. $a \propto x \;\Rightarrow\; a = -kx$
$$\frac{vdv}{dx} = -kx$$
$$\Rightarrow \quad \int_{v_1}^{v_2} vdv = -\int_0^x kx\,dx$$
$$\frac{v_2^2 - v_1^2}{2} = -\frac{kx^2}{2}$$
$$\Delta k \propto x^2$$

60. $F = -\frac{dU}{dx}$

At $r = r_1$, the slope is negative, so force is positive and hence, repulsive.

61. $E = U(x) + KE(x) = \text{constant}$
$$KE(x) = E - U(x)$$

So, the graph (a) represents the best variation of kinetic energy with x.

62. Given, $U = 20 + (x - 2)^2$

At $x = 2$ m, $U_{min} = 20$ J

So, maximum kinetic energy, $K_{max} = 36 - 20 = 16$ J

63. Given, $U(x) = \left(\frac{x^2}{4} - \frac{x}{2}\right)$ J

For minimum potential energy, $F = -\frac{dU}{dx} = \frac{x}{2} - \frac{1}{2} = 0$
$$x = 0 \quad \text{or} \quad \pm 1$$

So, $U_{min} = -\frac{1}{4}$

We have, $K_{max} + U_{min} = 2$
$$K_{max} = 2 + U_{min} = 2 + \frac{1}{4}$$
$$K_{max} = \frac{9}{4}$$
or $$\frac{1}{2}mv_{max}^2 = \frac{9}{4}$$
or $$v_{max} = \frac{3}{\sqrt{2}} \quad (\because m = 1 \text{ kg})$$

64. Given, $U = \frac{x^3}{3} - 4x + 12$

$$F = -\frac{dU}{dx} = -x^2 + 4$$

$$\Rightarrow \quad \frac{dU}{dx} = x^2 - 4$$

$$\frac{d^2U}{dx^2} = +2x$$

At equilibrium F,

or $\quad -\frac{dU}{dx} \quad$ or $\quad \frac{dU}{dx} = 0$

$$x^2 - 4 = 0$$

$$\therefore \quad x = \pm 2\,\text{m}$$

At $x = +2$ m, $\frac{d^2U}{dx^2} = +4$ m

$\therefore$ U is minimum

or $x = +2$ m is the position of stable equilibrium.

Similarly, $x = -2$ m is the position of unstable equilibrium.

65. The tension is given as,

$$T = \left(\frac{2m_1m_2}{m_1 + m_2}\right)g$$

For spring block systems A, B and C, we get

$$T_1 = 1.5\,g,\ T_2 = 2.4\,g,\ T_3 = 1.3\,g$$

As $x \propto T$, $x_2 > x_1 > x_3$

66. Power is the dot product of force and the velocity, it will be zero when they are perpendicular to each other.

At $t = 6$ s, $P = 0$

Hence, angle $= 90°$

67. The mass of water raised,

$$m = 1000 \times 9 = 9000\,\text{kg}$$

So, $\quad \text{power} = \frac{mgh}{t} = \frac{9000 \times 10 \times 10}{5 \times 60}$

$$= 3000\,\text{W} = 3\,\text{kW}$$

$\therefore$ Efficiency $= \frac{3\,\text{kW}}{10\,\text{kW}} = 0.3$ or $30\,\%$

68. Applying conservation of mechanical energy,

$$K_L + U_L = K_f + U_f$$

$$0 + mg\frac{L}{2}\sin 30° = \frac{1}{2}mv^2 - \frac{mg}{2}\left(\frac{L}{4}\right) + \frac{mg}{2}\left(\frac{L}{8}\right)$$

$$\frac{1}{2}mv^2 = \frac{5mgL}{16}$$

$$\Rightarrow \quad v = \sqrt{\frac{5gL}{8}} = \sqrt{\frac{5(10)(4/25)}{8}}$$

$$= 1\,\text{m/s}$$

69. The work done by frictional force = change in potential energy

$$mgh - mgh' = \mu mgL$$

$$h' = h - \mu L = 5 - 0.5\,(8) = 1\,\text{m}$$

So, the block climb to the smooth curved surface through 1 m.

70. We have, $P = \frac{1/2\,mv^2}{t}$

To obtain n times water from the same pipe in the same time, the velocity also becomes n times,

$$P' = \frac{1/2\,(n)\,m\,(nv)^2}{t} = n^3P$$

71. $F - mg\sin\theta = ma$

$$F = 2 \times 10 \times \sin 30° + 2 \times 1 = 12\,\text{N}$$

The velocity after 4 s, $v = at$

$$v = 1\,(4) = 4\,\text{m/s}$$

So, the power delivered at $t = 4$ s

$$= Fv = 12(4) = 48\,\text{W}$$

72. 2 kg block leaves the ground

when $T = kx = mg = 20$ N or $40\,x = 20$

$\therefore$ Extension, $x = \frac{1}{2}$ m

Now let us come to 5 kg block.

Decrease in its potential energy = Increase in its kinetic energy and spring potential energy.

$$\therefore \quad 5 \times 10 \times \frac{1}{2} = \frac{1}{2} \times 5 \times v^2 + \frac{1}{2} \times 40 \times \left(\frac{1}{2}\right)^2$$

$$\therefore \quad v = 2\sqrt{2}\,\text{m/s}$$

73. Given, $F = k_1x_1 = k_2x_2$

$$\therefore \quad x \propto \frac{1}{k}$$

$$\frac{1}{2}k_1x_1^2 > \frac{1}{2}k_2x_2^2$$

or $\quad x_1 > x_2 \qquad (k_1x_1 = k_2x_2)$

or $\quad k_1 < k_2 \qquad \left(\text{as } x \propto \frac{1}{k}\right)$

If stretched by same amount, then

$$\frac{W_1}{W_2} = \frac{\frac{1}{2}k_1x^2}{\frac{1}{2}k_2x^2} = \frac{k_1}{k_2} < 1 \qquad (\text{as } k_1 < k_2)$$

$$\therefore \quad W_1 < W_2$$

74. At point b, potential energy is minimum. So, it is stable equilibrium.

At point 2, $F \neq 0$. So, it is not an equilibrium.

75. For uniform acceleration,

$$v^2 = u^2 \pm 2ax$$

So, v^2 *versus* x, or kinetic energy *versus* x graph will be a straight line.

Since, the given KE *versus* x graph is not a straight line, therefore the acceleration of particle is non-uniform.

76. The area under P - x graph

$$= \int_1^v P \cdot dx \qquad \left(\because \frac{dx}{dt} = v\right)$$

$$= \int_1^v Fvdx = \int_1^v mv^2 \cdot dv = \frac{m\,(v^3 - 1)}{3}$$

$$30 = \frac{m}{3}[v^3 - 1] = \frac{10\,[v^3 - 1]}{7 \times 3}$$

$v = 4$ m/s

77. The total distance covered by the block on the horizontal surface before coming to rest is given as,

$$mgh = \mu\, mgd \qquad (\because \mu = 0.1, h = 4\,\text{m})$$

$$d = 40 \text{ m}$$

So, $BC = \frac{40}{2} = 20$ m

$AB = 19$ m

Distance $= 20(2) + 19 = 59$ m

78. $KE = \frac{1}{2}mv^2$ or $v^2 \propto s$

$\therefore \quad v \propto s^{\frac{1}{2}}$

$$\frac{dv}{ds} \propto s^{-\frac{1}{2}}$$

$$a \text{ or } v \cdot \frac{dv}{ds} \propto s^{-\frac{1}{2}} \cdot s^{\frac{1}{2}}$$

$\therefore \quad a$ or $v \cdot \frac{dv}{ds} \propto s^0$

or $\quad F$ or $ma \propto s^0$

79. Constant velocity means net force = 0

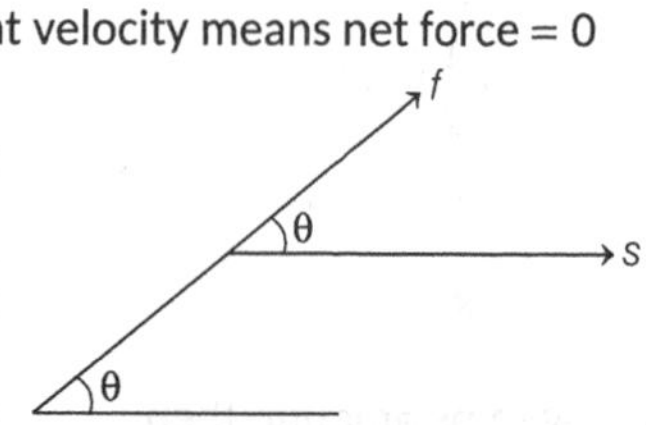

$\therefore \quad f = mg \sin\theta$

$s = vt$

$\therefore \quad W_f = fs\cos\theta = (mgvt)\sin\theta \cdot \cos\theta$

LEVEL 03 Previous Years Questions

80. Let the particle at height S from the surface of the earth is as shown.

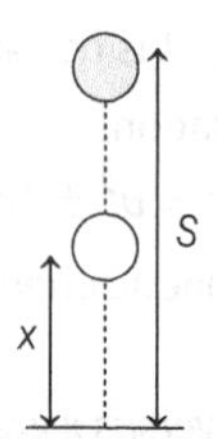

Given, at height x, kinetic energy

= 3 (potential energy)

$\Rightarrow \quad KE = 3\,mgx$

At height S, total energy, $TE = PE + KE$

$TE = mgS + 0 = mgS$

At height x, total energy,

$TE = PE + KE$

$TE = mgx + 3mgx$

$mgS = 4mgx$

$\Rightarrow \quad x = \frac{S}{4}$ ...(i)

Now, we shall determine the speed of the particle at this height.

As, $\quad KE = 3 \times mgx$

$KE = 3 \times mg\,\frac{S}{4}$ [from Eq. (i)]

$$\frac{1}{2}mv^2 = \frac{3}{4}mgS$$

$\Rightarrow \quad v = \sqrt{\frac{3}{2}gS}$

81. Given, the flow rate of the water, $\frac{m}{t} = 15$ kg/s

The height of the water fall, $h = 60$ m

Loss due to frictional force = 10%

The power used in the turbine

= (100 – 10) % = 90%

The acceleration due to gravity,

$g = 10\ \text{m/s}^2$

We know that, power generated by the turbine

= change in potential energy $= 0.90\left(\frac{mgh}{t}\right)$

$= 0.90 \times 15 \times 10 \times 60$

$= 8100$ W $= 8.1$ kW

82. Given, $E = 10^{-20}$ J

In terms of eV, we get

$$E = \frac{10^{-20}}{1.6 \times 10^{-19}} \text{ eV} = 0.06 \text{ eV}$$

Hence, correct option is (b).

83. Velocity of point mass in a vertical circle at lowest point,

$$V_l = \sqrt{7gr}$$

$\therefore \quad V_l = \sqrt{7gr} > \sqrt{5gr}$

Hence, point mass will have completed the vertical circular path.

We know that,

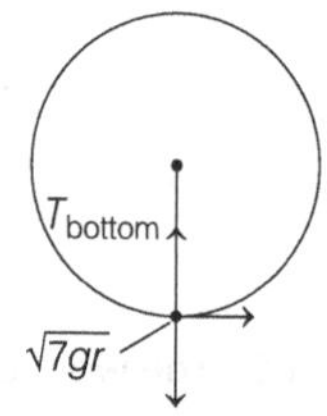

$$T_{bottom} - mg = \frac{mv^2}{r} = \frac{m}{r}(\sqrt{7gr})^2$$

$\Rightarrow \quad T_{bottom} - mg = 7mg$

$\Rightarrow \quad T_{bottom} = 8\,mg$

84. Work done by a force F, which is variable in nature in moving a particle from y_1 to y_2 is given by

$$W = \int_{y_1}^{y_2} F \cdot dy \quad \text{... (i)}$$

Here, force, $F = 20 + 10y, y_1 = 0$ and $y_2 = 1$ m

Substituting the given values in Eq. (i), we get

$$\Rightarrow \quad W = \int_0^1 (20 + 10y)dy = \left[20y + \frac{10y^2}{2}\right]_0^1$$

$$= (20 + 5) - (0) = 25 \text{ J}$$

$\therefore$ Work done will be 25 J.

85. The area under the force-displacement curve give the amount of work done.

From work-energy theorem,

$$W = \Delta KE \quad \text{...(i)}$$

$\therefore$ At $x = 8$ m,

W = Area $ABDO$ + Area $CEFD$

$= 20 \times 5 + 10 \times 3 = 130$ J

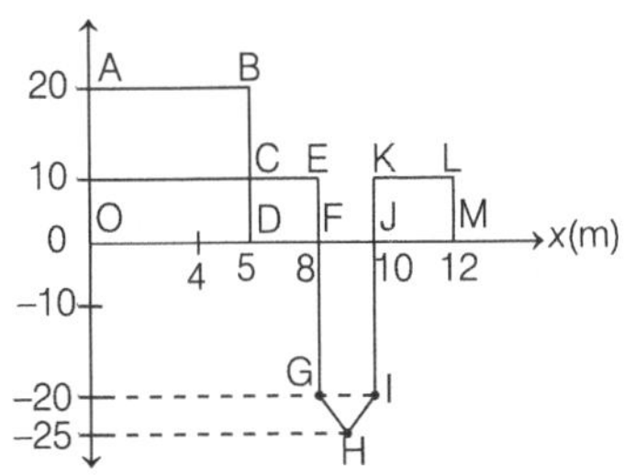

Using Eq. (i), we get

$$\Rightarrow \quad 130 = \frac{1}{2}mv^2 = \frac{1}{2} \times \frac{500}{1000} v^2$$

$$\Rightarrow \quad v = 2\sqrt{130} = 22.8 \text{ ms}^{-1} \approx 23 \text{ ms}^{-1}$$

At $x = 12$ m

W = Area $ABDO$ + Area $CEFD$ + Area $FGHIJ$ + Area $KLMJ$

$$W = 20 \times 5 + 10 \times 3 + (-20 \times 2) + \left(\frac{1}{2} \times -5 \times 2\right) + 10 \times 2$$

[$\because$ Area FGHIJ = Area FGIJ + Area GHI]

$= 100 + 30 - 40 - 5 + 20 = 105$ J

Using Eq. (i), we get

$$\therefore \quad 105 = \frac{1}{2} \times \frac{1}{2} \times v^2 \Rightarrow v = 2\sqrt{105} \simeq 20.6 \text{ ms}^{-1}$$

86. At bottom most point,

$$T = mg + \frac{mv^2}{r}$$

and it is maximum.

87. If a body is moving on a frictionless surface, then its total mechanical energy remains conserved.

According to the conservation of mechanical energy,

$$(\text{TE})_{\text{initial}} (\text{TE})_{\text{initial}} = (\text{TE})_{\text{final}}$$

$$\Rightarrow \quad (\text{KE})_i + (\text{PE})_i = (\text{KE})_f + (\text{PE})_f$$

$$0 + mgh = \frac{1}{2}mv_A^2 + 0$$

$$\Rightarrow \quad gh = \frac{v_A^2}{2} \text{ or } h = \frac{v_A^2}{2g} \quad \text{...(i)}$$

In order to complete the vertical circle, the velocity of the body at point A should be

$$v_A = v_{\min} = \sqrt{5gR}$$

where, R is the radius of the body.

Here, $$R = \frac{AB}{2} = \frac{D}{2}$$

$$\Rightarrow \quad v_{\min} = v_A = \sqrt{\frac{5}{2}gD}$$

Substituting the value of v_A in Eq. (i), we get

$$h = \frac{\left(\sqrt{\left(\frac{5}{2}gD\right)}\right)^2}{2g} = \frac{5gD}{2 \times 2g} = \frac{5}{4}D$$

88. By work-energy theorem, we have change in KE = work done by all of the forces.

Work done by gravitational force,

$$W_g = mgh = 10^{-3} \times 10 \times 1 \times 10^3 = 10 \text{ J}$$

Now, from work-energy theorem, we have

$$\Delta K = W_{\text{gravity}} + W_{\text{air resistance}}$$

$$\Rightarrow \quad \frac{1}{2} \times mv^2 = mgh + W_{\text{air resistance}}$$

$$\Rightarrow W_{\text{air resistance}} = \frac{1}{2}mv^2 - mgh$$

$$= 10^{-3}\left(\frac{1}{2} \times 50 \times 50 - 10 \times 10^3\right)$$

$$= -8.75 \text{ J}$$

89. Given, mass of particle $m = 0.01$ kg.

Radius of circle along which particle is moving,

$$r = 6.4 \text{ cm}$$

$\because$ Kinetic energy of particle,

$$\text{KE} = 8 \times 10^{-4} \text{ J}$$

$$\Rightarrow \quad \frac{1}{2}mv^2 = 8 \times 10^{-4} \text{ J}$$

$$\Rightarrow \quad v^2 = \frac{16 \times 10^{-4}}{0.01} = 16 \times 10^{-2} \quad \text{...(i)}$$

As it is given that KE of particle is equal to 8×10^{-4} J by the end of second revolution after the beginning of motion of particle. It means, its initial velocity (u) is 0 m/s at this moment.

$\because$ By Newton's 3rd equation of motion,

$$v^2 = u^2 + 2a_t s$$

$$\Rightarrow v^2 = 2a_t S \text{ or } v^2 = 2a_t (4\pi r)$$

($\because$ particle covers 2 revolutions)

$$\Rightarrow a_t = \frac{v^2}{8\pi r} = \frac{16 \times 10^{-2}}{8 \times 3.14 \times 6.4 \times 10^{-2}}$$

$(\because$ from Eq. (i), $v^2 = 16 \times 10^{-2})$

$$\therefore \; a_t = 0.1 \text{ m/s}^2$$

90. Position vectors of the particles are $\mathbf{r}_1 = -2\hat{\mathbf{i}} + 5\hat{\mathbf{j}}$ and $\mathbf{r}_2 = 4\hat{\mathbf{j}} + 3\hat{\mathbf{k}}$

$\therefore$ Displacement of the particle, $\Delta\mathbf{s} = \mathbf{r}_2 - \mathbf{r}_1$

$$= 4\hat{\mathbf{j}} + 3\hat{\mathbf{k}} - (-2\hat{\mathbf{i}} + 5\hat{\mathbf{j}}) = 2\hat{\mathbf{i}} - \hat{\mathbf{j}} + 3\hat{\mathbf{k}}$$

Force on the particle, $\mathbf{F} = 4\hat{\mathbf{i}} + 3\hat{\mathbf{j}}$ N

$\therefore$ Work done, $W = \mathbf{F} \cdot \Delta\mathbf{s}$

$$= (4\hat{\mathbf{i}} + 3\hat{\mathbf{j}}) \cdot (2\hat{\mathbf{i}} - \hat{\mathbf{j}} + 3\hat{\mathbf{k}})$$

$$= 8 - 3 = 5 \text{ J}$$

91. According to question, a body of mass 1 kg begins to move under the action of time dependent force,

$$\mathbf{F} = (2t\hat{\mathbf{i}} + 3t^2\hat{\mathbf{j}}) \text{ N}$$

where $\hat{\mathbf{i}}$ and $\hat{\mathbf{j}}$ are unit vectors along X and Y-axes.

$$\because \quad \mathbf{F} = m\mathbf{a}$$

$$\Rightarrow \quad \mathbf{a} = \frac{\mathbf{F}}{m}$$

$$\Rightarrow \quad \mathbf{a} = \frac{(2t\hat{\mathbf{i}} + 3t^2\hat{\mathbf{j}})}{1} \qquad (\because m = 1 \text{ kg})$$

$$\Rightarrow \quad \mathbf{a} = (2t\hat{\mathbf{i}} + 3t^2\hat{\mathbf{j}}) \text{ m/s}^2$$

$\because$ acceleration, $a = \dfrac{dv}{dt}$

$$\Rightarrow \quad dv = a\,dt \qquad \text{...(i)}$$

Integrating both sides, we get

$$\int dv = \int a\,dt = \int (2t\hat{\mathbf{i}} + 3t^2\hat{\mathbf{j}})\,dt$$

$$\mathbf{v} = t^2\hat{\mathbf{i}} + t^3\hat{\mathbf{j}}$$

$\because$ Power developed by the force at the time t will be given as

$$P = \mathbf{F} \cdot \mathbf{v} = (2t\hat{\mathbf{i}} + 3t^2\hat{\mathbf{j}}) \cdot (t^2\hat{\mathbf{i}} + t^3\hat{\mathbf{j}})$$

$$= (2t \cdot t^2 + 3t^2 \cdot t^3)$$

$$P = (2t^3 + 3t^5) \text{ W}$$

92. Let the tension at point A be T_A. So, from Newton's second law,

$$T_A - mg = \frac{mv_c^2}{R}$$

Energy at point $A = \dfrac{1}{2}mv_0^2$ ...(i)

Energy at point C is

$$= \frac{1}{2}mv_c^2 + mg \times 2R \qquad \text{...(ii)}$$

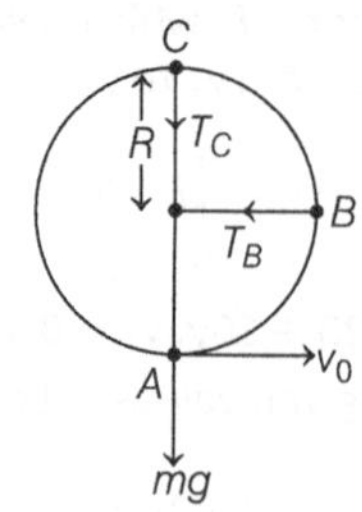

Applying Newton's 2nd law at point C,

$$T_c + mg = \frac{mv_c^2}{R}$$

To complete the loop $T_c \geq 0$

So, $$mg = \frac{mv_c^2}{R} \Rightarrow v_c = \sqrt{gR} \qquad \text{...(iii)}$$

From Eqs. (i) and (ii) by conservation of energy

$$\frac{1}{2}mv_0^2 = \frac{1}{2}mv_c^2 + 2mgR$$

$$\Rightarrow \quad \frac{1}{2}mv_0^2 = \frac{1}{2}mgR + 2mgR \qquad \text{[using Eq.(iii)]}$$

$$\Rightarrow \quad v_0^2 = gR + 4gR$$

$$\Rightarrow \quad v_0 = \sqrt{5gR}$$

93. Given, $K_P > K_Q$

In case (a), the elongation is same

i.e. $x_1 = x_2 = x$

So, $W_P = \dfrac{1}{2}K_P x^2$ and $W_Q = \dfrac{1}{2}K_Q x^2$

$$\therefore \quad \frac{W_P}{W_Q} = \frac{K_P}{K_Q} > 1 \Rightarrow W_P > W_Q$$

In case (b), the spring force is same

i.e. $F_1 = F_2 = F$

So, $x_1 = \dfrac{F}{K_P}, x_2 = \dfrac{F}{K_Q}$

$$\therefore \quad W_P = \frac{1}{2}K_P x_1^2 = \frac{1}{2}K_P \frac{F^2}{K_P^2} = \frac{1}{2}\frac{F^2}{K_P}$$

and $$W_Q = \frac{1}{2}K_Q x_2^2 = \frac{1}{2}K_Q \cdot \frac{F^2}{K_Q^2} = \frac{1}{2}\frac{F^2}{K_Q}$$

$$\therefore \quad \frac{W_P}{W_Q} = \frac{K_Q}{K_P} < 1$$

$$\Rightarrow \quad W_P < W_Q$$

94. From work-energy theorem,

Work done = Change in KE

$$\Rightarrow \quad W = K_f - K_i$$

$$\Rightarrow \quad K_f = W + K_i = \int_{x_1}^{x_2} Fx\,dx + \frac{1}{2}mv^2$$

$$= \int_{20}^{30} -0.1.x\,dx + \frac{1}{2}\times 10\times 10^2$$

$$= -0.1\left[\frac{x^2}{2}\right]_{20}^{30} + 500$$

$$= -0.05[30^2 - 20^2] + 500$$

$$= -0.05[900 - 400] + 500$$

$$\Rightarrow \quad K_f = -25 + 500 = 475\text{ J}$$

95. Total initial energy $= \frac{1}{2}m_1u_1^2 + \frac{1}{2}m_2u_2^2$

Since, after collision one particle absorb energy ε.

$\therefore$ Total final energy $= \frac{1}{2}m_1v_1^2 + \frac{1}{2}m_2v_1^2 + \varepsilon$

From conservation of energy,

$$\frac{1}{2}m_1u_1^2 + \frac{1}{2}m_2u_2^2 = \frac{1}{2}m_1v_1^2 + \frac{1}{2}m_2v_2^2 + \varepsilon$$

$$\Rightarrow \quad \frac{1}{2}m_1u_1^2 + \frac{1}{2}m_2u_2^2 - \varepsilon = \frac{1}{2}m_1v_1^2 + \frac{1}{2}m_2v_2^2$$

96. Conserving angular momentum,

$$L_i = L_f$$

$$\Rightarrow \quad mv_0R_0 = mv'\left(\frac{R_0}{2}\right) \Rightarrow v' = 2v_0$$

So, final kinetic energy of the particle is

$$K_f = \frac{1}{2}mv'^2 = \frac{1}{2}m(2v_0)^2$$

$$= 4\left(\frac{1}{2}mv_0^2\right) = 2mv_0^2$$

97. Suppose a ball rebounds with speed v,

$$v = \sqrt{2gh} = \sqrt{2\times 10\times 20}$$

$$= 20\text{ m/s}$$

Energy of a ball just after rebound,

$$E = \frac{1}{2}mv^2 = \frac{1}{2}\times m\times(20)^2 = 200\,m$$

As, 50% of energy loses in collision means just before collision, energy is 400 m.

According to law of conservation of energy, we have

$$\frac{1}{2}mv_0^2 + mgh = 400\,m$$

$$\Rightarrow \quad \frac{1}{2}mv_0^2 + m\times 10\times 20 = 400\,m$$

$$\Rightarrow \quad v_0 = 20\text{ m/s}$$

98. As the machine delivers a constant power

So F.v = constant = k (watts)

$$\Rightarrow \quad m\frac{dv}{dt}\cdot v = k \quad \Rightarrow \int v\,dv = \frac{k}{m}\int dt$$

$$\Rightarrow \quad \frac{v^2}{2} = \frac{k}{m}t \Rightarrow v = \sqrt{\frac{2k}{m}t}$$

Now, force on the particle is given by

$$F = m\frac{dv}{dt} = m\frac{d}{dt}\left(\frac{2kt}{m}\right)^{\frac{1}{2}}$$

$$= \sqrt{2km}\cdot\left(\frac{1}{2}t^{-\frac{1}{2}}\right) = \sqrt{\frac{mk}{2}}\cdot t^{-\frac{1}{2}}$$

99. Given, pressure = 150 mm of Hg

Pumping rate of heart of a man

$$= \frac{dV}{dt} = \frac{5\times 10^{-3}}{60}\text{ m}^2/\text{s}$$

Power of heart $= P\cdot\frac{dV}{dt} = \rho gh\cdot\frac{dV}{dt}$ $\quad [\because p = \rho gh]$

$$\frac{(13.6\times 10^3)(10\times 0.15\times 5\times 10^{-3})}{60} = 1.70\text{ W}$$

100. Given force, $F = 3\hat{\mathbf{i}} + \hat{\mathbf{j}}$

$$\mathbf{r}_1 = (2\hat{\mathbf{i}} + \hat{\mathbf{k}})\text{ m and } \mathbf{r}_2 = (4\hat{\mathbf{i}} + 3\hat{\mathbf{j}} - \hat{\mathbf{k}})\text{ m}$$

$$\therefore \quad \mathbf{s} = \mathbf{r}_2 - \mathbf{r}_1 = (4\hat{\mathbf{i}} + 3\hat{\mathbf{j}} - \hat{\mathbf{k}}) - (2\hat{\mathbf{i}} + \hat{\mathbf{k}})$$

$$= (2\hat{\mathbf{i}} + 3\hat{\mathbf{j}} - 2\hat{\mathbf{k}})\text{ m}$$

$$\therefore \quad W = \mathbf{F}\cdot\mathbf{s} = (3\hat{\mathbf{i}} + \hat{\mathbf{j}})\cdot(2\hat{\mathbf{i}} + 3\hat{\mathbf{j}} - 2\hat{\mathbf{k}})$$

$$= 3\times 2 + 3 + 0$$

$$= 6 + 3 = 9\text{ J}$$

101. Given, the potential energy of a particle in a force field,

$$U = \frac{A}{r^2} - \frac{B}{r}$$

For stable equilibrium, $F = -\frac{dU}{dr} = 0$

$$= \frac{dU}{dr} = -2Ar^{-3} + Br^{-2}$$

$$0 = -\frac{2A}{r^3} + \frac{B}{r^2} \qquad \left(\text{as } \frac{-dU}{dr} = 0\right)$$

or $\quad \frac{2A}{r} = B$

The distance of particle from the centre of the field

$$r = \frac{2A}{B}$$

102. Concept Work done is equal to area under the curve in *F*-*d* graph.

Work done = Area under (*F*-*d*) graph

$$= 2\times(7-3) + \frac{1}{2}\times 2\times(12-7)$$

$$= 8 + \frac{1}{2}\times 10 = 8 + 5 = 13\text{ J}$$

103. Given, velocity of water $v = 2$ m/s

Mass per unit length of water in the pipe

$= 100$ kg/m

So, power = (mass per unit length of water in pipe) $\times v^3$

$$= \frac{m}{l}\times v^3 = 100\times 2\times 2\times 2 = 800\text{ W}$$

Chapter 08

CIRCULAR MOTION

LEVEL 01 Objective Questions

1 If the equation for the displacement of a particle moving on a circular path is given by $\theta = 2t^3 + 0.5$, where θ is in radians and t in seconds, then the angular velocity of the particle after 2 s from its start is

(a) 8 rad/s (b) 12 rad/s
(c) 24 rad/s (d) 36 rad/s

2 The ratio of angular speeds of minute hand and hour hand of a watch is

(a) 1 : 12 (b) 6 : 1
(c) 12 : 1 (d) 1 : 6

3 A disc rotates about its axis with a uniform acceleration of 10 rad s^{-2}. The radial and tangential accelerations of a particle at a distance 2 cm from the axis, two seconds after the disc starts rotating are

(a) 8 ms^{-2}, 0.1 ms^{-2} (b) 2 ms^{-2}, 0.1 ms^{-2}
(c) 8 ms^{-2}, 0.2 ms^{-2} (d) 4 ms^{-2}, 0.2 ms^{-2}

4 A particle starting from rest moves with a constant angular acceleration α in a circular path. The time at which the magnitudes of tangential and radial accelerations are equal, is

(a) $\frac{1}{\alpha}$ (b) α (c) $\frac{1}{\sqrt{\alpha}}$ (d) $\sqrt{\alpha}$

5 A man is running with constant speed along a circular path of radius $2\sqrt{2}$ m. He completes 1 round in 10 s. Find instantaneous speed at 2.5 s.

(a) $\frac{\sqrt{2}\pi}{5}$ m/s (b) $\frac{2\sqrt{2}\pi}{5}$ m/s
(c) $\frac{2\sqrt{3}\pi}{5}$ m/s (d) $\frac{5\sqrt{2}\pi}{2}$ m/s

6 A particle of mass m describes a circle of radius r. The centripetal acceleration of the particle is $4/r^2$. What will be the momentum of the particle?

(a) $2mr$ (b) $2m/\sqrt{r}$ (c) $4m/\sqrt{r}$ (d) $4m/r$

7 A particle is moving on a circular path of radius r with uniform speed v. The magnitude of change in velocity, when the particle moves from P to Q is

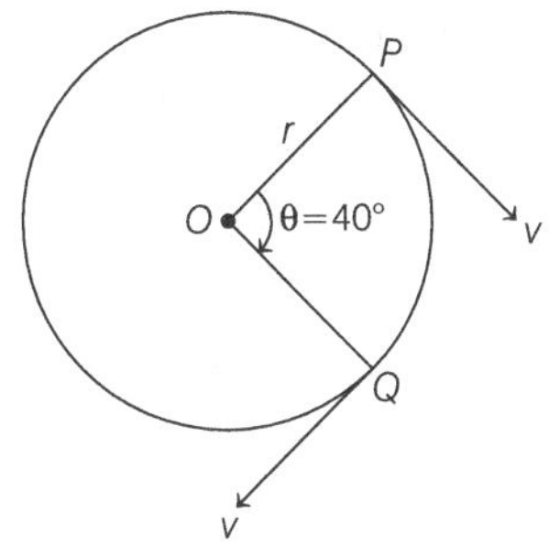

(a) $2v \cos 40°$ (b) $2v \sin 40°$ (c) $2v \sin 20°$ (d) $2v \cos 20°$

8 A particle moves in a circle of radius 0.5 m at a speed that uniformly increases. If its speed changes from 2 m/s to 4 m/s in 4 s, then angular acceleration of the particle is
(a) 0.5 rad/s^2 (b) 1 rad/s^2 (c) 2 rad/s^2 (d) 0.25 rad/s^2

9 An object moves at a constant speed along a circular path in horizontal xy-plane with the centre at origin when the object is at $x = -2$ m, its velocity is -4 m/s$\hat{\mathbf{j}}$. Acceleration of the object, when it is at $y = 2$ m is
(a) -8 m/s^2 $\hat{\mathbf{j}}$ (b) -8 m/s^2 $\hat{\mathbf{i}}$ (c) -4 m/s^2 $\hat{\mathbf{j}}$ (d) -4m/s^2 $\hat{\mathbf{i}}$

10 A particle starts from rest on a circle of radius $\frac{25}{\pi}$ m with uniformly increasing speed. If it completes 5 revolutions after 5 s of motion, then tangential acceleration will be
(a) 5 m/s^2 (b) 10 m/s^2 (c) 15 m/s^2 (d) 20 m/s^2

11 A particle is moving on a circle of radius 2 m. Its speed varies with time t as $v = \alpha t^2$, where α is a constant. The ratio of radial acceleration to tangential acceleration of the particle at $t = 2$ s is
(a) α (b) 8 (c) 2α (d) 4

12 A particle revolving in a circular path completes first one-third of circumference in 3 s, while next one-third in 2 s. The average angular velocity of the particle is
(a) $\frac{3\pi}{5}$ rad/s (b) $\frac{\pi}{15}$ rad/s (c) $\frac{2\pi}{15}$ rad/s (d) $\frac{4\pi}{15}$ rad/s

13 A particle moves in a circle of radius 30 cm. Its linear speed is given by $v = 2t$, where t is in second and v in m/s. Find out its radial and tangential acceleration at $t = 3$ s.
(a) 220 m/s^2, 50 m/s^2 (b) 100 m/s^2, 5 m/s^2
(c) 120 m/s^2, 2 m/s^2 (d) 110 m/s^2, 10 m/s^2

14 Two bodies of masses 10 kg and 5 kg are moving in concentric orbits of radii R and r, such that their periods are same. Then, the ratio between their centripetal acceleration is
(a) R/r (b) r/R (c) R^2/r^2 (d) r^2/R^2

15 Find the maximum speed at which a car can turn round a curve of 30 m radius on a level road, if the coefficient of friction between the tyres and the road is 0.4.
(a) 12 ms^{-1} (b) 10 ms^{-1}
(c) 11 ms^{-1} (d) 15 ms^{-1}

16 A man standing on a rough horizontal disc rotating with constant angular velocity of 10 rad/s. The distance from the centre of disc, where he should stand, so that he does not slip on the disc is (Take, coefficient of friction between disc and man's feet is 0.4)
(a) $R \geq 0.04$ m (b) $R \leq 0.02$ m
(c) $R \leq 0.04$ m (d) $R \geq 0.02$ m

17 Three identical particles are joined together by a thread as shown in the following figure. Assembly of particles is rotated in a horizontal plane about point O. The velocity on the outermost particle is v_0, then the ratio of tensions in the three sections of the string is

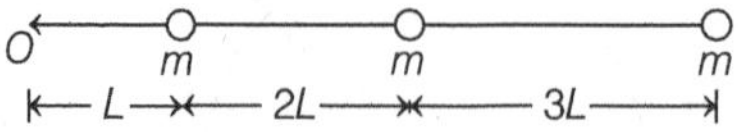

(a) 3 : 5 : 6 (b) 3 : 4 : 5 (c) 6 : 9 : 10 (d) 7 : 6 : 11

18 In the whole set-up shown in the following figure, particles (m_1 and m_2) are rotating with constant angular velocity ω on a horizontal frictionless table. If T_1 and T_2 are tensions in given strings, then ratio of tensions $\frac{T_1}{T_2}$ is $\left(\text{Take, } \frac{l_2}{l_1} = \frac{2}{1}\right)$

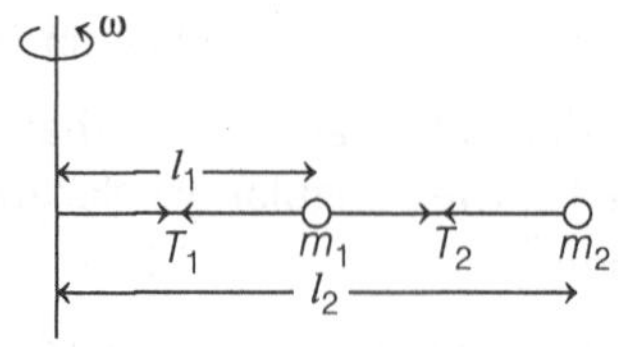

(a) $\frac{m_1}{m_2}$ (b) $\frac{m_1 + 2m_2}{2m_2}$ (c) $\frac{m_2}{m_1}$ (d) $\frac{m_2 + m_1}{m_1}$

19 Consider a car moving on a horizontal road, convex bridge and concave bridge of same radius R. The speed of the car in each situation is same and equal to v, the mass of the car is m, then

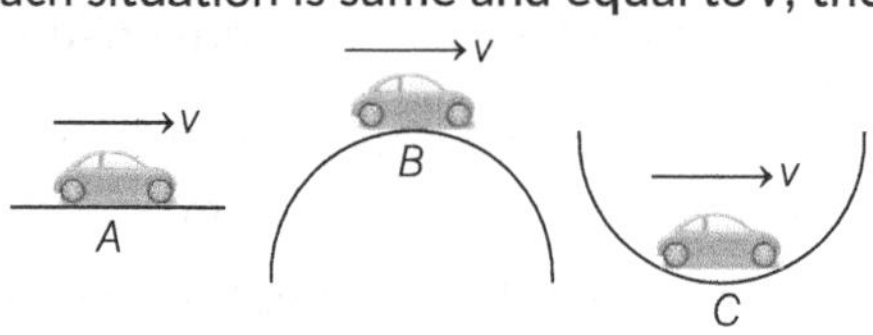

(i) $N_A = mg$ (ii) $N_B = mg - \frac{mv^2}{R}$ (iii) $N_C = mg + \frac{mv^2}{R}$ (iv) $N_C > N_A > N_B$

(a) (i), (ii) (b) (ii), (iii)
(c) (i), (ii), (iii) (d) All options are correct

20 A particle of mass 1 kg moves in a circular path of radius 1 m, such that its speed varies with time as $v = 3t^2$ m/s, where time t is in seconds. The power delivered by the force acting on the particle at $t = 1$s, is

(a) 36 W (b) 20 W (c) 18 W (d) zero

21 A motor car is travelling at 60 m/s on a circular road of radius 1200 m. It is increasing its speed at the rate of 4 m/s^2. The acceleration of the car is

(a) 3 m/s^2 (b) 4 m/s^2 (c) 5 m/s^2 (d) 7 m/s^2

22 A road is banked at an angle of 30° to the horizontal for negotiating a curve of radius $10\sqrt{3}$ m. At what velocity will a car experience no friction while negotiating the curve?

(a) 54 km/h (b) 72 km/h (c) 36 km/h (d) 18 km/h

23 Keeping the angle of banking unchanged, if the radius of curvature is made four times, the percentage increase in the maximum speed with which a vehicle can travel on a circular road is

(a) 25% (b) 50%
(c) 75% (d) 100%

24 A car is moving on a circular road of radius 100 m. At some instant, its speed is 20 m/s and is increasing at the rate of 3 m/s^2. The magnitude of its acceleration is

(a) 2 m/s^2 (b) 3 m/s^2
(c) 5 m/s^2 (d) 4 m/s^2

25 A simple pendulum of mass m swings about point B between extreme positions A and C. Net force acting on the bob at these three points is correctly shown by

(a) 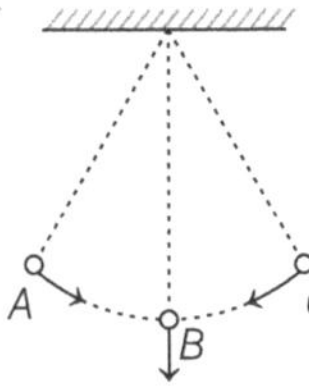

(b) 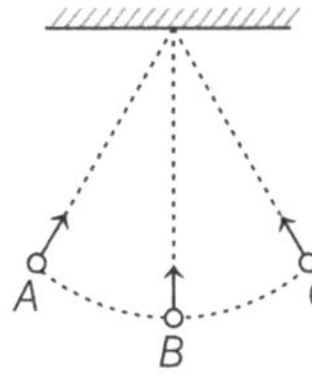

(c) 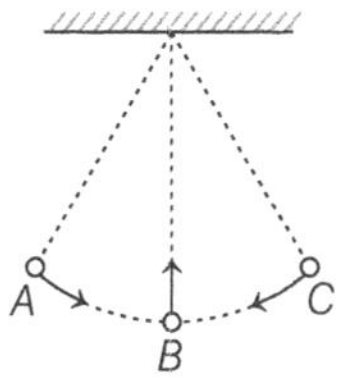

(d) 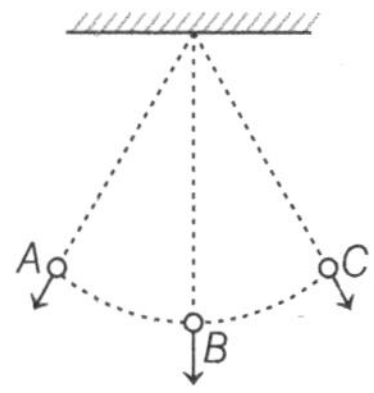

26 If a simple pendulum of length l has maximum angular displacement θ, then the maximum kinetic energy of the bob of mass m is

(a) $\left(\frac{1}{2}\right)m\sqrt{\frac{l}{g}}$ (b) $\left(\frac{1}{2}\right)\left(\frac{mg}{l}\right)$ (c) $mgl(1-\cos\theta)$ (d) $\left(\frac{1}{2}\right)mgl\sin\theta$

27 A point mass m is suspended from a light thread of length l, fixed at O is whirled in a horizontal circle at constant speed as shown in the following figure. With respect to frame of reference, which is fixed to rotating mass, the forces on the mass are

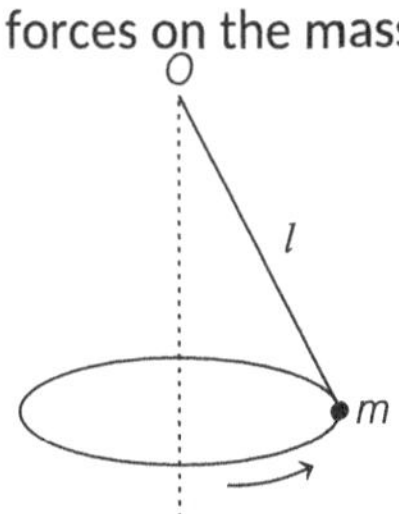

(a) 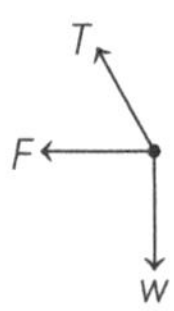

(b) 

(c) 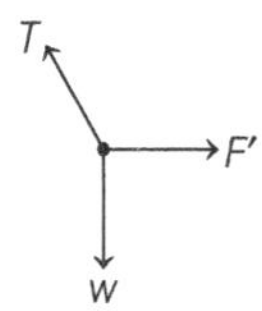

(d) 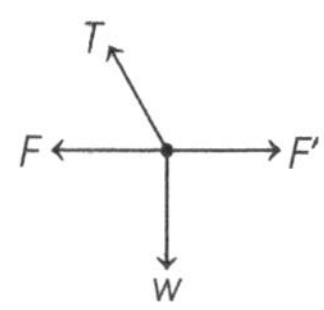

28 A stone of mass 1 kg tied to a light inextensible string of length $L = \frac{10}{3}$ m, is rotating in a circular path in a vertical plane. The ratio of maximum tension to the minimum tension in the string is 4, the speed of the stone at the highest point of the circle is (Take, $g = 10\ \text{m/s}^2$)

(a) 10 m/s (b) $5\sqrt{2}$ m/s (c) $10\sqrt{3}$ m/s (d) 20 m/s

29 A particle of mass m is attached with light string of length l. The string is fixed from other end at O as shown in the given figure. The ball is given horizontal velocity $v_0 = \sqrt{gl}$. So, the angular amplitude during the oscillations will be

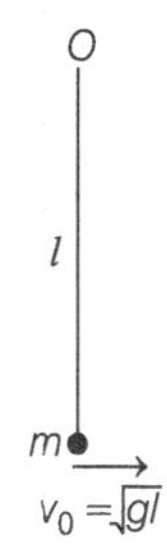

(a) 30°
(b) 60°
(c) 90°
(d) None of the above

30 A particle of mass m is rotating by means of a string in a vertical circle. The difference in the tension at the bottom and top would be

(a) 6 mg (b) 4 mg
(c) 3 mg (d) 2 mg

31 The bob of a simple pendulum is given a sharp hit to impart it a horizontal speed of $\sqrt{3gL}$. Find an angle made by the string with the upper vertical before it becomes slack. (L is length of the string)

(a) $\cos^{-1}\left(\frac{1}{3}\right)$ (b) $\cos^{-1}\left(\frac{2}{3}\right)$ (c) $\frac{\pi}{3}$ (d) $\frac{\pi}{6}$

32 A particle of mass m is attached to a string of length L and given a velocity $\sqrt{10gL}$ in the horizontal direction at the lowest point. Find the tension in the string when the particle is at the highest position.

(a) 4 mg (b) 5 mg (c) 11 mg (d) 8 mg

33 A particle is tied to a 20 cm long string. It performs circular motion in a vertical plane. What is the angular speed of particle at topmost point, when the tension in the string at the topmost point is zero?

(a) 5 rad/s (b) 2 rad/s (c) 7.5 rad/s (d) 7 rad/s

34 Block shown in the following figure is taken towards left, so that spring is compressed. Now, the block is released from rest. Block moves on smooth horizontal surface after leaving spring, then on circular path R. By how much distance spring should be compressed, so that block can complete circular motion?

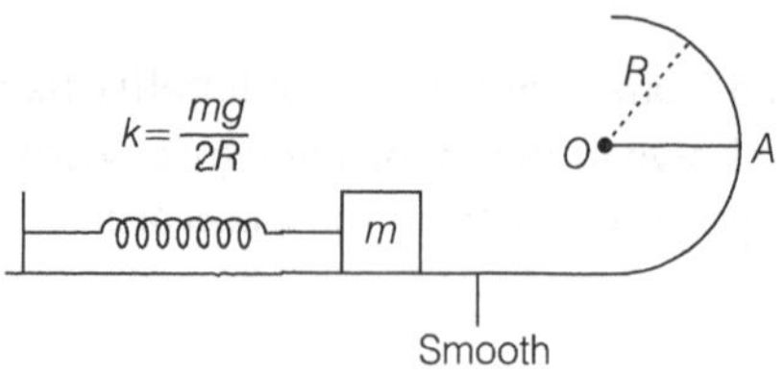

(a) $R\sqrt{3}$ (b) $R\sqrt{10}$ (c) $5R$ (d) $\frac{5}{2}R$

35 A small mass of 10 g, lies in a hemispherical smooth bowl of radius 0.5 m at a height of 0.2 m from the bottom of the bowl. The mass will be in equilibrium, if the bowl rotates at an angular speed of (Take, $g = 10\text{ m/s}^2$)

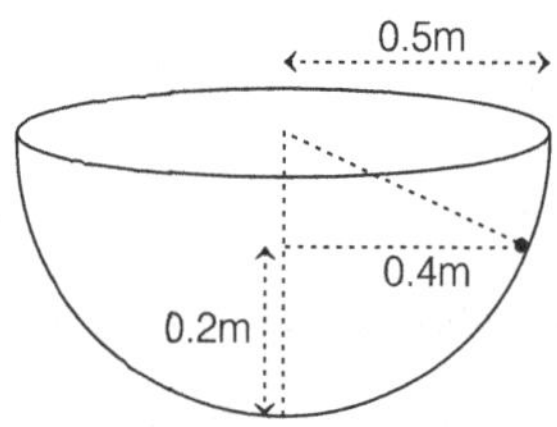

(a) $\frac{10}{\sqrt{3}}$ rad/s (b) $10\sqrt{3}$ rad/s (c) 10 rad/s (d) $\sqrt{20}$ rad/s

LEVEL 02 Objective Questions

36 The given figure shows a small mass connected to a string. If the mass is released from rest, when the string is horizontal, the magnitude of the total acceleration of the mass as a function of the angle θ is

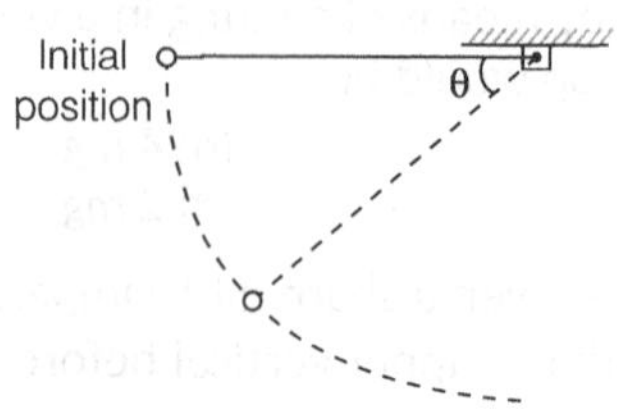

(a) $2g\sin\theta$ (b) $2g\cos\theta$ (c) $g\sqrt{3\cos^2\theta+1}$ (d) $g\sqrt{3\sin^2\theta+1}$

37 Two very small balls *A* and *B* of masses 4.0 kg and 5.0 kg are fixed to the ends of a light inextensible cord that passes through a frictionless ring of radius negligible compared to the length of the cord. The ring is fixed at some height above the ground. Ball *A* is pulled aside and given a horizontal velocity, so that it starts moving on a circular path parallel to the ground, keeping ball *B* in equilibrium as shown below. Speed of the ball *A* is closest to

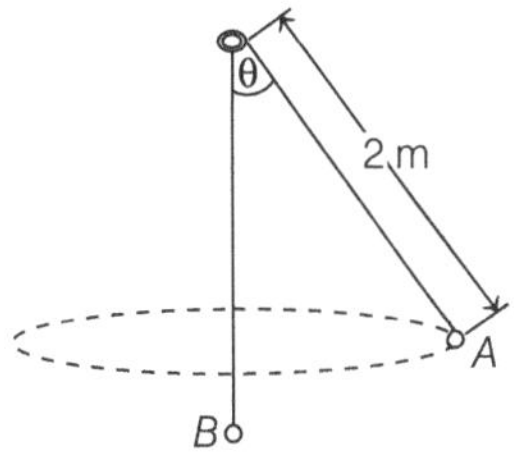

(a) 3.0 m/s
(b) 2.5 m/s
(c) 4.0 m/s
(d) Data insufficient

38 A simple pendulum is oscillating with an angular amplitude of 90° as shown in the following figure. For what value of α, is the acceleration directed?

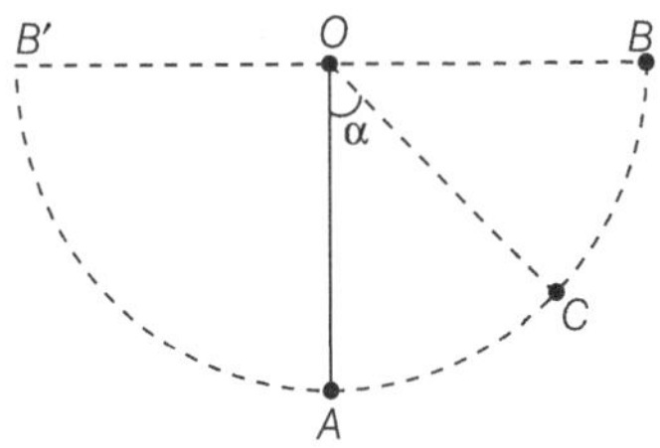

(i) Vertically upwards
(ii) Horizontally
(iii) Vertically downwards

(a) $0°, \cos^{-1}\left(\frac{1}{\sqrt{3}}\right), 90°$
(b) $90°, \cos^{-1}\left(\frac{1}{\sqrt{3}}\right), 0°$
(c) $\cos^{-1}\left(\frac{1}{\sqrt{3}}\right), 0°, 90°$
(d) $\cos^{-1}\left(\frac{1}{\sqrt{3}}\right), 90°, 0°$

39 A point *P* moves in counter clockwise direction on a circular path as shown in the following figure. The movement of *P* is such that it sweeps out a length $s = t^3 + 5$, where *s* is in metres and *t* in seconds. The radius of the path is 20 m. The acceleration of *P*, when $t = 2$ s is nearly

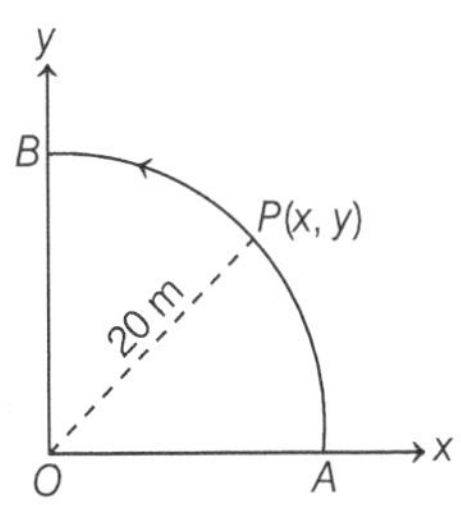

(a) 13 m/s^2
(b) 12 m/s^2
(c) 7.2 m/s^2
(d) 14 m/s^2

40 A particle moves in a circle with a uniform speed, when it goes from a point *A* to a diametrically opposite point *B*. The momentum of the particle changes by $\mathbf{p}_A - \mathbf{p}_B = 2$ kg-m/s $\hat{\mathbf{j}}$ and the centripetal force acting on it changes by $\mathbf{F}_A - \mathbf{F}_B = 8\text{N}\hat{\mathbf{i}}$, where $\hat{\mathbf{i}}, \hat{\mathbf{j}}$ are unit vectors. Then, the angular velocity of the particle is

(a) dependent on its mass
(b) 4 rad/s
(c) $2/\pi$ rad/s
(d) $16/\pi$ rad/s

41 A small ball is rolled with speed u from point A along a smooth circular track as shown in the following figure. If $x = 3R$, then determine the required speed u, so that the ball returns to A, the point of projection after passing through C, the highest point.

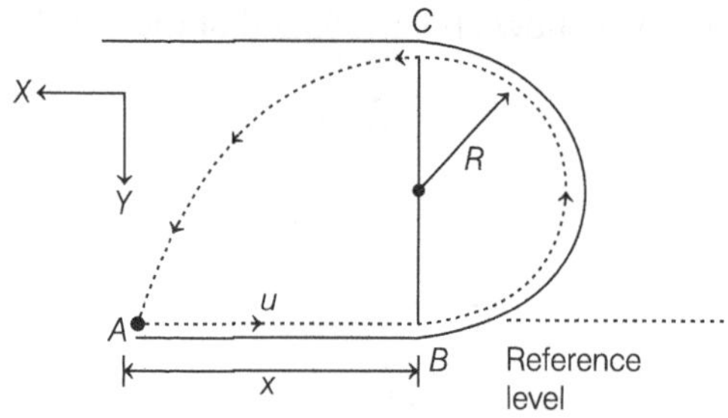

(a) $\frac{3}{2}\sqrt{gR}$ (b) $\frac{1}{2}\sqrt{gR}$ (c) $\frac{5}{3}\sqrt{gR}$ (d) $\frac{5}{2}\sqrt{gR}$

42 A wire which passes through the hole in a small bead, is bent in the form of quarter of a circle. The wire is fixed vertically on ground as shown in the figure. The bead is released from near the top of the wire and it slides along the wire without friction. As the bead moves from A to B, the force it applies on the wire is

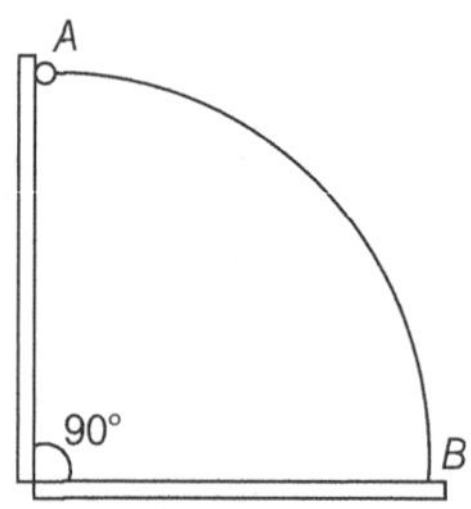

(a) always radially outwards
(b) always radially inwards
(c) radially outwards initially and radially inwards later
(d) radially inwards initially and radially outwards later

43 A particle P of mass m is attached to a vertical axis by two strings AP and BP of length l each. The separation $AB = l$. P rotates around the axis with an angular velocity ω. The tensions in the two strings are T_1 and T_2, then choose **wrong** option.

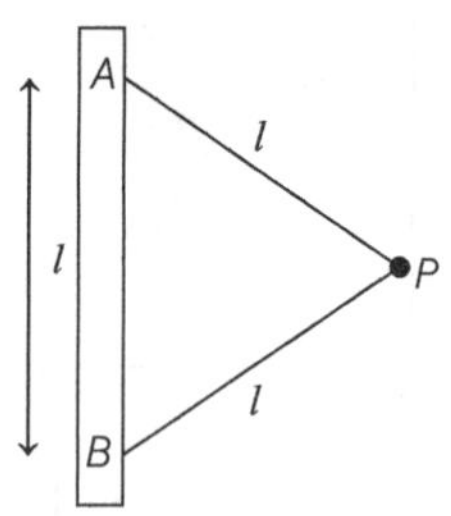

(a) $T_1 \neq T_2$ (b) $T_1 + T_2 = m\omega^2 l$
(c) $T_1 - T_2 = 2mg$ (d) $\frac{T_1}{T_2} = \frac{\omega^2 l}{g}$

44 A turn of radius 600 m is banked for a vehicle of mass 200 kg going with a speed of 180 km h^{-1}. Find the banking angle of its path.
(a) $\theta = 22.6°$ (b) $\theta = 30°$
(c) $\theta = 20.6°$ (d) $\theta = 30.6°$

45 The radius of the curved road on a national highway is R. The width of the road is b. The outer edge of the road is raised by h with respect to the inner edge, so that a car with speed v can pass safely over it. The value of h is

(a) $\frac{v^2 b}{Rg}$ (b) $\frac{v}{Rgb}$

(c) $\frac{v^2 R}{g}$ (d) $\frac{v^2 b}{R}$

46 A 2 kg block is gently pushed from rest at A and it slides down along the fixed smooth circular surface. If the attached spring has a force constant $k = 20$ N/m. What is unstretched length of spring, so that it does not allow the block to leave the surface until angle with the vertical is $\theta = 60°$?

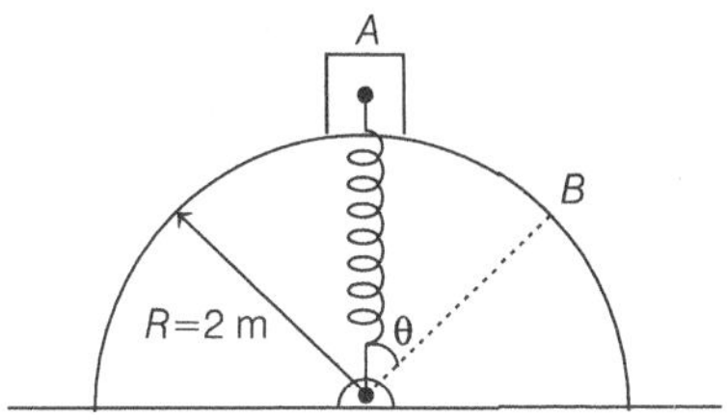

(a) 1 m (b) 1.5 m
(c) 0.5 m (d) 0.8 m

47 A particle is moving in a circular path. The acceleration and momentum of the particle at a certain moment are $\mathbf{a} = (4\hat{\mathbf{i}} + 3\hat{\mathbf{j}})$ m/s^2 and $\mathbf{p} = (8\hat{\mathbf{i}} - 6\hat{\mathbf{j}})$ kg-m/s. The motion of the particle is

(a) uniform circular motion (b) accelerated circular motion
(c) de-accelerated circular motion (d) we cannot say anything with **a** and **p** only

LEVEL 03 Previous Years Questions

48 A block of mass 10 kg is in contact against the inner wall of a hollow cylindrical drum of radius 1 m. The coefficient of friction between the block and the inner wall of the cylinder is 0.1. The minimum angular velocity needed for the cylinder to keep the block stationary when the cylinder is vertical and rotating about its axis, will be (Take, $g = 10$ m/s^2) [NEET 2019]

(a) $\frac{10}{2\pi}$ rad/s (b) 10 rad/s
(c) 10π rad/s (d) $\sqrt{10}$ rad/s

49 One end of the string of length l is connected to a particle of mass m and the other end is connected to a small peg on a smooth horizontal table. If the particle moves in circle with speed v, the net force on the particle (directed towards centre) will be (T represents the tension in the string) [NEET 2017]

(a) T (b) $T + \frac{mv^2}{l}$

(c) $T - \frac{mv^2}{l}$ (d) zero

50 A uniform circular disc of radius 50 cm at rest is free to turn about an axis, which is perpendicular to its plane and passes through its centre. It is subjected to a torque which produces a constant angular acceleration of 2.0 rad s^{-2}. Its net acceleration in ms^{-2} at the end of 2.0 s is approximately [NEET 2016]

(a) 7.0 (b) 6.0 (c) 3.0 (d) 8.0

51 A car is negotiating a curved road of radius *R*. The road is banked at angle θ. The coefficient of friction between the tyres of the car and the road is μ_s. The maximum safe velocity on this road is **[NEET 2016]**

(a) $\sqrt{gR\left(\frac{\mu_s + \tan\theta}{1 - \mu_s \tan\theta}\right)}$ (b) $\sqrt{\frac{g}{R}\left(\frac{\mu_s + \tan\theta}{1 - \mu_s \tan\theta}\right)}$

(c) $\sqrt{\frac{g}{R^2}\left(\frac{\mu_s + \tan\theta}{1 - \mu_s \tan\theta}\right)}$ (d) $\sqrt{gR^2\left(\frac{\mu_s + \tan\theta}{1 - \mu_s \tan\theta}\right)}$

52 A car of mass 1000 kg negotiates a banked curve of radius 90 m on a frictionless road. If the banking angle is 45°, then speed of the car is **[CBSE AIPMT 2012]**

(a) 20 ms^{-1} (b) 30 ms^{-1}

(c) 5 ms^{-1} (d) 10 ms^{-1}

53. A gramphone record is revolving with an angular velocity ω. A coin is placed at a distance *R* from the centre of the record. The static coefficient of friction is μ. The coin will revolve with the record, if **[CBSE AIPMT 2010]**

(a) $R > \frac{\mu g}{\omega^2}$ (b) $R = \frac{\mu g}{\omega^2}$ only

(c) $R < \frac{\mu g}{\omega^2}$ (d) $R \le \frac{\mu g}{\omega^2}$

ANSWER KEY

Level 1 Objective Questions

1 (c)	2 (c)	3 (c)	4 (c)	5 (b)	6 (b)	7 (c)	8 (b)	9 (a)	10 (d)
11 (c)	12 (d)	13 (c)	14 (a)	15 (c)	16 (c)	17 (c)	18 (b)	19 (d)	20 (c)
21 (c)	22 (c)	23 (d)	24 (c)	25 (c)	26 (c)	27 (c)	28 (a)	29 (b)	30 (a)
31 (a)	32 (b)	33 (d)	34 (b)	35 (a)					

Level 2 Objective Questions

36 (d)	37 (a)	38 (a)	39 (d)	40 (b)	41 (d)	42 (d)	43 (d)	44 (a)	45 (a)
46 (b)	47 (b)								

Level 3 Previous Years Questions

48 (b)	49 (a)	50 (d)	51 (a)	52 (b)	53 (d)

Solutions

LEVEL 01 Objective Questions

1. Given, $\theta = 2t^3 + 0.5$

$$\omega = \frac{d\theta}{dt} = 6t^2$$

At $t = 2$ s,

$$\omega = 6\,(2)^2 = 24 \text{ rad/s}$$

2. $\omega_{min} = \dfrac{2\pi}{60}$

$$\omega_{hr} = \frac{2\pi}{12 \times 60}$$

$$\frac{\omega_{min}}{\omega_{hr}} = 12:1$$

3. $a_t = R\alpha$

$$= 0.02\,(10) = 0.2 \text{ m/s}^2$$

Now, $\omega_f = \omega_i + \alpha t$

$$\omega_f = 0 + (10)(2) = 20 \text{ rad/s}$$

$\therefore$ $a_c = R\omega^2 = 0.02(20)^2 = 8 \text{ m/s}^2$

4. We know that,

$$\omega = \omega_0 + \alpha t$$

$$\omega = \alpha t \qquad (\because\ \omega_0 = 0)$$

$$\frac{v}{r} = \alpha t$$

$$v = \alpha t r \qquad \text{...(i)}$$

Now, tangential acceleration = centripetal acceleration

$$\alpha r = \frac{v^2}{r} \qquad \text{...(ii)}$$

From Eqs. (i) and (ii), we get

$$\alpha r = \frac{\alpha^2 t^2 r^2}{r} \Rightarrow t = \frac{1}{\sqrt{\alpha}}$$

5. We have, $r = 2\sqrt{2}$ m

So, time taken in completing 1 round = 10 s

$$v = \frac{2\pi r}{10} = \frac{2\pi(2\sqrt{2})}{10}$$

$$= \frac{2\sqrt{2}\pi}{5} \text{ m/s}$$

6. $a_c = \dfrac{4}{r^2}$

$$\Rightarrow \frac{v^2}{r} = \frac{4}{r^2}$$

$$\Rightarrow v = \frac{2}{\sqrt{r}}$$

$\therefore$ Momentum, $p = \dfrac{2m}{\sqrt{r}}$

7. Let the velocity at point P be $\mathbf{v}_1 = v\hat{\mathbf{i}}$

So, the velocity at point Q be

$$\mathbf{v}_2 = v\cos 40^\circ\ \hat{\mathbf{i}} - v\sin 40^\circ\ \hat{\mathbf{j}}$$

$\therefore$ Change in velocity $= \mathbf{v}_2 - \mathbf{v}_1$

$$= (v\cos 40^\circ\ \hat{\mathbf{i}} - v\sin 40^\circ\ \hat{\mathbf{j}}) - v\hat{\mathbf{i}}$$

Its magnitude,

$$= \sqrt{(v\cos 40^\circ - v)^2 + (-v\sin 40^\circ)^2}$$

So, we get

$$\Delta v = 2v\sin 20^\circ$$

8. $a_t = \dfrac{dv}{dt} = \dfrac{4-2}{4} = 0.5 \text{ m/s}^2$

$$a_t = R\alpha$$

$$\Rightarrow \alpha = \frac{a_t}{R} = \frac{0.5}{0.5} = 1 \text{ rad/s}^2$$

9. $a_c = \dfrac{v^2}{R} = \dfrac{(4)^2}{2} = 8 \text{ m/s}^2$

$$a_c = -8 \text{ m/s}^2\ \hat{\mathbf{j}}$$

10. $\theta = \dfrac{1}{2}\alpha t^2 = \dfrac{1}{2}\left(\dfrac{a_t}{R}\right)t^2$

$$\therefore\ a_t = \frac{2R\theta}{t^2} = \frac{2 \times \frac{25}{\pi} \times 10\pi}{(5)^2} = 20 \text{ m/s}^2$$

11. $v = \alpha t^2$

$$a_t = \frac{dv}{dt} = 2\alpha t$$

At $t = 2$ s,

$$a_t = 4\,\alpha$$

Now, $a_r = \dfrac{v^2}{R} = \dfrac{\alpha^2 t^4}{R}$

At $t = 2$ s,

$$a_r = \frac{\alpha^2 \times 16}{R} = \frac{16\,\alpha^2}{2} = 8\alpha^2$$

$$\therefore\ \frac{a_r}{a_t} = 2\alpha$$

12. We know that, $\omega_{avg} = \dfrac{\text{total angular displacement}}{\text{total time}}$

Angular displacement θ, for first one-third part of circle,

$$\theta_1 = \frac{2\pi r/3}{r} = \frac{2\pi}{3}$$

Similarly, for next one-third part of circle,

$$\theta_2 = \frac{2\pi r/3}{r} = \frac{2\pi}{3}$$

Total time taken = 2 + 3 = 5 s

$$\therefore \quad \omega_{avg} = \frac{\theta_1 + \theta_2}{t_1 + t_2}$$

$$= \frac{4\pi/3}{5} = \frac{4\pi}{15} \text{ rad/s}$$

13. Given, $v = 2t$,

$$r = 30 \text{ cm} = 0.3 \text{ m}$$

$$a_t = \frac{dv}{dt} = 2 \text{ m/s}^2$$

At $t = 3$ s,

$$a_c = \frac{v^2}{r} = \frac{(2t)^2}{0.3} = \frac{4t^2}{0.3}$$

$$= \frac{4\,(3)^2}{0.3} = 120 \text{ m/s}^2$$

14. ω is same, as T is same.

$$a = \omega^2 r$$

$\therefore \quad a \propto r$ or radius.

15. We know that for turning, centripetal force is provided by friction, so

$$\frac{mv^2}{r} \le f_L$$

As $f_L = \mu N = \mu mg$ (where, $N = mg$)

$$\frac{mv^2}{r} \le \mu mg$$

i.e., $v \le \sqrt{\mu gr}$, so that $v_{max} = \sqrt{\mu gr}$

Here, $\mu = 0.4$, $r = 30$ m, $g = 10 \text{ ms}^{-2}$

$$\therefore \quad v_{max} = \sqrt{0.4 \times 30 \times 10}$$

$$= 11 \text{ ms}^{-1}$$

16. $\mu mg \ge mR\omega^2$

$$\therefore \quad R \le \frac{\mu g}{\omega^2}$$

or $\quad R \le \dfrac{0.4 \times 10}{100}$

or $\quad R \le 0.04$ m

17.

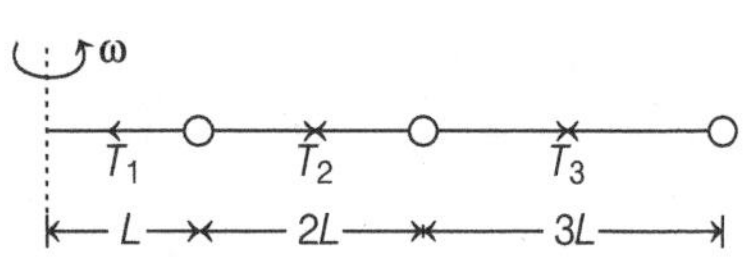

$$T_3 = m\,(6L)\,\omega^2 \quad \text{...(i)}$$

$$T_2 - T_3 = m\,(3L)\,\omega^2 \quad \text{...(ii)}$$

$$T_1 - T_2 = m\,(L)\,\omega^2 \quad \text{...(iii)}$$

On solving Eqs. (i), (ii) and (iii), we get

$$T_1 = 6\,mL\omega^2$$

$$T_2 = 9\,mL\omega^2$$

$$T_3 = 10\,mL\omega^2$$

So, the ratio is 6 : 9 : 10.

18. $T_1 = m_1\omega^2 l_1 + m_2\omega^2 l_2 \quad$ ...(i)

$T_2 = m_2\omega^2 l_2 \quad$ ...(ii)

$$\frac{T_1}{T_2} = \frac{m_1\omega^2 l_1 + m_2\omega^2 l_2}{m_2\omega^2 l_2} = \frac{\left[m_1 + m_2 \dfrac{l_2}{l_1}\right]}{m_2 \dfrac{l_2}{l_1}}$$

Here, $\dfrac{l_2}{l_1} = 2$, we get

$$\frac{T_1}{T_2} = \frac{m_1 + 2m_2}{2m_2}$$

19. In A, $\quad N_A = mg$

In B, $\quad mg - N_B = \dfrac{mv^2}{R} \Rightarrow N_B = mg - \dfrac{mv^2}{R}$

In C, $\quad N_C - mg = \dfrac{mv^2}{R} \Rightarrow N_C = mg + \dfrac{mv^2}{R}$

So, $N_C > N_A > N_B$

20. We have, $v = 3t^2$

$$a = \frac{dv}{dt} = 6t$$

At $t = 1$ s, $a = 6 \text{ m/s}^2$

So, the tangential force, $F = ma$

$$= 1\,(6) = 6 \text{ N}$$

$\therefore$ Power delivered by force at $t = 1$ s,

$$P = Fv$$

$$= 6 \times 3\,(1)^2$$

$$= 18 \text{ W}$$

21. Given, $v = 60$ m/s, $R = 1200$ m

Centripetal acceleration,

$$a_c = \frac{v^2}{R} = \frac{(60)^2}{1200} = 3 \text{ m/s}^2$$

$$a_t = 4 \text{ m/s}^2$$

So, the acceleration, $a = \sqrt{a_t^2 + a_c^2} = 5 \text{ m/s}^2$

22. $\tan\theta = \dfrac{v^2}{Rg}$

$$\frac{1}{\sqrt{3}} = \frac{v^2}{10\sqrt{3}\,(10)}$$

$$\Rightarrow \quad v = 10 \text{ m/s} \quad \text{or} \quad v = 36 \text{ km/h}$$

23. We know that, $\tan\theta = \dfrac{v^2}{rg}$

If the radius of curvature is made 4 times.

$$\tan\theta = \frac{u^2}{4rg}$$

We get, $u = 2v$

As speed is doubled.

So, % increase = 100 %

24. Given, radius of circular road = 100 m

Speed = 20 m/s

$$a_c = \frac{v^2}{R} = \frac{(20)^2}{100} = 4 \text{ m/s}^2$$

$$a_t = 3 \text{ m/s}^2$$

$$\therefore \quad a = \sqrt{4^2 + 3^2} = 5 \text{ m/s}^2$$

25. At extreme position, there is only tangential acceleration and in lowest position, there is only radial acceleration.

Further,

$$\mathbf{F}_{net} = m\mathbf{a}$$

So, net force is in the direction of **a**.

26. For a pendulum,

$$v^2 = 2gh = 2gl\,(1 - \cos\theta)$$

$$\therefore \quad \text{Maximum KE} = \frac{1}{2}mv^2 = mgl\,(1 - \cos\theta)$$

27. Three forces will act

(i) Weight, w

(ii) Tension, T

(iii) Centrifugal force F', which is outwards.

28. Let T_2 be the maximum tension in the string and T_1 be the minimum tension.

$$\frac{T_2}{T_1} = 4$$

$$\Rightarrow \quad T_2 = 4T_1$$

The difference between the two tensions must be 6 *mg*,

$$T_2 - T_1 = 6\,mg$$

$$4T_1 - T_1 = 6\,mg$$

$$\Rightarrow \quad T_1 = 2\,mg$$

Now, the tension T at top of a circle,

$$T_1 + mg = \frac{mv^2}{L}$$

$$2mg + mg = \frac{mv^2}{L}$$

$$\therefore \quad v = 10 \text{ m/s}$$

29. Increase in potential energy = Decrease in kinetic energy

$$\therefore \quad mgl\,(1 - \cos\theta) = \frac{1}{2}mv_0^2$$

Substituting the value of v_0 and then solving it, we get

$$\theta = 60°$$

30. At top point,

$$T_t + w = \frac{mv_t^2}{r} \quad \text{...(i)}$$

At bottom point,

$$T_b - w = \frac{mv_b^2}{r} \quad \text{...(ii)}$$

Applying conservation of energy,

$$\frac{1}{2}mv_b^2 = \frac{1}{2}mv_t^2 + mg\,(2l)$$

$$v_b^2 - v_t^2 = 4\,gl$$

From Eqs. (i) and (ii), we get

$$T_b - T_t = 2w + (v_b^2 - v_t^2)\frac{m}{l}$$

$$T_b - T_t = 2mg + 4gl\frac{m}{l}$$

$$T_b - T_t = 6\,mg$$

31. Applying conservation of energy,

$$\frac{1}{2}m\,(3gL) = \frac{1}{2}mv^2 + mg\,(L + L\cos\theta)$$

We get, $\quad v^2 = gL - 2gL\cos\theta \quad$ ...(i)

If the string slacks, then tension will become zero.

$$\therefore \quad mg\cos\theta = \frac{mv^2}{L}$$

$$v^2 = gL\cos\theta \quad \text{...(ii)}$$

Now, equating Eqs. (i) and (ii), we get

$$gL - 2gL\cos\theta = gL\cos\theta$$

So, we get $\quad \theta = \cos^{-1}\left(\frac{1}{3}\right)$

32. At highest point,

$$v^2 = u^2 - 2gh$$

$$v^2 = 10\,gL - 2g\,(2L)$$

$$= 6\,gL$$

Now, tension,

$$T = \frac{mv^2}{L} - mg$$

$$= \frac{m\,(6gL)}{L} - mg = 5\,mg$$

33. $mr\omega^2 = mg$

$$\Rightarrow \quad \omega = \sqrt{\frac{g}{r}} = \sqrt{\frac{9.8}{0.2}} = 7 \text{ rad/s}$$

34. Applying law of conservation of mechanical energy,

$$\frac{1}{2}mv^2 + mg\,(2R) = \frac{1}{2}kx^2$$

$$\frac{1}{2}mv^2 + mg\,(2R) = \frac{1}{2}\left(\frac{mg}{2R}\right)x^2$$

$$\Rightarrow \quad x = \sqrt{10}\,R$$

35. For the mass to be in equilibrium,

$$N\sin\theta = mg$$

$$N\cos\theta = mr\omega^2$$

$$\tan\theta = \frac{g}{r\omega^2}$$

From figure, $\quad \tan\theta = \frac{3}{4}$

$$\omega^2 = \frac{10}{0.4 \times 3/4}$$

$$\Rightarrow \quad \omega = \frac{10}{\sqrt{3}} \text{ rad/s}$$

LEVEL 02 Objective Questions

36. We have, $a_t = g \cos\theta$

$$v^2 = 2gh = 2gl \sin\theta$$

So, $$a_c = \frac{v^2}{R} = \frac{2gl\sin\theta}{l} = 2g \sin\theta$$

$$a = \sqrt{a_t^2 + a_c^2}$$

$$= g\sqrt{3 \sin^2\theta + 1}$$

37. $$T = m_B g = 50 \text{ N} \quad \text{...(i)}$$

$$T \cos\theta = m_A g = 40 \text{ N} \quad \text{...(ii)}$$

On dividing Eq. (ii) by Eq. (i), we get

$$\cos\theta = \frac{m_A}{m_B} = \frac{4}{5}$$

$$\theta = 37°$$

Now, $$T \sin\theta = \frac{mv^2}{r}$$

Here, $r = 2 \sin\theta$, $T = 50$

$$v^2 = \frac{2T \sin^2\theta}{m_A} = \frac{2 \times 50 (\sin^2 37°)}{4}$$

So, $v = 3$ m/s

38. The velocity of simple pendulum is given as

$$v^2 = 2gL \cos\theta$$

$$\therefore \quad a_c = \frac{v^2}{L} = 2g \cos\theta$$

$$a_t = g \sin\theta$$

For acceleration to be directed vertically upwards,

$$\theta = 0°$$

For the acceleration to be in horizontal direction,

$$a_c \cos\theta = a_t \sin\theta$$

So, we get

$$\theta = \cos^{-1}\left(\frac{1}{\sqrt{3}}\right)$$

Acceleration to be in downwards direction,

$$\theta = 90°$$

39. Given, $s = t^3 + 5$

$$v = \frac{ds}{dt} = 3t^2$$

$$a_t = \frac{dv}{dt} = 6t$$

At $t = 2$ s, $a_t = 12 \text{ m/s}^2$

Now, $$a_c = \frac{v^2}{R} = \frac{9t^4}{20} = \frac{9(2)^4}{20}$$

$$= 7.2 \text{ m/s}^2$$

$$\therefore \quad a = \sqrt{a_t^2 + a_c^2}$$

$$= \sqrt{12^2 + 7.2^2}$$

$$\approx 14 \text{ m/s}^2$$

40. We have given,

$$\mathbf{p}_A - \mathbf{p}_B = 2 \text{ kg-m/s } \hat{\mathbf{j}}$$

$$\mathbf{F}_A - \mathbf{F}_B = 8 \text{ N} \hat{\mathbf{i}}$$

$$\mathbf{v}_A = v\hat{\mathbf{j}} \Rightarrow \mathbf{v}_B = v(-\hat{\mathbf{j}})$$

Now, $m\mathbf{v}_A - m\mathbf{v}_B = m[v\hat{\mathbf{j}} - (-v\hat{\mathbf{j}})]$

$$2mv\,\hat{\mathbf{j}} = 2\hat{\mathbf{j}}$$

$$mv = 1 \text{ kg m/s}$$

We have, $\mathbf{F}_a = \frac{mv^2}{R}\hat{\mathbf{i}}$, $\mathbf{F}_B = \frac{mv^2}{R}(-\hat{\mathbf{i}})$

$$\mathbf{F}_A - \mathbf{F}_B = 8 \text{ N}\hat{\mathbf{i}}$$

$$\frac{2mv^2}{R}\hat{\mathbf{i}} = 8 \text{ N}\hat{\mathbf{i}}$$

$$\frac{mv^2}{R} = 4 \text{ N}$$

$$mv\left(\frac{v}{R}\right) = 4$$

$$\because \quad \frac{v}{R} = \frac{4}{1} = 4$$

$$\therefore \quad \omega = \frac{v}{R} = 4 \text{ rad/s}$$

41. We have, $x = 3R$

The ball has some velocity at point C and it is falling towards A.

So, time taken by ball to reach point A,

$$t = \sqrt{\frac{2h}{g}} = \sqrt{\frac{2(2R)}{g}} = 2\sqrt{\frac{R}{g}}$$

Now, the horizontal range $= v_c t$

$$3R = v_c \times 2\sqrt{\frac{R}{g}}$$

$$v_c = \frac{3\sqrt{gR}}{2}$$

Now, conserving energy at points A and C,

$$\frac{1}{2}mu^2 = \frac{1}{2}m\left(\frac{3}{2}\sqrt{gR}\right)^2 + mg(2R)$$

$$u = \frac{5\sqrt{gR}}{2}$$

42. Upto $\cos\theta = \frac{2}{3}$, force on bead is outwards, so force on wire is inwards. After that, force on wire is outwards.

43.

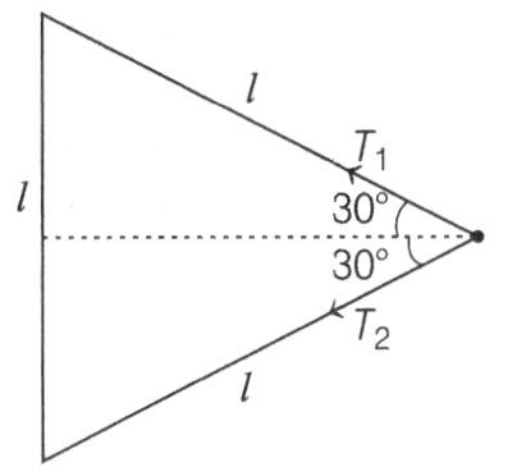

Radius of circle, $r = l\cos 30°$

Now, $T_1 \sin 30° - T_2 \sin 30° = mg$

$\therefore \quad T_1 - T_2 = 2mg \quad \text{...(i)}$

Further,

$$T_1 \cos 30° + T_2 \cos 30° = mr\omega^2$$

Substituting the value of r, we get

$$T_1 + T_2 = ml\omega^2 \quad \text{...(ii)}$$

From Eqs. (i) and (ii), we get

$$T_2 = \frac{ml\omega^2}{2} - mg$$

Now, $T_2 \geq 0$

$\therefore \quad \frac{l\omega^2}{2} \geq g$

or $\quad \omega \geq \sqrt{\frac{2g}{l}}$

44. The turn is banked for speed

$$v = 180 \text{ kmh}^{-1}$$

$$= 180 \times \frac{5}{18} \text{ ms}^{-1}$$

$$= 50 \text{ ms}^{-1}$$

$\because \quad \tan\theta = \frac{v^2}{Rg} = \frac{50 \times 50}{600 \times 10}$

$$= \frac{25}{60} = \frac{5}{12} = 0.417$$

$\Rightarrow \quad \tan\theta = 0.417$

$\Rightarrow \quad \theta = \tan^{-1}(0.417)$

$\Rightarrow \quad \theta = 22.6°$

45. In equilibrium,

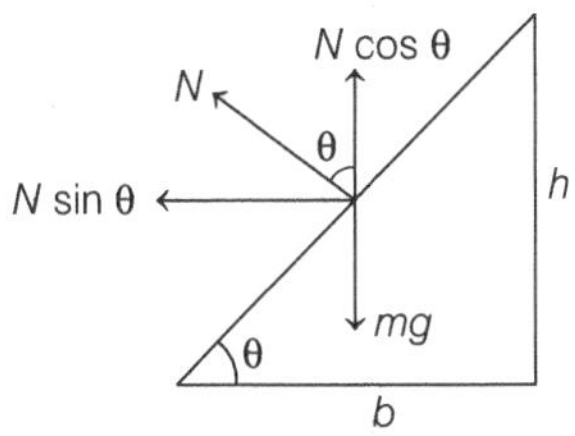

$$N\sin\theta = \frac{mv^2}{R} \quad \text{...(i)}$$

$$N\cos\theta = mg \quad \text{...(ii)}$$

Dividing Eq. (i) by Eq. (ii), we get

So, $\quad \tan\theta = \frac{v^2}{Rg}$

$$\frac{h}{b} = \frac{v^2}{Rg}$$

$\Rightarrow \quad h = \frac{v^2 b}{Rg}$

46. $mg\cos\theta + kx = \frac{mv^2}{R}$

$$2(10)\frac{1}{2} + 20x = \frac{2v^2}{2}$$

$$10 + 20x = v^2 \quad \text{...(i)}$$

Now, from conservation of energy,

$$0 + \frac{1}{2}kx^2 + 0 = \frac{1}{2}mv^2 - mgR(1 - \cos\theta) + \frac{1}{2}kx^2$$

$$v^2 = 20 \text{ m/s}$$

Now, substituting the value of v^2 in Eq. (i), we get

$$x = 0.5 \text{ m}$$

So, the unstretched length of spring $= R - x$

$$= 2 - 0.5 = 1.5 \text{ m}$$

47. Given, $\mathbf{a} = (4\hat{\mathbf{i}} + 3\hat{\mathbf{j}}) \text{ m/s}^2$

$$\mathbf{p} = (8\hat{\mathbf{i}} - 6\hat{\mathbf{j}}) \text{ kg-m/s}$$

The dot product of **a** and **p** is

$$\mathbf{a} \cdot \mathbf{p} = (4\hat{\mathbf{i}} + 3\hat{\mathbf{j}}) \cdot (8\hat{\mathbf{i}} - 6\hat{\mathbf{j}}) = 14$$

Here, **a** is not perpendicular to **p**. So, this is not uniform circular motion.

Here, angle is positive. So, it is uniform accelerated motion.

LEVEL 03 Previous Years Questions

48. Given, mass of cylinder, $m = 10$ kg

Radius of cylinder, $r = 1$ m

Coefficient of friction, $\mu = 0.1$

The given situation can be as shown in the figure given below.

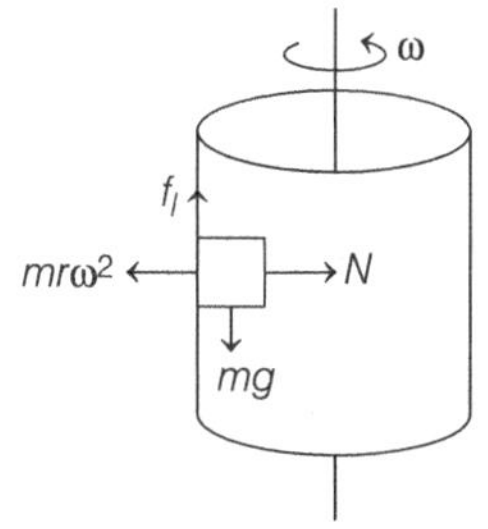

From the above figure, it can be concluded that the block will be stationary, when the limiting friction (f_l) is equal to or greater than the downward force or weight of block, i.e.

$$f_l \geq mg \quad \text{...(i)}$$

Also, the magnitude of limiting friction between two bodies is directly proportional to the normal reaction (N) between them, i.e.

$$f_l \propto N \quad \text{or} \quad f_l = \mu N \qquad \text{...(ii)}$$

From Eqs. (i) and (ii), we get

$$\mu N \geq mg \text{ or } \mu(mr\omega^2) \geq mg \qquad [\because N = mr\omega^2]$$

$$\Rightarrow \qquad \omega \geq \sqrt{\frac{g}{r\mu}}$$

Thus, the minimum angular velocity is

$$\omega_{\min} = \sqrt{\frac{g}{r\mu}} = \sqrt{\frac{10}{1 \times 0.1}} = 10 \text{ rad/s}$$

49. Consider the string of length l is connected to a particle as shown in the following figure.

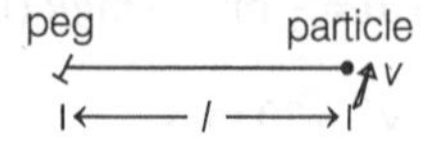

Speed of the particle is v. As the particle is in uniform circular motion, net force on the particle must be equal to centripetal force, which is provided by the tension (T).

$\therefore$ Net force = Centripetal force

$$\Rightarrow \qquad \frac{mv^2}{l} = T$$

50. According to given question, a uniform circular disc of radius 50 cm at rest is free to turn about an axis having perpendicular to its plane and passes through its centre. This situation can be shown by the figure given below

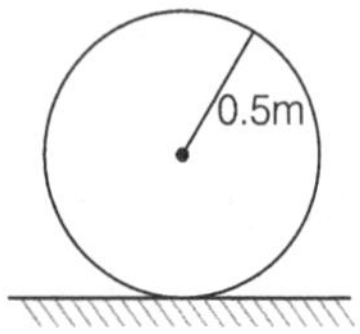

$\therefore$ Angular acceleration, $\alpha = 2 \text{ rad s}^{-2}$ (given)

Angular speed, $\omega = \alpha t = 4 \text{ rad s}^{-1}$

$\because$ Centripetal acceleration, $a_c = \omega^2 r$

$$= (4)^2 \times 0.5$$

$$= 16 \times 0.5$$

$$a_c = 8 \text{ m/s}^2$$

$\because$ Linear acceleration at the end of 2s,

$$a_t = \alpha r = 2 \times 0.5$$

$$\Rightarrow \qquad a_t = 1 \text{ m/s}^2$$

Therefore, the net acceleration at the end of 2.0 s is given by

$$a = \sqrt{a_c^2 + a_t^2}$$

$$a = \sqrt{(8)^2 + (1)^2} = \sqrt{65}$$

$$\Rightarrow \qquad a \approx 8 \text{ m/s}^2$$

51. According to question, a car is negotiating a curved road of radius R. The road is banked at an angle θ and the coefficient of friction between the tyres of car and the road is μ_s. So, this given situation can be drawn as shown in figure below.

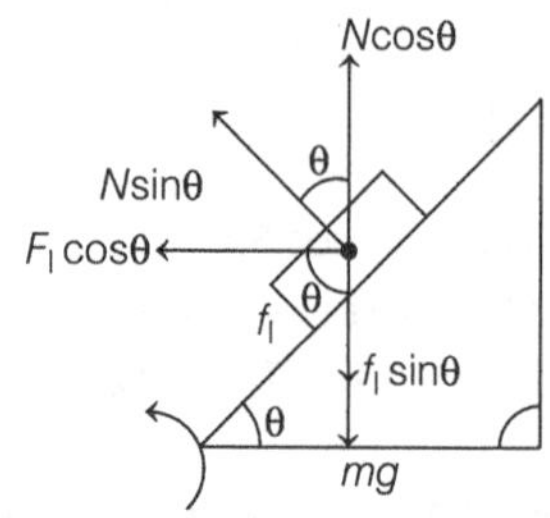

Considering the case of vertical equilibrium,

$$N\cos\theta = mg + f_l \sin\theta$$

$$\Rightarrow \qquad mg = N\cos\theta - f_l \sin\theta \qquad \text{...(i)}$$

Considering the case of horizontal equilibrium,

$$N\sin\theta + f_l \cos\theta = \frac{mv^2}{R} \qquad \text{...(ii)}$$

From Eqs. (i) and (ii), we get

$$\frac{v^2}{Rg} = \frac{\sin\theta + \mu_s \cos\theta}{\cos\theta - \mu_s \sin\theta} \qquad [f_l \propto \mu_s]$$

$$\Rightarrow \qquad v = \sqrt{Rg\left(\frac{\sin\theta + \mu_s \cos\theta}{\cos\theta - \mu_s \sin\theta}\right)}$$

$$\Rightarrow \qquad v = \sqrt{Rg\left(\frac{\tan\theta + \mu_s}{1 - \mu_s \tan\theta}\right)}$$

52. The angle of banking,

$$\tan\theta = \frac{v^2}{rg}$$

Given, $\theta = 45°$

Radius of banked curve road,

$$r = 90 \text{ m}$$

and $g = 10 \text{ m/s}^2$

$$\Rightarrow \qquad \tan 45° = \frac{v^2}{90 \times 10}$$

$$v = \sqrt{90 \times 10 \times \tan 45°}$$

$$= \sqrt{90 \times 10 \times 1}$$

$$= 30 \text{ m/s}$$

53. Frictional force will provide the centripetal force,

$$f = mR\omega^2$$

But $f \leq \mu\, mg$

$$mR\omega^2 \leq \mu\, mg$$

$$R \leq \frac{\mu g}{\omega^2}$$

Chapter 09

CENTRE OF MASS, COLLISION & IMPULSE

LEVEL 01 Objective Questions

1 A bomb at rest explodes into 3 parts, the momentum of the 2 parts are $-2p\hat{\mathbf{i}}$ and $p\hat{\mathbf{j}}$. The momentum of the third part will have a magnitude of

(a) p
(b) $\sqrt{3}\,p$
(c) $p\sqrt{5}$
(d) zero

2 At a certain instant of time, the mass of a rocket going up vertically is 100 kg. If it is ejecting 5 kg of gas per second at a speed of 400 m/s, the acceleration of the rocket would be ($g = 10\ \text{m/s}^2$)

(a) $20\ \text{m/s}^2$
(b) $10\ \text{m/s}^2$
(c) $2\ \text{m/s}^2$
(d) $1\ \text{m/s}^2$

3 A machine gun fires a bullet of mass 40 g with a velocity 1200 m/s. The man holding it can exert a maximum force of 144 N on the gun. How many bullets can be fired per second at the most?

(a) 1
(b) 4
(c) 2
(d) 3

4 Which of the following graphs represent a collision of two bodies in air ?

(a)
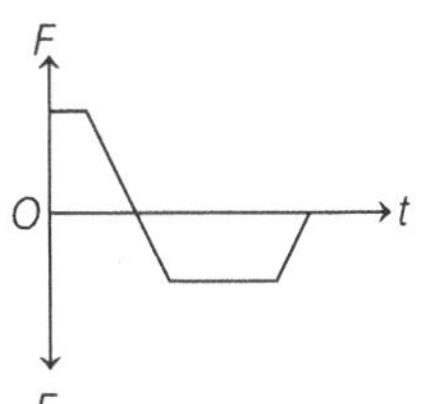

(b)
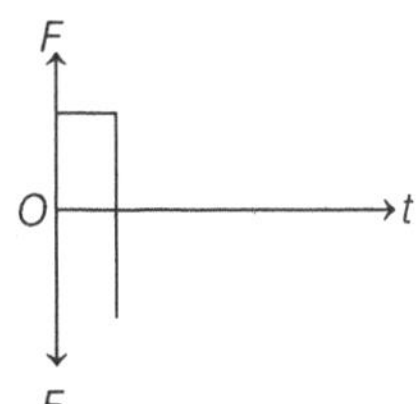

(c)
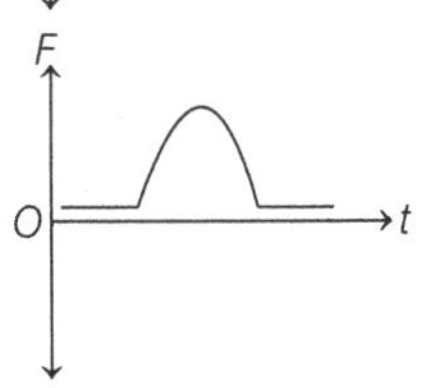

(d)
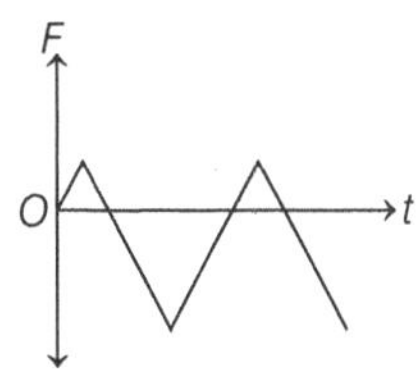

5 Figures I, II, III and IV depict variation of force with time. In which of the following situations, impulse depicted is maximum?

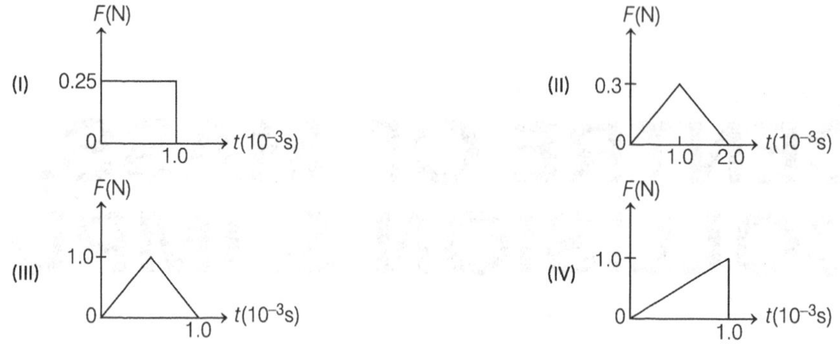

(a) I and II (b) III and I (c) III and IV (d) IV only

6 A rocket of initial mass 1500 kg ejects gas at a constant rate of 10 kg/s with a relative speed of 5 km/s. What is the acceleration of the rocket 50 s after the blast, neglecting gravity?

(a) 10 ms^{-2} (b) 25 ms^{-2} (c) 50 ms^{-2} (d) 100 ms^{-2}

7 A body of mass 40 kg is moving with a velocity of $\mathbf{v}_1$. If an impulse **J** is imparted to the body such that its velocity becomes $\mathbf{v}_2$, then

(a) $\mathbf{J} = 20\mathbf{v}_1 + 20\mathbf{v}_2$ (b) $\mathbf{J} = 40(\mathbf{v}_2 - \mathbf{v}_1)$ (c) $\mathbf{J} = 40(\mathbf{v}_1 + \mathbf{v}_2)$ (d) $\mathbf{J} = 20(\mathbf{v}_2 - \mathbf{v}_1)$

8 The figure below depicts the paths of two colliding steel balls *A* and *B*.

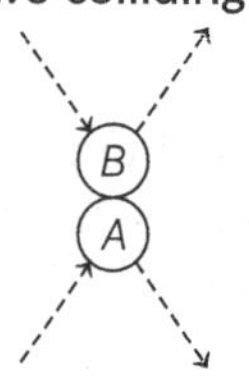

Which of the arrows 1-4 best represents the impulse applied to ball *A* during the collision?

(a)  (b)  (c)  (d) 

9 Find the position of centre of mass of given figure.

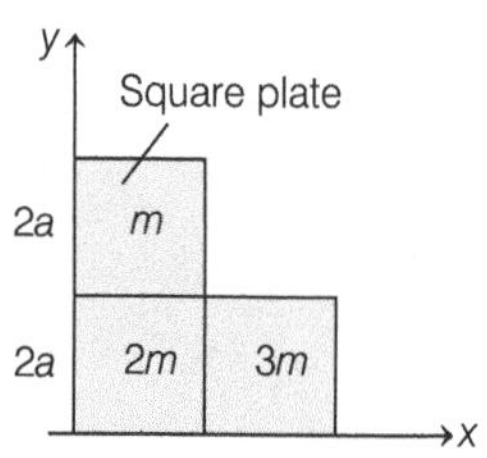

(a) $a\hat{\mathbf{i}} + a\hat{\mathbf{j}}$

(b) $\frac{3a}{2}\hat{\mathbf{i}} + \frac{3a}{2}\hat{\mathbf{j}}$

(c) $2a\hat{\mathbf{i}} + \frac{4a}{3}\hat{\mathbf{j}}$

(d) $2a\hat{\mathbf{i}} + 2a\hat{\mathbf{j}}$

10 There are hundred identical blocks equally spaced on a frictionless track as shown. Initially all the blocks are at rest. Block 1 is pushed with velocity *v* towards block 2. All collisions are perfectly inelastic. The final velocity of the set of hundred stucked blocks will be

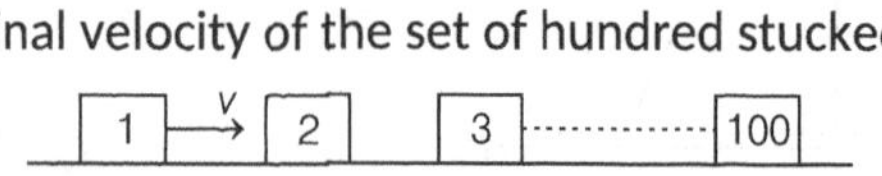

(a) $\frac{v}{99}$ (b) $\frac{v}{100}$ (c) *v* (d) zero

11 Two billiard balls *A* and *B*, each of mass 50 g and moving in opposite directions with speed of 5 ms^{-1} each, collide and rebound with the same speed. The impulse imparted to each ball is

(a) $0.5\ \frac{\text{kgm}}{\text{s}}$ (b) $-0.5\ \frac{\text{kgm}}{\text{s}}$ (c) $0.1\ \frac{\text{kgm}}{\text{s}}$ (d) $0.125\ \frac{\text{kgm}}{\text{s}}$

12 A bullet of mass m is fired from a gun of mass M. The recoiling gun compresses a spring of force constant k by a distance d. Then, the velocity of the bullet is

(a) $kd\sqrt{M/m}$ (b) $\frac{d}{M}\sqrt{km}$ (c) $\frac{d}{m}\sqrt{kM}$ (d) $\frac{kM}{m}\sqrt{d}$

13 The momentum of a particle is given by $p = at - bt^2$, where a and b are constants, then find the time at which force on the particle becomes zero.

(a) $\frac{b}{2a}$ (b) $\frac{a}{2b}$ (c) $\frac{b}{a}$ (d) None of these

14 Consider the following two statements :

(i) Linear momentum of a system of particles is zero.

(ii) Kinetic energy of a system of particles is zero.

Then,

(a) (i) implies (ii) but (ii) does not imply (i)
(b) (i) does not imply (ii) but (ii) implies (i)
(c) (i) implies (ii) and (ii) implies (i)
(d) (i) does not imply (ii) and (ii) does not imply (i)

15 In a two block system, an initial velocity v_0 (with respect to ground) is given to block A.

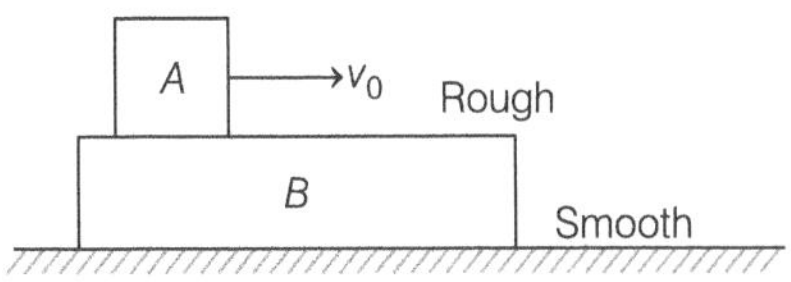

(a) The momentum of block A is not conserved
(b) The momentum of system of blocks A and B is conserved
(c) The increase in momentum of B is equal to the decrease in momentum of block A
(d) All of the above

16 If **J** is the impulse of a particular force, what is $\frac{dJ}{dt}$?

(a) Momentum
(b) Change in momentum
(c) Force
(d) Change in force

17 A mass of 20 kg moving with a speed of 10 m/s collides with another stationary mass of 5 kg. As a result of the collision, both masses stick together. The kinetic energy of the composite mass will be

(a) 600 J (b) 800 J (c) 1000 J (d) 1200 J

18 Three different projectiles, each with the same mass, are fired with speed v at a wall. In case A, the projectile bounces straight back with speed v. In case B, the projectile sticks to the wall. In case C, the projectile crashes through the wall and emerges with half its original speed. These three cases are shown here.

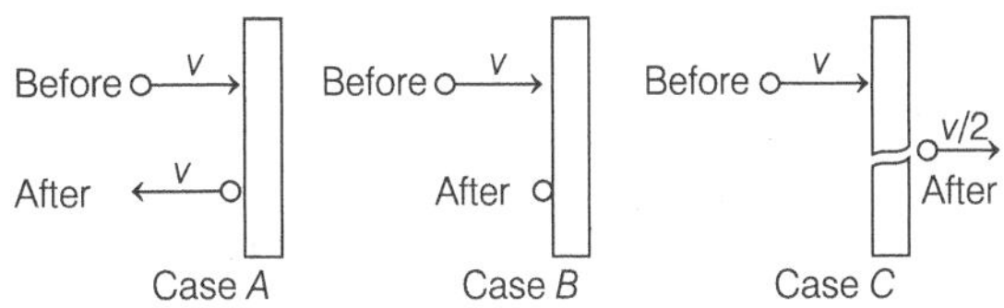

Place the impulse exerted by the wall on the projectile in each of these three cases in the correct order.

(a) $A > C > B$
(b) $C > B > A$
(c) $B > A > C$
(d) $A > B > C$

19 The correct statement about the centre of mass of a system of two particles, among the following is :

(a) The centre of mass lies on the line joining the two particles midway between them.
(b) The centre of mass lies on the line joining them at a point whose distance from each particle is inversely proportional to the mass of the particle.
(c) The centre of mass lies on the line joining them at a point whose distance from each particle is proportional to the square of the mass of that particle.
(d) The centre of mass is on the line joining them at a point whose distance from each particle is proportional to the mass of that particle.

20 Masses 8 kg, 2 kg, 4 kg and 2 kg are placed at the corners *A*, *B*, *C* and *D* respectively of a square *ABCD* of diagonal 80 cm. The distance of centre of mass from *A* will be

(a) 20 cm (b) 30 cm (c) 40 cm (d) 60 cm

21 Six rods of the same mass *m* and length *l* are arranged as shown in figure. Calculate the coordinate of the centre of mass of the system.

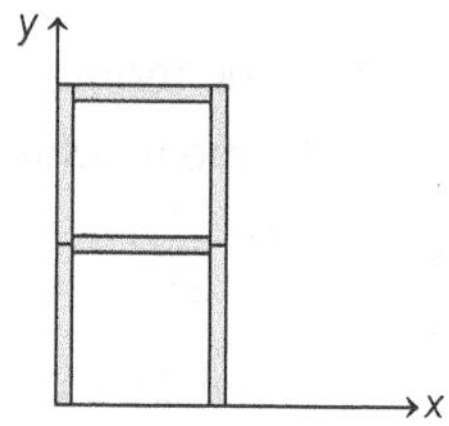

(a) $\left(\frac{l}{2}, \frac{7l}{6}\right)$ (b) $\left(\frac{l}{3}, l\right)$

(c) $\left(\frac{l}{2}, \frac{4l}{3}\right)$ (d) $\left(\frac{l}{2}, \frac{5l}{3}\right)$

22 In the given figure, the distance travelled by the plank when *A* and *B* exchange their positions.

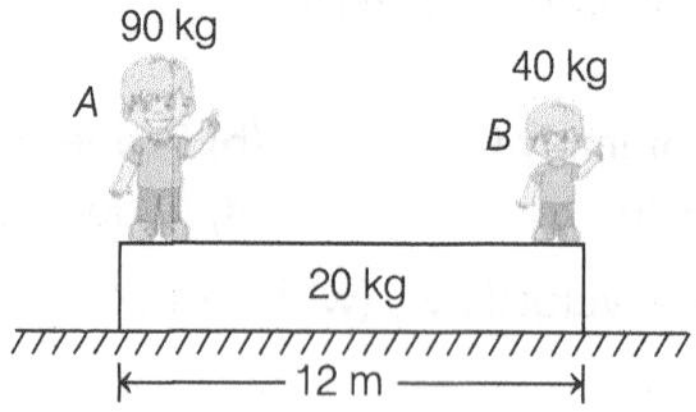

(a) 2 m (b) 4 m (c) 4.8 m (d) 5.4 m

23 Two particles are at rest as shown. Suddenly a force $F = 6\text{N}$ starts acting on the 3 kg particle. Find the velocity of the centre of mass of these particles at $t = 5\text{s}$.

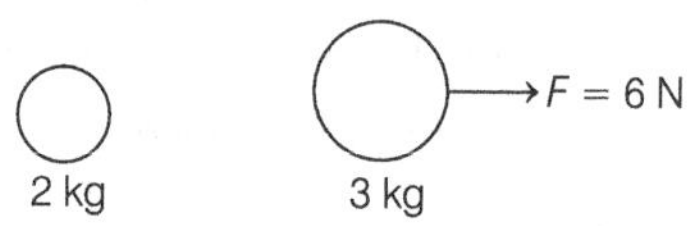

(a) 6 m/s (b) 5 m/s (c) 4 m/s (d) 10 m/s

24 The position-time graph of a body of mass 12 kg is given in the figure. What is the impulse on the body at $t = 4$ s.

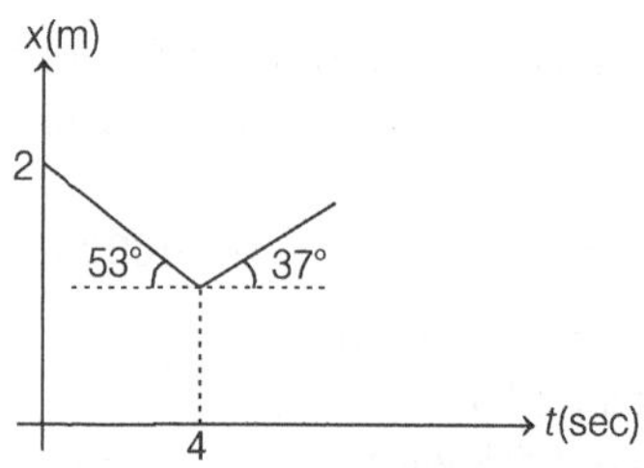

(a) $34 \frac{\text{kg-m}}{\text{s}}$ (b) $28 \frac{\text{kg-m}}{\text{s}}$ (c) $12 \frac{\text{kg-m}}{\text{s}}$ (d) $25 \frac{\text{kg-m}}{\text{s}}$

LEVEL 02 Objective Questions

25 Three men Anoop, Ashu and Ashish of masses 40 kg, 50 kg and 60 kg are standing on a plank of mass 90 kg, which is kept on a smooth horizontal plane. If Anoop and Ashish exchange their positions, then mass Ashu will shift (with respect to ground)

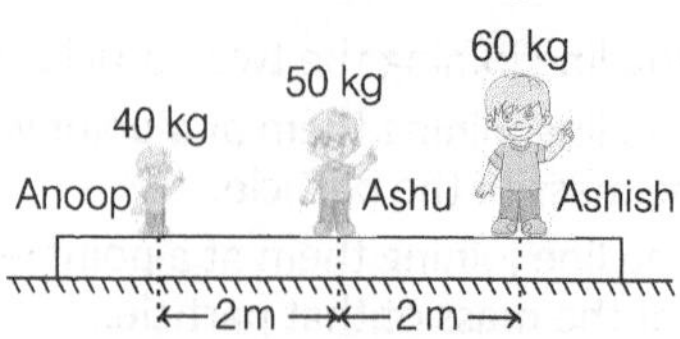

(a) (1/3) m towards left (b) (1/3) m towards right

(c) will not move with respect to ground (d) (5/3) m towards left

26 A wooden block of mass M is suspended by a cord and is at rest. A bullet of mass m, moving with a velocity v passes through the block and comes out with a velocity $v/2$ in the same direction. Upto what height, the block will rise?

(a) $m^2v^2/2M^2g$ (b) $m^2v^2/8M^2g$ (c) $mv^2/4Mg$ (d) $mv^2/2Mg$

27 A ball of mass 1 kg moving with a velocity of 4 m/s, collides with a stationary ball. The collision is oblique. After the collision, the first ball moves at right angle to its initial direction with a velocity of 3 m/s. The momentum of the second ball (in kg-m/s) after collision would be nearly

(a) 5 (b) 7 (c) 1 (d) 3

28 Two blocks each of mass m are connected to a spring of spring constant k. If both are given velocity v in opposite directions, then the maximum elongation of the spring is

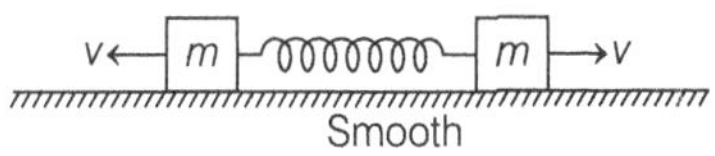

(a) $\sqrt{\frac{mv^2}{k}}$ (b) $\sqrt{\frac{2mv^2}{k}}$ (c) $\sqrt{\frac{mv^2}{2k}}$ (d) $2\sqrt{\frac{2mv^2}{k}}$

29 A thin rod of length 6 m is lying along the x- axis with its ends at $x = 0$ and $x = 6$ m. Its linear density (mass/length) varies with x as kx^4. Find the position of centre of mass of rod in metres.

(a) 5 m (b) 3 m (c) 2 m (d) 4 m

30 A circular disc of radius R is removed from a bigger circular disc of radius $2R$, such that the circumference of the discs coincide. The centre of mass of the new disc is αR from the centre of the bigger disc. The value of α is

(a) $\frac{1}{3}$ (b) $\frac{1}{2}$ (c) $\frac{1}{6}$ (d) $\frac{1}{4}$

31 A bullet fired into a wooden block loses half of its velocity after penetrating 40 cm. It comes to rest after penetrating a further distance of

(a) $\frac{22}{3}$ cm (b) $\frac{40}{3}$ cm (c) $\frac{20}{3}$ cm (d) $\frac{22}{5}$ cm

32 A 100 g iron ball having velocity 10 m/s collides with a wall at an angle 30° and rebounds with same speed at same angle. If the time of contact between ball and wall is 0.1 second, then the force experienced by the wall is

(a) 10 N (b) 100 N
(c) 1 N (d) 0.1 N

33 A stream of water flowing horizontally with a speed of 15 m/s gushes out of a tube of cross-sectional area 10^{-2} m^2 and hits a vertical wall normally. Assuming that it does not rebound from the wall, force exerted on the wall by the impact of water is

(a) 1.25×10^3 N (b) 2.25×10^3 N
(c) 3.25×10^3 N (d) 4.25×10^3 N

34 If initial momentum is 4 kg-m/s, then which of the following is **incorrect**?

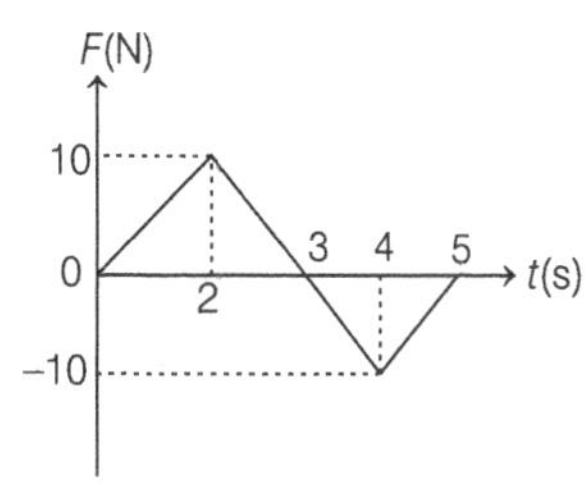

(a) Momentum at $t = 3$ s is 19 kg-m/s.
(b) Change in momentum during 3 s to 5 s is –10 kg-m/s.
(c) Change in momentum from $t = 0$ to $t = 5$ s will be zero.
(d) Momentum at $t = 2$ s is 14 kg-m/s.

35 Velocity of a particle of mass 2 kg varies with time t according to the equation $\mathbf{v} = (2t\hat{\mathbf{i}} - t\hat{\mathbf{j}})$ m/s. Here, t is in seconds. The impulse imparted to the particle in the time interval $0 \le t \le 2$ s is

(a) $4\hat{\mathbf{i}}$ kg-m/s (b) $(8\hat{\mathbf{j}} - 4\hat{\mathbf{i}})$ kg-m/s (c) $(8\hat{\mathbf{j}} + 4\hat{\mathbf{i}})$ kg-m/s (d) $(8\hat{\mathbf{i}} - 4\hat{\mathbf{j}})$ kg-m/s

36 A wedge of mass M is placed on a horizontal frictionless surface and a block of mass m is released on its frictionless slope. As the block slides down by a length L on this slope of inclination θ, the wedge would slide by a distance of

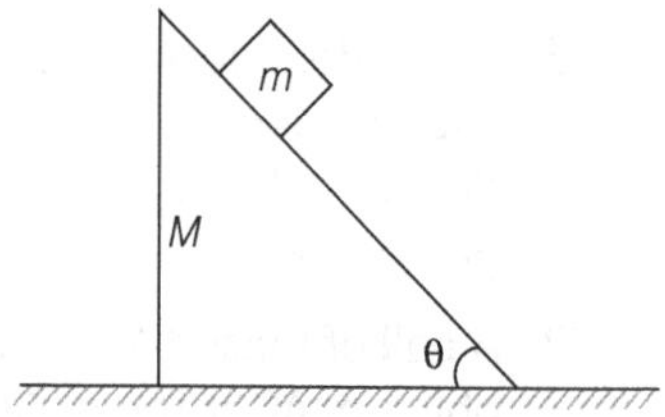

(a) $\frac{m}{M}L\cos\theta$
(b) $\frac{m}{L}(M+m)$
(c) $\frac{M+m}{mL\cos\theta}$
(d) $\frac{mL\cos\theta}{M+m}$

37 A particle of mass 200 g is dropped from a height of 50 m and another particle of mass 100 g is simultaneously projected up from the ground along the same line with a speed of 100 m/s, then velocity of centre of mass after 1 s will be

(a) $\frac{20}{3}$ m/s vertically down
(b) $\frac{20}{3}$ m/s vertically up
(c) $\frac{70}{3}$ m/s vertically down
(d) $\frac{70}{3}$ m/s vertically up

38 A block of mass m moving at a velocity v collides with another block of mass $2m$ at rest. The lighter block comes to rest after collision. Find the coefficient of restitution.

(a) $\frac{1}{2}$ (b) 1 (c) $\frac{1}{3}$ (d) $\frac{1}{4}$

39 A metal ball falls from a height of 32 m on a steel plate. If the coefficient of restitution is 0.5, to what height will the ball rise after second bounce?

(a) 2 m (b) 4 m (c) 8 m (d) 16 m

40 Particle x of mass 4 kg and y of mass 9 kg move directly towards each other, collide and then separate. If Δv_x and Δv_y are change in velocity of x and y respectively, then the magnitude of $\frac{\Delta v_x}{\Delta v_y}$ is

(a) $\frac{9}{4}$ (b) $\frac{3}{2}$ (c) $\frac{2}{3}$ (d) $\frac{4}{9}$

41 A bird is sitting in a large closed cage which is placed on a spring balance. It records a weight of 25 N. The bird of mass 0.5 kg fly upwards in the cage with an acceleration of 2 m/s^2. The spring balance will now record a weight of

(a) 24 N (b) 25 N (c) 25.5 N (d) 27 N

42 In figure, determine the character of the collision. The masses of the blocks and the velocities before and after are given. The collision is

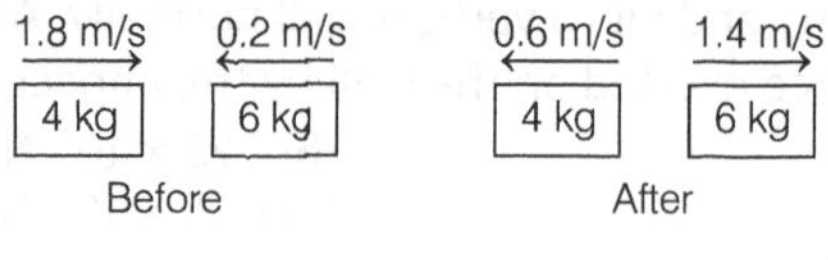

(a) completely inelastic
(b) partially inelastic
(c) perfectly elastic
(d) this collision is not possible

43 Two pendulum bobs of masses M and $2M$ are released from same height H. If they collide perfectly inelastically at mean position, then the height upto which the composite mass would rise is

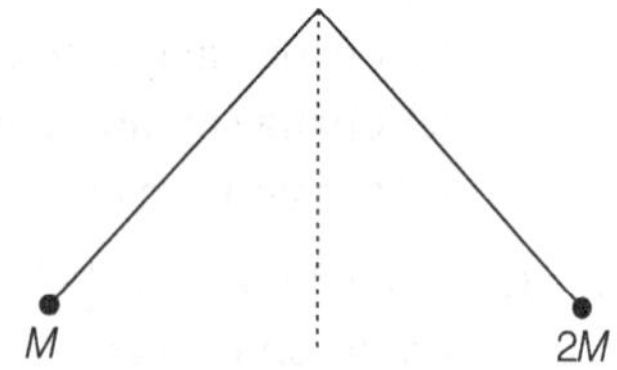

(a) $\frac{H}{3}$
(b) $\frac{H}{9}$
(c) $\frac{2H}{3}$
(d) $\frac{4H}{9}$

44 A plate of uniform thickness is shown in figure. It's centre of mass is at origin O. Now, if the shaded area is removed, the new position of the centre of mass is

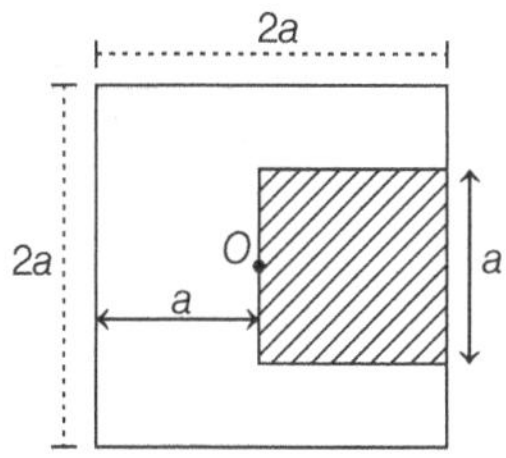

(a) $\frac{a}{6}\hat{i}$
(b) $\frac{a}{3}\hat{i}$
(c) $\frac{-a}{3}\hat{i}$
(d) $\frac{-a}{6}\hat{i}$

45 A balloon has 8 g of air. A small hole is pierced into it. The air escapes at a uniform rate of 7 cms^{-1}. If the balloon shrinks in 5.6 s, then the average force acting on the balloon is
(a) 10^{-4} N
(b) 10^{-2} dyne
(c) 56 dyne
(d) 10^{-6} N

46 A machine gun is mounted on a 2000 kg vehicle on a horizontal smooth road (friction negligible). The gun fires 10 bullets per sec with a velocity of 500 m/s. If the mass of each bullet be 10 g, what is the acceleration produced in the vehicle?
(a) 2.5 cm/s^2
(b) 25 cm/s^2
(c) 50 cm/s^2
(d) 25 m/s^2

47 Two men of masses 60 kg and 80 kg stand on a plank of mass 20 kg. Both of them can jump with a velocity of 1 m/s relative to the plank. Ashish jumps to left and after that Anoop jumps to right. Find final velocity of plank.

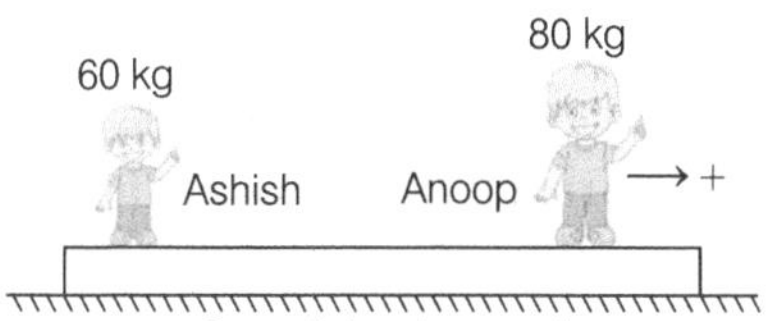

(a) $-\frac{17}{40}$ m/s
(b) $-\frac{1}{2}$ m/s
(c) $\frac{3}{8}$ m/s
(d) $-\frac{1}{8}$ m/s

48 Two particles of equal masses have velocities $\mathbf{v}_1 = a\hat{\mathbf{i}}$ and $\mathbf{v}_2 = a\hat{\mathbf{j}}$. The acceleration of first particle is $\mathbf{a}_1 = b(\hat{\mathbf{i}} + \hat{\mathbf{j}})$, where a and b are constants. If the acceleration of the second particle is zero, the centre of mass of the two particles moves along a
(a) straight line
(b) circle
(c) ellipse
(d) parabola

49 The velocities of two steel balls before impact are shown. If after head on impact, the velocity of ball B is observed to be 3 m/s to the right, the coefficient of restitution is

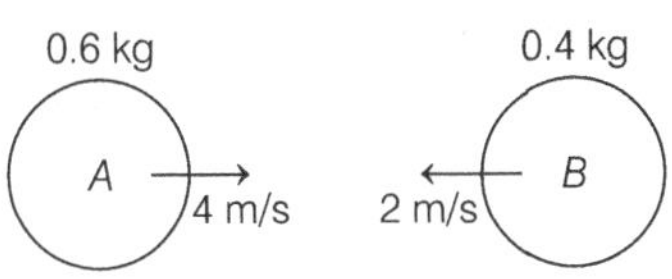

(a) $\frac{7}{18}$
(b) $\frac{6}{7}$
(c) $\frac{5}{18}$
(d) None of these

50 Find the centre of mass (x, y, z) of the following structure of four identical cubes, if the length of each side of a cube is 1 unit.

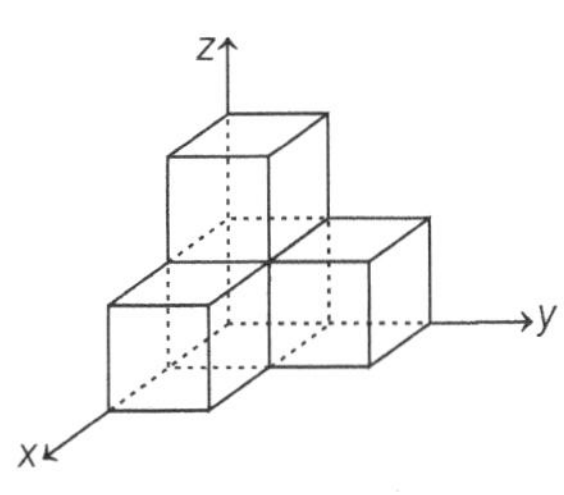

(a) (1/2, 1/2, 1/2)
(b) (1/3, 1/3, 1/3)
(c) (3/4, 3/4, 3/4)
(d) (1/2, 3/4, 1/2)

51 Figure shows a position-time graph for a body of mass 0.2 kg in one-dimensional motion, the magnitude of each impulse will be

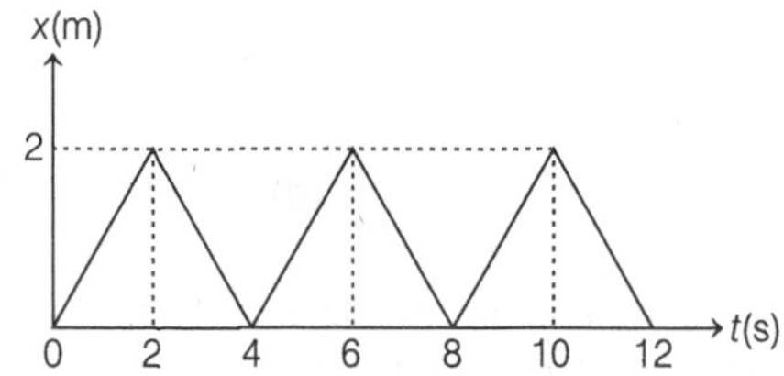

(a) 0.2 N-s (b) 0.4 N-s
(c) 0.8 N-s (d) 1.6 N-s

52 According to the given figure, find distance of centre of mass from A. Mass of each rod is same.

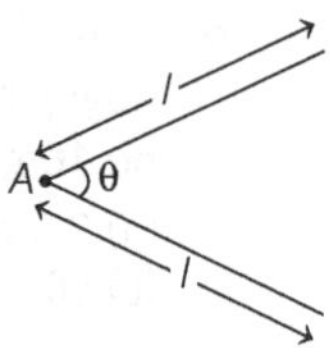

(a) $\frac{l}{2}\cos\theta$ (b) $\frac{l}{2}\sin\theta$
(c) $\frac{l}{2}\cos\frac{\theta}{2}$ (d) $\frac{l}{2}\sin\frac{\theta}{2}$

53 A square piece of side a is removed from a uniform circular disc of radius R as shown in figure. The centre of bigger disc is taken as origin. The shift of centre of mass from origin is

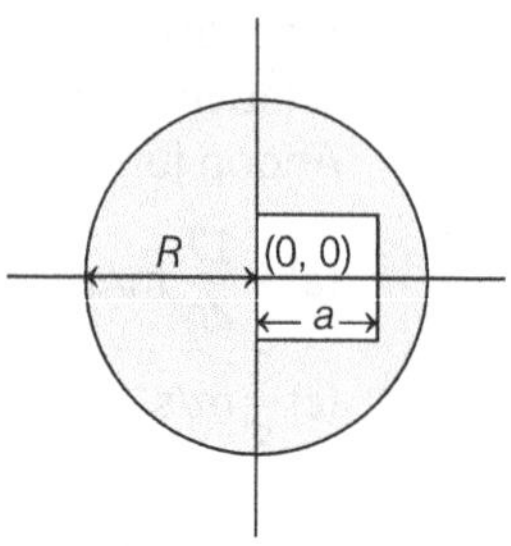

(a) $\frac{a^2}{\pi R^2 - a^2}$ (b) $\frac{a^3}{2[\pi R^2 - a^2]}$
(c) $\frac{a^4}{\frac{4}{3}\pi R^3 - a^3}$ (d) $\frac{a^3}{2\pi R^2 - a^2}$

54 A disc of mass 100 g is kept floating horizontally in air by firing bullets, each of mass 5 g with the same velocity at a constant rate of 10 bullets per second. The bullet rebound with the same speed in opposite direction. The velocity of each bullet at the time of impact is
(a) 196 cm^{-1} (b) 9.8 cm s^{-1}
(c) 98 cm s^{-1} (d) 980 cm s^{-1}

55 Find the distance of centre of mass of two square plates system as shown from point O. If masses of plates are $2m$ and m (their edges are a and $2a$, respectively).

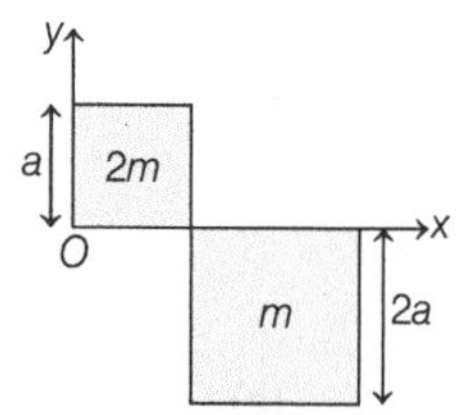

(a) $\frac{a}{2}$ (b) a
(c) $\frac{3a}{2}$ (d) $\frac{2a}{3}$

56 A man of mass 80 kg is riding on a small cart of mass 40 kg which is rolling along a level floor at a speed of 2 m/s. He is running on the cart, so that his velocity relative to the cart is 3 m/s in the direction opposite to the motion of the cart. What is the speed of the centre of mass of the system?
(a) 1.5 m/s (b) 1 m/s
(c) 3 m/s (d) zero

57 A mass m moves with a velocity v and collides inelastically with another identical mass initially at rest. After collision, the first mass moves with velocity $\frac{v}{\sqrt{3}}$ in a direction perpendicular to its initial direction of motion. The speed of second mass after collision is
(a) $\frac{2}{\sqrt{3}}v$ (b) $\frac{v}{\sqrt{3}}$ (c) v (d) $\sqrt{3}$ v

58 A ball of mass 0.2 kg is thrown vertically upwards by applying a force by hand. If the hand moves 0.2 m while applying the force and the ball goes upto 2 m height further, find the magnitude of the force. (Take, $g = 10 \text{ m/s}^2$)

(a) 4 N (b) 16 N (c) 20 N (d) 22 N

59 A tennis ball is dropped on the floor from a height of 4 m. It rebounds to a height of 2 m. If the ball is in contact with the floor for 12×10^{-3} s, its average acceleration during the contact is (Take, $g = 9.8 \text{ ms}^{-2}$)

(a) 0 (b) 1260 ms^{-2} (c) 980 ms^{-2} (d) 600 ms^{-2}

60 A ball is thrown with a velocity of 6 m/s vertically downwards from a height $H = 3.2$ m above a horizontal floor. If it rebounds back to same height, then coefficient of restitution e is (Take, $g = 10 \text{ m/s}^2$)

(a) 0.5 (b) 0.6 (c) 0.7 (d) 0.8

61 From a circular disc of radius R, a square is cut out with a radius as its diagonal. The centre of mass of the remaining part is at a distance (from the centre)

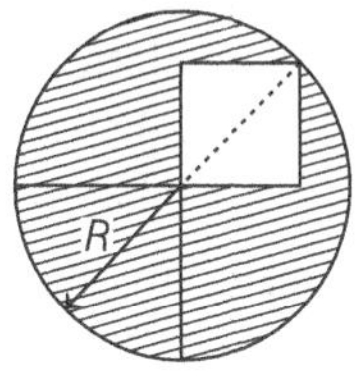

(a) $\frac{R}{(4\pi - 2)}$ (b) $\frac{R}{2\pi}$

(c) $\frac{R}{(\pi - 2)}$ (d) $\frac{R}{(2\pi - 2)}$

62 A man of mass M stands at one end of a plank of length L which lies at rest on a smooth surface. The man walks to the other end of the plank. If mass of the plank is $3M$, the distance that man moves relative to the ground is

(a) $\frac{L}{4}$ (b) $\frac{3L}{4}$ (c) $\frac{2L}{3}$ (d) $\frac{L}{3}$

63 In the given figure, if collision between blocks A and B is elastic, then maximum compression in the spring will be

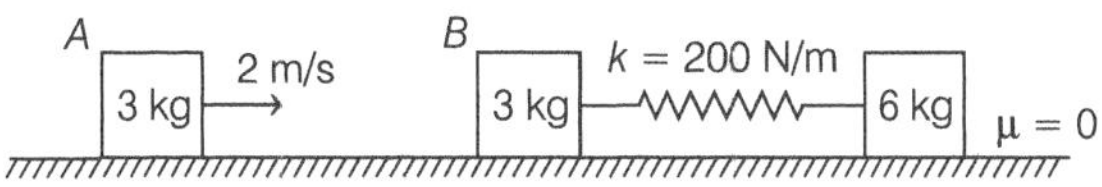

(a) 10 cm (b) 20 cm (c) 30 cm (d) None of these

64 The distance of the centre of mass of the T-shaped plate from O is

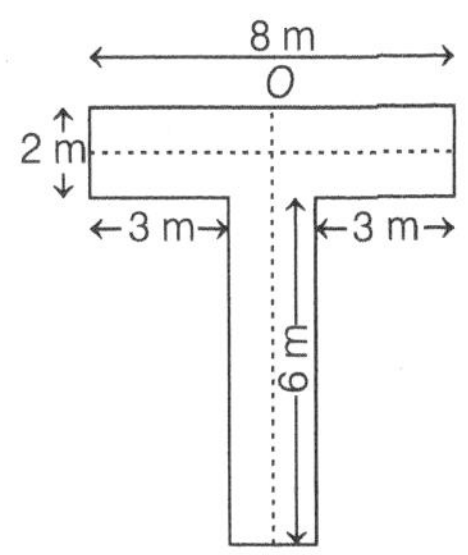

(a) 7 m (b) 2.7 m

(c) 4 m (d) 1 m

65 Three objects A, B and C are kept in a straight line on a smooth horizontal surface. These have masses m, $2m$ and $3m$ respectively, the object A is given velocity v_0 towards B. The head-on-elastic collision takes place between A and B and then B makes completely inelastic collision with C. All motions occur on the same straight line. The final speed of C will be

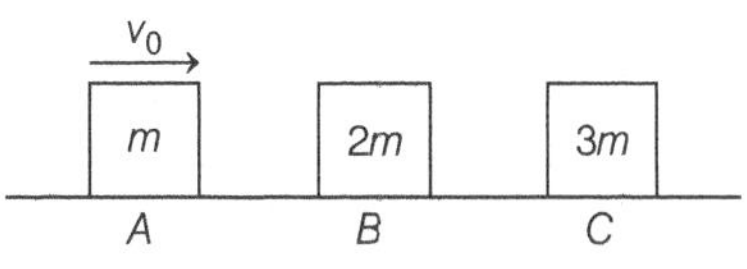

(a) $\frac{v_0}{15}$ (b) $\frac{2v_0}{15}$ (c) $\frac{3v_0}{15}$ (d) $\frac{4v_0}{15}$

66 Two masses of 1 kg and 2 kg respectively are connected by a massless spring as shown in figure. A force of 20 N acts on the 2 kg mass at the instant when the 1 kg mass has an acceleration of 10 ms^{-2} towards right, the acceleration of the 2 kg mass is

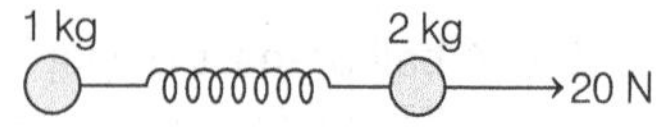

(a) 2 ms^{-2}
(b) 5 ms^{-2}
(c) 10 ms^{-2}
(d) 20 ms^{-2}

67 A man of 50 kg mass is standing in a gravity free space at a height of 10 m above the floor. He throws a stone of 0.5 kg mass downwards with a speed of 2 m/s. When the stone reaches the floor, the distance of the man above the floor will be

(a) 20 m
(b) 9.9 m
(c) 10.1 m
(d) 10 m

68 A ball of mass 1 kg strikes a heavy platform elastically, moving upwards with a velocity of 5 m/s. The speed of the ball just before the collision is 10 m/s downwards. Then, the impulse imparted by the platform on the ball is

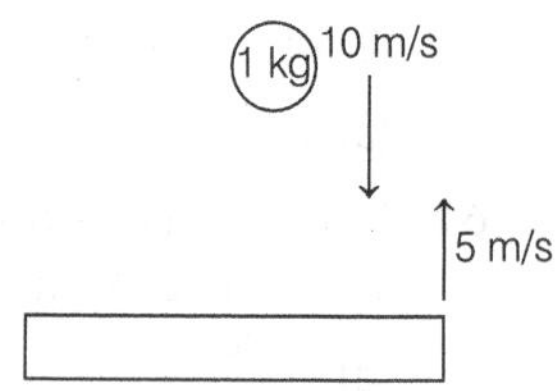

(a) 15 N-s
(b) 10 N-s
(c) 20 N-s
(d) 30 N-s

69 An object of mass $2m$ is projected with a velocity v at angle θ with the horizontal. At the highest point of its path, it explodes into two fragments of equal mass such that speed of one of the fragments immediately after explosion is zero. Distance from the point of projection at which the centre of mass hits the horizontal is

(a) $\frac{v^2 \sin\theta}{g}$
(b) $\frac{v^2 \sin\theta}{2g}$
(c) $\frac{v^2 \sin 2\theta}{g}$
(d) $\frac{v \sin\theta}{2g}$

70 A stone is projected from ground and hits a smooth vertical wall after 3 s and again falls back on the ground. The time taken by stone to reach the ground after the collision is 5 s. The maximum height reached by the same stone, if the vertical wall were not to be present is $(g = 10 \text{ m/s}^2)$

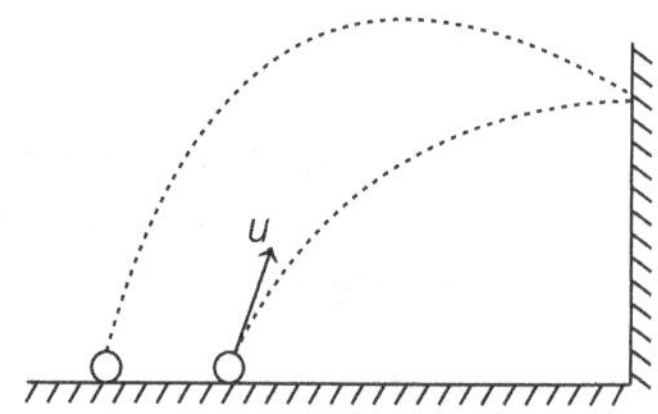

(a) 100 m
(b) 80 m
(c) 60 m
(d) 50 m

71 Three identical blocks A, B and C are placed on a horizontal frictionless surface. The blocks B and C are at rest. A is approaching towards B with a speed of 10 m/s. If coefficient of restitution for all collisions is 0.5, then speed of block C just after collision is approximately

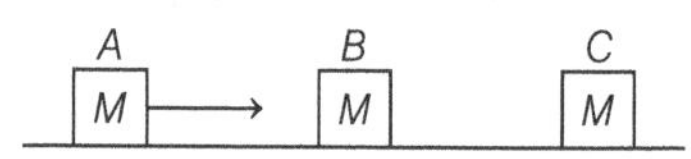

(a) 5.6 m/s
(b) 6.4 m/s
(c) 3.2 m/s
(d) 4.6 m/s

72 Two small spheres of equal masses start moving in opposite directions from a point A in a horizontal smooth circular tube with tangential velocities v and $2v$, respectively. The number of elastic collisions the spheres will make before they reach point A again is

(a) 4
(b) 3
(c) 1
(d) 2

73 Seven homogeneous bricks each of length l are arranged as shown. Each brick is displaced with respect to the one in contact by $l/10$. The x-coordinate of centre of mass of the system with respect to point O is

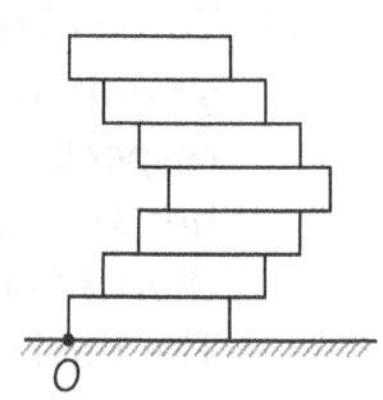

(a) $\frac{22l}{35}$
(b) $\frac{11l}{17}$
(c) $\frac{31l}{45}$
(d) $\frac{l}{2}$

74 Two masses $m_1 = 1$ kg and $m_2 = 2$ kg are connected by a light inextensible string and suspended by means of a weightless pulley as shown in the figure. Assuming that both the masses start from rest, the distance travelled by the centre of mass in two seconds is (Take, $g = 10$ ms^{-2})

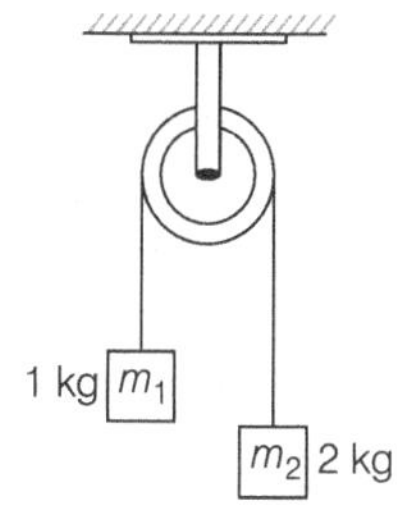

(a) $\frac{20}{9}$ m
(b) $\frac{40}{9}$ m
(c) $\frac{2}{3}$ m
(d) $\frac{1}{3}$ m

75 Four balls A, B, C and D are placed on a smooth horizontal surface. Ball B collides with ball C with an initial velocity u as shown in the figure. If all collisions are elastic, find the total number of collisions between the balls.

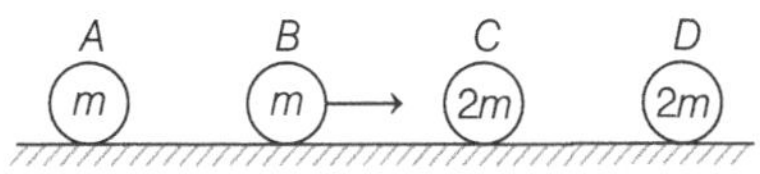

(a) 1
(b) 2
(c) 3
(d) 4

76 Block A is hanging from a vertical spring and is at rest. Block B strikes the block A with velocity v and sticks to it. Then, the value of v for which the spring just attains natural length is

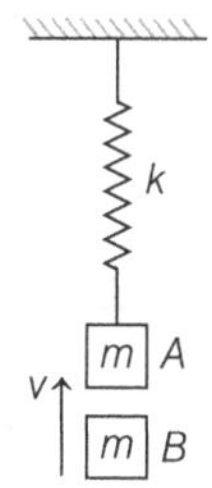

(a) $\sqrt{\frac{60mg^2}{k}}$
(b) $\sqrt{\frac{6mg^2}{k}}$
(c) $\sqrt{\frac{10mg^2}{k}}$
(d) $\sqrt{\frac{8mg^2}{k}}$

77 Two identical balls moving in opposite directions in free space collide head on. How they will move after the impact, is decided by coefficient of restitution. For different coefficients of restitution, few possible outcomes are shown in the following figures.

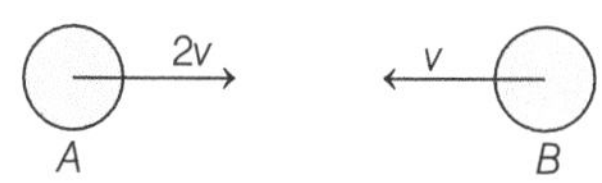

Which of the above outcomes are physically possible for a collision?

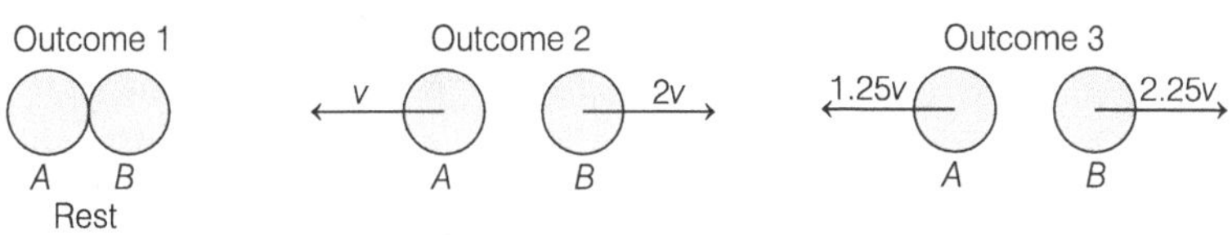

(a) Only 1
(b) Only 2
(c) 1 and 2 but not 3
(d) 1, 2 and 3 all

78 A particle of mass 1 kg is projected from a tower of height 375 m with initial velocity 100 ms^{-1} at an angle 30° with the horizontal. Find out its kinetic energy (in joule) just after collision with ground, if collision is inelastic with $e = \frac{1}{2}$. (Take, $g = 10$ ms^{-2})

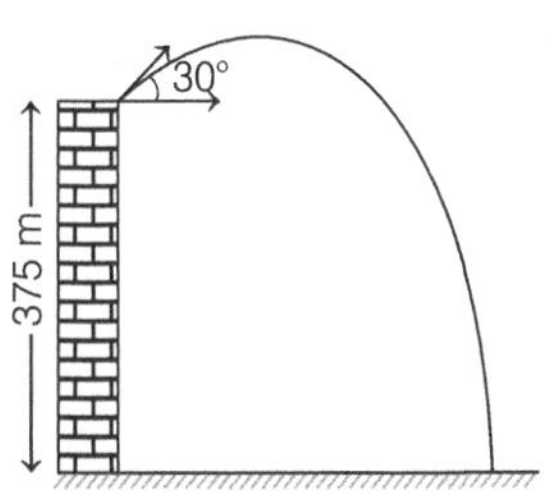

(a) 5000 J
(b) 10000 J
(c) 2500 J
(d) 6000 J

79 A 1 kg block moves towards a light spring with a velocity of 8 m/s. When the spring is compressed by 3 m, its momentum becomes half of the original momentum. Spring constant of the spring is

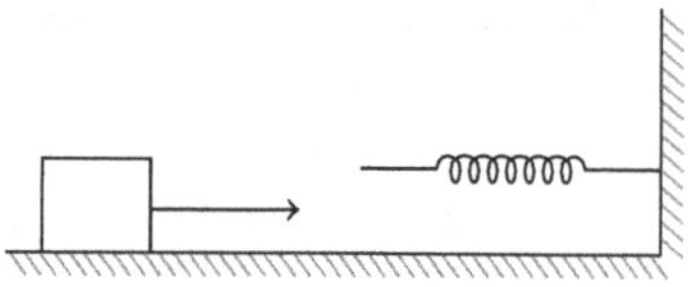

(a) (18/3) N/m (b) (16/3) N/m
(c) 3 N/m (d) 8 N/m

80 A block of mass m is pushed towards a movable wedge of mass $2m$ and height h with a velocity u. All surfaces are smooth. The minimum value of u for which the block will reach the top of the wedge is

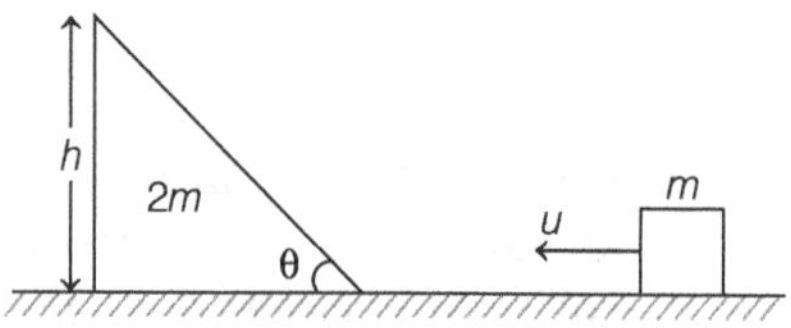

(a) $2\sqrt{gh}$ (b) $\sqrt{3gh}$
(c) $\sqrt{6gh}$ (d) $\sqrt{\frac{3}{2}gh}$

81 Infinite rods of uniform mass density and length $L, L/2, L/4$ .... are placed one over the another as shown in the figure. Find the x-coordinate of centre of mass.

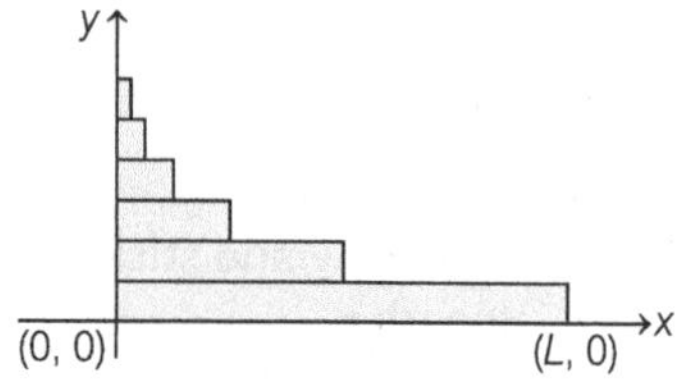

(a) $L/6$ (b) $4L/3$
(c) $L/2$ (d) $2L/3$

82 Two identical balls A and B are released from the positions shown in figure. They collide elastically on horizontal portion MN. The ratio of the heights attained by A and B after collision will be (neglect friction)

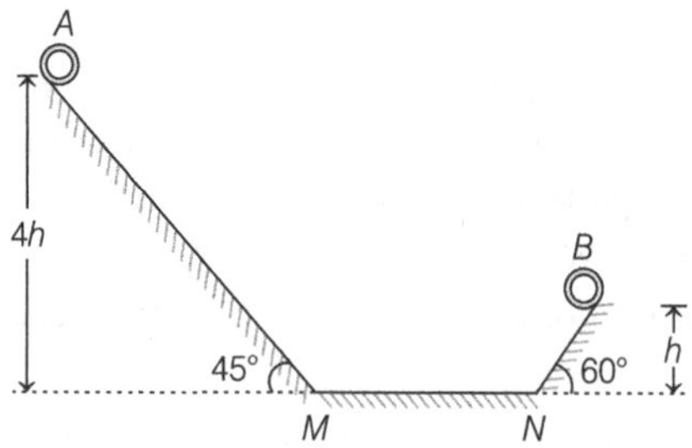

(a) 1 : 4 (b) 2 : 1 (c) 4 : 13 (d) 2 : 5

83 A ball falls vertically onto a floor with momentum p and then bounces repeatedly. If the coefficient of restitution is e, then the total momentum imparted by the ball on the floor is

(a) $p(1+e)$ (b) $\frac{p}{1-e}$ (c) $p\left(1+\frac{1}{e}\right)$ (d) $p\left(\frac{1+e}{1-e}\right)$

84 Two identical balls A and B lie on a smooth horizontal surface, which gradually merges into a curve to a height 3.2 m. Ball A is given a velocity of 10 m/s. It collides head on with ball B, which then moves on the curved path. The minimum coefficient of restitution e for the collision between A and B, in order that ball B reaches the highest point C of the curve is
(Take, $g = 10\ \text{m/s}^2$)

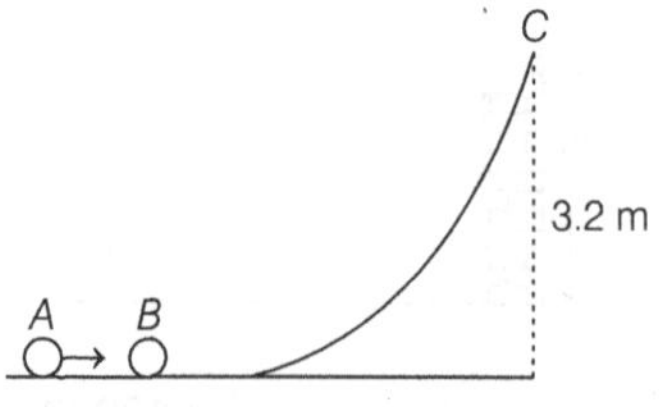

(a) $\frac{1}{4}$ (b) $\frac{3}{5}$ (c) $\frac{1}{2}$ (d) $\frac{3}{4}$

85 Consider a gravity free hall in which a tray of mass M, carrying a cubical block of ice of mass m and edge L, is at rest in the middle. If ice melts by what distance does the centre of mass of the tray and ice system descend?

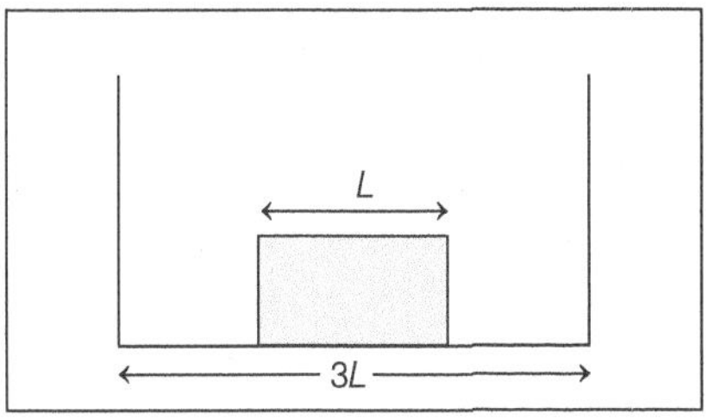

(a) $\frac{L}{3}$ (b) $\frac{L}{4}$ (c) $\frac{L}{2}$ (d) Zero

86 A man of mass 50 kg is in a gravity free space at a height of 20 m above the floor. He throws a stone of 0.5 kg downwards with a speed of 36 km/h at an angle of 60° with vertical. When the stone reaches the floor, the distance of man from the floor is

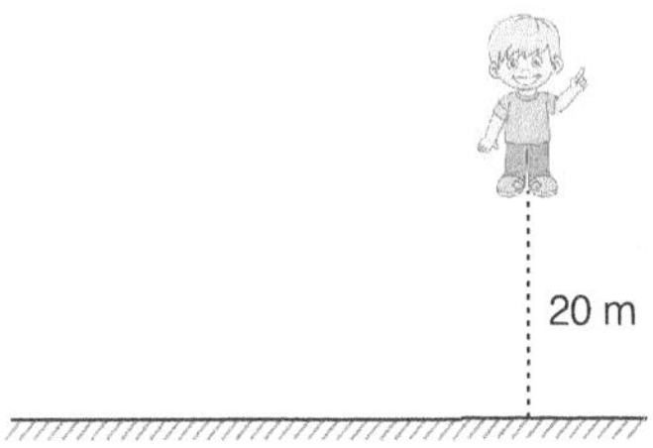

(a) 20.2 m (b) 20.1 m
(c) 20.0 m (d) 19.9 m

87 A set of n identical cubical blocks lies at rest parallel to each other along a line on a smooth horizontal surface. The separation between the near surfaces of any two adjacent blocks is L. The block at one end is given a speed v towards the next one at time $t = 0$. All collisions are completely inelastic, then

(a) the last block starts moving at $t = \frac{(n-1)L}{v}$

(b) the last block starts moving at $t = \frac{n(n-1)L}{2v}$

(c) the centre of mass of the system will have a final speed v

(d) the centre of mass of the system will have a final speed zero

88 Half of the rectangular plate shown in figure is made of a material of density ρ_1 and other half of density ρ_2. The length of the plate is L. The distance of centre of mass of the plate from origin O is

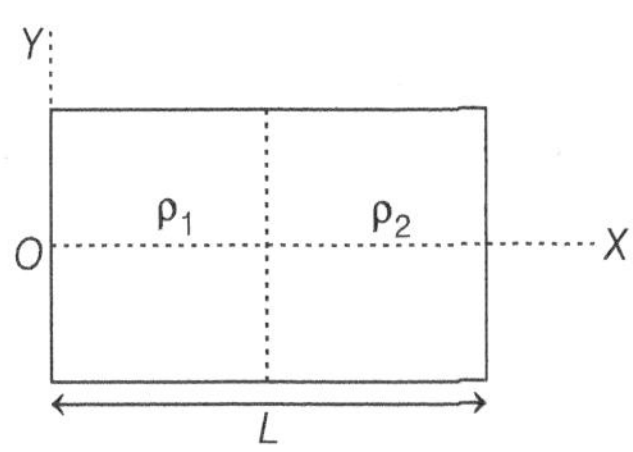

(a) $\frac{L}{2}$ (b) $\frac{(\rho_1 + 3\rho_2)}{4(\rho_1 + \rho_2)}L$ (c) $\frac{(3\rho_1 + \rho_2)}{(\rho_1 + \rho_2)}L$ (d) $\left[\frac{\rho_1 + 2\rho_2}{3(\rho_1 + \rho_2)}\right]L$

LEVEL 03 Previous Years Questions

89 A uniform rod of length 200 cm and mass 500 g is balanced on a wedge placed at 40 cm mark. A mass of 2 kg is suspended from the rod at 20 cm and another unknown mass m is suspended from the rod at 160 cm mark as shown in the figure. Find the value of m such that the rod is in equilibrium. (Take, $g = 10 \text{ m/s}^2$) [NEET 2021]

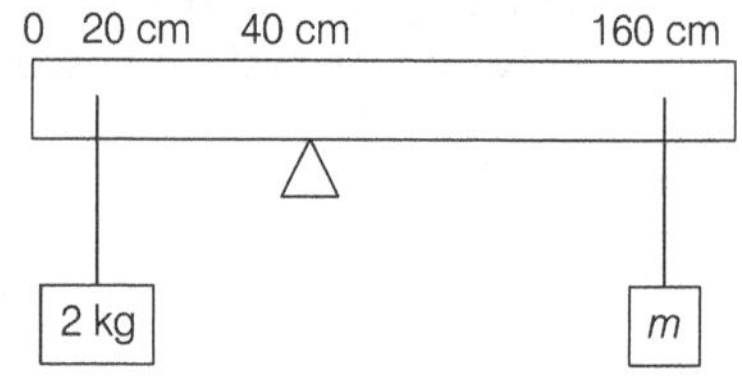

(a) $\frac{1}{2}$ kg (b) $\frac{1}{3}$ kg (c) $\frac{1}{6}$ kg (d) $\frac{1}{12}$ kg

90 Three identical spheres, each of mass M, are placed at the corners of a right angle triangle with the mutually perpendicular sides equal to 2 m (see figure). Taking the point of intersection of the two mutually perpendicular sides as the origin, find the position vector of centre of mass. [NEET 2020]

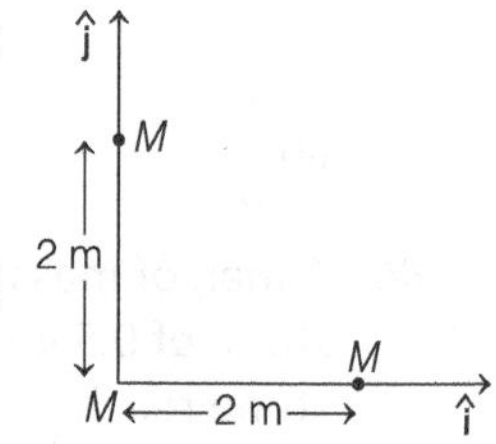

(a) $2(\hat{i} + \hat{j})$ (b) $(\hat{i} + \hat{j})$
(c) $\frac{2}{3}(\hat{i} + \hat{j})$ (d) $\frac{4}{3}(\hat{i} + \hat{j})$

91 Two particles of mass 5 kg and 10 kg respectively are attached to the two ends of a rigid rod of length 1 m with negligible mass.
The centre of mass of the system from the 5 kg particle is nearly at a distance of [NEET 2020]
(a) 50 cm (b) 67 cm
(c) 80 cm (d) 33 cm

92 An object flying in air with velocity $(20\hat{\mathbf{i}} + 25\hat{\mathbf{j}} - 12\hat{\mathbf{k}})$ suddenly breaks in two pieces whose masses are in the ratio 1 : 5. The smaller mass flies off with a velocity $(100\hat{\mathbf{i}} + 35\hat{\mathbf{j}} + 8\hat{\mathbf{k}})$. The velocity of the larger piece will be [NEET 2019]
(a) $4\hat{\mathbf{i}} + 23\hat{\mathbf{j}} - 16\hat{\mathbf{k}}$ (b) $-100\hat{\mathbf{i}} - 35\hat{\mathbf{j}} - 8\hat{\mathbf{k}}$
(c) $20\hat{\mathbf{i}} + 15\hat{\mathbf{j}} - 80\hat{\mathbf{k}}$ (d) $-20\hat{\mathbf{i}} - 15\hat{\mathbf{j}} - 80\hat{\mathbf{k}}$

93 A particle of mass $5m$ at rest suddenly breaks on its own into three fragments. Two fragments of mass m each move along mutually perpendicular direction with each speed v. The energy released during the process is [NEET 2019]
(a) $\frac{3}{5}mv^2$ (b) $\frac{5}{3}mv^2$ (c) $\frac{3}{2}mv^2$ (d) $\frac{4}{3}mv^2$

94 Body A of mass $4m$ moving with speed u collides with another body B of mass $2m$, at rest. The collision is head on and elastic in nature. After the collision, the fraction of energy lost by the colliding body A is [NEET 2019]
(a) $\frac{8}{9}$ (b) $\frac{4}{9}$ (c) $\frac{5}{9}$ (d) $\frac{1}{9}$

95 A moving block having mass m, collides with another stationary block having mass $4m$. The lighter block comes to rest after collision. When the initial velocity of the lighter block is v, the value of coefficient of restitution (e) will be [NEET 2018]
(a) 0.8 (b) 0.25
(c) 0.5 (d) 0.4

96 Two identical balls A and B having velocities of 0.5 m/s and −0.3 m/s respectively collide elastically in one dimension. The velocities of B and A after the collision respectively will be [NEET 2016]
(a) −0.5 m/s and 0.3 m/s (b) 0.5 m/s and −0.3 m/s
(c) −0.3 m/s and 0.5 m/s (d) 0.3 m/s and 0.5 m/s

97 On a frictionless surface, a block of mass M moving at speed v collides elastically with another block of same mass M which is initially at rest. After collision, the first block moves at an angle θ to its initial direction and has a speed $v/3$. The second block's speed after the collision is **[CBSE AIPMT 2015]**

(a) $\frac{2\sqrt{2}}{3}v$ (b) $\frac{3}{4}v$

(c) $\frac{3}{\sqrt{2}}v$ (d) $\frac{\sqrt{3}}{2}v$

98 A rod of weight w is supported by two parallel knife edges A and B and is in equilibrium in a horizontal position. The knives are at a distance d from each other. The centre of mass of the rod is at distance x from A. The normal reaction on A is **[CBSE AIPMT 2015]**

(a) $\frac{wx}{d}$ (b) $\frac{wd}{x}$

(c) $\frac{w(d-x)}{x}$ (d) $\frac{w(d-x)}{d}$

99 A body of mass $(4m)$ is lying in xy-plane at rest. It suddenly explodes into three pieces. Two pieces each of mass (m) move perpendicular to each other with equal speeds (v). The total kinetic energy generated due to explosion is **[CBSE AIPMT 2014]**

(a) mv^2 (b) $\frac{3}{2}mv^2$

(c) $2mv^2$ (d) $4\,mv^2$

100 A ball moving with velocity 2 ms^{-1} collides head on with another stationary ball of double the mass. If the coefficient of restitution is 0.5, then their velocities (in ms^{-1}) after collision will be **[CBSE AIPMT 2010]**

(a) 0,1 (b) 1, 1 (c) 1, 0.5 (d) 0, 2

ANSWER KEY

Level 1 Objective Questions

1 (c)	2 (b)	3 (d)	4 (c)	5 (c)	6 (c)	7 (b)	8 (b)	9 (c)	10 (c)
11 (b)	12 (c)	13 (b)	14 (b)	15 (d)	16 (c)	17 (b)	18 (d)	19 (b)	20 (b)
21 (a)	22 (b)	23 (a)	24 (d)						

Level 2 Objective Questions

25 (b)	26 (b)	27 (a)	28 (b)	29 (a)	30 (a)	31 (b)	32 (a)	33 (b)	34 (c)
35 (d)	36 (d)	37 (d)	38 (a)	39 (a)	40 (a)	41 (c)	42 (c)	43 (b)	44 (d)
45 (a)	46 (a)	47 (a)	48 (a)	49 (a)	50 (c)	51 (b)	52 (c)	53 (b)	54 (c)
55 (b)	56 (d)	57 (a)	58 (d)	59 (b)	60 (d)	61 (a)	62 (b)	63 (b)	64 (b)
65 (d)	66 (b)	67 (c)	68 (d)	69 (c)	70 (b)	71 (a)	72 (d)	73 (a)	74 (a)
75 (c)	76 (b)	77 (b)	78 (a)	79 (b)	80 (b)	81 (b)	82 (c)	83 (d)	84 (b)
85 (d)	86 (a)	87 (b)	88 (b)						

Level 3 Previous Years Questions

89 (d)	90 (b)	91 (b)	92 (a)	93 (d)	94 (a)	95 (b)	96 (c)	97 (a)	98 (d)
99 (b)	100 (a)								

Solutions

LEVEL 01 Objective Questions

1. Let $\mathbf{p}_3$ be the momentum of third part.

Applying conservation of linear momentum,

$$\mathbf{p}_{final} = \mathbf{p}_{initial}$$

$$-2p\hat{\mathbf{i}} + p\hat{\mathbf{j}} + \mathbf{p}_3 = 0$$

$$\Rightarrow \quad \mathbf{p}_3 = 2p\hat{\mathbf{i}} - p\hat{\mathbf{j}}$$

$$\Rightarrow \quad p_3 = p\sqrt{5}$$

2. Force acting upward in rocket,

$$F = \frac{dp}{dt} = \frac{dmv}{dt}$$

$$= 5 \times 400 = 2000 \text{ N}$$

Gravitational force acting downward $= mg$

$$= 100 \times 10 = 1000 \text{ N}$$

So, acceleration of rocket

$$= \frac{2000 - 1000}{100}$$

$$= 10 \text{ m/s}^2$$

3. Let n be the total number of bullets fired.

So, $p = nmv$

Now, $F = \frac{dp}{dt} = \frac{d(nmv)}{dt}$

$$\frac{dn}{dt} = \frac{F}{mv}$$

$$= \frac{144}{40 \times 10^{-3} \times 1200} = 3$$

So, the number of bullets fired per second = 3

4. Force must be slightly up than the zero line, which shows the gravity force is acting constantly. After the collision, both the bodies will experience an unexpected force, which is not linear.

5. $\mathbf{J} = \int F \cdot dt =$ Area under the F-t curve

In situations III and IV, maximum impulse is depicted.

6. Acceleration of a rocket at any instant,

$$a = \frac{v_r dm/dt}{m - t\frac{dm}{dt}} = \frac{5000 \text{ m/s} \times 10 \text{ kg/s}}{1500 - 50 \times 10}$$

$$= 50 \text{ m/s}^2$$

7. $\mathbf{J} = \Delta I = 40\,(\mathbf{v}_2 - \mathbf{v}_1)$

8. We know after the collision, the horizontal component remains unchanged but vertical component changes from upwards to downwards.

So, downward impulse is applied to A.

9. $X = \frac{3m\,(a) + 3m\,(3a)}{6m} = 2a$

$$Y = \frac{5m\,(a) + m\,(3a)}{6m} = \frac{4}{3}a$$

So, the position of centre of mass is $2\,a\hat{\mathbf{i}} + \frac{4a}{3}\hat{\mathbf{j}}$.

10. From momentum conservation,

Collisions are elastic and no friction is present. When block I strikes block II, they simply exchange velocities. So, when all 100 collisions are complete, we have 99 blocks at rest and 100 blocks are moving with speed v.

11. Impulse $= p_f - p_i = (-mv) - (mv)$

$$= -2\,mv = -0.5 \frac{\text{kg-m}}{\text{s}}$$

12. Let V and v be the recoiling speed of gun and velocity of the ball, respectively.

From conservation of linear momentum,

$$mv - MV = 0 \Rightarrow V = \frac{mv}{M} \quad \text{...(i)}$$

Now, from energy conservation,

$$\frac{1}{2}kd^2 = \frac{1}{2}MV^2 \quad \text{...(ii)}$$

From Eqs. (i) and (ii), we get

$$\frac{1}{2}kd^2 = \frac{1}{2}M\left(\frac{mv}{M}\right)^2$$

$$kd^2 = M\frac{m^2v^2}{M^2}$$

$$kd^2 = \frac{m^2v^2}{M}$$

$$\Rightarrow \quad v^2 = \frac{d^2}{m^2}kM$$

$$\Rightarrow \quad v = \frac{d}{m}\sqrt{kM}$$

13. Given, $\mathbf{p} = at - bt^2$

$$\mathbf{F} = \frac{dp}{dt} = 0 \Rightarrow a - 2bt = 0 \Rightarrow t = \frac{a}{2b}$$

14. If we consider a system of two particles in which its linear momentum is zero, it does not imply that momentum of each particle is zero.

Hence, statement (i) does not imply (ii).

Now, if the kinetic energy of a system is zero, then the kinetic energy and therefore momentum of each particle will be zero.

Thus, statement (i) implies (ii).

15. On block A, force of friction is in the opposite direction from block B. So, the net force on the system is zero. Hence, total momentum of the system remains conserved.

Therefore, the increase in momentum of B must be equal to the decrease in the momentum of A.

16. $\mathbf{J} = \mathbf{F} \cdot dt \Rightarrow \frac{d\mathbf{J}}{dt} = \mathbf{F}$

17. $p_{\text{initial}} = p_{\text{final}}$

$$m_1v_1 + m_2v_2 = (m_1 + m_2)v$$
$$20 \times 10 + 5 \times 0 = (20 + 5)v$$
$$v = 8 \text{ m/s}$$

KE of composite mass

$$= \frac{1}{2}(20 + 5)(8)^2 = 800 \text{ J}$$

18. Impulse $= p_f - p_i$

$$I_A = mv - (-mv) = 2mv$$
$$I_B = mv$$
$$I_C = mv - \frac{mv}{2} = \frac{mv}{2}$$

Hence, $I_A > I_B > I_C$

19. $m_1r_1 = m_2r_2$

So, the centre of mass of a particle is inversely proportional to the distance.

20.

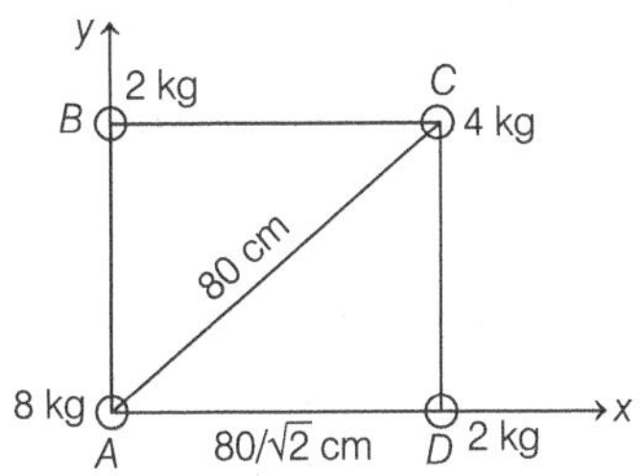

$$x_{CM} = \frac{0 + 0 + 4\left(\frac{80}{\sqrt{2}}\right) + 2\left(\frac{80}{\sqrt{2}}\right)}{16}$$
$$= 15\sqrt{2} \text{ cm}$$
$$y_{CM} = \frac{0 + 2\left(\frac{80}{\sqrt{2}}\right) + 4\left(\frac{80}{\sqrt{2}}\right) + 0}{16}$$
$$= 15\sqrt{2} \text{ cm}$$

Distance of CM from point A

$$= \sqrt{x_{CM}^2 + y_{CM}^2} = 30 \text{ cm}$$

21. $$x_{CM} = \frac{l}{2}$$
$$y_{CM} = \frac{2l(m) + 3l/2(m+m) + l/2(m+m) + l(m)}{6m}$$
$$y = \frac{7l}{6}$$

22. Let plank moves x towards left. Then,

$$90\,(12 - x) = (20)\,(x) + (40)\,(12 + x)$$

Solving this equation, we get

$$x = 4 \text{ m}$$

23. $$v_{CM} = \frac{m_1v_1 + m_2v_2}{m_1 + m_2} = \frac{0 + m_2 \times \left(\frac{F}{m_2} \times t\right)}{m_1 + m_2}$$
$$= \frac{30}{5} = 6 \text{ m/s}$$

24. Initial velocity of particle,

$$v_1 = \tan 53° = -\frac{4}{3} \text{ m/s}$$

Final velocity of particle,

$$v_2 = \tan 37° = \frac{3}{4} \text{ m/s}$$

Impulse $= p_f - p_i$

$$= 12\left[\frac{3}{4} - \left(\frac{-4}{3}\right)\right] = 25 \text{ kg-m/s}$$

LEVEL 02 *Objective Questions*

25. We know,

$$x_{CM} = \frac{m_1x_1 + m_2x_2 + m_3x_3}{m_1 + m_2 + m_3 + M}$$
$$x_{CM_1} = \frac{40(-2) + 50(0) + 60(2)}{40 + 50 + 60 + 90}$$
$$= \frac{1}{6} \text{ m}$$

Now, x_{CM} when Anoop and Ashish exchanges their positions.

$$x_{CM_2} = \frac{60(-2) + 50(0) + 40(2)}{40 + 50 + 60 + 90}$$
$$= \frac{-1}{6} \text{m}$$

So, Ashu will shift by, $x_{CM_1} - x_{CM_2}$

$$= \frac{1}{6} - \left(-\frac{1}{6}\right) = \frac{1}{6} + \frac{1}{6} = \frac{2}{6}$$
$$= \frac{1}{3} \text{m towards right side}$$

26. From conservation of linear momentum,

$$mv + 0 = Mv' + \frac{mv}{2}$$
$$\Rightarrow \quad v' = \frac{mv}{2M}$$
$$\because \quad v' = \sqrt{2gh}$$
$$h = \frac{m^2v^2}{8M^2g}$$

27. From conservation of linear momentum,

$$4\hat{i} = 3\hat{j} + \mathbf{p}$$

$$\Rightarrow \quad \mathbf{p} = 4\hat{i} - 3\hat{j}$$

$$p = \sqrt{4^2 + 3^2}$$

$$= 5 \text{ kg-m/s}$$

28. $\frac{1}{2}mv^2 + \frac{1}{2}mv^2 = \frac{1}{2}kx^2$

$$\Rightarrow \quad x^2 = \frac{2mv^2}{k}$$

$$\Rightarrow \quad x = \sqrt{\frac{2mv^2}{k}}$$

29. We know,

$$= \frac{\int x\, dm}{\int dm} = \frac{\int_0^6 x(kx^4)dx}{\int_0^6 kx^4 dx} \quad [\because dm = \lambda dx]$$

$$= 5\text{m}$$

30. $x_{CM} = \dfrac{\sigma\pi(2R)^2 \times 0 - \sigma(\pi R^2)R}{\sigma\pi(2R)^2 - \sigma\pi R^2}$

$$\frac{R}{3} = \alpha R \Rightarrow \alpha = \frac{1}{3}$$

31. Let v be the initial velocity of bullet and x be the distance travelled after penetrating 40 cm.

From equation $v^2 = u^2 + 2as$,

$$\left(\frac{v}{2}\right)^2 = v^2 - 2a\,(40)$$

$$a = \frac{3v^2}{320}$$

Now the distance x,

$$0 = \left(\frac{v}{2}\right)^2 - 2\left(\frac{3v^2}{320}\right)x$$

$$x = \frac{40}{3}\text{cm}$$

32. $\mathbf{F} = \dfrac{d\mathbf{p}}{dt} = \dfrac{m(v_f - v_i)}{t}$

$$= \frac{0.1\,[v\sin 30^\circ - (-v\sin 30^\circ)]}{0.1} = 10 \text{ N}$$

33. Mass of the water hitting the wall per second

$$= \rho \times A \times v = 10^3 \times 10^{-2} \times 15$$

$$= 150 \text{ kg/s}$$

So, force exerted $= \dfrac{p_f - p_i}{t} = \dfrac{m\,(v_f - v_i)}{t}$

$$= \frac{150(15)}{0.1} = 2.25 \times 10^3 \text{N}$$

34. Area under the F-t graph = Change in momentum

At 3 s, $\quad p_f - p_i = \frac{1}{2} \times 2 \times 10 + \frac{1}{2} \times 1 \times 10$

$$p_f = 4 + 15 = 19 \text{ kg-m/s}$$

At 2 s, force is 10 N.

Now, change in momentum in interval of $t = 3$ s to 5 s,

$$\Delta p = -\frac{1}{2} \times 10 \times 1 + \frac{1}{2} \times (-10) \times 1$$

$$= -10 \text{ kg-m/s}$$

Now, change in momentum from $t = 0$ to $t = 5$ s,

$$\Delta p = \frac{1}{2} \times 3 \times 10 - \frac{1}{2} \times 2 \times 10$$

$$= \frac{1}{2} \times 10 = 5 \frac{\text{kg-m}}{\text{s}}$$

Hence, the option (c) is incorrect.

35. Impulse = Change in momentum

$$= m\Delta v = 2(v_{t=2} - v_{t=0})$$

$$= 2\,(4\hat{i} - 2\hat{j} - 0) = (8\hat{i} - 4\hat{j}) \text{ kg-m/s}$$

36. Centre of mass of a system remains unchanged in the horizontal direction.

Hence, $(M + m)x = mL\cos\theta$

$$\Rightarrow \quad x = \frac{mL\cos\theta}{m + M}$$

37. Velocity of mass having 200 g is,

$$v_1 = u + at$$

$$= 0 + 10\,(1) = 10 \text{ m/s}$$

Similarly, the velocity of mass 100 g is,

$$v_2 = u - gt$$

$$= 100 - 10\,(1) = 90 \text{ m/s}$$

Velocity of centre of mass,

$$v_{CM} = \frac{m_1 v_1 + m_2 v_2}{m_1 + m_2}$$

$$v_{CM} = \frac{200 \times (-10) + 100 \times 90}{300}$$

$$= \frac{70}{3} \text{ m/s} \quad \text{(upwards)}$$

38. From conservation of linear momentum,

$$mv = 2mv_1$$

$$v_1 = \frac{v}{2}$$

So, $e = \dfrac{\text{velocity of separation}}{\text{velocity of approach}} = \dfrac{v_1 - 0}{v} = \dfrac{1}{2}$

39. Velocity of ball just before colliding,

$$v = \sqrt{2gh}$$

Now, $e = 0.5 = \dfrac{\text{velocity of separation}}{\text{velocity of approach}}$

$$0.5 = \frac{v'}{\sqrt{2gh}}$$

$$v' = 0.5\sqrt{2gh}$$

Now, h' be the height attained after the collision.

So, $$h' = \frac{v'^2}{2g} = \frac{(0.5\sqrt{2gh})^2}{2g}$$

$$h' = (0.5)^2 h \quad ...(i)$$

Similarly, the height attained after the 2nd collision,

$$h'' = (0.5)^2 h' \quad ...(ii)$$

From Eqs. (i) and (ii), we get

$$h'' = (0.5)^2 (0.5)^2 h$$

We get,

$$h'' = (0.5)^4 h = (0.5)^4 \times 32$$

$$h'' = 2\text{m}$$

40. $$\Delta v_x = v_{x_2} - v_{x_1} \quad ...(i)$$

$$\Delta v_y = v_{y_2} + v_{y_1} \quad ...(ii)$$

From momentum conservation,

$$4v_{x_1} - 9v_{y_1} = 4v_{x_2} + 9v_{y_2} \quad ...(iii)$$

Solving Eqs. (i), (ii) and (iii), we get

$$\frac{\Delta v_x}{\Delta v_y} = \frac{9}{4}$$

41. Force due to bird $= ma = 1$ N

Since, we know for every action, there is an equal and opposite reaction.

So, reaction force of 1 N will act in downward direction.

The weight recorded by spring balance

$$= 24.5 + 1 = 25.5 \text{ N}$$

42. $$e = \frac{\text{Relative velocity of separation}}{\text{Relative velocity of approach}}$$

$$= \frac{1.4 - (-0.6)}{1.8 - (-0.2)} = 1$$

$e = 1$, so the collision is perfectly elastic.

43. Let h be the height to which composite mass will rise.

Velocity of masses M and 2M $= \sqrt{2gH}$

From momentum conservation,

$$Mv - 2Mv = (M + 2M)\, v'$$

$$\Rightarrow \quad v' = -\frac{v}{3}$$

Now, from energy conservation,

$$\frac{1}{2} M v'^2 = Mgh$$

$$\frac{1}{2} M \left(-\frac{v}{3}\right)^2 = Mgh$$

$$\frac{1}{2} M \left(\frac{v^2}{9}\right) = Mgh$$

$$\frac{1}{2} M \frac{2gH}{9} = Mgh$$

$$\therefore \quad h = \frac{H}{9}$$

44. $$X = \frac{A_1 X_1 - A_2 X_2}{A_1 - A_2}$$

$$= \frac{(4a^2)(0) - (a^2)\left(\frac{a}{2}\right)}{4a^2 - a^2} = -\frac{a}{6}$$

45. $$F = v\frac{dm}{dt} = 7 \times 10^{-2}\left(\frac{8 \times 10^{-3}}{5.6}\right) = 10^{-4}\text{ N}$$

46. Momentum of each bullet $= 0.01 \times 500 = 5$ kg-m/s

Force on car $= 5 \times 10 = 50$ N

So, acceleration of car

$$= \frac{\text{force}}{\text{mass of car}} = \frac{50}{2000} = 0.025 \text{ m/s}^2$$

$$= 2.5 \text{ cm/s}^2$$

47. From momentum conservation,

$$60(-1 + v) = -100v$$

$$\Rightarrow \quad v = \frac{3}{8}$$

Now, when Ashish jumps to left and Anoop jumps to the right.

$$(80 + 20) \times \frac{3}{8} = 80(1 + v) + 20v$$

$$\Rightarrow \quad v = -\frac{17}{40} \text{ m/s}$$

48. $$x_1 = at + \frac{b}{2}t^2$$

$$x_2 = 0$$

$$x_{CM} = \frac{1}{2}\left(at + \frac{bt^2}{2}\right)$$

Similarly, $y_1 = at + \dfrac{bt^2}{2}$

$$y_{CM} = \frac{1}{2}\left(at + \frac{bt^2}{2}\right)$$

Since, $x_{CM} = y_{CM}$

So, centre of mass moves along a straight line.

49. From conservation of linear momentum,

$$m_a v_a + m_b v_b = m_b v_b' + m_a v$$

$$0.6 \times 4 - 0.4 \times 2 = 0.4 \times 3 + 0.6v$$

$$v = \frac{2}{3}$$

Now, $$e = \frac{\text{velocity of separation}}{\text{velocity of approach}} = \frac{3 - \frac{2}{3}}{6} = \frac{7}{18}$$

50. The coordinates of centre of mass for individual cubes x, y, z are

Cube 1 $\frac{1}{2},\frac{1}{2},\frac{1}{2}$, **Cube 2** $\frac{3}{2},\frac{1}{2},\frac{1}{2}$

Cube 3 $\frac{1}{2},\frac{3}{2},\frac{1}{2}$ **Cube 4** $\frac{1}{2},\frac{1}{2},\frac{3}{2}$

Now, $x_{CM} = \frac{mx_1 + mx_2 + mx_3 + mx_4}{4m}$

$= \frac{1/2 + 3/2 + 1/2 + 1/2}{4} = \frac{3}{4}$

Similarly, we get

$$y_{CM} = \frac{3}{4},\ z_{CM} = \frac{3}{4}$$

So, centre of mass x,y,z are $\frac{3}{4},\frac{3}{4},\frac{3}{4}$.

51. Position - time graph gives velocity as slope.

Initial velocity, $v_1 = \frac{2}{2} = 1$ m/s

Final velocity, $v_2 = \frac{-2}{2} = -1$ m/s

Impulse = Change in momentum

$= 0.2\,(2) = 0.4$ N-s

52.

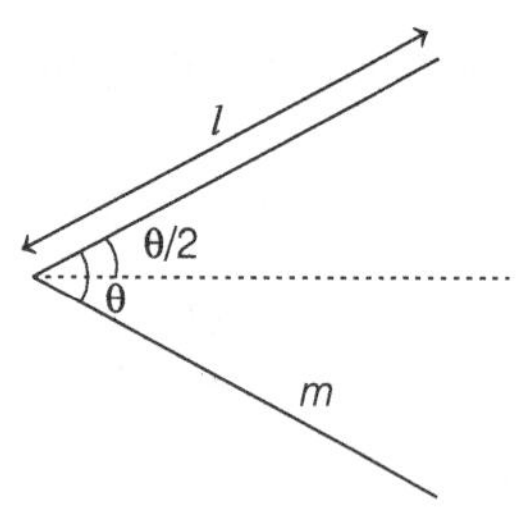

$$x_{CM} = \frac{\frac{l}{2}\cos\frac{\theta}{2}\,(m) + \frac{l}{2}\cos\frac{\theta}{2}\,(m)}{2\,m}$$

$$= \frac{l}{2}\cos\frac{\theta}{2}$$

$$y_{CM} = 0$$

53. Let the total mass of disc be m and m' be the mass of square plate.

$$\frac{m}{\pi R^2} = \frac{m'}{a^2}$$

$\Rightarrow$ $m' = \frac{ma^2}{\pi R^2}$ ...(i)

Now, $x_{CM} = 0$

$m'\left(\frac{a}{2}\right) + (m - m')\,x = 0$ ...(ii)

$$x = \frac{m'}{m - m'}\left(\frac{a}{2}\right)$$

From Eqs. (i) and (ii), we get

$$= \frac{ma^2/\pi R^2}{m - \frac{ma^2}{\pi R^2}} \times \frac{a}{2}$$

$$= \frac{-a^3}{2\,(\pi R^2 - a^2)}$$

So, the shift of centre of mass from the origin is $\frac{a^3}{2\,[\pi R^2 - a^2]}$.

54. We have given

Mass of disc = 100 g

Mass of bullet = 5 g

Number of bullets striking per second with the velocity, $v = 10$ m/s

So, total change in momentum when they rebound

$$= 2mv\,(10) = 20 \times 5v = 100\,v$$

$$F = \frac{\Delta p}{\Delta t} = 100\,v$$

For the disc to remain at rest,

$$100\,v = mg$$
$$100\,v = 10 \times 980$$
$$v = 98 \text{ cm/s}$$

55. $x_{CM} = \frac{2m(a/2) + m(2a)}{3m} = a$

$y_{CM} = \frac{2m(a/2) + m(-a)}{3m} = 0$

So, distance $= \sqrt{x_{CM}^2 + y_{CM}^2} = a$

56. $v_M = v_{M_m} - v_m$

$= 3 - 2 = 1$ m/s

$$v_{CM} = \frac{m_1v_1 + m_2v_2}{m_1 + m_2}$$

$$= \frac{80(-1) + 40(2)}{80 + 40} = 0$$

So, the speed of centre of mass is zero.

57. From conservation of linear momentum in x-direction,

$$mv + 0 = 0 + mv_x$$

$\Rightarrow$ $v_x = v$

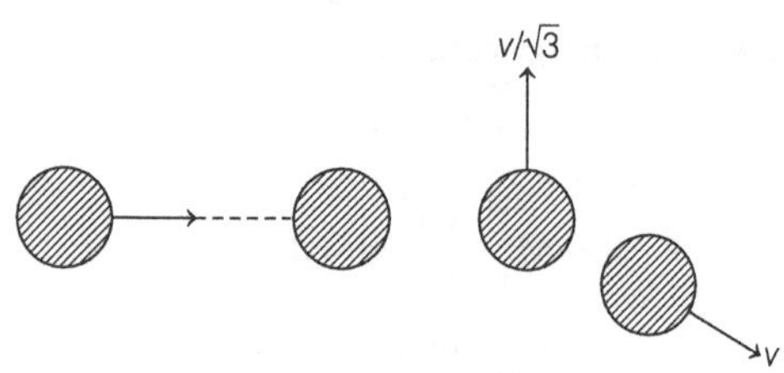

In y-direction,

$$0+0=\frac{mv}{\sqrt{3}}+mv_y$$

$$\Rightarrow \quad v_y=-\frac{v}{\sqrt{3}}$$

So, the velocity of second mass after collision

$$=\sqrt{v^2+\left(\frac{v}{\sqrt{3}}\right)^2}=\frac{2v}{\sqrt{3}}$$

58. Work done = Change in potential energy

$F\cdot s=mgh$

$$F=\frac{0.2(10)(0.2+2)}{0.2}=22\text{ N}$$

59. Velocity attained by the tennis ball when dropped from a height of 2m,

$$v=\sqrt{2gh}=\sqrt{2(9.8)(4)}=\sqrt{78.4}\text{ m/s}$$

Velocity attained by the ball when it rebounds to a height of 2 m,

$$v'^2=u'^2-2gh$$

$$v'^2=2gh$$

$$v'=\sqrt{2gh}=\sqrt{2(9.8)(2)}$$

$$=\sqrt{39.2}\text{ m/s}$$

So, average acceleration, $\frac{\Delta v}{\Delta t}=\frac{v-(-v')}{\Delta t}$

$$=\frac{\sqrt{78.4}+\sqrt{39.2}}{12\times10^{-3}}=1260\text{ m/s}^2$$

60. $v_1=\sqrt{6^2+2\times10\times3.2}=10$ m/s

$v_2=\sqrt{2\times10\times3.2}=8$ m/s

$$\therefore \quad e=\frac{v_2}{v_1}=\frac{8}{10}=0.8$$

61. Mass of remaining portion $=\left(\pi R^2-\frac{R^2}{2}\right)\sigma$

Now, $m_1x_1=m_2x_2$

$$\sigma\left(\pi R^2-\frac{R^2}{2}\right)x=\frac{R^2}{2}\sigma\frac{R}{2}$$

$$x=\frac{R}{2(2\pi-1)}=\frac{R}{(4\pi-2)}$$

62.

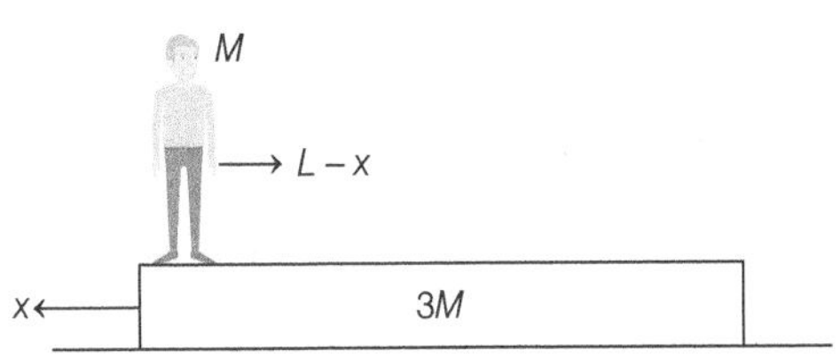

$$M(L-x)=(3M)x$$

$$\therefore \quad x=\frac{L}{4}$$

Displacement of man relative to ground

$$=L-x=\frac{3L}{4}$$

63. From momentum conservation,

$$3(2)=(3+6)v_f$$

$$\Rightarrow \quad v_f=\frac{2}{3}\text{ m/s}$$

From work-energy theorem,

$$\frac{1}{2}mv_i^2-\frac{1}{2}mv_f^2=\frac{1}{2}kx^2$$

$$\frac{1(3)}{2}(2)^2-\frac{1}{2}\times9\left(\frac{2}{3}\right)^2=\frac{1(200)}{2}x^2$$

$$x=0.2\text{ m}$$

$$\Rightarrow \quad x=0.2\times100$$

$$\Rightarrow \quad x=20\text{ cm}$$

64. Coordinates of part 1 = (0, 1 m)

Coordinates of part 2 = (0, 5 m)

Now, mass $\propto$ area

So, $x_{CM}=\frac{A_1x_1+A_2x_2}{A_1+A_2}$

$$=\frac{16(1)+12(5)}{16+12}=2.7\text{ m}$$

65. Let v_A and v_B be the velocity of A and B after the collision.

Since, collision is elastic,

$$e=1=\frac{v_B-v_A}{v_0}$$

$$v_B=v_A+v_0 \quad \text{...(i)}$$

Now, from momentum conservation,

$$mv_0=2mv_B+2mv_A$$

$$v_0=2v_B+v_A$$

$$v_0=2(v_A+v_0)+v_A \quad [\because \text{using Eq. (i)}]$$

$$3v_A+2v_0=v_0$$

$$v_A=-\frac{v_0}{3}$$

So, $v_B=\frac{2v_0}{3}$

Now, for the 2nd collision,

B makes perfectly inelastic collision with C.

So, let u be the velocity of both B and C.

Now, from momentum conservation,

$$2m\times\frac{2v_0}{3}=(2m+3m)u$$

$$\Rightarrow \quad \frac{4v_0}{3}=5u$$

$$\Rightarrow \quad u=\frac{4v_0}{15}$$

So, the final speed of C is $\frac{4v_0}{15}$.

66. FBD of 1 kg mass

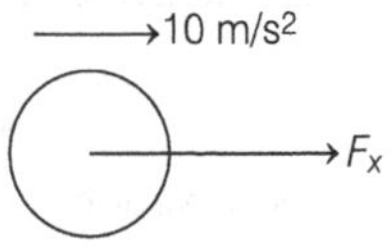

$$F_x = \text{spring balance}$$
$$F = 1 \times 10 = 10 \text{ N}$$

FBD of 2 kg mass

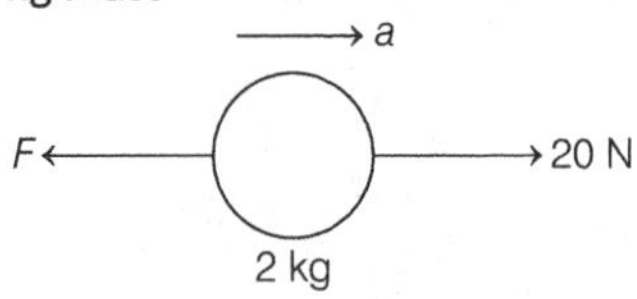

$$20 - F = ma$$
$$20 - 10 = 2a$$
$$a = 5 \text{ m/s}^2$$

So, the acceleration of the 2 kg mass is 5 m/s^2.

67.

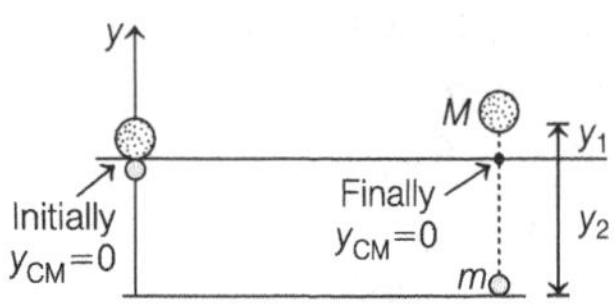

As man throws stone, force is internal to system.

Hence, $(y_{CM})_{\text{initial}} = (y_{CM})_{\text{final}}$

$\Rightarrow \quad 0 = My_1 + my_2$

$\Rightarrow \quad My_1 = -my_2$

$\Rightarrow \quad y_1 = -\dfrac{m}{M} \cdot y_2 = \dfrac{-0.5 \times (-10)}{50}$

$= 0.1 \text{ m}$

Final height of man above ground = 10 + 0.1 = 10.1 m

68. Speed of the block with respect to plank,

$$v_{bp} = v_b - v_p = 10 - (-5)$$
$$= 15 \text{ m/s}$$

Impulse, $J = p_f - p_i$

$= m(v_f - v_i) = 1[(15) - (-15)]$

$= 30 \text{ N-s}$

69. Since, no external force is acting on the system.

So, centre of mass will follow the same path.

So, $\quad \text{range} = v\cos\theta \times T$

$$= v\cos\theta \times \frac{2v\sin\theta}{g} \quad \left[\because T = \frac{2v\sin\theta}{g}\right]$$

$$= \frac{v^2 \sin 2\theta}{g}$$

70. We know, the time of flight only depends upon the vertical component. The collision of the stone with the vertical wall only changes the horizontal component.

So, time of flight in absence of wall

$= 3 + 5 = 8\text{s}$

$\therefore \quad \dfrac{2u_y}{g} = 8$

$\Rightarrow \quad u_y = 40 \text{ m/s}$

So, maximum height reached by stone

$$= \frac{u_y^2}{2g} = 80 \text{ m}$$

71. From conservation of linear momentum,

$$10 = v_A + v_B \quad \text{...(i)}$$

Now, $e = \dfrac{v_A - v_B}{10} = 0.5$

$$5 = v_A - v_B \quad \text{...(ii)}$$

From Eqs. (i) and (ii), we get

$$v_A = 7.5 \text{ m/s}$$

and $\quad v_B = 2.5 \text{ m/s}$

Now, applying momentum conservation for both B and C,

$$v_A = v_B' + v_C$$
$$7.5 = v_B' + v_C \quad \text{...(iii)}$$

Now, $e = \dfrac{v_C - v_B'}{v_A} = 0.5$

$$v_C - v_B' = 3.75 \quad \text{...(iv)}$$

From Eqs. (iii) and (iv), we get

$$v_C = 5.6 \text{ m/s}$$

72. We know in an elastic collisions, bodies exchange their velocities.

After Ist collision, $mA = 2v$, $mB = v$

Similarly, after collision, $mA = v$, $mB = 2v$

So, the total number of elastic collisions is 2.

73. Let the mass of each brick be m.

Centre of mass of each brick will be assumed to be concentrated at centre.

Now, x-coordinate of centre of 1st and 7th brick, 2nd and 6th brick, 3rd and 5th brick and 4th brick are

$$\frac{l}{2}, \frac{3l}{5}, \frac{7l}{10}, \frac{8l}{10}$$

So, the x-coordinate of centre of mass

$$= \frac{mx_1 + mx_2 + mx_3 + mx_4 + \ldots}{7m}$$

We get,

$$x_{CM} = \frac{22l}{35}$$

74. Acceleration of system, a

$$= \left(\frac{m_2 - m_1}{m_1 + m_2}\right) g = \left(\frac{2-1}{1+2}\right) g = \frac{g}{3}$$

Now, acceleration of centre of mass,

$$a_{CM} = \frac{m_1 a_1 + m_2 a_2}{m_1 + m_2}$$

$$= \frac{1(-a)+2a}{1+2} = \frac{a}{3} = \frac{g}{9}$$

So, the distance travelled by centre of mass in 2 s,

$$s = \frac{1}{2}at^2 = \frac{1}{2} \times \frac{10}{9}(2)^2 = \frac{20}{9} \text{m}$$

75. Between B and C

$$v'_{2m} = \left(\frac{2m-m}{3m}\right)(0) + \frac{2\times m}{3m}(u) = \frac{2}{3}u$$

$$v'_m = \left(\frac{m-2m}{3m}\right)(u) + \frac{2\times 2m}{3m}(0)$$

$$= -\frac{u}{3}$$

Now, C and D will interchange their velocities and similarly, B and A will also interchange their velocities.

So, after three collisions :

(i) Between B and C

(ii) Between C and D

(iii) Between B and A

Situation is as shown below

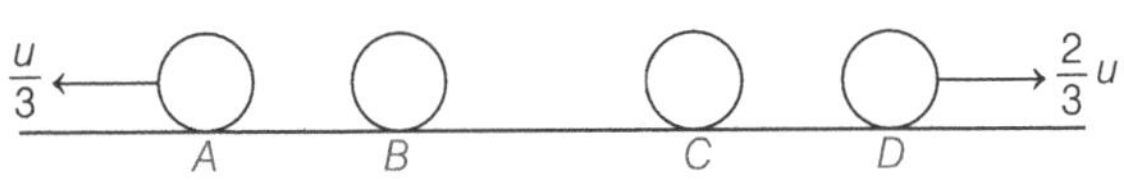

So, total collisions are three.

76. Initial elongation of block,

$$mg = kx_0 \Rightarrow x_0 = \frac{mg}{k} \quad \text{... (i)}$$

From momentum conservation,

$$mv = 2mv_1$$

$$\Rightarrow v_1 = \frac{v}{2} \quad \text{... (ii)}$$

Now, from conservation of energy,

$$\frac{1}{2}(2m)v_1^2 + \frac{1}{2}kx_0^2 = 2mgx_0 \quad \text{... (iii)}$$

From Eqns. (i), (ii) and (iii), we get

$$v = \sqrt{\frac{6mg^2}{k}}$$

77. We know in a collision, the identical balls will exchange their velocities. So, outcome 2 is physically possible in a collision.

78. We have, $u_x = 100 \cos 30° = 50\sqrt{3}$ m/s and

$u_y = 100 \cos 60° = 50$ m/s

Now, velocity at the ground,

$$v_y^2 = u_y^2 + 2as$$

$$v_y^2 = (50)^2 + 2(10)(375) = 2500 + 7500$$

$$v = 100 \text{ m/s}$$

After collision with ground, velocity

$$= ev = \frac{100}{2} = 50 \text{ m/s}$$

So, $\text{KE} = \frac{1}{2}(1)(50)^2 + \frac{1}{2}(1)(50\sqrt{3})^2$

$$= 5000 \text{ J}$$

79. $p_2 = \frac{1}{2}p_1$

$$\frac{1\times 8}{2} = mv_2 \Rightarrow v_2 = \frac{8}{2} = 4 \text{ m/s}$$

From conservation of mechanical energy, we get

$$\frac{1}{2}mv_1^2 = \frac{1}{2}mv_2^2 + \frac{1}{2}kx^2$$

$$k = \frac{16}{3} \text{ N/m}$$

80. From conservation of linear momentum,

$$mu = (m+2m)v \Rightarrow v = \frac{u}{3}$$

Now, from the energy conservation,

$$\frac{1}{2}mu^2 = \frac{1}{2}(m+2m)v^2 + mgh$$

$$\frac{1}{2}mu^2 = \frac{1}{2}(3m)\left(\frac{u}{3}\right)^2 + mgh$$

$$\Rightarrow u = \sqrt{3gh}$$

81. $x_{CM} = \frac{\Sigma m_i x_i}{\Sigma m} = \frac{m\frac{L}{2} + \frac{m}{2}\frac{L}{4} + \dots\dots\dots}{m + m/2 + m/4 + \dots\dots\dots}$

$$= \frac{m}{m}\frac{(L/2 + L/8 + \frac{L}{32} + \dots\dots\dots)}{(1 + 1/2 + 1/4 + \dots\dots\dots)}$$

Now, we know sum of infinite GP

$$= \frac{a}{1-r} = \frac{L\left(\frac{1/2}{1-1/4}\right)}{1-\frac{1}{2}} = \frac{4L}{3}$$

82. Since, we know after the collision, balls exchange their velocities,

$$v_A = \sqrt{2gh}$$

and $$v_B = \sqrt{2g(4h)} = 2\sqrt{2gh}$$

Height attained by ball $A = \frac{v_A^2}{2g} = h_A$

Now, using energy conservation for ball B,

$$\frac{1}{2}mv_B^2 = \frac{1}{2}mv^2 + mgh$$

$$\Rightarrow v = \sqrt{6gh}$$

Now, height attained,

$$h_B = h + \frac{v^2 \sin^2 60°}{2g} = \frac{13h}{4}$$

$$\therefore \frac{h_A}{h_B} = \frac{4}{13}$$

83. Let the momentum of ball after each successive collision be p_1, p_2 and p_3 and so on till the ball comes to rest and p be initial momentum.

For the 1st collision,

$$\frac{p_1 - 0}{-p - 0} = -e \Rightarrow p_1 = ep$$

So, the net momentum imparted by 1st collision

$$= p_1 - (-p) = ep + p = p(1+e)$$

Similarly, for 2nd collision,

$$\frac{p_2 - 0}{-p_1 - 0} = -e$$

$$\Rightarrow \quad p_2 = ep_1 = e^2 p$$

So, net momentum imparted $= p_2 - (-p_1) = e^2 p + ep$

$$= pe(1+e)$$

Momentum imparted to the floor by each other successive collisions,

$$pe^2 (1+e), pe^3(1+e) \text{ and so on.}$$

Total momentum imparted

$$= p(1+e) + pe(1+e) + pe^2(1+e) + pe^3(1+e) \ldots \infty$$

$$= p(1+e)[1 + e + e^2 + e^3 + \ldots \infty] = \frac{p(1+e)}{1-e}$$

84. $\frac{1}{2} mv^2 = mgh$

$$\Rightarrow \quad v_2 = \sqrt{2gh}$$

$$= \sqrt{2 \times 10 \times 3.2} = 8 \text{ m/s} \quad \ldots \text{(i)}$$

Now, from momentum conservation,

$$10 \times m = mv_1 + mv_2 \quad \ldots \text{(ii)}$$

From Eqs. (i) and (ii), we get

$$\Rightarrow \quad v_1 = 2 \text{ m/s}$$

$$\therefore \quad e = \frac{v_2 - v_1}{10} \Rightarrow e = \frac{8-2}{10}$$

$$= 0.6 = \frac{3}{5}$$

85. As the hall is gravity free space and there is no external force acting on the system. So, the centre of mass will not move.

86. Let v be the velocity of man in upward direction.

From conservation of linear momentum, in vertical direction

$$m_{man} v - m_{stone} v_{stone} = 0$$

$$50 v - 0.5 v_{stone} = 0$$

$$v = \frac{0.5 \times 5}{50}$$

$$\left(\because v_{stone} = 36 \times \frac{5}{18} \times \cos 60° = 5 \text{ m/s}\right)$$

$$= 0.05 \text{ m/s}$$

Time taken by wall to reach the ground

$$= \frac{20}{5} = 4 \text{ s}$$

So, the distance transversed in 4 s

$$= vt = 0.05 \times 4 = 0.2 \text{ m}$$

So, the distance of man from floor

$$= 20 + 0.2 = 20.2 \text{ m}$$

87. Since, the collision is perfectly elastic, so all the blocks will stick together.

Now, time required to cover a distance L by first block $= \frac{L}{v}$

Now, the first and second block will move together.

So, its velocity becomes $\frac{v}{2}$.

So, combined system will take time

$$= \frac{2L}{v}$$

So, total time $= \frac{L}{v} + \frac{2L}{3v} + \ldots \frac{(n-1)L}{v} = \frac{n(n-1)L}{2v}$

88. Let the width of rectangular plate be d.

So, mass of left part of plate,

$$m = \frac{Ld\rho_1}{2}$$

Mass of right part of rectangular plate, m'

$$= \frac{Ld\rho_2}{2} = \frac{m\left(\frac{L}{4}\right) + m'\left(\frac{3L}{4}\right)}{\frac{Ld}{4}\rho_1 + \frac{Ld}{2}\rho_2} = L\frac{(\rho_1 + 3\rho_2)}{4(\rho_1 + \rho_2)}$$

LEVEL 03 Previous Years Questions

89. Given, the length of a uniform rod,

$L = 200$ cm

The wedge is placed at the mark = 40 cm

The balanced mass placed at 100 cm mark on wedge = 500 g = 0.5 kg

The mass suspended from the rod at 20 cm distance from the end, M = 2 kg

Another unknown mass suspended from the rod at 160 cm distance from the end = m

Let's draw the diagram of the uniform rod suspended with mass.

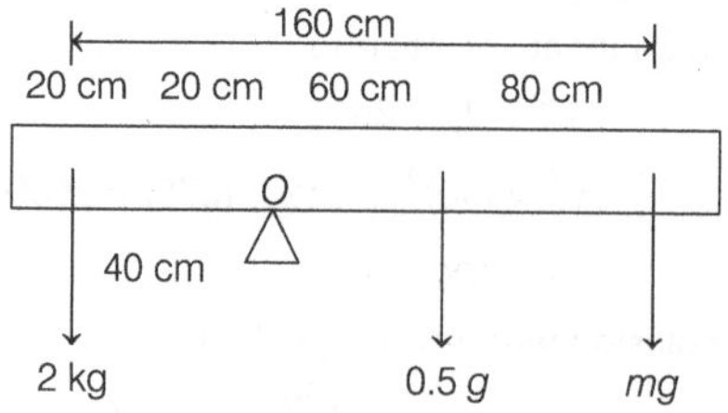

As we know in equilibrium, net moment of force is equals to zero.

$\Rightarrow 0.5\,g\,(0.60) + mg\,(1.20) - 2g\,(0.20) = 0$

$\Rightarrow 0.3 + 1.20\,m - 0.4 = 0 \Rightarrow 1.20m = 0.1$

$$\Rightarrow \quad m = \frac{1}{12}\text{ kg}$$

90. The given situation is shown in the figure.

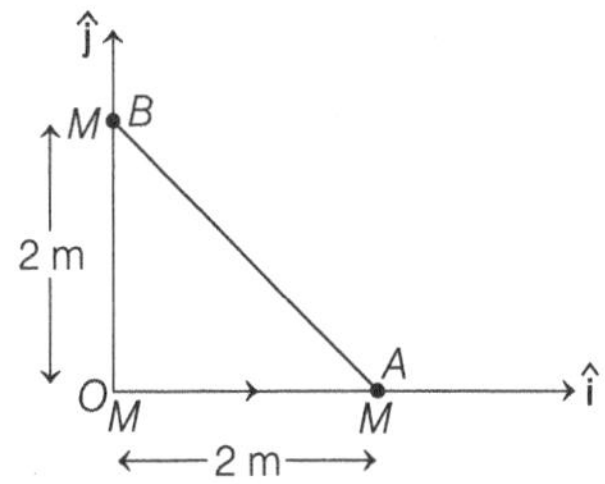

$$\mathbf{OA} = 2\hat{\mathbf{i}}$$

$$\mathbf{OB} = 2\hat{\mathbf{j}}$$

Position vector of centre of mass,

$$R_{CM} = \frac{m_1\mathbf{r}_1 + m_2\mathbf{r}_2}{M_1 + M_2} = \frac{M\,\mathbf{OA} + M\,\mathbf{OB}}{M + M}$$

$$= \frac{M \times 2\hat{\mathbf{i}} + M \times 2\hat{\mathbf{j}}}{2M} = \hat{\mathbf{i}} + \hat{\mathbf{j}}$$

91. Given, $m_1 = 5$ kg, $m_2 = 10$ kg and $r = 1$ m $= 100$ cm

Let the centre of mass lies at origin O.

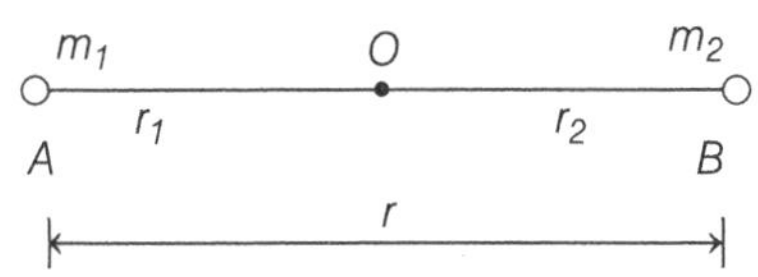

$$\therefore \quad \frac{m_1r_1 - m_2r_2}{m_1 + m_2} = 0$$

$$\Rightarrow \quad 5r_1 - 10r_2 = 0$$

$$\Rightarrow \quad r_2 = \frac{r_1}{2}$$

Also, $\quad r_1 + r_2 = 100$

$$\Rightarrow \quad r_1 + \frac{r_1}{2} = 100$$

$$\Rightarrow \quad 3r_1 = 200$$

$$\Rightarrow \quad r_1 = \frac{200}{3} \simeq 67\text{ cm}$$

92. Let m be the mass of an object flying with velocity v in air. When it splits into two pieces of masses in ratio 1 : 5, the mass of smaller piece is $m/6$ and of bigger piece is $\frac{5m}{6}$.

This situation can be interpreted diagrammatically as below

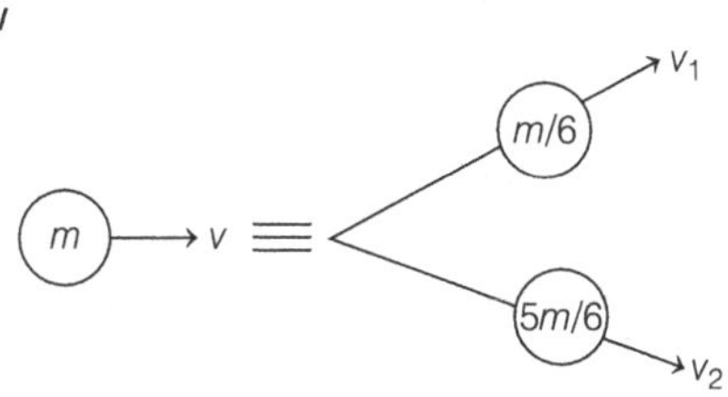

As the object breaks in two pieces, so the momentum of the system will remain conserved i.e. the total momentum (before breaking) = total momentum (after breaking)

$$m\mathbf{v} = \frac{m}{6}\mathbf{v}_1 + \frac{5m}{6}\mathbf{v}_2$$

$$\Rightarrow \quad \mathbf{v} = \frac{\mathbf{v}_1}{6} + \frac{5\mathbf{v}_2}{6}$$

Here, $\quad \mathbf{v} = 20\hat{\mathbf{i}} + 25\hat{\mathbf{j}} - 12\hat{\mathbf{k}}$

$$\mathbf{v}_1 = 100\hat{\mathbf{i}} + 35\hat{\mathbf{j}} + 8\hat{\mathbf{k}}$$

$$\Rightarrow \quad 20\hat{\mathbf{i}} + 25\hat{\mathbf{j}} - 12\hat{\mathbf{k}} = \frac{(100\hat{\mathbf{i}} + 35\hat{\mathbf{j}} + 8\hat{\mathbf{k}})}{6} + \frac{5\mathbf{v}_2}{6}$$

$$\Rightarrow \quad (120\hat{\mathbf{i}} + 150\hat{\mathbf{j}} - 72\hat{\mathbf{k}}) = (100\hat{\mathbf{i}} + 35\hat{\mathbf{j}} + 8\hat{\mathbf{k}}) + 5\mathbf{v}_2$$

$$\Rightarrow \quad \mathbf{v}_2 = \frac{1}{5}(20\hat{\mathbf{i}} + 115\hat{\mathbf{j}} - 80\hat{\mathbf{k}})$$

$$= 4\hat{\mathbf{i}} + 23\hat{\mathbf{j}} - 16\hat{\mathbf{k}}$$

93. The particle of mass $5m$ breaks in three fragments of mass m, m and $3m$ respectively. Two fragments of mass m each, move in perpendicular direction with velocity v and the left fragment will move in a direction with velocity $\mathbf{v}'$ such that the total momentum of the system must remain conserved.

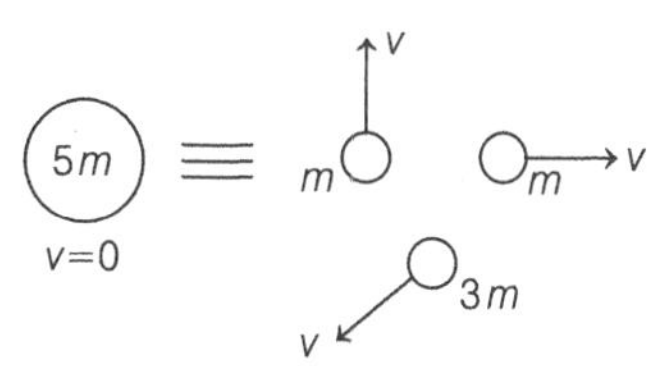

By law of conservation of momentum,

$$5m \times 0 = mv\hat{\mathbf{i}} + mv\hat{\mathbf{j}} + 3m\mathbf{v}'$$

$$\Rightarrow \quad \mathbf{v}' = -\frac{v}{3}\hat{\mathbf{i}} - \frac{v}{3}\hat{\mathbf{j}}$$

$$\therefore \quad |\mathbf{v}'| = \sqrt{\left(-\frac{v}{3}\right)^2 + \left(-\frac{v}{3}\right)^2} = \frac{v\sqrt{2}}{3}$$

$\therefore$ Energy released

$$E = \frac{1}{2}mv^2 + \frac{1}{2}mv^2 + \frac{1}{2} \times 3m\left(\frac{v\sqrt{2}}{3}\right)^2$$

$$= mv^2 + \frac{mv^2}{3} = \frac{4}{3}mv^2$$

94. In head-on elastic collision, momentum and kinetic energy before and after the collision is conserved. The given situation of collision can be drawn as

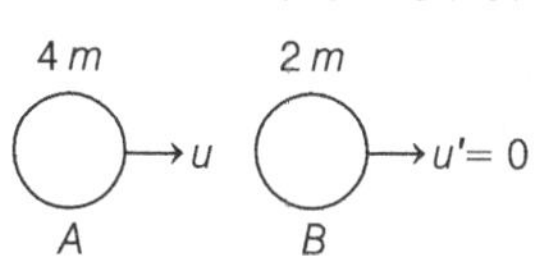

Before collision

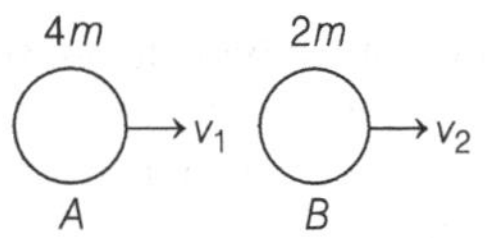

After collision

Applying conservation of linear momentum,
Initial momentum of system = Final momentum of system

$$\Rightarrow \quad (4m)u + (2m)u' = (4m)v_1 + (2m)v_2$$

$$4mu = 4mv_1 + 2mv_2$$

or $$2u = 2v_1 + v_2 \quad \text{...(i)}$$

As, $e = 1$,

So, $v_2 - v_1 = u_1 - u_2$

$$\Rightarrow \quad v_2 - v_1 = u \quad \text{...(ii)}$$

Now, using Eqs. (i) and (ii), we have

$$2u = 2v_1 + v_2 \quad \text{...(iii)}$$

$$-2u = 2v_1 - 2v_2 \quad \text{...(iv)}$$

Subtracting Eq. (iv) from Eq. (iii) we get

$$4u = 3v_2, v_2 = \frac{4}{3}u$$

So, $v_1 = v_2 - u = \frac{4}{3}u - u$

or $$v_1 = \frac{1}{3}u$$

Now, fractional loss of kinetic energy of A

$$= \frac{\Delta K}{K} = \frac{\frac{1}{2}m_1\,(u_1^2 - v_1^2)}{\frac{1}{2}m_1\,u_1^2}$$

or $$\frac{\Delta K}{K} = 1 - \left(\frac{v_1}{u_1}\right)^2 = 1 - \frac{1}{9} = \frac{8}{9} \qquad [\because u_1 = u]$$

95. Since, the collision mentioned is an elastic head-on collision. Thus, according to the law of conservation of linear momentum, we get

$$m_1u_1 + m_2u_2 = m_1v_1 + m_2v_2$$

where, m_1 and m_2 are the masses of the two blocks, respectively and u_1 and u_2 are their initial velocities and v_1 and v_2 are their final velocities, respectively.

Here, $m_1 = m, m_2 = 4m$

$$u_1 = v, u_2 = 0 \text{ and } v_1 = 0$$

$$mv + 4m \times 0 = 0 + 4mv_2$$

$$\Rightarrow \quad mv = 4mv_2 \text{ or } v_2 = \frac{v}{4} \quad \text{...(i)}$$

As, the coefficient of restitution is given as,

$$e = \frac{\text{Relative velocity of separation after collision}}{\text{Relative velocity of approach}}$$

$$= \frac{v_2 - v_1}{u_2 - u_1} = \frac{\frac{v}{4} - 0}{0 - v} \qquad \text{[from Eq. (i)]}$$

$$= \frac{1}{4}$$

$\therefore \quad e = 0.25$

96. In elastic collision, kinetic energy of the system remains unchanged and momentum is also conserved.
It is given that mass of balls are same and collision is perfectly elastic ($e = 1$), so their velocities will be interchanged.

Thus, $$v'_A = v_B = -0.3 \text{ m/s},$$
$$v'_B = v_A = 0.5 \text{ m/s}$$

97.

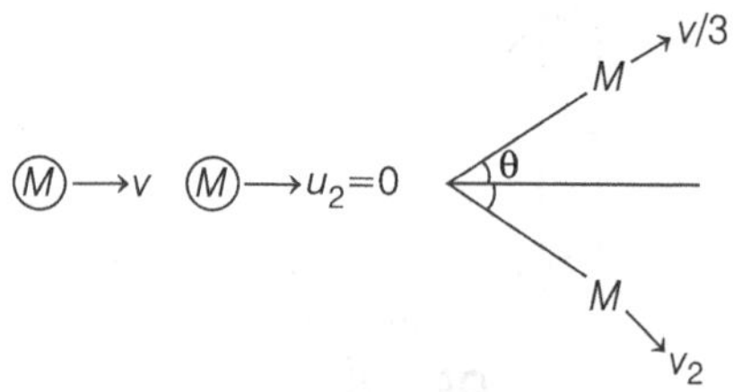

According to law of conservation of kinetic energy, we have

$$\frac{1}{2}Mv^2 + 0 = \frac{1}{2}M\left(\frac{v}{3}\right)^2 + \frac{1}{2}Mv_2^2$$

$$\Rightarrow \quad v^2 = \frac{v^2}{9} + v_2^2$$

$$\Rightarrow \quad v^2 - \frac{v^2}{9} = v_2^2$$

$$\Rightarrow \quad \frac{8v^2}{9} = v_2^2$$

$\therefore$ Velocity of second block after collision,

$$v_2 = \frac{2\sqrt{2}}{3}v$$

98. As the weight w balances the normal reactions.

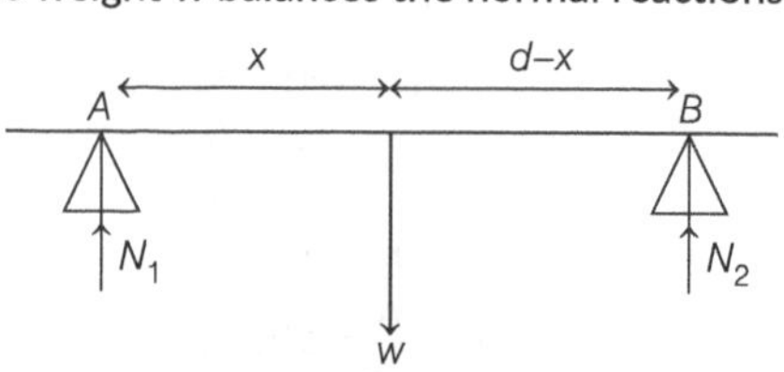

So, $$w = N_1 + N_2 \quad \text{...(i)}$$

Now balancing torque about the centre of mass, i.e. anti-clockwise momentum
= clockwise momentum

$$\Rightarrow \quad N_1x = N_2(d - x)$$

Putting the value of N_2 from equation (i), we get

$$N_1x = (w - N_1)(d - x)$$

$$\Rightarrow \quad N_1x = wd - wx - N_1d + N_1x$$

$$\Rightarrow \quad N_1d = w(d - x)$$

$$\Rightarrow \quad N_1 = \frac{w(d - x)}{d}$$

99. **Problem Solving Strategy** Conserve the momentum of third mass with the resultant momentum of 1st and 2nd masses. After getting velocity of third mass, calculate total kinetic energy.

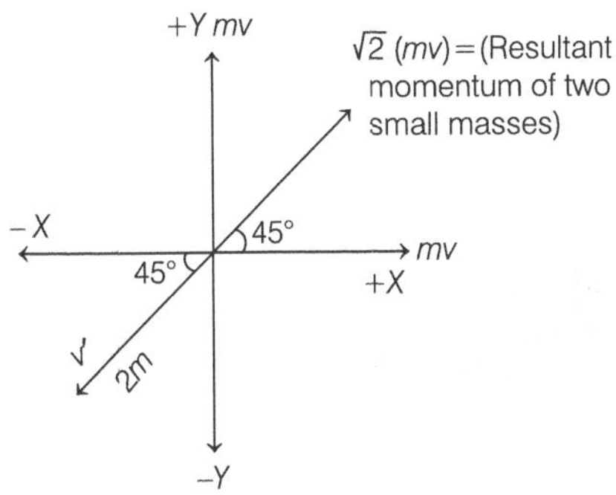

According to question, the third part of mass $2m$ will move as shown in the figure, because the total momentum of the system after explosion must remain zero. Let the velocity of third part be v'.

From the conservation of momentum

$$\sqrt{2}\,(mv) = (2m) \times v'$$

$$\Rightarrow \quad v' = \frac{v}{\sqrt{2}}$$

So, total kinetic energy generated by the explosion

$$= \frac{1}{2}mv^2 + \frac{1}{2}mv^2 + \frac{1}{2}(2m)\,v'^2$$

$$= mv^2 + m \times \left(\frac{v}{\sqrt{2}}\right)^2$$

$$= mv^2 + \frac{mv^2}{2}$$

$$= \frac{3}{2}mv^2$$

100. If two bodies collide head on with coefficient of restitution

$$e = \frac{v_2 - v_1}{u_1 - u_2} \qquad \text{...(i)}$$

From, the law of conservation of linear momentum

$$m_1u_1 + m_2u_2 = m_1v_1 + m_2v_2$$

$$\Rightarrow \quad v_1 = \left[\frac{m_1 - em_2}{m_1 + m_2}\right]u_1 + \left[\frac{(1+e)\,m_2}{m_1 + m_2}\right]u_2$$

Substituting $u_1 = 2\text{ ms}^{-1}$, $u_2 = 0$, $m_1 = m$ and $m_2 = 2m$, $e = 0.5$

we get, $v_1 = \left[\frac{m - m}{m + 2m}\right] \times 2$

$$\Rightarrow \quad v_1 = 0$$

Similarly, $v_2 = \left[\frac{(1+e)\,m_1}{m_1 + m_2}\right]u_1 + \left[\frac{m_2 - em_1}{m_1 + m_2}\right]u_2$

$$= \left[\frac{1.5 \times m}{3m}\right] \times 2 = 1\text{ ms}^{-1}$$

Chapter 10

ROTATIONAL MOTION

LEVEL 01 Objective Questions

1 What is the value of linear velocity, if $\omega = 3\hat{i} - 4\hat{j} + \hat{k}$ and $\mathbf{r} = 5\hat{i} - 6\hat{j} + 6\hat{k}$?

(a) $18\hat{i} + 13\hat{j} + 2\hat{k}$ (b) $-18\hat{i} - 13\hat{j} + 2\hat{k}$ (c) $4\hat{i} - 13\hat{j} + 6\hat{k}$ (d) $6\hat{i} - 2\hat{j} + 3\hat{k}$

2 A wheel is subjected to uniform angular acceleration about its axis. Initially its angular velocity is zero. In the first 2 s, it rotates through an angle θ_1, in the next 2 s, it rotates through an angle θ_2. The ratio of $\frac{\theta_2}{\theta_1}$ is

(a) 1 (b) 2 (c) 3 (d) 5

3 A uniform cube of side a and mass m rests on a rough horizontal table. A horizontal force F is applied normal to one of the faces at a point directly above the centre of face, at a height $\frac{3a}{4}$ above the base. The minimum value of F for which the cube begins to tip about an edge is

(a) $\frac{2}{3}mg$ (b) $\frac{1}{3}mg$ (c) $\frac{3}{4}mg$ (d) mg

4 Three parallel forces are acting on a rod at distances of 40 cm, 60 cm and 80 cm respectively, from one end of the rod (as shown in the following figure). Assuming no other forces, the rod under the influence of these forces will

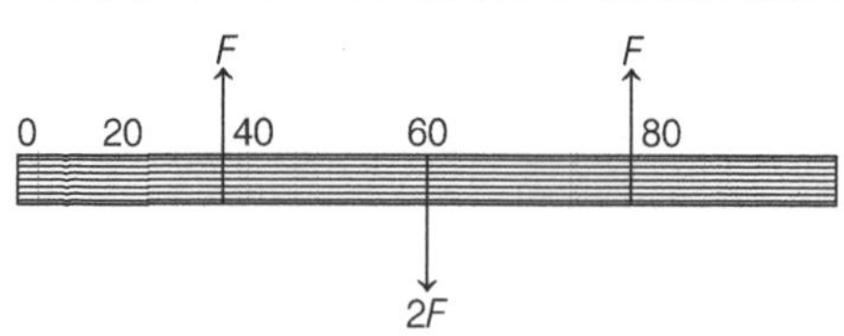

(a) be at rest
(b) exhibit rotational motion only
(c) exhibit translational motion
(d) exhibit combined rotational and translational motion

5 A circular disc A of radius r is made from an iron plate of thickness t and another circular disc B of radius $4r$ is made from an iron plate of thickness $t/4$. Find the relation between the moment of inertia I_A and I_B.

(a) $I_B = 4I_A$
(b) $I_A = 64I_B$
(c) $I_B = 64I_A$
(d) None of these

6 A rod has non-uniform mass distribution. The mass increases from left end A to right end B. C is the centre of the rod. It is easiest to rotate the rod about an axis

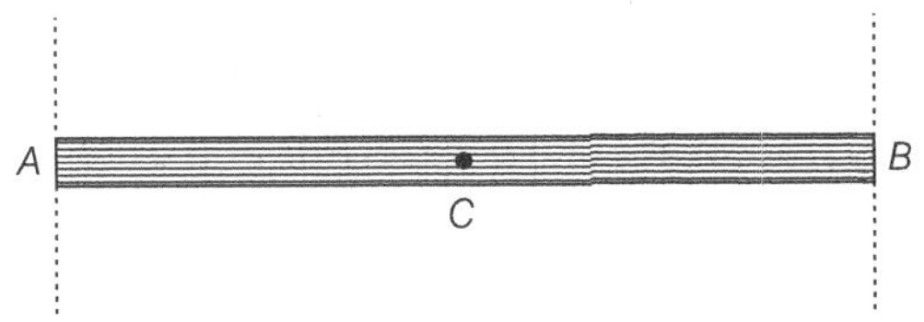

(a) passing through C
(b) passing through a point between B and C
(c) passing through a point between A and C
(d) passing through A

7 Two discs of same material and same thickness rotating about their own axis. If ratio of their radii are 1 : 3, then ratio of their moment of inertia about their axis will be

(a) 1 : 9 (b) 1 : 81 (c) 1 : 27 (d) 1 : 1

8 A spherical shell has mass one-fourth of mass of a solid sphere and both have same moment of inertia about their respective diameters. The ratio of their radii will be

(a) 1 2
(b) 2 : 1
(c) $\sqrt{3} : \sqrt{20}$
(d) $\sqrt{5} : \sqrt{20}$

9 Two rings having the same radius and mass are placed, such that their centres are at a common point and their planes are perpendicular to each other. Find the moment of inertia of the system about an axis passing through the centre and perpendicular to the plane of one of the rings. (M = mass of each ring and R = radius)

(a) MR^2 (b) $\frac{3}{2}MR^2$ (c) $2MR^2$ (d) None of these

10 Four thin rods of same mass m and same length L form a square as shown in the following figure. Moment of inertia of this system about an axis through centre O and perpendicular to its plane is

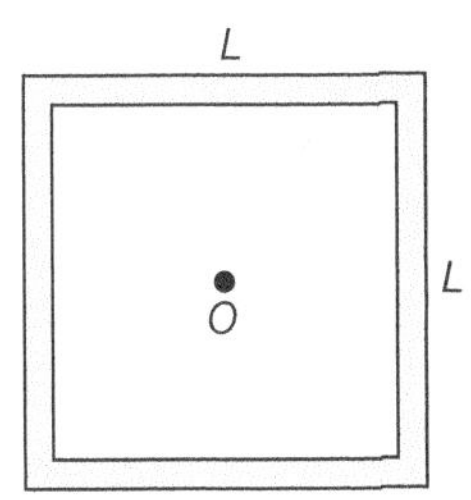

(a) $\frac{4mL^2}{3}$ (b) $\frac{mL^2}{3}$ (c) $\frac{mL^2}{6}$ (d) $\frac{2mL^2}{3}$

11 Three identical rods, each of length L, are joined to form a rigid equilateral triangle. Its radius of gyration about an axis passing through a corner and perpendicular to the plane of triangle is

(a) $\frac{L}{2}$ (b) $L\sqrt{\frac{3}{2}}$ (c) $\frac{L}{\sqrt{2}}$ (d) $\frac{L}{\sqrt{3}}$

12 We have two spheres one of which is hollow and the other is solid. They have identical masses and moment of inertia about their respective diameters. The ratio of their radius is given by

(a) 5 : 7
(b) 3 : 5
(c) $\sqrt{3} : \sqrt{5}$
(d) $\sqrt{3} : \sqrt{7}$

13 Two identical hollow spheres of mass M and radius R are joined together, and the combination is rotated about an axis tangent to one sphere and perpendicular to the line connecting them. The rotational inertia of the combination is

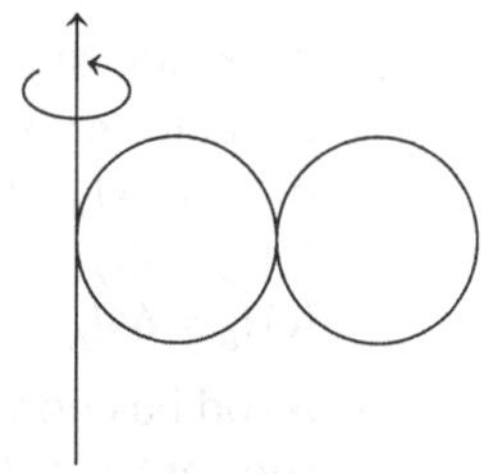

(a) $10\ MR^2$
(b) $\frac{4}{3}MR^2$
(c) $\frac{32}{3}MR^2$
(d) $\frac{54}{5}MR^2$

14 The moment of inertia of a solid sphere about an axis passing through centre of gravity is $\frac{2}{5}MR^2$, then its radius of gyration about a parallel axis at a distance $2R$ from first axis is

(a) $5R$
(b) $\sqrt{22/5}\,R$
(c) $\frac{5}{2}R$
(d) $\sqrt{12/5}\,R$

15 A wire of mass m and length l is bent in the form of a quarter circle. The moment of inertia of this wire about an axis passing through the centre of the quarter circle and perpendicular to the plane of the quarter circle is approximately

(a) $0.6\,ml^2$
(b) ml^2
(c) $0.2\,ml^2$
(d) $0.4\,ml^2$

16 Three identical thin rods each of length l and mass m are joined together to form a letter H. What is the moment of inertia of the system about one of the side of H?

(a) $\frac{ml^2}{4}$
(b) $\frac{ml^2}{3}$
(c) $\frac{2ml^2}{3}$
(d) $\frac{4ml^2}{3}$

17 Four identical rods each of mass m and length l make a square frame in xy-plane as shown in the adjoining figure. Calculate its moment of inertia about the x-axis.

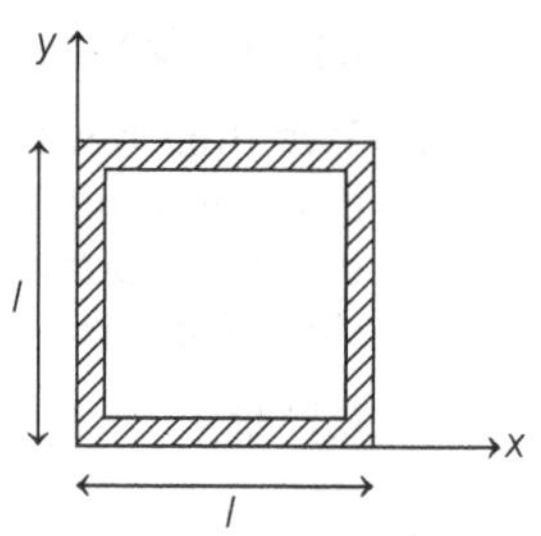

(a) $\frac{5}{3}ml^2$
(b) $\frac{4}{3}ml^2$
(c) ml^2
(d) None of these

18 Two thin uniform rings made of wire of same material and of radii R and $4R$ are joined as shown in the following figure. The mass of smaller ring is m. Find the moment of inertia of the system about an axis passing through the centre of mass of bigger ring and perpendicular to the plane.

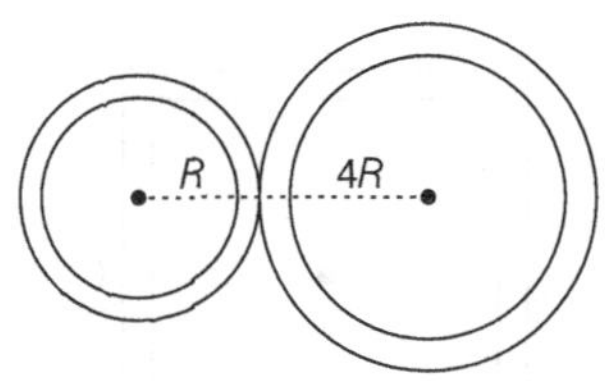

(a) $85\,mR^2$
(b) $90\,mR^2$
(c) $70\,mR^2$
(d) $35\,mR^2$

19 The moment of inertia about the axis AB for system of solid spheres shown will be ($M = 5$ kg, $R = 1$m)

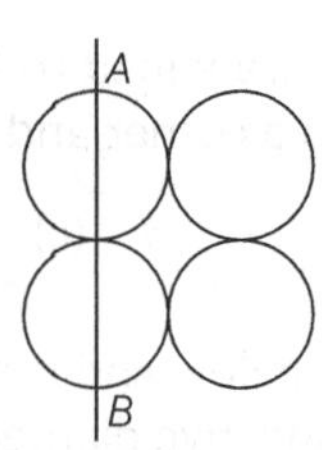

(a) 48 kg-m^2
(b) 60 kg-m^2
(c) 70 kg-m^2
(d) 50 kg-m^2

20 A thin wire of mass M and length L is turned into a ring. The moment of inertia about its axis will be

(a) $\frac{1}{4\pi^2}ML^2$ (b) $\frac{1}{12}ML^2$ (c) $\frac{1}{3\pi^2}ML^2$ (d) $\frac{1}{\pi^2}ML^2$

21 Find radius of gyration about an axis shown in the following figure.

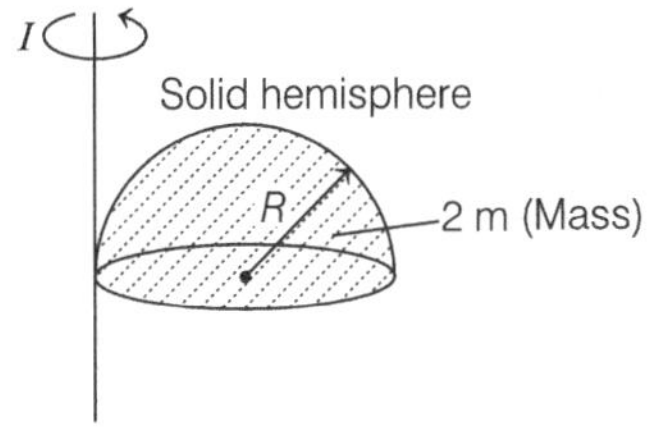

(a) $\sqrt{0.67}R$ (b) $\sqrt{0.4}R$ (c) $\sqrt{1.4}R$ (d) $\sqrt{1.33}R$

22 Moment of inertia of a uniform circular disc about a diameter is I. Its moment of inertia about an axis perpendicular to its plane and passing through a point on its rim will be

(a) $5I$ (b) $3I$ (c) $6I$ (d) $4I$

23 The radius of gyration of a uniform rod of length L about an axis passing through its centre of mass is

(a) $\frac{L}{\sqrt{12}}$ (b) $\frac{L^2}{12}$ (c) $\frac{L}{\sqrt{3}}$ (d) $\frac{L}{\sqrt{2}}$

24 A wheel having moment of inertia 2 kg-m^2 about its vertical axis, rotates at the rate of 60 rpm about this axis. The torque which can stop the wheel's rotation in 1 min would be

(a) $\frac{2\pi}{15}$ N-m (b) $\frac{\pi}{12}$ N-m (c) $\frac{\pi}{15}$ N-m (d) $\frac{\pi}{18}$ N-m

25 A child is standing on a platform with folded hands and the centre of the platform is rotating about its central axis. The kinetic energy of the system is K. The child now stretches his arms, so that the moment of inertia of the system doubles. The kinetic energy of the system now is

(a) $2K$ (b) $K/2$ (c) $K/4$ (d) $4K$

26 If the angular momentum of any rigid rotating body increases by 200%, then the increase in its kinetic energy is

(a) 400% (b) 800% (c) 200% (d) 100%

27 A uniform rod AB of length l is free to rotate about point A. The rod is released from rest in the horizontal position. The initial angular acceleration of the rod will be

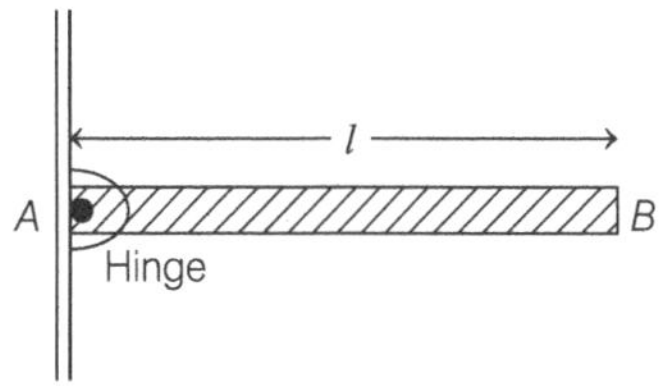

(a) $\frac{3g}{2l}$ (b) $\frac{2g}{3l}$

(c) $\frac{gl}{2}$ (d) $\frac{3}{2}gl$

28 A pulley has the shape of a uniform solid disc of mass 2 kg and radius 0.5 m. A string is wrapped over its rim and is pulled by a force of 2.5 N. The pulley is free to rotate about its axis. Initially, the pulley is at rest. Find the angular velocity and angle rotated by pulley in 10 s.

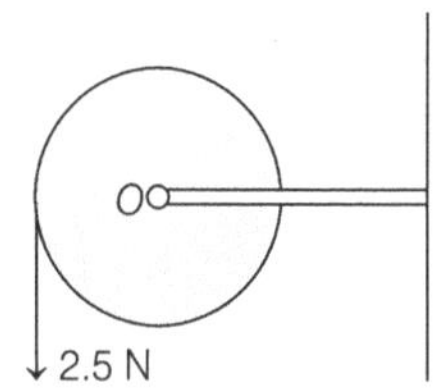

(a) 25 rad/s, 100 rad (b) 25 rad/s, 200 rad

(c) 10 rad/s, 200 rad (d) 50 rad/s, 250 rad

29 The moment of inertia of a body about a given axis is 1.2 kg-m^2. Initially, the body is at rest. In order to produce a rotational kinetic energy of 1500 J, an angular acceleration of 25 rad/s^2 must be applied about that axis for a duration of

(a) 4 s (b) 2 s (c) 8 s (d) 10 s

30 A sphere is rolling without slipping on a fixed horizontal plane surface. A is the point of contact, B is the centre of the sphere and C is top most point. Then,

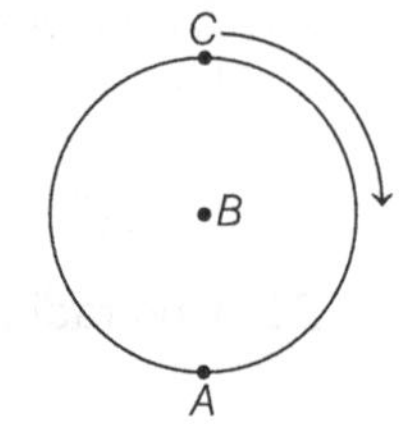

(a) $\mathbf{v}_C - \mathbf{v}_A = 2(\mathbf{v}_B - \mathbf{v}_C)$
(b) $\mathbf{v}_C - \mathbf{v}_B = \mathbf{v}_B - \mathbf{v}_A$
(c) $|\mathbf{v}_C - \mathbf{v}_A| = 2(\mathbf{v}_B - \mathbf{v}_C)$
(d) $|\mathbf{v}_C - \mathbf{v}_A| = 4|\mathbf{v}_B|$

31 A hoop of radius 2 m weighs 100 kg. It rolls along a horizontal floor, so that its centre of mass has a speed of 20 cm s^{-1}. How much work has to be done to stop it?
(a) 2 J (b) 4 J (c) 6 J (d) 8 J

32 A drum of radius R and mass M, rolls down without slipping along an inclined plane of angle θ. The frictional force
(a) decreases the rotational and translational motion
(b) dissipates energy as heat
(c) decreases the rotational motion
(d) converts translational energy to rotational energy

33 A solid sphere rolls down two different inclined planes of the same height but of different inclinations.
(a) In both cases, the speeds and time of descent will be same
(b) The speeds will be different but time of descent will be same
(c) Speeds will be same but time of descent will be different
(d) The speeds and time of descent will be different

34 A body is rolling down an inclined plane. If KE of rotation is 40% of KE in translatory state, then the body is a
(a) ring (b) cylinder (c) hollow ball (d) solid ball

35 The moment of inertia of two freely rotating bodies A and B are I_A and I_B, respectively. $I_A > I_B$ and their angular momenta are equal. If K_A and K_B are their kinetic energies, then
(a) $K_A = K_B$ (b) $K_A > K_B$
(c) $K_A < K_B$ (d) $K_A = 2K_B$

36 An inclined plane makes an angle of 30° with the horizontal. A solid sphere rolling down this inclined plane from rest without slipping has a linear acceleration equal to
(a) $\frac{g}{3}$ (b) $\frac{2g}{3}$ (c) $\frac{5g}{7}$ (d) $\frac{5g}{14}$

LEVEL 02 Objective Questions

37 A rigid lamina is rotating about an axis passing perpendicular to its plane through point O as shown in the following figure. The angular velocity of point B with respect to A is

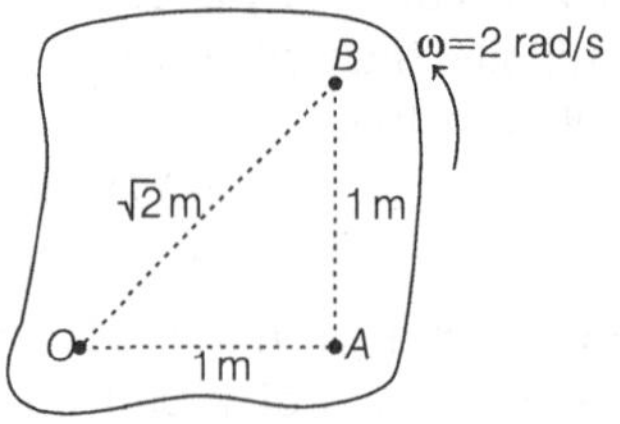

(a) 1 rad/s (b) 2 rad/s
(c) 3 rad/s (d) 4 rad/s

38 A table fan, rotating at a speed of 2400 rpm, is switched off and the resulting variation of the rpm with time is shown in the figure. The total number of revolutions of the fan before it comes to rest is

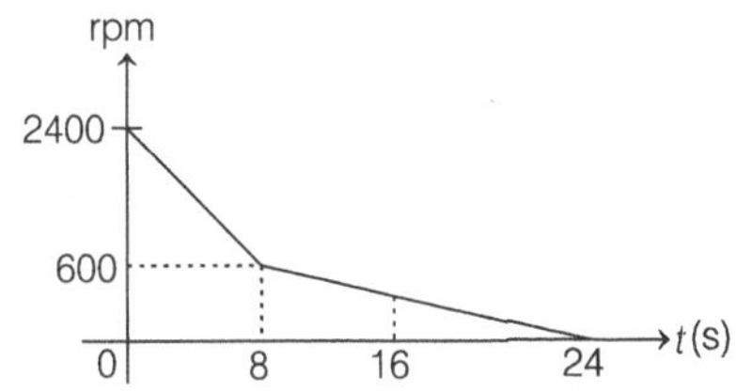

(a) 420 (b) 190 (c) 280 (d) 380

39 A wheel is rolling straight on ground without slipping. If the centre of mass of the wheel has speed v, the instantaneous velocity of a point P on the rim, relative to the ground will be

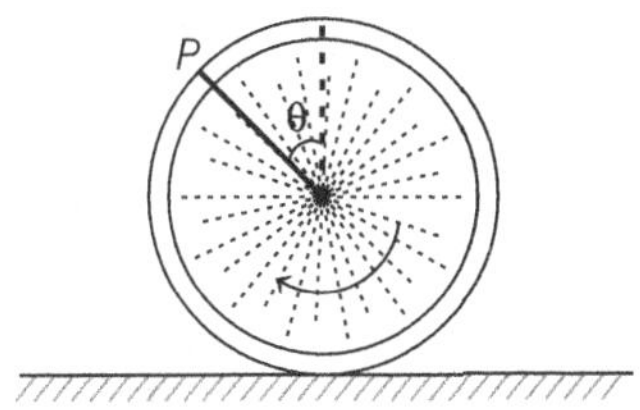

(a) $v\cos\left(\frac{1}{2}\theta\right)$ (b) $2v\cos\left(\frac{\theta}{2}\right)$ (c) $v\sin\left(\frac{\theta}{2}\right)$ (d) $v(\sin\theta)$

40 Four hollow spheres, each with a mass of 1 kg and a radius $R = 10$ cm are connected with massless rods to form a square with side of length $L = 50$ cm. In case-1, the masses rotate about an axis that bisects two sides of the square. In case-2, the masses rotate about an axis that passes through the diagonal of the square, as shown in the figure. Compute the ratio of the moments of inertia I_1/I_2 for the two cases.

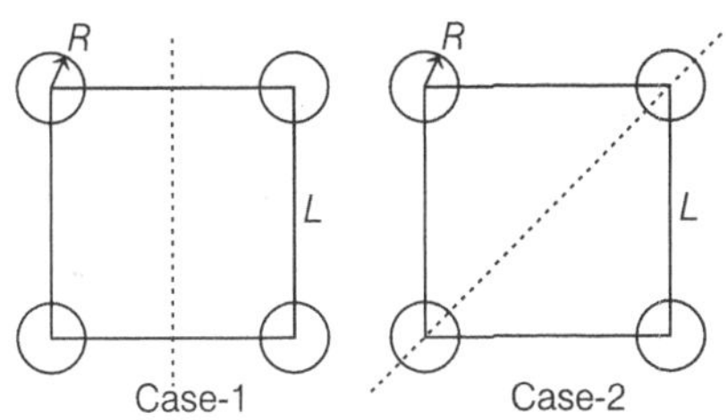

(a) $I_1/I_2 = 8$ (b) $I_1/I_2 = 4$ (c) $I_1/I_2 = 2$ (d) $I_1/I_2 = 1$

41 A uniform disc of radius R lies in the xy-plane with its centre at the origin. Its moment of inertia about z-axis is equal to its moment of inertia about line $y = x + c$. The value of c is

(a) $R/\sqrt{2}$ (b) $-R/2$ (c) $R/4$ (d) $-R$

42 A uniform thin stick of length l and mass m is held horizontally with its end B hinged at a point B on the edge of a table. Point A is suddenly released. The acceleration of the centre of mass of the stick at the time of release, is

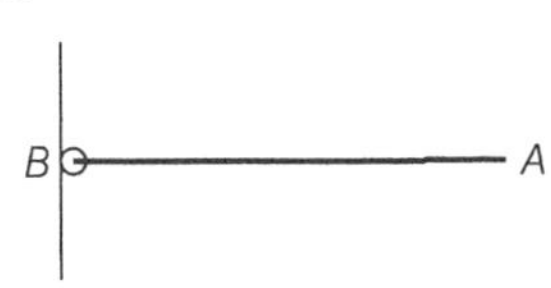

(a) $\frac{3}{4}g$ (b) $\frac{3}{7}g$ (c) $\frac{2}{7}g$ (d) $\frac{1}{7}g$

43 A rod of mass M and length L is hinged about one end and a particle of mass M is attached to its other end. If it is kept horizontal with the help of a spring constant k as shown in figure, find the extension in spring.

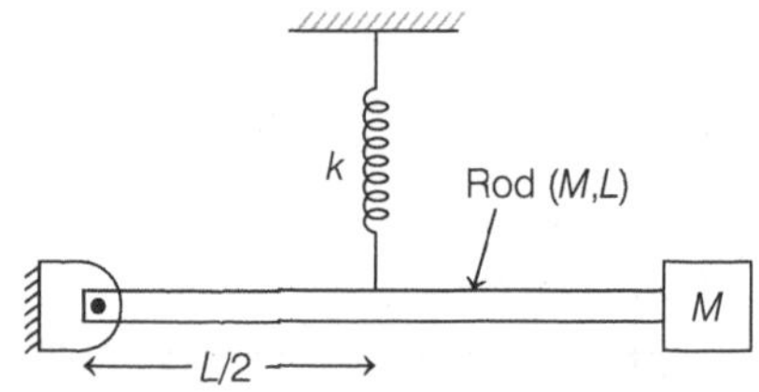

(a) $\frac{3Mg}{k}$ (b) $\frac{Mg}{k}$ (c) $\frac{2Mg}{k}$ (d) $\frac{4Mg}{k}$

44 A weightless rod of length $2l$ carries two equal masses m, one tied at lower end A and the other at the middle of the rod at B. The rod can rotate in vertical plane about a fixed horizontal axis passing through C. The rod is released from rest in horizontal position. The speed of the mass B at the instant rod becomes vertical is

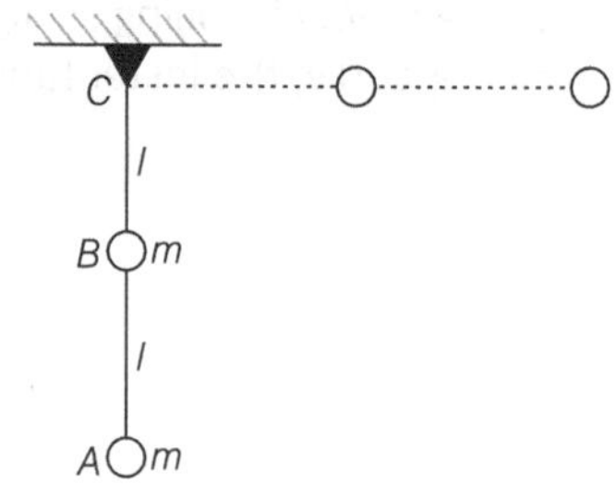

(a) $\sqrt{\frac{3gl}{5}}$ (b) $\sqrt{\frac{4gl}{5}}$

(c) $\sqrt{\frac{6gl}{5}}$ (d) $\sqrt{\frac{7gl}{5}}$

45 A pulley of radius 2 m is rotated about its axis by a force $F = (20t - 5t^2)$ N (where, t is measured in seconds) applied tangentially. If the moment of inertia of the pulley about its axis of rotation is 10 kg-m^2, the number of rotations made by the pulley before its direction of motion is reversed, is

(a) more than 6 but less than 9 (b) more than 9

(c) less than 3 (d) more than 3 but less than 6

46 A uniform rod of mass m and length L is at rest on a smooth horizontal surface. A horizontal impulse J is applied to the end B perpendicular to the rod. The time taken by the rod to turn through a right angle is

(a) $\frac{\pi mL}{J}$ (b) $\frac{\pi mL}{2J}$ (c) $\frac{\pi mL}{6J}$ (d) $\frac{\pi mL}{12J}$

47 A rod of mass m and length l is hinged at one of its end A as shown in the adjoining figure. A force F is applied at a distance x from A. The acceleration of centre of mass a varies with x as

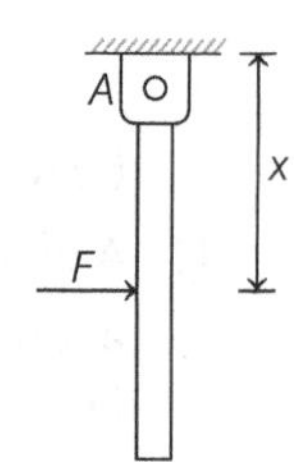

(a)
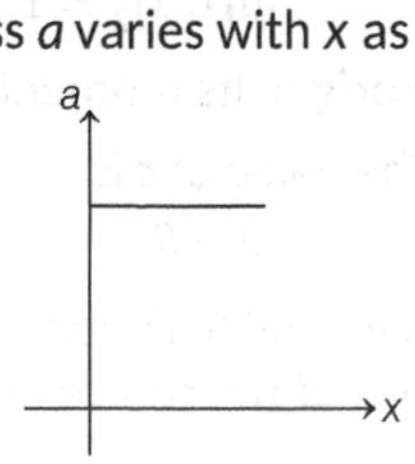

(b)
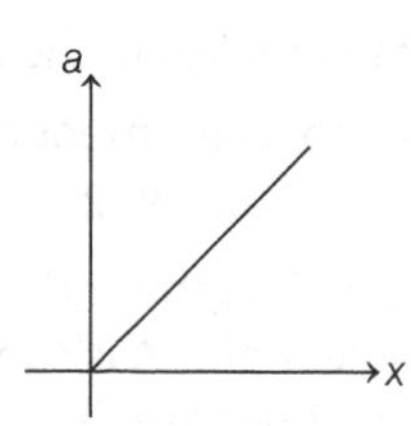

(c)
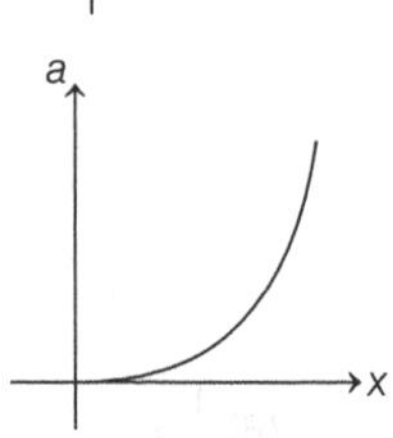

(d)
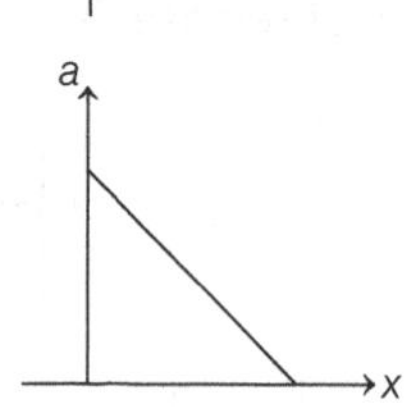

48 A uniform rod of mass m and length l is hanging vertically at one end as shown in the adjoining figure. A bullet of mass $\frac{m}{2}$ fired with velocity v strikes the rod in the middle and sticks to it, then angular velocity of the rod about hing just after being hit is

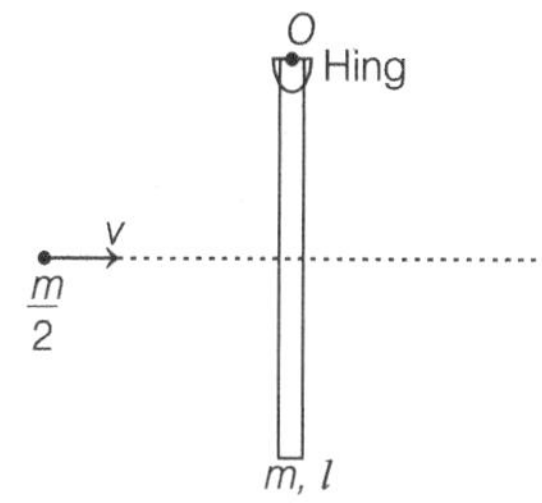

(a) $\frac{2v}{3l}$ (b) $\frac{6v}{11l}$
(c) $\frac{3v}{7l}$ (d) $\frac{4v}{3l}$

49 A stick of length L and mass M lies on a frictionless horizontal surface on which it is free to move in any way. A ball of mass m moving with speed v collides elastically with the stick as shown in the adjoining figure. If after the collision ball comes to rest, then what should be the mass of the ball?

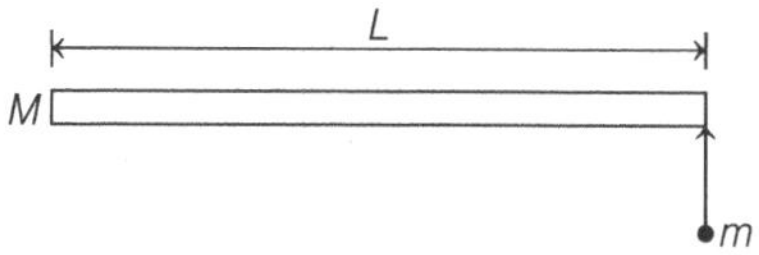

(a) $m = 2M$ (b) $m = M$ (c) $m = M/2$ (d) $m = M/4$

50 A body is rolling without slipping on a horizontal plane. The radius of body is R. If the ratio of rotational to translational kinetic energy is $\frac{9}{25}$, then the radius of gyration of the body is

(a) $\frac{5R}{3}$ (b) $\frac{9R}{25}$ (c) $\frac{3R}{5}$ (d) $\frac{25R}{9}$

51 A solid sphere of mass 2 kg rolls on a horizontal surface at 10 m/s. It then rolls up a rough enough inclined plane of inclination 30° with the horizontal. The height attained by the sphere before it stops is

(a) 1.7 m (b) 4.5 m
(c) 5.4 m (d) 7.1 m

52 A ring of mass m is rolling without slipping with linear velocity v as shown in the adjoining figure. A rod of identical mass is fixed along one of its diameter. The total kinetic energy of the system is

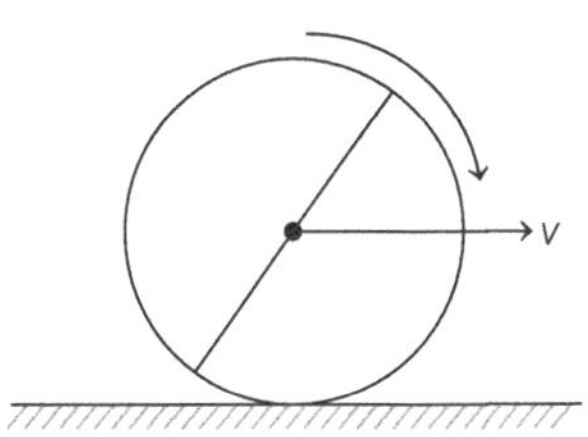

(a) $\frac{7}{5}mv^2$ (b) $\frac{2}{5}mv^2$ (c) $\frac{15}{8}mv^2$ (d) $\frac{5}{4}mv^2$

53 A wheel of radius 0.1 m (wheel A) is attached by a non-stretching belt to a wheel of radius 0.2 m (wheel B). The belt does not slip. By the time, wheel B turns through one revolution, wheel A will rotate through x revolutions. Find the value of x.

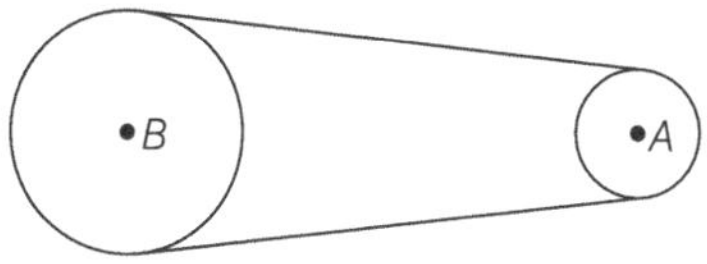

(a) 6 (b) 4
(c) 2 (d) 0

54 A solid sphere of mass 10 kg and radius 0.1 m is placed on a rough horizontal surface. A 20 N force is applied on the sphere horizontally passing through centre of mass. For pure rolling, the force of friction is

(a) 40/7 N in forward direction (b) 40/7 N in backward direction
(c) 20/7 N in forward direction (d) 20/7 N in backward direction

55 A spherical shell of mass m and radius R is pulled on a horizontal plane by a horizontal force F acting on its centre of mass. If shell is rolling without slipping, then angular acceleration of the spherical shell will be

(a) $\frac{3F}{2mR}$ (b) $\frac{3F}{5mR}$ (c) $\frac{5F}{7mR}$ (d) $\frac{5F}{2mR}$

56 A force F acts tangentially at the highest point of a solid cylinder of mass m kept on a rough horizontal plane. If the cylinder rolls without slipping, find the acceleration of the centre of cylinder.

(a) $\frac{4F}{3m}$ (b) $\frac{3F}{4m}$ (c) $\frac{5F}{4m}$ (d) $\frac{3F}{5m}$

57 A solid cylinder has a thin string wrapped several times around its circumference. The string is fixed at one end and the cylinder is released. Find the downward acceleration of cylinder and tension in the string.

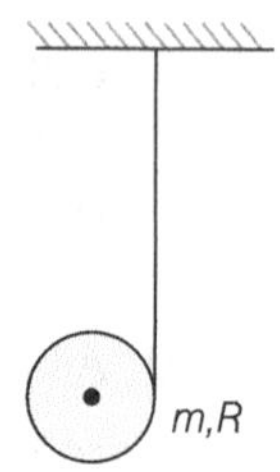

(a) $\frac{2g}{3}, \frac{mg}{3}$ (b) $\frac{g}{3}, \frac{mg}{3}$

(c) $\frac{g}{3}, \frac{2mg}{3}$ (d) $\frac{g}{3}, \frac{4mg}{3}$

58 A ring is rolling down an inclined plane which is inclined at an angle of 30° to the horizontal. Then, the speed of ring after travelling a distance of 10 m is

(a) 10 ms^{-1} (b) $\sqrt{50} \text{ ms}^{-1}$ (c) $\sqrt{30} \text{ ms}^{-1}$ (d) 2 ms^{-1}

59 A solid cylinder of mass M and radius R rolls down an inclined plane of height h. The angular velocity of the cylinder, when it reaches the bottom of the plane is

(a) $\frac{1}{R}\sqrt{\frac{gh}{2}}$ (b) $\frac{2}{R}\sqrt{\frac{gh}{3}}$

(c) $\frac{1}{R}\sqrt{\frac{2gh}{3}}$ (d) $\frac{3}{R}\sqrt{\frac{gh}{2}}$

60 A solid sphere is in pure rolling motion on an inclined surface having inclination θ,

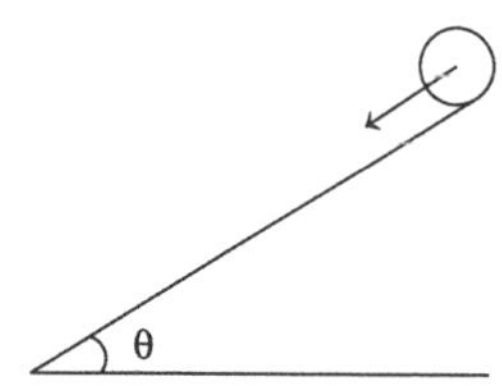

(i) frictional force acting on sphere is $f = \mu mg \cos\theta$

(ii) f is dissipative force

(iii) friction will increase its angular velocity and decreases its linear velocity

(iv) if θ decrease, friction will decrease

(a) (i, ii) (b) (i, iii)

(c) (ii, iii) (d) (iii, iv)

61 A mass m hangs with the help of a string wrapped around a pulley. The pulley has mass m and radius R. Assuming the pulley to be a perfect uniform circular disc, the acceleration of the mass m, if the string does not slip on the pulley, is

(a) $\frac{3}{2}g$ (b) g (c) $\frac{2}{3}g$ (d) $\frac{g}{3}$

62 A disc of mass m and radius R is hinged at its centre. Light string is wrapped on its rim and the free end of the string is being pulled by a force F. Assuming that the string does not slip over the disc, the acceleration of the free end of the string is

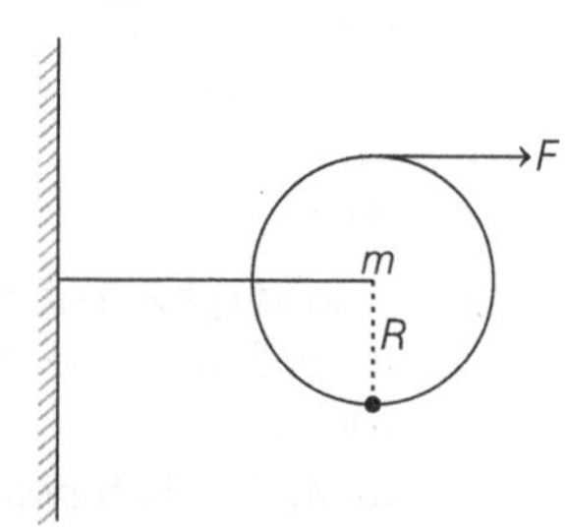

(a) $\frac{F}{m}$ (b) $\frac{F}{3m}$

(c) $\frac{2F}{m}$ (d) $\frac{F}{2m}$

63 A ring of mass m is thrown in horizontal direction on a rough horizontal surface with a speed v_0. The speed of the ring, when it starts pure rolling is

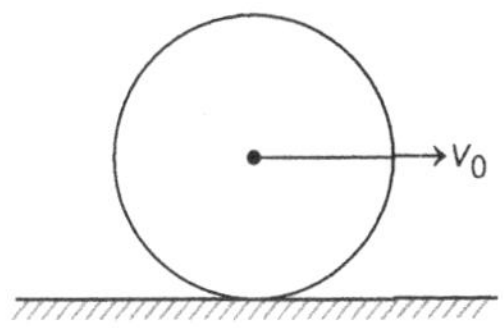

(a) $\frac{v_0}{2}$ (b) $\frac{v_0}{3}$ (c) $\frac{2v_0}{3}$ (d) $\frac{3v_0}{3}$

64 The total kinetic energy of a body of mass 10 kg and radius 0.5 m moving with a velocity of 2 m/s without slipping is 32.8 J. The radius of gyration of the body is

(a) 0.25 m (b) 0.2 m
(c) 0.5 m (d) 0.4 m

65 What will be the value of maximum acceleration of the truck in forward direction, so that the block kept on the back does not topple ?

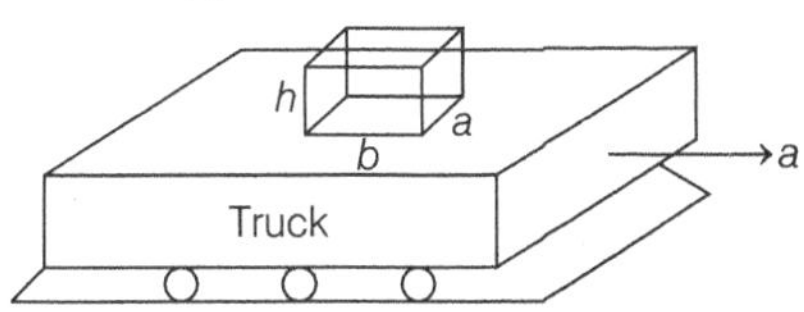

(a) ag/h (b) hg/a
(c) hg/b (d) bg/h

66 A sphere of mass m is being pulled over a rough step of height $R/2$ as shown below. The minimum force F required to just lift the sphere off the ground is

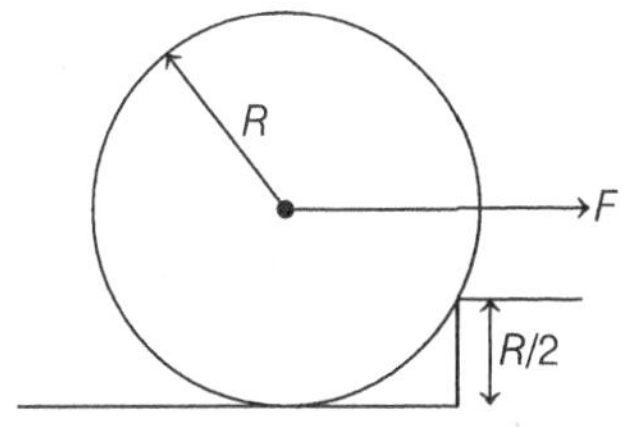

(a) mg (b) $\frac{mg}{2}$ (c) $\sqrt{3}\,mg$ (d) $\frac{mg}{\sqrt{3}}$

67 A T-shaped object with dimensions shown in the adjoining figure, is lying on a smooth floor. A force **F** is applied at the point P parallel to AB, such that the object has only the translational motion without rotation. Find the location of P with respect to C.

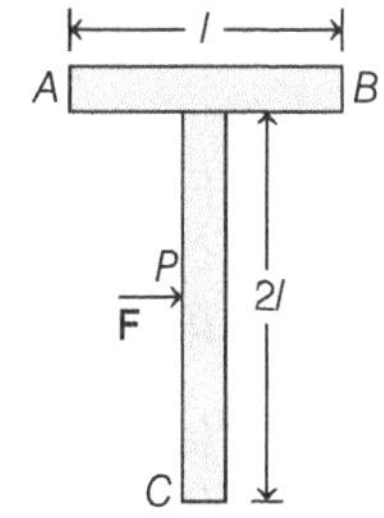

(a) $\frac{4l}{3}$ (b) l
(c) $\frac{2l}{3}$ (d) $\frac{3l}{2}$

68 A uniform rod AB of mass m, length $2a$ is allowed to fall under gravity with AB in horizontal position. When the speed of rod is v, suddenly the end A is fixed. The angular velocity of rod with which it begins to rotate will be

(a) $\frac{3v}{2a}$ (b) $\frac{5v}{3a}$
(c) $\frac{3v}{4a}$ (d) $\frac{5v}{2a}$

LEVEL 03 Previous Years Questions

69 From a circular ring of mass M and radius R, an arc corresponding to a 90° sector is removed. The moment of inertia of the remaining part of the ring about an axis passing through the centre of the ring and perpendicular to the plane of the ring is K times MR^2. Then, the value of K is **[NEET 2021]**

(a) $\frac{3}{4}$ (b) $\frac{7}{8}$
(c) $\frac{1}{4}$ (d) $\frac{1}{8}$

70 The angular speed of the wheel of a vehicle is increased from 360 rpm to 1200 rpm in 14 s. Its angular acceleration is **[NEET 2020]**

(a) 2π rad/s^2 (b) 28π rad/s^2
(c) 120π rad/s^2 (d) 1 rad/s^2

71 A disc of radius 2 m and mass 100 kg rolls on a horizontal floor. Its centre of mass has speed of 20 cm/s. How much work is needed to stop it? **[NEET 2019]**

(a) 30 kJ (b) 2 J
(c) 1 J (d) 3 J

72 A solid cylinder of mass 2 kg and radius 4 cm is rotating about its axis at the rate of 3 rpm. The torque required to stop after 2π revolutions is **[NEET 2019]**

(a) 2×10^{-3} N-m (b) 12×10^{-4} N-m
(c) 2×10^{6} N-m (d) 2×10^{-6} N-m

73 A solid sphere is rotating freely about its symmetry axis in free space. The radius of the sphere is increased keeping its mass same. Which of the following physical quantities would remain constant for the sphere? **[NEET 2018]**

(a) Rotational kinetic energy (b) Moment of inertia
(c) Angular velocity (d) Angular momentum

74 Two discs of same moment of inertia rotating about their regular axis passing through centre and perpendicular to the plane of disc with angular velocities ω_1 and ω_2. They are brought into contact face to face coinciding the axis of rotation. The expression for loss of energy during this process is **[NEET 2017]**

(a) $\frac{1}{2} I(\omega_1 + \omega_2)^2$ (b) $\frac{1}{4} I(\omega_1 - \omega_2)^2$
(c) $I(\omega_1 - \omega_2)^2$ (d) $\frac{I}{8} (\omega_1 - \omega_2)^2$

75 From a disc of radius R and mass M, a circular hole of diameter R, whose rim passes through the centre is cut. What is the moment of inertia of the remaining part of the disc about a perpendicular axis, passing through the centre ? **[NEET 2016]**

(a) $13\,MR^2/32$ (b) $11\,MR^2/32$
(c) $9\,MR^2/32$ (d) $15\,MR^2/32$

76 A light rod of length l has two masses m_1 and m_2 attached to its two ends. The moment of inertia of the system about an axis perpendicular to the rod and passing through the centre of mass is **[NEET 2016]**

(a) $\frac{m_1 m_2}{m_1 + m_2} l^2$ (b) $\frac{m_1 + m_2}{m_1 m_2} l^2$
(c) $(m_1 + m_2) l^2$ (d) $\sqrt{m_1 m_2} l^2$

77 Three identical spherical shells, each of mass m and radius r are placed as shown in figure. Consider an axis XX', which is touching to two shells and passing through diameter of third shell.

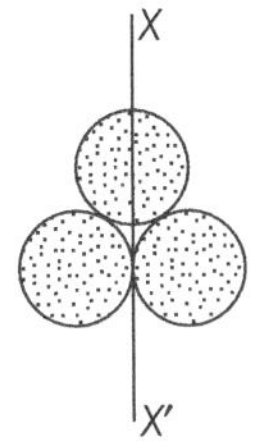

Moment of inertia of the system consisting of these three spherical shells about XX' axis is [CBSE AIPMT 2015]

(a) $\frac{11}{5}mr^2$ (b) $3\,mr^2$

(c) $\frac{16}{5}mr^2$ (d) $4\,mr^2$

78 The moment of inertia of a thin uniform rod of mass M and length L about an axis passing through its mid-point and perpendicular to its length is I_0. Its moment of inertia about an axis passing through one of its ends and perpendicular to its length is [CBSE AIPMT 2011]

(a) $I_0 + ML^2/4$ (b) $I_0 + 2ML^2$

(c) $I_0 + ML^2$ (d) $I_0 + ML^2/2$

79 A circular disc of moment of inertia I_t is rotating in a horizontal plane, about its symmetry axis, with a constant angular speed ω_i. Another disc of moment of inertia I_b is dropped coaxially onto the rotating disc. Initially the second disk has zero angular speed. Eventually both the discs rotate with a constant angular speed ω_f. The energy lost by initially rotating disc due to friction is [CBSE AIPMT 2010]

(a) $\frac{1}{2}\frac{I_b^2}{(I_t + I_b)}\omega_i^2$ (b) $\frac{1}{2}\frac{I_t^2}{(I_t + I_b)}\omega_i^2$

(c) $\frac{1}{2}\frac{I_b - I_t}{(I_t + I_b)}\omega_i^2$ (d) $\frac{1}{2}\frac{I_b I_t}{(I_t + I_b)}\omega_i^2$

80 A ball rolls without slipping. The radius of gyration of the ball about an axis passing through its centre of mass is k. If radius of the ball be R, then the fraction of total energy associated with its rotational energy will be [CBSE AIPMT 2003]

(a) $\frac{k^2}{k^2 + R^2}$ (b) $\frac{R^2}{k^2 + R^2}$

(c) $\frac{k^2 + R^2}{R^2}$ (d) $\frac{k^2}{R^2}$

81 A spherical ball rolls on a table without slipping. Then, the fraction of its total energy associated with rotation is [CBSE AIPMT 1994]

(a) $\frac{2}{5}$ (b) $\frac{2}{7}$

(c) $\frac{3}{5}$ (d) $\frac{3}{7}$

82 A thin uniform circular ring is rolling down an inclined plane of inclination 30° without slipping. Its linear acceleration along the inclined plane will be [CBSE AIPMT 1994]

(a) $\frac{g}{2}$ (b) $\frac{g}{3}$

(c) $\frac{g}{4}$ (d) $\frac{2g}{3}$

83 A flywheel rotating about a fixed axis has a kinetic energy of 360 J, when its angular speed is 30 rad/s. The moment of inertia of the wheel about the axis of rotation is **[CBSE AIPMT 1990]**

(a) 0.6 kg-m^2 (b) 0.15 kg-m^2
(c) 0.8 kg-m^2 (d) 0.75 kg-m^2

84 At any instant, a rolling body may be considered to be in pure rotation about an axis through the point of contact. This axis is translating forward with speed **[CBSE AIPMT 1989]**

(a) equal to centre of mass (b) zero
(c) twice of centre of mass (d) None of these

85 A solid homogeneous sphere of mass M and radius R is moving on a rough horizontal surface, partly rolling and partly sliding. During this kind of motion of the sphere, **[CBSE AIPMT 1988]**

(a) total kinetic energy is conserved
(b) the angular momentum of the sphere about the point of contact with the plane is conserved
(c) only the rotational kinetic energy about the centre of mass is conserved
(d) angular momentum about the centre of mass is conserved

ANSWER KEY

Level 1 Objective Questions

1 (b)	2 (c)	3 (a)	4 (a)	5 (c)	6 (a)	7 (b)	8 (b)	9 (b)	10 (a)
11 (b)	12 (c)	13 (d)	14 (b)	15 (d)	16 (d)	17 (a)	18 (b)	19 (a)	20 (a)
21 (c)	22 (c)	23 (a)	24 (c)	25 (b)	26 (b)	27 (a)	28 (d)	29 (b)	30 (b)
31 (b)	32 (d)	33 (c)	34 (d)	35 (c)	36 (d)				

Level 2 Objective Questions

37 (b)	38 (c)	39 (b)	40 (d)	41 (a)	42 (a)	43 (a)	44 (c)	45 (d)	46 (d)
47 (b)	48 (b)	49 (d)	50 (c)	51 (d)	52 (c)	53 (c)	54 (b)	55 (b)	56 (a)
57 (a)	58 (b)	59 (b)	60 (d)	61 (c)	62 (c)	63 (a)	64 (d)	65 (d)	66 (c)
67 (a)	68 (c)								

Level 3 Previous Years Questions

69 (a)	70 (a)	71 (d)	72 (d)	73 (d)	74 (b)	75 (a)	76 (a)	77 (d)	78 (a)
79 (d)	80 (a)	81 (b)	82 (c)	83 (c)	84 (a)	85. (b)			

Solutions

LEVEL 01 Objective Questions

1. $\mathbf{v} = \omega \times \mathbf{r}$

$$\Rightarrow \quad \mathbf{v} = \begin{vmatrix} \hat{\mathbf{i}} & \hat{\mathbf{j}} & \hat{\mathbf{k}} \\ 3 & -4 & 1 \\ 5 & -6 & 6 \end{vmatrix}$$

$$\Rightarrow \quad \mathbf{v} = -18\hat{\mathbf{i}} - 13\hat{\mathbf{j}} + 2\hat{\mathbf{k}}$$

2. Using $\theta = \omega_0 t + \frac{1}{2}\alpha t^2$

$$\theta_1 = 0 + \frac{1}{2}\alpha(2)^2 = 2\alpha$$

Now, the angular velocity changes to ω,

$$\omega = \omega_0 + \alpha t$$
$$\omega = 0 + 2\alpha = 2\alpha$$

Now, $\theta_2 = \omega t + \frac{1}{2}\alpha t^2$

$$= 2\alpha(2) + \frac{1}{2}\alpha(2)^2 = 6\alpha$$

$$\therefore \quad \frac{\theta_2}{\theta_1} = \frac{6\alpha}{2\alpha} = 3$$

3. For the force to be minimum, the torque due to F at a point should be equal to the torque due to weight,

$$F\left(\frac{3a}{4}\right) = mg\left(\frac{a}{2}\right)$$

$$\Rightarrow \quad F = \frac{2}{3}mg$$

4. Net force acting on the system

$$= 2F - 2F = 0$$

So, there will be no translational motion.

Now,

$$\tau_1 = F \times 10\text{ cm} = 0.1\,F \ \otimes$$
$$\tau_2 = 2F \times 0.1 = 0.2\,F \ \otimes$$
$$\tau_3 = 0.3 \times F = 0.3\,F \ \odot$$

So, net torque $= \tau_3 - (\tau_1 + \tau_2) = 0$

So, there will be no rotational motion as well.

5. Mass of disc, $M = (\pi R^2 t)\sigma$

$$I_A = \frac{MR^2}{2} = \frac{(\pi R^2 t\sigma)R^2}{2} = \frac{\pi R^4 \sigma t}{2}$$

Similarly, $I_B = 32\pi R^4 \sigma t$

$$\because \quad \frac{I_B}{I_A} = 64$$

$$\Rightarrow \quad I_B = 64 I_A$$

6. Moment of inertia about an axis passing through centre of mass is minimum.

7. Mass $\propto$ (Radius)^2

$$I = \frac{1}{2}mR^2 \quad \text{or} \quad I \propto mR^2$$

or $$I \propto R^4$$

$$\frac{I_1}{I_2} = \left(\frac{R_1}{R_2}\right)^4 = \left(\frac{1}{3}\right)^4 = \frac{1}{81}$$

R_1 and R_2 are the radii of two discs and I_1 and I_2 are their respective moment of inertia.

8. $I_{\text{solid}} = I_{\text{hollow}}$

$$\frac{2}{3}mR_1^2 = \frac{2}{3}\left(\frac{m}{4}\right)R_2^2$$

$$\Rightarrow \quad \frac{R_2}{R_1} = \frac{2}{1}$$

9. Moment of inertia of a given system

$$= I_1 + I_2 = MR^2 + \frac{MR^2}{2}$$
$$= \frac{3}{2}MR^2$$

10. Moment of inertia of rod AB about O

$$= \frac{mL^2}{12} + m\left(\frac{L}{2}\right)^2 = \frac{mL^2}{3}$$

Since, the system consists of four rods,

$$I_{\text{system}} = \frac{4mL^2}{3}$$

11. Moment of inertia of given system of rods about any corner

$$= 2 \cdot \frac{ML^2}{3} + \left(\frac{5}{6}ML^2\right)$$
$$= \left(\frac{2}{3} + \frac{5}{6}\right)ML^2 = \frac{3}{2}ML^2$$

Radius of gyration

$$= k = \sqrt{\frac{3}{2}} \cdot L$$

12. As the moment of inertia is equal,

$$I_{\text{hollow sphere}} = I_{\text{solid sphere}}$$
$$\frac{2}{3}mr_1^2 = \frac{2}{5}mr_2^2$$

$$\Rightarrow \quad \frac{r_1}{r_2} = \frac{\sqrt{3}}{\sqrt{5}}$$

13. Moment of inertia of combination

$$= I_1 + I_2$$
$$= I_{CM} + MR^2 + I_{CM} + M(R + 2R)^2$$
$$= 2I_{CM} + MR^2 + M(3R)^2$$
$$= \left(\frac{4}{5} + 10\right) MR^2 = \frac{54}{5} MR^2$$

14. From parallel axes theorem,

$$I \text{ at } 2R = \frac{2}{5} MR^2 + M(2R)^2 = \frac{22}{5} MR^2$$

Now, radius of gyration,

$$Mk^2 = \frac{22}{5} MR^2$$

$$\Rightarrow \quad k = \sqrt{\frac{22}{5}} R$$

15. $I = mr^2$

$$\because \quad r = \frac{2l}{\pi}$$

$$\therefore \quad I = m\left(\frac{2l}{\pi}\right)^2 = 0.4\, ml^2$$

16. Let I_1, I_2, I_3 be the moment of inertia.

So, $I_1 + I_2 + I_3 = 0 + \frac{ml^2}{3} + ml^2$

$$= \frac{4}{3} ml^2$$

17. Moment of inertia about x-axis

$$= \left(\frac{ml^2}{3}\right) \times 2 + ml^2 = \frac{5}{3} ml^2$$

18. Mass ∝ Length of ring ∝ Radius

∴ Mass of bigger ring = 4 m

$$I = [(4m)(4R)^2] + [mR^2 + m(5R)^2]$$
$$= 90\, mR^2$$

19. $I = 4\left[\frac{2}{5} MR^2\right] + 2[M(2R)^2] = \frac{48}{5} MR^2$

$$\because \quad M = 5 \text{ kg}$$
$$R = 1 \text{ m}$$
$$\Rightarrow \quad I = \frac{48}{5} \times 5 \times 1$$
$$I = 48 \text{ kg-m}^2$$

20. We know, $L = 2\pi R$

$$\Rightarrow \quad R = L/2\pi$$

So, moment of inertia of ring

$$= MR^2 = M\left(\frac{L}{2\pi}\right)^2 = \frac{ML^2}{4\pi^2}$$

21.

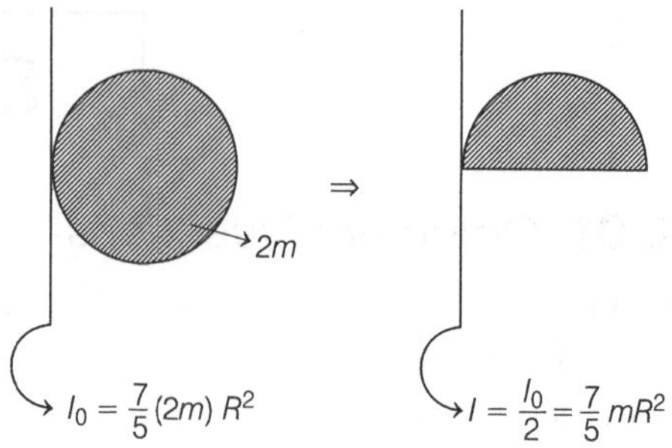

Radius of gyration $= \sqrt{\frac{I}{m}} = \sqrt{\frac{7}{5}} R = \sqrt{1.4}\, R$

22. Moment of inertia of disc about axis,

$$2I = \frac{mR^2}{2}$$

Now, from parallel axes theorem, moment of inertia of disc about the given axis

$$= 2I + mR^2 = 2I + 4I = 6I$$

23. $\frac{mL^2}{12} = mk^2$

So, $k = \frac{L}{2\sqrt{3}} = \frac{L}{\sqrt{12}}$

24. $\tau = I\alpha = I\left(\frac{\omega_0 - \omega}{t}\right)$

$$= \frac{2 \times 2\pi \times 60}{60 \times 60} = \frac{\pi}{15} \text{ N-m}$$

25. From angular momentum conservation,

$$I_1\omega_1 = I_2\omega_2$$

$$\because \quad I_2 = 2I, \omega_2 = \frac{\omega_1}{2}$$

Initial KE $= \frac{I_1\omega_1^2}{2} = K$

Final KE $= \frac{I_2\omega_2^2}{2} = \frac{2I\left(\frac{\omega_1}{2}\right)^2}{2} = \frac{K}{2}$

26. KE $= \frac{L^2}{2I}$

$$\text{KE} \propto L^2$$
$$\frac{E_2}{E_1} = \left(\frac{L_2}{L_1}\right)^2$$
$$L_2 = L_1 + \frac{200}{100} L_1 = 3L_1$$
$$\frac{E_2}{E_1} = 9$$

So, % change in kinetic energy

$$= \frac{E_2 - E_1}{E_1} = \frac{9E_1 - E_1}{E_1} \times 100\% = 800\%$$

27. $\tau = I\alpha$

$$mg\frac{l}{2} = \frac{ml^2}{3}\alpha \Rightarrow \alpha = \frac{3g}{2l}$$

So, initial acceleration of rod is $\frac{3g}{2l}$.

28. Moment of inertia of disc, I

$$= \frac{mR^2}{2} = \frac{2\times(0.5)^2}{2} = 0.25\ \text{kg/m}^2$$

Now, $\tau = F \times r$

$$= 2.5 \times 0.5$$
$$= 1.25\ \text{N-m}^2$$

$\therefore \quad \tau = I\alpha$

$$1.25 = 0.25\alpha$$

$\Rightarrow \quad \alpha = 5\ \text{rad/s}^2$

Using $\omega = \omega_0 + \alpha t$

$$= 0 + 5(10) = 50\ \text{rad/s}$$
$$\omega^2 = \omega_0^2 + 2\alpha\theta$$
$$(50)^2 = 0 + 2\times5\times\theta$$
$$\theta = 250\ \text{rad}$$

29. $KE_{rot} = \frac{1}{2}I\omega^2 = 1500\ \text{J}$

$\Rightarrow \quad 1500 = \frac{1}{2}\times1.2\times\omega^2$

$\Rightarrow \quad \omega = 50\ \text{rad/s}$

Now, $\omega = \omega_0 + \alpha t$

$$50 = 0 + 25t$$

$\Rightarrow \quad t = 2\ \text{s}$

30. $\mathbf{v}_A = 0$, $\mathbf{v}_B = \mathbf{v}$ (say). Then, $\mathbf{v}_C = 2\mathbf{v}$

31. Work done to stop the hoop

$$= \frac{1}{2}mv^2 + \frac{1}{2}I\omega^2$$

Here, $v = R\omega$ and $I = mR^2$

On substituting values, we get

$$W = 4\ \text{J}$$

32. Frictional force converts translational energy to the rotational energy. As the energy remains conserved.

33. From work-energy theorem,

$$mgh = \frac{1}{2}mv^2 + \frac{1}{2}I\omega^2$$

Here, $I = \frac{2}{5}mR^2$

So, $v^2 = \frac{10gh}{7}$

Hence, the speed is same in both the cases.

Now, $\quad v = at$

$$t = \frac{v\times1+I/mR^2}{g\sin\theta}$$

So, for the different inclination, time of descent will be different.

34. $$\frac{KE_{rot}}{KE_{trans}} = \frac{\frac{1}{2}mv^2\left(\frac{k^2}{R^2}\right)}{\frac{1}{2}mv^2} = \frac{2}{5}$$

$$k^2 = \frac{2}{5}R^2$$
$$mk^2 = \frac{2}{5}mR^2$$

So, the body is a solid sphere.

35. $KE = \frac{L^2}{2I} \Rightarrow KE \propto \frac{1}{I}$

So, $\quad I_A > I_B, K_A < K_B$

36. $$a = \frac{mg\sin\theta - F}{m} \quad \text{...(i)}$$

Now, $\alpha = \frac{FR}{I} = \frac{5FR}{2mR^2} = \frac{5F}{2mR}$

For no slipping, $a = R\alpha$

$$\frac{5F}{2m} = g\sin\theta - \frac{F}{m}$$
$$\frac{F}{m} = \frac{2}{7}g\sin\theta$$

On substituting F/m in equation (i), we get

$$a = \frac{5}{7}g\sin30° = \frac{5}{14}g$$

LEVEL 02 Objective Questions

37.

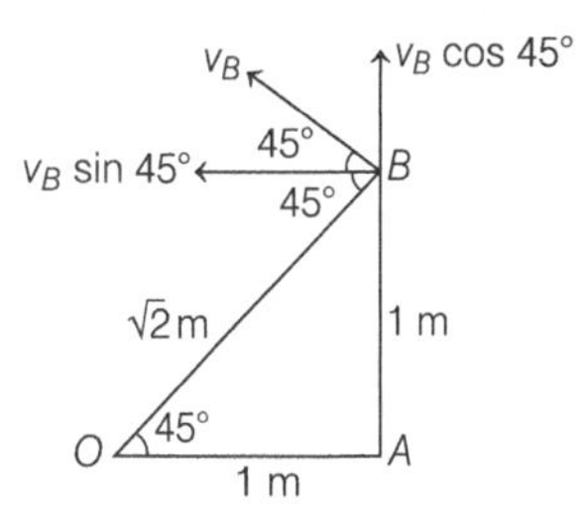

$$\omega_{BA} = \frac{|v_B - v_A|}{r_\perp} = \frac{\omega r\sin45° - 0}{1}$$
$$= \frac{2\sqrt{2}\times1}{\sqrt{2}} = 2\ \text{rad/s}$$

38. Total number of revolutions = Area under graph

$$= \frac{1}{2}\times\frac{8}{60}\times1800 + \frac{8}{60}\times600 + \frac{1}{2}\times\frac{16}{60}\times600$$
$$= 280$$

So, after 280 revolutions, fan will come in rest.

39. $v_R = \sqrt{v^2 + v^2 + 2v^2\cos\theta} = 2v\cos\frac{\theta}{2}$

40. For case I,

$$I_1 = 4\left(\frac{2}{3}mR^2\right) + m\left(\frac{L}{2}\right)^2 = \frac{8\times10^{-2}}{300} + \frac{1}{4}$$

For case II,

$$I_2 = 2\left(\frac{2}{3}mR^2\right) + m\left(\frac{L}{\sqrt{2}}\right)^2$$

$$= \frac{8\times10^{-2}}{300} + \frac{1}{4}$$

So, $I_1 = I_2$

41. Moment of inertia along line perpendicular to $y = x + c$ is $\frac{mR^2}{4}$.

Line $y = x + c$ is at $\frac{c}{\sqrt{2}}$ from centre.

So, moment of inertia,

$$\frac{mR^2}{4} + m\left(\frac{c}{\sqrt{2}}\right)^2 = \frac{mR^2}{2}$$

$$c = \pm\frac{R}{\sqrt{2}}$$

42. $\tau = I\alpha$

$$\Rightarrow \quad F \times r = I\alpha$$

$$\Rightarrow \quad mg\frac{l}{2} = \frac{ml^2}{3}\alpha \Rightarrow \alpha = \frac{3g}{2l}$$

Now, $\quad a = \frac{l}{2}\alpha = \frac{l}{2}\frac{3g}{2l}$

$$a = \frac{3g}{4}$$

43. Net torque about hinge = 0

$$(kx)\frac{L}{2} = (Mg)L + (Mg)\left(\frac{L}{2}\right)$$

$$\therefore \quad x = \frac{3Mg}{k}$$

44. Let $v_A = 2v$ and $v_B = v$

From energy conservation,

$$mgl + mg(2l) = \frac{1}{2}mv^2 + \frac{1}{2}m(2v)^2$$

$$v = \sqrt{\frac{6gl}{5}}$$

45. $\tau = FR$

$$\alpha = \frac{FR}{I} = \frac{(20t - 5t^2)\times 2}{10} = 4t - t^2$$

Now, $\quad \alpha = \frac{d\omega}{dt}$

$$d\omega = (4t - t^2)dt$$

On integrating, we get

$$\omega = 2t^2 - \frac{t^3}{3}$$

$$\omega = \frac{d\theta}{dt}$$

$$d\theta = \left(2t^2 - \frac{t^3}{3}\right)dt$$

Again on integrating, we get

$$\theta = \frac{2t^3}{3} - \frac{t^4}{12}$$

At $t = 6$ s, $\theta = 36$ rad

So, $n = \frac{36}{2\pi} < 6$

So, it is more than 3 but less than 6.

46. Impulse = Change in momentum

$$\frac{JL}{2} = \frac{mL^2}{12}\omega \Rightarrow \omega = \frac{6J}{mL}$$

For $\theta = \pi/2, t = \frac{\pi}{2\omega} = \frac{\pi mL}{12J}$

47. $\tau = I\alpha = \frac{ml^2}{3}\alpha$ ...(i)

Now, the centre of mass is at a distance $\frac{l}{2}$ from hinged end,

$$a = \frac{l}{2}\alpha \Rightarrow \alpha = \frac{2a}{l}$$

Substituting α in equation (i), we get

$$F_x = \frac{ml^2}{3}\alpha = \frac{ml^2}{3}\frac{2a}{l}$$

$$a = \frac{3F}{2ml}x$$

The above equation is of a straight line.

48. Angular momentum about hinge point should remain conserved.

$$L_i = L_f$$

$$\left(\frac{m}{2}\right)(v)\left(\frac{l}{2}\right) = I\omega = \left[\frac{ml^2}{3} + \frac{m}{2}\left(\frac{l}{2}\right)^2\right]\cdot\omega$$

$$\therefore \quad \omega = \frac{6}{11}\cdot\left(\frac{v}{l}\right)$$

49. From conservation of linear momentum,

$$mv = MV \qquad \text{...(i)}$$

Now, from conservation of angular momentum,

$$\frac{mvL}{2} = \frac{ML^2}{12}\omega \qquad \text{...(ii)}$$

Since, it is given that the collision is elastic, so

$$\frac{1}{2}mv^2 = \frac{1}{2}MV^2 + \frac{1}{2}I\omega^2 \qquad \text{...(iii)}$$

From Eqs. (i), (ii) and (iii), we get

$$m = \frac{M}{4}$$

50. $\dfrac{K_{rot}}{K_{trans}} = \dfrac{\frac{1}{2}I\omega^2}{\frac{1}{2}mv^2} = \dfrac{9}{25}$

Here, $v = R\omega$

So, we get $I = \dfrac{9m}{25}R^2$

Now, the radius of gyration, $K = \sqrt{\dfrac{I}{m}} = \dfrac{3R}{5}$

51. From conservation of energy,

$$KE_{trans} + KE_{rot} = PE$$

$$\frac{1}{2}mv^2 + \frac{1}{2}\cdot\frac{2}{5}mR^2\omega^2 = mgh$$

On solving and substituting the given values, we get

$$h = 7.1\text{ m}$$

Hence, height attained before it stops is 7.1 m.

52. Total kinetic energy of ring

$$= \frac{1}{2}mv^2 + \frac{1}{2}I\omega^2$$

$$= \frac{1}{2}mv^2 + \frac{1}{2}mr^2\left(\frac{v}{r}\right)^2$$

$$= mv^2$$

Total kinetic energy of rod

$$= \frac{1}{2}mv^2 + \frac{1}{2}I\omega^2$$

$$= \frac{1}{2}mv^2 + \frac{1}{2}m\frac{(3r)^2}{12}\times\left(\frac{v}{r}\right)^2 \qquad (\because\ l = 3r)$$

$$= \frac{7}{8}mv^2$$

So, total kinetic energy of system

$$= mv^2 + \frac{7}{8}mv^2 = \frac{15}{8}mv^2$$

53. Since, the distance travelled by the wheel remains the same,

$\therefore \quad v_A t = v_B t$

$\Rightarrow \quad v_A = v_B$

$\Rightarrow \quad \omega_A r_A = \omega_B r_B$

r_A is half

$\therefore\ \omega_A$ will be double.

54.

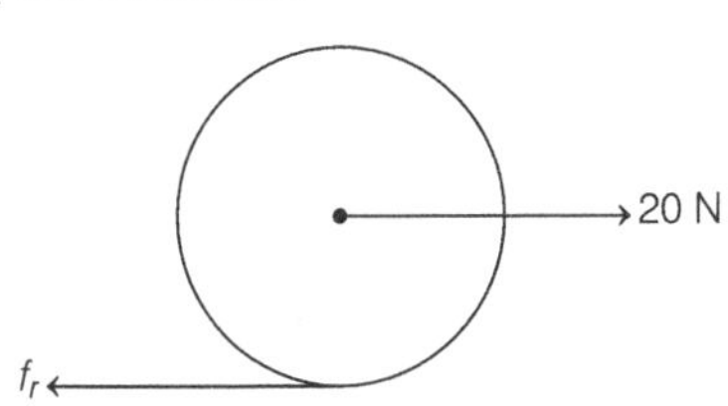

$$20 - f_r = ma \qquad \text{...(i)}$$

Now, equation of torque,

$$20\times 0 + f_r \times r = I\alpha$$

$$f_r \times r = \frac{2}{5}mr^2\alpha$$

$$f_r = \frac{2}{5}mr\alpha$$

On substituting in Eq. (i), we get

$$20 - \frac{2}{5}mr\alpha = ma$$

$$20 = \frac{2}{5}mr\alpha + ma$$

$$20 = \frac{2}{5}mr\alpha + mr\alpha \qquad (\because\ a = \alpha r)$$

$$20 = \frac{7}{5}mr\alpha \Rightarrow 100 = 7mr\alpha$$

$$\alpha = \frac{100}{7mr}$$

$$f_r = I\alpha$$

$$f_r = \frac{2mr}{5}\left(\frac{100}{7mr}\right) = \frac{40}{7}\text{ N}$$

So, $f_r = \dfrac{40}{7}$ N in the backward direction.

55. In case of pure rolling, angular acceleration can be obtained about bottom most point,

$$\alpha = \frac{\tau}{I} = \frac{FR}{\frac{5}{3}mR^2} = \frac{3F}{5mR}$$

56. As the force F moves the sphere rightwards, so friction will act towards left.

$$F - f = ma \qquad \text{...(i)}$$

Now, torque about centre,

$$\tau = F_r + f_r$$

$$\Rightarrow \quad I\alpha = F_r + f_r = \frac{mR^2}{2}\frac{a}{R}$$

$$\Rightarrow \quad F + f = \frac{mR^2}{2}\cdot\frac{a}{R}$$

$$F + f = ma/2 \qquad \text{...(ii)}$$

So, from Eqs. (i) and (ii), we get

$$a = 4F/3m$$

57. $\tau = I\alpha$

$$T\times R = \frac{mR^2}{2}\times\frac{a}{R}$$

$$T = \frac{ma}{2}$$

Now, $mg - T = ma$

$$\Rightarrow \quad mg - \frac{ma}{2} = ma \Rightarrow a = \frac{2g}{3}$$

So, $T = \dfrac{m}{2}\left(\dfrac{2g}{3}\right) = \dfrac{mg}{3}$

Tension and acceleration in the string are $\dfrac{mg}{3}$ and $\dfrac{2g}{3}$.

58. $v = \sqrt{2as} = \sqrt{2\left[\dfrac{g\sin\theta}{1 + I/mR^2}\right]s}$

59. From work-energy theorem,

$$Mgh = \frac{1}{2}Mv^2 + \frac{1}{2}I\omega^2$$

Here, $I = \frac{2}{5}MR^2$ and $v = \omega R$

So, we get $\omega = \frac{2}{R}\sqrt{\frac{gh}{3}}$

60. Frictional force opposes the motion of body. So, it decreases the linear velocity and increases the angular velocity, as the friction supports the rotational motion of sphere.

In pure rolling,

frictional force, $f = \dfrac{mg\sin\theta}{1 + \dfrac{MR^2}{I}}$

$\therefore \quad f \propto \sin\theta$

So, as θ decreases, frictional force also decreases.

61. $mg - T = ma$ ...(i)

$FR = I\alpha$

$\Rightarrow TR = I\alpha$

$\Rightarrow TR = \frac{mR^2}{2}\cdot\alpha$

$TR = \frac{mR^2 a}{2R}$ $\left[\because \alpha = \frac{a}{R}\right]$

$\Rightarrow T = \frac{ma}{2}$ ...(ii)

From Eqs. (i) and (ii), we get

$\frac{3ma}{2} = mg$

$\Rightarrow a = \frac{2g}{3}$

62. $\tau = FR$

$I\alpha = FR$

$\frac{mR^2}{2}\frac{a}{R} = FR \Rightarrow a = \frac{2F}{m}$

63. From angular momentum conservation, $mv_0R = I\omega$

Now, the moment of inertia at point P,

$I = mR^2 + mR^2 = 2mR^2$

$mv_0R = 2mR^2\left(\frac{v}{R}\right)$

$\Rightarrow v_0 = 2v$

$\Rightarrow v = \frac{v_0}{2}$

64. $KE_{trans} + KE_{rot} = 32.8\text{ J}$

$\frac{1}{2}mv^2 + \frac{1}{2}I\omega^2 = 32.8$

$\frac{1}{2}\times 10\times(2)^2 + \frac{1}{2}I\left(\frac{2}{0.5}\right)^2 = 32.8$ $\left[\because \omega = \frac{v}{R}\right]$

$\Rightarrow 20 + 8I = 32.8 \Rightarrow 8I = 12.8$

$\Rightarrow I = \frac{12.8}{8} \Rightarrow I = 1.6\text{ kgm}^2$

So, radius of gyration, $k = \sqrt{\frac{I}{m}}$

$\Rightarrow k = \sqrt{\frac{1.6}{10}} = \frac{4}{10} = 0.4\text{ m}$

65.

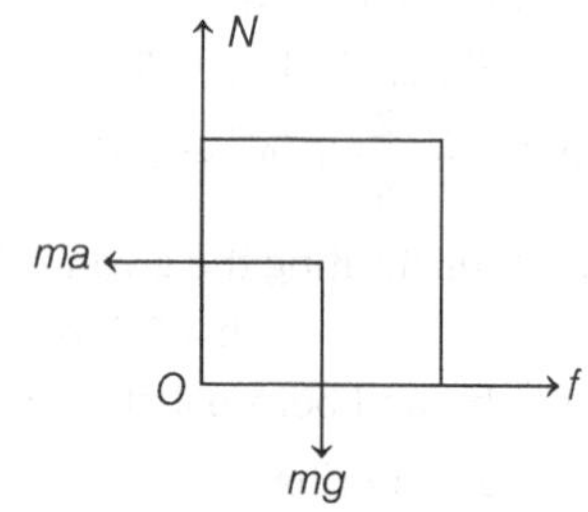

Here, N = Normal reaction,

f = friction,

ma = pseudo force.

Block will topple about an axis passing through O and perpendicular to plane of paper.

In critical case, N will pass through O.

Torque of ma about O = Torque of mg about O

$\therefore \quad (ma)\left(\frac{h}{2}\right) = (mg)\left(\frac{b}{2}\right)$ or $a = \frac{bg}{h}$

66.

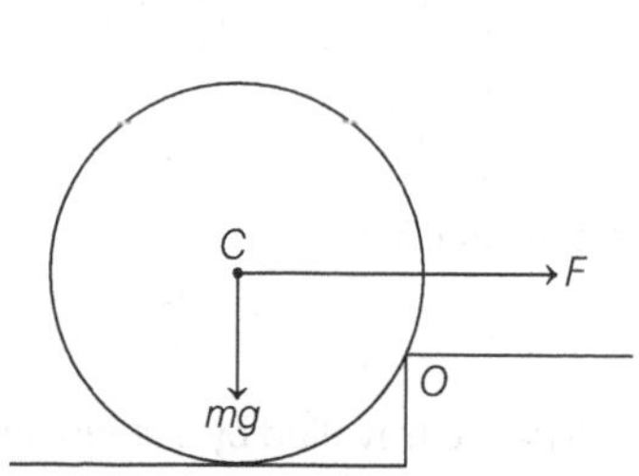

About O,

$\tau_F > \tau_{mg}$

$\therefore \quad F\left(\frac{R}{2}\right) > (mg)\sqrt{R^2 - \left(\frac{R}{2}\right)^2}$

or $F > \sqrt{3}\,mg$

67. For only translational motion, force should be applied at centre of mass. If mass of AB is m, then mass of CD is $2m$. Let distance of centre of mass from C be x. Then,

$$x = \frac{[(2m)(l) + (m)(2l)]}{(m + 2m)} = \frac{4l}{3}$$

68. From angular momentum conservation,

$mva = I\omega$

$mva = \frac{m(2a)^2}{3}\omega \Rightarrow \omega = \frac{3v}{4a}$

LEVEL 03 *Previous Years Questions*

69. Given, the mass of the circular ring = M

The radius of the circular ring = R

We know that, the moment of inertia of the circular ring, $I = MR^2$

An arc corresponding to a 90° sector in the circular ring is removed, means one-fourth part of circular ring is removed. Then, the remaining mass of the circular ring, $M' = \frac{M}{4}$

The new moment of inertia, $I' = M'R^2 = \frac{MR^2}{4}$

Now, the moment of inertia of the remaining part,

$$I'' = I - I' = MR^2 - \frac{MR^2}{4} \Rightarrow I'' = \frac{3MR^2}{4}$$

Comparing with $I'' = KMR^2$

The value of $K = 3/4$.

70. Initial angular speed of wheel,

$$\omega_0 = 2\pi f_0 = 2\pi \times \frac{360}{60} \text{ rad/s}$$
$$= 12\pi \text{ rad/s}$$

Final angular speed of wheel, $\omega = 2\pi f$

$$= 2\pi \times \frac{1200}{60} \text{ rad/s} = 40\pi \text{ rad/s}$$

$$t = 14 \text{ s}$$

From the equation of rotational motion,

$$\omega = \omega_0 + \alpha t \Rightarrow \alpha = \frac{\omega - \omega_0}{t}$$
$$= \frac{40\pi - 12\pi}{14} = \frac{28\pi}{14} = 2\pi \text{ rad/s}^2$$

71. Given, radius, $R = 2$ m,

mass, $m = 100$ kg and

v_{CM} (velocity centre of mass) = v = 20 cm/s

$= 20 \times 10^{-2}$ m/s

Then, according to work energy theorem, the work done in stopping the disc is equal to the change in its kinetic energy, i.e.

$$W = KE_f - KE_i$$

As, the disc stops at the end, so final velocity is zero.

Thus, $KE_f = 0$

Since, the disc is rolling, so its initial kinetic energy would have both rotational and translational kinetic energy component.

$$KE_i = KE_R + KE_T = \frac{1}{2}I\omega^2 + \frac{1}{2}mv^2$$
$$= \frac{1}{2}\left(\frac{1}{2}mR^2\right)\omega^2 + \frac{1}{2}mv^2 \quad \left[\because \text{for disc, } I = \frac{1}{2}mR^2\right]$$
$$= \frac{1}{4}mR^2\omega^2 + \frac{1}{2}mv^2$$
$$= \frac{1}{4}mR^2\frac{v^2}{R^2} + \frac{1}{2}mv^2 \quad [\because v = r\omega]$$
$$= \frac{3}{4}mv^2$$

$$\therefore W = |KE_i| = \frac{3}{4}mv^2$$

Substituting the given values, we get

$$W = \frac{3}{4} \times 100 \times (20 \times 10^{-2})^2$$
$$= \frac{3}{4} \times 400 \times 100 \times 10^{-4} = 3 \text{ J}$$

72. Key Idea *According to work-energy theorem, the change in kinetic energy of a particle is the amount of work done on the particle to move, i.e.*

$$W = -\Delta KE = KE_f - KE_i$$

Given, mass of cylinder, $m = 2$ kg

Radius of cylinder, $r = 4$ cm $= 4 \times 10^{-2}$ m

Rotational velocity, $\omega = 3$ rpm $= 3 \times \frac{2\pi}{60} = \frac{\pi}{10}$ rad/s

and $\theta = 2\pi$ revolution $= 2\pi \times 2\pi = 4\pi^2$ rad.

The work done in rotating an object by an angle θ from rest is given by $W = \tau\theta$

As the cylinder is brought to rest, so the work done will be negative.

According to work-energy theorem,

Work done = Change in rotational kinetic energy

$$-\tau\theta = \frac{1}{2}I\omega_f^2 - \frac{1}{2}I\omega_i^2 = \frac{1}{2}I(\omega_f^2 - \omega_i^2)$$

$$\Rightarrow \quad \tau = \frac{I(\omega_i^2)}{2\theta} \quad [\because \omega_f = 0]$$
$$= \frac{1}{2}\left(\frac{1}{2}mr^2\right)\frac{\omega_i^2}{\theta} \quad \left[I = \frac{1}{2}MR^2 \text{ (for cylinder)}\right]$$
$$= \frac{1}{4}mr^2\frac{\omega^2}{\theta} \quad [\because \omega_i = \omega]$$
$$= \frac{1}{4} \times 2 \times (4 \times 10^{-2})^2 \times \left(\frac{\pi}{10}\right)^2 \times \frac{1}{4\pi^2}$$
$$= \frac{1}{4} \times 2 \times 16 \times 10^{-4} \times \frac{\pi^2}{100} \times \frac{1}{4\pi^2}$$
$$= \frac{2}{100} \times 10^{-4} = 2 \times 10^{-6} \text{ N-m}$$

73. Moment of inertia of a rotating solid sphere about its symmetrical (diametric) axis is given as, $I = \frac{2}{5}mR^2$

Rotational kinetic energy of solid sphere is

$$K_r = \frac{1}{2}I\omega^2 = \frac{1}{2} \times \frac{2}{5}mR^2\omega^2 = \frac{1}{5}mR^2\omega^2$$

Angular velocity, $\omega = V_{CM}R$

As, we know that external torque, $\tau_{ext} = \frac{dL}{dt}$

where, L is the angular momentum.

Since, in the given condition, $\tau_{ext} = 0$

$$\Rightarrow \quad \frac{dL}{dt} = 0 \text{ or } L = \text{constant}$$

Hence, when the radius of the sphere is increased keeping its mass same, only the angular momentum remains constant. But other quantities like moment of inertia, rotational kinetic energy and angular velocity changes.

74. **Thinking Process** When no external torque acts on system, angular momentum of system remains constant.

Angular momentum before contact

$$= I_1\omega_1 + I_2\omega_2$$

Angular momentum after the discs brought into contact

$$= I_{net}\omega = (I_1 + I_2)\,\omega$$

So, final angular speed of system = ω

$$= \frac{I_1\omega_1 + I_2\omega_2}{I_1 + I_2}$$

Now, to calculate loss of energy, we subtract initial and final energies of system.

$$\Rightarrow \text{Loss of energy} = \frac{1}{2} I\omega_1^2 + \frac{1}{2} I\omega_2^2 - \frac{1}{2}(2I)\,\omega^2$$

$$= \frac{1}{4} I(\omega_1 - \omega_2)^2$$

75. Considering the information given in the question, let us draw the figure

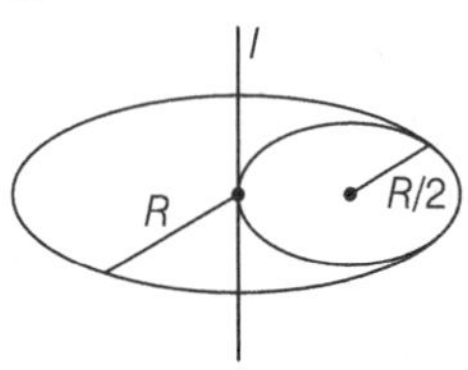

If the above figure is considered, then moment of inertia of disc will be given as

$$I = I_{remain} + I_{(R/2)} \Rightarrow I_{remain} = I - I_{(R/2)}$$

Putting the values, we get

$$= \frac{MR^2}{2} - \left[\frac{\frac{M}{4}\left(\frac{R}{2}\right)^2}{2} + \frac{M}{4}\left(\frac{R}{2}\right)^2\right]$$

$$= \frac{MR^2}{2} - \left[\frac{MR^2}{32} + \frac{MR^2}{16}\right]$$

$$= \frac{MR^2}{2} - \left[\frac{MR^2 + 2MR^2}{32}\right]$$

$$= \frac{MR^2}{2} - \frac{3MR^2}{32} = \frac{16MR^2 - 3MR^2}{32}$$

$$I_{remain} = \frac{13MR^2}{32}$$

76. COM of m_1 and m_2 masses lies at

$$r = \frac{m_1r_1 + m_2r_2}{m_1 + m_2}$$

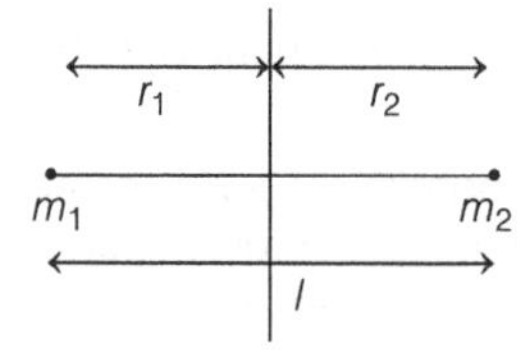

$\therefore$ Moment of inertia of the point masses about the given axis is

$$I = \Sigma m_i r_i^2 \Rightarrow I = m_1r_1^2 + m_2r_2^2$$

$$= m_1\left(\frac{m_2 l}{m_1 + m_2}\right)^2 + m_2\left(\frac{m_1 l}{m_1 + m_2}\right)^2$$

$$= \frac{m_1m_2l^2}{(m_1 + m_2)^2}(m_2 + m_1) = \frac{m_1m_2l^2}{(m_1 + m_2)}$$

77. The total moment of inertia of the system is

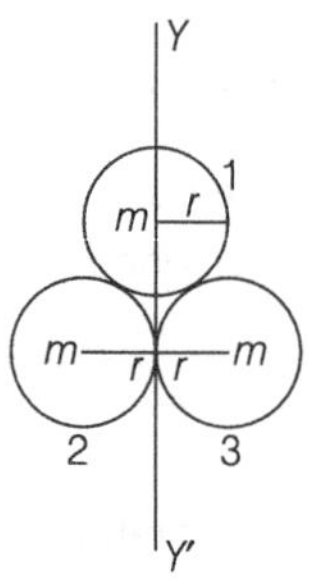

$$I = I_1 + I_2 + I_3 \quad \ldots\text{(i)}$$

Here, $\quad I_1 = \frac{2}{3}mr^2$

$$I_2 = I_3 = \frac{2}{3}mr^2 + mr^2$$

[From parallel axis theorem]

$$= \frac{5}{3}mr^2$$

From Eq. (i), $I = \frac{2}{3}mr^2 + 2 \times \frac{5}{3}mr^2$

$$= mr^2\left(\frac{2}{3} + \frac{10}{3}\right)$$

$$I = 4mr^2$$

78. **Concept** Apply parallel axes theorem of moment of inertia.

According to parallel axes theorem of moment of inertia,

$$I = I_{CM} + Mh^2$$

So, $\quad I = I_0 + M\left(\frac{L}{2}\right)^2$

$$\Rightarrow \quad I = I_0 + \frac{ML^2}{4}$$

79. Loss of energy is given by

$$\Delta E = \frac{1}{2} I_t \omega_i^2 - \frac{1}{2}\frac{I_t^2 \omega_i^2}{(I_t + I_b)}$$

$$= \frac{1}{2}\frac{I_b I_t \omega_i^2}{(I_t + I_b)}$$

80. Kinetic energy of rotation is

$$K_{rot} = \frac{1}{2} I\omega^2 = \frac{1}{2} Mk^2 \frac{v^2}{R^2} \qquad \left[\because K = \sqrt{\frac{I}{M}}\right]$$

where, k is radius of gyration.

Kinetic energy of translation is

$$K_{trans} = \frac{1}{2} Mv^2$$

Thus, total energy

$$E = K_{rot} + K_{trans} = \frac{1}{2} Mk^2 \frac{v^2}{R^2} + \frac{1}{2} Mv^2$$

$$= \frac{1}{2} Mv^2 \left(\frac{k^2}{R^2} + 1\right)$$

$$= \frac{1}{2}\frac{Mv^2}{R^2}(k^2 + R^2)$$

Hence, $$\frac{K_{rot}}{E} = \frac{\frac{1}{2} Mk^2 \frac{v^2}{R^2}}{\frac{1}{2}\frac{Mv^2}{R^2}(k^2 + R^2)} = \frac{k^2}{k^2 + R^2}$$

81. Concept The total kinetic energy of the ball rolling on a table without slipping is equal to its rotational kinetic energy and translational kinetic energy. Total kinetic energy of spherical ball is given by

K = Kinetic energy rotational (K_{rot}) + Kinetic energy translational (K_{trans})

$$= \frac{1}{2} I\omega^2 + \frac{1}{2} mv^2$$

For sphere, moment of inertia about its diameter

$$I = \frac{2}{5} mr^2$$

$$\therefore \quad K = \frac{1}{2}\left(\frac{2}{5} mr^2\right)\omega^2 + \frac{1}{2} mv^2$$

$$= \frac{1}{5} mr^2 \omega^2 + \frac{1}{2} mv^2$$

$$= \frac{1}{5} mv^2 + \frac{1}{2} mv^2 \qquad (\text{as } v = r\omega)$$

$$= \frac{7}{10} mv^2$$

$$\therefore \quad \frac{K_r}{K} = \frac{\frac{1}{5} mv^2}{\frac{7}{10} mv^2} = \frac{2}{7}$$

82. Acceleration of the centre of mass of the rolling body is given by

$$a = \frac{g\sin\theta}{1 + \left(\frac{I}{MR^2}\right)}$$

Moment of inertia of the ring about an axis perpendicular to the plane of the ring and passing through its centre is given by

$$I = MR^2$$

$$\therefore \quad a = \frac{g\sin\theta}{1 + MR^2/MR^2}$$

$$= \frac{g\sin 30^\circ}{1 + 1} = \frac{g}{4}$$

83. A flywheel is a large heavy wheel with a long cylindrical axle supported on ball bearings. Its centre of mass lies on its axis of rotation, so that it remains at rest in any position. Rotational kinetic energy of flywheel is given by

$$K_{rot} = \frac{1}{2} I\omega^2$$

where, I = moment of inertia of the wheel about the axis of rotation,

ω = angular velocity of flywheel

Given, rotational kinetic energy, K_r = 360 J

Angular velocity, ω = 30 rad/s

$$\therefore \quad I = \frac{2K_r}{\omega^2} = \frac{2 \times 360}{(30)^2} = 0.8 \text{ kg-m}^2$$

84.

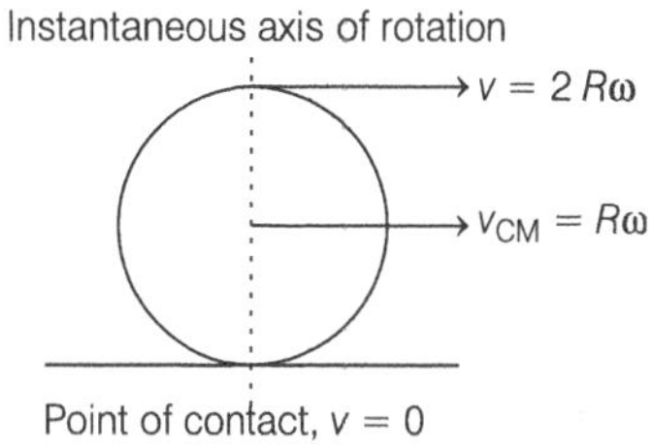

Velocity of instantaneous axis of rotation

$$= \frac{2\pi R}{T} = \omega R = v_{CM}$$

85. Angular momentum about the point of contact, for solid homogeneous sphere of mass M and radius R is conserved.

Chapter 11

GRAVITATION

LEVEL 01 Objective Questions

1 A planet moving around sun sweeps area A_1 in 2 days, A_2 in 3 days and A_3 in 6 days. Then, the relation between A_1, A_2 and A_3 is

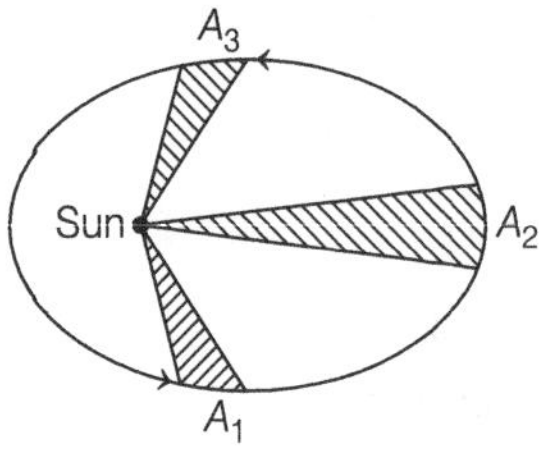

(a) $3A_1 = 2A_2 = A_3$
(b) $2A_1 = 3A_2 = 6A_3$
(c) $3A_1 = 2A_2 = 6A_3$
(d) $6A_1 = 3A_2 = 2A_3$

2 If a graph is plotted between T^2 and r^3 for a planet, then its slope will be

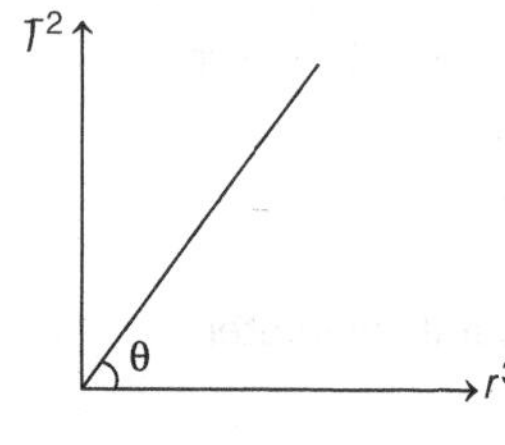

(a) $\frac{4\pi^2}{GM}$
(b) $\frac{GM}{4\pi^2}$
(c) $4\pi GM$
(d) $\frac{3\pi^2}{GM}$

3 If A is areal velocity of a planet of mass M, its angular momentum is

(a) $\frac{M}{A}$
(b) $2MA$
(c) A^2M
(d) MA

4 Which of the following quantities does not depend on the orbital radius of the satellite?

(a) $\frac{T}{R}$
(b) $\frac{T^2}{R^2}$
(c) $\frac{T^2}{R}$
(d) $\frac{T^2}{R^3}$

5 A small satellite is in elliptical orbit around earth as shown below. If L denotes the magnitude of its angular momentum and K denotes kinetic energy, then

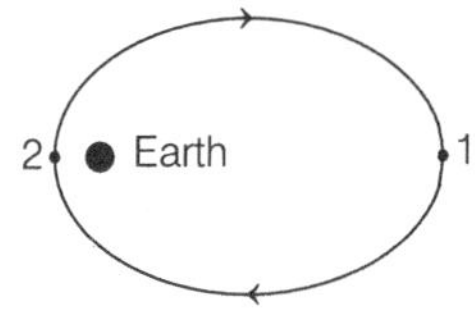

(a) $L_2 > L_1$ and $K_2 > K_1$
(b) $L_2 > L_1$ and $K_2 = K_1$
(c) $L_2 < L_1$ and $K_2 = K_1$
(d) $L_2 = L_1$ and $K_2 > K_1$

6 The dependence of acceleration due to gravity g on the distance r from the centre of the earth, assumed to be a sphere of radius R of uniform density is as shown in figures given below. The correct figure is

(a) 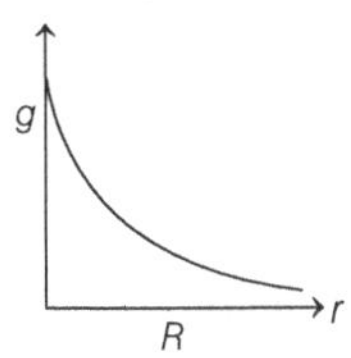

(b) 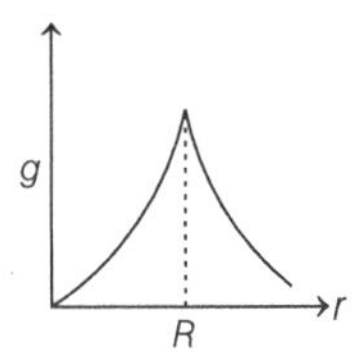

(c) 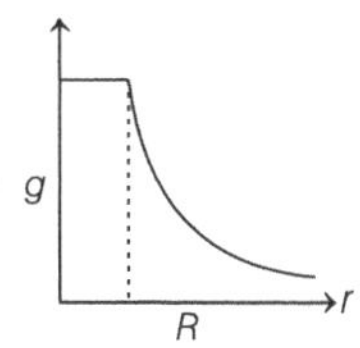

(d) 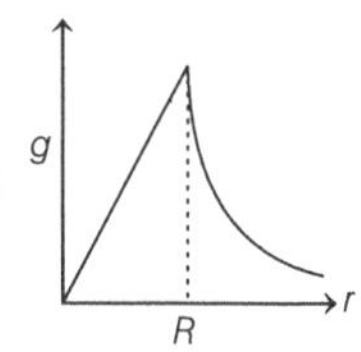

7 A body weighs 250 N on the surface of the earth. How much will it weigh half way down to the centre of the earth?

(a) 125 N
(b) 150 N
(c) 175 N
(d) 250 N

8 A planet has twice the density of earth but the acceleration due to gravity on its surface is exactly the same as that on the surface of earth. Its radius in terms of earth's radius R will be

(a) $\frac{R}{4}$
(b) $\frac{R}{2}$
(c) $\frac{R}{3}$
(d) $\frac{R}{8}$

9 The fractional change in the value of free-fall acceleration g for a particle, when it is lifted from the surface to an elevation $h (h << R)$ is

(a) $\frac{h}{R}$
(b) $\frac{2h}{R}$
(c) $-\frac{2h}{R}$
(d) $-\frac{h}{R}$

10 If the density of the earth is doubled keeping its radius constant, then acceleration due to gravity will be (Take, $g = 9.8 \text{ m/s}^2$)

(a) 19.6 m/s^2
(b) 9.8 m/s^2
(c) 4.9 m/s^2
(d) 2.45 m/s^2

11 The rotation of the earth having R radius about its axis speeds up to a value such that a man at latitude angle 60° feels weightlessness. The duration of the day in such a case is

(a) $2\pi\sqrt{\frac{R}{g}}$
(b) $\pi\sqrt{\frac{R}{g}}$
(c) $2\pi\sqrt{\frac{g}{R}}$
(d) $4\pi\sqrt{\frac{R}{g}}$

12 If a man at the equator would weigh (3/5)th of his weight, the angular speed of the earth is

(a) $\sqrt{\frac{2g}{5R}}$
(b) $\sqrt{\frac{g}{R}}$
(c) $\sqrt{\frac{R}{g}}$
(d) $\sqrt{\frac{2R}{5g}}$

13 Two identical particles each of mass m are separated by a distance $2R$. Find the gravitational field strength at P.

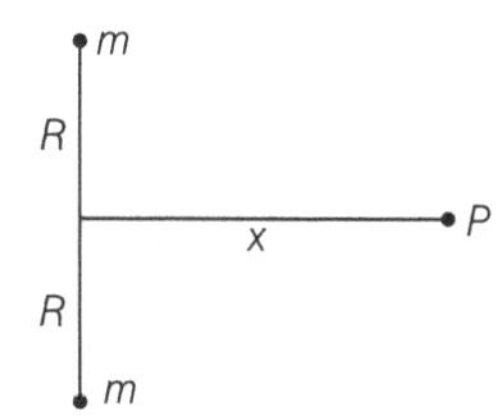

(a) $\frac{2Gmx}{(R^2 + x^2)^{3/2}}$
(b) $\frac{Gmx}{(R^2 + x^2)^3}$
(c) $\frac{Gmx}{(R^2 + x^2)^2}$
(d) None of these

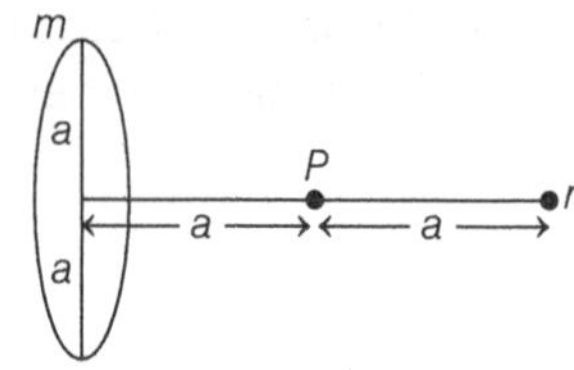

14 A particle of mass m is placed at a distance $2a$ from the centre of a ring of mass m on its axis. Gravitational field intensity at point P is

(a) $\frac{Gm}{a^2}\left(1-\frac{1}{2\sqrt{2}}\right)$ (b) $\frac{Gm}{a^2}\left(\frac{1}{\sqrt{2}}-1\right)$

(c) $\frac{Gm}{a^2}(1-2\sqrt{2})$ (d) None of these

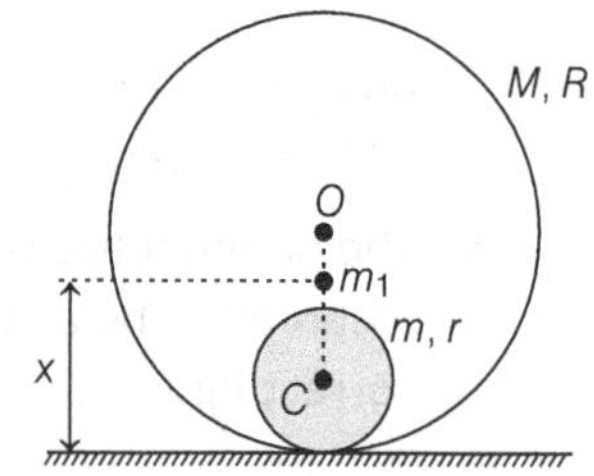

15 A solid sphere of mass m and radius r is placed inside a hollow thin spherical shell of mass M and radius R as shown in the adjoining figure. A particle of mass m_1 is placed on the line joining the two centres at a distance x from the point of contact of the sphere and the shell. If $2r < x < 2R$, then the magnitude of gravitational force due to the sphere and shell on this particle is

(a) $\frac{Gmm_1x}{r^3}$ (b) $\frac{Gmm_1(x-r)}{r^3}$

(c) $\frac{Gmm_1}{(x-r)^2}$ (d) $\frac{GMm_1}{(x-r)^2}$

16 A uniform solid sphere of mass M and radius a is surrounded symmetrically by a uniform thin spherical shell of equal mass and radius $2a$. The gravitational field at a distance $\frac{3}{2}a$ from the centre is

(a) $\frac{4}{9}\frac{GM}{a^2}$ (b) $\frac{1}{4}\frac{GM}{a^2}$ (c) $\frac{2}{3}\frac{GM}{a^2}$ (d) $\frac{21}{25}\frac{GM}{a^2}$

17 A uniform spherical shell gradually expands maintaining shape. Then, gravitational potential at centre will (mass remains constant)

(a) increases (b) decreases

(c) remains constant (d) oscillates

18 In a certain region of space, the gravitational field is given by $-k/r$, where r is the distance and k is a constant. If the gravitational potential at $r=r_0$ be V_0, then what is the expression for the gravitational potential (V)?

(a) $V_0+k\ln\left(\frac{r}{r_0}\right)$ (b) $V_0+k\ln\left(\frac{r_0}{r}\right)$ (c) $V_0+k\log\left(\frac{r}{r_0}\right)$ (d) $V_0+k\log\left(\frac{r_0}{r}\right)$

19 A particle of mass M is placed at the centre of a uniform spherical shell of mass $2M$ and radius R. The gravitational potential on the surface of the shell is

(a) $-\frac{GM}{R}$ (b) $-\frac{3GM}{R}$ (c) $-\frac{2GM}{R}$ (d) zero

20 A particle of mass M is situated at the centre of a spherical shell of same mass and radius a. The gravitational potential at a point situated at $\frac{a}{2}$ distance from the centre, will be

(a) $-\frac{4GM}{a}$ (b) $-\frac{3GM}{a}$

(c) $\frac{2GM}{a}$ (d) $-\frac{GM}{a}$

21 A thin wire of mass m and length l is bent to form a semicircle. The gravitational potential at the centre of semicircle is

(a) $-\frac{\pi Gm}{l}$ (b) $-\frac{2\pi Gm}{l}$

(c) $-\frac{3\pi Gm}{l}$ (d) zero

22 The additional kinetic energy to be provided to a satellite of mass m revolving around a planet of mass M, to transfer it from a circular orbit of radius R_1 to another of radius R_2 $(R_2 > R_1)$ is

(a) $GmM\left(\frac{1}{R_1^2}-\frac{1}{R_2^2}\right)$ (b) $GmM\left(\frac{1}{R_1}-\frac{1}{R_2}\right)$ (c) $2GmM\left(\frac{1}{R_1}-\frac{1}{R_2}\right)$ (d) $\frac{1}{2}GmM\left(\frac{1}{R_1}-\frac{1}{R_2}\right)$

23 Two satellites of earth S_1 and S_2 are moving in the same orbit. The mass of S_1 is four times the mass of S_2. Which one of the following statements is true ?

(a) The potential energies of earth and satellite in the two cases are equal.
(b) S_1 and S_2 are moving with the same speed.
(c) The kinetic energies of the two satellites are equal.
(d) The time period of S_1 is four times that of S_2.

24 Which pair of PE and KE is possible for a polar satellite? [Assume PE = 0 at infinity]

(a) KE = 1000 J, PE = – 500 J (b) KE = 500 J, PE = – 1000 J
(c) KE = 0, PE = – 1000 J (d) KE = 1000 J, PE = 1000 J

25 A satellite is orbiting in a circular path near earth surface. How much times its kinetic energy make to escape it from its orbit?

(a) 2 times (b) Half
(c) 4 times (d) 3 times

26 The time period of an artificial satellite in a circular orbit of radius R is 2 days and its orbital velocity is v_o. If time period of another satellite in a circular orbit is 16 days, then

(a) its radius of orbit is $4R$ and orbital velocity is v_o
(b) its radius of orbit is $4R$ and orbital velocity is $\frac{v_o}{2}$
(c) its radius of orbit is $2R$ and orbital velocity is v_o
(d) its radius of orbit is $2R$ and orbital velocity is $\frac{v_o}{2}$

27 Three planets of same density have radii R_1, R_2 and R_3, such that $R_1 = 2R_2 = 3R_3$. The gravitational field at their respective surfaces are g_1, g_2 and g_3 and escape velocities from their surfaces are v_1, v_2 and v_3, then

(1) $\frac{g_1}{g_2}=\frac{1}{2}$ (2) $\frac{g_1}{g_3}=3$ (3) $\frac{v_1}{v_2}=\frac{1}{4}$ (4) $\frac{v_1}{v_3}=3$

(a) 1 and 4 are correct (b) 1 and 3 are correct
(c) 2 and 3 are correct (d) 2 and 4 are correct

28 A satellite which is geostationary in a particular orbit is taken to another orbit. Its distance from the centre of earth in the new orbit is 2 times that of the earlier orbit. The time period in the new orbit is

(a) 48 h (b) $48\sqrt{2}$ h
(c) 24 h (d) $24\sqrt{2}$ h

29 A satellite of mass m is circulating around the earth with constant angular velocity. If radius of the orbit is R_o and mass of the earth is M, the ratio of angular momentum of the satellite about the centre of the earth and its kinetic energy will be

(a) $\frac{2}{M}\sqrt{\frac{R_o^2}{GM}}$ (b) $\sqrt{\frac{R_o}{GM}}$ (c) $\frac{2}{m}\sqrt{\frac{R_o}{GM}}$ (d) $2\sqrt{\frac{R_o^3}{GM}}$

30 The escape speed of a body on the earth's surface is 11.2 km/s. A body is projected with thrice of this speed. The speed of the body when it escapes the gravitational pull of earth is

(a) 11.2 km/s (b) $22.4\sqrt{2}$ km/s (c) $\frac{22.4}{\sqrt{2}}$ km/s (d) $22.4\sqrt{3}$ km/s

31 The orbital velocity of an artificial satellite in a circular orbit just above the earth's surface is v_o. For a satellite orbiting at an altitude of half the earth's radius above surface, the orbital velocity is
(a) $(3/2)\,v_o$ (b) $\sqrt{(3/2)}\,v_o$ (c) $(\sqrt{2/3})v_o$ (d) $(2/3)\,v_o$

32 The radius (in km) to which the present radius of the earth ($R = 6400$ km) to be compressed, so that the escape velocity from the surface of earth is increased 10 times, is
(a) 6.4 (b) 64 (c) 640 (d) 4800

33 If gravitational force between two masses varies as $F \propto \frac{1}{r^3}$, then orbital speed of satellite varies with radius of orbit as
(a) $v \propto \frac{1}{r}$ (b) $v \propto r^0$
(c) $v \propto \frac{1}{r^2}$ (d) $v \propto r$

34 In a vertical spring-mass system, if stretching in spring at equilibrium near earth surface is x_0, then the extension in spring at a height 1600 km from earth surface is (radius of earth, $R = 6400$ km)
(a) $\frac{5}{16}x_0$ (b) $\frac{5x_0}{4}$ (c) $\frac{16x_0}{25}$ (d) $\frac{x_0}{4}$

LEVEL 02 Objective Questions

35 A satellite of mass m is in a circular orbit of radius $2R_E$ about the earth. The energy required to transfer it to a circular orbit of radius $4R_E$ is (where, M_E and R_E is the mass and radius of the earth, respectively)
(a) $\frac{GM_Em}{2R_E}$ (b) $\frac{GM_Em}{4R_E}$ (c) $\frac{GM_Em}{8R_E}$ (d) $\frac{GM_Em}{16R_E}$

36 A satellite is orbiting the earth in a circular orbit of radius r. Its
(a) kinetic energy varies as r (b) angular momentum varies as $\frac{1}{\sqrt{r}}$
(c) linear momentum varies as $\frac{1}{r}$ (d) frequency of revolution varies as $\frac{1}{r^{3/2}}$

37 The height from the surface of earth at which the gravitational potential energy of a ball of mass m is half of that at the centre of earth is (where, R is the radius of earth)
(a) $\frac{R}{4}$ (b) $\frac{R}{3}$ (c) $\frac{3R}{4}$ (d) $\frac{4R}{3}$

38 A tunnel is dug along the diameter of the earth. There is a particle of mass m at the centre of the tunnel. Find the minimum velocity given to the particle, so that it just reaches to the surface of the earth. (M = mass of earth and R = radius of earth)
(a) $\sqrt{\frac{GM}{R}}$ (b) $\sqrt{\frac{GM}{2R}}$ (c) $\sqrt{\frac{2GM}{R}}$ (d) $\sqrt{\frac{3GM}{R}}$

39 A satellite is revolving around earth in a circular orbit of radius $2R$. If the satellite is stopped suddenly and is then released, then the velocity with which it strikes the surface of earth is (g = acceleration due to gravity on earth's surface and R = radius of earth)
(a) $\sqrt{2gR}$ (b) $\sqrt{gR}$
(c) $\sqrt{gR/2}$ (d) $\sqrt{4gR}$

40 A narrow tunnel is dug along the diameter of earth and a particle of mass m_0 is placed at $\frac{R}{2}$ distance from centre. Find the escape speed of the particle from that place.

(a) $\sqrt{\frac{11}{8}\frac{GM}{R}}$ (b) $\sqrt{\frac{11}{4}\frac{GM}{R}}$ (c) $\sqrt{\frac{4GM}{R}}$ (d) $\sqrt{\frac{GM}{2R}}$

41 A research satellite of mass 200 kg circles the earth in an orbit of average radius $3R/2$, where R is the radius of earth. Assuming the gravitational pull on the mass of 1 kg on the earth's surface to be 10 N, the pull on the satellite will be

(a) 1212 N (b) 889 N
(c) 1280 N (d) 960 N

42 A particle is thrown vertically upwards from the surface of earth and it reaches to a maximum height equal to the radius of earth. The ratio of the velocity of projection to the escape velocity on surface of earth is

(a) $\frac{1}{\sqrt{2}}$ (b) $\frac{1}{2}$
(c) $\frac{1}{4}$ (d) $\frac{1}{2\sqrt{2}}$

43 Two massive particles of masses M and m $(M > m)$ are separated by a distance l. They rotate with equal angular velocity under their gravitational attraction about a common axis through centre of mass. The linear speed of the heavier particle is

(a) $\sqrt{\frac{GMm}{(M+m)l}}$ (b) $\sqrt{\frac{GM^2}{(M+m)l}}$ (c) $\sqrt{\frac{Gm}{2l}}$ (d) $\sqrt{\frac{2Gm^2}{(M+m)l}}$

44 A body of mass m is kept at a small height h above the ground. If the radius of the earth is R and its mass is M, the potential energy of the body and earth system (with $h = \infty$ being the reference position) is

(a) $\frac{GMm}{R} + mgh$ (b) $\frac{-GMm}{R} + mgh$ (c) $\frac{GMm}{R} - mgh$ (d) $-\frac{GMm}{R} - mgh$

45 A satellite is seen after every 8 h over equator at a place on the earth when its sense of rotation is opposite to the earth. The time interval after which it can be seen at the same place when the sense of rotation of earth and satellite is same will be

(a) 8 h (b) 12 h
(c) 24 h (d) 6 h

46 Two particles of equal mass m each move in a circle of radius r under the action of their mutual gravitational attraction. Time period of each particle is

(a) $\frac{2\pi r^{3/2}}{\sqrt{Gm}}$ (b) $\frac{4\pi r^{3/2}}{\sqrt{Gm}}$ (c) $\frac{4\pi^2 r^{3/2}}{\sqrt{Gm}}$ (d) None of these

47 The escape velocity for a planet is v_e. A particle starts from rest at a large distance from the planet, reaches the planet only under gravitational attraction and passes through a smooth tunnel through its centre. Its speed at the centre of the planet will be

(a) $\sqrt{1.5}\, v_e$ (b) $\frac{v_e}{\sqrt{2}}$
(c) v_e (d) zero

48 A satellite is projected with a velocity $\sqrt{1.5}$ times the orbital velocity just above the surface of earth. Initial velocity of the satellite is perpendicular to the surface of earth. The maximum distance of the satellite from surface of earth will be

(a) $2R_e$ (b) $8R_e$
(c) $4R_e$ (d) $3R_e$

49 A spherical part of radius $R/2$ is excavated from the asteroid of mass M as shown in the adjoining figure. The gravitational acceleration at a point P on the surface of the asteroid just above the excavation is

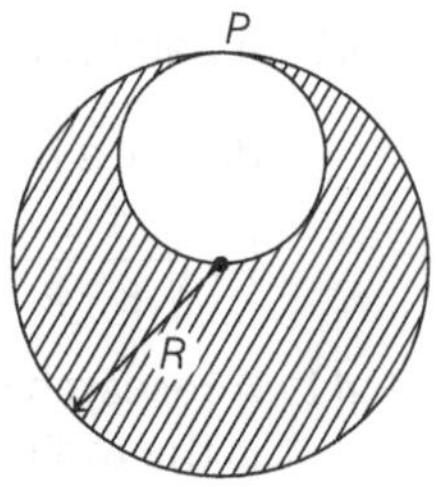

(a) $\frac{GM}{R^2}$ (b) $\frac{GM}{2R^2}$
(c) $\frac{GM}{8R^2}$ (d) $\frac{7GM}{8R^2}$

50 A spherical hole is made in a solid sphere of radius R. The mass of the sphere before hollowing was M. The gravitational field at the centre of the hole due to the remaining mass is

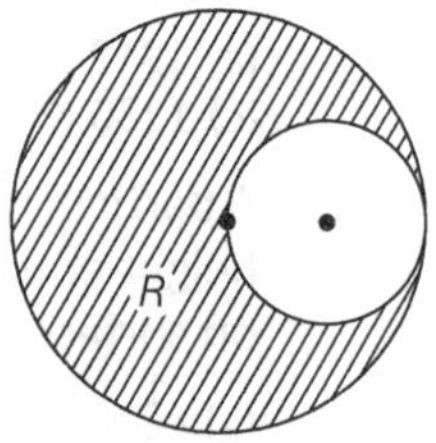

(a) zero (b) $\frac{GM}{8R^2}$
(c) $\frac{GM}{2R^2}$ (d) $\frac{GM}{R^2}$

51 A point P lies on the axis of a fixed ring of mass M and radius R at a distance $R\sqrt{3}$ from centre O. The particle starts from P and reach at O under gravitational attraction between ring and particle. The speed of particle at distance R from centre is

(a) $\sqrt{\frac{GM}{R}}$ (b) $\sqrt{\frac{2GM}{R}}$ (c) $\sqrt{\frac{GM}{2R}(\sqrt{2}-1)}$ (d) $\sqrt{\frac{GM}{R}(\sqrt{2}-1)}$

LEVEL 03 Previous Years Questions

52 The escape velocity from the earth's surface is v. The escape velocity from the surface of another planet having a radius, four times that of earth and same mass density is **[NEET 2021]**

(a) v (b) $2v$
(c) $3v$ (d) $4v$

53 A particle of mass m is projected with a velocity $v = kv_e (k < 1)$ from the surface of the earth. (Here, v_e = escape velocity). The maximum height above the surface reached by the particle is **[NEET 2021]**

(a) $R\left(\frac{k}{1-k}\right)^2$ (b) $R\left(\frac{k}{1+k}\right)^2$ (c) $\frac{R^2k}{1+k}$ (d) $\frac{Rk^2}{1-k^2}$

54 What is the depth at which the value of acceleration due to gravity becomes $1/n$ times the value to that at the surface of earth? (Radius of earth $=R$) **[NEET 2020]**

(a) R/n^2 (b) $R(n-1)/n$
(c) $Rn/(n-1)$ (d) R/n

55 A body weighs 72 N on the surface of the earth. What is the gravitational force on it, at a height equal to half of radius of the earth? **[NEET 2020]**

(a) 32 N (b) 30 N (c) 24 N (d) 48 N

56 Assuming that the gravitational potential energy of an object at infinity is zero, the change in potential energy (final – initial) of an object of mass m, when taken to a height h from the surface of earth (of radius R) is given by, **[NEET 2019]**

(a) $-\frac{GMm}{R+h}$ (b) $\frac{GMmh}{R(R+h)}$
(c) mgh (d) $\frac{GMm}{R+h}$

57 A body weighs 200 N on the surface of the earth. How much will it weigh half way down to the centre of the earth ? **[NEET 2019]**

(a) 200 N (b) 250 N
(c) 100 N (d) 150 N

58 The work done to raise a mass m from the surface of the earth to a height h, which is equal to the radius of the earth, is **[NEET 2019]**

(a) $2mgR$ (b) $\frac{1}{2}mgR$
(c) $\frac{3}{2}mgR$ (d) mgR

59 The time period of a geo stationary satellite is 24 h, at a height $6R_E$ (R_E is the radius of earth) from surface of earth. The time period of another satellite, whose height is 2.5 R_E from surface will be **[NEET 2019; CBSE AIPMT 2012]**

(a) $6\sqrt{2}$ h (b) $12\sqrt{2}$ h
(c) $\frac{24}{2.5}$ h (d) $\frac{12}{2.5}$ h

60 The kinetic energies of a planet in an elliptical orbit about the sun, at positions A, B and C are K_A, K_B and K_C, respectively. AC is the major axis and SB is perpendicular to AC at the position of the sun S as shown in the following figure. Then, **[NEET 2018]**

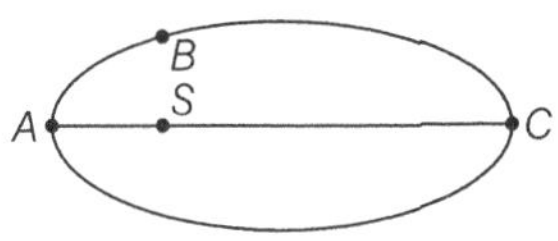

(a) $K_B < K_A < K_C$ (b) $K_A > K_B > K_C$
(c) $K_A < K_B < K_C$ (d) $K_B > K_A > K_C$

61 If the mass of the sun were ten times smaller and the universal gravitational constant were ten times larger in magnitude, which of the following is not correct? **[NEET 2018]**

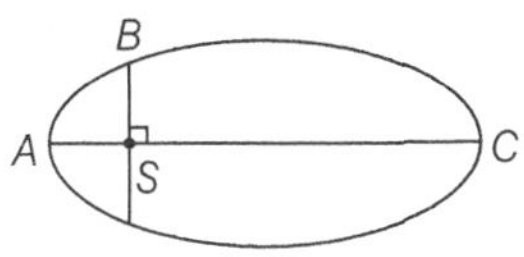

(a) Time period of a simple pendulum on the earth would decrease.
(b) Walking on the ground would become more difficult.
(c) Raindrops will fall faster.
(d) g on the earth will not change.

62 The acceleration due to gravity at a height 1 km above the earth is the same as at a depth d below the surface of earth. Then, **[NEET 2017]**

(a) $d = \frac{1}{2}$ km (b) $d = 1$ km
(c) $d = \frac{3}{2}$ km (d) $d = 2$ km

63 Two astronauts are floating in gravitational free space after having lost contact with their spaceship. The two will **[NEET 2017]**

(a) keep floating at the same distance between them
(b) move towards each other
(c) move away from each other
(d) become stationary

64 At what height from the surface of earth the gravitational potential and the value of g are -5.4×10^7 J kg^{-2} and 6.0 ms^{-2}, respectively? (Take, the radius of earth as 6400 km). [NEET 2016]

(a) 1600 km (b) 1400 km
(c) 2000 km (d) 2600 km

65 Starting from the centre of the earth having radius R, the variation of g (acceleration due to gravity) is shown by [NEET 2016]

(a) 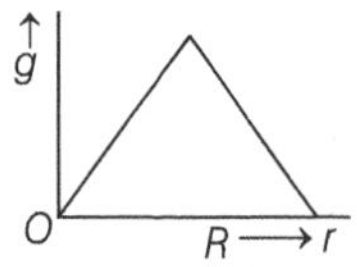

(b) 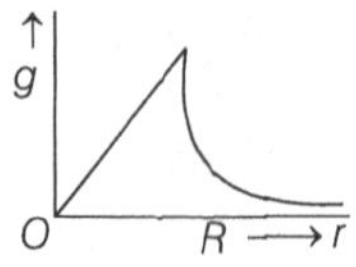

(c) 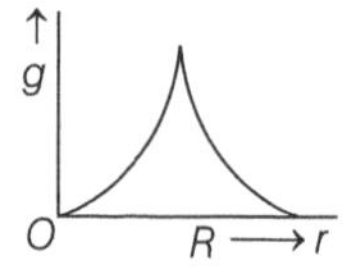

(d) 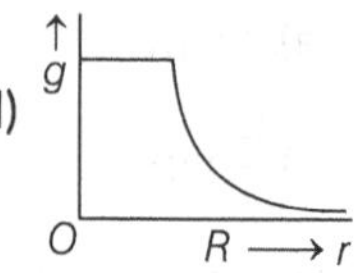

66 The ratio of escape velocity at earth (v_e) to the escape velocity at a planet (v_p) whose radius and mean density are twice as that of earth is [NEET 2016]

(a) $1:2\sqrt{2}$ (b) 1: 4
(c) $1:\sqrt{2}$ (d) 1: 2

67 A satellite of mass m is orbiting the earth (of radius R) at a height h from its surface. The total energy of the satellite in terms of g_0, the value of acceleration due to gravity at the earth's surface is [NEET 2016]

(a) $\dfrac{mg_0R^2}{2(R+h)}$ (b) $-\dfrac{mg_0R^2}{2(R+h)}$
(c) $\dfrac{2mg_0R^2}{R+h}$ (d) $-\dfrac{2mg_0R^2}{R+h}$

68 Kepler's third law states that square of period of revolution (T) of a planet around the sun is proportional to third power of average distance r between the sun and planet, i.e. $T^2 = Kr^3$, here K is constant. If the masses of the sun and planet are M and m respectively, then as per Newton's law of gravitational force of attraction between them is $F = \dfrac{GMm}{r^2}$, here G is gravitational constant. The relation between G and K is described as [CBSE AIPMT 2015]

(a) $GK = 4\pi^2$ (b) $GMK = 4\pi^2$
(c) $K = G$ (d) $K = \dfrac{I}{G}$

69 Two spherical bodies of masses M and 5M and radii R and $2R$ are released in free space with initial separation between their centres equal to 12 R. If they attract each other due to gravitational force only, then the distance covered by the smaller body before collision is [CBSE AIPMT 2015]

(a) 2.5 R (b) 4.5 R (c) 7.5 R d) 1.5 R

70 A remote sensing satellite of earth revolves in a circular orbit at a height of 0.25×10^6 m above the surface of earth. If earth's radius is 6.38×10^6 m and $g = 9.8$ ms^{-2}, then the orbital speed of the satellite is [CBSE AIPMT 2015]

(a) 7.76 kms^{-1} (b) 8.56 kms^{-1}
(c) 9.13 kms^{-1} (d) 6.67 kms^{-1}

71 A satellite S is moving in an elliptical orbit around the earth. The mass of the satellite is very small as compared to the mass of the earth. Then, [CBSE AIPMT 2015]

(a) angular momentum of S about the centre of the earth changes in direction but its magnitude remains constant
(b) total mechanical energy of S varies periodically with time
(c) linear momentum of S remains constant in magnitude
(d) acceleration of S is always directed towards the centre of the earth

72 A black hole is an object whose gravitational field is so strong that even light cannot escape from it. To what approximate radius would earth (mass = 5.98×10^{24} kg) have to be compressed to be a black hole? [CBSE AIPMT 2014]

(a) 10^{-9} m (b) 10^{-6} m
(c) 10^{-2} m (d) 10^{-7} m

73 Infinite number of bodies, each of mass 2 kg are situated on x-axis at distances 1 m, 2 m, 4 m and 8 m respectively, from the origin. The resulting gravitational potential due to this system at the origin will be [NEET 2013]

(a) $-G$ (b) $-\frac{8}{3}G$
(c) $-\frac{4}{3}G$ (d) $-4G$

74 A body of mass m taken from the earth's surface to the height equal to twice the radius (R) of the earth. The change in potential energy of body will be [NEET 2013]

(a) $mg2R$ (b) $\frac{2}{3}mgR$
(c) $3mgR$ (d) $\frac{1}{3}mgR$

75 The height at which the weight of a body becomes 1/16th, its weight on the surface of the earth (radius R), is [CBSE AIPMT 2012]

(a) 5 R (b) 15 R
(c) 3 R (d) 4 R

76 A compass needle which is allowed to move in a horizontal plane is taken to a geomagnetic pole. It [CBSE AIPMT 2012]

(a) will become rigid showing no movement
(b) will stay in any position
(c) will stay in North-South direction only
(d) will stay in East-West direction only

77 A spherical planet has a mass M_p and diameter D_p. A particle of mass m falling freely near the surface of this planet will experience an acceleration due to gravity, equal to [CBSE AIPMT 2012]

(a) $4GM_p/D_p^2$
(b) GM_pm/D_p^2
(c) GM_p/D_p^2
(d) $4GM_pm/D_p^2$

78 A planet moving along an elliptical orbit is closest to the sun at a distance r_1 and farthest away at a distance of r_2. If v_1 and v_2 are the linear velocities at these points respectively, then the ratio $\frac{v_1}{v_2}$ is [CBSE AIPMT 2011]

(a) r_2/r_1 (b) $(r_2/r_1)^2$
(c) r_1/r_2 (d) $(r_1/r_2)^2$

79 A body projected vertically from the earth reaches a height equal to earth's radius before returning to the earth. The power exerted by the gravitational force is greatest [CBSE AIPMT 2011]

(a) at the instant just before the body hits the earth
(b) it remains constant all through
(c) at the instant just after the body is projected
(d) at the highest position of the body

80 A particle of mass M is situated at the centre of a spherical shell of same mass and radius a. The gravitational potential at a point situated at $a/2$ distance from the centre, will be **[CBSE AIPMT 2010]**

(a) $-\frac{3\,GM}{a}$ (b) $-\frac{2\,GM}{a}$

(c) $-\frac{GM}{a}$ (d) $-\frac{4GM}{a}$

81 The radii of circular orbits of two satellites A and B of the earth are $4R$ and R, respectively. If the speed of satellite A is $3v$, then the speed of satellite B will be **[CBSE AIPMT 2010]**

(a) $3v/4$ (b) $6\,v$

(c) $12\,v$ (d) $3v/2$

ANSWER KEY

Level 1 Objective Questions

1 (a)	2 (a)	3 (b)	4 (d)	5 (d)	6 (d)	7 (a)	8 (b)	9 (c)	10 (a)
11 (b)	12 (a)	13 (a)	14 (a)	15 (c)	16 (a)	17 (a)	18 (a)	19 (b)	20 (b)
21 (a)	22 (d)	23 (b)	24 (b)	25 (a)	26 (b)	27 (d)	28 (b)	29 (d)	30 (b)
31 (c)	32 (b)	33 (a)	34 (c)						

Level 2 Objective Questions

35 (c)	36 (d)	37 (b)	38 (a)	39 (b)	40 (b)	41 (b)	42 (a)	43 (b)	44 (b)
45 (c)	46 (b)	47 (a)	48 (d)	49 (b)	50 (c)	51 (d)			

Level 3 Previous Years Questions

52 (d)	53 (d)	54 (b)	55 (a)	56 (b)	57 (c)	58 (b)	59 (a)	60 (b)	61 (d)
62 (d)	63 (b)	64 (d)	65 (b)	66 (a)	67 (b)	68 (b)	69 (c)	70 (a)	71 (d)
72 (c)	73 (d)	74 (b)	75 (c)	76 (c)	77 (a)	78 (a)	79 (a)	80 (a)	81 (b)

Solutions

LEVEL 01 Objective Questions

1. According to the Kepler's second law, equal area is swept in equal interval of time.

$$\therefore \quad \frac{\Delta A}{\Delta t} = \text{constant}$$

$$\therefore \quad \frac{A_1}{t_1} = \frac{A_2}{t_2} = \frac{A_3}{t_3}$$

$$\frac{A_1}{2} = \frac{A_2}{3} = \frac{A_3}{6}$$

$$3A_1 = 2A_2 = A_3$$

2. Time period of a planet is given as,

$$T^2 = \frac{4\pi^2}{GM} R^3$$

So, it is of the form $y = mx$

Hence, the slope of graph is $\frac{4\pi^2}{GM}$.

3. $$\text{Areal velocity} = \frac{\text{Area swept by radius vector}}{\text{Time taken}}$$

$$A = \frac{\pi R^2}{T}$$

Time period, $T = \frac{2\pi}{\omega}$

$$A = \frac{R^2\omega}{2}$$

$$\Rightarrow \quad \omega = \frac{2A}{R^2}$$

Now, the angular momentum, $L = I\omega$

$$= MR^2 \times \frac{2A}{R^2} = 2MA$$

4. From Kepler's law, $T^2 \propto R^3$

Thus, $\frac{T^2}{R^3}$ does not depend on radius.

5. By Kepler's law, when the satellite moves in the elliptical path, its angular momentum remains constant.

$$mv_1r_1 = mv_2r_2$$

$$\frac{r_1}{r_2} = \frac{v_2}{v_1}$$

Here, $r_1 > r$

$$\frac{r_1}{r_2} > 1$$

So, $\frac{v_2}{v_1}$ must be > 1

$$v_2 > v_1$$

Then, $\frac{1}{2}mv_2^2 > \frac{1}{2}mv_1^2$

$$\therefore \quad K_2 > K_1$$

6. Acceleration due to gravity at a depth below surface of earth,

$$g' = g\left(1 - \frac{d}{R}\right)$$

At $d = R$, $g' = 0$

Now for $r > R$,

$$g' = \frac{g}{\left(1 + \frac{h}{R}\right)^2} = \frac{gR^2}{r^2}$$

Here, $R + h = r$

$$g' \propto \frac{1}{r^2}$$

For $r < R$,

$$g' = g\left(1 - \frac{d}{R}\right) = \frac{gr}{R}$$

Here, $R - d = r$

So, $g' \propto r$

Therefore, the graph (d) represents the variation of g with distance from centre of earth.

7. Weight of the body at depth d is,

$$w' = mg' = mg\left(1 - \frac{d}{R_E}\right)$$

Here, $d = \frac{R_E}{2}$

So, we get

$$w' = 125 \text{ N}$$

8. $$g = \frac{GM}{R^2} = \frac{G\left(\frac{4}{3}\pi R^3\rho\right)}{R^2}$$

$$g \propto \rho R$$

$$g_1 = g_2$$

$$\rho_1 R_1 = \rho_2 R_2$$

$$\rho_1 R_1 = 2\rho_1 R_2 \qquad (\because \rho_2 = 2\rho_1)$$

$$\Rightarrow \quad R_2 = \frac{R_1}{2}$$

So, $R_2 = \frac{R}{2}$ $\quad \left(\because R_1 = \frac{R}{2}\right)$

9. $$g = \frac{Gm}{R^2} \qquad \text{...(i)}$$

$$\frac{dg}{dR} = -\frac{2Gm}{R^3}$$

Now, substituting $dR = h$,

$$\frac{dg}{h} = -\frac{2Gm}{R^2}\frac{1}{R} \quad ...(ii)$$

On dividing Eq. (ii) by E. (i), we get

$$\frac{dg}{g} = -\frac{2h}{R}$$

10. $g = \frac{Gm}{R^2} = \frac{G}{R^2}\cdot\frac{4}{3}\pi R^3 \times \rho$

$$= \frac{4}{3}\pi GR \times \rho$$

As the radius is kept constant,

$$g \propto \rho$$

Since, the density is doubled, $\rho_2 = 2\rho_1$

$$\frac{g_2}{g_1} = \frac{\rho_2}{\rho_1}$$

$$\Rightarrow \quad g_2 = g_1\left(\frac{2\rho_1}{\rho_1}\right) = 9.8 \times 2$$

$$= 19.6 \text{ m/s}^2$$

11. $g - R\omega^2 \cos^2 60° = 0$

$$\Rightarrow \quad \omega = 2\sqrt{\frac{g}{R}}$$

So, $\quad T = \pi\sqrt{\frac{R}{g}} \quad \left[\because T = \frac{2\pi}{\omega}\right]$

12. $\frac{3}{5}mg = mg - mR\omega^2 \Rightarrow \omega = \sqrt{\frac{2}{5}\frac{g}{R}}$

13.

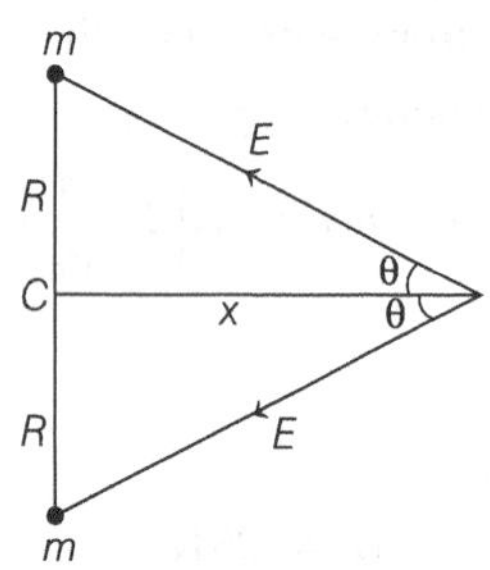

Net field strength $= 2E\cos\theta$ (towards C)

$$= 2\left[\frac{Gm}{R^2 + x^2}\right]\left[\frac{x}{\sqrt{R^2 + x^2}}\right]$$

$$= \frac{2\,Gmx}{(R^2 + x^2)^{3/2}}$$

14. $E = \frac{Gm}{a^2} - \frac{Gmx}{(a^2 + x^2)^{3/2}}$

Substituting $x = a$, we get

$$E = \frac{Gm}{a^2}\left[1 - \frac{1}{2\sqrt{2}}\right]$$

15. E due to a shell $= 0$

$\therefore$ Force due to $M = 0$

16. $E_P = \frac{GM}{\left(\frac{3}{2}a\right)^2} = \frac{4\,GM}{9a^2}$

17. Gravitational potential at the centre of spherical shell

$$= -\frac{Gm}{R}$$

As the spherical shell is expanding R increases, $\frac{Gm}{R}$ decreases.

So, $\frac{-Gm}{R}$ increases.

18. We have given,

$$E = -\frac{k}{r}$$

$$dV = -E \cdot dr$$

$$\int_{V_0}^{V} dV = k\int_{r_0}^{r}\frac{dr}{r} \quad \left(\because E = -\frac{k}{r}\right)$$

On integrating this, we get

$$V = V_0 + k\ln\left(\frac{r}{r_0}\right)$$

19. $v = v_1$ due to $M_1 + v_2$ due to M_2

$$= \frac{-GM}{R} + \left(\frac{-2GM}{R}\right) = \frac{-3GM}{R}$$

20. $V_p = V_{sphere} + V_{particle}$

$$= -\left(\frac{GM}{a} + \frac{GM}{a/2}\right) = -\frac{3GM}{a}$$

21. $l = \pi R$

$$\Rightarrow \quad R = \frac{l}{\pi}$$

So, potential at centre,

$$V_C = -\frac{Gm}{R} = -\frac{\pi Gm}{l}$$

22. $\text{KE} = \left(-\frac{GmM}{2R_2} + \frac{GmM}{2R_1}\right)$

$$= \frac{1}{2}GmM\left(\frac{1}{R_1} - \frac{1}{R_2}\right)$$

23. Velocity of satellite, $v = \sqrt{\frac{GM}{R}}$

Since, it is independent of mass of satellite, so S_1 and S_2 will be moving with same speed.

24. $\text{KE} = \frac{|\text{PE}|}{2}$

As, $\quad \text{PE} = -\frac{GMm}{r}$

and $\quad \text{KE} = \frac{GMm}{2r}$

25. Orbital velocity near the surface of earth, $v_o = \sqrt{gR}$

Escape velocity, $v_e = \sqrt{2gR} = \sqrt{2}v_0$

So, speed has to increase $\sqrt{2}$ times

or kinetic energy 2 times.

26. According to Kepler's law,

$$T^2 \propto R^3 \Rightarrow \frac{T_1}{T_2} = \left(\frac{R_1}{R_2}\right)^{3/2}$$

$$\Rightarrow \quad R_2 = R_1\left(\frac{T_2}{T_1}\right)^{2/3}$$

$$\Rightarrow \quad R_2 = R_1\left(\frac{16}{2}\right)^{2/3}$$

$$\Rightarrow \quad R_2 = 4R_1$$

Now, orbital velocity, $v_o = \sqrt{\frac{GM}{R}}$

$$\frac{v_{o2}}{v_{o1}} = \sqrt{\frac{R_1}{R_2}} = \sqrt{\frac{R_1}{4R_1}} = \frac{1}{2}$$

So, $v_{o2} = \frac{v_{o1}}{2} = \frac{v_o}{2}$

27. $g = \frac{GM}{R^2} = \frac{G\left(\frac{4}{3}\pi R^3\rho\right)}{R^2}$

or $\quad g \propto R \quad$ (as ρ = constant)

Escape velocity, $\quad v_e = \sqrt{2gR}$

Since, $g \propto R$

$\therefore \quad v \propto R$

28. In first time, $T^2 \propto R^3$

In second time, $T'^2 \propto (2R)^3$

$$\frac{T'}{T} = \left(\frac{2R}{R}\right)^{3/2} = 2^{3/2}$$

$$T' = 2^{3/2}T = 2^{3/2} \times 24$$

$$= 48\sqrt{2}\text{ h}$$

29. $L = mvR_o$

$$K = \frac{1}{2}mv^2$$

$$\therefore \quad \frac{L}{K} = \frac{2R_o}{v} = \frac{2R_o}{\sqrt{GM/R_o}} = 2\sqrt{\frac{R_o^3}{GM}}$$

30. From conservation of mechanical energy,

$$\frac{1}{2}mu^2 - \frac{GM_Em}{R_E} = \frac{1}{2}mv^2 - 0$$

$$v^2 = u^2 - 2\frac{GM_E}{R_E}$$

Here, $\frac{2GM_E}{R_E} = v_e^2$

$$v = \sqrt{(3v_e)^2 - v_e^2} = \sqrt{8}v_e$$

$$= 22.4\sqrt{2}\text{ km/s} \quad [\because v_e = 11.2\text{ km/s}]$$

31. Orbital velocity, $v_o = \sqrt{\frac{Gm}{R}}$

$$r = R + h$$

If $h = \frac{R}{2}$, then $r = \frac{3R}{2}$

So, orbital velocity becomes

$$v' = \sqrt{\frac{2}{3}} \cdot \sqrt{\frac{Gm}{R}} = \sqrt{\frac{2}{3}}v_o$$

32. We know, escape speed, $v = \sqrt{\frac{2Gm}{R}}$

Now, assuming the mass to be constant while compressing the radius.

$$v' = 10v$$

$$\sqrt{\frac{2Gm}{R'}} = 10 \times \sqrt{\frac{2Gm}{R}}$$

$$\sqrt{\frac{1}{R'}} = 10 \times \sqrt{\frac{1}{R}}$$

$$\sqrt{R'} = \frac{1}{10}\sqrt{R}$$

$$\Rightarrow \quad R' = \frac{1}{100}R \quad \text{(squaring both sides)}$$

$$\Rightarrow \quad R' = \frac{1}{100} \times 6400$$

So, we get

$$R' = 64\text{ km}$$

33. $\frac{Gm_1m_2}{r^3} = \frac{m_1v^2}{r}$

$$\Rightarrow \quad v = \sqrt{\frac{Gm_2}{r^2}} = \frac{\sqrt{Gm_2}}{r}$$

So, $v \propto \frac{1}{r}$

34. In the first case,

$$Mg = kx_0$$

In the second case,

$$Mg' = kx$$

$$\because \quad \frac{Mg}{Mg'} = \frac{kx_0}{kx}$$

$$\Rightarrow \quad \frac{g}{g'} = \frac{x_0}{x}$$

$$\frac{g}{g\left(1 + \frac{h}{R}\right)^2} = \frac{x_0}{x}$$

On substituting $h = 1600$ km and $R = 6400$ km, we get

$$x = \frac{16x_0}{25}$$

LEVEL 02 Objective Questions

35. Energy required to transfer the satellite to the desired orbit = Change in potential energy

$$= E_f - E_i$$

$$= -\frac{GM_E m}{8R_E} - \left(-\frac{GM_E m}{4R_E}\right) = \frac{GM_E m}{8R_E}$$

36. Frequency of revolution, $\nu = \frac{1}{r} = \frac{1}{2\pi}\sqrt{\frac{GM_E}{r^3}}$

So, $\nu \propto \frac{1}{r^{3/2}}$

37. $\frac{3}{2}\frac{GMm}{R}\left(\frac{1}{2}\right) = \frac{GMm}{x}$

$\Rightarrow$ $$x = \frac{4R}{3}$$

So, height from surface of earth

$$= \frac{4R}{3} - R = \frac{R}{3}$$

38. Applying conservation of energy,

$$\frac{1}{2}mv^2 + \left(\frac{-3GMm}{2R}\right) = 0 + \left(\frac{-GMm}{R}\right)$$

$\Rightarrow$ $$v = \sqrt{\frac{GM}{R}}$$

39. Increase in kinetic energy = Decrease in potential energy

$\therefore$ $$\frac{1}{2}mv^2 = \frac{mgh}{1 + \frac{h}{R}}$$

Substituting $h = (2R - R) = R$, we get

$$v = \sqrt{gR}$$

40. Applying conservation of energy,

$$\frac{1}{2}m_0 v_e^2 + \frac{Gm_0 M}{2R^3}\left[3R^2 - \left(\frac{R}{2}\right)^2\right] = 0$$

So, $v_e = \sqrt{\frac{11GM}{4R}}$

41. Force between earth and mass of 1 kg,

$$F = \frac{GM_E}{R_E^2} = 10 \text{ N}$$

Force between earth and satellite,

$$F' = \frac{GM_E \times 200}{\left(\frac{3}{2}R_E\right)^2}$$

$\because$ $$\frac{F'}{F} = \frac{800}{9}$$

$\Rightarrow$ $$F' = \frac{800}{9} \times 10$$

$$\simeq 889 \text{ N}$$

42. From energy conservation,

$$\frac{1}{2}mv^2 = -\frac{GMm}{R+R} - \left(\frac{-GMm}{R}\right) = \frac{GMm}{2R}$$

$$v = \sqrt{\frac{Gm}{R}} = \frac{\sqrt{\left(\frac{2Gm}{R}\right)}}{\sqrt{2}} = \frac{v_e}{\sqrt{2}}$$

$\therefore$ $$\frac{v}{v_e} = \frac{1}{\sqrt{2}}$$

43. Radius of circular path, $r = \frac{Ml}{M+m}$

$$F = \frac{mv^2}{r} = \frac{GMm}{l^2}$$

$$\frac{mv^2}{\frac{Ml}{M+m}} = \frac{GMm}{l^2}$$

So, we get

$$v = \sqrt{\frac{GM^2}{(M+m)l}}$$

44. Potential energy on the surface of earth,

$$U = -\frac{GMm}{R}$$

At $h << R$, increase in potential energy is mgh.

So, the potential energy of body and earth is given by,

$$U_h = -\frac{GMm}{R} + mgh$$

45. Relative angular velocity,

$$8 = \frac{2\pi}{\omega_1 + \omega_2} = \frac{2\pi}{\frac{2\pi}{T_1} + \frac{2\pi}{T_2}}$$

Here, $T_1 = 24\text{h}$ for earth, $T_2 = 12\text{h}$ for satellite, as it will remain the same as distance from centre of earth remains same.

Now, when the sense of rotation of earth and satellite is same,

$$T = \frac{2\pi}{\omega_1 - \omega_2} = \frac{2\pi}{\frac{2\pi}{T_1} - \frac{2\pi}{T_2}} = 24\text{h}$$

46. $\frac{Gmm}{(2r)^2} = \frac{mv^2}{r}$

$\Rightarrow$ $$v = \frac{1}{2}\sqrt{\frac{Gm}{r}}$$

Time period $= \frac{2\pi r}{v}$

$$= \frac{4\pi r^{3/2}}{\sqrt{Gm}}$$

47. $K_i + U_i = K_f + U_f$

$$0 + 0 = \frac{1}{2}mv^2 + mV_C \quad \text{...(i)}$$

V_C = gravitational potential at centre

$$= -\frac{3}{2}\left(\frac{GM}{R}\right)$$

Substituting in Eq. (i), we get

$$v = \sqrt{\frac{3GM}{R}} = \sqrt{1.5}\, v_e$$

As, $v_e = \sqrt{2gR}$ or $\sqrt{\frac{2GM}{R}}$

48. From conservation of energy,

$$-\frac{GMm}{R_e} + \frac{1}{2}mv^2 = -\frac{GMm}{R_e + h}$$

$$-\frac{GMm}{R_e} + \frac{1m}{2}\left(\sqrt{1.5}\sqrt{\frac{Gm}{R_e}}\right)^2 = -\frac{GMm}{R_e + h}$$

So, we get

$$h = 3R_e$$

49. Mass of removed hole, $M' = \frac{M}{\frac{4}{3}\pi R^3} \times \frac{4}{3}\pi\left(\frac{R}{2}\right)^3$

$$M' = \frac{M}{8}$$

So, gravitational acceleration at a point on the surface above the excavation is

$$= \frac{GM}{R^2} - \frac{GM'}{\left(\frac{R}{2}\right)^2} = \frac{GM}{R^2} - \frac{G\left(\frac{M}{8}\right)}{\left(\frac{R}{2}\right)^2} = \frac{GM}{2R^2}$$

50. From principle of superposition,

$$\mathbf{E} = \mathbf{E}_1 + \mathbf{E}_2$$

Here, **E** is net field at centre due to entire mass and $\mathbf{E}_1$ due to remaining mass.

$\mathbf{E}_2$ is the field due to mass in hole = 0

So, $E = E_1 = \left(\frac{GM}{R^3}\right)r$

Here, $r = \frac{R}{2}$

So, $E = \frac{GM}{2R^2}$

51. Potential due to the axis of a ring,

$$V = -\frac{GM}{\sqrt{a^2 + R^2}}$$

Now, from conservation of mechanical energy,

$$\frac{-M(-GM)}{\sqrt{R^2 + (R\sqrt{3})^2}} = \frac{1}{2}\frac{Mv^2}{} - \frac{GMM}{(R\sqrt{3})^2}$$

We get,

$$v = \sqrt{\frac{GM}{R}(\sqrt{2} - 1)}$$

LEVEL 03 Previous Years Questions

52. We know that, escape velocity from earth's surface,

$$v_e = \sqrt{\frac{2GM_e}{R_e}}$$

where, G is the gravitational constant,
R_e is the radius of the earth,
M_e is the mass of the earth.

Also, density $= \frac{\text{mass}}{\text{volume}}$

$\Rightarrow$ Mass = Density × Volume

or $M_e = \rho \times V$

or $M_e = \rho \times \frac{4}{3}\pi R_e^3$ $\left(\because V = \frac{4}{3}\pi R_e^3\right)$

Substituting the values in the escape velocity expression, we get

$$v_e = \sqrt{\frac{2G\rho \times \frac{4}{3}\pi R_e^3}{R_e}}$$

$$v_e = \sqrt{2G\rho \times \frac{4}{3}\pi R_e^2} \quad \text{...(i)}$$

Now, we shall determine the escape velocity for another planet.

Given, radius of another planet is 4 times the radius of the earth,

$$\Rightarrow v_p = \sqrt{2G\rho \times \frac{4}{3}\pi R_p^2}$$

$$v_p = \sqrt{2G\rho \times \frac{4}{3}\pi(4R_e)^2} \quad (\because R_p = 4R_e)$$

$$v_p = 4\sqrt{2G\rho \times \frac{4}{3}\pi(R_e)^2} \quad \text{...(ii)}$$

On dividing Eq. (ii) by Eq. (i), we get

$$\frac{v_p}{v_e} = \frac{4\sqrt{2G\rho \times \frac{4}{3}\pi(R_e)^2}}{\sqrt{2G\rho \times \frac{4}{3}\pi R_e^2}}$$

$$\Rightarrow v_p = 4v_e = 4v$$

53. Using the law of conservation of energy,

initial energy at the time of projection = final energy at the maximum height h from the surface of the earth

$$\frac{1}{2}mv^2 - \frac{GMm}{R} = -\frac{GMm}{R + h}$$

Substituting the values in the above equation, we get

$$\frac{1}{2}m(kv_e)^2 - \frac{GMm}{R} = -\frac{GMm}{R + h}$$

$$\Rightarrow \frac{1}{2}mk^2\left(2\frac{GM}{R}\right) - \frac{GMm}{R} = -\frac{GMm}{R + h} \quad \left(\because v_e = \sqrt{\frac{2GM}{R}}\right)$$

$$\Rightarrow \quad \frac{k^2}{R} - \frac{1}{R} = -\frac{1}{R+h}$$

$$\Rightarrow \quad \frac{1}{R+h} = \frac{1}{R} - \frac{k^2}{R}$$

$$\Rightarrow \quad \frac{R}{R+h} = (1-k^2)$$

$$\Rightarrow \quad \frac{1}{1-k^2} = 1 + \frac{h}{R}$$

$$\Rightarrow \quad h = R\left(\frac{k^2}{1-k^2}\right)$$

54. Radius of earth $= R$

Let at depth d, gravitational acceleration becomes $\frac{g}{n}$.

i.e. $$g_d = \frac{g}{n} \Rightarrow g\left(1 - \frac{d}{R}\right) = \frac{g}{n}$$

$$\Rightarrow \quad 1 - \frac{d}{R} = \frac{1}{n} \Rightarrow 1 - \frac{1}{n} = \frac{d}{R}$$

$$\Rightarrow \quad \frac{n-1}{n} = \frac{d}{R} \Rightarrow d = \left(\frac{n-1}{n}\right)R$$

55. Given, $w = mg = 72$ N (on the surface of earth).

At height equal to half of radius of the earth, $\left(\text{i.e. } h = \frac{R}{2}\right)$

Acceleration due to gravity,

$$g' = g\left(\frac{R}{R+h}\right)^2 = g\left(\frac{R}{R+R/2}\right)^2 = g\left(\frac{4R^2}{9R^2}\right)$$

$$\Rightarrow \quad g' = \frac{4}{9}g \quad \Rightarrow \quad mg' = \frac{4}{9}mg = \frac{4}{9} \times 72 = 32 \text{ N}$$

$$w' = 32 \text{ N}$$

56. The gravitational potential energy of an object placed at earth's surface is

$$U_1 = -\frac{GMm}{R} \quad \text{...(i)}$$

where, $G =$ gravitational constant,

$M =$ mass of earth, $m =$ mass of object and $R =$ radius of the earth.

The negative sign in the above relation indicates that it is the work done in bringing the object from infinity to a distance R.

The gravitational potential energy of object at a height h above the surface of earth is

$$U_2 = -\frac{GMm}{(R+h)} \quad \text{...(ii)}$$

So, the change in potential energy is

$$\Delta U = U_2 - U_1 = -\frac{GMm}{R+h} - \left(-\frac{GMm}{R}\right)$$

[from Eqs. (i) and (ii)]

$$= -GMm\left[\frac{1}{(R+h)} - \frac{1}{R}\right]$$

$$= -GMm\left(-\frac{h}{R(R+h)}\right)$$

$$= \frac{GMmh}{R(R+h)}$$

57. Given, weight of the body, $w = 200$ N

As we know,

$w = mg$, where m is the mass of the body and g (≈ 10 m/s^2) is acceleration due to gravity of the body at the surface of the earth.

Since, mass m remains constant irrespective of the position of the body on the earth. However, g is not constant and its value at a depth d below the earth's surface is given as

$$g' = g\left(1 - \frac{d}{R}\right) \quad \text{... (i)}$$

where, R is the radius of the earth.

Multiplying m on the both sides of Eq. (i), we get

$$mg' = mg\left(1 - \frac{d}{R}\right)$$

Thus, the weight of the body at half way down $\left(\text{i.e. } d = \frac{R}{2}\right)$ to the centre of the earth is,

$$mg' = 200 \times \left(1 - \frac{R/2}{R}\right)$$

$$= 200\left(1 - \frac{1}{2}\right) = 200 \times \frac{1}{2}$$

$$= 100 \text{ N}$$

$\therefore$ The body will weigh 100 N half way down to the centre of the earth.

58. Key Idea Amount of work done in moving the given body from one point to another against the gravitational force is equal to the change in potential energy of the body.

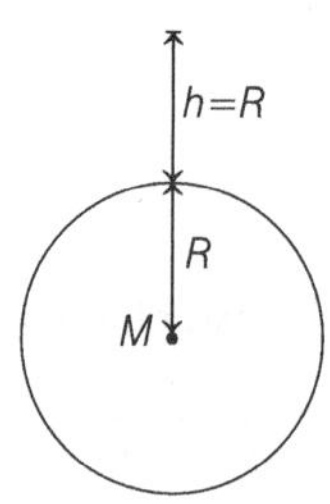

As we know, the potential energy of body of mass m of the surface of earth is,

$$U_1 = -\frac{GMm}{R} \quad \text{... (i)}$$

where, $G =$ gravitational constant,

$M =$ mass of earth and $R =$ radius of earth.

When the mass is raised to a height h from the surface of the earth, the potential energy of the body becomes

$$U_2 = -\frac{GMm}{(R+h)}$$

Here, $h = R$ (given)

$$\Rightarrow \quad U_2 = -\frac{GMm}{2R} \quad \text{... (ii)}$$

Thus, the change in potential energy, $\Delta U = U_2 - U_1$

Substituting the values from Eqs. (i) and (ii), we get

$$\Delta U = -\frac{GMm}{2R} + \frac{GMm}{R}$$

$$= \frac{GMm}{2R} = \frac{gR^2m}{2R} \quad \left[\because g = \frac{GM}{R^2}\right]$$

$$= \frac{mgR}{2}$$

Thus, the work done in raising the mass to a height R is equals to $\frac{mgR}{2}$.

59. From Kepler's third law, the time period of revolution of satellite around earth is

$$T^2 \propto r^3 \quad \text{or} \quad T \propto r^{3/2} \quad \text{...(i)}$$

where, r is the radius of satellite's orbit.

Here, $r_1 = 6R_E + R_E, T_1 = 24$ h,

$r_2 = 2.5R_E + R_E, T_2 = ?$

where, R_E = radius of earth.

So, from Eq. (i), we get

$$\frac{T_1}{T_2} = \left(\frac{r_1}{r_2}\right)^{3/2}$$

$$\frac{24}{T_2} = \left(\frac{6R_E + R_E}{2.5R_E + R_E}\right)^{3/2} = \left(\frac{7}{3.5}\right)^{3/2} = 2^{3/2}$$

$$\Rightarrow \quad T_2 = \frac{24}{(2)^{3/2}} = \frac{24}{2\sqrt{2}} = \frac{12}{\sqrt{2}} = 6\sqrt{2}\text{ h}$$

60. According to the question,

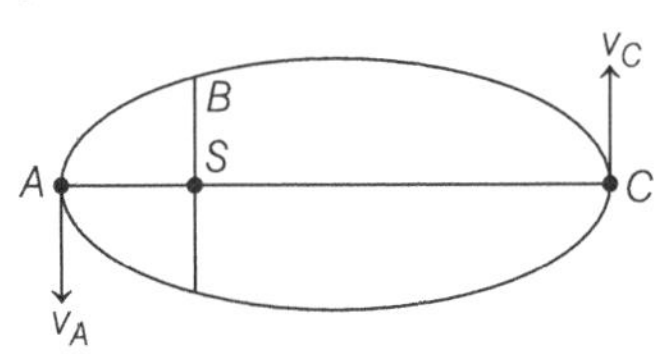

The above figure shows an ellipse traced by a planet around the sun S. The closed point A is known as perihelion (perigee) and the farthest point C is known as aphelion (apogee).

Since, as per the result, the Kepler's second law of area, that the planet will move slowly (v_{min}) only when it is farthest from the sun and more rapidly (v_{max}) when it is nearest to the sun.

Thus, $\quad v_A = v_{max}, v_C = v_{min}$

Therefore, we can write

$$v_A > v_B > v_C \quad \text{...(i)}$$

Kinetic energy of the planet at any point is given as,

$$K = \frac{1}{2}mv^2$$

Thus, at A, $K_A = \frac{1}{2}mv_A^2$

At B, $\quad K_B = \frac{1}{2}mv_B^2$

At C, $\quad K_C = \frac{1}{2}mv_C^2$

From Eq. (i), we can write

$$K_A > K_B > K_C$$

61. Let the original mass of sun was M_s and gravitational constant G'.

According to the question,

New mass of sun, $M_s' = \frac{M_s}{10}$

New gravitational constant, $G' = 10G$

As, the acceleration due to gravity is given as

$$g = \frac{GM_E}{R^2} \quad \text{...(i)}$$

where, M_E is the mass of earth and R is the radius of the earth.

Now, new acceleration due to gravity,

$$g' = \frac{G'M_e}{R^2} = \frac{10M_eG}{R^2} \quad \text{...(ii)}$$

$\therefore \quad g' = 10g$ [from Eqs. (i) and (ii)]

This means the acceleration due to gravity has been increased. Hence, force of gravity acting on a body placed on or surface of the earth increases.

Due to this, rain drops will fall faster and walking on ground would become more difficult.

As, time period of the simple pendulum is,

$$T = 2\pi\sqrt{\frac{l}{g}} \quad \text{or} \quad T \propto \frac{1}{\sqrt{g}}$$

Thus, time period of the pendulum also decreases with the increase in g.

62. Thinking Process g_h = Acceleration due to gravity at height h above earth's surface

$$= g\left(\frac{R}{R+h}\right)^2 = g\left(1 - \frac{2h}{R}\right)$$

g_d = Acceleration at depth d below earth's surface

$$= g\left(1 - \frac{d}{R}\right)$$

Given, when $h = 1$ km, $g_d = g_h$

or $\quad g\left(1 - \frac{d}{R}\right) = g\left(1 - \frac{2h}{R}\right)$

$\Rightarrow \quad d = 2h$

or $\quad d = 2$ km

63. In the space, there is no external gravity. Due to masses of the astronauts, there will be small gravitational attractive force between them. Thus, these astronauts will move towards each other.

64. Gravitational potential at some height h from the surface of the earth is given by

$$V = -\frac{GM}{R+h} \quad \text{...(i)}$$

And acceleration due to gravity at some height h from the earth surface can be given as

$$g' = \frac{GM}{(R+h)^2} \quad \text{...(ii)}$$

From Eqs. (i) and (ii), we get

$$\frac{|V|}{g'} = \frac{GM}{(R+h)} \times \frac{(R+h)^2}{GM}$$

$$\Rightarrow \quad \frac{|V|}{g'} = R + h \quad \text{(iii)}$$

$\because \; V = -5.4 \times 10^7 \text{ J kg}^{-2}$ and $g' = 6.0 \text{ ms}^{-2}$

Radius of earth, $R = 6400$ km.

Substitute these values in Eq. (iii), we get

$$\frac{5.4 \times 10^7}{6.0} = R + h$$

$$\Rightarrow \quad 9 \times 10^6 = R + h$$

$$\Rightarrow \quad h = (9 - 6.4) \times 10^6$$

$$= 2.6 \times 10^6 \text{ m}$$

$$\Rightarrow \quad h = 2600 \text{ km}$$

65. Acceleration due to gravity at a depth d below the surface of the earth is given by

$$g_{\text{depth}} = g_{\text{surface}}\left(1 - \frac{d}{R}\right)$$

Also, for a point at height h above the surface,

$$g_{\text{height}} = g_{\text{surface}}\left[\frac{R^2}{(R+h)^2}\right]$$

Therefore, we can say that value of g increases from centre of maximum at the surface and then decreases as depicted in graph (b).

66. As, $v_{\text{escape}} \propto R\sqrt{\rho}$

So, $$\frac{v_e}{v_p} = \frac{R\sqrt{\rho}}{2R\sqrt{2\rho}}$$

$$\Rightarrow \quad \frac{v_e}{v_p} = \frac{1}{2\sqrt{2}} = 1 : 2\sqrt{2}$$

67. $\because$ Total energy of a satellite at height h is

$$= \text{KE} + \text{PE} = \frac{GMm}{2(R+h)} - \frac{GMm}{(R+h)}$$

$$= \frac{-GMm}{2(R+h)} = \frac{-mg_0R^2}{2(R+h)} \quad \left(\because \; g_0 = \frac{GM}{R^2}\right)$$

68. The gravitational force of attraction between the planet and sun provide the centripetal force.

i.e. $$\frac{GMm}{r^2} = \frac{mv^2}{r} \Rightarrow v = \sqrt{\frac{GM}{r}}$$

The time period of planet will be

$$T = \frac{2\pi r}{v}$$

$$\Rightarrow \quad T^2 = \frac{4\pi^2 r^2}{\frac{GM}{r}} = \frac{4\pi^2 r^3}{GM} \quad \text{...(i)}$$

Also, from Kepler's third law,

$$T^2 = Kr^3 \quad \text{...(ii)}$$

From Eqs. (i) and (ii), we get

$$\frac{4\pi^2 r^3}{GM} = Kr^3 \Rightarrow GMK = 4\pi^2$$

69. Suppose, the smaller body cover a distance x before collision, then

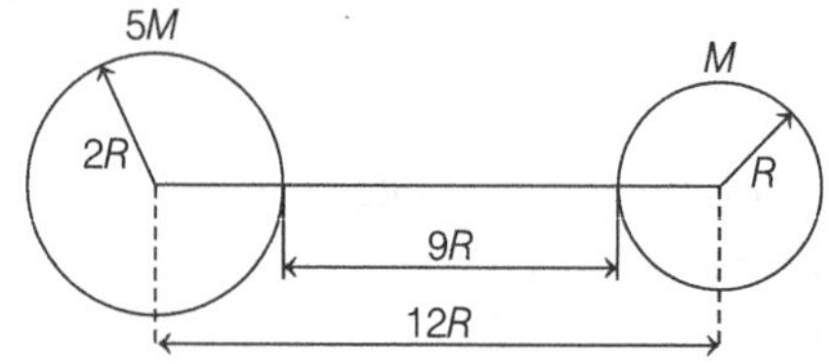

$$Mx = 5M(9R - x)$$

or $$x = 45R - 5x$$

or $$x = \frac{45R}{6} = 7.5\,R$$

70. Given, height of a satellite,

$$h = 0.25 \times 10^6 \text{ m}$$

Earth's radius, $R_e = 6.38 \times 10^6$ m

For the satellite revolving around the earth, orbital velocity of the satellite,

$$v_o = \sqrt{\frac{GM_e}{R_e}} = \sqrt{\frac{GM_e}{R_e\left[1 + \frac{h}{R_e}\right]}} \Rightarrow v_o = \sqrt{\frac{gR_e}{1 + \frac{h}{R_e}}}$$

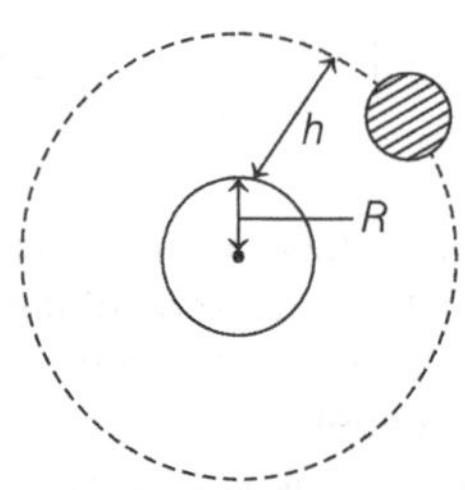

Substitutes the values of g, R_e and h, we get

$$v_o = \sqrt{60 \times 10^6} \text{ m/s}$$

$$v_o = 7.76 \times 10^3$$

$$= 7.76 \text{ km/s}$$

71. As we know that, force on satellite is only gravitational force which will always be towards the centre of earth. Thus, the acceleration of S is always directed towards the centre of the earth.

72. Problem Solving Strategy For the black hole, the escape speed is more than c (speed of light). We should compare the escape speed with the c (Note that the escape speed should be at least just greater than c).

$$v_e = \sqrt{\frac{2GM}{R'}} \quad [R' = \text{New radius of the earth}]$$

$$c = \sqrt{\frac{2GM}{R'}} \quad [v_e \approx c]$$

$$\Rightarrow \quad c^2 = 2\frac{GM}{R'}$$

$$R' = \frac{2GM}{c^2}$$

$$= \frac{2 \times 6.67 \times 10^{-11} \times 6 \times 10^{24}}{9 \times 10^{16}}$$

$$= \frac{4 \times 6.67}{3} \times 10^{-3}$$

$$= 8.89 \times 10^{-3}$$

$$= 0.889 \times 10^{-2} \simeq 10^{-2} \text{ m}$$

73. The resulting gravitational potential,

$$V = -2G\left[\frac{1}{1} + \frac{1}{2} + \frac{1}{4} + \frac{1}{8} + \ldots\right]$$

$$\Rightarrow \quad V = -2G\left[1 + \frac{1}{2} + \frac{1}{2^2} + \frac{1}{2^3}\ldots\right]$$

$$\Rightarrow \quad V = -2G\left(1 + \frac{1}{2}\right)^{-1}$$

$$\Rightarrow \quad V = -\frac{2G}{\left(1 - \frac{1}{2}\right)} = -\frac{2G}{\left(\frac{1}{2}\right)} = -4G$$

74. Change in potential energy,

$$\Delta U = -\frac{GMm}{R + 2R} - \left(-\frac{GMm}{R}\right)$$

where, $U_{\text{final}} = \frac{GMm}{R + 2R}$

$$U_{\text{initial}} = -\frac{GMm}{R}$$

$$\therefore \quad \Delta U = -\frac{GMm}{3R} + \frac{GMm}{R}$$

$$= \frac{2GMm}{3R} = \frac{2}{3}mgR \quad \left[\because g = \frac{GM}{R^2}\right]$$

75. According to the question,

$$\frac{GMm}{(R + h)^2} = \frac{1}{16}\frac{GMm}{R^2}$$

where, m = mass of the body

and $\frac{GM}{R^2}$ = gravitational acceleration

$$\frac{1}{(R + h)^2} = \frac{1}{16R^2} \text{ or } \frac{R}{R + h} = \frac{1}{4} \text{ or } \frac{R + h}{R} = 4$$

$$h = 3R$$

76. It will stay in North-South direction only at geomagnetic North and South poles.

77. According to Newton's law of gravitation force,

$$F = \frac{GMm}{R^2}$$

Force on planet of mass M_P and body of mass m is given by

$$F = \frac{GM_Pm}{(D_P/2)^2}$$

$$\left[\begin{array}{l}\text{where, } D_P = \text{diameter of planet} \\ \text{and } R_P = \text{radius of planet} = \frac{D_P}{2}\end{array}\right]$$

$$F = \frac{4GM_Pm}{D_P^2}$$

As we know that, $F = ma$

So, acceleration due to gravity

$$a = \frac{F}{m} = \frac{4GM_P}{D_P^2}$$

78. Concept Apply conservation of angular momentum.

From the law of conservation of angular momentum,

$L_1 = L_2$

So, $mr_1v_1 = mr_2v_2$

$$\left[\begin{array}{l}\text{where, } m = \text{mass of the planet} \\ r = \text{radius of orbit} \\ v = \text{velocity of the planet}\end{array}\right]$$

$$r_1v_1 = r_2v_2$$

$$\Rightarrow \quad \frac{v_1}{v_2} = \frac{r_2}{r_1}$$

79. We know that, Power, $P = \mathbf{F} \cdot \mathbf{v} = Fv\cos\theta$ So, just before hitting, θ is zero, power will be maximum.

80. Gravitational potential at point $a/2$ distance from centre is given by,

$$V = -\frac{GM}{a} - \frac{GM}{a/2} = -\frac{3GM}{a}$$

81. Orbital velocity of satellite is given by,

$$v = \sqrt{\frac{GM}{r}}$$

Ratio of orbital velocities of A and B is given by,

$$\Rightarrow \quad \frac{v_A}{v_B} = \sqrt{\frac{r_B}{r_A}} = \sqrt{\frac{R}{4R}} = \frac{1}{2}$$

$$\therefore \quad \frac{v_A}{v_B} = \frac{3v}{v_B} = \frac{1}{2}$$

$$\therefore \quad v_B = 6v$$

Chapter 12

PROPERTIES OF MATTER

LEVEL 01 Objective Questions

1 The stress *versus* strain graphs for wires of two materials A and B is as shown in the following figure. If Y_A and Y_B are the Young's modulus of the materials, then

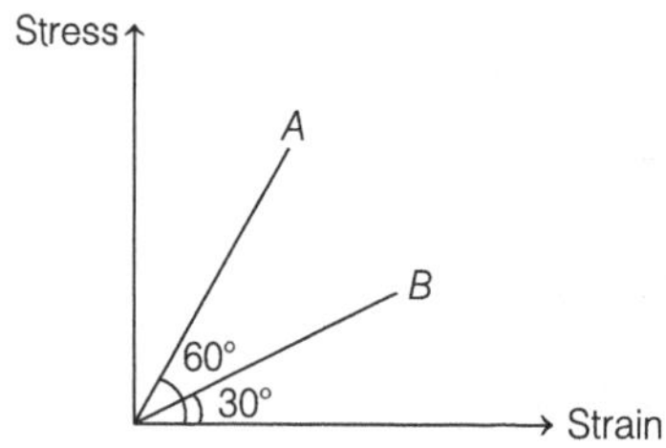

(a) $Y_B = 2Y_A$ (b) $Y_A = Y_B$
(c) $Y_B = 3Y_A$ (d) $Y_A = 3Y_B$

2 In determination of Young's modulus of elasticity of a wire, a force is applied and extension is recorded. Initial length of wire is 1 m. The curve between extension and stress is depicted, then Young's modulus of wire will be

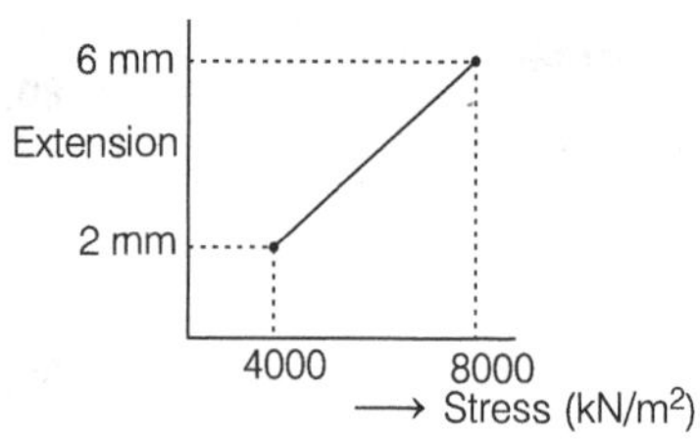

(a) 2×10^9 N/m^2 (b) 1×10^9 N/m^2
(c) 2×10^{10} N/m^2 (d) 1×10^{10} N/m^2

3 Two wires of diameter 0.25 cm, one made of steel and other made of brass are loaded as shown in the following figure. The unloaded length of the steel wire is 1.5 m and that of brass is 1.0 m. Young's modulus of steel is 2.0×10^{11} Nm^{-2} and that of brass is 1.0×10^{11} Nm^{-2}. The ratio of elongations of steel and brass wires $\left(\frac{\Delta l_{steel}}{\Delta l_{brass}} = ?\right)$

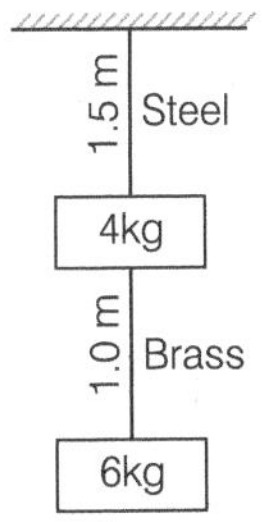

(a) 1.25 (b) 1.5 (c) 2.0 (d) 0.5

4 The diameter of a brass rod is 4 mm and Young's modulus of brass is 9×10^{10} N/m^2. The force required to stretch by 0.1% of its length is

(a) $36\,\pi \times 10^5$ N (b) $144\,\pi \times 10^3$ N
(c) 36 N (d) $360\,\pi$ N

5 Two wires, one of steel ($Y_{steel} = 2 \times 10^{11}$ N/m^2) and the other of copper ($Y_{Cu} = 10^{11}$ N/m^2) have same length and radii. If on applying a force F, the length of first wire increases by l_0, then force required to increase the length of second wire by same amount is

(a) $4F$ (b) $\frac{F}{4}$ (c) $\frac{F}{2}$ (d) $2F$

6 When a certain weight is suspended from a long uniform wire, its length increases by 1 cm. If the same weight is suspended from another wire of the same material and length but having a diameter half of the first one, the increase in length will be

(a) 0.5 cm (b) 2 cm
(c) 4 cm (d) 8 cm

7 A wire of length 2 m is made from 10 cm^3 of copper. A force F is applied, so that its length increases by 2 mm. Another wire of length 8 m is made from the same volume of copper. If the force F is applied to it, its length will increase by

(a) 0.8 cm (b) 1.6 cm
(c) 2.4 cm (d) 3.2 cm

8 One end of a horizontal thick copper wire of length $2L$ and radius $2R$ is welded to an end of another horizontal thin copper wire of length L and radius R. When the arrangement is stretched by applying forces at two ends, the ratio of the elongation in the thin wire to that in the thick wire is

(a) 0.25 (b) 0.50
(c) 2.00 (d) 4.00

9 Four identical hollow cylindrical columns, support a big structure of mass M. The inner and outer radii of a column are R_1 and R_2, respectively. Assuming the load distribution to be uniform, the compressional strain of each column is (where, Y is Young's modulus of the column)

(a) $\frac{Mg}{\pi(R_2^2 - R_1^2)Y}$ (b) $\frac{Mg}{4\pi(R_2^2 - R_1^2)Y}$

(c) $\frac{Mg}{\pi(R_1^2 - R_2^2)Y}$ (d) $\frac{Mg}{4\pi(R_1^2 - R_2^2)Y}$

10 Two bodies of masses 1 kg and 2 kg are connected by a metal wire as shown in the following figure. A force of 10 N is applied on the body of mass 2 kg. The breaking stress of metal wire is 2×10^9 N/m^2. What should be the minimum radius of the wire used, if it is not to break?

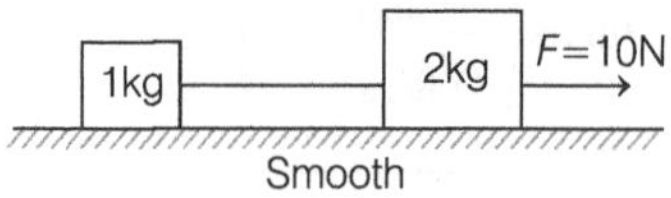

(a) 0.23×10^{-4} m
(b) 4×10^{-4} m
(c) 5×10^{-4} m
(d) 5.2×10^{-4} m

11 An increase in pressure required to decrease the 200 L volume of a liquid by 0.004% in a container is (Take, bulk modulus of the liquid = 2100 MPa)

(a) 188 kPa
(b) 8.4 kPa
(c) 18.8 kPa
(d) 84 kPa

12 The compressibility of water is 4×10^{-5} per unit atmospheric pressure. The decrease in volume of 100 cc of water under a pressure of 100 atmosphere will be

(a) 0.4 cc
(b) 4×10^{-5} cc
(c) 0.025 cc
(d) 0.004 cc

13 The energy stored per unit volume in a stretched wire is given by

(a) $\frac{1}{2}\frac{(\text{Stress})}{Y}$
(b) $\frac{1}{2}\frac{(\text{Stress})^2}{Y}$
(c) $\frac{1}{2}\frac{(\text{Stress})^2}{Y^2}$
(d) $\frac{1}{2}\frac{(\text{Stress})}{Y^2}$

14 The work done in increasing the length of one metre long wire of cross-sectional area 1 mm^2 through 1 mm will be (Take, $Y = 2 \times 10^{11}$ Nm^{-2})

(a) 0.1 J
(b) 5 J
(c) 10 J
(d) 250 J

15 If the work done in stretching a wire by 1 mm is 2 J, the work necessary for stretching another wire of same material but with double radius of cross-section and half the length by 1 mm is

(a) 16 J
(b) 8 J
(c) 4 J
(d) $\frac{1}{4}$ J

16 The stress-strain graph for a metal wire is as shown in the following figure. In the graph, the region in which Hooke's law is obeyed, the ultimate strength and fracture point are represented by

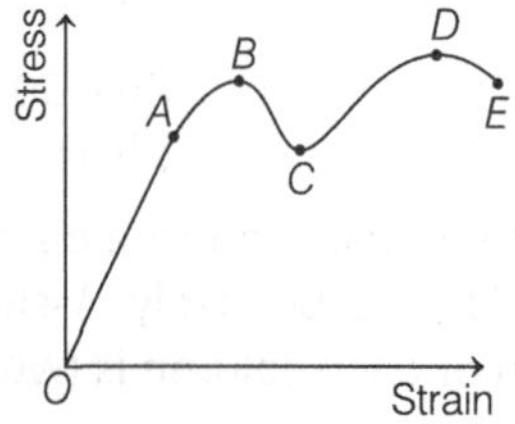

(a) *OA*, *C*, *D*
(b) *OB*, *D*, *E*
(c) *OA*, *D*, *E*
(d) *OB*, *C*, *D*

LEVEL 02 Objective Questions

17 A uniform cylindrical rod of length $3l$, cross-sectional area A and Young's modulus Y is acted upon by the forces as shown in the following figure. The elongation of the rod is

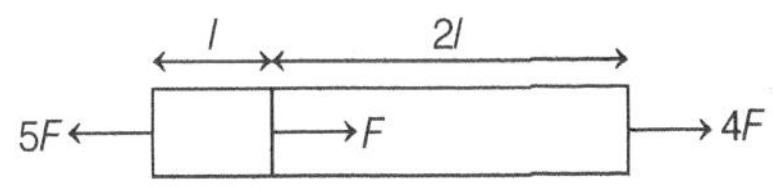

(a) $\frac{12}{5}\frac{Fl}{AY}$ (b) $\frac{13}{5}\frac{Fl}{AY}$ (c) $\frac{13}{8}\frac{Fl}{AY}$ (d) $\frac{13Fl}{AY}$

18 If ρ is the density of the material of a wire and σ is breaking stress, the greatest length of the wire that can hang freely without breaking is

(a) $\frac{2\sigma}{\rho g}$ (b) $\frac{\rho}{\sigma g}$ (c) $\frac{\rho g}{2\sigma}$ (d) $\frac{\sigma}{\rho g}$

19 The density of water at surface of ocean is ρ. If the bulk modulus of water is β, then the density of ocean water at depth, where the pressure is αp_0 will be (p_0 = atmospheric pressure)

(a) $\frac{\rho\beta}{\beta-(\alpha-1)p_0}$ (b) $\frac{\rho\beta}{\beta+(\alpha-1)p_0}$

(c) $\frac{\rho\beta}{\beta-\alpha p_0}$ (d) $\frac{\rho\beta}{\beta+\alpha p_0}$

20 Find the depth of lake at which density of water is 1% greater than at the surface. (Take, compressibility, $K = 50\times10^{-6}$ /atm)

(a) 3 km (b) 2 km

(c) 4 km (d) 1 km

21 Given : σ is the compressibility of water, ρ is the density of water and K is the bulk modulus of water. What is the energy density of water at the bottom of a lake h metre deep?

(a) $\frac{1}{2}\sigma(h\rho g)^2$ (b) $\frac{1}{2}\sigma(h\rho g)$ (c) $\frac{1}{2}\frac{h\rho g}{\sigma}$ (d) $\frac{h\rho g}{\sigma}$

22 A steel wire of mass 3.16 kg is stretched to a tensile strain of 1×10^{-3}. The elastic deformation energy is (Take, density, $\rho = 7.9$ g cc^{-1} and $Y = 2\times10^{11}$ Nm^{-2})

(a) 4 kJ (b) 0.4 kJ (c) 0.04 kJ (d) 4 J

LEVEL 03 Previous Years Questions

23 A wire of length L, area of cross-section A is hanging from a fixed support. The length of the wire changes to L_1 when mass M is suspended from its free end. The expression for Young's modulus is **[NEET 2020]**

(a) $\frac{Mg(L_1-L)}{AL}$ (b) $\frac{MgL}{AL_1}$ (c) $\frac{MgL}{A(L_1-L)}$ (d) $\frac{MgL_1}{AL}$

24 The stress-strain curves are drawn for two different materials X and Y. It is observed that the ultimate strength point and the fracture point are close to each other for material X but are far apart for material Y. We can say that materials X and Y are likely to be (respectively) **[NEET 2019]**

(a) ductile and brittle (b) brittle and ductile

(c) brittle and plastic (d) plastic and ductile

25 When a block of mass M is suspended by a long wire of length L, the length of the wire becomes $(L + l)$. The elastic potential energy stored in the extended wire is **[NEET 2019]**

(a) MgL (b) $\frac{1}{2}Mgl$ (c) $\frac{1}{2}MgL$ (d) Mgl

26 Two wires are made of the same material and have the same volume. The first wire has cross-sectional area A and the second wire has cross-sectional area $3A$. If the length of the first wire is increased by Δl on applying a force F, how much force is needed to stretch the second wire by the same amount? **[NEET 2018]**

(a) $4F$ (b) $6F$ (c) $9F$ (d) F

27 The bulk modulus of a spherical object is B. If it is subjected to uniform pressure p, the fractional decrease in radius is **[NEET 2017]**

(a) $\frac{p}{B}$ (b) $\frac{B}{3p}$

(c) $\frac{3p}{B}$ (d) $\frac{p}{3B}$

28 Copper of fixed volume V is drawn into wire of length l. When this wire is subjected to a constant force F, the extension produced in the wire is Δl. Which of the following graphs is a straight line? **[CBSE AIPMT 2014]**

(a) Δl *versus* $\frac{1}{l}$ (b) Δl *versus* l^2

(c) Δl *versus* $\frac{1}{l^2}$ (d) Δl *versus* l

29 The following four wires are made of same material. Which of these will have the largest extension, when the same tension is applied? **[NEET 2013]**

(a) Length = 50 cm, diameter = 0.5 mm
(b) Length = 100 cm, diameter = 1 mm
(c) Length = 200 cm, diameter = 2 mm
(d) Length = 300 cm, diameter = 3 mm

ANSWER KEY

Level 1 Objective Questions

1 (d)	**2** (b)	**3** (a)	**4** (d)	**5** (c)	**6** (c)	**7** (d)	**8** (c)	**9** (b)	**10** (a)
11 (d)	**12** (a)	**13** (b)	**14** (a)	**15** (a)	**16** (c)				

Level 2 Objective Questions

17 (d)	**18** (d)	**19** (a)	**20** (b)	**21** (a)	**22** (c)

Level 3 Previous Years Questions

23 (c)	**24** (b)	**25** (b)	**26** (c)	**27** (d)	**28** (b)	**29** (a)

Solutions

LEVEL 01 Objective Questions

1. $Y = \frac{\text{Stress}}{\text{Strain}}$ = Slope of graph

$$\frac{Y_A}{Y_B} = \frac{\tan 60^\circ}{\tan 30^\circ} = 3$$

$\therefore \quad Y_A = 3Y_B$

2. $\Delta L = \frac{F}{A} \times \frac{L}{Y}$

Here, $\frac{L}{Y}$ = slope of curve.

$$\frac{L}{Y} = \frac{(6-2)\times 10^{-3}}{4000 \times 10^3}$$

Here, $L = 1$ m

So, $Y = \frac{4000\times 10^3}{4\times 10^{-3}} = 1\times 10^9 \text{ N/m}^2$

3. We have, $Y_1 = \frac{F_1 \times L_1}{A_1 \times \Delta L_1}$

$\Rightarrow \quad \Delta L_1 = \frac{F_1 L_1}{A_1 Y_1}$

Similarly, $\Delta L_2 = \frac{F_2 L_2}{A_2 Y_2}$

So, $\frac{\Delta L_1}{\Delta L_2} = \frac{F_1}{F_2} \cdot \frac{L_1}{L_2} \cdot \frac{A_2}{A_1} \cdot \frac{Y_2}{Y_1}$

$$= \frac{10}{6} \times \frac{1.5}{1.0} \times 1 \times \frac{1}{2}$$

$\Rightarrow \quad \frac{\Delta L_1}{\Delta L_2} = \frac{\Delta L_{steel}}{\Delta L_{brass}} \approx 1.25$

4. $Y = \frac{FL}{A\Delta L}$

So $F = \frac{YA\Delta L}{L} = 9\times 10^{10} \times \pi \times \left(\frac{4\times 10^{-3}}{2}\right)^2 \times \frac{0.1}{100}$

$= 360\ \pi$ N

5. $Y = \frac{\text{Stress}}{\text{Strain}} = \frac{\frac{F}{A}}{\text{Strain}}$

$$F \cdot \frac{YA}{\frac{\Delta l}{l}} = \frac{Y\pi R^2}{\frac{\Delta l}{l}}$$

$$F \propto \frac{Y}{\Delta l / l}$$

$\because \quad \frac{F'}{F} = \frac{Y_{Cu}}{Y_{steel}} \times \frac{l}{l} \times \frac{\Delta l_{steel}}{\Delta l_{Cu}}$

($\because$ length of both the wires are same)

$$\frac{F'}{F} = \frac{10^{11}}{2\times 10^{11}} \times \frac{l_0}{l_0}$$

$$F' = \frac{F}{2}$$

6. $Y = \frac{FL}{A\Delta L}$

$$\Delta L \propto \frac{1}{A} \propto \frac{1}{D^2}$$

$\because \quad \frac{\Delta L_2}{\Delta L_1} = \frac{D_1^2}{D_2^2} = 4 \quad \left(\because D_2 = \frac{1}{2} D_1\right)$

$$\Delta L_2 = 4\Delta L_1 = 4 \text{ cm}$$

7. $l = \frac{FL}{AY} = \frac{FL^2}{(AL)Y} = \frac{FL^2}{VY}$

$\therefore \quad l \propto L^2$

$\because \quad \frac{l_2}{l_1} = \left(\frac{L_2}{L_1}\right)^2 = \left(\frac{8}{2}\right)^2 = 16$

$\Rightarrow \quad l_2 = 16 \times l_1 = 16 \times 2$

$= 32$ mm

$= 3.2$ cm

8. Change in length, $\Delta L = \frac{F\cdot L}{YA}$

$\Rightarrow \quad \Delta L = \frac{F\cdot L}{Y\pi R^2}$

$\therefore \quad \Delta L \propto \frac{L}{R^2}$

$\because \quad \frac{\Delta L_1}{\Delta L_2} = \frac{L_1}{L_2} \times \frac{R_2^2}{R_1^2} = \frac{2L_2}{L_2} \times \frac{R_1^2}{4R_1^2}$

or $\quad \frac{\Delta L_1}{\Delta L_2} = \frac{1}{2}$ or $\frac{\Delta L_2}{\Delta L_1} = 2$

9. $\Delta l = \frac{Fl}{AY}$

$\Rightarrow \quad \frac{\Delta l}{l} = \frac{F}{AY} \quad$ ...(i)

Here, $F = \frac{Mg}{4}$ and $A = \pi (R_2^2 - R_1^2)$

Substituting in Eq. (i), we have
Compressional strain,

$$\frac{\Delta l}{l} = \frac{Mg}{4\pi (R_2^2 - R_1^2)\, Y}$$

10. From Newton's second law,

$$F - T = 2a \quad \text{...(i)}$$

and $\quad T = 1\times a = a$

On substituting T in Eq. (i), we get

$$a = \frac{F}{3} = \frac{10}{3}\ \text{m/s}^2$$

$$T = \frac{10}{3}\text{N}$$

So, stress $= \frac{T}{\pi r^2}$

$$r = \sqrt{\frac{10}{\pi \times 2 \times 10^9 \times 3}} = 0.23 \times 10^{-4}\ \text{m}$$

So, the minimum radius of wire used is 0.23×10^{-4} m.

11. Bulk modulus $(K) = \dfrac{p}{\dfrac{\Delta V}{V}}$

$\because$ $$K = 2100 \times 10^6\ \text{Pa}$$

$$\frac{\Delta V}{V} = \frac{0.004}{100}$$

So, $p = K \times \frac{\Delta V}{V} = 2100 \times 10^6 \times \frac{0.004}{100} = 84$ kPa

12. $K = \left(\frac{\Delta V}{V}\right) \times \Delta p$

$$4 \times 10^{-5} = \frac{\Delta V}{100} \times 100$$

$\Rightarrow$ $\Delta V = 4 \times 10^{-5}$ $\Rightarrow$ $\Delta V = 0.4$ cc

13. Energy density,

$$u = \frac{1}{2} \times \text{Stress} \times \text{Strain}$$

$$\text{Strain} = \frac{\text{Stress}}{Y}$$

$\therefore$ $$u = \frac{1}{2}\frac{(\text{Stress})^2}{Y}$$

14. $W = \frac{1}{2} \times Y \times \left(\frac{\Delta L}{L}\right)^2 \times AL$

$$= \frac{YA(\Delta L)^2}{2L} = \frac{2 \times 10^{11} \times 10^{-6} \times (10^{-3})^2}{2L}$$

$$= \frac{2 \times 10^{11} \times 10^{-6} \times 10^{-6}}{2 \times 1} = 0.1\ \text{J}$$

15. Work done $= \frac{AY}{2L} = \frac{\pi r^2 Y}{2L}$

So, work done $\propto \frac{r^2}{L}$

$$\frac{W_1}{W_2} = \frac{r_1^2}{L_1} \times \frac{L_2}{r_2^2} = \frac{r_1^2}{L_1} \times \frac{\frac{1}{2}L_1}{(2r_1)^2} = \frac{1}{8}$$

$\Rightarrow$ $\frac{W_1}{W_2} = \frac{1}{8}$ $\Rightarrow$ $W_2 = 8W_1 = 16$ J

16. In region OA, graph is linear showing that stress is proportional to the strain. Thus, in this region, Hooke's law is obeyed. Point D on the graph is tensile strength and point E on the graph is known as fracture point.

LEVEL 02 *Objective Questions*

17. Elongation of rod,

$$\Delta l = \Delta l_1 + \Delta l_2$$

$$= \frac{5F \times l}{AY} + \frac{4F \times 2l}{AY} = \frac{13Fl}{AY}$$

18. Stress $= \frac{\text{Weight}}{\text{Area}}$

$$\sigma = \frac{AL\rho g}{A} \quad \therefore \quad L = \frac{\sigma}{\rho g}$$

19. Variation of density with pressure is given by

$$\rho' = \frac{\rho}{\left[1 - \frac{\Delta p}{\beta}\right]}$$

Here, $$\Delta p = \alpha p_0 - p_0 = p_0(\alpha - 1)$$

So, $\rho' = \dfrac{\rho}{1 - \dfrac{p_0(\alpha-1)}{\beta}} = \dfrac{\rho\beta}{\beta - (\alpha - 1)p_0}$

20. $\beta = \dfrac{\Delta p}{-\dfrac{\Delta V}{V}} \Rightarrow \dfrac{\Delta V}{V} = -\dfrac{\Delta p}{\beta}$ ...(i)

Now, $m = \rho V =$ constant

$$d\rho V + dV\rho = 0$$

$$\frac{d\rho}{\rho} = -\frac{dV}{V} \quad \text{...(ii)}$$

From Eqs. (i) and (ii), we get

So, $\frac{\Delta\rho}{\rho} = \frac{\Delta p}{\beta} = \frac{1}{100}$

$$\frac{1}{100} = \frac{h\rho g}{\beta} \quad \text{(assuming } \rho = \text{constant)}$$

$\Rightarrow$ $$h\rho g = \frac{\beta}{100} = \frac{1}{100\,K}$$

$\Rightarrow$ $$h\rho g = \frac{1 \times 10^5}{100 \times 50 \times 10^{-6}}$$

$$h = \frac{10^5}{5000 \times 10^{-6} \times 1000 \times 10}$$

$$= \frac{100 \times 10^3}{50} = 2\ \text{km}$$

21. Bulk modulus, $K = \frac{1}{\sigma} = -\left(\frac{\Delta p}{\Delta V}\right)V$

$\therefore$ $$\left(-\frac{\Delta V}{V}\right) = (\sigma)(\Delta p) = \sigma(\rho g h) = \text{Strain}$$

Energy density, $u = \frac{1}{2} \times \text{Stress} \times \text{Strain}$

$$= \frac{1}{2}(\Delta p)(\sigma)(\rho g h) = \frac{1}{2}(\rho g h)(\sigma)(\rho g h)$$

$$= \frac{1}{2}\sigma(\rho g h)^2$$

22. Elastic deformation energy

$$= \frac{1}{2} \times Y \times \text{Strain} \times \text{Strain} \times \text{Volume}$$

Here, volume $= \dfrac{\text{mass}}{\rho}$

$$= \frac{1}{2} \times 2 \times 10^{11} \times 1 \times 10^{-3} \times 1 \times 10^{-3} \times \frac{3.16}{7.9 \times 1000}$$

$= 0.04$ kJ

LEVEL 03 *Previous Years Questions*

23. Here, change in length, $\Delta L = (L_1 - L)$

Area $= A$, Force, $F = Mg$

Young's modulus, $Y = \dfrac{\text{Normal stress}}{\text{Longitudinal strain}}$

$$\Rightarrow \quad Y = \frac{(F/A)}{\left(\frac{\Delta L}{L}\right)} = \frac{\frac{Mg}{A}}{\left(\frac{L_1 - L}{L}\right)} = \frac{MgL}{A(L_1 - L)}$$

24. The stress-strain curve for a material is shown below

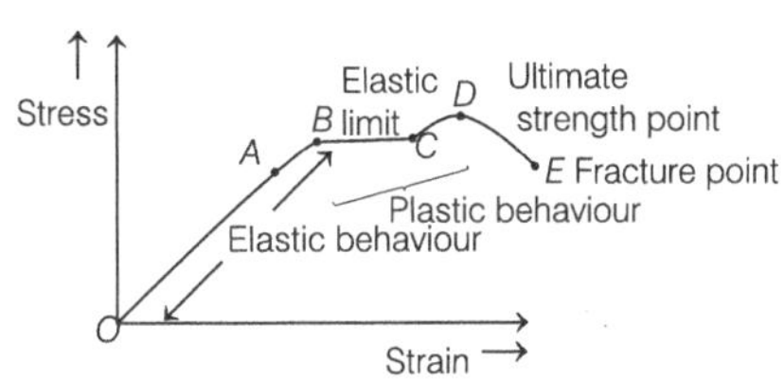

This curve specifies the behaviour of material.

For the material X, if the distance between strength point and fracture point is small, it is brittle and will break easily on the application of some extra stress after point D.

For material Y, if the distance between strength point and fracture point is large, so it is a ductile material and can withstand for some extra stress beyond point D.

25.

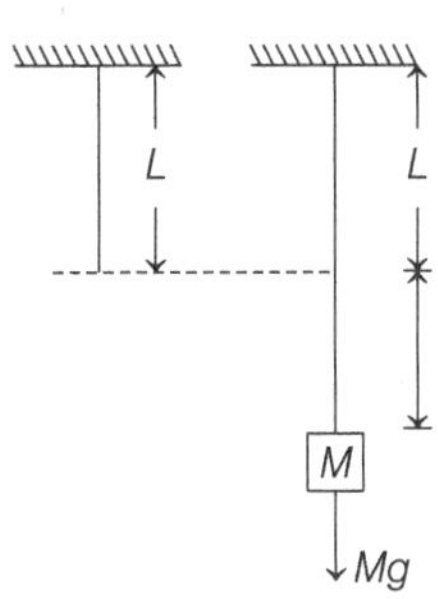

$$U = W = \frac{1}{2} \times \text{Force }(F) \times \text{Elongation }(l)$$

$$= \frac{1}{2} Fl = \frac{1}{2} \times Mg \times l = \frac{1}{2} Mgl$$

26. According to the question,

For wire 1

$$\Rightarrow \quad Y_1 = \frac{F_1 l_1}{A_1 \Delta l} \quad \text{...(i)}$$

For wire 2

$$Y_2 = \frac{F_2 l_2}{A_2 \Delta l} \quad \text{...(ii)}$$

$\because$ Volume, $V = Al$ or $l = \dfrac{V}{A}$

Substituting the value of l in Eqs. (i) and (ii), we get

$$Y_1 = \frac{F_1 V}{A_1^2 \Delta l} \quad \text{and} \quad Y_2 = \frac{F_2 V}{A_2^2 \Delta l}$$

As it is given that the wires are made up of same material, i.e. $Y_1 = Y_2$

$$\Rightarrow \quad \frac{F_1 V}{A_1^2 \Delta l} = \frac{F_2 V}{A_2^2 \Delta l}$$

$$\Rightarrow \quad \frac{F_1}{F_2} = \frac{A_1^2}{A_2^2} = \frac{A^2}{9A^2} \quad (\because A_1 = A \text{ and } A_2 = 3A)$$

$$= \frac{1}{9} \quad \text{or} \quad F_2 = 9F_1 = 9F \quad (\text{given, } F_1 = F)$$

27. The object is spherical and the bulk modulus is represented by B. It is the ratio of normal stress to the volumetric strain.

Hence, $B = \dfrac{F/A}{\Delta V/V}$

$$\Rightarrow \quad \frac{\Delta V}{V} = \frac{p}{B} \Rightarrow \left|\frac{\Delta V}{V}\right| = \frac{p}{B}$$

Here, p is applied pressure on the object and $\dfrac{\Delta V}{V}$ is volume strain.

Fractional decreases in volume,

$$\frac{\Delta V}{V} = 3\frac{\Delta R}{R} \quad \left[\because V = \frac{4}{3}\pi R^3\right]$$

Volume of the sphere decreases due to the decrease in its radius.

Hence, $\dfrac{\Delta V}{V} = \dfrac{3\Delta R}{R} = \dfrac{p}{B} \Rightarrow \dfrac{\Delta R}{R} = \dfrac{p}{3B}$

28. Youngs' modulus is given by

$$Y = \frac{F \times l}{A \times \Delta l} \quad \text{...(i)}$$

As $\quad V = A \times l = \text{constant} \quad$ ...(ii)

From Eqs. (i) and (ii), we get

$$Y = \frac{F \times l^2}{V \times \Delta l} \Rightarrow \Delta l = \frac{F}{V \times Y} \times l^2$$

$$\Rightarrow \quad \Delta l \propto l^2$$

29. As, $\gamma = \dfrac{F \times L}{\Delta L \times A} = \dfrac{mg \cdot L}{\Delta L \cdot A}$

or $\quad \Delta L = \dfrac{mgL}{\gamma \cdot A}$

$\Rightarrow \Delta L \propto \dfrac{L}{A}$, which is maximum for option (a).

Chapter 13

FLUID MECHANICS

LEVEL 01 Objective Questions

1 Equal masses of three liquids are kept in three identical cylindrical vessels A, B and C. The densities are ρ_A, ρ_B and $\rho_C (\rho_A < \rho_B < \rho_C)$. The force on the base will be maximum in

(a) A (b) B
(c) C (d) same in all

2 The gate of a canal is 8 m wide. The level of water on one side is 30 m and on the other side is 15 m. The resultant force on the gate is

(a) 270×10^5 N (b) 270×10^6 N
(c) 540×10^5 N (d) 540×10^6 N

3 A liquid X of density 3.36 g/cm^3 is poured in right arm of a U-tube, containing mercury. Another liquid Y is poured in left arm with column of height 8 cm. Upper levels of X and Y are at same level as shown in the adjoining figure. If density of mercury is 13.6 g/cc, then the density of liquid Y is

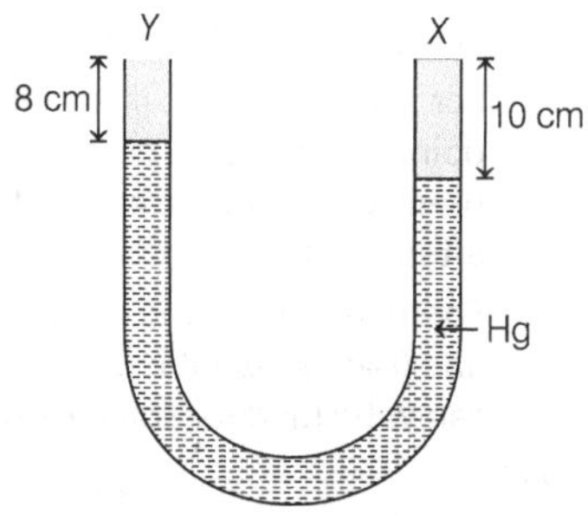

(a) 0.8 g/cc (b) 1.2 g/cc
(c) 1.4 g/cc (d) 1.6 g/cc

4 The area of cross-section of the wider tube shown in the following figure is 800 cm^2. If a mass of 12 kg is placed on the massless piston, the difference in the height h in the level of water in two tubes, is

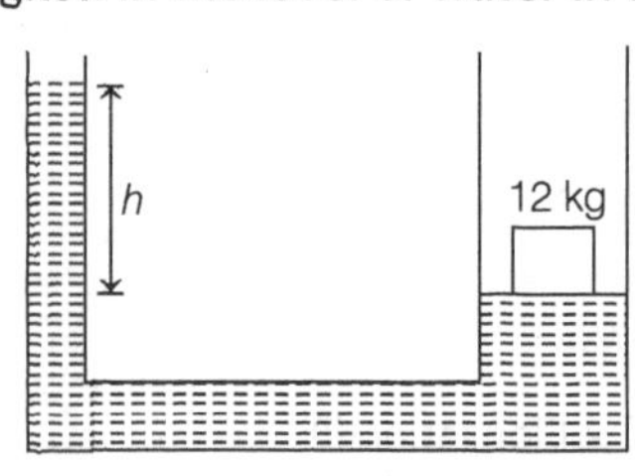

(a) 10 cm (b) 6 cm
(c) 15 cm (d) 2 cm

5 The liquids in U-tube of a uniform cross-section are water and mercury. If the difference of the height of the mercury columns is 2 cm, then the height h of the water column is

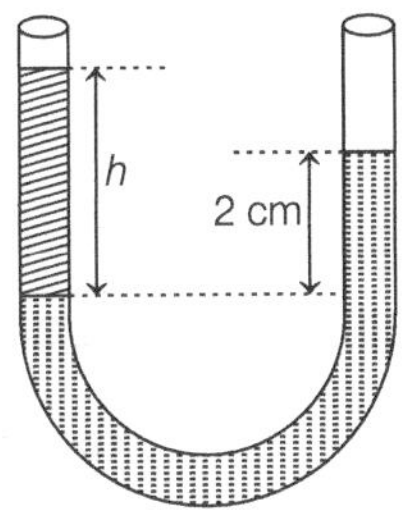

(a) 6.8 cm (b) 13.6 cm (c) 27.2 cm (d) 54 cm

6 A rubber balloon has 200 g of water in it. Its weight in water will be (neglect the weight of balloon)

(a) 100 g (b) 200 g (c) 50 g (d) zero

7 A tank with a square base of area 2 m^2 is divided into two compartments by a vertical partition in the middle. There is a small hinged door of face area 20 cm^2 at the bottom of the partition. Water is filled in one compartment and an acid of relative density 1.5 in the other, both to a height of 4 m. The force necessary to keep the door closed is (Take, $g = 10 \text{ ms}^{-2}$)

(a) 10 N (b) 20 N (c) 40 N (d) 80 N

8 A block of wood floats in water with $\left(\frac{4}{5}\right)$th of its volume submerged. If the same block just floats in a liquid, the density of the liquid (in kg m^{-3}) is

(a) 1250 (b) 600 (c) 400 (d) 800

9 A cube of wood supporting 200 g mass just floats in water. When the mass is removed, the cube rises by 2 cm. What is the side of cube?

(a) 5 cm (b) 10 cm
(c) 15 cm (d) 20 cm

10 A rectangular shape of soap having density 800 kg/m^3 floats in water of density 1000 kg/m^3. Oil of density 300 kg/m^3 is slowly added forming a layer that does not mix with the water. When the top surface of the oil is at the same level as the top surface of the soap. What is the ratio of the oil layer thickness to the soap's thickness $\frac{x}{L}$?

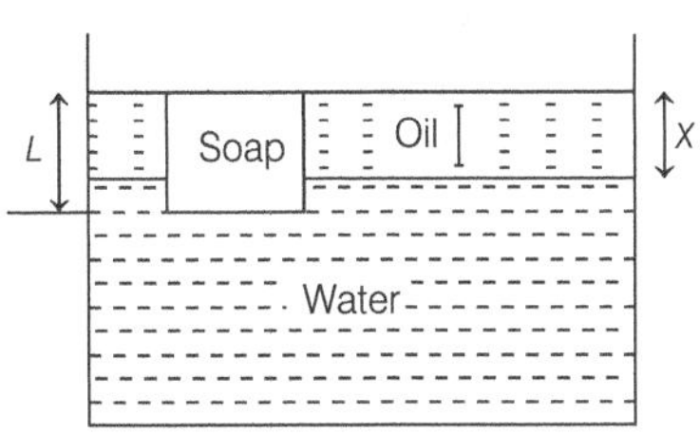

(a) $\frac{2}{10}$ (b) $\frac{2}{7}$ (c) $\frac{3}{10}$ (d) $\frac{3}{8}$

11 In the given figure, if mass m is removed from the top of cubical block, then block comes up by 3 cm, then value of m will be

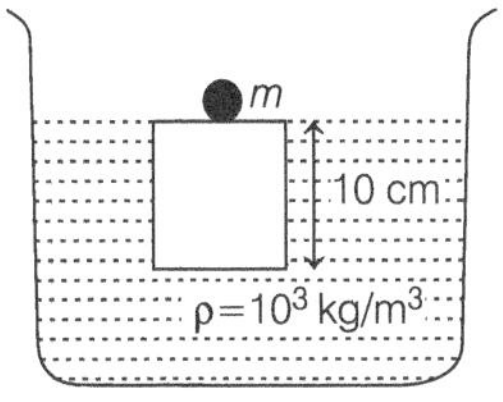

(a) 200 g
(b) 300 g
(c) 400 g
(d) None of the above

12 A metallic sphere floats in an immiscible mixture of water ($\rho_w = 10^3 \text{kg/m}^3$) and a liquid ($\rho_L = 13.5 \times 10^3 \text{ kg/m}^3$), such that its (4/5)th portion is in water and (1/5)th portion in the liquid. The density of metal is

(a) 4.5×10^3 kg/m^3 (b) 4.0×10^3 kg/m^3
(c) 3.5×10^3 kg/m^3 (d) 1.9×10^3 kg/m^3

13 The tube shown below is of uniform cross-section. Liquid flows through it at a constant speed in the direction shown by the arrows. The liquid exerts on the tube,

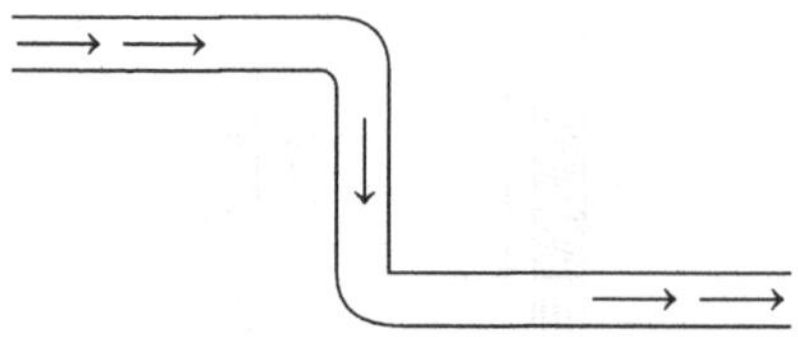

(a) a net force to the right
(b) a net force to the left
(c) a clockwise torque
(d) an anti-clockwise torque

14 An ideal fluid is flowing through a horizontal tube. The pressure at a point in the tube is p, where the velocity of flow is v. At another point, where the pressure is $\frac{p}{2}$, the velocity of flow is (density of fluid = ρ)

(a) $\sqrt{v^2 + \frac{p}{\rho}}$
(b) $\sqrt{v^2 - \frac{p}{\rho}}$
(c) $\sqrt{v^2 + \frac{2p}{\rho}}$
(d) $\sqrt{v^2 - \frac{2p}{\rho}}$

15 A water tank stands on the roof of a building as shown below. Then, the value of h for which the distance covered by the water (x) is greatest, is

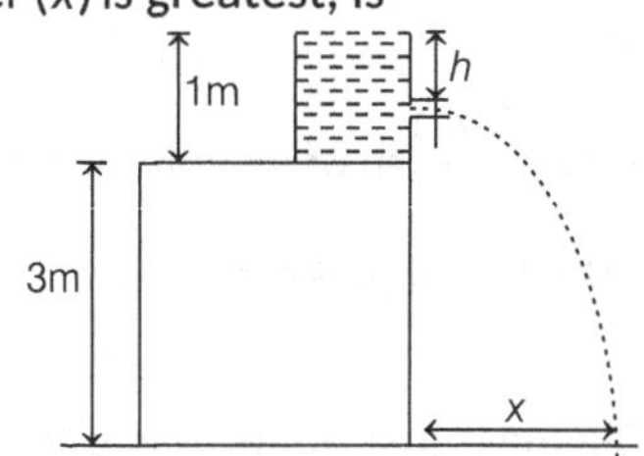

(a) 0.5 m
(b) 0.67 m
(c) 1 m
(d) None of these

16 Water flows through a horizontal tube as shown in the following figure. If the difference of heights of water column in the vertical tubes is $h = 0.02$ m and the areas of cross-section at A and B are 4×10^{-1} m^2 and 2×10^{-1} m^2 respectively, then the rate of flow of water across any section is

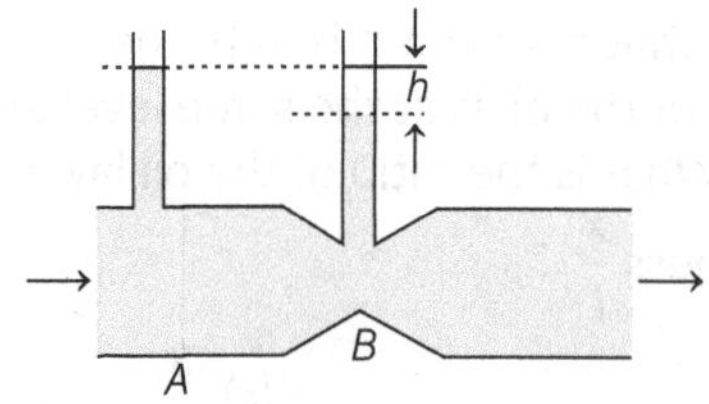

(a) 130×10^{-6} m^3/s
(b) 146×10^{-3} m^3/s
(c) 160×10^{-6} m^3/s
(d) 170×10^{-3} m^3/s

17 The following figure shows venturimeter through which water is flowing. The speed of water at X is 2 cm/s. Find the speed of water at Y. (Take, $g = 1000$ cm/s^2)

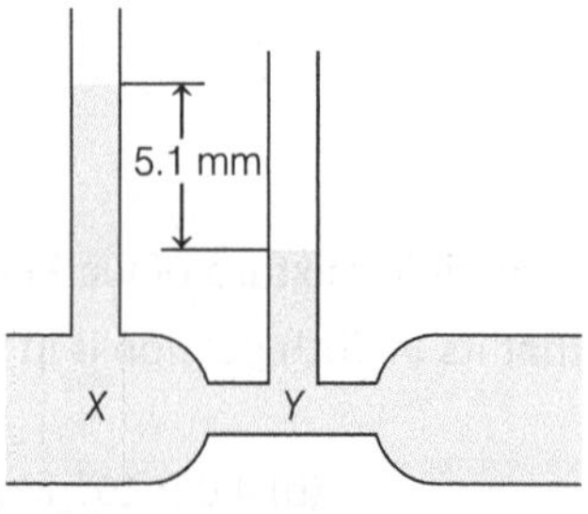

(a) 16 cm/s
(b) 32 cm/s
(c) 8 cm/s
(d) 4 cm/s

18 A cylinder containing water stands on a table of height H. A small hole is punched in the side of cylinder at its base. The stream of water strikes the ground at a horizontal distance R from the table. Then, the depth of water in the cylinder is

(a) H (b) R
(c) $\sqrt{RH}$ (d) $R^2/4H$

19 A thin liquid film formed between a U-shaped wire and a light slider supports a weight of 1.5×10^{-2} N. The length of the slider is 30 cm and its weight is negligible. The surface tension of the liquid film is

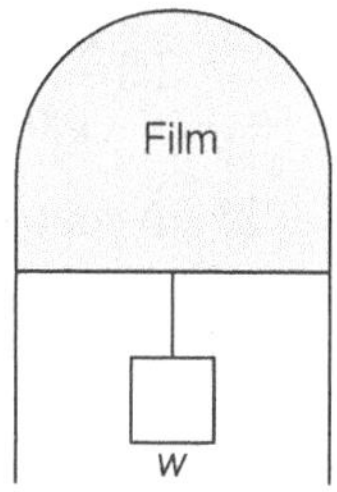

(a) 0.0125 Nm^{-1}
(b) 0.1 Nm^{-1}
(c) 0.05 Nm^{-1}
(d) 0.025 Nm^{-1}

20 A frame made of metallic wire enclosing a surface area A is covered with a soap film. If the area of the frame of metallic wire is reduced by 50%, the energy of the soap film will be changed by

(a) 100% (b) 75%
(c) 50% (d) 25%

21 Two very wide parallel glass plates are held vertically at a small separation d and dipped in water. Some water climbs up in the gap between the plates. Let S be the surface tension of water, p_0 = atmospheric pressure, p = pressure of water just below the water surface in the region between the plates. Then,

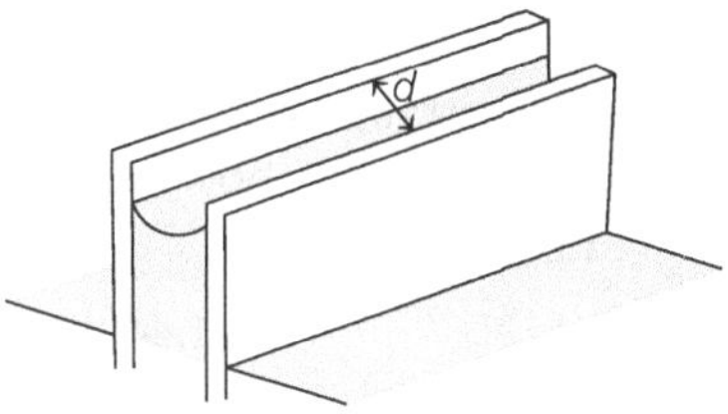

(a) $p = p_0 - \frac{2S}{d}$ (b) $p = p_0 + \frac{2S}{d}$ (c) $p = p_0 - \frac{4S}{d}$ (d) $p = p_0 + \frac{4S}{d}$

22 Water rises to a height of 2 cm in a capillary tube. If the tube is tilted 60° from the vertical, water will rise in the tube to a length

(a) 4.0 cm (b) 2.0 cm
(c) 1.0 cm (d) water will not rise at all

23 Liquid reaches an equilibrium position as shown, in a capillary tube of internal radius r. If the surface tension of the liquid is T, the angle of contact θ and density of liquid ρ, then the pressure difference between P and Q is

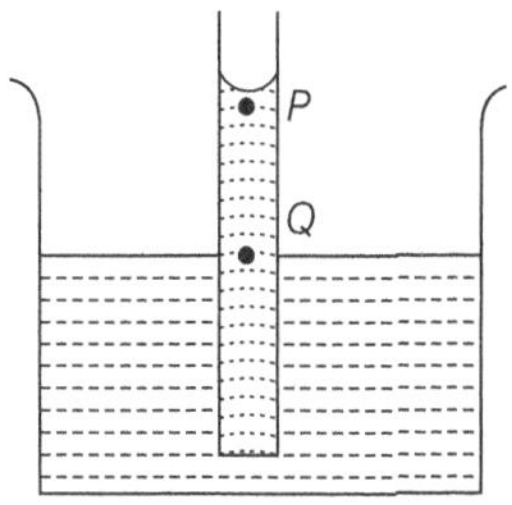

(a) $\left(\frac{2T}{r}\right)\cos\theta$ (b) $\frac{T}{r\cos\theta}$
(c) $\frac{2T}{r\cos\theta}$ (d) $\left(\frac{4T}{r}\right)\cos\theta$

24 The radii of the two columns in U-tube are r_1 and r_2. When a liquid of density ρ (angle of contact is 0°) is filled in it the level of difference of liquid in two arms is *h*. The surface tension of liquid is (g = acceleration due to gravity)

(a) $\frac{\rho g h r_1 r_2}{2(r_2 - r_1)}$
(b) $\frac{\rho g h (r_1 - r_2)}{2 r_1 r_2}$
(c) $\frac{2(r_2 - r_1)}{\rho g h r_1 r_2}$
(d) $\frac{\rho g h}{2(r_2 - r_1)}$

25 A soap bubble has radius *R* and thickness $d(<< R)$ as shown in the following figure. It collapses into a spherical drop. The ratio of excess pressure in the drop to the excess pressure inside the bubble is

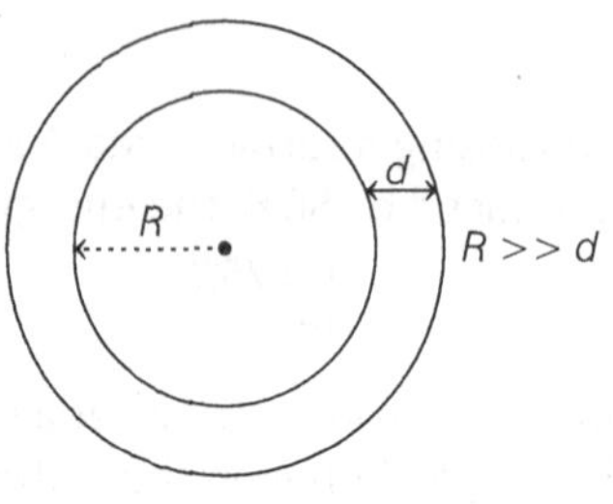

(a) $\left(\frac{R}{3d}\right)^{\frac{1}{3}}$
(b) $\left(\frac{R}{6d}\right)^{\frac{1}{3}}$
(c) $\left(\frac{R}{24d}\right)^{\frac{1}{3}}$
(d) None of these

26 A film of water is formed between two straight parallel wires of length 10 cm and separated by 0.5 cm. If their separation is increased by 1 mm, how much work will have to be done? (Take, surface tension of water $= 7.2 \times 10^{-2}$ N/m)

(a) 7.22×10^{-6} J
(b) 1.44×10^{-5} J
(c) 2.88×10^{-5} J
(d) 5.76×10^{-5} J

27 A square wire frame of length *l* is dipped in a liquid. When the frame is taken out, a liquid film is formed. What is the algebraic sum of all the forces acting on the frame due to surface tension of the liquid? (Here, σ = surface tension of the liquid)

(a) $4\sigma l$
(b) $8\sigma l$
(c) $10\sigma l$
(d) $12\sigma l$

28 A soap bubble of radius *r* is blown until radius becomes 2*r* under isothermal conditions. If *T* is the surface tension of soap solution, the energy spent in the blowing, is

(a) $3\pi T r^2$ joule
(b) $6\pi T r^2$ joule
(c) $12\pi T r^2$ joule
(d) $24\pi T r^2$ joule

29 There is an air bubble of radius 1 mm in a liquid of a surface tension 0.075 Nm^{-1} and density 10^3 kgm^{-3}. The bubble is at a depth 10 cm below the free surface. By what amount is the pressure inside the bubble greater than the atmospheric pressure. ?

(a) 1030 Nm^{-1}
(b) 1230 Nm^{-1}
(c) 1130 Nm^{-1}
(d) 1210 Nm^{-1}

30 Consider an ice cube of edge 1.0 cm kept in a gravity free space. Find the surface area of the water when the ice melts. Neglect the difference in densities of ice and water.

(a) $(36\pi)^{\frac{2}{3}}$ cm^2
(b) $(36\pi)^{\frac{1}{3}}$ cm^2
(c) 4π cm^2
(d) None of these

31 A spherical body of mass *m* is moving in a viscous liquid with constant velocity *v*. Calculate constant velocity of another spherical body of same material and mass 8*m* in same liquid.

(a) v
(b) 2v
(c) 4v
(d) 3v

32 A solid ball of volume V is dropped into a viscous liquid. It experiences a viscous force equal to F. If a solid ball of volume $2V$ and of the same material is dropped into the same liquid, the viscous force will be

(a) F (b) $2F$ (c) $4F$ (d) $8F$

33 The graph between terminal velocity (along y-axis) and square of radius (along x-axis) of spherical body of density ρ allowed to fall through a fluid of density σ is a

(a) straight line with positive slope
(b) straight line with negative slope
(c) straight line perpendicular to x-axis
(d) straight line perpendicular to y-axis

34 A metal block of area $0.10\ m^2$ is connected to a 0.01 kg mass *via* a string that passes over a massless and frictionless pulley as shown in the following figure. A liquid with a film thickness of 0.3 mm is placed between the block and the table. When released the block moves to the right with a constant speed of $0.085\ ms^{-1}$. The coefficient of viscosity of the liquid is (Take, $g = 10\ ms^{-2}$)

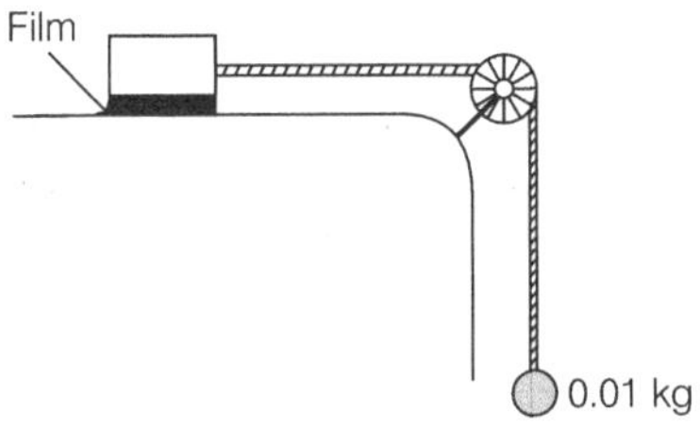

(a) 2.5×10^{-3} Pa s (b) 3.5×10^{-3} Pa s (c) 4.5×10^{-3} Pa s (d) 6.5×10^{-3} Pa s

35 Three liquids A, B and C are allowed to flow in identical capillaries under identical conditions but the volumes of liquids collected in equal time interval are $2\ cm^3$, $4\ cm^3$ and $8\ cm^3$. The ratio of coefficients of viscosity of liquids A, B and C is

(a) 1 : 2 : 4
(b) 1 : 1 : 1
(c) 1 : 4 : 16
(d) 4 : 2 : 1

36 Water flows in a streamlined manner through a capillary of radius a, the pressure difference being p and the rate of flow Q. If the radius is reduced to $a/2$ and the pressure increased to $4p$, the rate of flow becomes

(a) $4Q$
(b) Q
(c) $Q/4$
(d) $Q/8$

LEVEL 02 Objective Questions

37 A U-tube is partially filled with water. Oil which does not mix with water, is next poured into one side until water rises by 25 cm, on the other side. If the density of oil be $0.8\ g\ cm^{-3}$, the oil level will stand higher than water level by

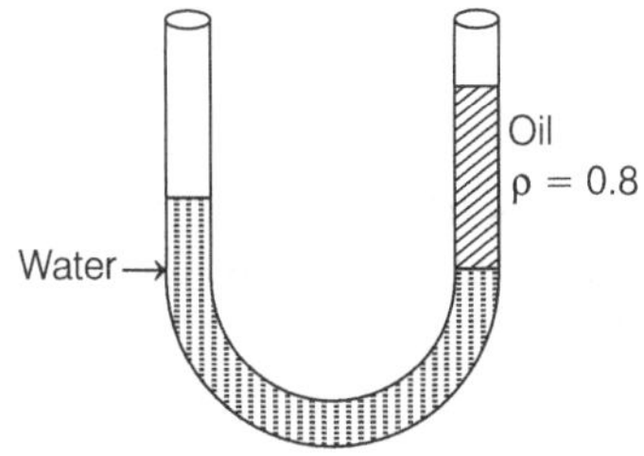

(a) 6.25 cm (b) 12.50 cm (c) 31.25 cm (d) 20 cm

38 There are three different liquids with densities ρ_1, ρ_2 and ρ_3 in a U-shaped container as shown in the following figure. The lengths shown are $H_1 = 15$ cm and $H_2 = 10$ cm. Which of the following equations give the correct relation between the densities of the fluids in the container?

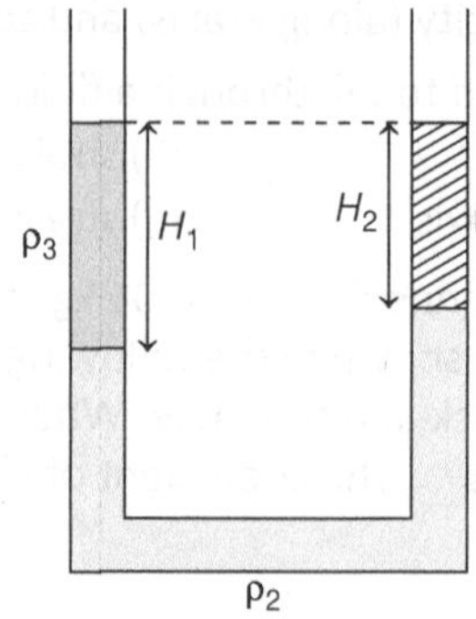

(a) $3\rho_3 = 2\rho_1 + \rho_2$ (b) $\rho_3 = 2\rho_1 + 3\rho_2$ (c) $2\rho_3 = 3\rho_1 + \rho_2$ (d) $\rho_3 = 3\rho_1 + 2\rho_2$

39 A U-tube closed at one end is filled with three immiscible liquids having densities ρ, 3ρ and 8ρ as shown in the following figure. The pressure at point C inside the tube is

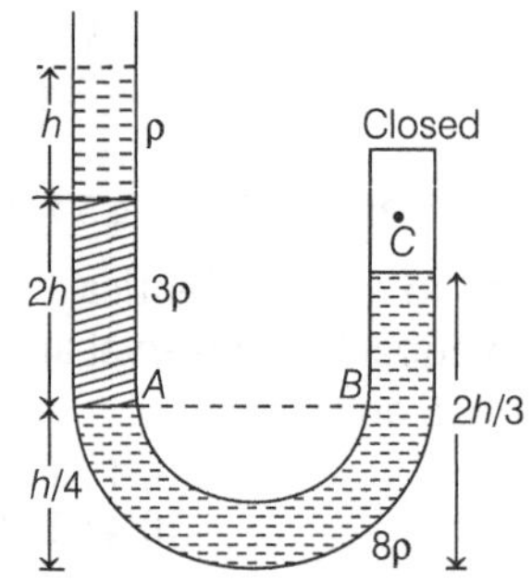

(a) $p_0 + \frac{h\rho g}{3}$ (b) $p_0 + \frac{11h\rho g}{3}$ (c) $p_0 + \frac{2h\rho g}{3}$ (d) $p_0 + h\rho g$

40 A cubical block of wood of specific gravity 0.5 and a chunk of concrete of specific gravity 2.5 are fastened together. The ratio of the mass of wood to the mass of concrete, which makes the combination to float with its entire volume submerged under water is

(a) 1/5 (b) 1/3
(c) 3/5 (d) 2/5

41 A cubical block of wood of edge a and density ρ floats in water of density 2ρ. The lower surface of the cube just touches the free end of a massless spring of force constant K fixed at the bottom of the vessel. The weight w put over the block, so that it is completely immersed in water without wetting the weight is

(a) $a(a^2\rho g + K)$ (b) $a(a\rho g + 2K)$
(c) $a\left(\frac{a\rho g}{2} + 2K\right)$ (d) $a\left(a^2\rho g + \frac{K}{2}\right)$

42 The pressure at the bottom of a tank of water is $4p$, where p is the atmospheric pressure. If the water is drawn out, such that the level of water is lowered by one-third, the pressure at the bottom of the tank will now be

(a) $\frac{2p}{3}$ (b) $3p$
(c) $4p$ (d) $\frac{5p}{3}$

43 An empty balloon weight 1 g. The balloon is filled with water to the neck and tied with a massless thread. The weight of balloon along with water is 101 g. The balloon filled with water is weighed, when fully immersed. Then, its weight in water is

(a) 1 g (b) 101 g (c) 201 g (d) 51 g

44 A piece of copper having an internal cavity weights 264 g in air and 221 g in water. Find the volume of cavity. (Take, density of Cu = 8.8 g/cc)

(a) 30 cc (b) 20 cc (c) 43 cc (d) 13 cc

45 A wooden block of mass 8 kg is tied to a string attached to the bottom of a tank. The block is completely inside the water. Relative density of wood is 0.8. Taking $g = 10 \text{ m/s}^2$, what is the tension in the string?

(a) 100 N (b) 80 N (c) 50 N (d) 20 N

46 A solid uniform ball having volume V and density ρ floats at the interface of two immiscible liquids as shown in the following figure.

The densities of the upper and the lower liquids are ρ_1 and ρ_2 respectively, such that $\rho_1 < \rho < \rho_2$. What fraction of the volume of the ball will be in the lower liquid?

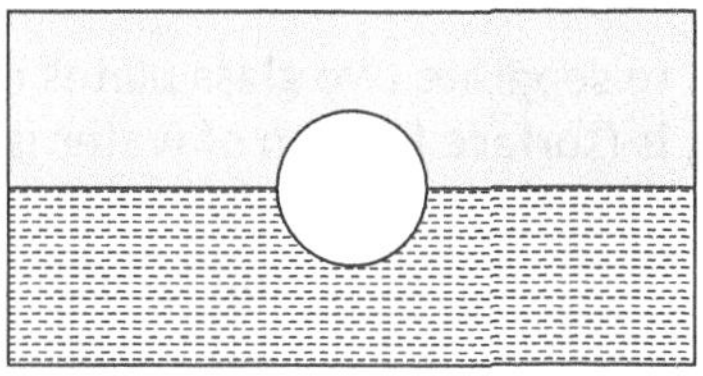

(a) $\frac{\rho - \rho_2}{\rho_1 - \rho_2}$ (b) $\frac{\rho_1}{\rho_1 - \rho_2}$ (c) $\frac{\rho_1 - \rho}{\rho_1 - \rho_2}$ (d) $\frac{\rho_1 - \rho_2}{\rho_2}$

47 A ball of mass m and radius r is gently released in a viscous liquid. The mass of the liquid displaced by it is m, such that $m > m'$. The terminal velocity is proportional to

(a) $\frac{m - m'}{r}$ (b) $\frac{m + m'}{r}$ (c) $\frac{m + m'}{r^2}$ (d) $(m - m')r^2$

48 A liquid of density 2ρ is filled in a cylindrical vessel, whose cross-sectional area is $2A$. A wooden cylinder of height H, cross-sectional area A and density ρ is floating in the liquid at equilibrium with its axis vertical. The cylinder is pushed down by a small distance x from its equilibrium position and released. Find its initial acceleration.

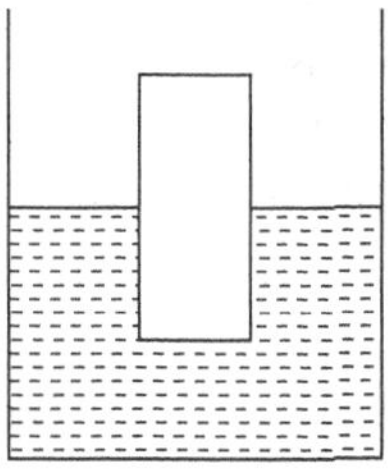

(a) $a = \left(\frac{g}{H}\right)x$ (b) $a = \left(\frac{2g}{H}\right)x$ (c) $a = \left(\frac{4g}{H}\right)x$ (d) Zero

49 A tank full of water has a small hole at its bottom. If $\frac{1}{4}$ th tank is emptied in time t_1 and remaining $\frac{3}{4}$ th tank is emptied in t_2, then ratio $t_1 : t_2$ is

(a) $\frac{\sqrt{2} - 1}{\sqrt{2}}$ (b) $\frac{2 - \sqrt{3}}{\sqrt{3}}$ (c) $\frac{\sqrt{2} + 1}{\sqrt{2}}$ (d) $\frac{2 + \sqrt{3}}{3}$

50 Equal volumes of two immiscible liquids of densities ρ and 2ρ are filled in a vessel as shown in the following figure. Two small holes are punched at depth $h/2$ and $3h/2$ from surface of lighter liquid. If v_1 and v_2 are the velocities of efflux at these two holes, then v_1/v_2 is

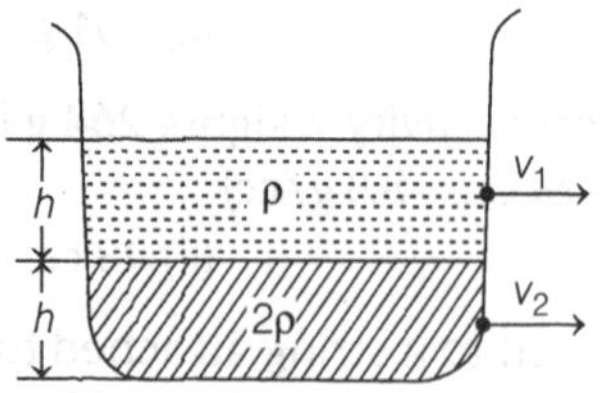

(a) $\frac{1}{2\sqrt{2}}$ (b) $\frac{1}{2}$ (c) $\frac{1}{4}$ (d) $\frac{1}{\sqrt{2}}$

51 Two parallel glass plates are dipped partly in the liquid of density d, keeping them vertical. If the distance between the plates is x, surface tension of liquid T and angle of contact is θ, then rise of liquid between plates will be

(a) $\frac{T\cos\theta}{xd}$ (b) $\frac{2T\cos\theta}{xdg}$ (c) $\frac{2T}{xdg\cos\theta}$ (d) $\frac{T\cos\theta}{xdg}$

52 The minimum force required to separate two glass plates of area 1000 cm^2 with film of water 0.1 mm thick between them, is (surface tension of water is 70×10^{-3} N/m and angle of contact is 0°)

(a) 70 N (b) 50 N
(c) 140 N (d) 100 N

53 A drop of liquid pressed between two glass plates spread into a circle of diameter 10 cm. Thickness of the liquid film is 0.5 mm and surface tension is 70×10^{-3} N/m. The force required to pull them apart

(a) 4.4 N (b) 1.1 N
(c) 2.2 N (d) 3.6 N

54 A bucket having area of cross-section A, has a small hole of area a at bottom, is placed under a tap of water. The rate of flow of water from tap is V m^3/s. The maximum height up to which water can be filled in bucket is

(a) $\frac{V^2}{2a^2g}$ (b) $\frac{V}{2ag}$ (c) $\frac{V^2}{a^2g}$ (d) $\frac{V^2}{ag}$

55 A non-uniform cylinder of mass m, length l and radius r is having its centre of mass at a distance $l/4$ from the centre and lying on the axis of the cylinder. The cylinder is kept in a liquid of uniform density ρ. The moment of inertia of the rod about the centre of mass is I. The angular acceleration of point A relative to point B just after the rod is released from the position shown in the following figure is

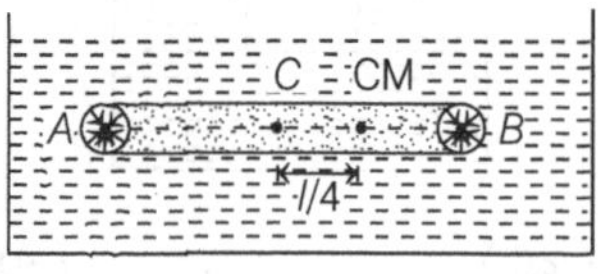

(a) $\frac{\pi\rho g l^2 r^2}{I}$ (b) $\frac{\pi\rho g l^2 r^2}{4I}$ (c) $\frac{\pi\rho g l^2 r^2}{2I}$ (d) $\frac{3\pi\rho g l^2 r^2}{4I}$

56 If the terminal speed of a sphere of gold (density $= 19.5$ kg m^{-3}) is 0.2 ms^{-1} in a viscous liquid (density $= 1.5$ kg m^{-3}), the terminal speed of a sphere of silver (density $= 10.5$ kg m^{-3}) of the same size in the same liquid is

(a) 0.1 ms^{-1} (b) 0.2 ms^{-1}
(c) 0.4 ms^{-1} (d) 0.133 ms^{-1}

LEVEL 03 Previous Years Questions

57 The velocity of a small ball of mass M and density d, when dropped in a container filled with glycerine becomes constant after some time. If the density of glycerine is $d/2$, then the viscous force acting on the ball will be [NEET 2021]

(a) $\frac{Mg}{2}$ (b) Mg (c) $\frac{3}{2}Mg$ (d) $2\,Mg$

58 A barometer is constructed using a liquid (density $= 760\ kg/m^3$). What would be the height of the liquid column, when a mercury barometer reads 76 cm?
(Take, density of mercury $= 13600\ kg/m^3$) [NEET 2020]

(a) 1.36 m (b) 13.6 m
(c) 136 m (d) 0.76 m

59 A liquid does not wet the solid surface, if angle of contact is [NEET 2020]

(a) equal to 45° (b) equal to 60°
(c) greater than 90° (d) zero

60 A capillary tube of radius r is immersed in water and water rises in it to a height h. The mass of the water in the capillary tube is 5 g. Another capillary tube of radius $2r$ is immersed in water. The mass of water that will rise in this tube is [NEET 2020]

(a) 5.0 g (b) 10.0 g
(c) 20.0 g (d) 2.5 g

61 In a U-tube as shown in a figure, water and oil are in the left side and right side of the tube, respectively. The heights from the bottom for water and oil columns are 15 cm and 20 cm, respectively. The density of the oil is (Take, $\rho_{water} = 1000\ kg/m^3$) [NEET 2019]

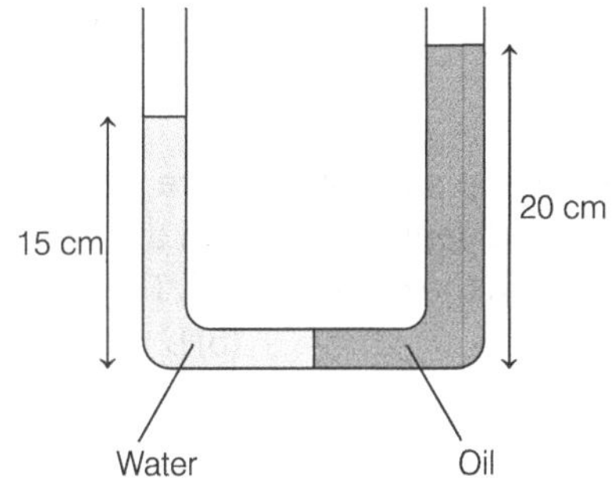

(a) $1200\ kg/m^3$ (b) $750\ kg/m^3$
(c) $1000\ kg/m^3$ (d) $1333\ kg/m^3$

62 Two small spherical metal balls, having equal masses are made from materials of densities ρ_1 and ρ_2 ($\rho_1 = 8\rho_2$) and have radii of 1 mm and 2 mm, respectively. They are made to fall vertically (from rest) in viscous medium whose coefficient of viscosity equals η and whose density is $0.1\,\rho_2$. The ratio of their terminal velocities would be [NEET 2019]

(a) $\frac{79}{72}$ (b) $\frac{19}{36}$
(c) $\frac{39}{72}$ (d) $\frac{79}{36}$

63 A small hole of area of cross-section $2\ mm^2$ is present near the bottom of a fully-filled open tank of height 2 m. Take, $g = 10\ m/s^2$, the rate of flow of water through the open hole would be nearly [NEET 2019]

(a) $8.9 \times 10^{-6}\ m^3/s$ (b) $2.23 \times 10^{-6}\ m^3/s$
(c) $6.4 \times 10^{-6}\ m^3/s$ (d) $12.6 \times 10^{-6}\ m^3/s$

64 A soap bubble, having radius of 1 mm, is blown from a detergent solution having a surface tension of 2.5×10^{-2} N/m. The pressure inside the bubble equals at a point Z_0 below the free surface of water in a container. Take, $g = 10$ m/s^2, density of water $= 10^3$ kg/m^3, the value of Z_0 is **[NEET 2019]**

(a) 10 cm (b) 1 cm
(c) 0.5 cm (d) 100 cm

65 A small sphere of radius r falls from rest in a viscous liquid. As a result, heat is produced due to viscous force. The rate of production of heat when the sphere attains its terminal velocity, is proportional to **[NEET 2018]**

(a) r^5 (b) r^2
(c) r^3 (d) r^4

66 A U tube with both ends open to the atmosphere, is partially filled with water. Oil, which is immiscible with water, is poured into one side until it stands at a distance of 10 mm above the water level on the other side. Meanwhile the water rises by 65 mm from its original level (see diagram). The density of the oil is **[NEET 2017]**

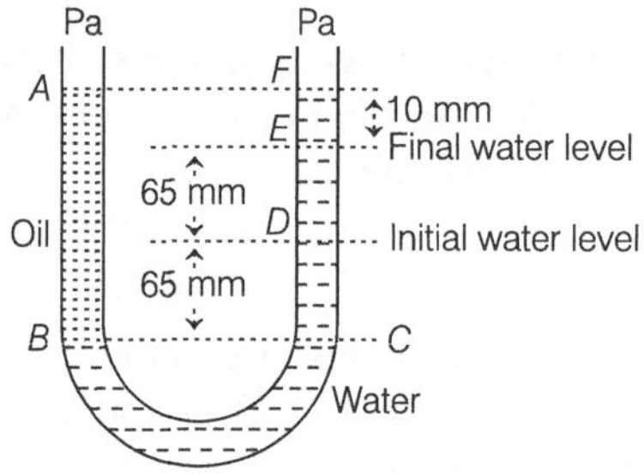

(a) 650 kg m^{-3} (b) 425 kg m^{-3}
(c) 800 kg m^{-3} (d) 928 kg m^{-3}

67 Two non-mixing liquids of densities ρ and $n\rho(n > 1)$ are put in a container. The height of each liquid is h. A solid cylinder of length L and density d is put in this container. The cylinder floats with its axis vertical and length pL $(p < 1)$ in the denser liquid. The density d is equal to **[NEET 2016]**

(a) $\{2 + (n + 1)p\}\rho$ (b) $\{2 + (n - 1)p\}\rho$
(c) $\{1 + (n - 1)p\}\rho$ (d) $\{1 + (n + 1)p\}\rho$

68 The approximate depth of an ocean is 2700 m. The compressibility of water is 45.4×10^{-11} Pa^{-1} and density of water is 10^3 kg/m^3. What fractional compression of water will be obtained at the bottom of the ocean? **[CBSE AIPMT 2015]**

(a) 0.8×10^{-2} (b) 1.0×10^{-2}
(c) 1.2×10^{-2} (d) 1.4×10^{-2}

69 A rectangular film of liquid is extended from (4 cm $\times$ 2 cm) to (5 cm $\times$ 4 cm). If the work done is 3×10^{-4} J, the value of the surface tension of the liquid is **[NEET 2016]**

(a) 0.250 Nm^{-1} (b) 0.125 Nm^{-1}
(c) 0.2 Nm^{-1} (d) 8.0 Nm^{-1}

70 Three liquids of densities ρ_1, ρ_2 and ρ_3 (with $\rho_1 > \rho_2 > \rho_3$), having the same value of surface tension T, rise to the same height in three identical capillaries. The angles of contact θ_1, θ_2 and θ_3 obey **[NEET 2016]**

(a) $\frac{\pi}{2} > \theta_1 > \theta_2 > \theta_3 \ge 0$ (b) $0 \le \theta_1 < \theta_2 < \theta_3 < \frac{\pi}{2}$

(c) $\frac{\pi}{2} < \theta_1 < \theta_2 < \theta_3 < \pi$ (d) $\pi > \theta_1 > \theta_2 > \theta_3 > \frac{\pi}{2}$

71 A wind with speed 40 m/s blows parallel to the roof of a house. The area of the roof is 250 m^2. Assuming that the pressure inside the house is atmospheric pressure, the force exerted by the wind on the roof and the direction of the force will be (Take, $\rho_{air} = 1.2\ kg/m^3$) [CBSE AIPMT 2015]

(a) 4.8×10^5 N, downwards
(b) 4.8×10^5 N, upwards
(c) 2.4×10^5 N, upwards
(d) 2.4×10^5 N, downwards

72 The cylindrical tube of a spray pump has radius R, one-end of which has n fine holes, each of radius r. If the speed of the liquid in the tube is v, the speed of the ejection of the liquid through the holes is [CBSE AIPMT 2015]

(a) $\frac{vR^2}{n^2r^2}$
(b) $\frac{vR^2}{nr^2}$
(c) $\frac{vR^2}{n^3r^2}$
(d) $\frac{v^2R}{nr}$

73 A certain number of spherical drops of a liquid of radius r coalesce to form a single drop of radius R and volume V. If T is the surface tension of the liquid, then [CBSE AIPMT 2014]

(a) energy $= 4VT\left(\frac{1}{r} - \frac{1}{R}\right)$ is released

(b) energy $= 3VT\left(\frac{1}{r} + \frac{1}{R}\right)$ is absorbed

(c) energy $= 3VT\left(\frac{1}{r} - \frac{1}{R}\right)$ is released

(d) energy is neither released nor absorbed

74 The wettability of a surface by a liquid depends primarily on [NEET 2013]

(a) viscosity
(b) surface tension
(c) density
(d) angle of contact between the surface and the liquid

ANSWER KEY

Level 1 Objective Questions

1 (d)	2 (a)	3 (a)	4 (c)	5 (c)	6 (d)	7 (c)	8 (d)	9 (b)	10 (b)
11 (b)	12 (c)	13 (c)	14 (a)	15 (c)	16 (b)	17 (b)	18 (d)	19 (d)	20 (c)
21 (a)	22 (a)	23 (a)	24 (a)	25 (c)	26 (b)	27 (b)	28 (d)	29 (c)	30 (b)
31 (c)	32 (b)	33 (a)	34 (b)	35 (d)	36 (c)				

Level 2 Objective Questions

37 (b)	38 (a)	39 (b)	40 (c)	41 (d)	42 (b)	43 (a)	44 (d)	45 (d)	46 (c)
47 (a)	48 (c)	49 (b)	50 (d)	51 (b)	52 (c)	53 (c)	54 (a)	55 (b)	56 (a)

Level 3 Previous Years Questions

57 (a)	58 (b)	59 (c)	60 (d)	61 (b)	62 (d)	63 (d)	64 (b)	65 (a)	66 (d)
67 (c)	68 (c)	69 (b)	70 (b)	71 (c)	72 (b)	73 (c)	74 (d)		

Solutions

LEVEL 01 Objective Questions

1. The force on the base will be same. This force $= (mg + p_0 A)$

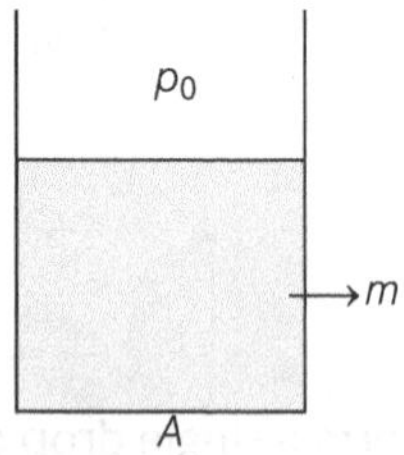

Free body diagram of liquid,

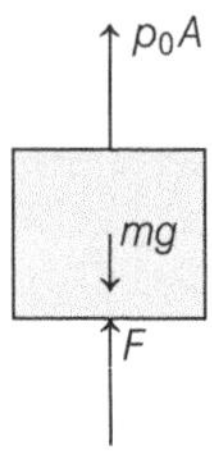

$$F = (p_0 A + mg)$$

2. Net force on the canal, $F = p\Delta A$

$$F = \rho g h \Delta A$$

So, $$F = \rho g \left(30 \times 15 \times 8 - 15 \times \frac{15}{2} \times 8 \right)$$

$$= 10^3 \times 10 \left(30 \times 15 \times 8 - 15 \times \frac{15}{2} \times 8 \right)$$

$$= 10^4 \times 8 \times 15 \times 15 \times \frac{3}{2}$$

$$= 270 \times 10^5 \text{ N}$$

3. As the pressure remains same in the two arms. Hence,

$$8\,\rho_Y + 2 \times 13.6 = 10 \times 3.36$$

$$\Rightarrow \quad \rho_Y = \frac{33.6 - 27.2}{8}$$

$$= 0.8 \text{ g/cc}$$

4. Pressure, $p = \rho g h = \dfrac{mg}{A}$

$$1 \times 980 \times h = \frac{12 \times 1000 \times 980}{800}$$

$$\Rightarrow \quad h = 15 \text{ cm}$$

5. At lower dotted line pressure from both sides should be same.

$$\therefore \quad p_0 + \rho_w g h = p_0 + \rho_{Hg} g \cdot 2$$

$$\therefore \quad h = (2)(\rho_{Hg}) \qquad (\text{as } \rho_w = 1 \text{ g/cm}^3)$$

$$= 2 \times 13.6$$

$$= 27.2 \text{ cm}$$

6. The weight of balloon in water will be,

$$\rho_w V g = 0.2\, g$$

Mass, $m = \rho_w V$

$$mg = 0.2\, g$$

As the net force is zero, so weight of balloon in water is zero.

7. Pressure due to water is given as,

$$p_1 = h_1 \rho_1 g = 4 \times 10^3 \times 10$$

$$p_1 = 4 \times 10^4 \text{ Pa}$$

Similarly, the pressure due to the acid,

$$p_2 = h_2 \rho_2 g = 4 \times 1.5 \times 10^3 \times 10$$

$$p_2 = 6 \times 10^4 \text{ Pa}$$

Now, the pressure difference between water and acid columns,

$$\Delta p = p_2 - p_1$$

$$= (6 \times 10^4 - 4 \times 10^4)$$

$$= 2 \times 10^4 \text{ Pa}$$

So, force necessary to keep door closed $= \Delta p A$

$$= 2 \times 10^4 \times 20 \times 10^{-4}$$

$$= 40 \text{ N}$$

8. $\dfrac{4}{5} V \rho_W g = V \rho_L g$

$$\Rightarrow \quad \rho_L = \frac{4}{5} \times \rho_W = \frac{4}{5} \times 1000$$

$$\rho_L = 800 \text{ kg/m}^3$$

9. Let L be the edge of a cube and ρ be the density. In initial condition,

$$\rho_{wood}\, g (L^3) + 200\, g = \rho_{water}\, g (L^3)$$

$$\rho_{wood} (L^3) = \rho_{water} (L^3) - 200 \qquad \text{... (i)}$$

Now, when the mass is removed,

$$\rho_{wood}\, g\, (L^3) = L^2\, (L - 2)\, g \qquad \text{... (ii)}$$

Now, from Eqs. (i) and (ii), we get

$$\rho_{water} (L^3) - 200 = L^2 (L - 2) g \qquad [\because \rho_{water} = 1]$$

So, we get

$$L = 10 \text{ cm}$$

10. As the soap is in equilibrium,

weight of soap = buoyant force

$$\rho_{oil}\, Ax + \rho_{water}\, A(L - x) = \rho_{soap} \cdot AL$$

$$300x + 1000L - 1000x = 800L$$

$$\Rightarrow \quad 200L = 700x$$

$$\Rightarrow \quad \frac{x}{L} = \frac{2}{7}$$

11. Initially, the block is in equilibrium, so

$$mg = \rho ghA \qquad [\because F = pA = \rho ghA]$$

$$m \times 10 = 10^3 \times 10 \times 0.03 \times (0.1)^2 = 0.3 \text{ kg}$$

$$= 300 \text{ g}$$

12. As the metallic sphere floats on immiscible mixture.

$$\therefore \quad Mg = M_1 g + M_2 g$$

$$V\rho g = \frac{4V}{5} \times 10^3 g + \frac{V}{5} \times 13.5 \times 10^3 g$$

On solving, we get

$$\rho = 3.5 \times 10^3 \text{ kg/m}^3$$

13. The forces are exerted only at the bends which forms the couple exerting a clockwise torque.

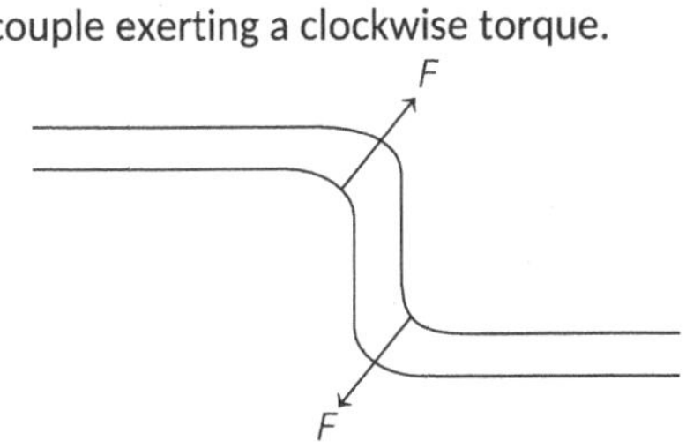

14. Applying Bernoulli's equation,

$$p + \frac{1}{2}\rho v^2 = \frac{p}{2} + \frac{1}{2}\rho v'^2$$

$$\Rightarrow \quad v' = \sqrt{v^2 + \frac{p}{\rho}}$$

15. Total height, $H = (3 + 1) = 4$ m

Maximum value of x comes from $h = \frac{H}{2} = 2$ m from top.

But this point is lying on table. Therefore, from $h = 1$ m, we will get maximum value of x.

From $h = 0$ to $h = 2$ m, value of x will first increase, then from $h = 2$ m to $h = 4$ m, value of x will decrease.

16. From equation of continuity

$$v_A A_A = v_B A_B$$

$$v_A \times 4 = v_B \times 2$$

$$v_B = 2v_A$$

Now, applying Bernoulli's equation, we get

$$\frac{1}{2}\rho v_A^2 + p_A = \frac{1}{2}\rho v_B^2 + p_B$$

$$p_A - p_B = \frac{1}{2}\rho (v_B^2 - v_A^2)$$

$$\frac{\rho gh}{\rho} = \frac{4v_A^2 - v_A^2}{2}$$

$$v_A = \sqrt{\frac{2gh}{3}} = \sqrt{\frac{2 \times 10 \times 0.02}{3}}$$

$$= 0.365 \text{ m/s}$$

Now, the rate of flow of water $= A_A v_A = A_B v_B$

$$= 0.365 \times 4 \times 10^{-1}$$

$$= 146 \times 10^{-3} \text{ m}^3/\text{s}$$

17. $\Delta\rho gh = \frac{\rho}{2}(v_y^2 - v_x^2)$

$$\Delta hg = \frac{1}{2}(v_y^2 - v_x^2)$$

$$0.51 \times 1000 = \frac{1}{2}[v_y^2 - (2)^2]$$

So, $v_y = 32$ cm/s

18. Let d be the depth of water.

Velocity of water ejecting out $= \sqrt{2gd}$

Time taken by water to strike the ground,

$$t = \sqrt{\frac{2H}{g}}$$

Now, $R = vt$

$$= \sqrt{2gd} \times \sqrt{\frac{2H}{g}}$$

$$d = \frac{R^2}{4H}$$

19. As the liquid film has two surfaces,

$$2TL = w$$

$$\Rightarrow \quad T = \frac{w}{2L}$$

$$= \frac{1.5 \times 10^{-2}}{2 \times 30 \times 10^{-2}}$$

$$= 0.025 \text{ Nm}^{-1}$$

20. We know,

surface energy, $E = T \times 2A$

$$E_1 = T \times 2\left(\frac{A}{2}\right) = T \times A$$

So, % decrease in surface energy

$$= \frac{E - E_1}{E} \times 100$$

$$= \frac{2TA - TA}{2TA} \times 100$$

$$= 50\%$$

21. We know, $F = S \times L$

$$F_t = S \times \frac{A}{d} \times 2$$

$$= \frac{2AS}{d}$$

Now, $F = p_0 A - F_t$

$$F = p_0 A - \frac{2AS}{d}$$

Pressure, $p = \frac{F}{A} = p_0 - \frac{2S}{d}$

22. As the water rises to a height of 2 cm in a capillary tube tilted at 60° from vertical.

So, height, $h = L\cos 60°$

$2 = L\cos 60°$

$L = 4$ cm

23. $R = \dfrac{r}{\cos\theta}$

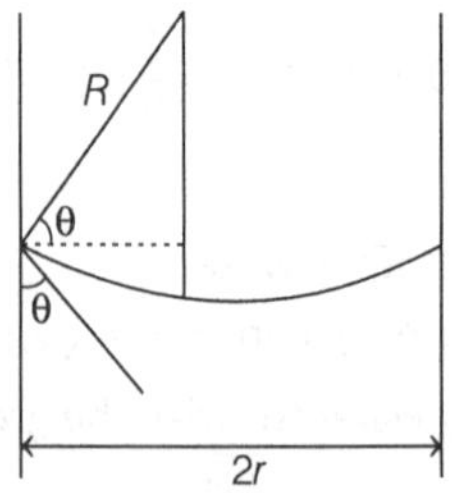

Difference in pressure, $\Delta p = \dfrac{2T}{R}$

$$\Delta p = \frac{2T\cos\theta}{r}$$

24. Rise of liquid in the capillary tube, $h = \dfrac{2T}{r\rho g}$

So, $h_1 = \dfrac{2T}{r_1\rho g}, h_2 = \dfrac{2T}{r_2\rho g}$

Now, $h = h_1 - h_2$

$$= \frac{2T}{\rho g}\left(\frac{1}{r_1} - \frac{1}{r_2}\right) = \frac{2T}{\rho g}\left[\frac{r_2 - r_1}{r_1 r_2}\right]$$

Hence, $T = \dfrac{h\rho g\, r_1 r_2}{2(r_2 - r_1)}$

25. Volume of the water forming bubble

$$= \frac{4}{3}\pi[R^3 - (R-d)^3]$$

$$= \frac{4}{3}\pi r^3$$

$d << R$ [so neglecting d^2 and d^3]

$r^3 = 3R^2 d$

Ratio of excess pressure in the drop to the excess pressure inside the bubble

$$= \frac{2\sigma/r}{4\sigma/R} = \frac{1}{2}\left(\frac{R}{r}\right)$$

On substituting the value of r, we get

$$= \left(\frac{R}{24d}\right)^{1/3}$$

26. Initial surface area = 2 × Length × Separation

$= 2 \times 10 \text{ cm} \times 0.5 \text{ cm}$

$= 10 \text{ cm}^2 = 10 \times 10^{-4} \text{m}^2$

Final surface area $= 2 \times 10 \text{ cm} \times (0.5 + 0.1)$ cm

$= 12 \times 10^{-4} \text{ m}^2$

So, the work done $= T\Delta A$

$= 7.2 \times 10^{-2}(12 \times 10^{-4} - 10 \times 10^{-4})$

$= 7.2 \times 10^{-2} \times 2 \times 10^{-4}$

$= 1.44 \times 10^{-5}$ J

27. Perimeter of frame $= 4l$

Since, the square film has two free surfaces,

$$F = 2\sigma \times 4l = 8\sigma l$$

28. Initial surface area of soap bubble, $A_1 = 4\pi r^2$

Under the isothermal condition when the radius becomes $2r$.

$\therefore$ Surface area $= 4\pi(2r)^2 = 16\pi r^2$

So, $\Delta A = 2(16\pi r^2 - 4\pi r^2)$

$= 24\pi r^2$

Energy spent,

$W = T \times \Delta A$

$= 24\,\pi T r^2$ joule

29. Given, $r = 1 \text{ mm} = 10^{-3}$ m

$S = 0.075 \text{ Nm}^{-1}$

$\rho = 10^3 \text{ kgm}^{-3}$

$h = 10$ cm

$\therefore$ Pressure inside the bubble, which is greater inside the atmospheric pressure is

$$\frac{2S}{r} + h\rho g = \frac{2 \times 0.075}{10^{-3}} + 0.10 \times 10^3 \times 9.8$$

$$= 150 + 980 = 1130 \text{ Nm}^{-2}$$

30. In the absence of gravity, the water tends to acquire spherical surface when the ice melts.

Volume remains same,

$$a^3 = \left(\frac{4}{3}\right)\pi r^3 \Rightarrow r = \left(\frac{3a^3}{4\pi}\right)^{1/3}$$

$\therefore$ Surface area $= 4\pi r^2$

$$= 4\pi\left(\frac{3a^3}{4\pi}\right)^{2/3} = 3^{2/3}\cdot 4^{1/3}\cdot \pi^{1/3}.\, a^2$$

$= (36\pi)^{1/3}$ cm² $\quad [\because a = 1 \text{ cm}]$

31. $v_t \propto r^2$

$m \propto r^3$

$\Rightarrow \quad r \propto m^{1/3}$

So, $v_t \propto (m^{1/3})^2$

$\because \quad \dfrac{v}{v'} = \left(\dfrac{m_1}{m_2}\right)^{2/3}$

$\Rightarrow \quad \dfrac{v}{v'} = \left(\dfrac{m}{8m}\right)^{2/3}$

$\Rightarrow \quad v' = 4v$

32. $V \propto R^3$, $F = 6\pi\eta R v_T \Rightarrow F \propto R v_T$

Further $v_T \propto R^2$

$\therefore \quad F \propto R^3$

or $\quad F \propto V$

It means when a solid ball of volume $2V$ and of the same material is dropped into the same liquid, the viscous force will be $2F$.

Therefore, answer will be (b).

33. $v_t = \frac{2}{9} r^2 \frac{(\rho - \sigma)}{n} g$

$\Rightarrow \quad v_t \propto r^2$

So, graph between v_t along y- axis and r^2 along x-axis is a straight line passing through origin with positive slope.

34. Shear force, $F = T = Mg = 0.01 \text{ kg} \times 10 \text{ m/s}^2$

$= 0.1 \text{kg-m/s}^2$

Shear stress $= \frac{F}{A} = \frac{0.1}{0.1} = 1 \text{kgm}^{-1}\text{s}^{-2}$

Strain rate $= \frac{v}{L} = \frac{0.085 \text{ m/s}}{0.3 \times 10^{-3} \text{ m}}$

Now, η coefficient of viscosity,

$\eta = \frac{\text{shear stress}}{\text{strain rate}} = \frac{1}{0.085} \times 0.3 \times 10^{-3}$

$= 3.5 \times 10^{-3}$ Pa s

35. We know, $Q = \frac{p\pi R^4}{8n^2}$, so $\eta \propto \frac{1}{\sqrt{Q}}$

$n_A : n_B : n_C = \frac{1}{\sqrt{2}} : \frac{1}{\sqrt{4}} : \frac{1}{\sqrt{8}} = 4 : 2 : 1$

36. We know, $Q = \frac{\pi p r^4}{8nL}$

$Q \propto r^4 \propto p$

So, $\frac{Q'}{Q} = \frac{4p \times (a/2)^4}{p \times a^4}$

$Q' = \frac{Q}{4}$

LEVEL 02 Objective Questions

37. Let h be the height of the oil level, at which pressure due to $(h + 50)$ cm is equal to the pressure due to 50 cm water.

$(h + 50) \times 0.8 \times g = 50 \times 1 \times g$

$h = 12.50$ cm of Hg

38. Let us consider points A and B at which $p_A = p_B$

$p_0 + \rho_3 g H_1 = p_0 + \rho_1 g H_2 + \rho_2 g (H_1 - H_2)$

$H_1\rho_3 = H_2\rho_1 + H_1\rho_2 - H_2\rho_2$

$15\rho_3 = 10\rho_1 + 5\rho_2$

$3\rho_3 = 2\rho_1 + \rho_2$

39. Pressure at dotted line from both sides should be same.

$\therefore \quad p_0 + \rho g h + (3\rho)(g)(2h) = p + (8\rho)(g)\left(\frac{2h}{3} - \frac{h}{4}\right)$

$\therefore \quad p = p_0 + \frac{11\rho g h}{3}$

40. We have, $\frac{\rho_1 V_1 + \rho_2 V_2}{V_1 + V_2} = 1 \Rightarrow \frac{0.5 V_1 + 2.5 V_2}{V_1 + V_2} = 1$

We get, $\quad \frac{V_1}{V_2} = 3$

Now, $\frac{\text{mass of wood}}{\text{mass of concrete}} = \frac{\rho_1 V_1}{\rho_2 V_2}$

$= \frac{0.5 \times 3}{2.5} = \frac{3}{5}$

41. $F_b = w$

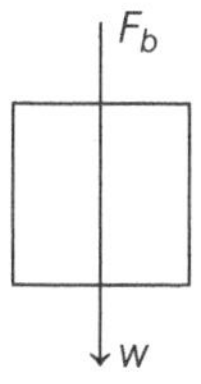

$\rho_L V g = Mg$

$\Rightarrow 2\rho \cdot a^2 \cdot x \cdot g = Mg \quad \left[\because \rho = \frac{M}{a^3}\right]$

$\frac{2M}{a^3} \cdot a^2 \cdot x \cdot g = Mg \Rightarrow x = \frac{a}{2}$

Now, when the weight is put over the block,

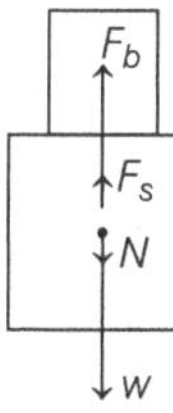

$F_b + F_s = w + N$

$F_b + F_s = w + w'$

$2\rho \cdot a^3 \cdot g + K\left(\frac{a}{2}\right) = a^3 \cdot \rho \cdot g + w'$

$w' = \rho a^3 g + \frac{Ka}{2} = a\left(\rho a^2 g + \frac{K}{2}\right)$

42. Pressure at the bottom of a tank is $3p$. If we ignore the atmospheric pressure p.

Now, we know, $\quad h\rho g = 3p$

If the height getting lowered by 1/3rd, then the height becomes 2/3.

$\therefore \quad \frac{2}{3} h\rho g = \frac{2}{3} \times (3p) = 2p$

If we include the atmospheric pressure, it becomes

$= 2p + p = 3p$

43. Apparent weight = Actual weight – Upthrust

$= (101)\, g - (\text{Volume immersed})\, \rho_w g \quad$ ...(i)

Volume of balloon is negligible.

So, volume immersed is actually volume of water only.

Substituting in Eq. (i), we get

Apparent weight $= (101)g - \left(\frac{m_w}{\rho_w}\right)\rho_w g$

$= (101)g - (100)g$

$= 1g$

or we can say its 1 g.

44. We have, mass of copper in air = 264 g

Mass of copper in water = 221 g

Apparent loss of mass in water = 264 − 221 = 43 g

So, volume displaced $= \frac{43}{1} = 43$ cc

Volume of copper piece + Volume of cavity = 43 cc

Volume of copper block $= \frac{264}{8.8} = 30$ cc

Therefore, volume of the cavity = 43 − 30 = 13 cc

45. $T + mg = F_b$

$$\Rightarrow \quad T = F_b - mg$$

$$T = V\rho_{water}\, g - mg$$

$$T = \frac{m}{\rho_{rel}\,\rho_{water}}\rho_{water}\, g - mg$$

$$T = mg\left(\frac{1}{\rho_{rel}} - 1\right)$$

$$= 8 \times 10\left(\frac{1}{0.8} - 1\right) = 20 \text{ N}$$

46. Let f fraction is in the lower liquid.

Weight = Upthrust from lower liquid + Upthrust from upper liquid

$$\therefore \quad V\rho g = (fV)\,\rho_2 g + (1-f)\,V\rho_1 g$$

Solving this equation, we get

$$f = \frac{\rho_1 - \rho}{\rho_1 - \rho_2}$$

47. $v_t = \frac{2}{9}r^2\frac{(\rho_{ball} - \rho_{liquid})}{n} \times g$

$$v_t \propto r^2\,(\rho_{ball} - \rho_{liquid})$$

Now, $\rho_{ball} = \frac{m}{\frac{4}{3}\pi r^3} = \frac{3m}{4\pi r^3}$

$$\rho_{liquid} = \frac{m'}{V} = \frac{3m'}{4\pi r^3}$$

Now, $v_t \propto r^2\left(\frac{3m}{4\pi r^3} - \frac{3m'}{4\pi r^3}\right)$

$$v_t \propto \frac{3}{4\pi}\left(\frac{m}{r} - \frac{m'}{r}\right) \Rightarrow v_t \propto \frac{m - m'}{r}$$

48. After the displacement of cylinder from the equilibrium position, its volume,

$$\Delta V = A \cdot x$$

Now, the water level rises for compensation,

$$x_{water} = \frac{\Delta V}{2A - A} = x$$

So, total volume under the water is $2Ax$.

Now, $F = \rho_{water} g(2Ax)$

$$\therefore \quad a = \frac{\rho_{water}\, g(2Ax)}{m} \quad [\because \rho_{water} = 2\rho]$$

$$= \frac{2\rho g \cdot 2Ax}{\rho AH} = 4g\frac{x}{H}$$

49. Let h be the total height.

Time taken for tank to empty $\frac{h}{4}$ is t_1.

So, $v = u + at$

$$t_1 = \frac{v - u}{a} = \frac{\sqrt{2g\frac{3h}{4}} - \sqrt{2gh}}{g}$$

$$t_2 = \sqrt{\frac{2g\frac{3h}{4}}{g}}$$

On solving for t_1 and t_2, we get

$$\frac{t_1}{t_2} = \frac{2 - \sqrt{3}}{\sqrt{3}}$$

50. We have,

$$v_1 = \sqrt{2g \times \frac{h}{2}} = \sqrt{gh}$$

Now, from Bernoulli's theorem,

$$\rho gh + 2\rho g\left(\frac{h}{2}\right) = \frac{1}{2}(2\rho)v_2^2$$

$$\Rightarrow \quad v_2 = \sqrt{2gh}$$

Now, $\frac{v_1}{v_2} = \frac{\sqrt{gh}}{\sqrt{2gh}} = \frac{1}{\sqrt{2}}$

51. $2T \times \cos\theta = x\,hdg$

$$\therefore \quad h = \frac{2T\cos\theta}{xdg}$$

52. Surface tension $= \frac{F}{L}$

Here, length, $L = \frac{2 \times \text{area}}{\text{thickness}} = \frac{2 \times 0.1}{0.0001} = 2 \times 10^3$ m

So, $F = 70 \times 10^{-3} \times 2 \times 10^3$

$$= 140 \text{ N}$$

53. Force required to pull the glass plates apart,

$$F = 2T \times L = 2T \times \frac{A}{d}$$

$$= \frac{2 \times 70 \times 10^{-3} \times \pi \times (5 \times 10^{-2})^2}{0.5 \times 10^{-3}}$$

$$= 2.2 \text{ N}$$

54. $V = a\sqrt{2gh}$

So, $h = \frac{V^2}{a^2 \times 2g} = \frac{V^2}{2a^2 g}$

55. Torque, $\tau = F_{buoyancy} \times \frac{l}{4}$

$$\tau = \rho\pi r^2 lg \times \frac{l}{4} = \frac{\rho\pi r^2 l^2 g}{4}$$

Now, $\tau = I\alpha$

$$\Rightarrow \quad \alpha = \frac{\tau}{I} = \frac{\pi\rho g l^2 r^2}{4I}$$

56. $v_t = \frac{2r^2}{9\eta}(\rho_1 - \rho_2)g \Rightarrow v_t \propto (\rho_1 - \rho_2)$

$$\frac{v_{t_2}}{0.2} = \frac{(10.5 - 1.5)}{(19.5 - 1.5)}$$

$$\Rightarrow \quad v_{t_2} = 0.2 \times \frac{9}{18} = 0.1 \text{ m/s}$$

LEVEL 03 *Previous Years Questions*

57. Given, the density of the small ball is d
The mass of the small ball is M
The density of the glycerine is $d/2$.
When ball reaches terminal speed,
Viscous force = Weight – Buoyant force
Viscous force = $Vd_1g - Vd_2g$
Here, V is the volume of submerged bodies,
g is the acceleration due to gravity,
d_1 is the density of the small ball,
d_2 is the density of the glycerine,

$$d_1 = d \text{ and } d_2 = d/2$$

Substituting the given values in the viscous force expression, we get

$$\text{Viscous force} = Vdg - V\frac{d}{2}g$$

$$\text{Viscous force} = \frac{Vdg}{2} = \frac{Mg}{2} \qquad (\because M = d \times V)$$

58. Density of liquid, $\rho_l = 760 \text{ kg/m}^3$
Density of mercury, $\rho_m = 13600 \text{ kg/m}^3$
Height of liquid column in mercury barometer,

$$h_m = 76 \text{ cm} = 0.76 \text{ m}$$

If height of liquid in liquid column be h_l, then

$$p_{\text{liquid}} = p_{\text{mercury}}$$

$$\Rightarrow \quad h_l\rho_l g = h_m\rho_m g$$

$$\Rightarrow \quad h_l = \frac{h_m\rho_m}{\rho_l} = \frac{0.76 \times 13600}{760} = 13.6 \text{ m}$$

59. A liquid does not wet the solid surface, if the angle of contact is obtuse, i.e. $\theta > 90°$.

60. Relation for height of water in capillary tube is

$$h = \frac{2S\cos\theta}{\rho g r}$$

$$\Rightarrow \quad h \propto \frac{1}{r}$$

$$\Rightarrow \quad \frac{h_1}{h_2} = \frac{r_2}{r_1} = \frac{2r}{r} = 2$$

As $m = A \cdot h \cdot \rho$

$$\therefore \quad \frac{m_2}{m_1} = \frac{Ah_2\rho}{Ah_1\rho} = \frac{h_2}{h_1} = \frac{1}{2}$$

$$\Rightarrow \quad m_2 = \frac{m_1}{2} = \frac{5}{2} = 2.5 \text{ g}$$

61. According to Pascal's law, "Pressure applied to an enclosed fluid is transmitted undiminished to every point of the fluid and the walls of the containing vessel."
In the given situation as shown in the figure below.

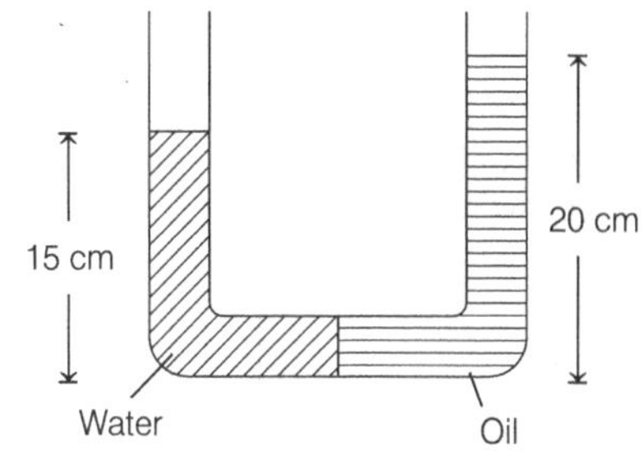

Pressure due to water column of height 15 cm = Pressure due to oil column of height 20 cm

$$\Rightarrow \quad h_w\rho_w g = h_0\rho_0 g$$

$$15\rho_w = 20\rho_0 \quad \Rightarrow \quad \rho_0 = \frac{15}{20}\rho_\omega$$

$$\rho_0 = \frac{15}{20} \times 1000 \qquad (\text{given, } \rho_w = 1000 \text{ kg m}^{-3})$$

$$= 750 \text{ kgm}^{-3}$$

62. The terminal velocity achieved by ball in a viscous fluid is

$$v_t = \frac{2(\rho - \sigma)r^2 g}{9\eta}$$

where, ρ = density of metal of ball,
σ = density of viscous medium,
r = radius of ball and
η = coefficient of viscosity of medium.

Terminal velocity of first ball,

$$v_{t_1} = \frac{2(\rho_1 - \sigma)r_1^2 g}{9\eta}$$

$$= \frac{2(8\rho_2 - \sigma)r_1^2 g}{9\eta} \qquad \text{...(i) } [\because \rho_1 = 8\rho_2]$$

Similarly, for second ball

$$v_{t_2} = \frac{2(\rho_2 - \sigma)r_2^2 g}{9\eta} \qquad \text{...(ii)}$$

From Eqs. (i) and (ii), we get

$$\frac{v_{t_1}}{v_{t_2}} = \frac{2(8\rho_2 - \sigma)r_1^2 g}{2(\rho_2 - \sigma)r_2^2 g} \times \frac{9\eta}{9\eta}$$

$$v_{t_1} = \left(\frac{8\rho_2 - 0.1\rho_2}{\rho_2 - 0.1\rho_2}\right)\left(\frac{r_1}{r_2}\right)^2 \quad [\because \sigma = 0.1\rho_2] \quad \text{...(iii)}$$

Here, $r_1 = 1\,\text{mm}$ and $r_2 = 2\,\text{mm}$

Substituting these values in Eq. (iii), we get

$$\Rightarrow \quad \frac{v_{t_1}}{v_{t_2}} = \left(\frac{7.9\rho_2}{0.9\rho_2}\right)\left(\frac{1}{2}\right)^2 = \frac{79}{36}$$

63. The rate of liquid flow moving with velocity v through an area a is given by

Rate (R) = Area (a) × Velocity (v)

Given, area of hole,

$$a = 2\,\text{mm}^2$$
$$= 2 \times 10^{-6}\,\text{m}^2$$

height of tank, $h = 2\,\text{m}$.

The given situation can also be depicted as shown in the figure given below.

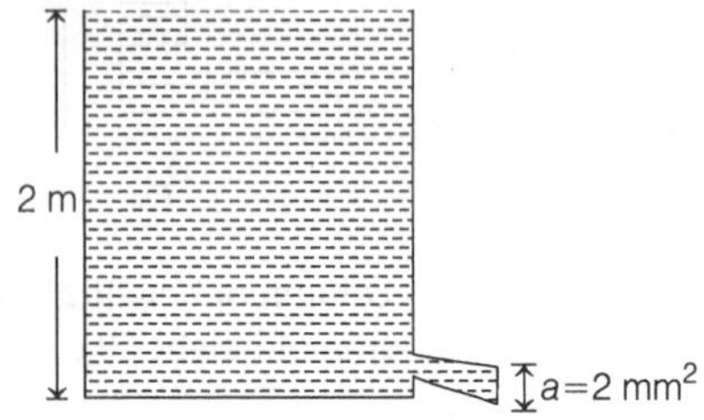

As the velocity of liquid flow is given as $v = \sqrt{2gh}$

$$\therefore \quad R = av = a\sqrt{2gh}$$

Substituting the given values, we get

$$R = 2 \times 10^{-6} \times \sqrt{2 \times 10 \times 2}$$
$$= 2 \times 10^{-6} \times 6.32$$
$$= 12.64 \times 10^{-6}\,\text{m}^3/\text{s}$$
$$\simeq 12.6 \times 10^{-6}\,\text{m}^3/\text{s}$$

64. The excess pressure inside a soap bubble of radius r is given by

$$p = \frac{4T}{r}$$

where, T = surface tension.

If p_0 be the outside pressure from the water, then total pressure inside the bubble becomes

$$p_1 = p_0 + \frac{4T}{r} \quad \text{... (i)}$$

The pressure at the depth Z_0 below the water surface is

$$p_2 = p_0 + Z_0\rho g \quad \text{... (ii)}$$

As it is given that the pressure inside the bubble is same as the pressure at depth Z_0, then equating Eqs. (i) and (ii), we get

$$p_0 + \frac{4T}{r} = p_0 + Z_0\rho g$$

$$\Rightarrow \quad Z_0 = \frac{4T}{r\rho g} \quad \text{... (iii)}$$

Here, $T = 2.5 \times 10^{-2}\,\text{N/m}$,

$$\rho = 10^3\,\text{kg/m}^3,\; g = 10\,\text{ms}^{-2}$$

and $r = 1\,\text{mm} = 1 \times 10^{-3}\,\text{m}$

Substituting these values in Eq. (iii), we get

$$Z_0 = \frac{4 \times 2.5 \times 10^{-2}}{1 \times 10^{-3} \times 10^3 \times 10}$$
$$= 10 \times 10^{-3}\,\text{m}$$
$$= 1\,\text{cm}$$

65. The rate of heat generation is equal to the rate of work done by the viscous force which in turn is equal to its power.

Rate of heat produced, $\dfrac{dQ}{dt} = F \times v_T$

where, F is the viscous force and v_T is the terminal velocity.

As, $\quad F = 6\pi\eta r v_T$

$$\Rightarrow \quad \frac{dQ}{dt} = 6\pi\eta r v_T \times v_T$$
$$= 6\pi\eta r v_T^2 \quad \text{...(i)}$$

From the relation for terminal velocity,

$$v_T = \frac{2}{9}\frac{r^2(\rho - \sigma)}{\eta}g, \text{ we get}$$

$$v_T \propto r^2 \quad \text{...(ii)}$$

From Eq. (ii), we can rewrite Eq. (i) as

$$\frac{dQ}{dt} \propto r \cdot (r^2)^2$$

or $$\frac{dQ}{dt} \propto r^5$$

66. **Thinking Process** Pressure of two points lie in the same horizontal level should be same and $p = hdg$

Both ends of the U tube are open, so the pressure on both the free surfaces must be equal.

i.e. $\quad p_1 = p_2$

$$h_{\text{oil}} \cdot S_{\text{oil}}\, g = h_{\text{water}} \cdot S_{\text{water}} \cdot g$$

S_{oil} = specific density of oil

$$S_{\text{oil}} = \frac{h_{\text{water}} \cdot S_{\text{water}} \cdot g}{h_{\text{oil}} \cdot g}$$

From figure, $$S_{\text{oil}} = \frac{(65 + 65) \times 1000}{(65 + 65 + 10)}$$
$$= 928\,\text{kgm}^{-3}$$

67. According to the question, the situation can be drawn as follows.

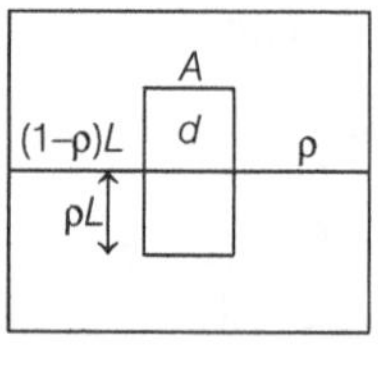

Applying Archemedes' principle,

Weight of cylinder = (upthrust)$_1$ + (upthrust)$_2$

i.e. $ALdg = (1-p)\,LA\rho g + (pLA)\,n\rho g$

$\Rightarrow \quad d = (1-p)\,\rho + pn\rho = \rho - p\rho + n\,p\rho$

$= \rho + (n-1)\,p\rho = \rho\,[1+(n-1)\,p]$

68. Given, $d = 2700$ m $\Rightarrow \rho = 10^3$ kg/m^3

Compressibility $= 45.4 \times 10^{-11}$ Pa^{-1}

The pressure at the bottom of ocean is given by

$$p = \rho g d$$
$$= 10^3 \times 10 \times 2700$$
$$= 27 \times 10^6 \text{ Pa}$$

So, fractional compression

= compressibility × pressure

$= 45.4 \times 10^{-11} \times 27 \times 10^6 = 1.2 \times 10^{-2}$

69. Increase in surface energy =

Work done in area × Surface tension

∵ Increase in surface area,

$\Delta A = (5\times4 - 4\times2)\times2$ (∵ film has two surfaces)

$= (20-8)\times 2$ cm^2

$= 24$ cm^2 $= 24 \times 10^{-4}$ m^2

So, work done, $W = T\cdot\Delta A$

$3\times10^{-4} = T \times 24 \times 10^{-4}$

$\therefore \quad T = \frac{1}{8} = 0.125$ N/m

70. According to ascent formula for capillary tube,

$$h = \frac{2T\cos\theta}{\rho g r}$$

$$\therefore \quad \frac{\cos\theta_1}{\rho_1} = \frac{\cos\theta_2}{\rho_2} = \frac{\cos\theta_3}{\rho_3}$$

Thus, $\cos\theta \propto \rho$

$\therefore \quad \rho_1 > \rho_2 > \rho_3$

$\therefore \quad \cos\theta_1 > \cos\theta_2 > \cos\theta_3$

$$0 \le \theta_1 < \theta_2 < \theta_3 < \frac{\pi}{2}$$

71. From Bernoulli's theorem,

$$p_1 + \frac{1}{2}\rho v_1^2 = p_2 + \frac{1}{2}\rho v_2^2$$

where, p_1, p_2 are pressure inside and outside the roof and v_1, v_2 are velocities of wind inside and outside the roof. Neglect the width of the roof.

Pressure difference is

$$p_1 - p_2 = \frac{1}{2}\rho\,(v_2^2 - v_1^2)$$
$$= \frac{1}{2} \times 1.2\,(40^2 - 0)$$
$$= 960 \text{ N/m}^2$$

Force acting on the roof is given by

$$F = (p_1 - p_2)A = 960 \times 250$$
$$= 24 \times 10^4 \text{N}$$
$$= 2.4 \times 10^5 \text{ N}$$

As the pressure inside the roof is more than outside to it. So, the force will act in the upward direction, i.e. $\mathbf{F} = 2.4 \times 10^5$N, upwards.

72.

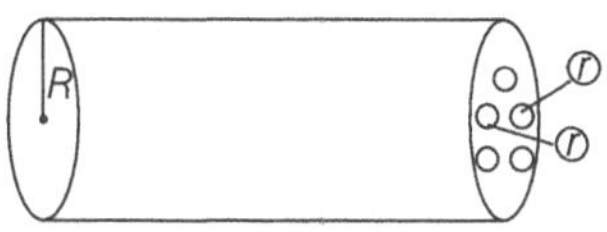

According to equation of continuity,

Volume in flow rate = Volume in out flow rate

$$\pi R^2 v = n\,\pi r^2 v' \Rightarrow v' = \frac{R^2 v}{nr^2}$$

Thus, speed of the ejection of the liquid through the holes is $\frac{R^2 v}{nr^2}$.

73. As the volume is same,

$$n\left(\frac{4}{3}\pi r^3\right) = \frac{4}{3}\pi R^3$$

$$\Rightarrow \quad n = \frac{R^3}{r^3}$$

Now, $\Delta U = T\Delta A$

$$= T\,[4\pi R^2 - n \times 4\pi r^2]$$
$$= 4\pi T\left[R^2 - \frac{R^3}{r^3}r^2\right]$$
$$= 4\pi R^3 T\left[\frac{1}{R} - \frac{1}{r}\right]$$
$$= 3\left(\frac{4}{3}\pi R^3\right)T\left(\frac{1}{R} - \frac{1}{r}\right)$$
$$= -3VT\left(\frac{1}{r} - \frac{1}{R}\right)$$

74. The wettability of a surface by a liquid depends primarily on angle of contact between the surface and the liquid.

Chapter 14

SIMPLE HARMONIC MOTION

LEVEL 01 Objective Questions

1 Shown is the position-time graph of a mass undergoing SHM. What can you say about the velocity and force at the instant indicated by dashed line? (Positive direction is to the right)

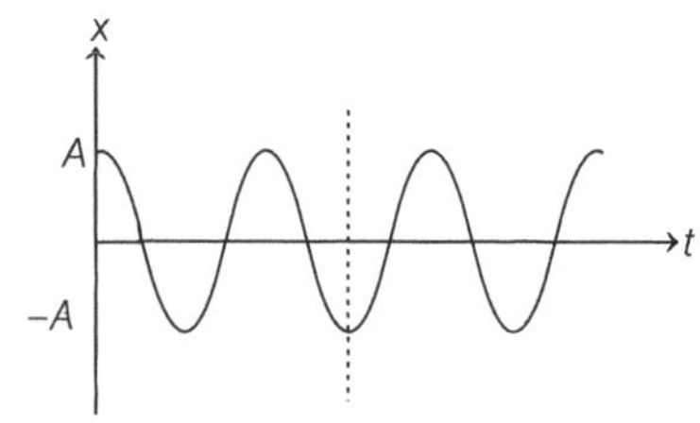

(a) Velocity is zero, force is to the right.
(b) Velocity is positive, force is to the right.
(c) Velocity is negative, force is to the right.
(d) Velocity is zero, force is to the left.

2 A particle is performing simple harmonic motion along x-axis with amplitude 4 cm and time period 1.2 s. The minimum time taken by the particle to move from $x = 2$ cm to $x = 4$ cm and back again is given by (Mean position at $x = 0$)
(a) 0.6 s (b) 0.4 s
(c) 0.3 s (d) 0.2 s

3 A simple harmonic motion of a particle is represented by an equation $x = 10 \sin\left(10t + \frac{\pi}{3}\right)$ and the motion starts at $t = 0$. When does the particle first come to rest?
(a) $\frac{\pi}{30}$ s (b) $\frac{\pi}{60}$ s
(c) $\frac{\pi}{40}$ s (d) $\frac{\pi}{80}$ s

4 A particle of mass 1 kg is undergoing SHM for which graph between force and displacement (from mean position) is shown. Its time period (in second) is

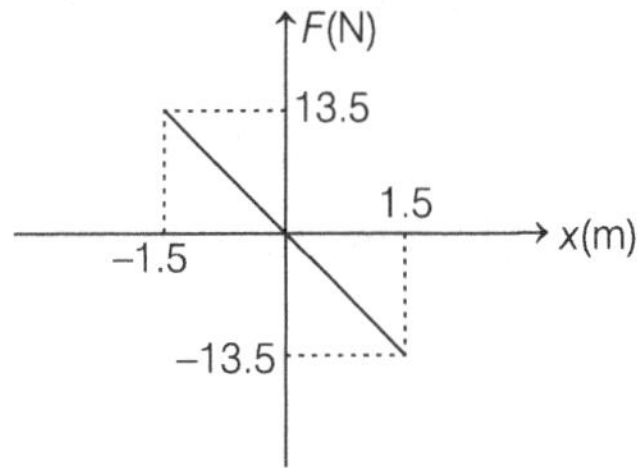

(a) $\frac{\pi}{3}$ (b) $\frac{2\pi}{3}$ (c) $\frac{\pi}{6}$ (d) $\frac{3}{\pi}$

5 Figure shows the position-time graph of an object in SHM. The correct equation representing this motion is

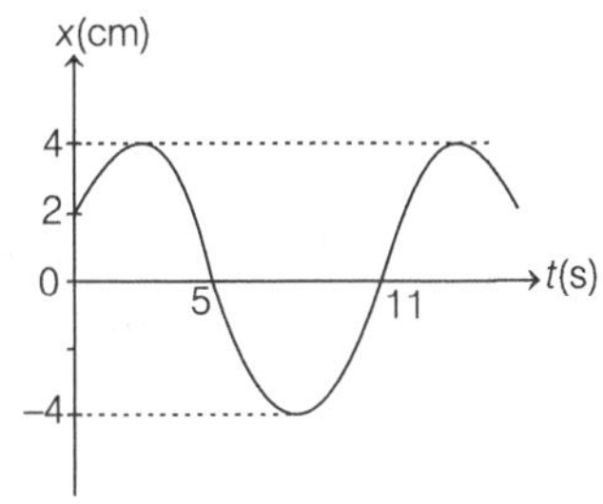

(a) $2\sin\left(\frac{2\pi}{5}t+\frac{\pi}{6}\right)$ (b) $4\sin\left(\frac{\pi}{5}t+\frac{\pi}{6}\right)$ (c) $4\sin\left(\frac{\pi}{6}t+\frac{\pi}{3}\right)$ (d) $4\sin\left(\frac{\pi}{6}t+\frac{\pi}{6}\right)$

6 The x - t graph of a particle undergoing simple harmonic motion is shown below.The acceleration of the particle at $t=\frac{4}{3}$ s is

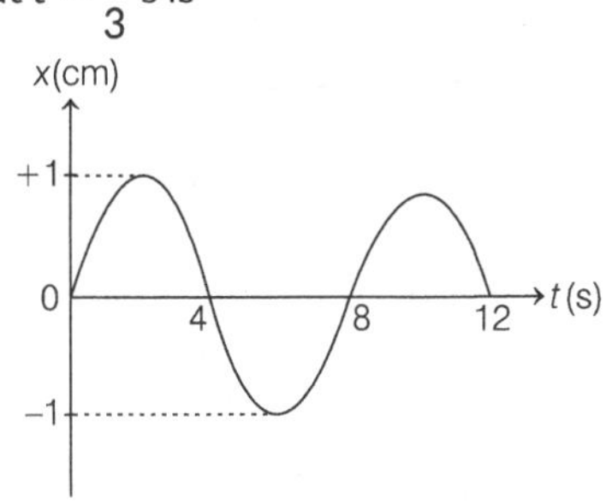

(a) $\frac{\sqrt{3}}{32}\pi^2\text{cm/s}^2$ (b) $\frac{-\pi^2}{32}\text{cm/s}^2$ (c) $\frac{\pi^2}{32}\text{cm/s}^2$ (d) $-\frac{\sqrt{3}}{32}\pi^2\text{cm/s}^2$

7 The acceleration-displacement graph of a particle executing simple harmonic motion is shown in figure. The time period of simple harmonic motion is

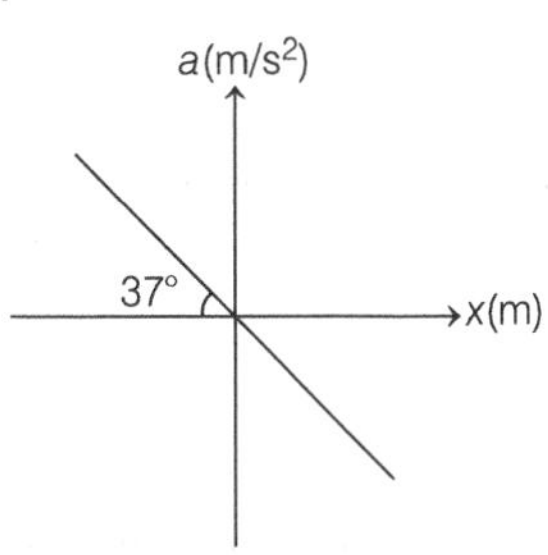

(a) $\frac{4\pi}{\sqrt{3}}$ s (b) $\frac{2\pi}{\sqrt{3}}$ s (c) 2π s (d) $\frac{4\pi}{3}$ s

8 A body of mass 36 g is undergoing SHM of amplitude $A = 13$ cm and time period $T = 12$ s. At time $t = 0$, the displacement x is 13 cm. The shortest time of passage from $x = 6.5$ cm to $x = -6.5$ cm is
(a) 4 s (b) 2 s (c) 6 s (d) 3 s

9 A particle performs simple harmonic motion with amplitude A. Its speed is tripled at the instant, when it is at a distance $\frac{2A}{3}$ from equilibrium position. The new amplitude of the motion is
(a) $3A$ (b) $A\sqrt{3}$ (c) $\frac{7A}{3}$ (d) $\frac{A}{3}\sqrt{41}$

10 A body performs simple harmonic motion about $x = 0$ with an amplitude a and time period T. The speed of the body at $x = a/2$ will be
(a) $\frac{\pi a\sqrt{3}}{T}$ (b) $\frac{\pi a\sqrt{3}}{2T}$ (c) $\frac{\pi a}{T}$ (d) $\frac{3\pi^2 a}{T}$

11 A particle of mass m is executing SHM. If amplitude is a and frequency n, the value of its force constant will be
(a) mn^2 (b) $4mn^2a^2$ (c) ma^2 (d) $4\pi^2 mn^2$

12 A particle is executing SHM along x-axis with origin as the mean position. If velocity v of the particle is related with position x as $9v^2 = 36 - x^2$, then time period of the oscillation will be
(a) 3 s (b) 3π s (c) 6 s (d) 6π s

13 A man measures the period of a simple pendulum inside a stationary lift and finds it to be T s. If the lift accelerates upwards with an acceleration $g/4$, then the period of the pendulum will be

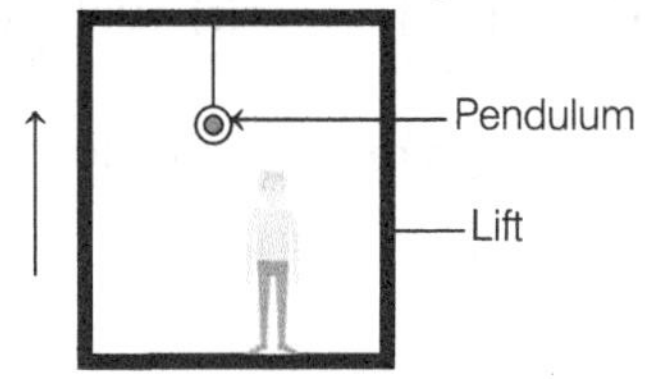

(a) T (b) $\frac{T}{4}$ (c) $\frac{2T}{\sqrt{5}}$ (d) $2T\sqrt{5}$

14 The amplitude of a damped oscillator becomes one-third in 2 s. If its amplitude after 6 s is $1/n$ times the original amplitude, then the value of n is
(a) 3^2 (b) $3\sqrt{2}$ (c) $3\sqrt{3}$ (d) 3^3

15 A rectangular block of mass m and area of cross-section A floats in a liquid of density ρ. If it is given a small vertical displacement from equilibrium, it oscillates with a time period T, then
(a) $T \propto \frac{1}{\sqrt{m}}$ (b) $T \propto \sqrt{\rho}$ (c) $T \propto \frac{1}{\sqrt{A}}$ (d) $T \propto \frac{1}{\rho}$

16 The potential energy of a particle executing SHM changes from maximum to minimum in 5 s. Then, the time period of SHM is
(a) 5 s (b) 10 s (c) 15 s (d) 20 s

17 A simple pendulum with a metallic bob has a time period T. The bob is now immersed in a non-viscous liquid. If the density of the liquid is 1/4 that of the metal, the time period of the same pendulum will be
(a) $\frac{T}{\sqrt{3}}$ (b) $\frac{2T}{\sqrt{3}}$ (c) $\frac{4}{3}T$ (d) $\frac{2}{3}T$

18 The kinetic energy and potential energy of a particle executing SHM with amplitude A will be equal, when its displacement is
(a) $A\sqrt{2}$ (b) $A/2$ (c) $A/\sqrt{2}$ (d) $A\sqrt{2/3}$

19 Five identical springs are used in the following three configurations. The time periods of vertical oscillations in configurations (i), (ii) and (iii) are in the ratio

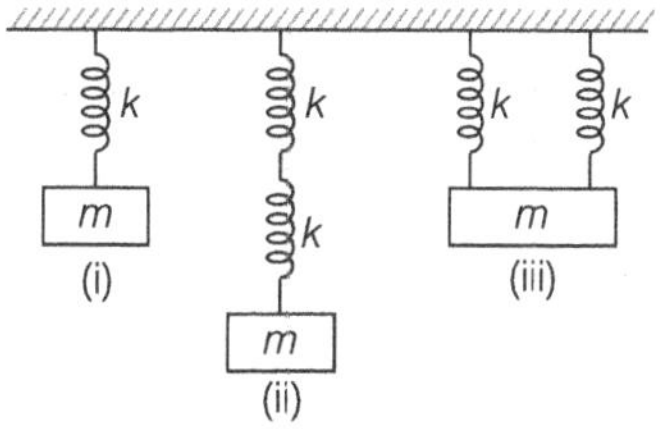

(a) $1 : \sqrt{2} : \frac{1}{\sqrt{2}}$ (b) $2 : \sqrt{2} : \frac{1}{\sqrt{2}}$

(c) $\frac{1}{\sqrt{2}} : 2 : 1$ (d) $2 : \frac{1}{\sqrt{2}} : 1$

20 On a smooth inclined plane, a body of mass M is attached between springs as shown. The other ends of the springs are fixed to firm supports. If each spring has force constant k, the period of oscillation of the body is (Assuming the springs as massless)

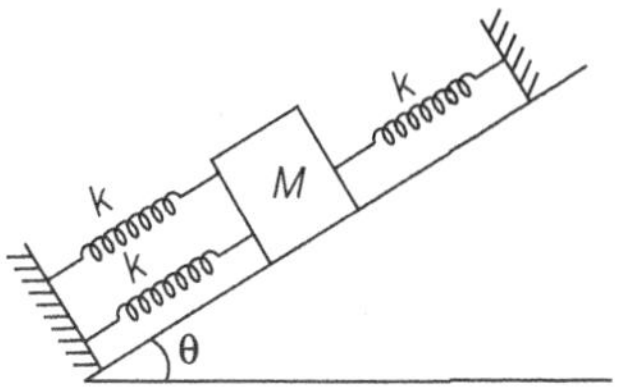

(a) $2\pi\left(\frac{M}{3k}\right)^{1/2}$ (b) $2\pi\left(\frac{2M}{k}\right)^{1/2}$ (c) $2\pi\frac{Mg\sin\theta}{2k}$ (d) $2\pi\left(\frac{2Mg}{k}\right)^{1/2}$

21 A spring of spring constant k is divided in 3 parts of length $l/4, l/4$ and $l/2$, where l is its natural length. Then, they are connected as shown. What is the equivalent spring constant of the combination?

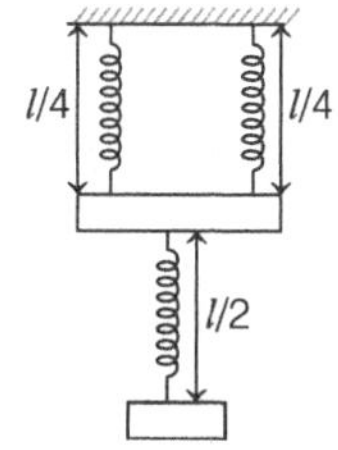

(a) 2.4 k
(b) 1.6 k
(c) 4 k
(d) 10 k

22 A plank of mass 12 kg is supported by two identical springs as shown in figure. The plank always remains horizontal. When the plank is pressed down and released, it performs simple harmonic motion with time period 3 s. When a block of mass m is attached to the plank, the time period changes to 6 s. The mass of the block is

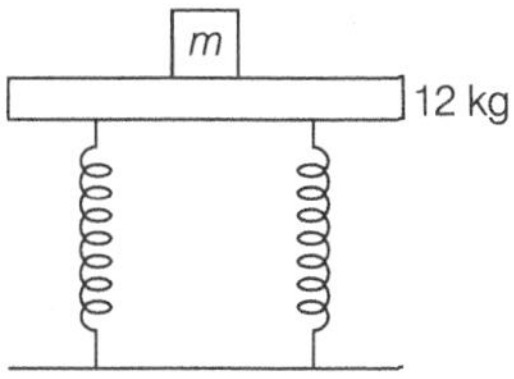

(a) 48 kg (b) 36 kg
(c) 24 kg (d) 12 kg

23 When a mass m is connected individually to two springs S_1 and S_2 separately, the oscillation frequencies are f_1 and f_2, respectively. If the same mass is connected with spring combination as shown in figure, then the oscillation frequency will be

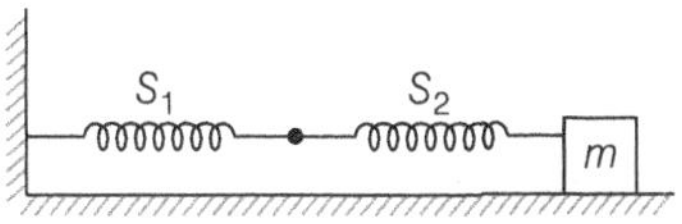

(a) $\frac{f_1^2 f_2^2}{f_1^2 + f_2^2}$ (b) $\frac{f_1 f_2}{f_1 + f_2}$ (c) $\frac{f_1 f_2}{\sqrt{f_1 + f_2}}$ (d) $\frac{f_1 f_2}{\sqrt{f_1^2 + f_2^2}}$

24 One end of a spring of force constant k is fixed to a vertical wall and the other to a block of mass m resting on a smooth horizontal surface. There is another wall at a distance x_0 from the block. The spring is then compressed by $2x_0$ and released. The time taken to strike the wall is

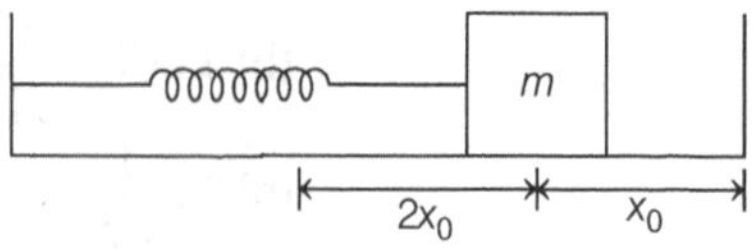

(a) $\frac{1}{6}\pi\sqrt{\frac{k}{m}}$ (b) $\sqrt{\frac{k}{m}}$ (c) $\frac{2\pi}{3}\sqrt{\frac{m}{k}}$ (d) $\frac{\pi}{4}\sqrt{\frac{k}{m}}$

25 The ratio of time period of three identical springs, if they are first joined in parallel and then in series is (Assume same mass is suspended from them)

(a) 1 : 3 (b) 2 : 1
(c) 2 : 3 (d) 9 : 1

26 The potential energy of a particle of mass m situated in a unidimensional potential field varies as $U(x) = U_0[1 - \cos ax]$, where U_0 and a are constants. The time period of small oscillations of the particle about the mean position is

(a) $2\pi\sqrt{\frac{m}{aU_0}}$ (b) $2\pi\sqrt{\frac{am}{U_0}}$ (c) $2\pi\sqrt{\frac{m}{a^2U_0}}$ (d) $2\pi\sqrt{\frac{a^2m}{U_0}}$

LEVEL 02 Objective Questions

27 A mass of 2.0 kg is put on a flat pan attached to a vertical spring fixed on the ground as shown in the figure. The mass of the spring and the pan is negligible. When pressed slightly and released, the mass executes a simple harmonic motion. The spring constant is 200 N/m. What should be the minimum amplitude of the motion, so that the mass gets detached from the pan (Take, $g = 10\ \text{m/s}^2$)?

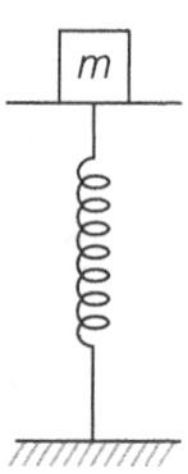

(a) 10 cm (b) any value less than 12.0 cm
(c) 4.0 cm (d) 8.0 cm

28 The displacement of a particle varies according to the relation $x = 3\sin 100t + 8\cos^2 50t$. Which of the following is **incorrect** about this motion?

(a) The motion of the particle is SHM.
(b) The amplitude of the SHM of the particle is 5 units.
(c) The minimum displacement of the particle from the origin is 2 units.
(d) The maximum displacement of the particle from the origin is 9 units.

29 The velocity of particle is given by $v = v_0 \sin \omega t$, where v_0 is constant and $\omega = \frac{2\pi}{T}$, find the average velocity in time interval $t = 0$ to $t = \frac{1}{2}$ s.

(a) v_0 (b) $\frac{2v_0}{\pi}$ (c) $\frac{v_0}{\pi}$ (d) πv_0

30 A particle of mass m is performing linear simple harmonic motion. Its equilibrium is at $x = 0$, force constant is k and amplitude of SHM is A. The maximum power supplied by the restoring force to the particle during SHM will be

(a) $\frac{k^{\frac{3}{2}}A^2}{\sqrt{m}}$ (b) $\frac{2k^{\frac{3}{2}}A^2}{\sqrt{m}}$ (c) $\frac{k^{\frac{3}{2}}A^2}{3\sqrt{m}}$ (d) $\frac{k^{\frac{3}{2}}A^2}{2\sqrt{m}}$

31 Acceleration of a particle which is at rest at $x = 0$ is $\mathbf{a} = (4 - 2x)\,\hat{\mathbf{i}}$ cm/s^2. Select the correct alternatives(s).
(a) Particle further comes to rest at $x = 4$ cm
(b) Particle oscillates about $x = 2$ cm
(c) Maximum speed of particle is $2\sqrt{2}$ cm/s
(d) All of these

32 A person normally weighing 60 kg stands on a platform which oscillates up and down harmonically at a frequency of 2 Hz and an amplitude 5 cm. A machine on the platform gives the person's weight. The maximum reading it will show is (Take, $g = 10$ ms^{-2})
(a) 1290 N
(b) 1500 N
(c) 1080 N
(d) None of these

33 A particle is executing SHM with period T and amplitude 13 cm. Its equilibrium position being O. Its velocity at a distance of x cm from O is 96 cm/s. Then, T in seconds and x in cm may be

	T	**x**		**T**	**x**
(a)	$5\pi/48$	12	(b)	$\pi/4$	6
(c)	$2\pi/15$	8	(d)	$3\pi/20$	10

34 Kinetic energy of a particle executing simple harmonic motion in a straight line is pv^2 and potential energy is qx^2, where v is speed at distance x from the mean position. It's time period is given by the expression
(a) $2\pi\sqrt{\frac{q}{p}}$
(b) $2\pi\sqrt{\frac{p}{q}}$
(c) $2\pi\sqrt{\frac{q}{p+q}}$
(d) $2\pi\sqrt{\frac{p}{p+q}}$

35 Two particles are oscillating along two close parallel straight lines side by side, with the same frequency and amplitudes. They pass each other moving in opposite directions when their displacement is half of the amplitude. The phase difference between the particles is
(a) $\frac{\pi}{6}$
(b) 0
(c) $\frac{2\pi}{3}$
(d) π

36 While a particle executes simple harmonic motion, the rate of change of acceleration is maximum and minimum, respectively at the
(a) mean position and extreme position
(b) extreme positions and mean position
(c) mean position alternatively
(d) extreme positions alternatively

37 A particle is performing SHM according to the equation $x = (3\text{ cm})\sin\left(\frac{2\pi}{18}t + \frac{\pi}{6}\right)$, where t is in seconds. The distance travelled by the particle in 39 s is
(a) 24 cm
(b) 1.5 cm
(c) 25.5 cm
(d) 12 cm

38 A particle performing simple harmonic motion having time period 3 s is in phase with another particle which also undergoes simple harmonic motion at $t = 0$. The time period of second particle is T (less than 3 s). If they are again in the same phase for the third time after 45 s, then the value of T will be
(a) 2.8 s
(b) 2.7 s
(c) 2.5 s
(d) 1.8 s

39 A disc of radius R and mass M is pivoted at the rim and is set for small oscillations. If simple pendulum has to have the same period as that of the disc, the length of a simple pendulum should be
(a) (5/4) R
(b) (2/3) R
(c) (3/4) R
(d) (3/2) R

40 In damped oscillations, the amplitude after 50 oscillations is 0.8 a_0, where a_0 is the initial amplitude, then the amplitude after 150 oscillations is
(a) 0.512 a_0
(b) 0.280 a_0
(c) zero
(d) a_0

41 An air chamber of volume V, has cross-sectional area A. A ball of mass m is fitted smoothly in the neck. The bulk modulus of air is B. If the ball is pressed down slightly and released, the time period of its oscillations is

(a) $2\pi\sqrt{\frac{mV}{2BA^2}}$ (b) $\pi\sqrt{\frac{2mV}{BA^2}}$ (c) $2\pi\sqrt{\frac{mV}{BA^2}}$ (d) $\frac{\pi}{2}\sqrt{\frac{m}{BA^2}}$

42 A simple pendulum of length l is suspended in a car. The car starts moving on a horizontal road according to the equation $x = \frac{g}{2}\sqrt{3}t^2$. Find the time period of oscillations of the pendulum.

(a) $2\pi\sqrt{\frac{l}{g}}$ (b) $\pi\sqrt{\frac{l}{2g}}$ (c) $2\pi\sqrt{\frac{l}{2g}}$ (d) $2\pi\sqrt{\frac{l}{g\sqrt{3}}}$

43 A solid disk of radius R is suspended from a spring of linear spring constant k and torsional constant c, as shown in figure. In terms of k and c, what value of R will give the same period for the vertical and torsional oscillations of this system?

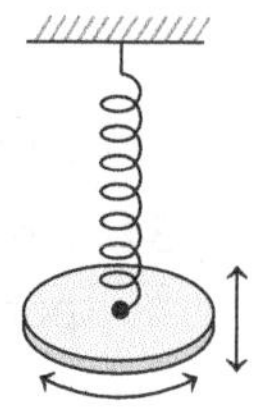

(a) $\sqrt{\frac{2c}{k}}$ (b) $\sqrt{\frac{c}{2k}}$ (c) $2\sqrt{\frac{c}{k}}$ (d) $\frac{1}{2}\sqrt{\frac{c}{k}}$

44 A simple pendulum attached to the ceiling of a stationary lift has a time period T. The distance y covered by the lift moving upwards varies with time t as $y = t^2$, where y is in metres and t in second. If $g = 10\ m/s^2$, the time period of pendulum will be

(a) $\sqrt{\frac{4}{5}}T$ (b) $\sqrt{\frac{5}{6}}T$ (c) $\sqrt{\frac{5}{4}}T$ (d) $\sqrt{\frac{6}{5}}T$

45 The bob of a simple pendulum of length L is released at time $t = 0$ from a position of small angular displacement θ_0. Its angular displacement at time t is given by

(a) $\theta = \theta_0 \sin 2\pi\sqrt{(L/g)}t$ (b) $\theta = \theta_0 \cos 2\pi\sqrt{(g/L)}t$

(c) $\theta = \theta_0 \sin\sqrt{(g/L)}t$ (d) $\theta = \theta_0 \cos\sqrt{(g/L)}t$

46 The coordinates of a particle moving in a plane are given by $x = a\cos\omega t$ and $y = a\sin\omega t$. The distance travelled by the particle in a time interval from $t = 0$ to $t = \frac{\pi}{2\omega}$ is

(a) π (b) $a\pi$ (c) $\frac{a}{2}\pi$ (d) $3a\pi$

47 A thin uniform vertical rod of mass m and length l is pivoted at point O as shown in the figure. Both springs has force constant k. Angular frequency of small oscillations of rod will be

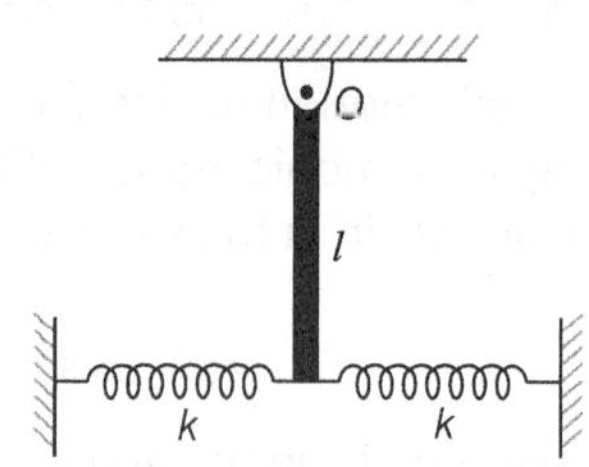

(a) $\sqrt{\frac{3k}{m}+\frac{2g}{3l}}$ (b) $\sqrt{\frac{6k}{m}+\frac{3g}{2l}}$ (c) $\sqrt{\frac{3k}{2m}+\frac{3g}{l}}$ (d) $\sqrt{\frac{6k}{2m}+\frac{g}{l}}$

48 A particle executing SHM has velocities u and v and acceleration a and b in two of its positions. Find the distance between these positions.

(a) $\frac{u^2 - v^2}{a+b}$ (b) $\frac{v^2 + u^2}{a-b}$ (c) $\frac{v^2 + u^2}{a+b}$ (d) $\frac{v^2 - u^2}{a-b}$

49 In a certain oscillatory system (particle is performing SHM), the amplitude of motion is 5 m and the time period is 4 s. The minimum time taken by the particle for passing between two points, which are at a distance of 4 m and 3 m from the centre and on the same side of it will approximately be

(a) $\frac{16}{45}$ s (b) $\frac{7}{45}$ s (c) $\frac{8}{45}$ s (d) $\frac{13}{45}$ s

50 The string of a simple pendulum is replaced by a uniform rod of length L and mass M while the bob has a mass m. It is allowed to make small oscillations. Its time period is

(a) $2\pi\sqrt{\left(\frac{2M}{3m}\right)\frac{L}{g}}$ (b) $2\pi\sqrt{\frac{2(M+3m)}{3(M+2m)}\frac{L}{g}}$

(c) $2\pi\sqrt{\left(\frac{M+m}{M+3m}\right)\frac{L}{g}}$ (d) $2\pi\sqrt{\left(\frac{2m+M}{3(M+2m)}\right)\frac{L}{g}}$

51 The coefficient of friction between block of masses m and $2m$ is $\mu = 2\tan\theta$. There is no friction between block of mass $2m$ and inclined plane. The maximum amplitude of the two block system for which there is no relative motion between both the blocks is

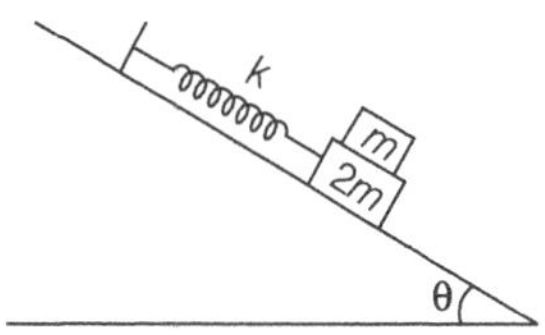

(a) $g\sin\theta\sqrt{\frac{k}{m}}$ (b) $\frac{mg\sin\theta}{k}$

(c) $\frac{3mg\sin\theta}{k}$ (d) None of these

52 One end of a spring of force constant k is fixed to a vertical wall and the other to a body of mass m resting on a smooth horizontal surface. There is another wall at a distance x_0 from the body. The spring is then compressed by $3x_0$ and released. The time taken to strike the wall from the instant of release is [Take, $\sin^{-1}(1/3) = (\pi/9)$]

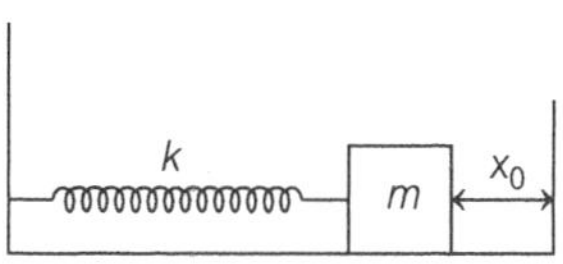

(a) $\frac{\pi}{6}\sqrt{\frac{m}{k}}$ (b) $\frac{2\pi}{3}\sqrt{\frac{m}{k}}$ (c) $\frac{\pi}{4}\sqrt{\frac{m}{k}}$ (d) $\frac{11\pi}{9}\sqrt{\frac{m}{k}}$

53 In an angular SHM, angular amplitude of oscillation is π rad and time period is 0.4 s, then calculate its angular velocity at angular displacement $\pi/2$ rad.

(a) 34.3 rad/s (b) 42.7 rad/s
(c) 22.3 rad/s (d) 50.3 rad/s

54 A block of mass M attached to a horizontal spring of spring constant k is performing simple harmonic motion with amplitude A_0. At the instant, the block is passing through its equilibrium position, a lump of putty of mass m is dropped vertically onto the block from a small height, which sticks to it. The new amplitude and period of the block is

(a) $A_0\sqrt{\frac{m+m}{M}}$; $2\pi\sqrt{\frac{m}{k}}$ (b) $A_0\sqrt{\frac{M}{M+m}}$; $2\pi\sqrt{\frac{M+m}{k}}$

(c) $A_0\sqrt{\frac{M}{m}}$; $2\pi\sqrt{\frac{M}{k}}$ (d) $A_0\sqrt{\frac{m}{M}}$; $2\pi\sqrt{\frac{M-m}{k}}$

55 A particle of mass m is attached to three springs A, B and C of equal force constant k as shown in figure. If particle is pushed slightly against spring C and released, find the period of oscillation.

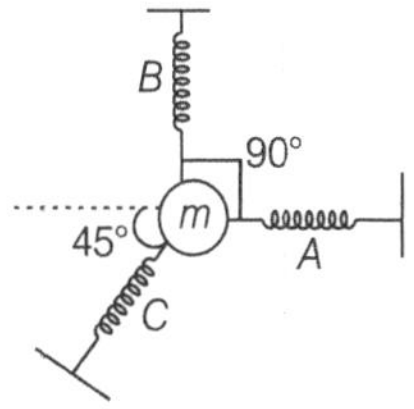

(a) $2\pi\sqrt{\frac{m}{2k}}$ (b) $2\pi\sqrt{\frac{2m}{k}}$

(c) $2\pi\sqrt{\frac{m}{4k}}$ (d) $2\pi\sqrt{\frac{m}{k}}$

56 A body moves along a straight line *OAB* simple harmonically. It has zero velocity at points *A* and *B* which are at a distance *a* and *b* respectively, from O and have maximum velocity *v*, when half way between *A* and *B*. Find the period of SHM.

(a) $\frac{2\pi(b-a)}{v}$ (b) $\frac{\pi(b-a)}{v}$ (c) $\frac{\pi(b+a)}{v}$ (d) $\frac{2\pi(a+b)}{v}$

57 A body is executing simple harmonic motion. At a displacement *x*, its potential energy is E_1 and at a displacement *y*, its potential energy is E_2. The potential energy *E* at displacement $(x+y)$ is

(a) $\sqrt{E} = \sqrt{E_1} - \sqrt{E_2}$ (b) $\sqrt{E} = \sqrt{E_1} + \sqrt{E_2}$

(c) $E = E_1 + E_2$ (d) $E = E_1 - E_2$

58 The following figure depicts a circular motion. The radius of the circle, the period of revolution, the initial position and the sense of revolution are indicated in the figure.

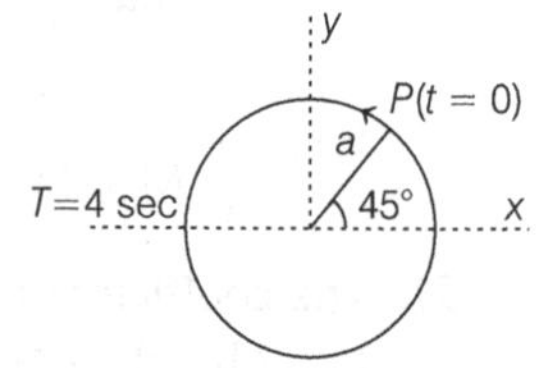

The simple harmonic motion of the *x*-projection of the radius vector of the rotating particle *P* can be shown as

(a) $x(t) = a\cos\left(\frac{2\pi t}{4} + \frac{\pi}{4}\right)$ (b) $x(t) = a\cos\left(\frac{\pi t}{4} + \frac{\pi}{4}\right)$

(c) $x(t) = a\sin\left(\frac{2\pi t}{4} + \frac{\pi}{4}\right)$ (d) $x(t) = a\cos\left(\frac{\pi t}{3} + \frac{\pi}{2}\right)$

59 A bent tube of uniform cross-sectional area *A* has a non-viscous liquid of density ρ. The mass of liquid in the tube is *m*. The time period of oscillation of the liquid is

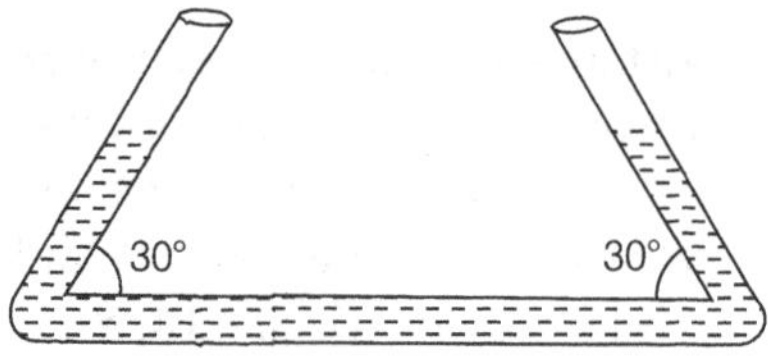

(a) $2\pi\sqrt{\frac{m}{\rho g A}}$ (b) $2\pi\sqrt{\frac{m}{2\rho g A}}$ (c) $2\pi\sqrt{\frac{2m}{\rho g A}}$ (d) $\pi\sqrt{\frac{m}{\rho g A}}$

60 The springs have been compressed from their equilibrium position at $x = 0$ cm. When released, they will start to oscillate. Rank in order from highest to lowest, the maximum speeds of the oscillators.

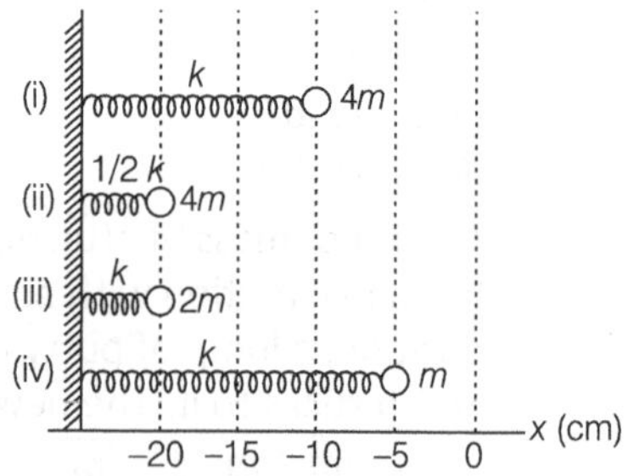

(a) (iii) = (ii) > (i) > (iv)

(b) (i) = (iv) > (ii) > (iii)

(c) (i) > (iv) > (ii) = (iii)

(d) (iii) > (ii) > (i) = (iv)

61 For a particle executing SHM with amplitude *A*, the position varies with time according to the following graph. The corresponding position equation with time is given by

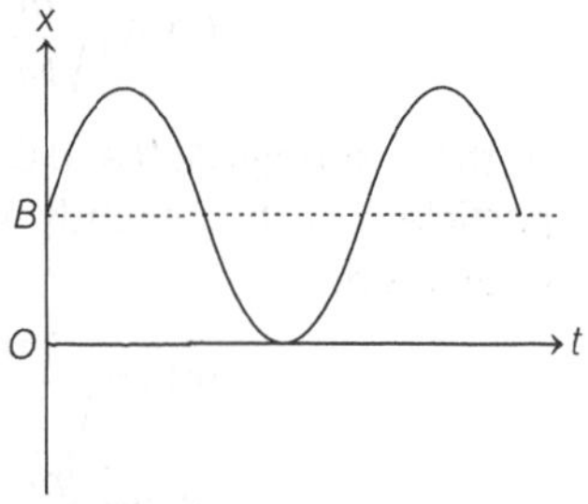

(a) $x = A\sin\omega t$ (b) $x = A + B\sin\omega t$ (c) $x = B + A\cos\omega t$ (d) $x = B + A\sin\omega t$

62 A particle moves along a straight line to follow the equation $ax^2 + bv^2 = k$, where a, b and k are constants and x and v are displacement and velocity of the particle, respectively. Find the amplitude.

(a) $\sqrt{\frac{k}{b}}$ (b) $\sqrt{\frac{b}{k}}$ (c) $\sqrt{\frac{a}{k}}$ (d) $\sqrt{\frac{k}{a}}$

63 A particle starts from a point P at a distance of $A/2$ from the mean position O and travel towards left as shown in the figure. If the time period of SHM executed about O is T and amplitude A, then the equation of motion of particle is

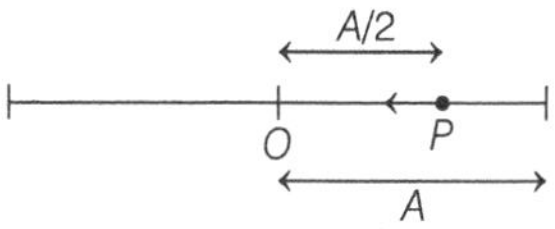

(a) $x = A\sin\left(\frac{2\pi}{T}t + \frac{\pi}{6}\right)$ (b) $x = A\sin\left(\frac{2\pi}{T}t + \frac{5\pi}{6}\right)$

(c) $x = A\cos\left(\frac{2\pi}{T}t + \frac{\pi}{6}\right)$ (d) $x = A\cos\left(\frac{2\pi}{T}t + \frac{\pi}{4}\right)$

64 After learning SHM, Guddu a curious student of XIth class, designed a spring pendulum with identical springs of spring constant k and a small block of mass m as shown in figure.

When block is at rest, it just touches the lower spring. So, Guddu forgot to connect lower spring with block. The time period of oscillation of the block, when it is slightly pushed downward and released is

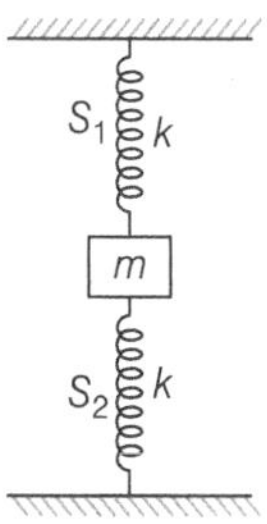

(a) $2\pi\sqrt{\frac{m}{2k}}$ (b) $2\pi\sqrt{\frac{2m}{k}}$

(c) $\pi\sqrt{\frac{m}{2k}}$ (d) $\pi\sqrt{\frac{m}{k}}\left[1 + \frac{1}{\sqrt{2}}\right]$

65 The displacement from mean position of a particle executing simple harmonic motion is given by $y = 5\sin\left(4t + \frac{\pi}{3}\right)$ cm. If T is time period and mass is 2 g, then kinetic energy of particle at $t = \frac{T}{4}$ s is

(a) 4×10^{-3} J (b) 5×10^{-5} J (c) 3×10^{-3} J (d) 3×10^{-5} J

66 A spring of spring constant k is connected to two masses m and $2m$. Forces of same magnitude are applied on both masses as shown in figure. The time period of oscillation of the system will be

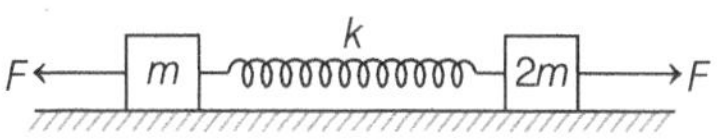

(a) $2\pi\sqrt{\frac{m}{k}}$ (b) $2\pi\sqrt{\frac{2m}{3k}}$ (c) $2\pi\sqrt{\frac{3m}{2k}}$ (d) $2\pi\sqrt{\frac{m}{2k}}$

67 A particle of mass M is performing SHM about origin along x-axis. If it's time period is 2π s, then find out correct relation.

(a) 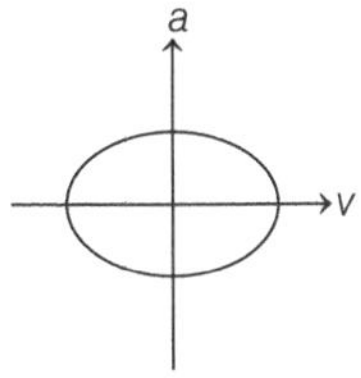

(b) 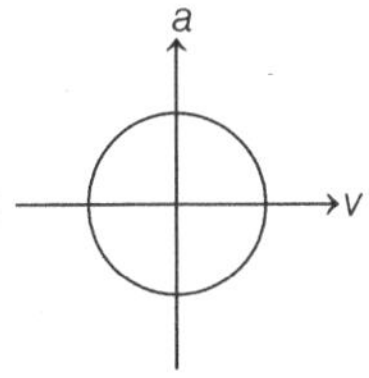

(c) 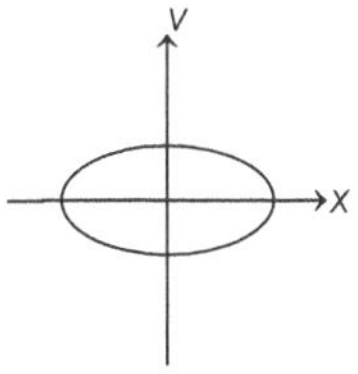

(d) 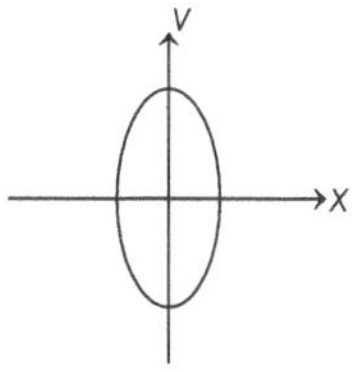

68 A particle performs SHM along x-axis with amplitude A and time period T. If at $t = 0$, particle starts from rest, then time taken by it to travel a distance $\frac{A}{4}$ will be

(a) $\frac{T}{16}$ (b) $\frac{T}{16}\cos^{-1}\left(\frac{3}{4}\right)$ (c) $\frac{T}{2\pi}\cos^{-1}\left(\frac{3}{4}\right)$ (d) $\frac{T}{2\pi}\cos^{-1}\left(\frac{1}{4}\right)$

LEVEL 03 Previous Years Questions

69 A body is executing simple harmonic motion with frequency n, the frequency of its potential energy is **[NEET 2021]**

(a) n (b) $2n$ (c) $3n$ (d) $4n$

70 A spring is stretched by 5 cm by a force 10 N. The time period of the oscillations, when a mass of 2 kg is suspended by it is **[NEET 2021]**

(a) 0.0628 s (b) 6.28 s (c) 3.14 s (d) 0.628 s

71 Identify the function which represents a periodic motion. **[NEET 2020]**

(a) $e^{\omega t}$ (b) $\log_e(\omega t)$

(c) $\sin\omega t + \cos\omega t$ (d) $e^{-\omega t}$

72 The phase difference between displacement and acceleration of a particle in a simple harmonic motion is **[NEET 2020]**

(a) $\frac{3\pi}{2}$ rad (b) $\frac{\pi}{2}$ rad

(c) zero (d) π rad

73 The distance covered by a particle undergoing SHM in one time period is (amplitude $= A$) **[NEET 2019]**

(a) zero (b) A

(c) $2A$ (d) $4A$

74 Average velocity of a particle executing SHM in one complete vibration is **[NEET 2019]**

(a) $A\omega$ (b) $\frac{A\omega^2}{2}$ (c) zero (d) $\frac{A\omega}{2}$

75 The displacement of a particle executing simple harmonic motion is given by

$$y = A_0 + A\sin\omega t + B\cos\omega t$$

Then the amplitude of its oscillation is given by **[NEET 2019]**

(a) $\sqrt{A^2+B^2}$ (b) $\sqrt{A_0^2+(A+B)^2}$

(c) $A+B$ (d) $A_0+\sqrt{A^2+B^2}$

76 The radius of circle, the period of revolution, initial position and sense of revolution are indicated in the figure. y-projection of the radius vector of rotating particle P is **[NEET 2019]**

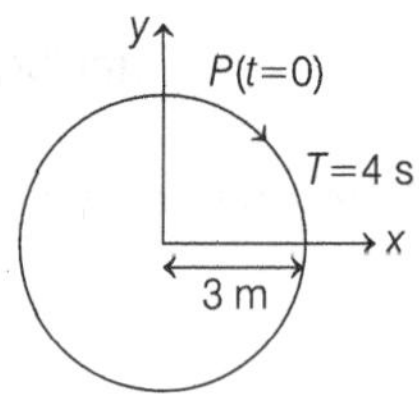

(a) $y(t) = 4\sin\left(\frac{\pi t}{2}\right)$, where y in m (b) $y(t) = 3\cos\left(\frac{3\pi t}{2}\right)$, where y in m

(c) $y(t) = 3\cos\left(\frac{\pi t}{2}\right)$, where y in m (d) $y(t) = -3\cos 2\pi t$, where y in m

77 A mass falls from a height 'h' and its time of fall 't' is recorded in terms of time period T of a simple pendulum. On the surface of earth it is found that $t = 2T$. The entire set up is taken on the surface of another planet whose mass is half of earth and radius the same. Same experiment is repeated and corresponding times noted as t' and T'. **[NEET 2019]**

(a) $t' = \sqrt{2}\,T'$ (b) $t' > 2T'$

(c) $t' < 2T'$ (d) $t' = 2T'$

78 A pendulum is hung from the roof of a sufficiently high building and is moving freely to and fro like a simple harmonic oscillator. The acceleration of the bob of the pendulum is 20 m/s^2 at a distance of 5 m from the mean position. The time period of oscillation is [NEET 2018]

(a) 2 s (b) π s (c) 2π s (d) 1 s

79 A particle executes linear simple harmonic motion with an amplitude of 3 cm. When the particle is at 2 cm from the mean position, the magnitude of its velocity is equal to that of its acceleration. Then, its time period (in seconds) is [NEET 2017]

(a) $\frac{\sqrt{5}}{\pi}$ (b) $\frac{\sqrt{5}}{2\pi}$ (c) $\frac{4\pi}{\sqrt{5}}$ (d) $\frac{2\pi}{\sqrt{3}}$

80 A spring of force constant k is cut into lengths of ratio 1 : 2 : 3. They are connected in series and the new force constant is k'. If they are connected in parallel and force constant is k'', then $1k' : k''$ is [NEET 2017]

(a) 1 : 6 (b) 1 : 9
(c) 1 : 11 (d) 1 : 14

81 A body of mass m is attached to the lower end of a spring whose upper end is fixed. The spring has negligible mass. When the mass m is slightly pulled down and released, it oscillates with a time period of 3 s. When the mass m is increased by 1 kg, the time period of oscillations becomes 5 s. The value of m in kg is [NEET 2016]

(a) $\frac{3}{4}$ (b) $\frac{4}{3}$
(c) $\frac{16}{9}$ (d) $\frac{9}{16}$

82 When two displacements represented by $y_1 = a\sin(\omega t)$ and $y_2 = b\cos(\omega t)$ are superimposed, the motion is [CBSE AIPMT 2015]

(a) not a simple harmonic
(b) simple harmonic with amplitude $\frac{a}{b}$
(c) simple harmonic with amplitude $\sqrt{a^2 + b^2}$
(d) simple harmonic with amplitude $\frac{(a + b)}{2}$

83 A particle is executing SHM along a straight line. Its velocities at distances x_1 and x_2 from the mean position are v_1 and v_2, respectively. Its time period is [CBSE AIPMT 2015]

(a) $2\pi\sqrt{\frac{x_1^2 + x_2^2}{v_1^2 + v_2^2}}$ (b) $2\pi\sqrt{\frac{x_2^2 - x_1^2}{v_1^2 - v_2^2}}$
(c) $2\pi\sqrt{\frac{v_1^2 + v_2^2}{x_1^2 + x_2^2}}$ (d) $2\pi\sqrt{\frac{v_1^2 - v_2^2}{x_1^2 - x_2^2}}$

84 A particle is executing a simple harmonic motion. Its maximum acceleration is α and maximum velocity is β. Then, its time period of vibration will be [CBSE AIPMT 2015]

(a) $\frac{\beta^2}{\alpha^2}$ (b) $\frac{\alpha}{\beta}$
(c) $\frac{\beta^2}{\alpha}$ (d) $\frac{2\pi\beta}{\alpha}$

85 The oscillation of a body on a smooth horizontal surface is represented by the equation,
$X = A\cos(\omega t)$ where, X = displacement at time t, ω = frequency of oscillation
Which one of the following graphs shows correctly the variation a with t? [CBSE AIPMT 2014]

(a) 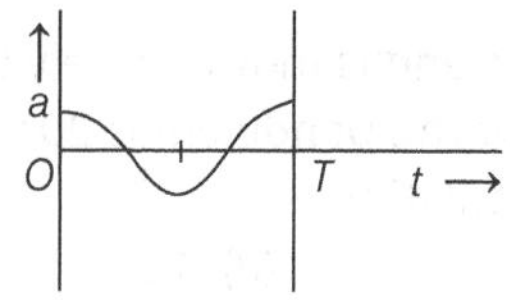

(b) 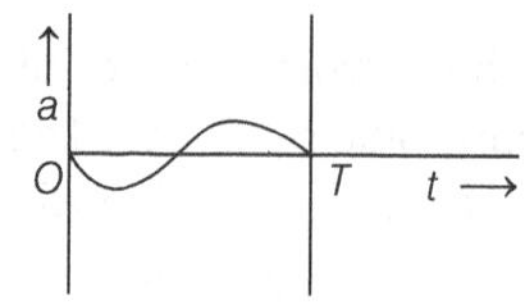

(c) 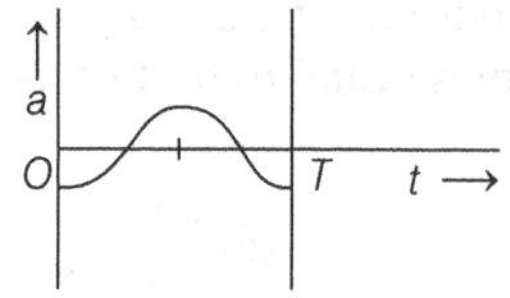

(d) 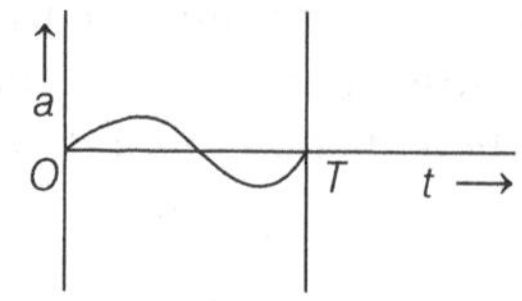

Here, a = acceleration at time t, T = time period.

86 Out of the following functions representing motion of a particle which represents SHM? **[CBSE AIPMT 2011]**

I. $y = \sin \omega t - \cos \omega t$

II. $y = \sin^3 \omega t$

III. $y = 5\cos\left(\frac{3\pi}{4} - 3\,\omega t\right)$

IV. $y = 1 + \omega t + \omega^2 t^2$

(a) Only (IV) does not represent SHM
(b) (I) and (III)
(c) (I) and (II)
(d) Only (I)

87 The displacement of a particle along the x-axis is given by $x = a\sin^2 \omega t$. The motion of the particle corresponds to **[CBSE AIPMT 2010]**

(a) simple harmonic motion of frequency ω/π
(b) simple harmonic motion of frequency $3\omega/2\pi$
(c) non-simple harmonic motion
(d) simple harmonic motion of frequency $\omega/2\pi$

88 The period of oscillation of a mass M suspended from a spring of negligible mass is T. If along with it another mass M is also suspended, the period of oscillation will now be **[CBSE AIPMT 2010]**

(a) T (b) $T/\sqrt{2}$ (c) $2T$ (d) $\sqrt{2}\,T$

ANSWER KEY

Level 1 Objective Questions

1 (a)	2 (b)	3 (b)	4 (b)	5 (d)	6 (d)	7 (a)	8 (b)	9 (c)	10 (a)
11 (d)	12 (d)	13 (c)	14 (d)	15 (c)	16 (d)	17 (b)	18 (c)	19 (a)	20 (a)
21 (b)	22 (b)	23 (d)	24 (c)	25 (a)	26 (c)				

Level 2 Objective Questions

27 (a)	28 (c)	29 (b)	30 (d)	31 (d)	32 (c)	33 (a)	34 (b)	35 (c)	36 (a)
37 (c)	38 (c)	39 (d)	40 (a)	41 (c)	42 (c)	43 (a)	44 (b)	45 (d)	46 (c)
47 (b)	48 (a)	49 (c)	50 (b)	51 (c)	52 (d)	53 (b)	54 (b)	55 (a)	56 (b)
57 (b)	58 (a)	59 (a)	60 (b)	61 (d)	62 (d)	63 (b)	64 (d)	65 (d)	66 (b)
67 (b)	68 (c)								

Level 3 Previous Years Questions

69 (b)	70 (d)	71 (c)	72 (d)	73 (d)	74 (c)	75 (a)	76 (c)	77 (d)	78 (b)
79 (c)	80 (c)	81 (d)	82 (c)	83 (b)	84 (d)	85 (c)	86 (b)	87 (c)	88 (d)

Solutions

LEVEL 01 Objective Questions

1. Velocity is zero, as the slope of position graph is zero. The negative value of x indicates that the particle is to the left. So, the restoring force is to the right.

2. Time taken by particle to move from $x = 0$ to $x = 4$ cm

$= \frac{T}{4} = \frac{1.2}{4} = 0.3$ s

Now, $x = a \sin \omega t$

$$2 = 4 \sin \frac{2\pi}{T} t$$

$$\frac{1}{2} = \sin \frac{2\pi}{1.2} t$$

$$\frac{\pi}{6} = \frac{2\pi}{1.2} t$$

$$\Rightarrow \quad t = 0.1 \text{ s}$$

So, time taken to move from $x = 2$ cm to $x = 4$ cm will be

$$= 0.3 - 0.1$$

$$= 0.2 \text{ s}$$

So, total time taken to move from $x = 2$ cm to $x = 4$ cm and back again $= 2 \times 0.2 = 0.4$ s

3. At the extreme position, the velocity becomes zero.

$$v = 100 \cos \left(10t + \frac{\pi}{3}\right)$$

$$\Rightarrow \quad v = 0$$

$$\Rightarrow \quad 10t + \frac{\pi}{3} = \frac{\pi}{2} \qquad \left(\because \cos \frac{\pi}{2} = 0\right)$$

$$t = \frac{\pi}{60} \text{ s}$$

At $t = \frac{\pi}{60}$ s, the particle comes to rest.

4. $F = -kx$

$$\Rightarrow \quad k = -\frac{F}{x} = -\frac{(-13.5)}{1.5} = 9 \text{ N/m}$$

So, $T = 2\pi\sqrt{\frac{m}{k}} = 2\pi\sqrt{\frac{1}{9}} = \frac{2\pi}{3}$ s

5. $\frac{T}{2} = 11 - 5 = 6$ s

Time period $= 2 \times 6 = 12$ s

Amplitude, $A = 4$

So, $\omega = \frac{2\pi}{T} = \frac{2\pi}{12}$

Now, $x = A \sin (\omega t + \theta)$

$$= 4 \sin\left(\frac{\pi}{6} t + \theta\right)$$

At $t = 0$, $x = 2$ cm, we get

$$\theta = \frac{\pi}{6}$$

$$\therefore \quad x = 4 \sin\left(\frac{\pi}{6} t + \frac{\pi}{6}\right)$$

6. $T = 8$ s, $\omega = \frac{2\pi}{T} = \frac{\pi}{4}$ rad s^{-1}

$$x = A \sin \omega t$$

$$\therefore \quad a = -\omega^2 x = -\frac{\pi^2}{16} \sin\left(\frac{\pi}{4} t\right) \qquad (\because A = 1)$$

At $t = \frac{4}{3}$ s, we get

$$a = -\frac{\sqrt{3}}{32} \pi^2 \text{cm/s}^2$$

7. $a = -\omega^2 x = -\frac{k}{m} x$

Slope, $m = -\frac{3}{4}$

$$\frac{m}{k} = \frac{4}{3}$$

Time period, $T = 2\pi\sqrt{\frac{m}{k}}$

$$= 2\pi\sqrt{\frac{4}{3}} = \frac{4\pi}{\sqrt{3}} \text{ s}$$

8. $y = A \cos \omega t$

where, $A = 13$ cm, $T = 12$ s and $\omega = \frac{2\pi}{T} = \frac{2\pi}{12} = \frac{\pi}{6}$ s

At $t = 0$, $\quad y = 13 \cos\left(\frac{\pi t}{6}\right)$

When the particle is at $y = 6.5$ cm,

$$6.5 = 13 \cos\left(\frac{\pi t}{6}\right)$$

$$\frac{6.5}{13} = \cos\left(\frac{\pi t}{6}\right)$$

$$\Rightarrow \quad \frac{65}{130} = \cos\left(\frac{\pi t}{6}\right)$$

$$\Rightarrow \quad \frac{1}{2} = \cos\left(\frac{\pi t}{6}\right)$$

$$\Rightarrow \quad \cos \frac{\pi}{3} = \cos\left(\frac{\pi t}{6}\right)$$

$$\Rightarrow \quad \frac{\pi t}{6} = \frac{\pi}{3}$$

or $\quad t = 2$ s

and when $y = -6.5$ cm, time t_2 is,

$$\cos\left(\frac{\pi t}{6}\right) = -\frac{1}{2}$$

$$\frac{\pi t}{6} = \frac{2\pi}{3} \qquad \left(\because \cos\frac{2\pi}{3} = \frac{-1}{2}\right)$$

$\Rightarrow$ $$t = 4\text{ s}$$

So, time difference, $\Delta t = t_2 - t_1 = (4 - 2)$
$= 2$ s

9. $$v = \sqrt{A^2 - x^2} = \sqrt{A^2 - \left(\frac{2A}{3}\right)^2} \quad \text{...(i)}$$

$$3v = \sqrt{A'^2 - \left(\frac{2A}{3}\right)^2} \quad \text{...(ii)}$$

From Eqs. (i) and (ii), we get

$$A' = \frac{7A}{3}$$

10. $v = \omega\sqrt{a^2 - x^2}$

At $x = \frac{a}{2}$, $$v = \frac{2\pi}{T}\sqrt{a^2 - \left(\frac{a}{2}\right)^2}$$

$$= \frac{2\pi}{T}\frac{\sqrt{3a^2}}{2} = \frac{\pi a\sqrt{3}}{T}$$

11. Frequency, $f = n$

$$\omega = 2\pi f = 2\pi n$$

$\therefore$ $$2\pi n = \sqrt{\frac{k}{m}}$$

$\Rightarrow$ $$k = 4\pi^2 n^2 m$$

12. $v = \pm\frac{1}{3}\sqrt{6^2 - x^2}$

Comparing with $v = \pm\,\omega\sqrt{A^2 - x^2}$, we have

$$\omega = \frac{1}{3} = \frac{2\pi}{T}$$

$\therefore$ $$T = 6\pi\text{ s}$$

13. Time period of pendulum in a stationary lift,

$$T = 2\pi\sqrt{\frac{L}{g}}$$

When the lift moves up with acceleration a. So, its time period,

$$T' = 2\pi\sqrt{\frac{L}{g + a}} = 2\pi\sqrt{\frac{L}{g + \frac{g}{4}}}$$

$\because$ $$\frac{T'}{T} = \frac{2}{\sqrt{5}}$$

$\Rightarrow$ $$T' = \frac{2T}{\sqrt{5}}$$

14. $A = A_0 e^{-kt}$

$$\frac{A_0}{3} = A_0 e^{-2k} \quad \text{...(i)}$$

$$\frac{A_0}{n} = A_0 e^{-6k} \quad \text{...(ii)}$$

On solving Eqs. (i) and (ii), we get

$$n = 3^3$$

15. Time period of SHM for small oscillations in a liquid is given as, $T = 2\pi\sqrt{\frac{L}{g}}$

Now, $mg = AL\rho g$

$\Rightarrow$ $$L = \frac{m}{A\rho}$$

$\therefore$ $$T = 2\pi\sqrt{\frac{m}{A\rho g}} \Rightarrow T \propto \frac{1}{\sqrt{A}}$$

16. Potential energy is maximum at extreme position and minimum at mean position.
So, time taken to go from extreme to the mean position, is $\frac{T}{4}$.

$$5\text{ s} = \frac{T}{4}$$

$\Rightarrow$ $$T = 20\text{ s}$$

17. In the initial situation, when the bob is not immersed in liquid, its time period,

$$T = 2\pi\sqrt{\frac{L}{g}}$$

Now, when the bob is immersed in the non-viscous liquid,

so $g_{\text{effective}} = \frac{\rho - \sigma}{\rho} g$

$$= \left(1 - \frac{\sigma}{\rho}\right)g = \left(1 - \frac{1}{4}\right)g = \frac{3g}{4}$$

So, time period, $T' = 2\pi\sqrt{\frac{4L}{3g}}$

$\therefore$ $$\frac{T'}{T} = \frac{2\pi\sqrt{\frac{4L}{3g}}}{2\pi\sqrt{\frac{L}{g}}}$$

$\Rightarrow$ $$T' = \frac{2T}{\sqrt{3}}$$

18. KE = PE

$$\frac{1}{2}k(A^2 - x^2) = \frac{1}{2}kx^2$$

$$x = \frac{A}{\sqrt{2}}$$

19. Time period of vertical oscillation in configuration (i),

$$T_1 = 2\pi\sqrt{\frac{m}{k}}$$

Similarly, $T_2 = 2\pi\sqrt{\frac{2m}{k}} \left[\because \text{series, } k_{eq} = \frac{k_1k_2}{k_1+k_2}\right]$

Similarly, $T_3 = 2\pi\sqrt{\frac{m}{2k}}$ $[\because \text{parallel}, k_{eq} = k_1 + k_2]$

$\therefore$ $T_1 : T_2 : T_3 = 1 : \sqrt{2} : \frac{1}{\sqrt{2}}$

20. $k_{eff} = k + k + k = 3k$

$\therefore$ $T = 2\pi\sqrt{\frac{M}{k_{eff}}} = 2\pi\sqrt{\frac{M}{3k}}$

21. Force constant, $k \propto \frac{1}{l}$

So, $k_{eq} = \frac{2 \times (4k) \times 2k}{8k + 2k} = \frac{8k \times 2k}{10k}$

$= 1.6\, k$

22. $k_{eq} = k_1 + k_2 = 2k$ [for parallel, $k_{eq} = k_1 + k_2$]

Time period, $T = 2\pi\sqrt{\frac{m}{k_{eq}}}$

$$3 = 2\pi\sqrt{\frac{12}{2k}} \quad \text{...(i)}$$

$$6 = 2\pi\sqrt{\frac{m+12}{2k}} \quad \text{...(ii)}$$

On dividing Eqs. (i) by (ii) we get

$$\frac{1}{2} = \sqrt{\frac{12}{12+m}}$$

$\Rightarrow$ $12 + m = 48$

$\Rightarrow$ $m = 36$ kg

23. $f = \frac{1}{2\pi}\sqrt{\frac{k}{m}} \Rightarrow k = cf^2$

Here, C is a constant.

As springs are in series

$$\frac{1}{k} = \frac{1}{k_1} + \frac{1}{k_2}$$

$$\Rightarrow \frac{1}{cf^2} = \frac{1}{cf_1^2} + \frac{1}{cf_2^2}$$

$$\Rightarrow f = \frac{f_1f_2}{\sqrt{f_1^2 + f_2^2}}$$

24. We know, time taken to move from extreme to mean position is $\frac{T}{4}$.

Now, $x_0 = 2x_0 \sin\left(\frac{2\pi}{T} \times t\right)$, we get

$$\frac{1}{2} = \sin\left(\frac{2\pi t}{T}\right)$$

$$\sin\left(\frac{\pi}{6}\right) = \sin\left(\frac{2\pi t}{T}\right)$$

$$\frac{\pi}{6} = \frac{2\pi t}{T} \Rightarrow t = \frac{T}{12}\text{ s}$$

Now, total time taken, $t = \frac{T}{4} + \frac{T}{12} = \frac{T}{3}$ s

Now, $T = 2\pi\sqrt{\frac{m}{k}} = \frac{2\pi}{3}\sqrt{\frac{m}{k}}$

25. In series, $k_{eq} = \frac{k}{3}$

In parallel, $k_{eq} = 3k$

In series, $T_1 = 2\pi\sqrt{\frac{3m}{k}}$

In parallel, $T_2 = 2\pi\sqrt{\frac{m}{3k}}$

$\therefore$ $\frac{T_2}{T_1} = \frac{1}{3}$

26. $F = -\frac{dU}{dx} = -a^2U_0x$ (for small amplitude)

$\Rightarrow$ $\omega^2 = \frac{a^2U_0}{m}$

So, $T = 2\pi\sqrt{\frac{m}{a^2U_0}}$

LEVEL 02 Objective Questions

27. To detach the mass from the pan, the force applied must be greater than its weight. Now, let a be the minimum amplitude, so that the mass get detached from the pan.

$$ka = mg$$

$$\Rightarrow a = \frac{mg}{k} = \frac{2 \times 10}{200} = 0.1\text{ m} = 10\text{ cm}$$

28. $x = 3 \sin 100\, t + 8 \cos^2 50t$

Using $2\cos^2 A = (1 + \cos 2A)$

$$x = 3 \sin 100\, t + 4 \cos 100\, t + 4$$
$$= 5 \sin(100t + 53^\circ) + 4$$

Motion of particle is simple harmonic motion of amplitude 5 units and mean position at $x = 4$ units

So, extreme positions are $x = 9$ and $x = -1$

$$A = \sqrt{3^2 + 4^2} = 5\text{ units}$$

29. $\bar{v} = \frac{\int_0^{T/2} v\,dt}{\int_0^{T/2} dt} = \frac{\int_0^{T/2} v_0 \sin(\omega t)\,dt}{\int_0^{T/2} dt}$

$$= \frac{2v_0}{T}\left[\frac{-\cos\omega t}{\omega}\right]_0^{T/2}$$

$$= \frac{2v_0}{\omega T}\left[1 - \cos\frac{\omega T}{2}\right] \quad ...(i)$$

We have given, $\omega = \frac{2\pi}{T}$

$$\omega \times \frac{T}{2} = \frac{2\pi}{T} \times \frac{T}{2}$$

$$\omega T = 2\pi \quad ...(ii)$$

$\therefore$ From Eqs. (i) and (ii), we get

$$\therefore \quad \frac{2v_0}{2\pi}(1 - \cos\pi) = \frac{2v_0}{2\pi}(2) = \frac{2v_0}{\pi}$$

30. Power, $P = Fv$

At displacement x,

$$P = [(kx)]\,[\omega\sqrt{A^2 - x^2}] \quad ...(i)$$

So, P is a function of x.

At maximum power,

$$\frac{dP}{dx} = 0$$

$$\therefore \quad \omega k\left[\sqrt{A^2 - x^2} + \frac{\left(\frac{1}{2}\right)(x)(-2x)}{\sqrt{A^2 - x^2}}\right] = 0$$

or

$$x = \frac{A}{\sqrt{2}}$$

Therefore, at $x = \frac{A}{\sqrt{2}}$, power is maximum.

Substituting $x = \frac{A}{\sqrt{2}}$ in Eq. (i), we get

$$P_{max} = \frac{k^{3/2}A^2}{2\sqrt{m}}$$

31. Given, $\mathbf{a} = (4 - 2x)\,\hat{\mathbf{i}}$

$$v\frac{dv}{dx} = 4 - 2x$$

$$\int v dv = \int (4 - 2x)\, dx$$

On integrating, we get

$$v^2 = 8x - 2x^2$$

$$v = \sqrt{8x - 2x^2} \quad ...(i)$$

Now, $\frac{dv}{da} = \frac{4 - 2x}{\sqrt{8x - 2x^2}} = 0$

$$\Rightarrow \quad x = 2$$

Now, substituting $x = 2$,

$$v = \sqrt{8 \times 2 - 2(2)^2}$$

$$= 2\sqrt{2} \text{ cm/s}$$

As the particle comes to rest,

$$v = \sqrt{8x - 2x^2} = 0$$

$$\Rightarrow \quad 8x - 2x^2 = 0$$

$$\Rightarrow \quad x = 4 \text{ cm}$$

Now, at the mean position, velocity will be maximum which is $2\sqrt{2}$ m/s.

Substituting this in Eq. (i), we get

$$8 = 8x - 2x^2$$

$$x = 2$$

So, the particle oscillates at $x = 2$ cm

32. We have given, $A = 5$ cm

and $\omega = 2\pi(2) = 4\pi$ rad/s

$$N = mg + ma$$

$$= mg + m(\omega^2 y)$$

$$= 60 \times 10 + 60 \times (4\pi)^2 \times 0.05$$

$$= 1073.2 \text{ N}$$

$$\approx 1080 \text{ N}$$

33. $\because \; v = \omega\sqrt{A^2 - x^2} = \frac{2\pi}{T}\sqrt{A^2 - x^2}$

Option (a) satisfy this equation.

34. Kinetic energy of a particle is given as,

$$KE = pv^2$$

$$\frac{1}{2}mv^2 = pv^2$$

$$\Rightarrow \quad m = 2p \quad ...(i)$$

Potential energy of a particle is given as,

$$PE = qx^2$$

$$\Rightarrow \quad \frac{1}{2}kx^2 = qx^2$$

$$\Rightarrow \quad k = 2q \quad ...(ii)$$

Time period of oscillation, $T = 2\pi\sqrt{\frac{m}{k}}$ ...(iii)

$$= 2\pi\sqrt{\frac{2p}{2q}}$$

Form Eqs. (i), (ii) and (iii), we get

$$T = 2\pi\sqrt{\frac{p}{q}}$$

35. Time taken by particle to travel from $x = 0$ to $\frac{A}{2}$ is $\frac{T}{12}$.

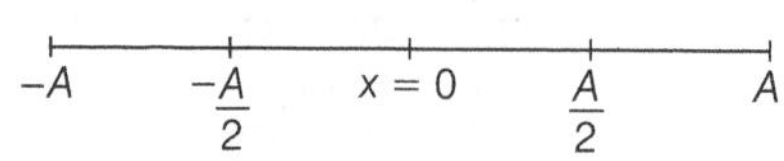

So, time taken by particle to travel from $x = A$ to $x = \frac{A}{2}$ is $\frac{T}{6}$.

Time difference $= \frac{T}{6} + \frac{T}{6} = \frac{T}{3}$

$\therefore$ Phase difference, $\phi = \frac{2\pi}{T} \times \frac{T}{3} = \frac{2\pi}{3}$

36. $a = -\omega^2 x = \omega^2 x$ (numerically)

$$\frac{da}{dt} = \omega^2 \frac{dx}{dt} = -\omega^2 v$$

$$\frac{da}{dt} = \omega^3 \sqrt{A^2 - x^2}$$

Now for $\frac{da}{dt}$ to be maximum, x^2 should be minimum, at $x = 0$.

For $\frac{da}{dt}$ to be minimum, x^2 should be maximum, i.e.

$$x = \pm a.$$

37. $T = \frac{2\pi}{\omega} = 18$ s

At $t = 0$, $x = 1.5$ cm and $\frac{dx}{dt}$ or v is positive.

Hence at $t = 0$, position of particle is as shown below :

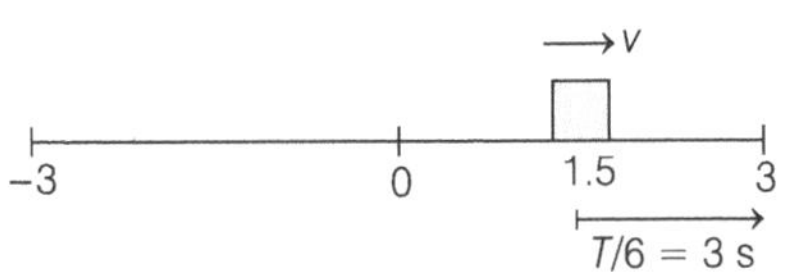

Given, time, $t = \frac{T}{6} + 2T$

$$\therefore \quad d = \frac{A}{2} + 2 \times 4A$$

$$= 1.5 + 2 \times 12 \qquad (\because A = 3\text{cm})$$

$$= 25.5 \text{ cm}$$

38. Let ω_1 and ω_2 be the angular frequencies of first and second particle and their phases are $\omega_1 t$ and $\omega_2 t$.

Now, $\omega_2 t - \omega_1 t = 3 \times 3\pi$ for $t = 45$ s

$$\frac{2\pi}{T} - \frac{2\pi}{3} = \frac{3 \times 2\pi}{45}$$

We get, $T = 2.5$ s

39. $\tau = -MgR \sin\theta = -Mg\, R\theta \qquad [\because \sin\theta \approx \theta]$

$$-MgR\,\theta = I\alpha \qquad (\because T = I\alpha)$$

$$-MgR\theta = \left(\frac{MR^2}{2} + MR^2\right)\alpha$$

$$-MgR\theta = \frac{3}{2} MR^2 \alpha$$

$$\Rightarrow \quad \alpha = -\frac{2g}{3R}\theta = -\omega^2 \theta$$

$$\Rightarrow \quad T = 2\pi\sqrt{\frac{3R}{2g}} \qquad \left(\because \omega = \frac{2\pi}{T}\right)$$

For simple pendulum, we know,

$$T = 2\pi\sqrt{\frac{l}{g}}$$

So, length of a simple pendulum should be $\left(\frac{3}{2}\right)R$.

40. Amplitude as a function of time is given as,

$$a = a_0 e^{-\alpha t}$$

$$a_{50} = a_0 e^{-\alpha \times 50t} = 0.80\, a_0$$

$$a_{150} = a_0 e^{-\alpha \times 150t} = a_0 (0.8)^3$$

$$= 0.512\, a_0$$

41. When the ball is pressed down a distance x, the change in volume $= V - \Delta V$

So, the pressure increases from p to $p + \Delta p$,

$$\Delta V = Ax$$

Now, $\Delta p = \frac{B\Delta V}{V}$

Restoring force, $F = \frac{B\Delta V}{V} A$

$$F = -\frac{BA^2 x}{V} \qquad (\because \Delta V = Ax)$$

$$\Rightarrow \quad F \propto -kx \qquad \left(\text{where, } k = -\frac{BA^2}{V}\right)$$

$$\therefore \quad T = 2\pi\sqrt{\frac{m}{k}} = 2\pi\sqrt{\frac{mV}{BA^2}}$$

42. Horizontal acceleration,

$$a = \frac{d^2x}{dt^2} = \sqrt{3}\, g$$

$$g_{\text{eff}} = \sqrt{a^2 + g^2} = 2g$$

$$\therefore \quad T = 2\pi\sqrt{\frac{l}{g_{\text{eff}}}} = 2\pi\sqrt{\frac{l}{2g}}$$

43. Time period for vertical oscillation,

$$T_v = 2\pi\sqrt{\frac{m}{k}}$$

Time period for torsional oscillation,

$$T_t = 2\pi\sqrt{\frac{I}{c}} = 2\pi R\sqrt{\frac{m}{2c}} \qquad \left[\because I = \frac{mR^2}{2}\right]$$

Now, $T_v = T_t$

$$2\pi\sqrt{\frac{m}{k}} = 2\pi R\sqrt{\frac{m}{2c}} \Rightarrow R = \sqrt{\frac{2c}{k}}$$

44. Acceleration of lift upwards,

$$a = \frac{d^2y}{dt^2} = 2 \text{ m/s}^2$$

$$a_{\text{relative}} = g + a = 10 + 2 = 12 \text{ m/s}^2$$

Time period when the lift was stationary,

$$T = 2\pi\sqrt{\frac{L}{g}} = 2\pi\sqrt{\frac{L}{10}}$$

Time period when the lift was moving with $a = 2$ m/s^2 upwards,

$$T' = 2\pi\sqrt{\frac{L}{g + a}} = 2\pi\sqrt{\frac{L}{12}}$$

So, $\frac{T'}{T} = \frac{\sqrt{10}}{\sqrt{12}} \Rightarrow T' = \sqrt{\frac{5}{6}}\, T$

45. Amplitude $= \theta_0$

It's angular frequency, $\omega = \sqrt{\frac{g}{L}}$

The angular displacement at any time is given as,

$$\theta = \theta_0 \cos \omega t = \theta_0 \cos \sqrt{\frac{g}{L}}\, t$$

46. Speed, $v = \sqrt{v_x^2 + v_y^2}$

$$= \sqrt{\left(\frac{dx}{dt}\right)^2 + \left(\frac{dy}{dt}\right)^2} = \omega a$$

Since, speed is constant.

$\therefore$ Distance = Speed $\times$ Time

$$= \omega a \times \frac{\pi}{2\omega} = \frac{a\pi}{2}$$

47.

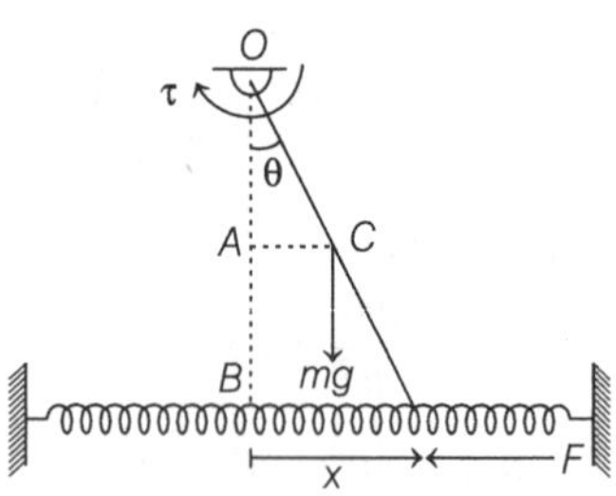

$$x = l\theta \Rightarrow F = 2kx$$

Restoring torque, $\tau = -[(mg)\, AC + F\,(OB)]$

$$\tau = -\left[(mg)\frac{l}{2}\sin\theta + 2\,kxl\right]$$

For small angles, $\sin\theta \approx \theta$

$\therefore$ $$\tau = -\left[\frac{mgl}{2} + 2\,kl^2\right]\theta \quad (\because x = l\theta) \quad \text{... (i)}$$

Angular acceleration,

$$\alpha = \frac{\tau}{I} = \frac{\tau}{\left(\frac{ml^2}{3}\right)} = \left(\frac{3}{ml^2}\right)\tau \quad \text{... (ii)}$$

On solving Eqs. (i) and (ii), we get

$$\alpha = -\left[\frac{3g}{2l} + \frac{6k}{m}\right]\theta$$

Comparing with $\alpha = -\omega^2\theta$, we get

$$\omega^2 = \left(\frac{3g}{2l} + \frac{6k}{m}\right) \Rightarrow \omega = \sqrt{\frac{3g}{2l} + \frac{6k}{m}}$$

48. Let x_1 and x_2 be the distances of the two positions from the mean position.

$$u^2 = \omega^2(A^2 - x_1^2) \quad \text{...(i)}$$

$$v^2 = \omega^2(A^2 - x_2^2) \quad \text{...(ii)}$$

$$a = \omega^2 x_1 \quad \text{...(iii)}$$

$$b = \omega^2 x_2 \quad \text{...(iv)}$$

On subtracting Eq. (i) from Eq. (ii), we get

$$u^2 - v^2 = \omega^2(x_2^2 - x_1^2) \quad \text{...(v)}$$

On adding Eqs. (iii) and (iv), we get

$$a + b = \omega^2(x_1 + x_2) \quad \text{...(vi)}$$

On dividing Eq. (v) by Eq. (vi), we get

$$x_2 - x_1 = \frac{u^2 - v^2}{a + b}$$

49.

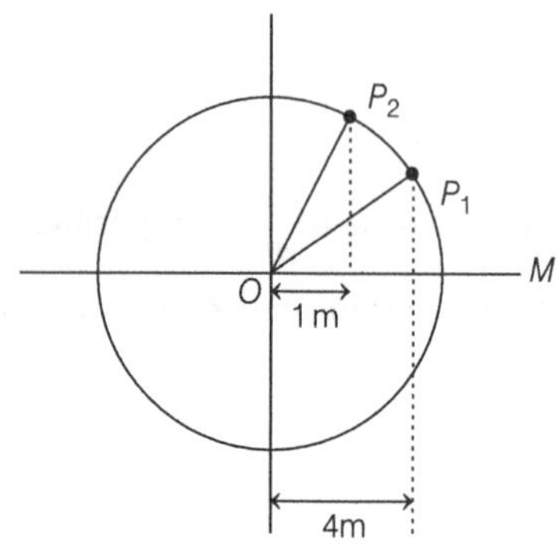

$OP_1 = OP_2 =$ Amplitude $= 5$ m

$\angle P_2OM = 53°$

$\angle P_1OM = 37°$

$\therefore$ $\angle P_2OP_1 = 16°$

$$t = \frac{T}{360°} \times 16° = \frac{4}{360°} \times 16° = \frac{8}{45}\ \text{s}$$

50. We know, the time period for a compound pendulum is given by,

$$T = 2\pi\sqrt{\frac{I}{mgd}} \quad \text{...(i)}$$

Now, moment of inertia I about point of suspension

$$= \frac{ML^2}{3} + mL^2 = \frac{(M + 3m)L^2}{3}$$

Now, d is the separation between point of suspension and centre of mass of the pendulum

$$= \frac{M\left(\frac{L}{2}\right) + m\,(L)}{M + m} = \frac{(M + 2m)L}{2\,(M + m)}$$

Substituting the values in Eq. (i), we get

$$T = 2\pi\sqrt{\frac{2(M + 3m)L}{3\,(M + 2m)\,g}}$$

51. In the two-block system, $\omega = \sqrt{\frac{k}{3m}}$ ...(i)

Maximum frictional force, $f_{max} = \mu\, mg \cos\theta$

$$= 2\tan\theta\, mg\cos\theta = 2mg\sin\theta$$

Now, from Newton's second law on the block at lower extreme position,

$$f - mg\sin\theta = m\omega^2 A$$

As $f \le f_{max}$ $\quad \omega^2 A = g \sin\theta$ ...(ii)

From Eqs. (i) and (ii), we get

$$A = \frac{3\,mg \sin\theta}{k}$$

52. $x_0 = 3x_0 \sin \omega t$

$$\therefore \quad \omega t = \sin^{-1}\left(\frac{1}{3}\right) \Rightarrow \left(\frac{2\pi}{T}\right)t = \frac{\pi}{9}$$

$\therefore$ $t = \frac{T}{18}$ s = time taken by the particle in moving from $x = 0$ to $x = x_0$.

$$\text{Total time} = \frac{T}{18} + \frac{T}{4} + \frac{T}{4} + \frac{T}{18} = \frac{11}{18}T$$

Here, $\quad T = 2\pi\sqrt{\frac{m}{k}}$

$$\therefore \text{Total time} = \frac{11\pi}{9}\sqrt{\frac{m}{k}}$$

53. $\theta = \theta_0 \sin \omega t$

$$\theta = \pi \sin\left(\frac{2\pi}{0.4}t\right) \quad \text{...(i)}$$

$$\omega = \frac{d\theta}{dt} = \pi \times \frac{2\pi}{0.4}\cos\left(\frac{2\pi}{0.4}t\right) \quad \text{...(ii)}$$

Substituting $\theta = \frac{\pi}{2}$ in Eq. (i), we get

$$\sin\left(\frac{2\pi}{0.4}t\right) = \frac{1}{2}$$

$$\sin\left(\frac{2\pi}{0.4}t\right) = \sin\frac{\pi}{6}$$

$$\Rightarrow \quad \frac{2\pi}{0.4}t = \frac{\pi}{6} \Rightarrow t = \frac{0.4}{12}\text{ s}$$

Putting t in Eq. (ii), we get

$$\omega = \frac{2\pi^2}{0.4}\cos\left(\frac{2\pi}{0.4} \times \frac{0.4}{12}\right) = 42.7 \text{ rad/s}$$

54. When a mass m is dropped onto the block, its combined mass is $M + m$.

Now, from momentum conservation,

$$Mv_0 = (m + M)v$$

$$\Rightarrow \quad v = \frac{Mv_0}{M+m} = \frac{M\omega A_0}{M+m}$$

where, $\omega = \sqrt{\frac{k}{M}}$

Now, KE = PE

$$\frac{1}{2}(M+m)v^2 = \frac{1}{2}kA^2$$

Substituting the values, we get

$$A = A_0\sqrt{\frac{M}{M+m}}$$

Further, $T = 2\pi\sqrt{\frac{M+m}{k}}$

55.

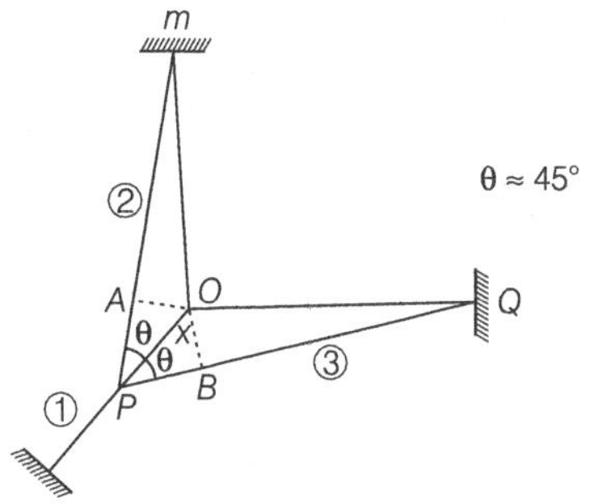

Compression in 1 = x

Elongation in 2 or 3 $\approx x \cos 45° = \frac{x}{\sqrt{2}}$

Net restoring force,

$$F = -\left[kx + 2 \times k\frac{x}{\sqrt{2}} \times \cos 45°\right]$$

$$F = -2kx$$

$$\therefore \quad T = 2\pi\sqrt{\frac{m}{2k}}$$

56. C is the mean position of SHM, where velocity is zero.

Now, amplitude, $A = \frac{b-a}{2}$

So, velocity, $v = \omega A = \omega\frac{(b-a)}{2}$

$$v = \frac{2\pi}{T}\frac{(b-a)}{2} \quad \left(\because \omega = \frac{2\pi}{T}\right)$$

$$\Rightarrow \quad v = \frac{\pi(b-a)}{T}$$

$$\Rightarrow \quad T = \frac{\pi(b-a)}{v}$$

57. At displacement x,

$$E_1 = \frac{1}{2}m\omega^2x^2$$

or $\quad \sqrt{E_1} = x\sqrt{\frac{1}{2}m\omega^2}$ ...(i)

At displacement y,

$$E_2 = \frac{1}{2}m\omega^2y^2$$

or $\quad \sqrt{E_2} = y\sqrt{\frac{1}{2}m\omega^2}$ ...(ii)

At displacement $x + y$,

$$E = \frac{1}{2}m(x+y)^2$$

or $\quad \sqrt{E} = x+y\sqrt{\frac{1}{2}m\omega^2}$ ...(iii)

From Eqs. (i), (ii) and (iii), we get

$$\sqrt{E} = \sqrt{E_1} + \sqrt{E_2}$$

58. We have, amplitude, $A = a$

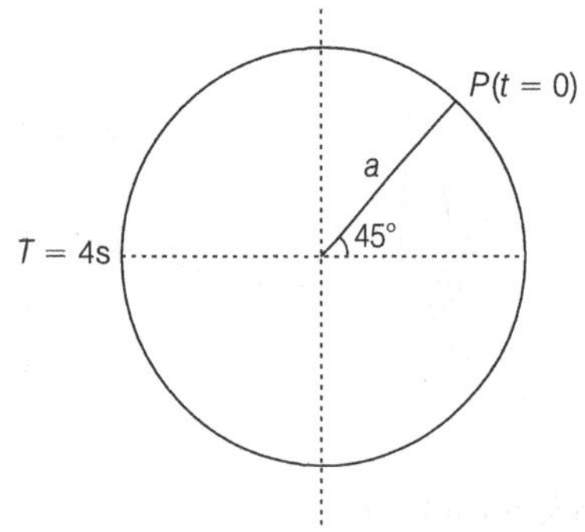

At $t = 0, \theta = \frac{\pi}{4}$

At time t, $\theta = \frac{2\pi}{T}t$

Now, $x = a\cos[\omega t + \theta] = a\cos\left[\frac{2\pi t}{T} + \frac{\pi}{4}\right]$

At $T = 4$ s,

$$x = a\cos\left[\frac{2\pi t}{4} + \frac{\pi}{4}\right]$$

59. Volume of a liquid $= \frac{\text{Mass}}{\text{Density}} = \frac{m}{\rho}$

$$\frac{m}{\rho} = (L_1 + L_2 + L_3)A$$

$$\therefore \quad L = L_1 + L_2 + L_3 = \frac{m}{\rho A}$$

$$T = 2\pi\sqrt{\frac{L}{g}} = 2\pi\sqrt{\frac{m/\rho A}{g}} = 2\pi\sqrt{\frac{m}{\rho g A}}$$

60. We know, $v_{max} = \omega A = \sqrt{\frac{k}{m}}A$

$$v_1 = \sqrt{\frac{k}{4m}} \times 10 = 5\sqrt{\frac{k}{m}}$$

$$v_2 = \sqrt{\frac{k}{2 \times 4m}} \times 20 = \frac{10}{\sqrt{2}}\sqrt{\frac{k}{m}}$$

$$v_3 = \sqrt{\frac{k}{2m}} \times 20 = 10\sqrt{2}\sqrt{\frac{k}{m}}$$

$$v_4 = \sqrt{\frac{k}{m}} \times 5 = 5\sqrt{\frac{k}{m}}$$

So, (i) = (iv) > (ii) > (iii)

61. At $t = 0$, $x = B$

At $t = \frac{T}{4}$ or $\omega t = 90°$, $x = A + B$

Option (d) satisfy both these conditions.

62. We know, $x = A\sin\omega t$ and $y = A\cos\omega t$

$$x^2 = A^2\sin^2\omega t \quad \text{...(i)}$$

$$y^2 = A^2\cos^2\omega t \quad \text{...(ii)}$$

Now, adding Eqs. (i) and (ii) and multiplying with ω, we get

$$\omega^2(x^2 + y^2) = A^2\omega^2(\sin^2\omega t + \cos^2\omega t)$$

Now, we have given,

$$ax^2 + bv^2 = k$$

$$\frac{a}{b}x^2 + v^2 = \frac{k}{b}$$

So, $A^2 = \frac{\frac{k}{b}}{\frac{a}{b}} \quad \Rightarrow A = \sqrt{\frac{k}{a}}$

63.
$$x = A\cos(\omega t + \theta)$$

$$\frac{dx}{dt} = v = \frac{d}{dt}[A\cos(\omega t + \theta)]$$

$$v = -\omega A\sin(\omega t + \theta)$$

At $t = 0$,

$$x = A\cos\theta \quad \text{and} \quad v = -\omega A\sin\theta$$

$$\Rightarrow \quad \frac{A}{2} = A\cos\theta$$

$$\cos\theta = \frac{1}{2}$$

$$\therefore \quad \theta = \frac{\pi}{3}$$

So, $x = A\cos\left(\omega t + \frac{\pi}{3}\right) = A\cos\left(\frac{2\pi t}{T} + \frac{\pi}{3}\right)$

$$= A\sin\left(\frac{2\pi t}{T} + \frac{\pi}{2} + \frac{\pi}{3}\right)$$

$$= A\sin\left(\frac{2\pi t}{T} + \frac{5\pi}{6}\right)$$

64. Half oscillation with only k and half oscillation with $2k$.

$$\therefore \quad T = \frac{2\pi\sqrt{\frac{m}{k}}}{2} + \frac{2\pi\sqrt{\frac{m}{2k}}}{2}$$

$$= \pi\sqrt{\frac{m}{k}}\left[1 + \frac{1}{\sqrt{2}}\right]$$

65. At $t = \frac{T}{4}$ s

$$\Rightarrow \quad t = \frac{\pi}{2\omega} = \frac{T}{2 \times 4} = \frac{\pi}{8}$$

$$y = 5\sin\left[4 \times \frac{\pi}{8} + \frac{\pi}{3}\right] = 2.5 \text{ cm}$$

Now, $\text{KE} = \frac{1}{2} \times m \times \omega^2(A^2 - y^2)$

$$= \frac{1}{2}(2 \times 10^{-3})(4)^2(5^2 - 2.5^2) \times 10^{-4}$$

$$= 3 \times 10^{-5} \text{ J}$$

66. When the two blocks are connected to the end of a spring, it's time period,

$$T = 2\pi\sqrt{\frac{\mu}{k}}$$

$$u = \text{reduced mass} = \frac{m_1 m_2}{m_1 + m_2}$$

So, $$T = 2\pi\sqrt{\frac{(m) \times (2m)}{3m \times k}} = 2\pi\sqrt{\frac{2m}{3k}}$$

67. $\omega = \frac{2\pi}{T} = \frac{2\pi}{2\pi} = 1 \text{ rad/s}$

***v versus x* :**

$$v^2 = \omega^2(A^2 - x^2)$$

or $$v^2 + x^2 = A^2$$

So, v *versus* x graph is a circle.

***a versus v* :**

$$a = -\omega^2 x$$

or $$a = -x \quad (\text{as } \omega = 1 \text{ rad/s})$$

$$v^2 = \omega^2 (A^2 - x^2)$$

$\therefore$ $$v^2 = A^2 - a^2$$

or $$v^2 + a^2 = A^2$$

So, v *versus* a graph is also a circle.

68. $x = A\cos(\omega t)$

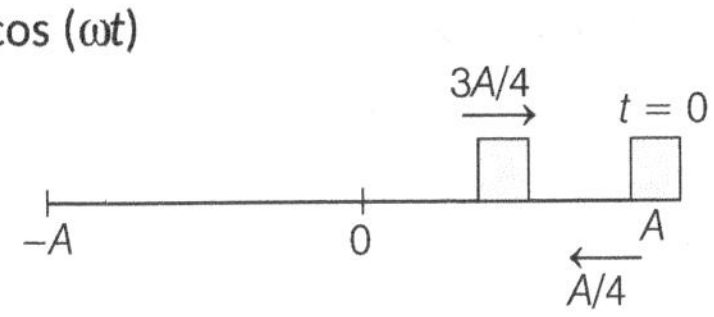

$$\frac{3A}{4} = A\cos\omega t$$

$\therefore$ $$\omega t = \cos^{-1}\left(\frac{3}{4}\right)$$

Putting the value of $\omega = \frac{2\pi}{T}$, we get

$$t = \frac{T}{2\pi}\cos^{-1}\left(\frac{3}{4}\right)$$

LEVEL 03 Previous Years Questions

69. In a simple harmonic motion, both kinetic energy and potential energy attains their maximum value two times in one complete oscillation. Hence, frequency of kinetic energy and potential energy is 2 for one complete oscillation. So, the frequency of the potential energy of a body executing SHM with frequency n is $2n$.

70. Given, the mass of suspended, $m = 2$ kg

The spring is stretched, $x = 5 \text{ cm} = 0.05$ m

The constant force applied on the spring, $F = 10$ N

As we know that, spring force,

$$F = kx \Rightarrow 10 \text{ N} = k\,(0.05 \text{ m})$$

$\Rightarrow$ $$k = 200 \text{ N/m}$$

Now, time period of the oscillation,

$$T = 2\pi\sqrt{\frac{m}{k}} \Rightarrow T = 2\pi\sqrt{\frac{2}{200}}$$

Time period, $T = 0.628$ s

71. $\sin\omega t$ and $\cos\omega t$, both are periodic function of period $\frac{2\pi}{\omega}$. We know that, sum of two periodic functions is also a periodic function, hence $(\sin\omega t + \cos\omega t)$ represents periodic motion.

72. In SHM, equation of displacement of a particle is $y = a\sin\omega t$

and equation of acceleration of a particle is

$$A = a\omega^2\sin\omega t = a\omega^2\sin(\omega t + \pi)$$

$\therefore$ Phase difference between displacement and acceleration of a particle is

$$= (\omega t + \pi) - \omega t = \pi \text{ rad}$$

73. In a simple harmonic motion, the particle oscillates about its mean position on a straight line.

The particle moves from its mean position (O) to an extreme position (P) and then return to its mean position covering same distance of A.

Then by the conservative force, it is moved in opposite direction to a point Q by distance A and then back to mean position covering a distance of A. This comprises of one time period as shown below

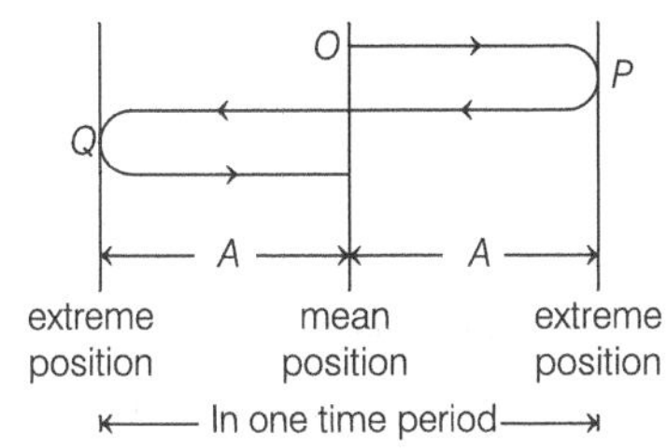

Hence, in one time period, it covers a distance of

$$x = OP + PO + OQ + QO$$
$$= A + A + A + A = 4A$$

74. The average velocity of a particle executing simple harmonic motion (SHM) is

$$v_{av} = \frac{\text{Total displacement}}{\text{Time interval}} = \frac{x_f - x_i}{T}$$

where, x_f and x_i are the initial and final position of the particle executing SHM.

As, in vibrational motion, the particle executes SHM about its mean position. So, after one complete vibration of the particle, it will reach its initial position, i.e.

Displacement, $x_f - x_i = 0$

$\therefore$ $$v_{av} = \frac{0}{T}$$

Hence, the average velocity is zero.

75. The displacement of given particle is,

$$y = A_0 + A\sin\omega t + B\cos\omega t \quad \text{... (i)}$$

The general equation of SHM can be given as,

$$x = a\sin\omega t + b\cos\omega t \quad \text{... (ii)}$$

So, from Eqs. (i) and (ii), we can say that A_0 be the value of mean position, at which $y = 0$.

$\therefore$ Amplitude, $R = \sqrt{A^2 + B^2 + 2AB\cos\theta}$

As two function sine and cosine have phase shift to 90°.

$$\therefore \quad R = \sqrt{A^2 + B^2} \quad [\because \cos 90^\circ = 0]$$

76. Let O be the centre of circle, then at $t = 0$, the displacement y is maximum and have value 3m.

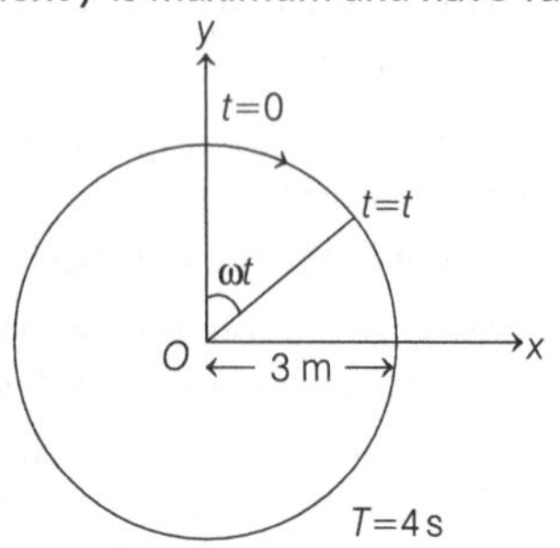

As, the general equation of displacement of a particle will be in the form,

$$y = A\cos\omega t$$

Here, $A = 3$ m

Then, $\omega = \frac{2\pi}{T} = \frac{2\pi}{4} = \frac{\pi}{2}$ [given, $T = 4$ s]

$$\therefore \quad y = 3\cos\left(\frac{\pi}{2}t\right) \text{(in m)}$$

77. The distance covered by the mass falling from height h during its time of fall t is given by,

$$s = h = ut + \frac{1}{2}gt^2$$

As, $u = 0 \Rightarrow h = \frac{1}{2}gt^2 \Rightarrow t = \sqrt{\frac{2h}{g}}$ ...(i)

The time period of simple pendulum is,

$$T = 2\pi\sqrt{\frac{l}{g}} \quad \text{...(ii)}$$

where, l is the length of the pendulum.

From equations (i) and (ii), since 'h' and 'l' are constant, so we can conclude that,

$$t \propto \frac{1}{\sqrt{g}} \text{ and } T \propto \frac{1}{\sqrt{g}}$$

$$\therefore \quad \frac{t}{T} = 1$$

Thus, the ratio of time of fall and time period of pendulum is independent of value of gravity (g) or any other parameter like mass and radius of the planet. Thus, the relation between t' and T' on another planet irrespective of its mass or radius will remain same as it was on earth, i.e.

$$t' = 2T'$$

78. The acceleration of particle/body executing SHM at any instant (at position x) is given as,

$$a = -\omega^2 x$$

where, ω is the angular frequency of the body.

$$\Rightarrow \quad |a| = \omega^2 x \quad \text{...(i)}$$

Here, $x = 5$ m, $|a| = 20$ ms^{-2}

Substituting the given values in Eq. (i), we get

$$20 = \omega^2 \times 5$$

$$\Rightarrow \quad \omega^2 = \frac{20}{5} = 4$$

or $\omega = 2$ rad s^{-1}

As, we know that,

Time period, $T = \frac{2\pi}{\omega}$ ...(ii)

$\therefore$ Substituting the value of ω in Eq. (ii), we get

$$T = \frac{2\pi}{2} = \pi \text{ s}$$

79. **Thinking Process** Magnitude of velocity of particle, when it is at displacement x from mean position

$$= \omega\sqrt{A^2 - x^2}$$

Also, magnitude of acceleration of particle in SHM

$$= \omega^2 x$$

Given, when $x = 2$ cm

$$|\mathbf{v}| = |\mathbf{a}|$$

$$\Rightarrow \quad \omega\sqrt{A^2 - x^2} = \omega^2 x$$

$$\Rightarrow \quad \omega = \frac{\sqrt{A^2 - x^2}}{x} = \frac{\sqrt{9-4}}{2}$$

$\Rightarrow$ Angular velocity, $\omega = \frac{\sqrt{5}}{2}$ rad/s

$\therefore$ Time period of motion,

$$T = \frac{2\pi}{\omega} = \frac{4\pi}{\sqrt{5}} \text{ s}$$

80. When the spring is cut into pieces, they will have the new force constant. The spring is divided into 1 : 2 : 3 ratio.

When the pieces are connected in series, the resultant force constant,

$$\frac{1}{k'} = \frac{1}{k_1} + \frac{1}{k_2} + \frac{1}{k_3}$$

$$\frac{1}{k'} = \frac{1}{x} + \frac{1}{2x} + \frac{1}{3x}$$

$$k' = \frac{6x}{11}$$

In parallel, the net force constant,

$$k'' = x + 2x + 3x = 6x = 11k$$

The required ratio, $\frac{k'}{k''} = \frac{6x/11}{6x}$

$$= 1 : 11$$

81. As we know that,

Time period, $T = 2\pi\sqrt{\frac{m}{k}}$

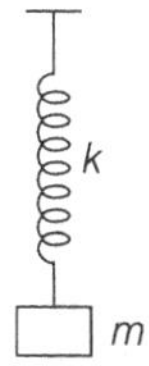

Case I $\quad T_1 = 2\pi\sqrt{\frac{m}{k}}$ ...(i)

Case II When the mass m is increased by 1 kg, then it becomes $m + 1$ ...(ii)

From Eqs. (ii) and (i), we get

$$\frac{T_2}{T_1} = \sqrt{\frac{m+1}{m}}$$

$$\Rightarrow \frac{5}{3} = \sqrt{\frac{m+1}{m}} \Rightarrow \frac{25}{9} = \frac{m+1}{m}$$

$$\Rightarrow \frac{25}{9} = 1 + \frac{1}{m}$$

$$\Rightarrow \frac{1}{m} = \frac{16}{9}$$

$$\therefore m = \frac{9}{16}\text{ kg}$$

82. Given, $y_1 = a\sin\omega t$

$$y_2 = b\cos\omega t = b\sin\left(\omega t + \frac{\pi}{2}\right)$$

The resultant displacement is given by

$$y = y_1 + y_2 = \sqrt{a^2 + b^2}\sin(\omega t + \phi)$$

Hence, the motion of superimposed wave is simple harmonic with amplitude $\sqrt{a^2+b^2}$.

83. Let A be the amplitude of oscillation, then

$$v_1^2 = \omega^2(A^2 - x_1^2) \quad \dots(i)$$

$$v_2^2 = \omega^2(A^2 - x_2^2) \quad \dots(ii)$$

Subtracting Eq. (ii) from Eq. (i), we get

$$v_1^2 - v_2^2 = \omega^2(x_2^2 - x_1^2)$$

$$\Rightarrow \omega = \sqrt{\frac{v_1^2 - v_2^2}{x_2^2 - x_1^2}} \Rightarrow \frac{2\pi}{T} = \sqrt{\frac{v_1^2 - v_2^2}{x_2^2 - x_1^2}}$$

$$\Rightarrow T = 2\pi\sqrt{\frac{x_2^2 - x_1^2}{v_1^2 - v_2^2}}$$

84. For a particle executing SHM, we have maximum acceleration,

$$\alpha = A\omega^2 \quad \dots(i)$$

where, A is maximum amplitude and ω is angular velocity of a particle.

Maximum velocity, $\beta = A\omega$ ...(ii)

Dividing Eq. (i) by Eq. (ii), we get

$$\frac{\alpha}{\beta} = \frac{A\omega^2}{A\omega} \Rightarrow \frac{\alpha}{\beta} = \omega = \frac{2\pi}{T}$$

i.e. $T = \frac{2\pi\beta}{\alpha}$

Thus, its time period of vibration, $T = \frac{2\pi\beta}{\alpha}$

85. As, $x = A\cos\omega t$

$$\therefore v = \frac{dx}{dt} = -A\omega\sin\omega t \quad \dots(i)$$

and $$a = \frac{d^2x}{dt^2} = -A\omega^2\cos\omega t \quad \dots(ii)$$

We can find the correct graph by putting different values of t in Eq. (ii).

At $t = 0$, $a = -A\omega^2$

At $t = \frac{T}{4}$, $a = -A\omega^2\cos\left(\frac{2\pi}{T}\times\frac{T}{4}\right) = 0$

At $t = \frac{T}{2}$, $a = -A\omega^2\cos\left(\frac{2\pi}{T}\times\frac{T}{2}\right)$

$$= -A\omega^2\cos\pi = +A\omega^2$$

At $t = \frac{3T}{4}$, $a = -A\omega^2\cos\left(\frac{2\pi}{T}\times\frac{3T}{4}\right) = 0$

At $t = T$, $a = -A\omega^2\cos\left(\frac{2\pi}{T}\times T\right) = -A\omega^2$

This condition is represented by graph in option (c).

86. For a simple harmonic motion,

$$a \propto \frac{d^2y}{dt^2} \propto -y$$

Hence, equations $y = \sin\omega t - \cos\omega t$ and $y = 5\cos\left(\frac{3\pi}{4} - 3\,\omega t\right)$ are satisfying this condition and equation $y = 1 + \omega t + \omega^2t^2$ is not periodic and $y = \sin^3\omega t$ is periodic but not simple hormonic motion.

87. For a particle executing SHM

Acceleration $(a) \propto -\omega^2$ displacement (x) ...(i)

Given $\quad x = a\sin^2\omega t$ ...(ii)

Differentiating the above equation w.r.t. t, we get

$$\frac{dx}{dt} = 2a\omega(\sin\omega t)(\cos\omega t)$$

Again differentiating, we get

$$\frac{d^2x}{dt^2} = a = 2a\omega^2[\cos^2\omega t - \sin^2\omega t]$$

$$= 2a\omega^2\cos 2\omega t$$

The given equation does not satisfy the condition for SHM [Eq. (i)]. Therefore, motion is not simple harmonic.

88. Time period of spring pendulum,

$$T = 2\pi\sqrt{\frac{M}{k}}.$$

If mass is doubled, then time period

$$T' = 2\pi\sqrt{\frac{2M}{k}} = \sqrt{2}T$$

Chapter 15

WAVE MOTION

LEVEL 01 Objective Questions

1 For the wave shown in figure given below, the frequency and wavelength, if its speed is 320 m/s are

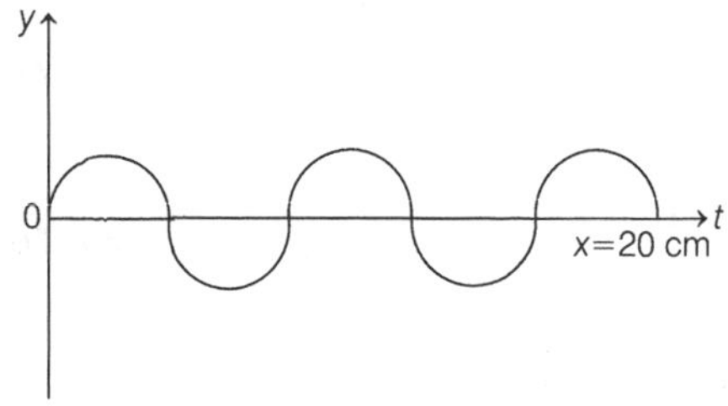

(a) 8 cm, 400 Hz (b) 80 cm, 40 Hz (c) 8 cm, 4000 Hz (d) 40 cm, 8000 Hz

2 A wave pulse is generated in a string that lies along x-axis. At the points A and B as shown in figure, if R_A and R_B are ratios of magnitudes of wave speed to the particle speed at point A and B respectively, then

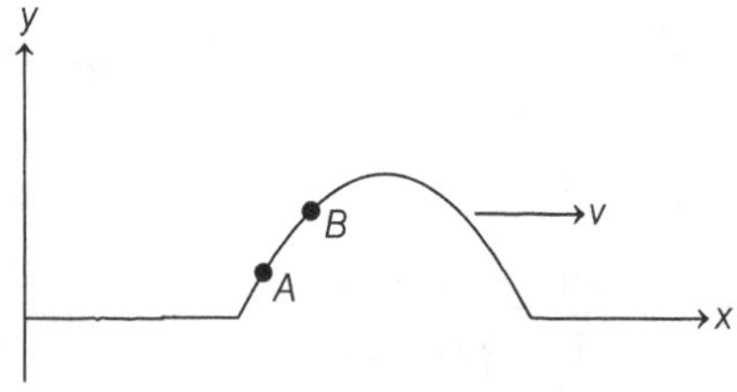

(a) $R_A > R_B$ (b) $R_B > R_A$ (c) $R_B = R_A$ (d) Data Insufficient

3 Figure represents the displacement y *versus* distance x along the direction of propagation of a longitudinal wave. The pressure is maximum at position marked

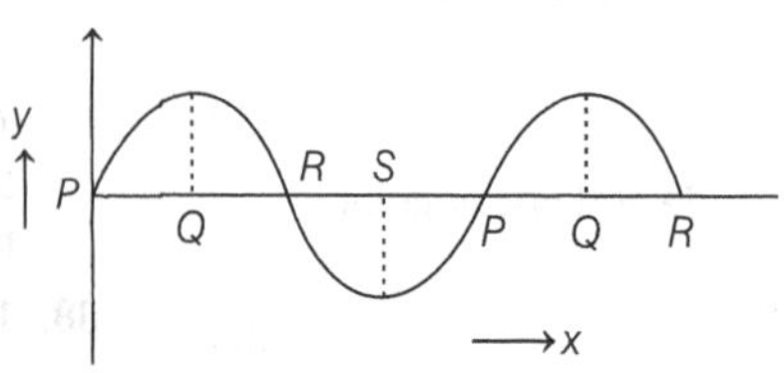

(a) P (b) Q (c) R (d) S

4 Figure given below shows a sinusoidal wave on a string. If the frequency of the wave is 150 Hz, what is the velocity and wavelength of the given wave?

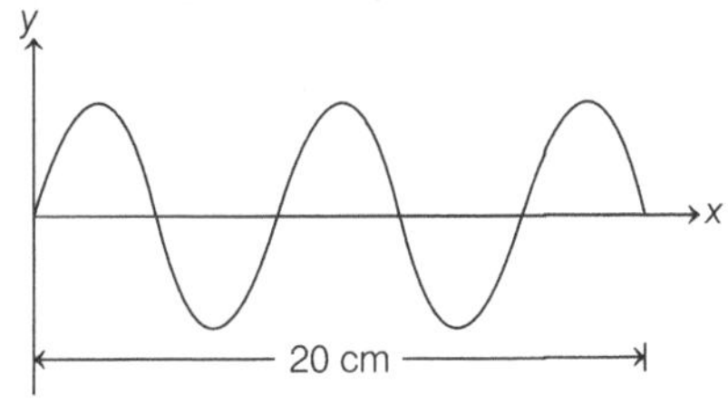

(a) 0.04 m, 10 ms^{-1} (b) 0.06 m, 12 ms^{-1} (c) 0.08 m, 10 ms^{-1} (d) 0.08 m, 12 ms^{-1}

5 Consider the snapshot of a wave travelling in positive x-direction.

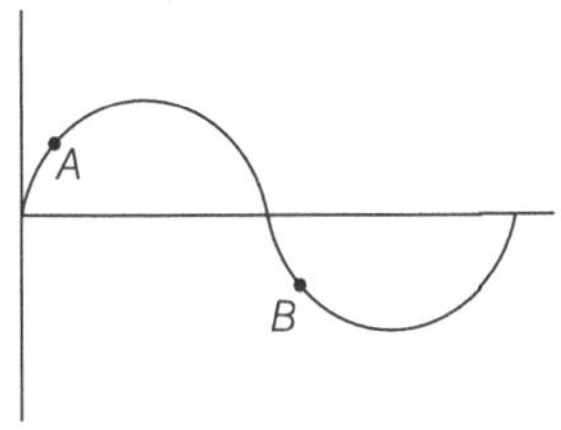

(a) The particle A is moving in $-$ve y-direction and particle B is moving in $+y$-direction
(b) The particle B is moving in $-$ve y-direction and particle A is moving in $+y$-direction
(c) Both are moving in the +ve y-direction
(d) Both are moving in the $-$ve y-direction

6 The speed of a wave in a certain medium is 960 m/s. If 3600 waves pass over a certain point of the medium in 1 min, the wavelength is
(a) 2 m (b) 4 m (c) 8 m (d) 16 m

7 A wave is represented by the equation $y = 10 \sin 2\pi(100t - 0.02x) + 10\sin 2\pi(100t + 0.02x)$. The maximum amplitude and loop length are respectively,
(a) 20 units and 30 units (b) 20 units and 25 units
(c) 30 units and 20 units (d) 25 units and 20 units

8 If $x = a \sin(\omega t + \pi/6)$ and $x' = a \cos \omega t$, then what is the phase difference between the two waves?
(a) $\pi/3$ (b) $\pi/6$ (c) $\pi/2$ (d) π

9 The phase difference between two points separated by 1 m in a wave of frequency 120 Hz is 90°. The wave velocity is
(a) 180 m/s (b) 240 m/s (c) 480 m/s (d) 720 m/s

10 Two graphs of the same harmonic wave are shown below. The graph(1) shows the displacement of wave y, as a function of position x for a given instant of time. The graph (2) shows the displacement of wave as a function of time t for a given position. The speed of the wave is

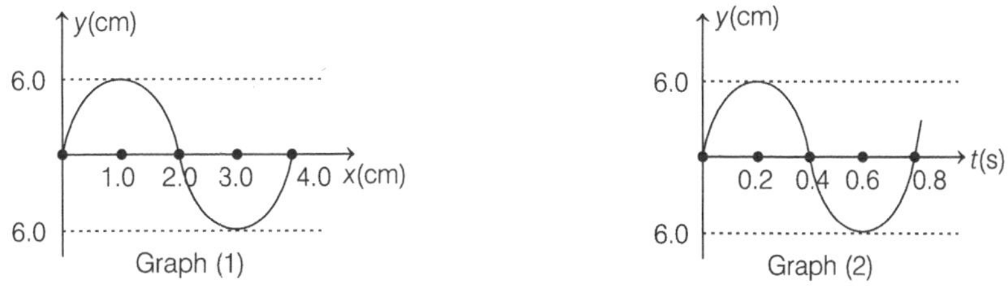

(a) 5.0 cms^{-1} (b) 0.5 cms^{-1} (c) 0.4 cms^{-1} (d) 4.0 cms^{-1}

11 A wave travelling in the +ve x-direction having maximum displacement along y-direction as 1 m, wavelength 2π m and frequency of $\frac{1}{\pi}$ Hz is represented by

(a) $y = \sin(10\pi x - 20\pi t)$
(b) $y = \sin(2\pi x + 2\pi t)$
(c) $y = \sin(x - 2t)$
(d) $y = \sin(2\pi x - 2\pi t)$

12 The vibrations of a string of length 60 cm fixed at both ends are represented by the equation $y = 4\sin(\pi x/15)\cos(96\pi t)$, where x and y are in cm and t in s. The maximum displacement at $x = 5$ cm is

(a) $2\sqrt{3}$ cm
(b) 4 cm
(c) zero
(d) $4\sqrt{2}$ cm

13 A wave travelling in positive x-direction with $a = 0.2$ m has a velocity of 360 m/s. If $\lambda = 60$ m, then correct expression for the wave is

(a) $y = 0.2\sin\left[2\pi\left(6t + \frac{x}{60}\right)\right]$
(b) $y = 0.2\sin\left[\pi\left(6t + \frac{x}{60}\right)\right]$
(c) $y = 0.2\sin\left[2\pi\left(6t - \frac{x}{60}\right)\right]$
(d) $y = 0.2\sin\left[\pi\left(6t - \frac{x}{60}\right)\right]$

14 Differential equation of wave motion is (symbols have their usual meanings)

(a) $\frac{\partial^2 y}{\partial t^2} = v^2\frac{\partial^2 y}{\partial x^2}$
(b) $\frac{\partial^2 y}{\partial x^2} = v^2\frac{\partial^2 y}{\partial t^2}$
(c) $\frac{\partial^2 y}{\partial t^2} = v\frac{\partial^2 y}{\partial x^2}$
(d) $\frac{\partial^2 y}{\partial x^2} = v\frac{\partial^2 y}{\partial t^2}$

15 The velocity of sound is generally greater in solids than in gases because

(a) the density of solid is high, but its elasticity is low
(b) the density of solid is high, but the elasticity of solid is very high
(c) both the density and elasticity of solids are low
(d) the density of solids is low but the elasticity is high

16 A 100 m long rod of density 10.0×10^4 kg/m^3 and having Young's modulus $Y = 10^{11}$ Pa, is clamped at one end. It is hammered at the other free end. The longitudinal pulse goes to right end, gets reflected and again returns to the left end. How much time the pulse take to go back to initial point?

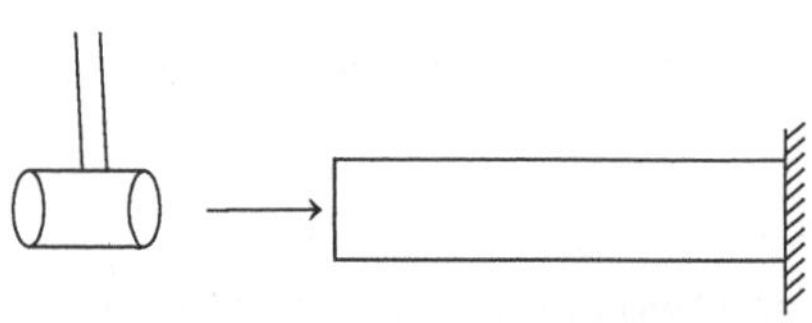

(a) 0.1 s
(b) 0.2 s
(c) 0.3 s
(d) 2 s

17 Four pendulums A, B, C and D are suspended from the same elastic support as shown in figure. A and C are of the same length, while B is smaller than A and D is larger than A. If A is given a transverse displacement,

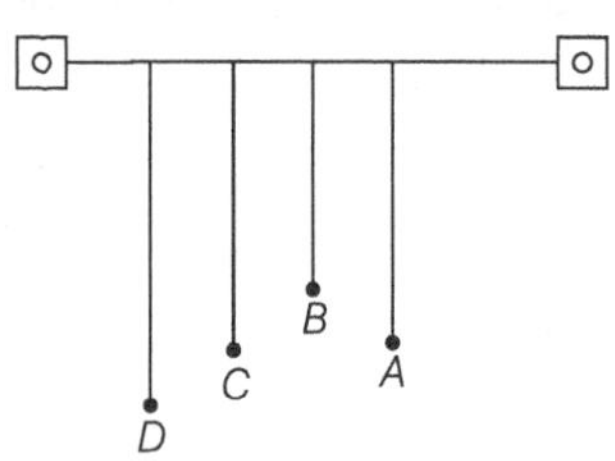

(a) D will vibrate with maximum amplitude
(b) C will vibrate with maximum amplitude
(c) B will vibrate with maximum amplitude
(d) All the four will oscillate with equal amplitude

18 A string fixed at both ends resonates at a certain fundamental frequency. Which of the following adjustments would not affect the fundamental frequency?
(a) Tension is made four times and length is doubled
(b) Tension is doubled and length is halved
(c) Both tension and length are halved
(d) Both length and tension are doubled

19 The stationary wave $y = 2a\sin kx \cos \omega t$ in a stretched string is the result of superposition of $y_1 = a\sin(kx - \omega t)$ and
(a) $y_2 = a\cos(kx + \omega t)$
(b) $y_2 = a\sin(kx + \omega t)$
(c) $y_2 = a\cos(kx - \omega t)$
(d) $y_2 = a\sin(kx - \omega t)$

20 When the string of a sonometer wire of length L, vibrates in the first overtone, the amplitude of vibration is maximum at
(a) $\frac{L}{8}, \frac{5L}{8}, \frac{7L}{8}$
(b) $\frac{L}{2}$
(c) $\frac{L}{4}, \frac{3L}{4}$
(d) $\frac{L}{6}, \frac{3L}{6}, \frac{5L}{6}$

21 Stationary waves are produced in 10 m long stretched string. If the string vibrates in 5 segments and wave velocity 20 m/s, the frequency is
(a) 10 Hz
(b) 5 Hz
(c) 4 Hz
(d) 2 Hz

22 The diagram below shows two pulses travelling towards each other in a uniform medium with same speed. Pulses in the figure are at the same distance from X and has same height width.

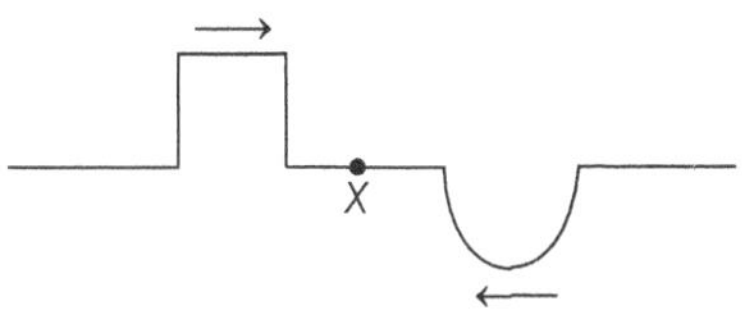

Which diagram best represents the medium, when the pulses meet at point X?

(a) 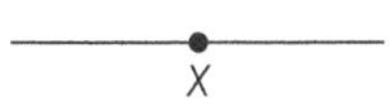

(b) 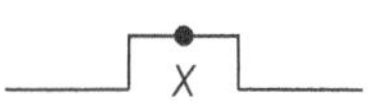

(c) 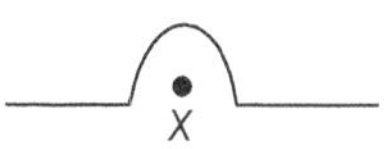

(d) 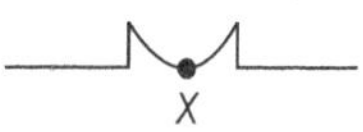

23 A source of sound of frequency 600 Hz is placed inside water. The speed of sound in water is 1500 m/s and in air, it is 300 m/s. The frequency of sound recorded by an observer who is standing in air is
(a) 200 Hz
(b) 3000 Hz
(c) 120 Hz
(d) 600 Hz

24 A string vibrates in 5 segments to a frequency of 480 Hz. The frequency that will cause it to vibrate in 2 segments will be
(a) 96 Hz
(b) 192 Hz
(c) 1200 Hz
(d) 2400 Hz

25 A string is stretched between two fixed points separated by 75 cm. It is observed to have resonant frequencies of 420 Hz and 315 Hz. There are no other resonant frequencies between these two. Then, the lowest resonant frequency for this string is
(a) 10.5 Hz
(b) 105 Hz
(c) 1.05 Hz
(d) 1050 Hz

26 Two pulses of same amplitude are moving in opposite directions along the string according to the figure. The speed of the wave is 2 cm/s. Initially, the pulses are 8 cm apart. What will be the position of pulses after two seconds?

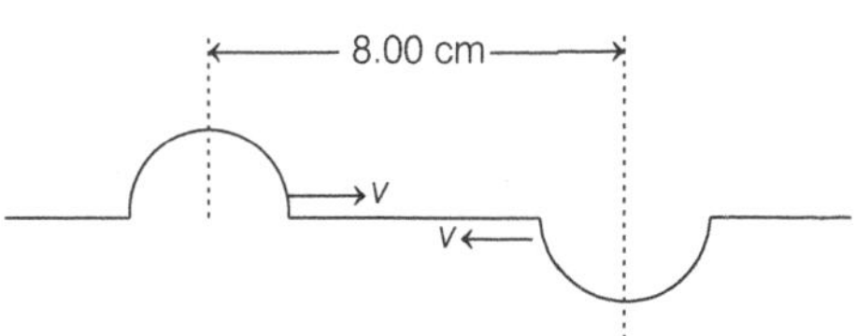

(a) 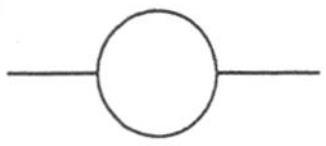

(b) 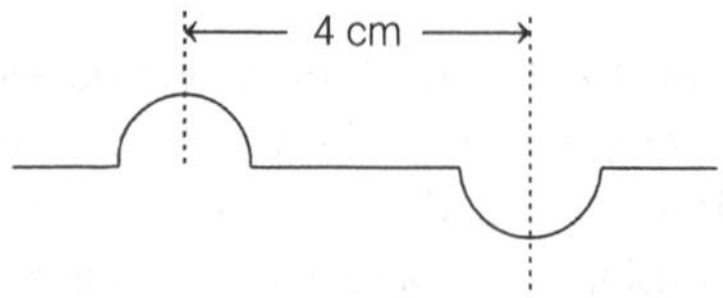

(c)

(d) 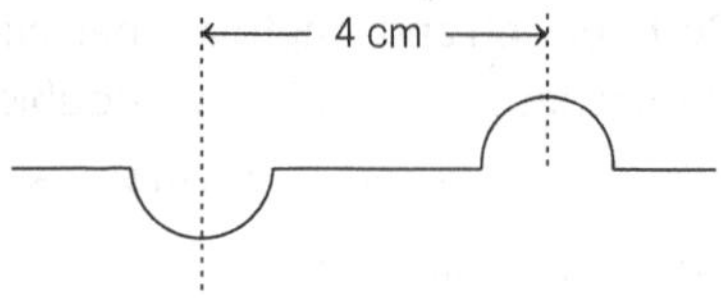

27 A stretched string is vibrating in its 5th harmonic as shown. Consider a particle 1 (figure). At an instant, this particle is at mean position and is moving towards its negative extreme. Which of the following set of particles are in same phase with particle 1?

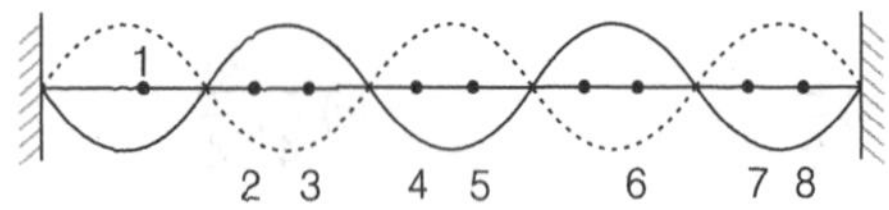

(a) 2, 4, 7 (b) 3, 5, 6 (c) 5, 8 (d) 2, 4, 6

28 At some point, four waves are represented by

(i) $y = a_1 \sin \omega t$ (ii) $y = a_2 \sin(\omega t + \theta)$

(iii) $y = a_1 \sin 2\omega t$ (iv) $y = a_2 \sin(2\omega t + \theta)$

Interference (x) may be observed due to superposition of

(a) (i) and (ii), (iii) and (iv) (b) (i) and (iii), (iii) and (iv)

(c) (ii) and (iv), (iii) and (iv) (d) (iii) and (iv) only

29 In a long cylindrical tube, the water level is adjusted and the air column above it is made to vibrate in unison with a vibrating tuning fork kept at the open end. Maximum sound is heard, when the air column lengths are equal to

(a) $\frac{\lambda}{4}, \frac{\lambda}{2}, \frac{3\lambda}{4}$ (b) $\frac{\lambda}{2}, \lambda, \frac{3\lambda}{2}$ (c) $\frac{\lambda}{2}, \frac{3\lambda}{2}, \frac{5\lambda}{2}$ (d) $\frac{\lambda}{4}, \frac{3\lambda}{4}, \frac{5\lambda}{4}$

30 An organ pipe of length L open at both ends is found to vibrate in its first harmonic when sounded with a tuning fork of 480 Hz. What should be the length of a pipe closed at one end, so that it also vibrates in its first harmonic with the same tuning fork?

(a) $\frac{L}{4}$ (b) $\frac{L}{2}$ (c) L (d) $2L$

31 A pipe closed at one end produces a fundamental note of 412 Hz. It is cut into two pieces of equal length, the fundamental notes produced by the two pieces are

(a) 824 Hz, 1648 Hz (b) 412 Hz, 824 Hz (c) 206 Hz, 412 Hz (d) 216 Hz, 824 Hz

32 A cylindrical tube ($L = 120$ cm) is in resonance with a tuning fork of frequency 220 Hz. If it is being filled by water, then to get resonance, minimum length of water column is (Take, $v_{air} = 330$ m/s)

(a) 37.5 cm (b) 82.5 cm

(c) 45 cm (d) 7.5 cm

33 In a resonance tube experiment, two consecutive resonances are observed when the length of the air columns are 16 cm and 49 cm. If the frequency of the tuning fork used is 500 Hz, the velocity of sound in air is

(a) 310 ms^{-1} (b) 320 ms^{-1} (c) 330 ms^{-1} (d) 340 ms^{-1}

34 An open pipe of length 33 cm resonates to a frequency of 1000 Hz. It is (Take, $v_{sound} = 330\ ms^{-1}$)

(a) fundamental frequency of pipe (b) first harmonic of pipe

(c) second harmonic of pipe (d) fourth harmonic of pipe

35 A pipe of length 20 cm is open at both ends. Which harmonic mode of the pipe may be in resonance with a 1700 Hz source? (Take, speed of sound = 340 ms^{-1})

(a) First harmonic (b) Second harmonic (c) Third harmonic (d) Fourth harmonic

36 The vibrations of four air columns of same length are represented in figure. The ratio of frequencies $n_p : n_q : n_r : n_s$ is

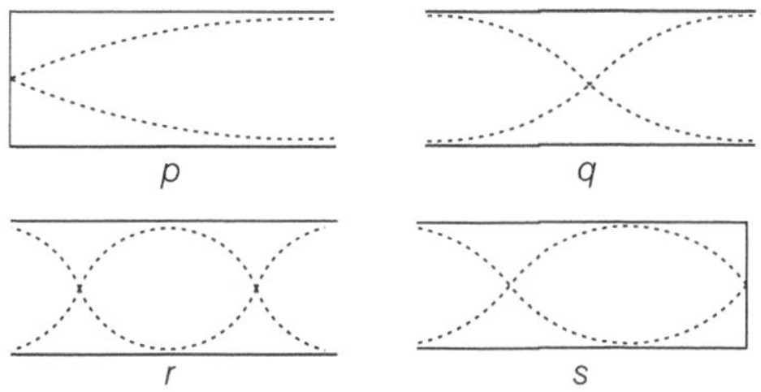

(a) 12 : 6 : 3 : 4 (b) 1 : 2 : 4 : 3 (c) 4 : 2 : 3 : 1 (d) 6 : 2 : 3 : 4

37 Two whistles *A* and *B* produce notes of frequencies 660 Hz and 596 Hz, respectively. There is a listener at the mid-point of the line joining them. Both the whistle *B* and the listener start moving with speed 30 m/s away from whistle *A*. If speed of sound be 330 m/s, how many beats per second will be heard by the listener?

(a) 2 (b) 4 (c) 6 (d) 8

38 If the velocity of sound in air is 320 m/s, then the maximum and minimum length of a pipe closed at one end, that would produce a just audible sound would be

(a) 2.6 m and 3.6 mm (b) 4 m and 4.2 mm
(c) 3 m and 3 mm (d) 4 m and 4 mm

39 When two tuning forks *A* and *B* are sounded together, *x* beats/s are heard. Frequency of *A* is *n*. Now, when one prong of fork *B* is loaded with a little wax, the number of beats/s decreases. The frequency of fork *B* is

(a) $n + x$ (b) $n - x$ (c) $n - x^2$ (d) $n - 2x$

40 A person is observing two trains one coming towards him and other leaving with the same speed 4 m/s. If their whistling frequencies are 240 Hz each, then the number of beats per second heard by the person will be (Take, velocity of sound is 320 m/s)

(a) 3 (b) 6
(c) 9 (d) zero

41 Two tuning forks of frequency 250 Hz and 256 Hz produce beats. If a maximum of intensity is observed just now, after how much time, the minimum is observed at the same place?

(a) $\frac{1}{18}$ s (b) $\frac{1}{4}$ s (c) $\frac{1}{3}$ s (d) $\frac{1}{12}$ s

42 A number of tuning forks are arranged in the order of increasing frequency and any two successive tuning forks produce 4 beats per second, when sounded together. If the last tuning fork has a frequency equal to octave (in octave one frequency is double) of the first tuning fork and the frequency of the first tuning fork is 256 Hz, then the number of tuning forks is

(a) 63 (b) 64
(c) 65 (d) 66

43 If the source is moving towards right, wavefront of sound waves get modified to

(a) 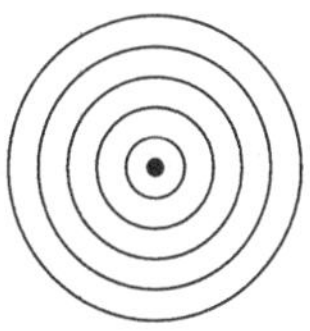 (b) 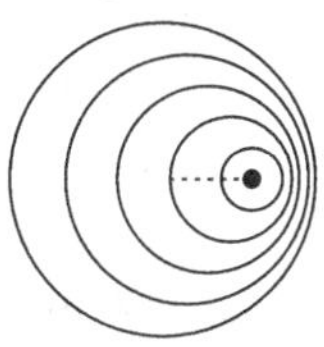 (c) 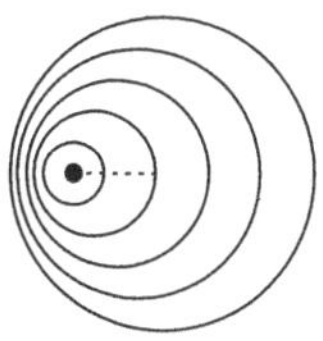 (d) None of these

44 A source is moving towards an observer with a speed of 20 m/s and having frequency of 240 Hz. The observer is also moving towards the source with a velocity of 20 m/s. Apparent frequency heard by the observer, if velocity of sound is 340 m/s, is

(a) 270 Hz (b) 240 Hz (c) 268 Hz (d) 360 Hz

45

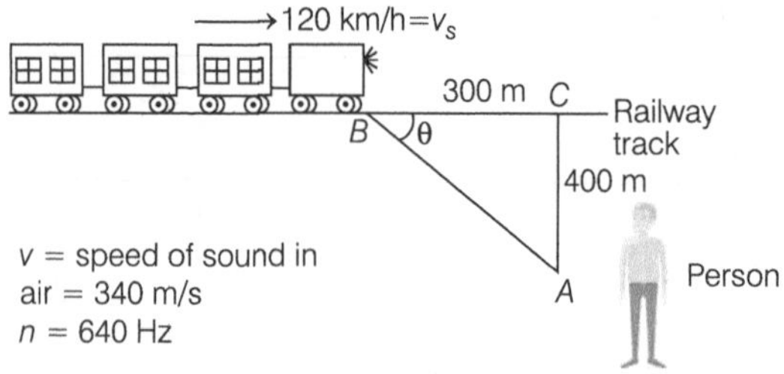

If engine of train produce horn at *B* point, then find apparent frequency observed by observer at *A* point.

(a) 720 Hz (b) 680 Hz (c) 600 Hz (d) 620 Hz

46 Two tuning forks *A* and *B*, produce notes of frequencies 258 Hz and 262 Hz. An unknown note sounded with *A* produces certain beats. When the same note is sounded with *B*, the beat frequency gets doubled. The unknown frequency is

(a) 250 Hz (b) 252 Hz (c) 254 Hz (d) 256 Hz

47 A whistle producing sound waves of frequencies 9500 Hz and above, is approaching a stationary person with a speed of v ms^{-1}. The velocity of sound in air is 300 ms^{-1}. If the person can hear frequencies up to a maximum of 10000 Hz, the maximum value of v up to which he can hear the whistle is

(a) $15\sqrt{2}$ ms^{-1} (b) $15/\sqrt{2}$ ms^{-1}
(c) 15 ms^{-1} (d) 30 ms^{-1}

48 A sound absorber attenuates the sound level by 20 dB. The intensity decreases by a factor of

(a) 100 (b) 1000 (c) 10000 (d) 10

49 Two cats in a house meow at each other with sound intensities 5×10^{-9} Wm^{-2} and 9×10^{-6} Wm^{-2}. By how many decibels (nearly) is the louder sound above the other?

(a) ~ 13 dB (b) ~ 23 dB
(c) ~ 33 dB (d) ~ 43 dB

50 There are 10 sound sources each producing intensity I at a point independently. They are incoherent. Average intensity of sound at that point will be

(a) I (b) $10I$ (c) $100\,I$ (d) 0

LEVEL 02 Objective Questions

51 A man is standing between two parallel cliffs and fires a gun. If he hears first and second echoes after 1.5 s and 3.5 s respectively, the distance between the cliffs is (Take, velocity of sound in air = 340 ms^{-1})

(a) 1190 m (b) 850 m
(c) 595 m (d) 510 m

52 A point sound source emits 200 π W power. The loudness of sound at a distance of 10 m from the source is nearly

(a) 113 dB (b) 117 dB
(c) 120 dB (d) 60 dB

53 A travelling wave represented by $y_1 = A\sin(\omega t - kx)$ is superposed with another travelling wave represented by $y_2 = -A\sin(\omega t + kx)$. The resultant is

(a) a wave travelling in $+x$-direction

(b) a wave travelling in $-x$-direction

(c) a standing wave having nodes at $x = \left(n + \frac{1}{2}\right)\frac{\lambda}{2}; n = 0, 1, 2, \ldots\ldots$

(d) a standing wave having nodes at $x = \frac{n\lambda}{2}; n = 0, 1, 2, \ldots\ldots$

54 The same progressive wave is represented by two graphs I and II. Graph I shows how the displacement y varies with the distance x along the wave at a given time. Graph II shows how y varies with time t at a given point on the wave. The ratio of measurements AB to CD, marked on the curves, represents

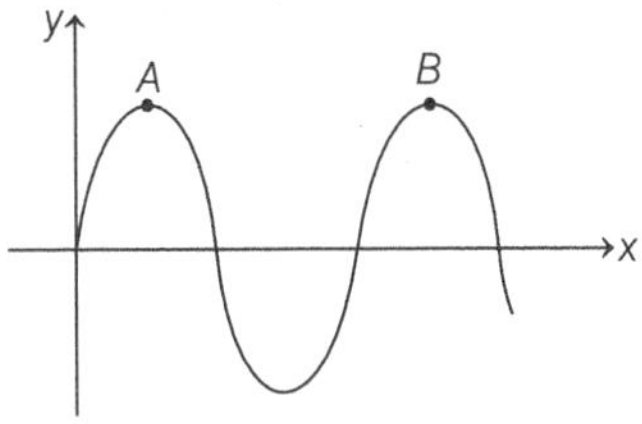

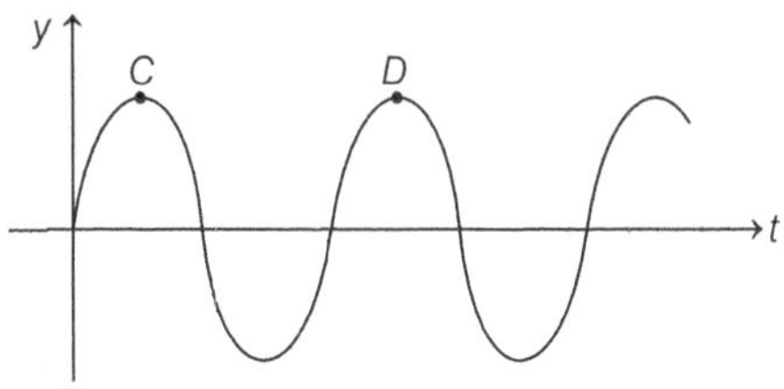

(a) wave number (K) (b) wave speed (v) (c) frequency (ν) (d) angular frequency (ω)

55 A stone is dropped into a quiet lake and waves move in circles spreading out radially at the speed of 5 cm/s. At the instant when the radius of the circular wave is 8 cm, how fast is the enclosed area increasing?

(a) 80π cm^2/s (b) 90π cm^2/s (c) 85π cm^2/s (d) 40π cm^2/s

56 The speed of sound through oxygen at T K is v ms^{-1}. As the temperature becomes $2T$ and oxygen gas dissociates into atomic oxygen, the speed of sound

(a) remains the same (b) becomes $2v$ (c) becomes $\sqrt{2}\,v$ (d) None of these

57 A uniform rope of length L and mass m_1 hangs vertically from a rigid support. A block of mass m_2 is attached to the free end of the rope. A transverse wave pulse of wavelength λ_1 is produced at the lower end of the rope. The wavelength of the pulse, when it reaches the top of the rope is λ_2. The ratio λ_2/λ_1 is

(a) $\sqrt{\frac{m_1}{m_2}}$ (b) $\sqrt{\frac{m_1 + m_2}{m_2}}$ (c) $\sqrt{\frac{m_2}{m_1}}$ (d) $\sqrt{\frac{m_1 + m_2}{m_1}}$

58 A small piece of cork in a ripple tank oscillates up and down as ripples pass through it. If the ripples travelling at 0.3 m/s have a wavelength of 1.5π cm and the cork vibrates with an amplitude of 5 mm, the maximum velocity of the cork is

(a) 20 cm/s (b) 20 m/s (c) 0.02 m/s (d) 200 cm/s

59 Disturbances of two waves in same medium are shown as a function of time in the following figure. The ratio of their intensities will be

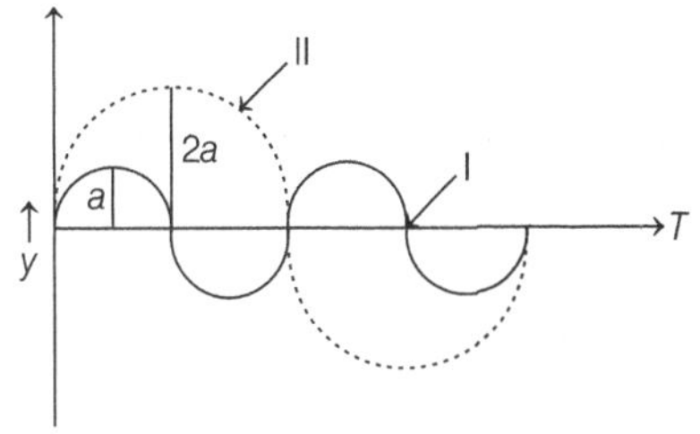

(a) 1 : 4 (b) 4 : 1 (c) 1 : 2 (d) 1 : 1

60 Two identical piano wires, kept under the same tension T have a fundamental frequency of 600 Hz. The fractional increase in the tension of one of the wires which will lead to occurrence of 6 beats when both the wires oscillate together would be

(a) 0.01
(b) 0.02
(c) 0.03
(d) 0.04

61 A sonometer wire has a total length of 2 m between the fixed ends. Two wooden bridges are placed below the wire at a distance of $\frac{2}{7}$ m from one end and $\frac{8}{7}$ m from the other end. The three segments of the wire have their first overtone frequencies in the ratio

(a) $1:\frac{1}{2}:\frac{1}{3}$
(b) 1 : 2 : 3
(c) 1 : 1 : 1
(d) 4 : 2 : 1

62 Three successive frequencies of a stretched wire are 90, 150 and 210 Hz. If the length of the string is 80 cm, the speed of the transverse wave in the string is ....

(a) 96 m/s
(b) 48 m/s
(c) 24 m/s
(d) 32 m/s

63 A wire having a linear mass density 5.0×10^{-3} kg/m is stretched between two rigid supports with tension of 450 N. The wire resonates at frequency 400 Hz. The next higher frequency at which the wire resonates is 500 Hz. The length of the wire will be

(a) 1 m
(b) 4 m
(c) 4.5 m
(d) 1.5 m

64 Two uniform strings A and B made of steel are made to vibrate under the same tension. If the first overtone of A is equal to the second overtone of B and if the radius of A is twice that of B, the ratio of lengths of the strings is

(a) 1 : 2
(b) 1 : 3
(c) 1 : 4
(d) 1 : 6

65 A tunning fork gives 4 beats with 50 cm length of a sonometer wire. If the length of the wire is shortened by 1 cm, the number of beats is still the same. The frequency of the fork is

(a) 404 Hz
(b) 400 Hz
(c) 396 Hz
(d) 384 Hz

66 A tuning fork of frequency 512 Hz makes 4 beats per second with the vibrating string of a piano. The beat frequency decreases to 2 beats per second, when the tension in the piano string is slightly increased. The frequency of the piano string before increasing the tension was

(a) 510 Hz
(b) 514 Hz
(c) 516 Hz
(d) 508 Hz

67 In a resonance-column experiment to measure the velocity of sound, the first resonance is obtained at a length l_1 and the second resonance at a length l_2. Then, if end correction is taken in account;

(a) $l_2 > 3l_1$
(b) $l_2 = 3l_1$
(c) $l_2 < 3l_1$
(d) may be any of the above, depending on the frequency of the tuning fork used

68 A hollow pipe of length 0.8 m is closed at one end. At its open end, a 0.5 m long uniform string is vibrating in its second harmonic and it resonates with the fundamental frequency of the pipe. If the tension in the wire is 50 N and the speed of sound is 320 ms^{-1}, the mass of the string is

(a) 5 g
(b) 10 g
(c) 20 g
(d) 40 g

69 In a resonance tube experiment, the first resonance is obtained for 10 cm of air column and the second resonance for 32 cm. The end correction for this pipe is

(a) 3 cm (b) 2 cm

(c) 1 cm (d) $\frac{1}{2}$ cm

70 A stone is hung in air from a wire which is stretched over a sonometer. The bridges of the sonometer are 40 cm apart, when the wire is in unison with a tuning fork of frequency 256 Hz. When the stone is completely immersed in water, the length between the bridges is 22 cm for re-establishing unison. The specific gravity of the material of the stone is

(a) $\frac{(40)^2}{(40)^2 + (22)^2}$ (b) $\frac{(40)^2}{(40)^2 - (22)^2}$

(c) $256 \times \frac{22}{40}$ (d) $256 \times \frac{40}{22}$

71 A closed organ pipe and an open organ pipe of same length produce 2 beats when they are set into vibration simultaneously in their fundamental mode. The length of the open organ pipe is now halved and of the closed organ pipe is doubled, the number of beats produced will be

(a) 8 (b) 7

(c) 4 (d) 2

72 The number of possible overtones of air column in a closed pipe of length 83.2 cm and of diameter 6 cm, whose frequencies lie below 1000 Hz, will be (Take, velocity of sound in air = 340 m/s)

(a) 5 (b) 4

(c) 10 (d) 7

73 An organ pipe filled with oxygen gas at 47°C resonates in its fundamental mode at frequency 300 Hz. If it is now filled with nitrogen gas, at what temperature will it resonate at the same frequency in the fundamental mode?

(a) 7°C (b) 27°C

(c) 87°C (d) 107°C

74 When a train approaches a stationary observer, the apparent frequency of the whistle is f_1 and when the same train recedes away from the observer, the apparent frequency is f_2. Then, the apparent frequency f, when the observer moves with the train is

(a) $f = \frac{f_1 + f_2}{2}$ (b) $f = \sqrt{f_1 f_2}$ (c) $f = \frac{2f_1 f_2}{f_1 + f_2}$ (d) $f = \frac{2f_1 f_2}{f_1 - f_2}$

75 In the figure, the intensity of waves arriving at D from two coherent sources S_1 and S_2 is I_0 each. The wavelength of the wave is $\lambda = 4$ m. Resultant intensity at D will be

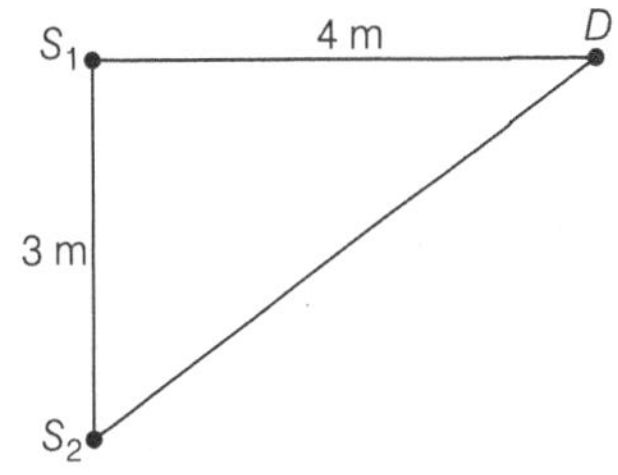

(a) $4I_0$ (b) I_0

(c) $2I_0$ (d) zero

76 A policeman on duty detects a drop of 10% in frequency of horn of a moving car as it crosses him. If the velocity of sound is 330 m/s, then the speed of a car will be

(a) 20 m/s (b) 17.3 m/s

(c) 25 m/s (d) 27 m/s

77 A train moving at a speed of 220 ms^{-1} towards a stationary object, emits a sound of frequency 1000 Hz. Some of the sound reaching the object gets reflected back to the train as echo. The frequency of the echo as detected by the driver of the train is (Take, speed of sound in air is 330 ms^{-1})

(a) 5000 Hz (b) 3000 Hz (c) 3500 Hz (d) 4000 Hz

78 A train is moving with a constant speed $7\sqrt{2}$ m/s along a circular track. The length of the train makes an angle 90° at the centre. A siren in its engine is emitting a sound of frequency 1.7 kHz. The apparent frequency of sound as observed by a passenger at the rear end of the train is (Take, velocity of sound as 333 m/s)

(a) 1.8 kHz (b) more than 1.8 kHz
(c) less than 1.8 kHz (d) data insufficient

79 A person carrying a whistle emitting continuously a note of 272 Hz is running towards a reflecting surface with a speed of 18 kmh^{-1}. The speed of sound in air is 345 ms^{-1}. The number of beats heard by him is

(a) 4 (b) 6
(c) 8 (d) 3

80 Two cars A and B are moving on two crossed roads with speed of 20 m/s and 10 m/s respectively, towards the crossing. Car B blows a horn of frequency 360 Hz, when line joining the cars make angle of 45° with the roads, then frequency as observed by the driver of car A will be (Take, velocity of sound in air = 330 m/s)

(a) 336 Hz (b) 384 Hz
(c) 298 Hz (d) 398 Hz

81 S_1 and S_2 are two coherent sources of sound with no initial phase difference. The velocity of sound is 330 m/s. No minima will be obtained on the line passing through S_2 and perpendicular to the line joining S_1 and S_2, if the frequency of both the sources is

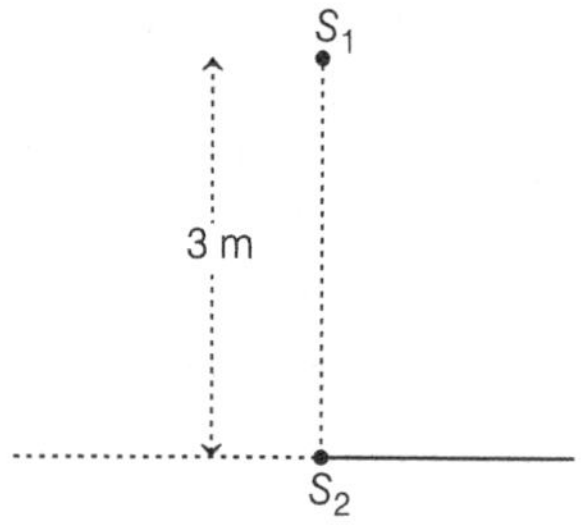

(a) 50 Hz
(b) 60 Hz
(c) 70 Hz
(d) 80 Hz

82 A source of sound placed at the open end of a resonance column sends an acoustic wave of pressure amplitude p_0 inside the tube. If the atmospheric pressure is p_A, then the ratio of maximum and minimum pressure at the closed end of the tube will be

(a) $\frac{(p_A + p_0)}{(p_A - p_0)}$ (b) $\frac{(p_A + 2p_0)}{(p_A - 2p_0)}$ (c) $\frac{p_A}{p_0}$ (d) $\frac{\left(p_A + \frac{1}{2}p_0\right)}{\left(p_A - \frac{1}{2}p_0\right)}$

83 Oxygen is 16 times heavier than hydrogen. Equal volumes of hydrogen and oxygen are mixed. The ratio of the velocity of sound in the mixture to that in oxygen is

(a) $\sqrt{\frac{1}{8}}$ (b) $\sqrt{\frac{32}{17}}$ (c) $\sqrt{\frac{17}{32}}$ (d) $\sqrt{8}$

84 A cylindrical tube, open at one end and closed at the other, is in unison with an external source held at the open end of the tube, in its fundamental note. Then,

(a) the displacement wave from the source gets reflected with a phase change of π at the closed end
(b) the pressure wave from the source gets reflected without a phase change at the closed end
(c) the wave reflected from the closed end again gets reflected at the open end
(d) All of the above

85 A sound source is moving on a circular path of radius R with constant angular speed ω in anti-clockwise direction and emits a frequency n. An observer performs simple harmonic motion along the path QPR with time period $T = \frac{2\pi}{\omega}$ as shown in the figure. If at $t = 0$, source is at A and observer is at Q. Assume that OP is very large as compared to radius R and QP, then

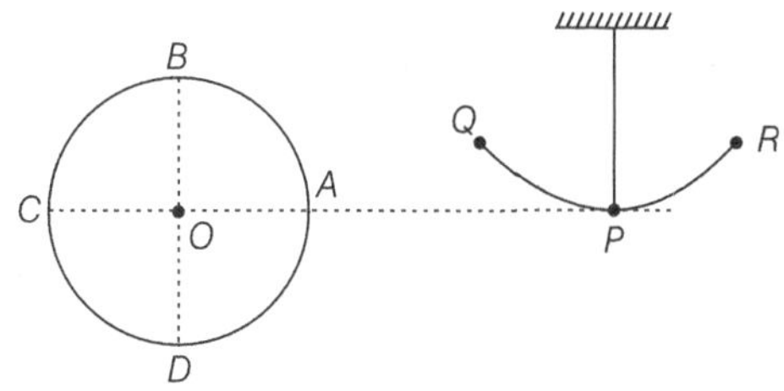

(a) observer will hear maximum frequency at P, if source is at B.
(b) observer will hear maximum frequency at P, if source is at D.
(c) observer will hear maximum frequency at Q, if source is at B
(d) observer will hear maximum frequency at Q, if source is at D

86 There are three sources of sound of equal intensity with frequencies 500 Hz, 502 Hz and 504 Hz, respectively. The number of beats heard per second is
(a) zero (b) two
(c) four (d) eight

87 The intensity of a plane progressive wave having frequency 1000 Hz is 10^{-10} W/m^2. Given that the speed of sound is 330 m/s and density of air is 1.293 kg/m^3, then the maximum change in pressure (in N/m^2) is
(a) 3×10^{-4} (b) 3×10^{-5}
(c) 3×10^{-3} (d) 3×10^{-2}

88 A cord fixed at one end while other end is attached to a vibrating fork oscillates in 8 loops, when tension in the cord is 49 N. The tension at which it will vibrate in 4 loops is
(a) 98 N (b) 124 N
(c) 36 N (d) 196 N

89 A whistle emitting a sound of frequency 440 Hz is tied to a string of 1.5 m length and rotated with an angular velocity of 20 rad/s in the horizontal plane. Then, the range of frequencies heard by an observer stationed at a large distance from the whistle will be (Take, $v = 330$ m/s)
(a) 400.0 Hz to 484.0 Hz (b) 403.3 Hz to 480.0 Hz
(c) 400.0 Hz to 480.0 Hz (d) 403.3 Hz to 484.0 Hz

90 Two stationary sources each emitting waves of wavelength λ. An observer moves from one source to other with velocity u. Then, number of beats heard by him
(a) $\frac{2u}{\lambda}$ (b) $\frac{u}{\lambda}$ (c) $\sqrt{u\lambda}$ (d) $\frac{u}{2\lambda}$

91 A source and an observer are located at the same point. The source starts moving away from the observer at $t = 0$, with a constant acceleration a. If natural frequency of the source is n_0 and speed of sound in air is v, then frequency received by the observer at time t will be
(a) equal to $\left(\frac{n_0 v}{v + at}\right)$ (b) equal to $\left(\frac{n_0 v}{v - at}\right)$
(c) greater than $\left(\frac{n_0 v}{v + at}\right)$ (d) None of these

92 A string of length l is fixed at both ends. It is vibrating in 3rd overtone with maximum amplitude a. The amplitude at a distance $l/3$ from one end is
(a) a (b) 0 (c) $\sqrt{3}a/2$ (d) $a/2$

93 A sonometer wire resonates with a given tuning fork forming standing waves with five antinodes between the two bridges, when a mass of 9 kg is suspended by the wire. When an additional mass m is attached, the wire resonates with the same tuning fork forming three antinodes for the same position of the bridges. The value of m is

(a) 25 kg (b) 20 kg
(c) 16 kg (d) 12 kg

94 Four open organ pipes of different lengths and different gases at same temperature as shown in figure. Let f_A, f_B, f_C and f_D be their fundamental frequencies, then (Take, $\gamma_{CO_2} = 7/5$)

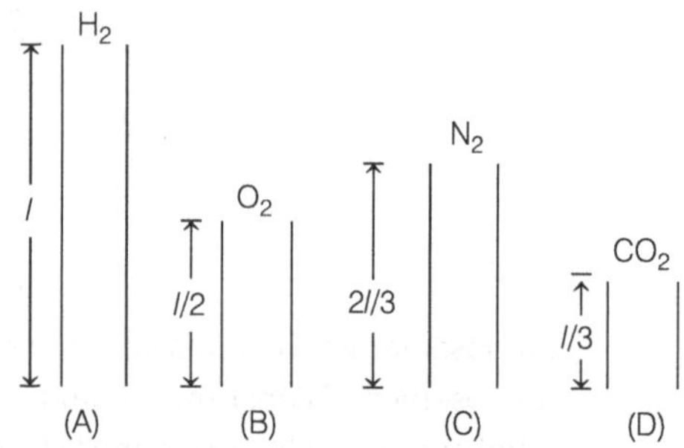

(a) $f_A / f_B = \sqrt{2}$
(b) $f_B / f_C = \sqrt{72/28}$
(c) $f_C / f_D = \sqrt{11/28}$
(d) $f_D / f_A = \sqrt{76/11}$

95 Two wires of same material and radii r and $2r$ respectively, are welded together end-to-end and the combination is used in the sonometer and is kept under tension T. The welded point is midway between the bridges in the sonometer. What should be the ratio of number of loops formed in wires, such that the joint is a node?

(a) 2 : 3 (b) 1 : 4
(c) 1 : 2 (d) 1 : 3

96 The equation of the standing wave in a string clamped at both ends, vibrating in its third harmonic is given by $y = 0.4 \sin (0.314x) \cos (600 \pi t)$, where x and y are in cm and t is in second.

(a) The frequency of vibration is 300 Hz
(b) The length of the string is 30 cm
(c) The nodes are located at $x = 0$, 10 cm, 20 cm and 30 cm
(d) All of the above

97 A person is producing wave in string by moving his hand first up and then down. If frequency is $\frac{1}{8}$ Hz, then find out time taken by particle which is at a distance of 9 m from source to move to lower extreme first time. (Take, $\lambda = 24$ m)

(a) 3 s (b) 5 s
(c) 9 s (d) 8 s

98 A piano string 1.5 m long is made of steel of density 7.7×10^3 kg/m^3 and Young's modulus 2×10^{11} N/m^2. It is maintained at a tension which produces an elastic strain of 1% in the string. The fundamental frequency of transverse vibrations of string is

(a) 85 Hz (b) 170 Hz
(c) 340 Hz (d) 310 Hz

99 A wave $y = 3 \text{ mm} \sin (2\pi x - 200 \pi t)$ is propagating in the left string. It strikes the boundary at $x = 0$ and is reflected and transmitted partially. Choose the **incorrect** statement.

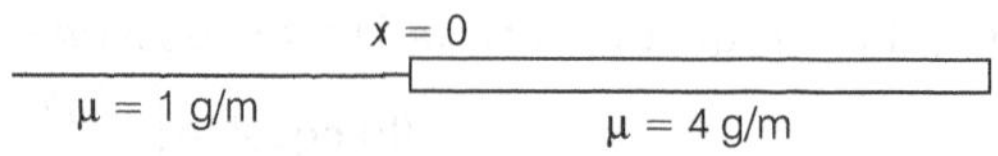

(a) The tension in strings is 20 N.
(b) The amplitude of transmitted wave is 2 mm.
(c) The wave undergoes a phase change of π on reflection.
(d) The wavelength of transmitted wave is $\frac{1}{2}$ m.

100 A massless rod is suspended by two identical strings *AB* and *CD* of equal length. A block of mass *m* is suspended from point *O*, such that *BO* is equal to *x*. Further, it is observed that the frequency of fundamental note in *AB* is equal to 2nd harmonic frequency in *CD*. Then, length of *BO* is

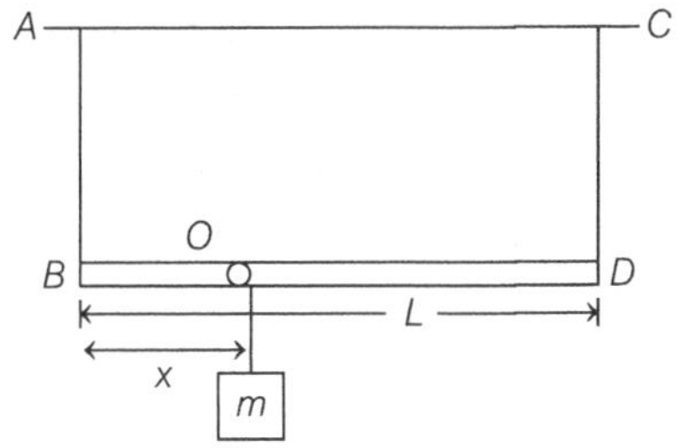

(a) $\frac{L}{5}$ (b) $\frac{4L}{5}$ (c) $\frac{3L}{4}$ (d) $\frac{L}{4}$

101 At $t = 0$, a transverse wave pulse travelling in the positive *x*-direction with a speed of 2 m/s in a wire is described by the function $y = \frac{6}{x^2}$. Given that $x \neq 0$. Transverse velocity of a particle at $x = 2$ m and $t = 2$ s is

(a) 3 m/s (b) – 3 m/s (c) 8 m/s (d) – 8 m/s

102 A source of sound *S* is travelling at $\frac{100}{3}$ m/s along a road, towards a point *A*. When the source is 3 m away from *A*, a person standing at a point *O* on a road perpendicular to *AS* hears a sound of frequency *n′*. The distance of *O* from *A* at that time is 4 m. If the original frequency is 640 Hz, then the value of *n′* is (Take, velocity of sound is 340 m/s)

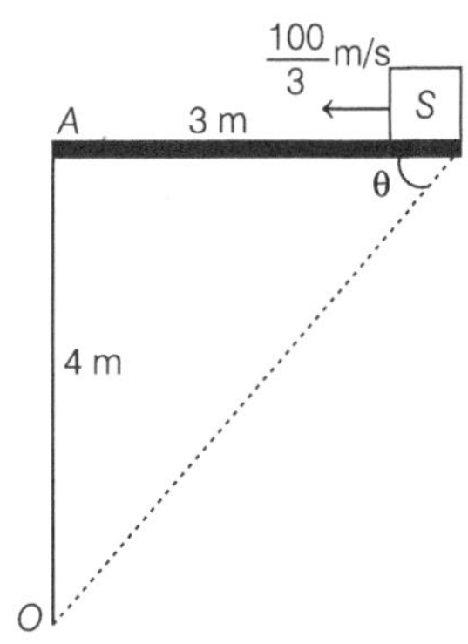

(a) 620 Hz (b) 680 Hz (c) 720 Hz (d) 840 Hz

LEVEL 03 Previous Years Questions

103 The length of the string of a musical instrument is 90 cm and has a fundamental frequency of 120 Hz. Where should it be pressed to produce fundamental frequency of 180 Hz? [NEET 2020]

(a) 75 cm (b) 60 cm
(c) 45 cm (d) 80 cm

104 In a guitar, two strings *A* and *B* made of same material are slightly out of tune and produce beats of frequency 6 Hz. When tension in *B* is slightly decreased, the beat frequency increases to 7 Hz. If the frequency of *A* is 530 Hz, the original frequency of *B* will be [NEET 2020]

(a) 524 Hz (b) 536 Hz (c) 537 Hz (d) 523 Hz

105 A tuning fork with frequency 800 Hz produces resonance in a resonance column tube with upper end open and lower end closed by water surface. Successive resonance are observed at length 9.75 cm, 31.25 cm and 52.75 cm. The speed of sound in air is **[NEET 2019]**

(a) 500 m/s (b) 156 m/s
(c) 344 m/s (d) 172 m/s

106 The fundamental frequency in an open organ pipe is equal to the third harmonic of a closed organ pipe. If the length of the closed organ pipe is 20 cm, the length of the open organ pipe is **[NEET 2018]**

(a) 12.5 cm (b) 8 cm
(c) 13.3 cm (d) 16 cm

107 A tuning fork is used to produce resonance in a glass tube. The length of the air column in this tube can be adjusted by a variable piston. At room temperature of 27°C, two successive resonances are produced at 20 cm and 73 cm of column length. If the frequency of the tuning fork is 320 Hz, the velocity of sound in air at 27°C is **[NEET 2018]**

(a) 350 m/s (b) 339 m/s
(c) 330 m/s (d) 300 m/s

108 The two nearest harmonics of a tube closed at one end and open at other end are 220 Hz and 260 Hz. What is the fundamental frequency of the system? **[NEET 2017]**

(a) 10 Hz (b) 20 Hz
(c) 30 Hz (d) 40 Hz

109 The second overtone of an open organ pipe has the same frequency as the first overtone of a closed pipe L metre long. The length of the open pipe will be **[NEET 2016]**

(a) L (b) $2L$
(c) $L/2$ (d) $4L$

110 Three sound waves of equal amplitudes have frequencies $(n-1), n, (n+1)$. They superimpose to give beats. The number of beats produced per second will be **[NEET 2016]**

(a) 1 (b) 4
(c) 3 (d) 2

111 A uniform rope of length L and mass m_1 hangs vertically from a rigid support. A block of mass m_2 is attached to the free end of the rope. A transverse pulse of wavelength λ_1 is produced at the lower end of the rope. The wavelength of the pulse, when it reaches the top of the rope is λ_2. The ratio λ_2/λ_1 is **[NEET 2016]**

(a) $\sqrt{\dfrac{m_1+m_2}{m_2}}$ (b) $\sqrt{\dfrac{m_2}{m_1}}$
(c) $\sqrt{\dfrac{m_1+m_2}{m_1}}$ (d) $\sqrt{\dfrac{m_1}{m_2}}$

112 The fundamental frequency of a closed organ pipe of length 20 cm is equal to the second overtone of an organ pipe open at both the ends. The length of organ pipe open at both the ends is **[CBSE AIPMT 2015]**

(a) 80 cm (b) 100 cm
(c) 120 cm (d) 140 cm

113 An air column, closed at one end and open at the other, resonates with a tunning fork when the smallest length of the column is 50 cm. The next larger length of the column resonating with the same tunning fork is **[CBSE AIPMT 2015]**

(a) 100 cm (b) 150 cm
(c) 200 cm (d) 66.7 cm

114 A string is stretched between fixed points separated by 75.0 cm. It is observed to have resonant frequencies of 420 Hz and 315 Hz. There are no other resonant frequencies between these two. The lowest resonant frequency for this strings is [CBSE AIPMT 2015]

(a) 155 Hz (b) 205 Hz
(c) 10.5 Hz (d) 105 Hz

115 If n_1, n_2 and n_3 are the fundamental frequencies of three segments into which a string is divided, then the original fundamental frequency n of the string is given by [CBSE AIPMT 2014]

(a) $\frac{1}{n}=\frac{1}{n_1}+\frac{1}{n_2}+\frac{1}{n_3}$ (b) $\frac{1}{\sqrt{n}}=\frac{1}{\sqrt{n_1}}+\frac{1}{\sqrt{n_2}}+\frac{1}{\sqrt{n_3}}$
(c) $\sqrt{n}=\sqrt{n_1}+\sqrt{n_2}+\sqrt{n_3}$ (d) $n=n_1+n_2+n_3$

116 The number of possible natural oscillations of air column in a pipe closed at one end of length 85 cm whose frequencies lie below 1250 Hz are (Take, velocity of sound = 340 ms^{-1}) [CBSE AIPMT 2014]

(a) 4 (b) 5
(c) 7 (d) 6

117 If we study the vibration of a pipe open at both ends, which of the following statements is not true? [NEET 2013]

(a) Open end will be antinode.
(b) Odd harmonics of the fundamental frequency will be generated.
(c) All harmonics of the fundamental frequency will be generated.
(d) Pressure change will be maximum at both ends.

118 A source of unknown frequency gives 4 beat/s when sounded with a source of known frequency 250 Hz. The second harmonic of the source of unknown frequency gives 5 beat/s when sounded with a source of frequency 513 Hz. The unknown frequency is [NEET 2013]

(a) 254 Hz (b) 246 Hz
(c) 240 Hz (d) 260 Hz

119 A wave travelling in the positive x-direction having displacement along y-direction as 1 m, wavelength 2π m and frequency of $\frac{1}{\pi}$ Hz is represented by [NEET 2013]

(a) $y=\sin(x-2t)$ (b) $y=\sin(2\pi x-2\pi t)$
(c) $y=\sin(10\pi x-20\pi t)$ (d) $y=\sin(2\pi x+2\pi t)$

120 When a string is divided into three segments of lengths l_1, l_2 and l_3, the fundamental frequencies of these three segments are ν_1, ν_2 and ν_3 respectively. The original fundamental frequency (ν) of the string is [CBSE AIPMT 2012]

(a) $\sqrt{\nu}=\sqrt{\nu_1}+\sqrt{\nu_2}+\sqrt{\nu_3}$ (b) $\nu=\nu_1+\nu_2+\nu_3$
(c) $\frac{1}{\nu}=\frac{1}{\nu_1}+\frac{1}{\nu_2}+\frac{1}{\nu_3}$ (d) $\frac{1}{\sqrt{\nu}}=\frac{1}{\sqrt{\nu_1}}+\frac{1}{\sqrt{\nu_2}}+\frac{1}{\sqrt{\nu_3}}$

121 Two sources of sound placed closed to each other, are emitting progressive waves given by $y_1=4\sin 600\pi t$ and $y_2=5\sin 608\pi t$
An observer located near these two sources of sound will hear [CBSE AIPMT 2012]

(a) 4 beat/s with intensity ratio 25 : 16 between waxing and waning
(b) 8 beat/s with intensity ratio 25 : 16 between waxing and waning
(c) 8 beat/s with intensity ratio 81 : 1 between waxing and waning
(d) 4 beat/s with intensity ratio 81 : 1 between waxing and waning

122 Two waves are represented by the equations $y_1 = a \sin(\omega t + kx + 0.57)$ m and $y_2 = a \cos(\omega t + kx)$ m, where x is in metre and t in second. The phase difference between them is **[CBSE AIPMT 2011]**

(a) 1.25 rad (b) 1.57 rad
(c) 0.57 rad (d) 1 rad

123 Sound waves travel at 350 m/s through a warm air and at 3500 m/s through brass. The wavelength of a 700 Hz acoustic wave as it enters brass from warm air **[CBSE AIPMT 2011]**

(a) increases by a factor 20 (b) increases by a factor 10
(c) decreases by a factor 20 (d) decreases by a factor 10

124 A transverse wave is represented by $y = A \sin(\omega t - kx)$. For what value of the wavelength is the wave velocity equal to the maximum particle velocity? **[CBSE AIPMT 2010]**

(a) $\pi A/2$ (b) πA
(c) $2\pi A$ (d) A

125 A tuning fork of frequency 512 Hz makes 4 beat/s with the vibrating string of a piano. The beat frequency decreases to 2 beat/s when the tension in the piano string is slightly increased. The frequency of the piano string before increasing the tension was **[CBSE AIPMT 2010]**

(a) 510 Hz (b) 514 Hz
(c) 516 Hz (d) 508 Hz

ANSWER KEY

Level 1 Objective Questions

1 (c)	2 (b)	3 (c)	4 (d)	5 (a)	6 (d)	7 (b)	8 (a)	9 (c)	10 (a)
11 (c)	12 (a)	13 (c)	14 (b)	15 (b)	16 (b)	17 (b)	18 (a)	19 (b)	20 (c)
21 (b)	22 (d)	23 (d)	24 (b)	25 (b)	26 (c)	27 (c)	28 (a)	29 (d)	30 (b)
31 (a)	32 (b)	33 (c)	34 (c)	35 (b)	36 (b)	37 (b)	38 (d)	39 (a)	40 (b)
41 (d)	42 (c)	43 (b)	44 (a)	45 (b)	46 (c)	47 (c)	48 (a)	49 (c)	50 (b)

Level 2 Objective Questions

51 (b)	52 (c)	53 (d)	54 (b)	55 (a)	56 (d)	57 (b)	58 (a)	59 (d)	60 (b)
61 (d)	62 (b)	63 (d)	64 (b)	65 (c)	66 (d)	67 (c)	68 (b)	69 (c)	70 (b)
71 (b)	72 (b)	73 (a)	74 (c)	75 (c)	76 (b)	77 (a)	78 (a)	79 (c)	80 (b)
81 (a)	82 (b)	83 (b)	84 (d)	85 (b)	86 (b)	87 (a)	88 (d)	89 (d)	90 (a)
91 (c)	92 (c)	93 (a)	94 (c)	95 (c)	96 (d)	97 (c)	98 (b)	99 (a)	100 (a)
101 (b)	102 (b)								

Level 3 Previous Years Questions

103 (b)	104 (a)	105 (c)	106 (c)	107 (b)	108 (b)	109 (b)	110 (a)	111 (a)	112 (c)
113 (b)	114 (d)	115 (a)	116 (d)	117 (d)	118 (a)	119 (a)	120 (c)	121 (d)	122 (d)
123 (b)	124 (c)	125 (d)							

Solutions

LEVEL 01 Objective Questions

1. 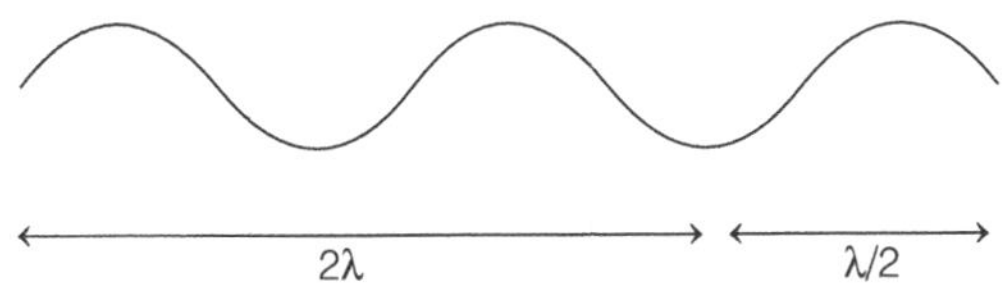

So, $2\lambda + \frac{\lambda}{2} = 20$

$\Rightarrow \lambda = 8$ cm or 0.08 m

$$f = \frac{v}{\lambda} = \frac{320}{0.08} = 4000 \text{ Hz}$$

2. $v_p = -v\left(\frac{\partial y}{\partial x}\right)$

$$\therefore \left|-\frac{v}{v_p}\right| = \frac{1}{\left|\frac{\partial y}{\partial x}\right|} = \frac{1}{|\text{Magnitude of slope}|}$$

Slope at B is less. Therefore, magnitude of ratio at B will be more.

3. $p = -B\frac{dy}{dx}$

At R, $\frac{dy}{dx}$ is most negative. Hence, the pressure is maximum.

4. From the figure, $\frac{5\lambda}{2} = 20$ cm

Wavelength of wave,

$$\lambda = \frac{40}{5} = 8 \text{ cm} = 0.08 \text{ m}$$

Velocity of the wave,

$$v = f\lambda = 150 \times 0.08 = 12 \text{ m/s}$$

5. $\frac{v_{\text{particle}}}{v_{\text{wave}}} = -$ Slope

At A, slope is positive, so v_p is negative.

At B, slope is negative, so v_p is positive.

6. Frequency, $f = \frac{3600}{60} = 60$ Hz

$$\therefore \lambda = \frac{v}{f} = \frac{960}{60} = 16 \text{ m}$$

7. Maximum amplitude $= 2A = 2(10) = 20$ units

$$k = \frac{2\pi}{\lambda} = 0.02$$

or $\lambda = 50$ units

Therefore, loop length $= \frac{\lambda}{2} = 25$ units

8. $x = a\sin\left(\omega t + \frac{\pi}{6}\right)$

$$y = a\sin\left(\omega t + \frac{\pi}{2}\right)$$

$$\phi_1 = \frac{\pi}{6} \text{ and } \phi_2 = \frac{\pi}{2}$$

$$\Delta\phi = \phi_2 - \phi_1 = \frac{\pi}{3}$$

9. $\Delta\phi = \frac{2\pi}{\lambda}\Delta x$

$$\frac{\pi}{2} = \frac{2\pi}{\lambda}(1) \Rightarrow \lambda = 4 \text{ m}$$

So, velocity, $v = f\lambda = 4 \times 120 = 480$ m/s

10. From graph 1,

$$\lambda = 4 \text{ cm}$$

From graph 2,

$$T = 0.8 \text{ s}$$

So, speed of the wave,

$$v = \frac{\lambda}{T} = \frac{4}{0.8} = 5.0 \text{ cms}^{-1}$$

11. $y = a\sin(kx - \omega t)$

$$y = (1)\sin\left(\frac{2\pi}{2\pi}x - 2\pi \times \frac{1}{\pi}t\right) = \sin(x - 2t)$$

12. Amplitude, $A = 4\sin\left(\frac{\pi x}{15}\right)$

At $x = 5$ cm,

$$A = 4\sin\left(\frac{\pi \times 5}{15}\right) = 2\sqrt{3} \text{ cm}$$

13. General equation of a travelling wave is given as,

$$y = a\sin(\omega t + kx)$$

As the wave is travelling along the positive x-axis, $-$ sign must be used.

$$y = a\sin(\omega t - kx)$$

Here, $\omega = \frac{2\pi v}{\lambda}$ and $k = \frac{2\pi}{\lambda}$

$$y = a\sin\frac{2\pi}{\lambda}(vt - x) \quad \text{...(i)}$$

Now, we have given, $a = 0.2$ m,

$v = 360$ m/s, $\lambda = 60$ m

On substituting them in the first equation, we get

$$y = 0.2\sin\left[2\pi\left(6t - \frac{x}{60}\right)\right]$$

14. This is a standard result. So, no solution is required.

15. $v = \sqrt{\frac{E}{\rho}}$ ρ is large but E is very high.

16. $v_s = \sqrt{\frac{Y}{\rho}} = \sqrt{\frac{10^{11}}{10 \times 10^4}} = 10^3$ m/s

$$\therefore \quad t = \frac{2l}{v} = \frac{2 \times 100}{10^3} = 0.2 \text{ s}$$

17. Time period for a simple pendulum,

$$T = 2\pi\sqrt{\frac{l}{g}}$$

So, time period of A and C must be same.

Since, the resonance occurs, pendulum C will vibrate with the maximum amplitude.

18. Fundamental frequency f is given as,

$$f = \frac{1}{2l}\sqrt{\frac{T}{\mu}}$$

If $T \to 4T$ and $l \to 2l$

Fundamental frequency will remain same.

19. $y = 2a \sin kx \cos \omega t$

On simplifying this, we get

$$y = a \sin(\omega t + kx) - a \sin(\omega t - kx)$$

$$\therefore \quad y_1 = a \sin(kx - \omega t)$$

$$y_2 = a \sin(kx + \omega t)$$

20.

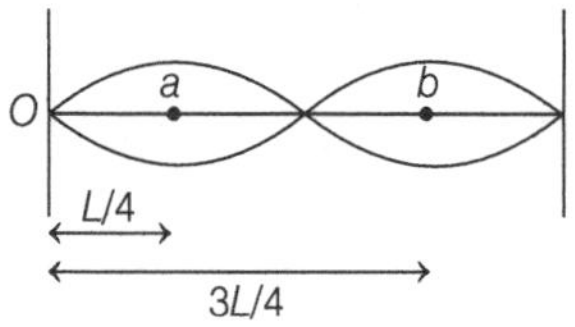

At a and b, amplitude of oscillations will be maximum.

21. Since, there are 5 loops in 10 m.

So, length of one loop = 2 m

Length of two loops = $\lambda = 2 \times 2 = 4$ m

Now, $v = f\lambda$

$$20 = f \times 4$$

$$f = 5 \text{ Hz}$$

22. $y = y_1 + y_2$

Here, y_1 is positive and constant for the wave pulse travelling towards right.

y_2 is negative and first increases, then decreases as we move from left to right on the wave pulse.

23. Frequency made under the water will be same, because the frequency do not depend on the propagation medium.

Hence, frequency of sound heard by an observer standing in air will be 600 ms^{-1}.

24. For the 5 segments, $\lambda = \frac{2l}{5}$

So, $v = \frac{f}{\lambda} \Rightarrow \lambda = \frac{f}{v} = \frac{5v}{2l} = 480$ Hz

For the 2 segments, $\lambda = l$

$$v = \frac{f}{\lambda} \Rightarrow \lambda = f/v = \frac{2}{5}\frac{5v}{2l}$$

$$= \frac{2}{5} \times 480 = 192 \text{ Hz}$$

25. $\frac{nv}{2l} = 315$ ...(i)

$$\frac{(n+1)v}{2l} = 420 \quad \text{...(ii)}$$

On solving Eqs. (i) and (ii), we get

$$\frac{v}{2l} = 105 \text{ Hz}$$

26. Both pulse covers 4 cm in the 2 s. So, after the 2 s, both the pulse overlaps completely. Hence, the resulting amplitude will be zero.

27. Particles 5 and 8 are in same phase with particle 1.

28. For the superposition of the two waves, its frequency must be same.

Since it can be seen that frequency in Eqs. (i) and (ii) are same and those given in Eqs. (iii) and (iv) are also same.

29. $l = \frac{(2n-1)\lambda}{4}$, $n = 1, 2, 3, \ldots$

Substituting $n = 1, 2, 3$, we get

Length as, $\frac{\lambda}{4}, \frac{3\lambda}{4}, \frac{5\lambda}{4}$

30. $\frac{v}{2L} = 480$ Hz ...(i)

$$\frac{v}{4L'} = 480 \text{ Hz} \quad \text{...(ii)}$$

From equations (i) and (ii), we get

$$L' = \frac{L}{2}$$

31. $f = \frac{v}{4l} = 412$ Hz

Now, when cut into the two pieces of equal length.

So, frequency of closed end $= \frac{v}{4\left(\frac{l}{2}\right)} = 2\left(\frac{v}{4l}\right)$

$$= 2 \times 412 = 824 \text{ Hz}$$

Frequency of open pipe $= \frac{v}{2\left(\frac{l}{2}\right)} = 4\frac{v}{(4l)} = 4 \times 412$

$$= 1648 \text{ Hz}$$

32. $\lambda = \frac{v}{f} = \frac{3}{2}$ m or 1.5 m

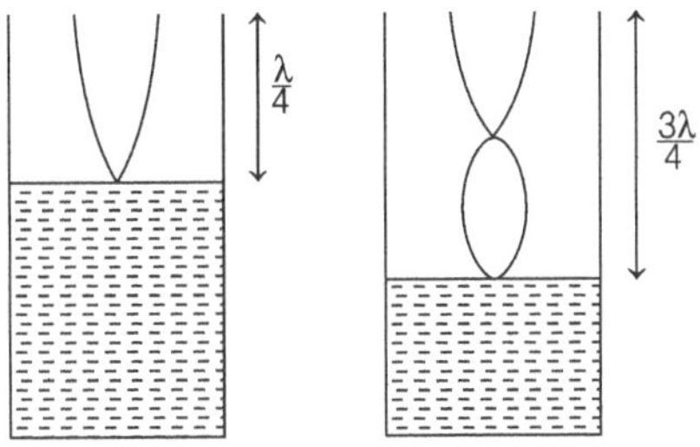

Lengths of air column are $\frac{\lambda}{4}, \frac{3\lambda}{4}, \frac{5\lambda}{4}$... and so.

or, 0.375 m, 1.125 m, 1.875 m and so on.

Since, length of tube is 1.2 m. Hence, maximum length of air column may be 0.375 m.

$\therefore$ Minimum length of water column is (1.2 – 0.375) m = 0.825 m = 82.5cm.

33.

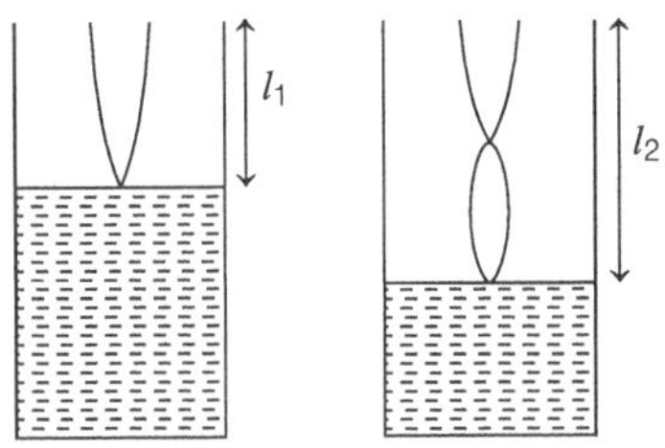

$$l_2 = \frac{3\lambda}{4} \quad \text{and} \quad l_1 = \frac{\lambda}{4}$$

$$l_2 - l_1 = \frac{\lambda}{2}$$

$$\therefore \quad \lambda = 2(l_2 - l_1)$$

$$v = f\lambda = 2f(l_2 - l_1)$$

$$= 2 \times 500\,(0.49 - 0.16)$$

$$= 330 \text{ m/s}$$

34. Frequency of the open pipe,

$$f = \frac{v}{2l} = \frac{330}{2 \times 33 \times 10^{-2}} = 500 \text{ Hz}$$

Since, the frequency (1000 Hz) is double of the fundamental frequency. Hence, the second harmonic will take place.

35. $1700 = \frac{nv}{2l} \Rightarrow 1700 = \frac{n \times 340}{2 \times \frac{1}{5}}$

$$n = 2 \text{ (second harmonic)}$$

36. $n_p = \frac{v}{4l}, n_q = \frac{v}{2l}, n_r = \frac{2 \cdot v}{2l}, n_s = \frac{3 \cdot v}{4l}$

$$n_p : n_q : n_r : n_s = 1 : 2 : 4 : 3$$

37. Number of the beats will be produced by the listener

$$= 600\left(\frac{330 - 30}{330}\right) \approx 600 \text{ Hz}$$

So, number of beats = 600 – 596 = 4

38. For the closed organ pipe,

$$f = \frac{v}{4l} \times (2n - 1)$$

For the minimum and the maximum length of pipe, fundamental frequency must be 20 kHz and 20 Hz.

$$20 = \frac{320}{4l_{max}}$$

$$\Rightarrow \quad l_{max} = 4 \text{ m}$$

$$20 \times 10^3 = \frac{320}{4l_{min}} \quad \Rightarrow \quad l_{min} = 4 \text{ mm}$$

39. $f_A - f_B = x$

So, f_B can be $x + n$ or $x - n$.

Now, when the B is loaded with the wax, its frequency decreases. So, its frequency is $(n + x)$.

40. Frequency observed by first train,

$$f_1 = \frac{v}{v - v_s} \times f = \frac{320}{320 - 4} \times 240$$

$$= 243 \text{ Hz}$$

Frequency observed by second train,

$$f_2 = \frac{v}{v + v_s} \times f = \frac{320}{320 + 4} \times 240$$

$$= 237 \text{ Hz}$$

$\therefore$ Number of beats = 243 – 237 = 6

41. Beat frequency, $f = 256 - 250 = 6$ Hz

Therefore, time taken from one maxima to minima,

$$T = \frac{1}{2f} = \frac{1}{2(6)} = \frac{1}{12} \text{ s}$$

42. Number of beats, $\Delta f = 4$

It may be $f_2 - f_1 = 4$

$\Rightarrow \quad f_2 = 4 + f_1$

Now, $f_n' = f_1 + (n - 1)\,4$

Here, $f_n' = 2f_1$

$$2f_1 = f_1 + (n - 1)\,4$$

$$2 \times 256 = 256 + (n - 1)\,4$$

$$\therefore \quad n = 65$$

43. Towards right wavelength gets compressed and towards left wavelength gets expanded.

44. $f' = \left(\frac{v + v_o}{v - v_s}\right) f = \left(\frac{340 + 20}{340 - 20}\right) \times 240$

$$= 270 \text{ Hz}$$

45. $n' = \frac{nv}{v - v_s \cos\theta}$

$$= \frac{(640)(340)}{340 - 120 \times \frac{5}{18} \times \frac{3}{5}}$$

$$= 680 \text{ Hz}$$

46. Let f be the frequency of the unknown tuning fork.

$$f_a - f = f_b \quad ...(i)$$
$$f_B - f = 2f_b \quad ...(ii)$$

From Eqs. (i) and (ii), we get

$$f_B - f_a = f_b$$
$$262 - 258 = f_b$$
$$\Rightarrow \quad f_b = 4 \text{ Hz}$$

From Eq. (i), we get

$$258 - 4 = f$$
$$\Rightarrow \quad f = 254 \text{ Hz}$$

47. From Doppler effect,

$$f_{app} = \frac{v_{air} \times f}{v_{air} \pm v_{source}}$$
$$10000 = \frac{300}{300 - v_{source}} \times 9500$$
$$\Rightarrow \quad v_{source} = 15 \text{ ms}^{-1}$$

48. $\Delta I = 10 \log_{10} \frac{I_1}{I_2}$

$$20 \text{ dB} = 10 \log_{10} \frac{I_1}{I_2}$$
$$\therefore \quad \frac{I_1}{I_2} = 100$$

49. $L_2 - L_1 = 10 \log_{10} \frac{I_2}{I_1}$

$$= 10 \log_{10} \left[\frac{9 \times 10^{-6}}{5 \times 10^{-9}}\right] = 10 \log_{10} (1800)$$
$$= 10 \log_{10} [2 \times 3^2 \times 10^2]$$
$$= 10 [\log_{10} 2 + 2\log_{10} 3 + 2\log_{10} 10]$$
$$= 10 [0.3 + 2 \times 0.477 + 2 \times 1]$$
$$= 32.5 \text{ dB}$$

50. Average intensity of the sound, $I_{avg} = I_1 + I_2 + \dots + I_{10}$

Hence, $I_{avg} = 10\, I$

LEVEL 02 Objective Questions

51. Let d_1 and d_2 be the distance from the man and the cliffs.

$$\therefore \quad 2d_1 + 2d_2 = v \times t_1 + v \times t_2$$
$$\text{So,} \quad d_1 + d_2 = \frac{v(t_1 + t_2)}{2}$$
$$= \frac{340 \times (1.5 + 3.5)}{2}$$
$$= 850 \text{ m}$$

52. Intensity of sound,

$$I = \frac{P}{A} = \frac{200\pi}{2\pi \times (10)^2} = 1 \text{ W/m}^2$$
$$I = 10 \log_{10} \frac{I}{I_0}$$
$$I = 10 \log_{10} \frac{1}{10^{-12}}$$
$$I = 120 \text{ dB}$$

53. $y = y_1 + y_2$

$$= A [\sin (\omega t - kx) - \sin (\omega t + kx)]$$
$$= -2A \cos \omega t \sin kx$$
$$= -A_x \cos \omega t$$

Here, $A_x = 2A \sin kx$

At node, $A_x = 0$

$$\therefore \quad kx = n\pi, \text{ where } n = 0, 1, 2, \dots$$
$$\text{or} \quad \frac{2\pi}{\lambda} \cdot x = n\pi$$
$$\text{or} \quad x = \frac{n\lambda}{2}$$

54. Graph 1 represents the wavelength and the graph 2 represents the time period.

$$\frac{\lambda}{T} = \text{wave speed } (v)$$

55. Area of a circle with radius r is given by, $A = \pi r^2$

$$\frac{dA}{dt} = \frac{d(\pi r^2)}{dt}$$
$$\frac{dA}{dt} = 2\pi r \frac{dr}{dt} = 2\pi(8)(5) \quad \left[\because \frac{dr}{dt} = 5\right]$$
$$\frac{dA}{dt} = 80\pi \text{ cm}^2/\text{s}$$

56. $v = \sqrt{\frac{\gamma RT}{M}}$ or $v \propto \sqrt{\frac{\gamma T}{M}}$

$$\because \quad \frac{v'}{v} = \left[\frac{\sqrt{\frac{5}{3}(2T)}}{\sqrt{\frac{7}{5}(T)}}\right] \left[\sqrt{\frac{32}{16}}\right] = \sqrt{\frac{100}{21}}$$
$$\therefore \quad v' = \left[\sqrt{\frac{100}{21}}\right] v$$

57. $\lambda \propto$ velocity, v

$$\frac{\lambda_2}{\lambda_1} = \sqrt{\frac{T_2}{T_1}} = \sqrt{\frac{(m_1 + m_2)g}{m_2 g}}$$
$$= \sqrt{\frac{m_1 + m_2}{m_2}}$$

58. Frequency of the ripple,

$$f = \frac{v}{\lambda} = \frac{30}{1.5\pi} = \frac{20}{\pi} \text{ Hz}$$

Now, the maximum velocity of cork,

$$v_{max} = \omega A$$
$$v_{max} = 2\pi f A$$
$$= 2\pi \left(\frac{20}{\pi}\right) \times 0.5 = 20 \text{ cm/s}$$

59. $I = \frac{1}{2}\rho\omega^2 A^2 v = 2\pi^2 \rho v \cdot \left(\frac{A}{T}\right)^2 \qquad \left\{\because \omega = \frac{2\pi}{T}\right\}$

$$\Rightarrow \quad I \propto \frac{A^2}{T^2} \Rightarrow \frac{I_1}{I_2} = \left(\frac{A_1}{A_2}\right)^2 \cdot \left(\frac{T_2}{T_1}\right)^2$$

$$= \left(\frac{a}{2a}\right)^2 \cdot \left(\frac{2T}{T}\right)^2 = 1$$

60. Frequency, $f = \frac{v}{2L}$ or $v = \frac{1}{2l}\sqrt{\frac{T}{\mu}}$

So, the fractional change in the frequency,

$$\frac{\Delta v}{v} = \frac{1}{2}\frac{\Delta T}{T}$$

$$\Rightarrow \quad \frac{6}{600} = \frac{1}{2}\frac{\Delta T}{T}$$

$$\therefore \quad \frac{\Delta T}{T} = 0.02$$

61. Three segments of the wire have the length ratio as 1 : 2 : 4.

Fundamental frequency, $f \propto \frac{1}{l}$

$\therefore$ Overtone frequencies ratio

$$= 4 : 2 : 1$$

62. The ratio of given frequencies is 3, 5, 7, etc.

So, 90 Hz is actually 3rd harmonic.

$$\therefore \quad 90 = 3\left(\frac{v}{2l}\right) \quad \text{or} \quad v = 60\,l$$

$$= 60 \times 0.8 = 48 \text{ m/s}$$

63. If the wire vibrates at 400 Hz in the nth harmonic and at the 500 Hz in its $(n+1)$ th harmonic, then

$$400 = \frac{n}{2l}\sqrt{\frac{T}{\mu}} \qquad \text{...(i)}$$

$$500 = \frac{n+1}{2l}\sqrt{\frac{T}{\mu}} \qquad \text{...(ii)}$$

So, we get $n = 4$

Putting $n = 4$ in the equation (i), we get

$$400 = \frac{4}{2l}\sqrt{\frac{450}{5 \times 10^{-3}}}$$

We get, $L = 1.5$ m

64. $f = \frac{1}{2l}\sqrt{\frac{T}{\mu}} = \frac{1}{2l}\sqrt{\frac{T}{\rho(A)}} = \frac{1}{2l}\sqrt{\frac{T}{\rho\pi r^2}}$

So, $f \propto \frac{1}{l} \propto \frac{1}{\sqrt{r^2}}$

First overtone is $2f$ and second overtone is $3f$,

$$3f_B = 2f_A$$

$$3 \times \frac{1}{l_B} \times \frac{1}{\sqrt{(r_B)^2}} = 2 \times \frac{1}{l_A} \times \frac{1}{\sqrt{(r_A)^2}}$$

$$\Rightarrow \quad \frac{l_A}{l_B} = \frac{2}{3}\left(\frac{r_B}{r_A}\right) = \frac{2}{3} \times \frac{1}{2} = \frac{1}{3} \qquad \{\because r_A = 2r_B\}$$

$$\Rightarrow \quad l_A : l_B = 1 : 3$$

65. $(f - 4)\,50 = (f + 4)\,49$

$$\Rightarrow \quad f = 396 \text{ Hz}$$

66. Let f_1 be the frequency of piano and f_2 be the frequency of tuning fork.

So, $f_1 = f_2 - x$

$$\Rightarrow \quad f_1 = 512 - 4$$

$$= 508 \text{ Hz}$$

67. $l_1 = \frac{\lambda}{4} + \delta$

$$l_2 = \frac{3\lambda}{4} + \delta$$

$$l_2 = (l_1 - \delta) + \delta = 3l_1 - 2\delta$$

So, $\quad l_2 < 3l_1$

68. $\frac{c}{4L} = \frac{2v}{L} = \frac{1}{5}\sqrt{\frac{T}{\mu}}$

$$\frac{320}{4 \times 0.8} = \frac{1}{5}\sqrt{\frac{50}{\mu}}$$

$$\Rightarrow \quad \mu = 0.02 \text{ kg m}^{-1}$$

$$\therefore \quad m = \mu L = 0.02 \times 0.5$$

$$= 10 \text{ g}$$

69. $l_1 + e = \frac{\lambda}{4}$ and $l_2 + e = \frac{3\lambda}{4}$

$$3l_1 + 3e = l_2 + e$$

or $\quad 2e = l_2 - 3l_1$

$$e = \frac{1}{2}(l_2 - 3l_1) = \frac{1}{2}[32 - 3(10)]$$

$$= 1.0 \text{ cm}$$

70. $f \propto \frac{v}{l} \propto \frac{\sqrt{T}}{l}$ f is same. Therefore,

$$\frac{\sqrt{W}}{40} = \frac{\sqrt{W - U}}{22}$$

$$\Rightarrow \quad \frac{\sqrt{d}}{40} = \frac{\sqrt{d-1}}{22}$$

$$\Rightarrow d = \text{specific gravity} = \frac{(40)^2}{(40)^2 - (22)^2}$$

71. Fundamental frequency of open pipe,

$$f_o = \frac{v}{2l}$$

Fundamental frequency of closed pipe,

$$f_c = \frac{v}{4l}$$

We have, $f_o - f_c = 2$

$$\frac{v}{2l} - \frac{v}{4l} = 2 \quad \text{or} \quad \frac{v}{l} = 8 \qquad \text{...(i)}$$

Now, when the length of open organ pipe is halved and that of closed pipe is doubled,

$$f_o' = \frac{v}{2\left(\frac{l}{2}\right)} = \frac{v}{l}$$

$$f_c' = \frac{v}{4(2l)} = \frac{v}{8l}$$

$$f_o' - f_c' = \frac{v}{l} - \frac{v}{8l} = \frac{7v}{8l}$$

Number of beats produced

$$= f_o' - f_c' = 7 \quad \left[\text{from Eq. (i), } \frac{v}{l} = 8\right]$$

72. $f = (2n-1)\dfrac{v}{4(l+0.6r)}$ (where, $n = 1, 2, 3, \dots$)

$$\therefore \frac{(2n-1)v}{4(l+0.6r)} \le 1000 \text{ Hz}$$

or $$n \le \frac{\frac{4000(l+0.6r)}{v} + 1}{2}$$

$$\le \frac{\frac{4000(0.832+0.6\times 0.03)}{340} + 1}{2}$$

$$\le 5.5$$

So, $n = 1, 2, 3, 4, 5$ is possible.

1 is fundamental, other four are overtones.

73. $f \propto \sqrt{\dfrac{T}{M}}$, if γ is same, f is same. Therefore,

$$\frac{T_{N_2}}{M_{N_2}} = \frac{T_{O_2}}{M_{O_2}}$$

$$\Rightarrow T_{N_2} = \left(\frac{28}{32}\right)(273 + 47)$$

$$= 280 \text{ K} = 7°\text{C}$$

74. When the train approaches stationary observer,

$$f_1 = \frac{fv}{v - v_s} \quad \text{...(i)}$$

When the train recedes away train,

$$f_2 = \frac{fv}{v + v_s} \quad \text{...(ii)}$$

From Eqs. (i) and (ii), we get

$$f = \frac{2f_1f_1}{f_1 + f_2}$$

75. $S_2D = \sqrt{S_1D^2 + S_1S_2^2}$

$$= \sqrt{4^2 + 3^2} = 5 \text{ m}$$

$$\Delta x = S_2D - S_1D = 5 - 4 = 1 \text{ m}$$

$$\Delta\phi = \frac{2\pi}{\lambda}\Delta x = \frac{2\pi}{4}(1) = \frac{\pi}{2}$$

$$I_0 = 4I_0 \cos^2\frac{\pi}{2} = 2I_0$$

76. $$\frac{f_1}{f_2} = \frac{\frac{c}{c-v}}{\frac{c}{c+v}} = \frac{c+v}{c-v} = \frac{100}{9}$$

So, $10c = 190v \Rightarrow v = \dfrac{10 \times 330}{190}$

$$= 17.3 \text{ m/s}$$

77. $f' = f\left(\dfrac{v + v_s}{v - v_s}\right) = 1000\left(\dfrac{330 + 220}{330 - 220}\right) = 5000 \text{ Hz}$

78. Velocity of source and observer,

$$v_o, v_s = 7\sqrt{2} \text{ m/s}$$

Original frequency, $f = 1.7$ kHz

Applying Doppler effect,

$$f' = \left(\frac{v + v_o}{v - v_s}\right)f$$

$$f' = \left(\frac{333 + 7\sqrt{2}}{333 - 7\sqrt{2}}\right) \times 1.7 \times 10^3$$

$$f' = 1.8 \text{ kHz}$$

79. $v' = \left(\dfrac{v + v_o}{v - v_s}\right)f = \left(\dfrac{345 + 5}{345 - 5}\right) \times 272 = 280 \text{ Hz}$

So, number of beats produced

$$= 280 - 272 = 8$$

80.

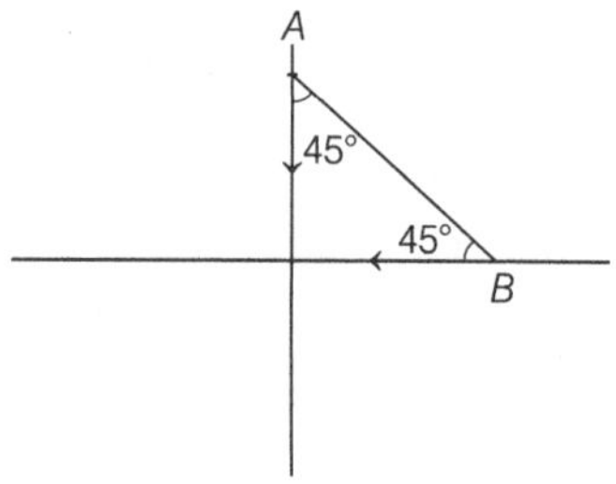

A is observer and B is source.

$$f' = f\left(\frac{v + v_A \cos 45°}{v - v_B \cos 45°}\right)$$

$$= 360\left[\frac{330 + 10/\sqrt{2}}{330 - 20/\sqrt{2}}\right]$$

$$\approx 384 \text{ Hz}$$

81. Maximum path difference on the said line is S_1S_2 or 3 m.

Now for minima, $\dfrac{\lambda}{2}$ or $\dfrac{v}{2f} > 3$

or $$f < \frac{v}{6} \text{ or } f < \frac{330}{6}$$

or $$f < 55 \text{ Hz}$$

82. Closed end is pressure antinode.

$$(\Delta p_x)_{max} = 2p_0$$

$\therefore$ $$p_{max} = p_A + 2p_0$$

and $$p_{min} = p_A - 2p_0$$

83. Density of mixture, $\rho_{mix} = \dfrac{\text{Total mass}}{\text{Total volume}}$

$$\rho_{mix} = \frac{v\rho_{H_2} + v\,(16\rho_{H_2})}{2v}$$

$$\rho_{mix} = \frac{17\rho_{H_2}}{2}$$

Now, $$v = \sqrt{\frac{rp}{\rho}} \propto \frac{1}{\sqrt{\rho}}$$

$\therefore$ $$\frac{v_{mix}}{v_{O_2}} = \sqrt{\frac{\rho_{O_2}}{\rho_{mix}}} = \sqrt{\frac{16\rho_{H_2}}{\rho_{mix}}}$$

$$= \sqrt{\frac{16\rho_{H_2} \times 2}{17\rho_{H_2}}} = \sqrt{\frac{32}{17}}$$

84. Theoretical question. So, no solution is required.

85. Time period of both is same. At $t = 3T/4$, source is at D, observer is at P and both are approaching towards each other.

86. Beat frequencies (say f) and beat time periods $\left(\text{say } T = \dfrac{1}{f}\right)$:

$f_{12} = f_2 - f_1 = 2\text{ Hz}$ $(\because\ T_{12} = 0.5\text{ s})$

$f_{23} = f_3 - f_2 = 2\text{ Hz}$ $(\because\ T_{23} = 0.5\text{ s})$

$f_{13} = f_3 - f_1 = 4\text{ Hz}$ $(\because\ T_{13} = 0.25\text{ s})$

Combined beat time will be 0.5 s. This is because, if we assume that all three are in same phase at $t = 0$. or, they produce maxima at $t = 0$, they are again in same phase at $t = 0.5$ s (not $t = 0.25$ s).

Therefore, beat time period is 0.5 s. So, beat frequency is 2 Hz.

87. $I = \dfrac{p_0^2}{2\rho v}$

$\Rightarrow$ $$p_0 = \sqrt{2I\rho v}$$

$$= \sqrt{2 \times 10^{-10} \times 1.293 \times 330}$$

$\Rightarrow$ $$= 3 \times 10^{-4}\ \text{N/m}^2$$

88. Fundamental frequency is given as,

$$f = \frac{n}{2l}\sqrt{\frac{T}{\mu}}$$

$\sqrt{T} \propto \dfrac{1}{n}$ (number of loops)

$$\frac{n_1^2}{n_2^2} = \frac{T}{49}$$

$\Rightarrow$ $$\frac{(8)^2}{(4)^2} = \frac{T}{49}$$

$\Rightarrow$ $$T = 196\text{ N}$$

89.

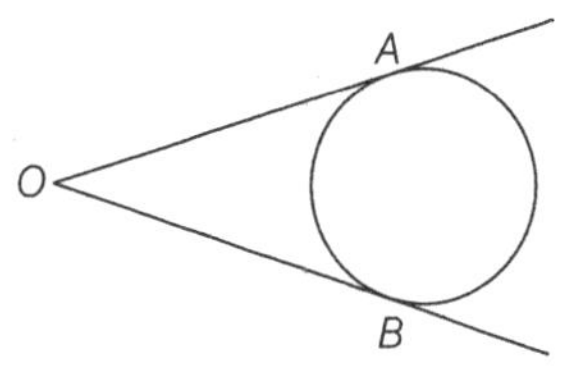

Speed of source, $v_s = R\omega = (1.5)\,(20)$
$= 30$ m/s

At position A, maximum frequency is heard.

$$f_{max} = f\left(\frac{v}{v - v_s}\right) = 440\left(\frac{330}{330 - 30}\right) = 484\text{ Hz}$$

At position B, minimum frequency is heard,

$$f_{min} = f\left(\frac{v}{v + v_s}\right) = 440\left(\frac{330}{330 + 30}\right)$$

$$= 403.33\text{ Hz}$$

90. $f_1 = \left(\dfrac{v - u}{v}\right) f$

$$f_2 = \left(\frac{v + u}{v}\right) f$$

So, beat frequency $= f_2 - f_1$

$$= \frac{2uf}{v} = \frac{2u}{\lambda}$$

91. Source is moving away from the observer, so

$$n = \frac{n_0 v}{v + v_s}$$

Frequency received by the observer at time t would have already been emitted before time t, say at time $t_0 (< t)$.

Hence, v_s at that time will be at_0

and $$n = \frac{n_0 v}{v + at_0}$$

Since $t_0 < t$, therefore $n > \dfrac{n_0 v}{v + at}$

92.

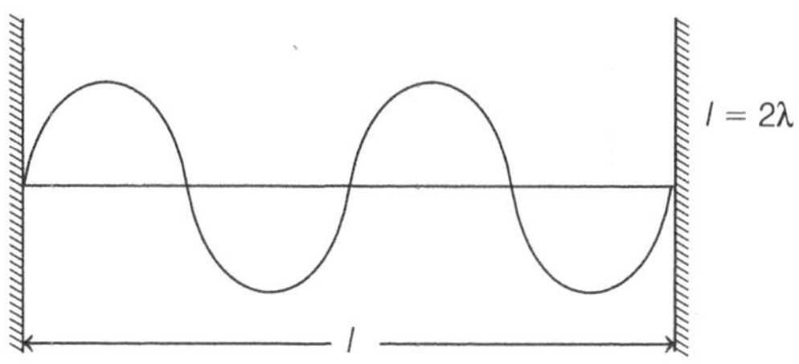

$$A = a \sin kx = a \sin\left(\frac{2\pi}{\lambda}\frac{l}{3}\right)$$

$$= a \sin\left(\frac{2\pi}{\lambda} \times \frac{2\lambda}{3}\right)$$

$$= a\sin(4\pi/3)$$
$$= a\sin(\pi + \pi/3)$$
$$= a\sin\frac{\pi}{3}$$
$$A = \frac{\sqrt{3}a}{2}$$

93. Same tuning fork is used. Therefore, frequency in two cases is same. In first case, there are five antinodes, so its fifth harmonic and in the second case, its third harmonic. Thus,

$$5\left(\frac{v_1}{2l}\right) = 3\left(\frac{v_2}{2l}\right)$$
$$v \propto \sqrt{T}$$
$$\therefore \quad 5\sqrt{T_1} = 3\sqrt{T_2}$$
or $$5\sqrt{9g} = 3\sqrt{mg}$$
or $$m = 25 \text{ kg}$$

94. Fundamental frequency of an open pipe is given by,

$$f = \frac{v}{2l} = \frac{\sqrt{\frac{\gamma RT}{M}}}{2l} \qquad \text{...(i)}$$

H_2, O_2 and N_2 are diatomic gases.
Therefore, value of γ is same for all gases. Because $\gamma = \frac{7}{5}$ for diatomic gases.
From equation (i), we get

$$f = \frac{\alpha}{l\sqrt{M}} \qquad (\alpha = \text{proportionality constant})$$
$$f_A = \frac{\alpha}{l\sqrt{2}}, f_B = \frac{\alpha}{2\sqrt{2}l}, f_C = \frac{3\alpha}{4\sqrt{7}l}, f_D = \frac{3\alpha}{2\sqrt{11}l}$$
$$\therefore \quad \frac{f_C}{f_D} = \sqrt{\frac{11}{28}}$$

95. If n_1 and n_2 be the number of loops formed in the wires of radii r and $2r$.

So, $$v_1 = \frac{n_1}{2l}\sqrt{\frac{T}{\mu_1}} \quad \text{and} \quad v_2 = \frac{n_2}{2l}\sqrt{\frac{T}{\mu_2}}$$

v_1 must be equal to the v_2.
So, we get

$$\frac{n_1}{n_2} = \sqrt{\frac{\mu_1}{\mu_2}} = \sqrt{\frac{\pi r^2 \times 1 \times \rho}{\pi (2r)^2 \times 1 \times \rho}} = \frac{1}{2}$$
$$n_1 : n_2 = 1 : 2$$

96. $\sin(0.314x) = 0$

$$0.314x = \pi k \Rightarrow x = 10\,k$$

Hence, the nodes are located at $x = 0$, 10 cm, 20 cm and 30 cm.
Now, $2\pi f = 600\pi \Rightarrow f = 300$ Hz
So, length of the string $= \frac{3\lambda}{2} = 3 \times 10 = 30$ cm

97. At the source, $t_1 = \frac{3T}{4} = 6$ s

$$v = f\lambda = 3 \text{ m/s}$$

Time taken for the wave to travel a distance of 9 m is,

$$t_2 = \frac{9}{3} = 3 \text{ s}$$
$$\therefore \quad t = t_1 + t_2 = 9 \text{ s}$$

98. Stress $= Y \times$ Strain

$$= Y \times \frac{1}{100} = \frac{2\times 10^{11}}{100} = 2 \times 10^9 \text{ N/m}^2$$

Tension, $T = \pi r^2 \times \text{Stress} = \pi r^2 \times 2 \times 10^9$ N
Mass, $m = \pi r^2 \rho = \pi r^2 \times 7700$

$$\therefore \quad f_o = \frac{1}{2l}\sqrt{\frac{T}{m}} = \frac{1}{2\,(1.5)}\sqrt{\frac{\pi r^2 \times 2 \times 10^9}{\pi r^2 \times 7700}}$$
$$= 170 \text{ Hz}$$

99. $v = \frac{\omega}{k} = \sqrt{\frac{T}{\mu}}$

$$\therefore \quad T = \left(\frac{\omega}{k}\right)^2(\mu) = \left(\frac{200\,\pi}{2\pi}\right)^2(10^{-3}) = 10 \text{ N}$$

100. $\frac{1}{2L}\sqrt{\frac{T_1}{\mu}} = \frac{1}{L}\sqrt{\frac{T_2}{\mu}} \Rightarrow T_2 = \frac{T_1}{4}$

$$\Sigma\tau_{net} = 0 \Rightarrow T_1 \times L = T_2 \times (L - x)$$
$$T_1 \times L = \frac{T_1}{4} \times (L - x)$$
$$x = \frac{L}{5}$$

101. $y = \frac{6}{x^2}$, then $y = \frac{6}{(x-2t)^2}$

$$\frac{\delta y}{\delta t} = \frac{24}{(x-2t)^3}$$

At $t = 2$ s, $x = 2$

$$v_y = -3 \text{ m/s}$$

102. $$v_s = \frac{100}{3}\cos\theta$$
$$v_s = \frac{100}{3} \times \frac{3}{5} = 20 \text{ m/s}$$

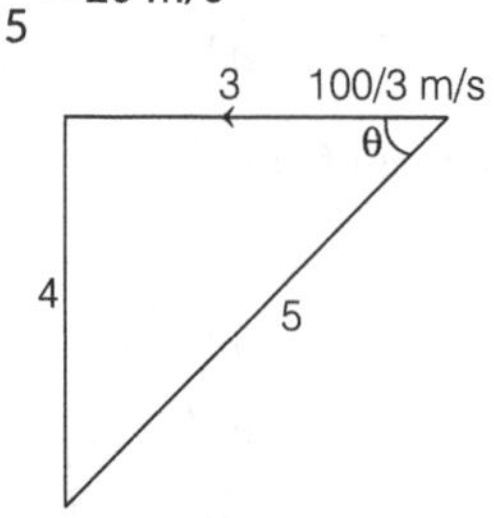

So, $$f' = \left(\frac{v}{v + v_s}\right) \times f$$
$$= \frac{340}{340 - 20} \times 640 = 680 \text{ Hz}$$

LEVEL 03 *Previous Years Questions*

103. Length of string of musical instrument,

$$l = 90 \text{ cm} = 0.9 \text{ m}$$

Fundamental frequency, $f_1 = 120$ Hz

$$f_2 = 180 \text{ Hz}$$

$\therefore$ We know that $f \propto \frac{1}{l}$

$$\Rightarrow \quad \frac{f_1}{f_2} = \frac{l_2}{l_1} \Rightarrow l_2 = \frac{f_1 l_1}{f_2} = \frac{120 \times 0.9}{180}$$

$$= \frac{2}{3} \times 0.9 = 0.6 \text{ m} = 60 \text{ cm}$$

104. In case of beats formation unknown frequency $(\nu_B) = \nu_A \pm$ beats

where, $\nu_A = 530$ Hz and beats $= 6$ Hz

$\Rightarrow \nu_B = 530 \pm 6 = 536$ or 524 Hz

When tension in B is slightly decreased, the beat frequency increases to 7 Hz. This is possible, if we take original frequency of B as 524 Hz.

Hence, correct option is (a).

105. For vibrating tunning fork over a resonance tube, the first resonance is obtained at the length

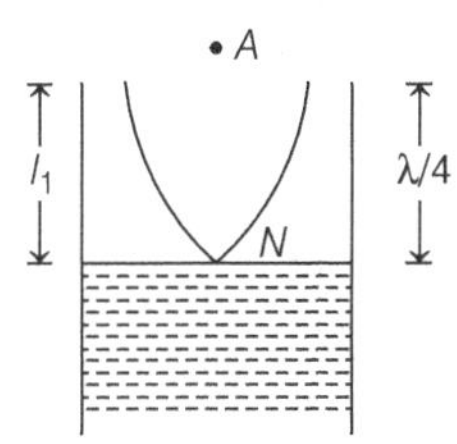

$$l_1 = \frac{\lambda}{4} \quad \text{...(i)}$$

and for second resonance,

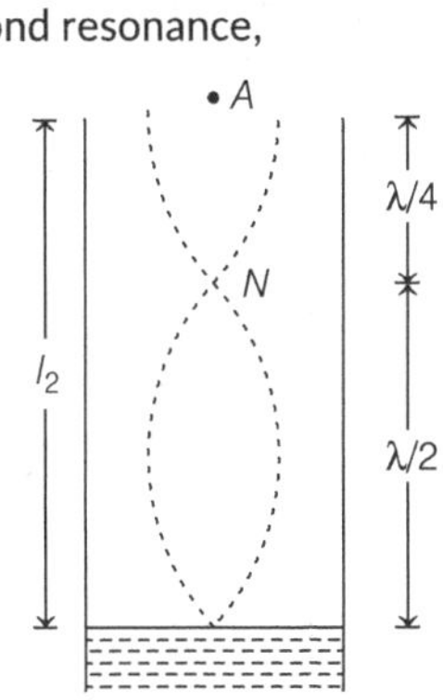

$$l_2 = \frac{\lambda}{4} + \frac{\lambda}{2} = \frac{3\lambda}{4} \quad \text{...(ii)}$$

Subtract Eq. (i) from Eq. (ii), we get

$$l_2 - l_1 = \frac{3\lambda}{4} - \frac{\lambda}{4} = \frac{\lambda}{2}$$

$$\Rightarrow \quad \lambda = 2(l_2 - l_1)$$

$$\Rightarrow \quad v = 2f(l_2 - l_1) \quad \text{...(iii)} \left(\because \lambda = \frac{v}{f}\right)$$

Here, $f = 800$ Hz, $l_1 = 9.75$ cm, $l_2 = 31.25$ cm

Substituting the given values in Eqs. (iii), we get

$$\Rightarrow \quad v = 2 \times 800(31.25 - 9.75)$$

$$= 34400 \text{ cm/s}$$

$$= 344 \text{ m/s}$$

106. Fundamental frequency for an open organ pipe is given as

$$f_1 = \frac{v}{2L}$$

where, L is the length of the open organ pipe.Third harmonic for a closed organ pipe is given as

$$f' = \frac{3v}{4L'}$$

where, L' is the length of closed organ pipe. According to the question,

$$f = f'$$

$$\frac{v}{2L} = \frac{3v}{4L'}$$

$$L = \frac{2}{3}L'$$

Given, $\quad L' = 20$ cm

$$\Rightarrow \quad L = \frac{2}{3} \times 20 \text{ cm} = \frac{40}{3} \text{ cm} = 13.3 \text{ cm}$$

107. For first resonance, $l_1 = \frac{\lambda}{4}$

For second resonance, $l_2 = \frac{3\lambda}{4}$

$$\therefore \quad (l_2 - l_1) = \frac{3\lambda}{4} - \frac{\lambda}{4}$$

$$\text{or} \quad \lambda = 2(l_2 - l_1) \quad \text{...(i)}$$

As, velocity of sound wave is given as,

$$v = f\lambda$$

where, f is the frequency.

$$\Rightarrow \quad v = f[2(l_2 - l_1)] \quad \text{[from equation (i)]}$$

Here, $f = 320$ Hz, $l_2 = 0.73$ m, $l_1 = 0.20$ m

$$\Rightarrow \quad v = 2[320(0.73 - 0.20)]$$

$$= 2 \times 320 \times 0.53$$

$$= 339.2 \text{ ms}^{-1} \simeq 339 \text{ ms}^{-1}$$

108. Thinking Process Frequency of nth harmonic in a closed end tube

$$\Rightarrow \quad f = \frac{(2n-1)v}{4l} \quad \text{where, } n = 1, 2, 3, \ldots$$

Also, only odd harmonics exists in a closed end tube.

Now, given two nearest harmonics are of frequency 220 Hz and 260 Hz.

$$\therefore \quad \frac{(2n-1)v}{4l} = 220 \text{ Hz} \quad \text{...(i)}$$

Next harmonic occurs at

$$\frac{(2n+1)v}{4l} = 260 \text{ Hz} \quad \text{...(ii)}$$

On subtracting equation (i) from equation (ii), we get

$$\frac{\{(2n+1)-(2n-1)\}v}{4l} = 260 - 220$$

$$2\left(\frac{v}{4l}\right) = 40 \Rightarrow \frac{v}{4l} = 20 \text{ Hz}$$

$\therefore$ Fundamental frequency of the system $= \dfrac{v}{4l} = 20$ Hz

109. For an open organ pipe,

$$\nu_n = \frac{n}{2L}v, \text{ where } n = 1, 2, 3, \ldots$$

For second overtone, $n = 3$, $\nu_{2_0} = \dfrac{3}{2L_1}v$

$$L_1 = \text{length of open organ pipe} \quad \ldots\text{(i)}$$

For closed organ pipe, $\nu_n = \left(\dfrac{2n+1}{4L}\right)v$

where, $n = 0, 1, 2, 3, \ldots$

Ist overtone for closed organ pipe, $n = 1$

$$\nu_{1_C} = \frac{3}{4L}v \quad \ldots\text{(ii)}$$

$\because \quad \nu_{2_0} = \nu_{1_C} \Rightarrow \dfrac{3v}{2L_1} = \dfrac{3}{4L}v$

$\Rightarrow \quad L_1 = 2L$

110. As we know that,

Beat frequency $= f_1 \sim f_2 = n - (n-1) = 1$

and similarly for n and $n + 1$

Beat frequency $= n + 1 - n = 1$

111. According to question, we have

Wavelength of transphase pulse, $\lambda = \dfrac{v}{f}$ ...(i)

(where, v = velocity of the wave; f = frequency of the wave)

As we know $\quad v = \sqrt{\dfrac{T}{\mu}}$ ...(ii)

(Here, T = tension in the spring, μ = mass per unit length of the rope)

From equations (I) and (II), we get

$$\lambda = \frac{1}{f}\sqrt{\frac{T}{\mu}}$$

As f and μ remains constant,

$\Rightarrow \quad \lambda \propto \sqrt{T}$

So, for two different case, we get

$$\frac{\lambda_2}{\lambda_1} = \sqrt{\frac{T_2}{T_1}} = \sqrt{\frac{m_1 + m_2}{m_2}}$$

112. The fundamental frequencies of closed and open organ pipe are given as

$$\nu_c = \frac{v}{4l} \Rightarrow \nu_o = \frac{v}{2l'}$$

Given the second overtone (i.e. third harmonic) of open pipe is equal to the fundamental frequency of closed pipe,

i.e. $\quad 3\nu_o = \nu_c \Rightarrow 3\dfrac{v}{2l'} = \dfrac{v}{4l}$

$\Rightarrow \quad l' = 6l = 6 \times 20 = 120$ cm

113. The smallest length of the air column is associated with fundamental mode of vibration of the air column as shown in the diagram.

$\because \quad L_{min} = \dfrac{\lambda}{4} \Rightarrow 50 \text{ cm} = \dfrac{\lambda}{4}$

$\Rightarrow \quad \lambda = 200$ cm

The next higher length of the air column is

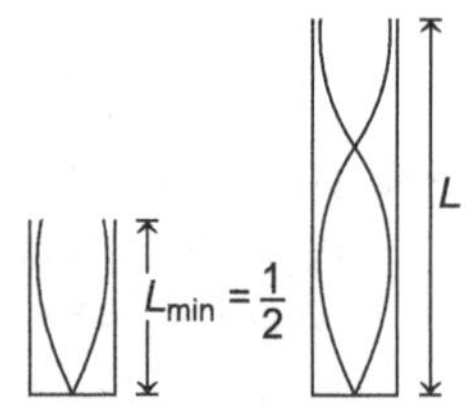

$$L = \frac{\lambda}{4} + \frac{\lambda}{2} = \frac{\lambda + 2\lambda}{4} = \frac{3\lambda}{4}$$

$$= \frac{3}{4} \times 200 = 150 \text{ cm}$$

114. Given, $l = 75$ cm, $f_1 = 420$ Hz and $f_2 = 315$ Hz

As, two consecutive resonant frequencies for a string fixed at both ends will be

$$f_1 = \frac{nv}{2l} \text{ and } f_2 = \frac{(n+1)v}{2l}$$

$\Rightarrow \quad f_2 - f_1 = 420 - 315$

$\Rightarrow \quad \dfrac{(n+1)v}{2l} - \dfrac{nv}{2l} = 105$ Hz

$\Rightarrow \quad \dfrac{v}{2l} = 105$ Hz

Thus, lowest resonant frequency of a string is 105 Hz.

115. **Problem Solving Strategy** In this problem, the fundamental frequencies of each part could be find. The fundamental frequency of the complete wire could be find. One should check each option for the given values.

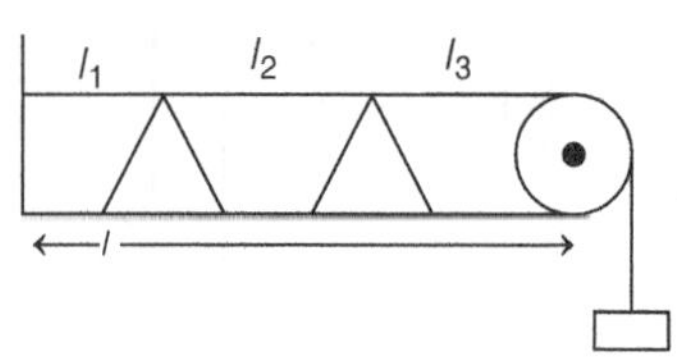

For Ist part, $n_1 = \dfrac{v}{2l_1} \Rightarrow l_1 = \dfrac{v}{2n_1}$

For IInd part, $n_2 = \dfrac{v}{2l_2} \Rightarrow l_2 = \dfrac{v}{2n_2}$

For IIIrd part, $n_3 = \dfrac{v}{2l_3} \Rightarrow l_3 = \dfrac{v}{2n_3}$

For the complete wire, $n = \dfrac{v}{2l}$

$\Rightarrow \quad l = \dfrac{v}{2n}$

We have, $l = l_1 + l_2 + l_3$

$$\frac{v}{2n} = \frac{v}{2n_1} + \frac{v}{2n_2} + \frac{v}{2n_3}$$

$$\frac{1}{n} = \frac{1}{n_1} + \frac{1}{n_2} + \frac{1}{n_3}$$

116. For pipe closed at one end,

$$f_n = n\left(\frac{v}{4l}\right), \text{ here } n \text{ is an odd number.}$$

$$= n\left[\frac{340}{4 \times 85 \times 10^{-2}}\right] = n\,[100]$$

Here, n is an odd number, so for the given condition, n can go upto $n = 11$,

i.e. $n = 1, 3, 5, 7, 9, 11$

So, number of possible natural oscillations could be 6. Which are below 1250 Hz.

117. Statement (d) is not true, because at the open ends, pressure change will be zero.

118. Given,

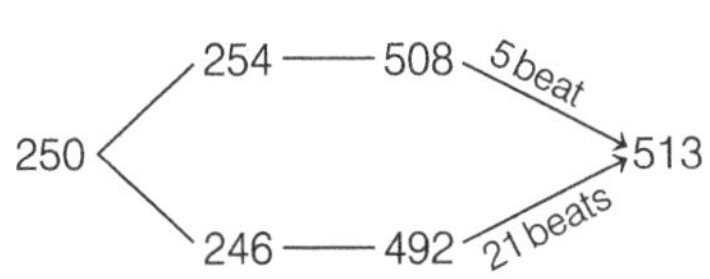

Hence, unknown frequency is 254 Hz.

119. Given, $a = 1$ m

As $y = a \sin(kx - \omega t)$

$$= \sin\left(\frac{2\pi}{2\pi}x - 2\pi \times \frac{1}{\pi}t\right) = \sin(x - 2t)$$

120. The fundamental frequency of string

$$\nu = \frac{1}{2l}\sqrt{\frac{T}{m}}$$

$$\therefore \quad \nu_1 l_1 = \nu_2 l_2 = \nu_2 l_3 = k \qquad \text{...(i)}$$

From Eq. (i),

$$l_1 = \frac{k}{\nu_1}, l_2 = \frac{k}{\nu_2}, l_3 = \frac{k}{\nu_3}$$

Original length, $l = \frac{k}{\nu}$

Here, $l = l_1 + l_2 + l_3$

$$\frac{k}{\nu} = \frac{k}{\nu_1} + \frac{k}{\nu_2} + \frac{k}{\nu_3}$$

$$\frac{1}{\nu} = \frac{1}{\nu_1} + \frac{1}{\nu_2} + \frac{1}{\nu_3}$$

121. Given, $y_1 = 4\sin 600\pi t$

and $y_2 = 5\sin 608\pi t$

Comparing with general equation

$$y = a\sin 2\pi f t$$

We get, $f_1 = 300$ Hz

and $f_2 = 304$ Hz

So,number of beats $= f_2 - f_1 = 4\,\text{s}^{-1}$

We know that, $\frac{I_{max}}{I_{min}} = \left(\frac{a_1 + a_2}{a_1 - a_2}\right)^2 = \left(\frac{4+5}{4-5}\right)^2 = 81$

122. According to question,

$$y_1 = a\sin(\omega t + kx + 0.57)$$

and $y_2 = a\cos(\omega t + kx)$

or $y_2 = a\sin\left(\frac{\pi}{2} + \omega t + kx\right)$

As phase difference,

$$\Delta\phi = \phi_2 - \phi_1 = \frac{\pi}{2} - 0.57$$

$$= 1.57 - 0.57 = 1 \text{ rad}$$

123. Velocity of a wave is given by

$$v = n\lambda \qquad \begin{bmatrix} n = \text{frequency of wave} \\ \lambda = \text{wavelength of wave} \end{bmatrix}$$

So, for two different cases,

$$v_1 = n_1\lambda_1$$

$$v_2 = n_2\lambda_2$$

$$\lambda_2 = \lambda_1 \frac{v_2}{v_1}$$

$$= \lambda_1 \times \frac{3500}{350} = \lambda_1 \times 10 \qquad [\because n_1 = n_2]$$

$$\lambda_2 = 10\,\lambda_1$$

124. Given, $y = A\sin(\omega t - kx)$

As we know that wave velocity is given by

$$v_w = \frac{\lambda}{T} = \frac{\omega\lambda}{2\pi} \qquad \text{...(i)} \left[T = \frac{2\pi}{\omega}\right]$$

and maximum particle velocity is given by

$$v_p = A\omega \qquad \text{...(ii)}$$

$$\begin{bmatrix} A = \text{amplitude} \\ \omega = \text{angular frequency} \end{bmatrix}$$

So, as Eq. (i) is equal to Eq. (ii),

$$A\omega = \frac{\omega\lambda}{2\pi}, \ \lambda = 2\pi A$$

125. Let n_p be the frequency of piano

As $(n_p \propto \sqrt{T})$

n_f = frequency of tuning fork = 512 Hz

x = Beat frequency = 4 beats/s, which is decreasing $(4 \to 2)$ after changing the tension of piano wire.

Also, tension of piano wire is increasing so $n_p \uparrow$

Hence, $n_p \uparrow - n_f = x \downarrow \longrightarrow$ wrong

$n_f - n_p \uparrow = x \downarrow \longrightarrow$ correct

$$n_p = n_f - x = 512 - 4 = 508 \text{ Hz}$$

Chapter 16

HEAT AND THERMODYNAMICS

LEVEL 01 Objective Questions

1 5 g of ice at 0°C is dropped in a beaker containing 20 g of water at 40°C. The final temperature will be

(a) 16°C (b) 8°C
(c) 4°C (d) 32°C

2 50 g of copper is heated to increase its temperature by 10°C. If the same quantity of heat is given to 10 g of water, the rise in its temperature is (Take, specific heat of copper $= 420\ \text{Jkg}^{-1}\,°\text{C}^{-1}$)

(a) 5°C (b) 6°C
(c) 7°C (d) 8°C

3 Two thermometers X and Y have ice points marked at 15° and 25° and steam points marked as 75° and 125°, respectively. When thermometer X measures the temperature of a bath as 60° on it, what would thermometer Y read when it is used to measure the temperature of the same bath?

(a) 60° (b) 75° (c) 100° (d) 90°

4 A liquid of mass m and specific heat C is heated to a temperature $2T$. Another liquid of mass $\frac{m}{2}$ and specific heat $2C$ is heated to a temperature T. If these two liquids are mixed. The resulting temperature of the mixture is

(a) $\left(\frac{2}{3}\right)T$ (b) $\left(\frac{8}{5}\right)T$ (c) $\left(\frac{3}{5}\right)T$ (d) $\left(\frac{3}{2}\right)T$

5 Steam is passed into 22 g of water at 20°C. The mass of water that will be present when the water acquires a temperature of 90°C (Latent heat of steam is 540 cal/g) is

(a) 24.8 g (b) 24 g (c) 36.6 g (d) 30 g

6 A vessel contains 110 g of water. The heat capacity of the vessel is equal to 10 g of water. The initial temperature of water in vessel is 10°C. If 220 g of hot water at 70°C is poured in the vessel, the final temperature neglecting radiation loss, will be

(a) 70°C (b) 80°C (c) 60°C (d) 50°C

7 Two tanks A and B contain water at 30°C and 80°C, respectively. Calculate the amount of water that must be taken from each tank respectively to prepare 40 kg of water at 50°C.
(a) 24 kg, 16 kg (b) 16 kg, 24 kg (c) 20 kg, 20 kg (d) 30 kg, 10 kg

8 10 g of ice at –20°C is dropped into a calorimeter containing 10 g of water at 10°C. The specific heat of water is twice that of ice. When equilibrium is reached, the calorimeter will contain
(a) 20 g of water (b) 20 g of ice
(c) 10 g ice and 10 g of water (d) 5 g ice and 15 g of water

9 A student takes 50 g wax (specific heat = 0.6 k cal/kg-°C) and heats it till it boils. The graph between temperature and time is shown below. Heat supplied to the wax per minute and boiling point are, respectively.

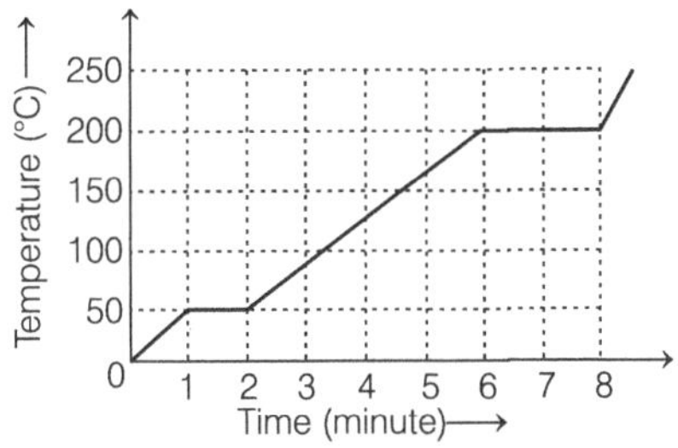

(a) 500 cal, 50°C (b) 1000 cal, 100°C (c) 1500 cal, 200°C (d) 1000 cal, 200°C

10 The molar heat capacity of water at constant pressure C is 75 J $K^{-1}mol^{-1}$. When 1.0 kJ of heat is supplied to 100 g of water which is free to expand, the increase in temperature of water is
(a) 1.2 K (b) 2.4 K (c) 4.8 K (d) 6.6 K

11 When x grams of steam at 100°C is mixed with y grams of ice at 0°C, we obtain $(x + y)$ grams of water at 100°C. What is the ratio y/x?
(a) 1 (b) 2 (c) 3 (d) 4

12 A rectangular metal plate have two holes as shown in the figure. Now, the plate is heated, then separation between the holes

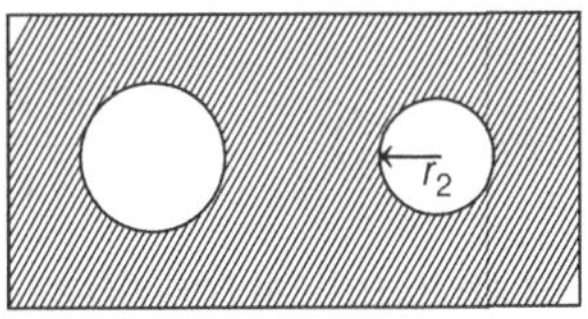

(a) decreases (b) increases (c) remain constant (d) Cannot say

13 A brass wire 1.8 m long at 27°C is held taut with little tension between two rigid supports. If the wire is cooled to a temperature of –39°C, what is the tension developed in the wire, if its diameter is 2.0 mm? (Coefficient of linear expansion of brass = 2.0×10^{-5} /°C, Young's modulus of brass = 0.91×10^{11} Pa)
(a) 2.4×10^2 N (b) 3.8×10^2 N (c) 1.8×10^2 N (d) 4.8×10^2 N

14 A composite bar of length $L = L_1 + L_2$ is made up from a rod of material 1 and of length L_1 attached to a rod of material 2 and of length L_2 as shown. If α_1 and α_2 are their respective coefficients of linear expansion, then equivalent coefficient of linear expansion for the composite rod is

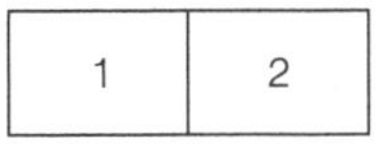

(a) $\frac{\alpha_1 L_2 + \alpha_2 L_1}{L}$ (b) $\frac{\alpha_1 L_2 + \alpha_2 L_2}{L}$ (c) $\frac{\alpha_1 L_1 + \alpha_2 L_2}{L}$ (d) $\frac{\alpha_1 \alpha_2 (L_1^2 + L_2)}{(\alpha_1 L_1 + \alpha_2 L_2)}$

15 What must be the lengths of steel and copper rods at 0°C for the difference in their lengths to be 10 cm at any common temperature?
(Take, $\alpha_{steel} = 1.2 \times 10^{-5} \, {}^\circ K^{-1}$ and $\alpha_{copper} = 1.8 \times 10^{-5} \, {}^\circ K^{-1}$)
(a) 30 cm for steel and 20 cm for copper
(b) 20 cm for steel and 30 cm for copper
(c) 40 cm for steel and 30 cm for copper
(d) 30 cm for steel and 40 cm for copper

16 A one litre flask contains some mercury. It is found that at different temperatures the volume of air inside the flask remains the same. What is the volume of mercury in the flask? Take, the coefficients of linear expansion of glass is 9×10^{-6} /°C and the coefficient of volume expansion of mercury is 1.8×10^{-4} /°C.
(a) 50 cm^3 (b) 100 cm^3 (c) 150 cm^3 (d) 200 cm^3

17 On heating one end of a rod, the temperature of whole rod will be uniform when
(a) $K = 1$ (b) $K = 0$
(c) $K = 100$ (d) $K = \infty$

18 A wall is made up of two layers A and B. The thickness of the two layers is the same but materials are different. The thermal conductivity of A is double than that of B. In thermal equilibrium, the temperature difference between the two ends is 36°C. Then, the difference of temperature at the two surfaces of A will be
(a) 6°C (b) 12°C
(c) 18°C (d) 24°C

19 The effective thermal conductivity of the combination shown in figure will be

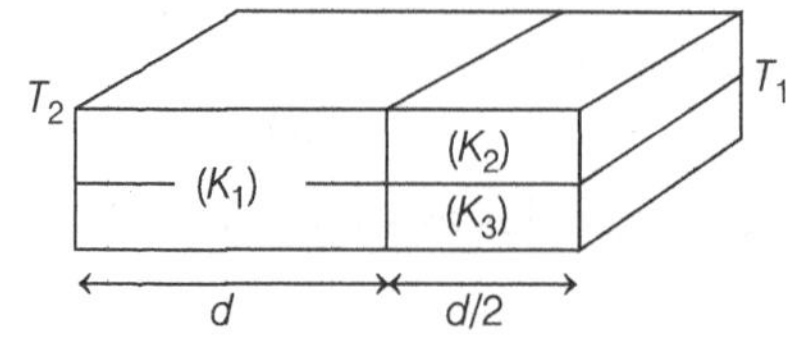

(a) $K_1 + \frac{K_2 + K_3}{2}$ (b) $\frac{2}{3}(K_1 + K_2 + K_3)$ (c) $\frac{K_1 K_2 K_3}{K_1 + K_2 + K_3}$ (d) $\frac{3}{2}\left(\frac{1}{K_1} + \frac{1}{K_2 + K_3}\right)^{-1}$

20 A rod CD is joined at the middle of an identical rod AB as shown in figure. The ends A, B and D are maintained at 100°C, 10°C and 20°C, respectively. The temperature at C is

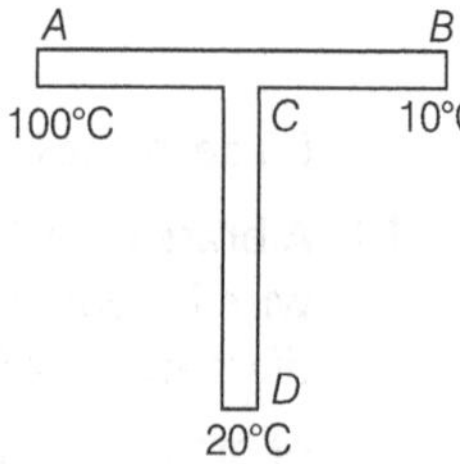

(a) 48°C
(b) 45°C
(c) 40°C
(d) 35°C

21 Heat is flowing through two cylindrical rods of same material. The diameters of the rods are in the ratio 1 : 2 and their lengths are in the ratio 2 : 1. If the temperature difference between their ends is the same, the ratio of rates of flow of heat through them will be
(a) 1 : 1 (b) 2 : 1
(c) 1 : 4 (d) 1 : 8

22 Which of the following statements is **incorrect**?
(a) Rough surfaces are better radiators than smooth surface.
(b) Highly polished mirror like surfaces are very good radiators.
(c) Black surfaces are better absorbers than white ones.
(d) Black surfaces are better radiators than white ones.

23 Energy radiated per second at temperature T is E for a body of diameter d. If temperature becomes $2T$ and diameter becomes $\frac{d}{4}$, then radiated energy will be

(a) $4E$ (b) $16E$ (c) E (d) $E/16$

24 The energy spectrum of a black body exhibits a maximum around a wavelength λ_0. The temperature of the black body is now changed such that the energy is maximum around a wavelength $\frac{3\lambda_0}{4}$. The power radiated by the black body will now increase by a factor of

(a) 256/81 (b) 64/27
(c) 16/9 (d) 4/3

25 A tungsten body of diameter 2.3 cm is at 2000°C. It radiates 30% of the energy radiated by a black body of same radius and temperature. Find radius of black body which will radiate energy at same rate at the same temperature?

(a) 2.32 cm (b) 1.49 cm
(c) 0.629 cm (d) 0.123 cm

26 Two samples A and B which are initially at the same temperature and pressure, are compressed from volume V to $\frac{V}{2}$ (A isothermally and B adiabatically). If their final pressures are respectively p_A and p_B, then

(a) $p_A < p_B$ (b) $p_A = p_B$
(c) $p_A > p_B$ (d) $p_A = 2p_B$

27 A balloon contains 1500 m^3 of helium at 27°C and 4 atmospheric pressure. The volume of helium at −3°C temperature and 2 atmospheric pressure will be

(a) 1500 m^3 (b) 1700 m^3 (c) 1900 m^3 (d) 2700 m^3

28 Under which of the following conditions is the law $pV = nRT$ obeyed most closely by a real gas?

(a) High pressure and high temperature (b) Low pressure and low temperature
(c) Low pressure and high temperature (d) High pressure and low temperature

29 Find the approximate number of molecules in a vessel of volume 7 L at 0°C and pressure 1.3×10^5 Pa.

(a) 2.4×10^{23} (b) 3×10^{23} (c) 6×10^{23} (d) 4.8×10^{23}

30 A gas is suddenly compressed to 1/9th of its initial volume. What is final temperature, if initial temperature was 127°C and $\gamma = 1.5$?

(a) 1200°C (b) 927°C
(c) 400°C (d) 900 K

31 According to Boyle's law $pV = C$, the value of C depends on

(i) mass of the gas (ii) type of gas (iii) temperature of gas

(a) (i), (ii) (b) (ii), (iii) (c) (i), (iii) (d) (i), (ii), (iii)

32 At pressure p and absolute temperature T, a mass M of an ideal gas fills a closed container of volume V. An additional mass $2M$ of the same gas is added into the container and the volume is then reduced to $\frac{V}{3}$ and the temperature to $\frac{T}{3}$. The pressure of the gas will now be

(a) $\frac{p}{3}$ (b) p (c) $3p$ (d) $9p$

33 Two gases A and B having same pressure p, volume V and temperature T are mixed. If mixture has volume and temperature as V and T respectively, then the pressure of the mixture will be

(a) $4p$ (b) $3p$
(c) $2p$ (d) p

34 An ideal gas has an initial pressure of 3 pressure units and an initial volume of 4 volume units. The table gives the final pressure and volume of the gas (in those same units) in four processes. Which processes start and end on the same isothermal?

	A	*B*	*C*	*D*
p	5	4	12	6
V	7	6	1	3

(a) *A* (b) *B* (c) *C* (d) *D*

35 Pressure *versus* temperature graph of an ideal gas is as shown in figure. Density of the gas at point *A* is ρ_0. Density at point *B* will be

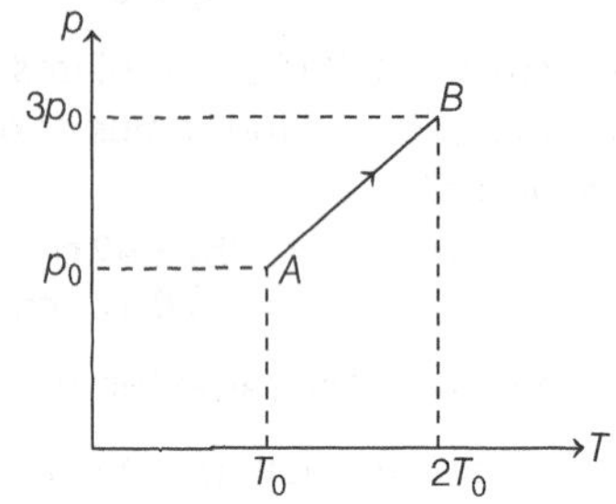

(a) $\frac{3}{4}\rho_0$ (b) $\frac{3}{2}\rho_0$ (c) $\frac{4}{3}\rho_0$ (d) $2\rho_0$

36 If the pressure of an ideal gas contained in a closed vessel is increased by 0.5%, the increase in temperature is 2K. The initial temperature of the gas is
(a) 27°C (b) 127°C (c) 300°C (d) 400°C

37 If 2 mol of an ideal monoatomic gas at temperature T_0 is mixed with 4 mol of another ideal monoatomic gas at temperature $2T_0$, then the temperature of the mixture is
(a) $\frac{5}{3}T_0$ (b) $\frac{3}{2}T_0$ (c) $\frac{4}{3}T_0$ (d) $\frac{5}{4}T_0$

38 If pressure of a gas contained in a closed vessel is increased by 0.4% when heated by 1°C, the initial temperature must be
(a) 250 K (b) 250°C (c) 2500 K (d) 25°C

39 If *p* is pressure, *V* is volume, *K* is Boltzmann constant and *T* is absolute temperature of gas, then the quantity $\frac{pV}{KT}$ gives
(a) molecular mass of the gas (b) kinetic energy of the gas
(c) number of molecules of the gas (d) number of moles of the gas

40 A gas is compressed at a constant pressure of 50 N/m^2 from a volume of 10 m^3 to a volume of 4 m^3. Energy of 100 J is then added to the gas by heating. Its internal energy is
(a) increased by 400 J (b) decreased by 400 J
(c) increased by 200 J (d) decreased by 200 J

41 Molar specific heat of oxygen at constant pressure is $C_p = 7.2$ cal/mol-°C and $R = 8.3$ J/mol-K. At constant volume, 5 mol of oxygen are heated from 10°C to 20°C. The quantity of heat required is approximately
(a) 26 cal (b) 360 cal
(c) 260 cal (d) 500 cal

42 Value of two principal specific heats of a gas in cal/mol-K determined by different students are given. Which is most reliable?
(a) 5, 2 (b) 6, 5 (c) 7, 5 (d) 7, 4

43 In an adiabatic process, $R = \frac{2}{3} C_V$, the pressure of the gas will be proportional to

(a) $T^{5/3}$ (b) $T^{5/2}$ (c) $T^{5/4}$ (d) $T^{5/6}$

44 If for an ideal gas $\frac{R}{C_V} = \frac{2}{5}$, then the degree of freedom of the gas will be

(a) 3 (b) 8
(c) 6 (d) 5

45 The number of molecules per unit volume (n) of a gas is given by

(a) $\frac{p}{kT}$ (b) $\frac{kT}{p}$ (c) $\frac{p}{RT}$ (d) $\frac{RT}{p}$

46 A gas of adiabatic exponent γ is supplied heat at a constant pressure. What is the ratio between ΔQ, ΔW and ΔU in the process? (Symbols have their usual meanings)

(a) $\gamma : (\gamma - 1) : \frac{1}{\gamma}$ (b) $1 : 1 : (\gamma - 1)$ (c) $\gamma : (\gamma - 1) : 1$ (d) $\gamma : 1 : (\gamma - 1)$

47 In a process, the molar heat capacity of diatomic gas is $\frac{10}{3}R$. Heat Q is supplied to the gas. Find the work done by the gas.

(a) $\frac{3Q}{4}$ (b) $\frac{Q}{3}$ (c) $\frac{Q}{4}$ (d) $\frac{2Q}{3}$

48 Volume *versus* temperature graph of two moles of helium gas is as shown in figure. The ratio of heat absorbed and the work done by the gas in process 1-2 is

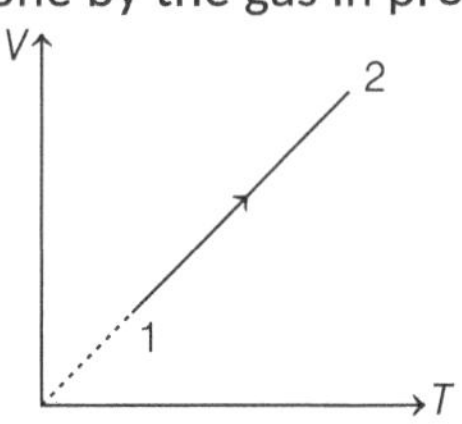

(a) 3 (b) $\frac{5}{2}$ (c) $\frac{5}{3}$ (d) $\frac{7}{2}$

49 $5n$, n and $5n$ moles of a monoatomic, diatomic and non-linear polyatomic gases (which do not react chemically with each other) are mixed at room temperature. The equivalent degree of freedom for the mixture is

(a) $\frac{25}{7}$ (b) $\frac{48}{11}$ (c) $\frac{52}{11}$ (d) $\frac{50}{11}$

50 An ideal gas in state (p_1, V_1) is isothermally expanded to state (p_2, V_2). Then, it is adiabatically compressed to initial volume V_1. Pressure being p_3 is in final state. If W is work done by the gas in the whole process, then

(a) $p_3 > p_1$ and $W < 0$ (b) $p_3 < p_1$ and $W > 0$
(c) $p_3 < p_1$ and $W = 0$ (d) $p_3 > p_1$ and $W > 0$

51 A fixed mass of an ideal gas undergoes the change represented by $XYZX$ below (shown in figure)

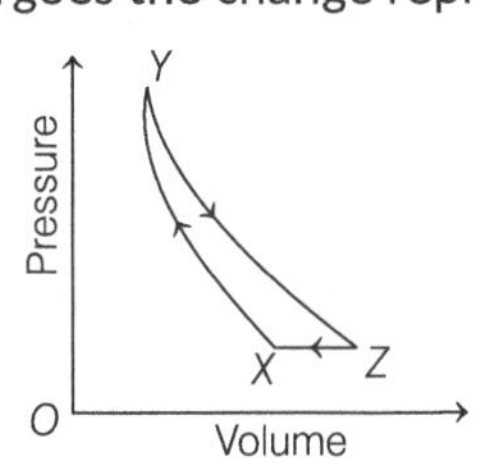

Which one of the following sets could describe this set of changes?

	XY	YZ	ZX
(a)	Isothermal expansion	adiabatic compression	compression at constant pressure
(b)	Adiabatic expansion	isothermal compression	pressure reduction at constant volume
(c)	Isothermal compression	adiabatic expansion	compression at constant pressure
(d)	Adiabatic compression	isothermal expansion	compression at constant pressure

52 Figure shows two flasks connected to each other. The volume of the flask 1 is twice that of flask 2. The system is filled with an ideal gas at temperature 100 K and 200 K respectively in the flasks. If the mass of the gas in 1 be m, then what is the mass of the gas in flask 2 in equilibrium?

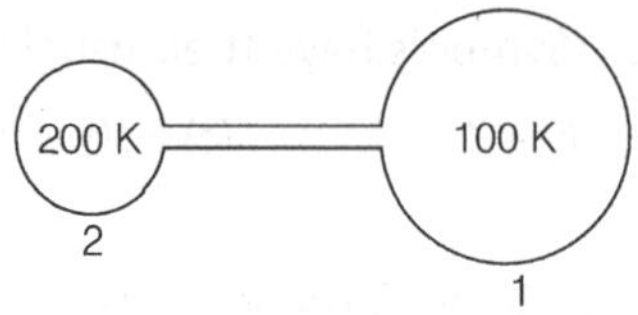

(a) m (b) $\frac{m}{2}$ (c) $\frac{m}{4}$ (d) $\frac{m}{8}$

53 One mole of an ideal monoatomic gas at temperature T_0 expands slowly according to the law $p = kV$ (k is a constant). If final temperature is $2T_0$, heat supplied to the gas is

(a) $2RT_0$ (b) $3/2\,RT_0$
(c) RT_0 (d) $RT_0/2$

54 If the ratio of specific heats of a gas at constant pressure to that at constant volume is γ, the change in internal energy of the given mass of gas when the volume changes from V to $2V$ at constant pressure p is

(a) $R/(\gamma - 1)$ (b) pV (c) $pV/(\gamma - 1)$ (d) $\gamma pV/(\gamma - 1)$

55 A refrigerator absorbs 2000 cal of heat from ice trays. If the coefficient of performance is 4, then work done by the motor is

(a) 2100 J (b) 8000 J (c) 33600 J (d) 500 J

56 The coefficient of performance of a refrigerator is 5. If the temperature inside freezer is $-20°$ C, the temperature of the surroundings to which it rejects heat is

(a) 21°C (b) 31°C
(c) 41°C (d) 11°C

57 Choose the **incorrect** statement from the following :

S1: The efficiency of a heat engine can be 1 but the coefficient of performance of a refrigerator can never be infinity.

S2: The first law of thermodynamics is basically the principle of conservation of energy.

S3: The second law of thermodynamics does not allow several phenomena consistent with the first law.

(a) S1 (b) S2
(c) S3 (d) None of these

58 A Carnot engine operating between temperatures T_1 and T_2 has efficiency $\frac{1}{6}$. When T_2 is lowered by 62 K, its efficiency increases to $\frac{1}{3}$. Then, T_1 and T_2 are respectively

(a) 330 K and 268 K (b) 310 K and 248 K
(c) 372 K and 310 K (d) 372 K and 330 K

59 Two jars A and B contain helium gas at pressure of 1 atm and 2 atm respectively at room temperature. They are mixed. If V_1 is the rms velocity of molecules in A, V_2 of those in B and V that of mixture, then check the correct statement.

(a) V is less than V_1 and V_2 (b) V is greater than V_1 and V_2
(c) $V_1 = V_2 = V$ (d) V is between V_1 and V_2

60 A refrigerator converts 100 g of water at 25°C into ice at −10°C in one hour and 50 min. The quantity of heat removed per minute is (Take, specific heat of ice = 0.5 cal/g°C, latent heat of fusion = 80 cal/g)

(a) 50 cal (b) 100 cal
(c) 200 cal (d) 75 cal

61 A Carnot engine, having an efficiency of $\eta = 1/10$ as heat engine is used as a refrigerator. If the work done on the system is 10 J, the amount of energy absorbed from the reservoir at lower temperature is

(a) 99 J (b) 90 J
(c) 1 J (d) 100 J

62 A heat engine is having a source at temperature 527°C and sink at temperature 127°C. If the useful work is required to be done by the engine at the rate of 750 W, then the amount of heat absorbed per second from the source in calories and the efficiency of heat engine are

(a) 482.2 cal/s, 50% (b) 482.2 cal/s, 25%
(c) 357.14 cal/s, 50% (d) 357.14 cal/s, 25%

63 The ratio of momentum of H_2 and O_2 gas at same temperature is

(a) 4 : 1 (b) 1 : 4
(c) 1 : 16 (d) $1 : 2\sqrt{2}$

64 The root mean square speed of the molecules of a diatomic gas is v. When the temperature is doubled, the molecules dissociate into two atoms. The new root mean square speed of the atom is

(a) v (b) $2v$ (c) $\sqrt{2}v$ (d) $4v$

65 The average translational kinetic energy of 10 g of H_2 at 27°C is

(a) 37250 J (b) 18700 J (c) 12450 J (d) 3737 J

66 The rms speed of air molecules in a vessel is 9×10^2 m/s. If half of the gas leaks out at constant temperature, the rms speed of the remaining air molecules will be

(a) 3×10^2 m/s (b) 30 m/s
(c) 9×10^2 m/s (d) 90 m/s

67 One kg of a diatomic gas is at a pressure of 8×10^4 N/m^2. The density of the gas is 4 kg/m^3. What is the energy of the gas due to its thermal motion?

(a) 3×10^4 J (b) 5×10^4 J
(c) 6×10^4 J (d) 7×10^4 J

68 One half mole each of nitrogen, oxygen and carbon dioxide are mixed together in 5 L enclosure at 27°C temperature. Pressure exerted by mixture is

(a) 7.5×10^5 Nm^{-2} (b) 5×10^5 Nm^{-2}
(c) 2×10^5 Nm^{-2} (d) 3×10^5 Nm^{-2}

69 If the internal energy of n_1 moles of He at $10T$ temperature is equal to the internal energy of n_2 moles of H_2 at $6T$ temperature, then the ratio of (n_1/n_2) is

(a) $\frac{3}{5}$ (b) 2 (c) 1 (d) $\frac{5}{3}$

70 How many times will the mean square speed of molecules of an ideal gas change by increasing its volume 2 times? The gas pressure increases 8 times and the mass remains unchanged.

(a) 2 times (b) 4 times
(c) 16 times (d) Decreases by a factor of 1/4

LEVEL 02 Objective Questions

71 On an X temperature scale, water freezes at $-125.0^\circ X$ and boils at $375.0^\circ X$. On a Y temperature scale, water freezes at $-70.0^\circ Y$ and boils at $-30.0^\circ Y$. The value of temperature on X-scale equal to the temperature of $50.0^\circ Y$ on Y-scale is

(a) $455.0^\circ X$ (b) $-125.0^\circ X$ (c) $1375.0^\circ X$ (d) $1500.0^\circ X$

72 Two temperature scales A and B are related by $\frac{A-42}{110} = \frac{B-72}{220}$

At which temperature, two scales have the same reading?

(a) $-42^\circ C$ (b) $-72^\circ C$ (c) $+12^\circ C$ (d) $-40^\circ C$

73 The table gives the initial length l_0, change in temperature ΔT and change in length Δl of four rods. Which rod has greatest coefficient of linear expansion?

Rod	l_0(m)	$\Delta T(^\circ C)$	$\Delta l(^\circ C)$
A_1	1	100	1
A_2	1	100	2
A_3	1.5	50	3
A_4	2.5	20	4

(a) A_1 (b) A_2 (c) A_3 (d) A_4

74 A pendulum clock keeps correct time at 20°C. The correction to be made during summer per day, where the average temperature is 40°C will be (Take, coefficient of linear expansion of the material of pendulum $= 10^{-5}/^\circ C$)

(a) 5.86 s (b) 7.82 s (c) 8.64 s (d) 9.28 s

75 A pendulum clock is 5 s fast each day at a temperature of 15°C and 10 s slow at a temperature of 30°C. The temperature at which it is designed to give correct time, is

(a) 18°C (b) 20°C (c) 24°C (d) 25°C

76 Two rods of different material having linear expansion coefficients α_1 and α_2 and Young's modulii Y_1 and Y_2 are fixed between two rigid walls. Both the rods are heated to the same temperature. If there is no bending of the rods, the thermal stress developed in them are equal provided that

(a) $\frac{Y_1}{Y_2} = \frac{\alpha_1}{\alpha_2}$ (b) $\frac{Y_1}{Y_2} = \frac{\alpha_1}{(\alpha_1 + \alpha_2)}$ (c) $\frac{Y_1}{Y_2} = \frac{\alpha_2}{(\alpha_1 + \alpha_2)}$ (d) $\frac{Y_1}{Y_2} = \frac{\alpha_2}{\alpha_1}$

77 The increase in moment of inertia I of a uniform rod, having thermal coefficient of linear expansion α, about an axis passing through its one end and perpendicular to its length when its temperature is slightly increased by ΔT, is

(a) $I\alpha\Delta T$ (b) $\frac{I\alpha\Delta T}{2}$ (c) $2I\alpha\Delta T$ (d) $3I\alpha\Delta T$

78 The specific heat of many solids at low temperature varies with absolute temperature T according to the relation $s = aT^3$, where a is a constant. The heat energy required to raise the temperature of mass m of such a solid from $T = 0K$ to $T = 20K$ is

(a) $4 \times 10^4\ ma$ (b) $2 \times 10^3\ ma$ (c) $8 \times 10^6\ ma$ (d) $2 \times 10^6\ ma$

79 Two spheres A and B have diameters in the ratio 1 : 2, densities in the ratio 2 : 1 and specific heats in the ratio 1 : 3, find the ratio of their thermal capacities.

(a) 1 : 6 (b) 1 : 12

(c) 1 : 3 (d) 1 : 4

80 Three containers of the same volume contain three different gases. The masses of the molecules are m_1, m_2 and m_3 and the number of molecules in their respective containers are N_1, N_2 and N_3. The gas pressure in the containers are p_1, p_2 and p_3, respectively. All the gases are now mixed and put in one of the containers. The pressure p of mixture will be

(a) $p < (p_1 + p_2 + p_3)$
(b) $p = \frac{p_1 + p_2 + p_3}{3}$
(c) $p = p_1 + p_2 + p_3$
(d) $p > (p_1 + p_2 + p_3)$

81 Three liquids with masses m_1, m_2 and m_3 are thoroughly mixed. If their specific heats are c_1, c_2, c_3 and their temperatures T_1, T_2, T_3 respectively, then the temperature of the mixture in equilibrium is

(a) $\frac{c_1T_1 + c_2T_2 + c_3T_3}{m_1c_1 + m_2c_2 + m_3c_3}$
(b) $\frac{m_1c_1T_1 + m_2c_2T_2 + m_3c_3T_3}{m_1c_1 + m_2c_2 + m_3c_3}$
(c) $\frac{m_1c_1T_1 + m_2c_2T_2 + m_3c_3T_3}{m_1T_1 + m_2T_2 + m_3T_3}$
(d) $\frac{m_1T_1 + m_2T_2 + m_3T_3}{c_1T_1 + c_2T_2 + c_3T_3}$

82 Two substances A and B of equal masses m are heated at uniform rate of 6 cal s^{-1} under similar conditions. A graph between temperature and time is shown in figure. Ratio of heats absorbed (H_A / H_B) by them for complete fusion is

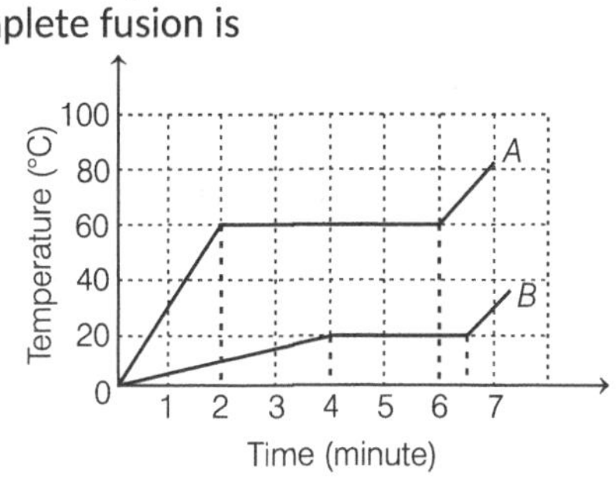

(a) 9/4
(b) 4/9
(c) 8/5
(d) 5/8

83 A small quantity, mass m of water at a temperature θ (in °C) is poured on a larger mass M of ice which is at its melting point. If c is the specific heat capacity of water, L the latent heat of fusion of ice, then the mass of ice melted is given by

(a) $\frac{ML}{mc\theta}$
(b) $\frac{Mc\theta}{L}$
(c) $\frac{mc\theta}{ML}$
(d) $\frac{mc\theta}{L}$

84 The emission spectrum of a black body at two different temperatures are shown by curves P and Q (as shown in figure). The ratio of the areas under the two curves P and Q will be

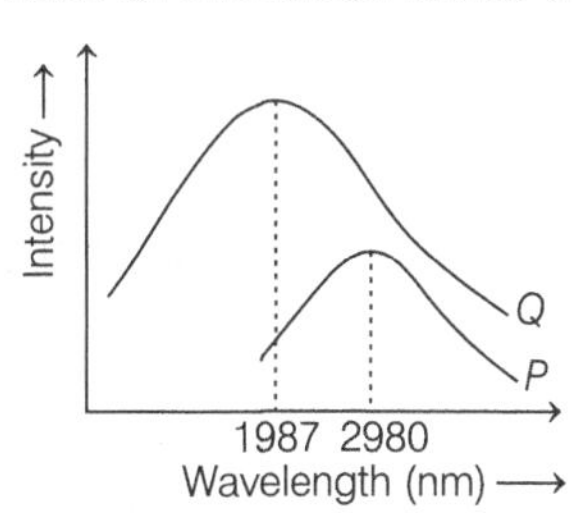

(a) 1 : 16
(b) 4 : 9
(c) 81 : 256
(d) 16 : 81

85 A cylindrical metallic rod in thermal contact with two reservoirs of heat at its two ends conduct an amount of heat Q in time t. The metallic rod is melted and the material is formed into a rod of one-third the radius of the original rod. What is the amount of heat conducted by the new rod, when placed in thermal contact with the two reservoirs in time t?

(a) $\frac{Q}{2}$
(b) $\frac{Q}{24}$
(c) $\frac{Q}{81}$
(d) $\frac{Q}{3}$

86 A lead bullet strikes against a steel plate with a velocity 200 ms^{-1}. If the impact is perfectly inelastic and the heat produced is equally shared between the bullet and the target, then the rise in temperature of the bullet is (Take, specific heat capacity of lead = 125 J $kg^{-1}K^{-1}$)

(a) 80°C (b) 60°C
(c) 160°C (d) 40°C

87 The equation of state for 5 g of nitrogen gas at a pressure p and temperature T, when occupying a volume V, will be

(a) $pV = 5\,RT$ (b) $pV = (5/2)\,RT$
(c) $pV = (5/14)\,RT$ (d) $pV = (5/28)\,RT$

88 During an adiabatic process, the pressure of a gas is found to be proportional to the cube of its absolute temperature. The ratio C_p/C_V for the gas is

(a) $\frac{3}{2}$ (b) $\frac{4}{3}$ (c) 2 (d) $\frac{5}{3}$

89 Find the work done by the gas in the process ABC.

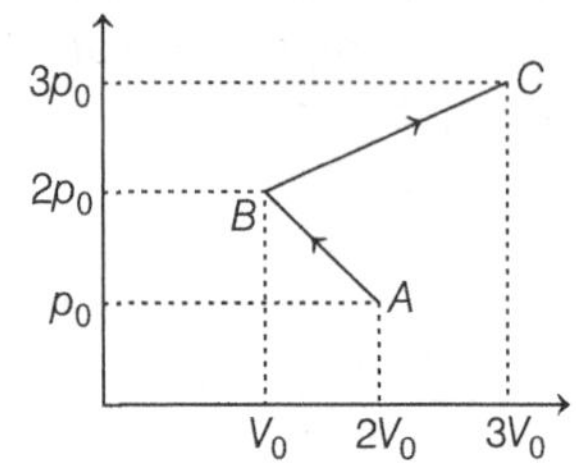

(a) $\frac{3}{2}p_0V_0$ (b) $\frac{5}{2}p_0V_0$ (c) $\frac{7}{2}p_0V_0$ (d) $4p_0V_0$

90 A cyclic process is shown in figure. Work done during isobaric expansion is

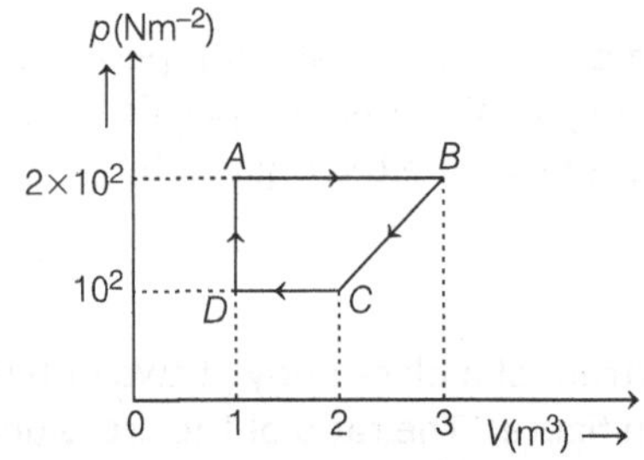

(a) 150 J (b) 100 J
(c) 400 J (d) 200 J

91 What is the nature of change in internal energy in the following three thermodynamical processes shown in figure.

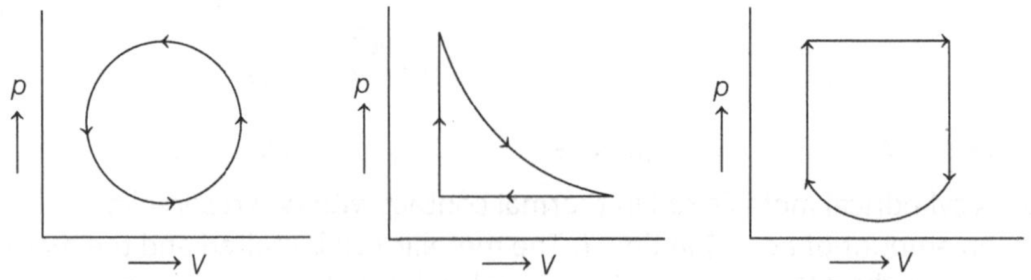

(a) ΔU is positive in all the three cases (b) ΔU is negative in all the three cases
(c) ΔU is + for (i), – for (ii), zero for (iii) (d) $\Delta U = 0$, in all the cases

92 A diatomic ideal gas undergoes a thermodynamic change according to the the p- V diagram shown in the figure. The heat given the gas in AB is

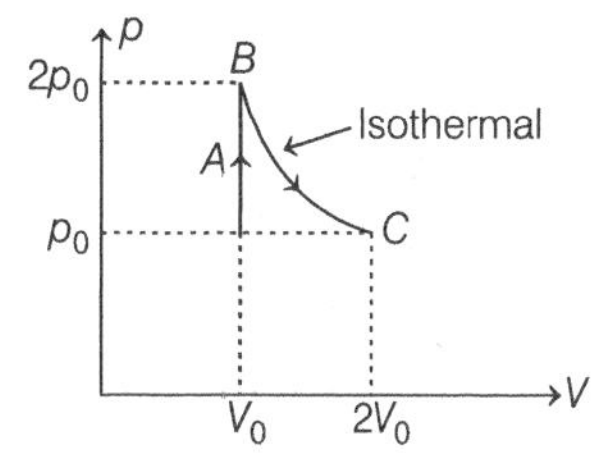

(a) $2.5\, p_0 V_0$
(b) $1.4\, p_0 V_0$
(c) $3.9\, p_0 V_0$
(d) $1.1\, p_0 V_0$

93 An ideal gas is taken along the path AB as shown in the figure. The work done by the gas is

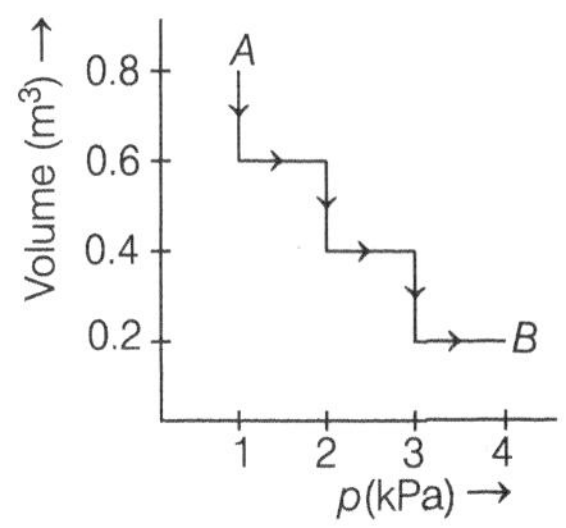

(a) 600 J
(b) 1200 J
(c) –600 J
(d) –1200 J

94 Consider two processes A and B on a system as shown in the figure. Let ΔW_1 and ΔW_2 be the work done by the system in the process A and B respectively, then

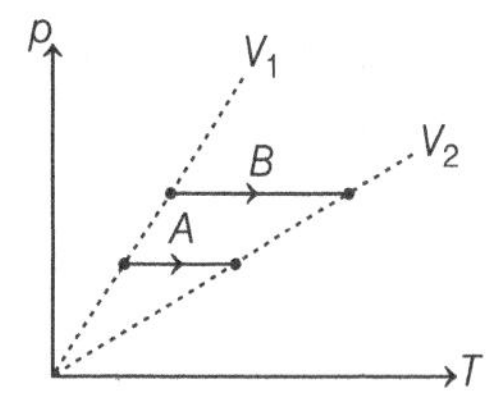

(a) $\Delta W_1 > \Delta W_2$
(b) $\Delta W_1 < \Delta W_2$
(c) $\Delta W_1 = \Delta W_2$
(d) Nothing can be said about the relation between ΔW_1 and ΔW_2

95 An ideal gas is taken from state 1 to state 2 through optional paths A, B, C and D as shown in the p-V diagram. Let Q, W and U represent the heat supplied, work done and change in internal energy of the gas, respectively. Then,

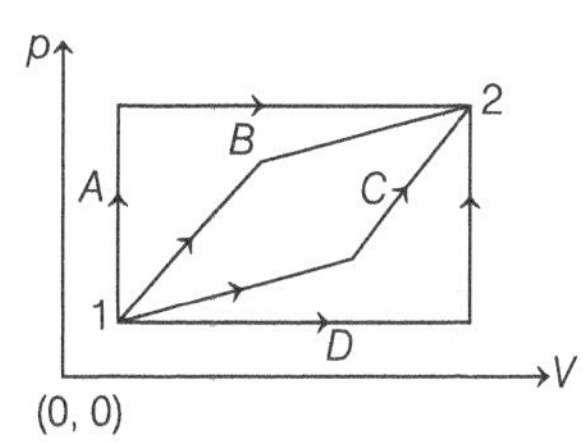

(a) $Q_A - Q_D = W_A - W_D$
(b) $Q_B - W_B > Q_C - W_C$
(c) $W_A < W_B < W_C - W_D$
(d) $Q_A < Q_B < Q_C < Q_D$

96 A system undergoes a process in which $\Delta U = +300$ J while absorbing 400 J of heat energy and undergoing an expansion against constant pressure of 0.5 bar. What is the change in the volume (in L) ?

(a) 4
(b) 5
(c) 2
(d) 3

97 Figure shows two paths that may be taken by a gas to go from a state A to a state C.

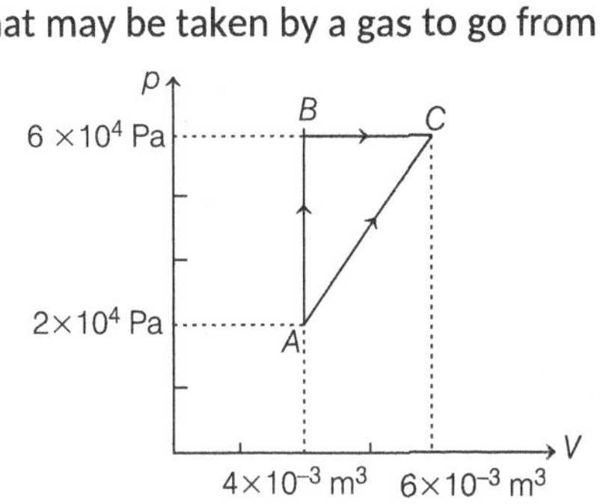

In process AB, 400 J of heat is added to the system and in process BC, 100 J of heat is added to the system. The heat absorbed by the system in the process AC will be

(a) 500 J (b) 460 J
(c) 300 J (d) 380 J

98 A cyclic process $ABCD$ is shown in the following p-V diagram. Which of the following curves represents the same process?

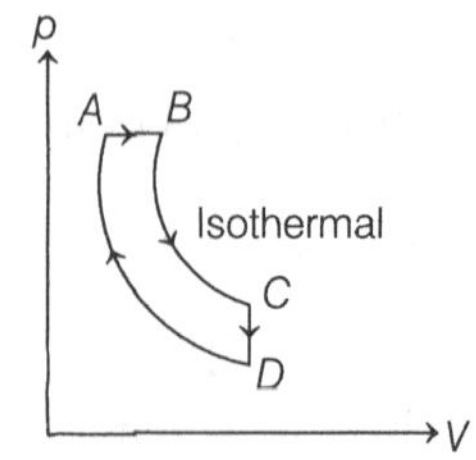

(a) 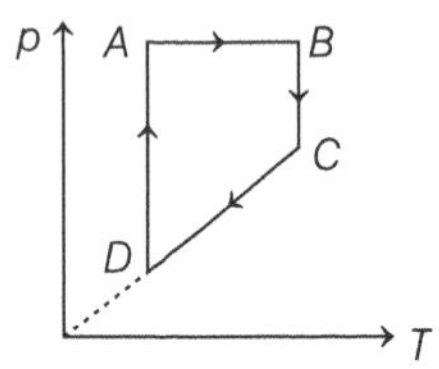

(b) 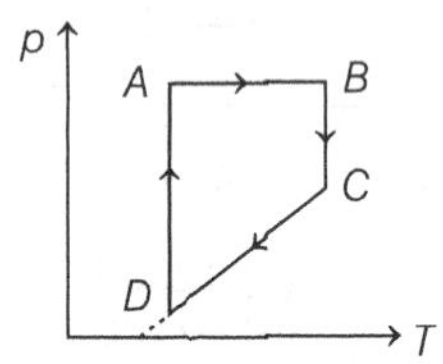

(c) 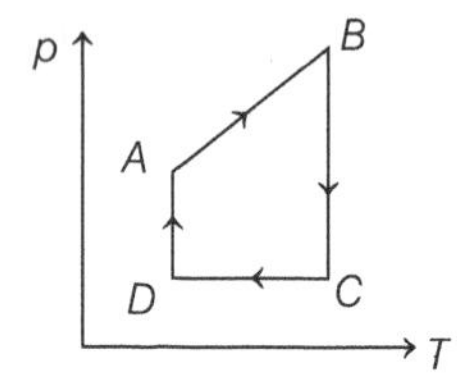

(d) 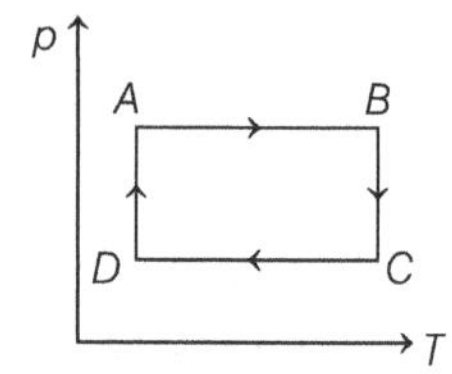

99 The equivalent p - V graph for the given p - U graph is

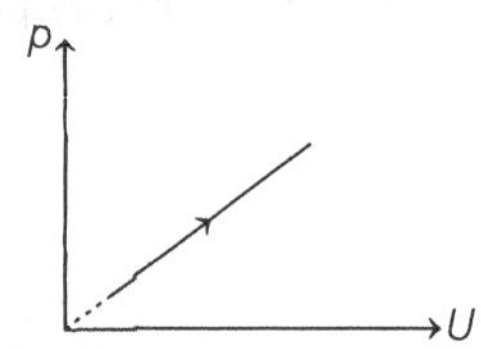

(a) 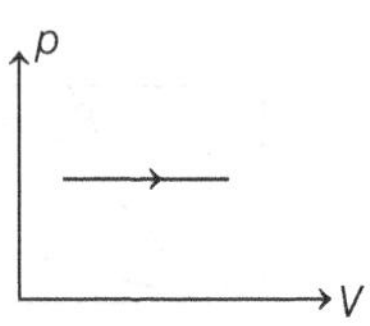

(b) 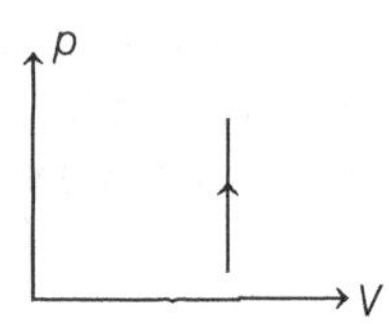

(c) 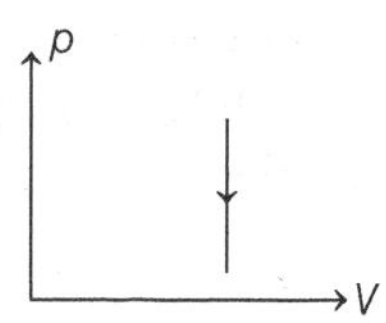

(d) 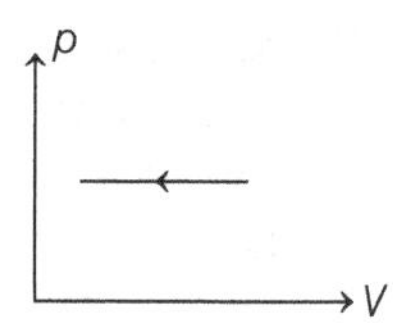

100 Consider p-V diagram for fixed mass of an ideal gas as shown in figure. Choose correct pressure-density (p-ρ) diagram.

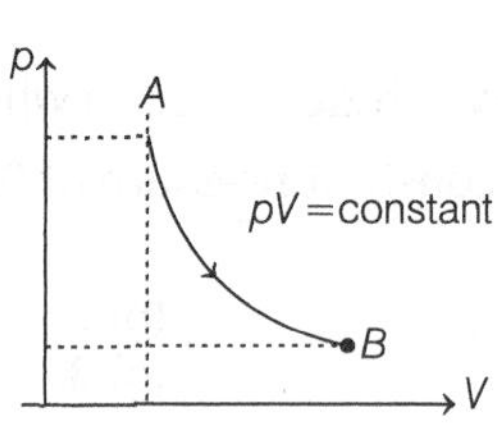

(a) 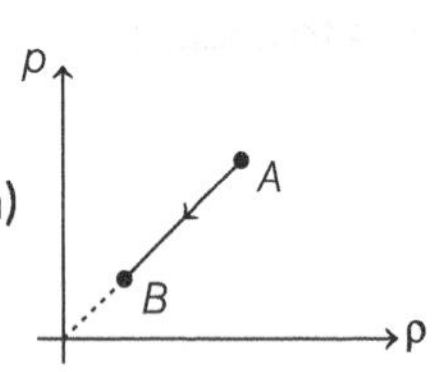

(b) 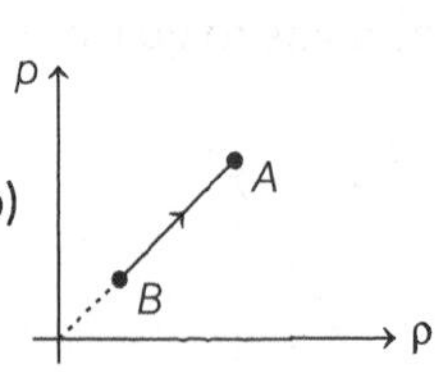

(c) 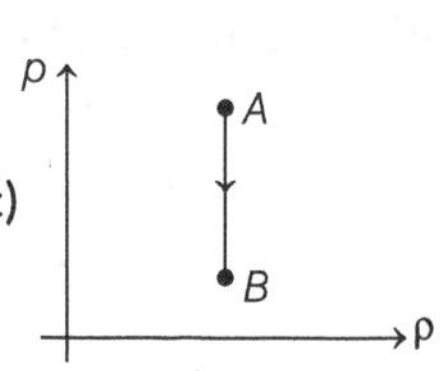

(d) 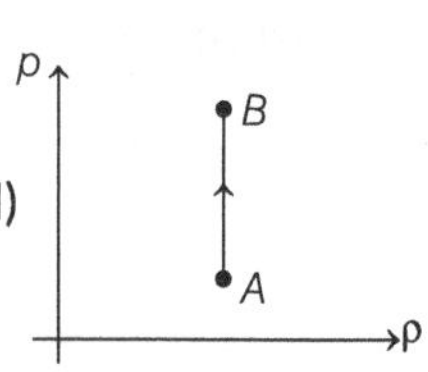

101 A gas has molar heat capacity $C = 4.5\,R$ in the process pT = constant. The number of degrees of freedom of molecules in the gas is

(a) 4 (b) 3
(c) 6 (d) 5

102 One mole of an ideal diatomic gas undergoes a transition from A to B along a path AB as shown in the figure. The change in internal energy of the gas during the transition is

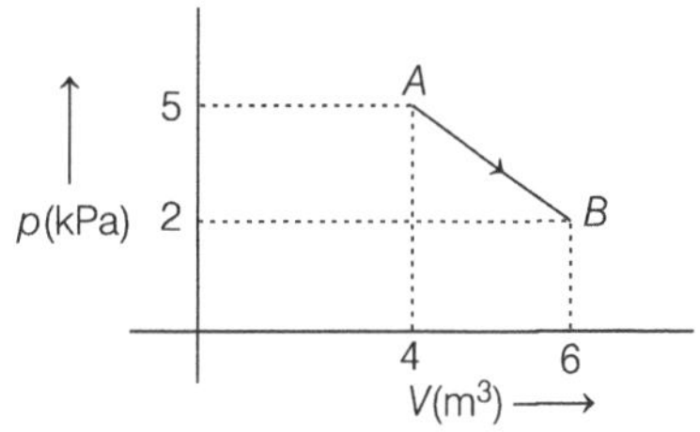

(a) – 20 kJ
(b) 20 J
(c) – 12 kJ
(d) 20 kJ

103 A polyatomic gas with n degrees of freedom has a mean energy per molecule given

(a) $\frac{nkT}{N}$
(b) $\frac{nkT}{2N}$
(c) $\frac{nkT}{2}$
(d) $\frac{3kT}{2}$

104 A gas mixture contains 1 g of H_2 and 1 g of He. If temperature of gas mixture is increased from 0°C to 100°C at isobaric process, then find heat given to the gas mixture.

$$\gamma_{He} = \frac{5}{3}; \gamma_{H_2} = \frac{7}{5}; R = \frac{2 \text{ cal}}{\text{mol-K}}$$

(a) 124 cal
(b) 327 cal
(c) 218 cal
(d) 457 cal

105 A thermodynamic process is shown in the figure. The pressures and volumes corresponding to some points in the figure are $p_A = 3 \times 10^4$ Pa, $p_B = 8 \times 10^4$ Pa and $V_A = 2 \times 10^{-3} \text{ m}^3$, $V_D = 5 \times 10^{-3} \text{ m}^3$. In process AB, 600 J of heat is added to the system and in process BC, 200 J of heat is added to the system. The change in internal energy of the system in process AC would be

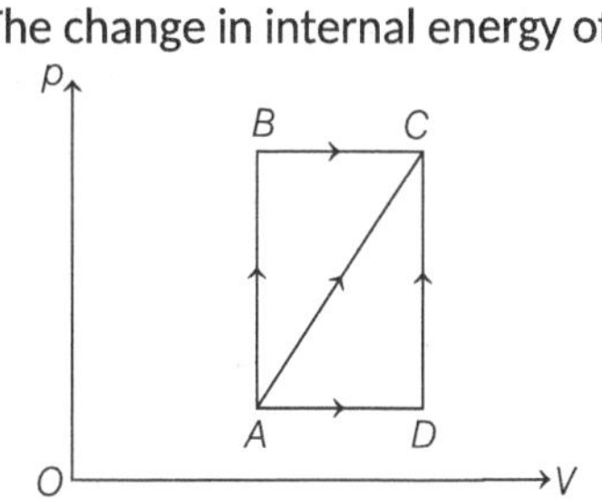

(a) 560 J
(b) 800 J
(c) 600 J
(d) 640 J

106 An ideal monoatomic gas undergoes a cyclic process $ABCA$ as shown in the figure. The ratio of heat absorbed during AB to the work done on the gas during BC is

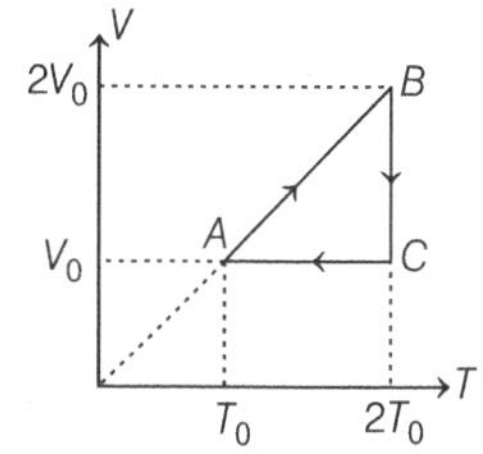

(a) $\frac{5}{2\ln 2}$
(b) $\frac{5}{3}$
(c) $\frac{5}{4\ln 2}$
(d) $\frac{5}{6}$

107 One mole of an ideal gas at an initial temperature of T K does $6R$ joules of work adiabatically. If the ratio of specific heats of this gas at constant pressure and at constant volume is 5/3, the final temperature of gas will be

(a) $(T + 2.4)$K
(b) $(T - 2.4)$K
(c) $(T + 4)$K
(d) $(T - 4)$K

108 A monoatomic ideal gas undergoes a process *ABC*. The heat given to the gas is

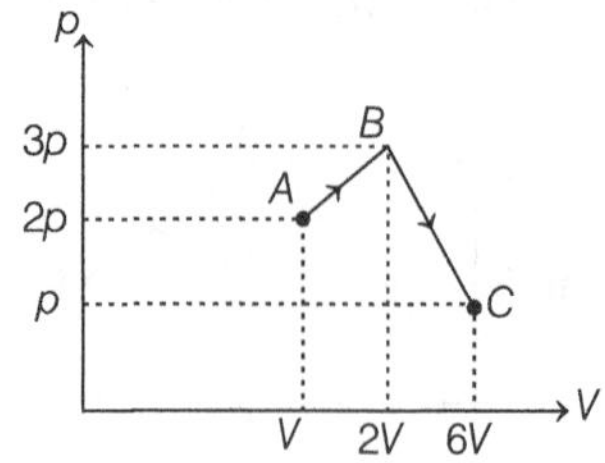

(a) 7.5 *pV* (b) 12.5 *pV*
(c) 16.5 *pV* (d) 20.5 *pV*

109 The *p*-*V* diagram of a system undergoing thermodynamic transformation is as shown in figure. The work done by the system in going from $A \rightarrow B \rightarrow C$ is 30 J and 40 J heat is given to the system. The change in internal energy between *A* and *C* is

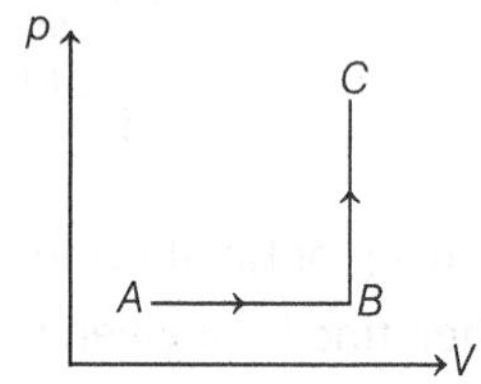

(a) 10 J (b) 70 J (c) 84 J (d) 134 J

110 *N* moles of an ideal diatomic gas are in a cylinder at temperature *T*. Suppose on supplying heat to the gas, its temperature remains constant but *n* moles get dissociated into atoms. Heat supplied to the gas is

(a) zero (b) $\frac{1}{2}nRT$ (c) $\frac{3}{2}nRT$ (d) $\frac{3}{2}(N-n)RT$

111 The graph shows logarithmic readings of pressure and volume for two ideal gases *A* and *B* undergoing adiabatic process. It can be concluded that (one of the gases *A* and *B* is monoatomic and one is diatomic)

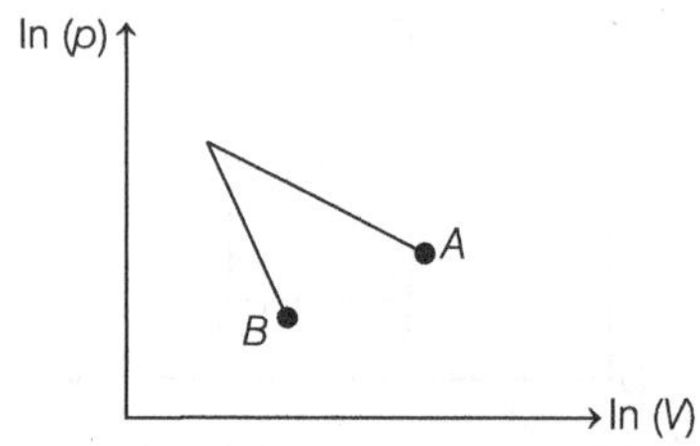

(a) *A* is diatomic (b) *B* is diatomic
(c) *B* is monoatomic (d) Both (a) and (c)

112 One mole of an ideal gas having initial volume *V*, pressure 2*p* and temperature *T* undergoes a cyclic process *ABCDA*. The net work done in the complete cycle is

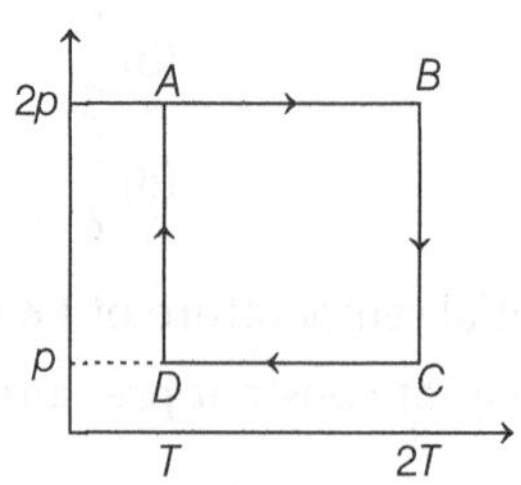

(a) zero (b) *RT* ln2
(c) 2*RT* ln2 (d) 1/2 *RT*ln2

113 The ratio of heat absorbed to work done by a monoatomic gas in a process is $\frac{5}{2}$. The process is shown in the figure. Find the parameters along x and y-axes.

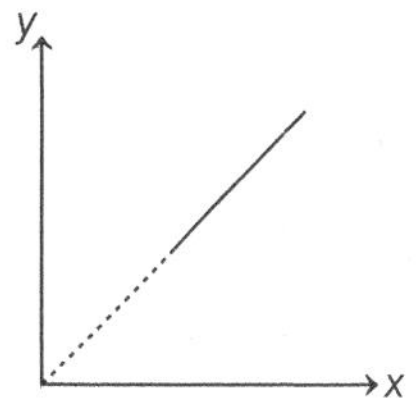

(a) Temperature and pressure
(b) Temperature and volume
(c) Internal energy and pressure
(d) Volume and density

114 The following graph shows two isotherms for a fixed mass of an ideal gas. The ratio of rms speed of the molecules at temperatures T_1 and T_2 is

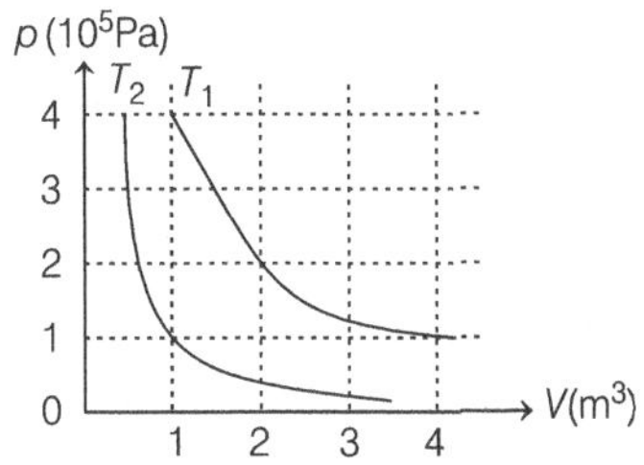

(a) $2\sqrt{2}$ (b) $\sqrt{2}$ (c) 2 (d) 4

115 One mole of an ideal gas undergoes a process $p = p_0\left[1+\left(\frac{2V_0}{V}\right)^2\right]^{-1}$, where p_0, V_0 are constants. Change in temperature of the gas when volume is changed from $V = V_0$ to $V = 2V_0$ is

(a) $\frac{4}{5}\frac{p_0V_0}{R}$ (b) $\frac{3}{4}\frac{p_0V_0}{R}$ (c) $\frac{2}{3}\frac{p_0V_0}{R}$ (d) $\frac{9}{7}\frac{p_0V_0}{R}$

116 Two identical vessels contain the same gas at pressures p_1 and p_2 at absolute temperatures T_1 and T_2, respectively. On joining the vessels with a small tube as shown in figure. The gas reaches a common temperature T and a common pressure p, determine the ratio p/T.

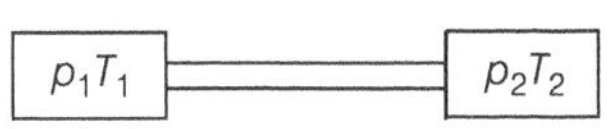

(a) $\left[\frac{p_1T_1 + p_2T_2}{T_1T_2}\right]$ (b) $\frac{1}{2}\left[\frac{p_1T_1 + p_2T_2}{T_1T_2}\right]$ (c) $\frac{1}{2}\left[\frac{p_1T_2 + p_2T_1}{T_1T_2}\right]$ (d) $\left[\frac{p_1T_2 + p_2T_1}{T_1T_2}\right]$

117 The quantity $\frac{2U}{fkT}$ represents (where, f = degree of freedom of the gas)

(a) mass of gas (b) kinetic energy of gas
(c) number of moles of gas (d) number of molecules in the gas

118 A Carnot engine whose low temperature reservoir is at 7°C has an efficiency of 50%. It is desired to increase the efficiency to 60%. By how many degrees should the temperature of the high temperature reservoir be increased?

(a) 140 K (b) 280 K
(c) 560 K (d) 380 K

119 In a Carnot engine, when $T_2 = 0°C$ and $T_1 = 200°C$, its efficiency is η_1 and when $T_1 = 0°C$ and $T_2 = -200°C$, its efficiency is η_2, then what is η_1/η_2?

(a) 0.577 (b) 0.733
(c) 0.638 (d) 0.427

120 The molar specific heat of the process $V \propto T^4$ for CH_4 gas at room temperature is

(a) $4R$ (b) $7R$ (c) $3R$ (d) $8R$

121 A cooking vessel on a slow burner contains 5 kg of water and unknown mass of ice in equilibrium at 0°C at time $t = 0$. The temperature of the mixture is measured at various time and the result is plotted as shown in given figure. During the first 50 min, the mixture remains at 0°C. From 50 min to 60 min, the temperature increases to 2°C. Neglecting the heat capacity of the vessel, the initial mass of the ice is

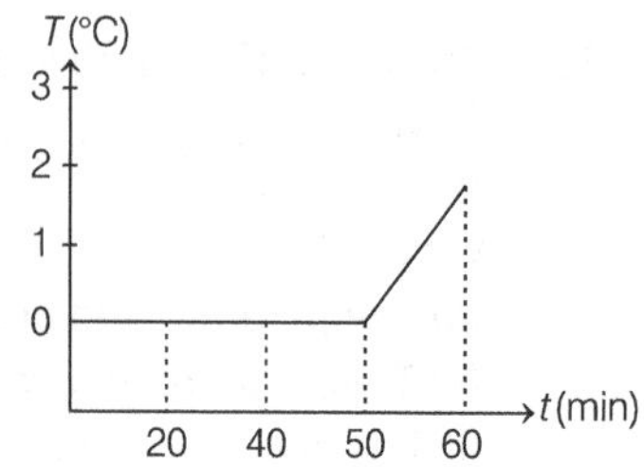

(a) $\frac{10}{7}$ kg (b) $\frac{5}{7}$ kg

(c) $\frac{5}{4}$ kg (d) $\frac{5}{8}$ kg

122 The internal energy of an ideal diatomic gas corresponding to volume V and pressure p is $U = 2.5\,pV$. The gas expands from 1 litre to 2 litre at a constant pressure of one atmosphere. The heat supplied to the gas is approximately

(a) 100 J (b) 250 J (c) 350 J (d) 50 J

123 20 g ice at −10°C is mixed with m g steam at 100°C. The minimum value of m, so that finally all ice and steam converts into water is (Use $s_{ice} = 0.5$ cal / g°C, $s_{water} = 1$ cal/g°C), L(melting) = 80 cal/g and L(vaporisation = 540 cal/g)

(a) $\frac{185}{27}$ g (b) $\frac{135}{17}$ g

(c) $\frac{85}{32}$ g (d) $\frac{113}{17}$ g

124 An ideal gas is expanding such that pT^2 = constant. The coefficient of volume expansion of the gas is

(a) $\frac{1}{T}$ (b) $\frac{2}{T}$ (c) $\frac{3}{T}$ (d) $\frac{4}{T}$

125 For an adiabatic expansion of a perfect gas, the value of $\Delta p/p$ is equal to

(a) $\frac{\Delta V}{V}$ (b) $\gamma \frac{\Delta V}{V}$ (c) $-\gamma \frac{\Delta V}{V}$ (d) $-\gamma^2 \frac{\Delta V}{V}$

126 The graph of pressure *versus* volume of a monoatomic ideal gas is as shown. What is the internal energy of the system in state 1 and the heat given to the system during the process 4 to 1? Use SI system.

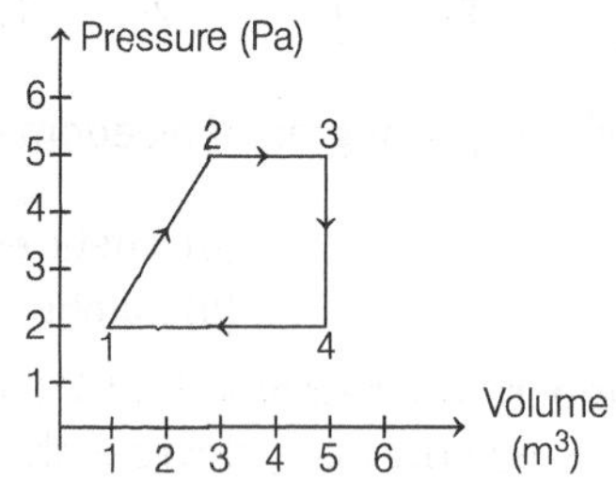

(a) 15, −5 (b) 3, −20 (c) 7, 9 (d) 7, −10

127 28 g of N_2 gas is contained in a flask at a pressure of 10 atm and at a temperature of 57°C. It is found that due to leakage in the flask, the pressure is reduced to half the initial value and the temperature to 27°C. The quantity of N_2 gas that leakes out is

(a) $\frac{5}{63}$ g (b) $\frac{63}{5}$ g

(c) $\frac{11}{20}$ g (d) $\frac{20}{11}$ g

128 Six moles of an ideal gas performs a cycle as shown in figure. If the temperature $T_A = 600$ K, $T_B = 800$ K, $T_C = 2200$ K and $T_D = 1200$ K, work done by the gas per cycle is

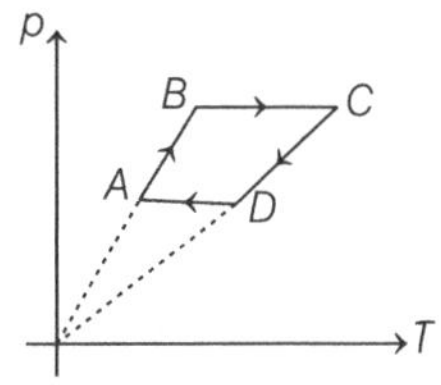

(a) 20 kJ (b) 30 kJ (c) 40 kJ (d) 60 kJ

129 1 mole of a gas expands with temperature according to the relation $V = kT^{2/3}$. When the temperature changes by 30°C, the work done will be

(a) $10R$ (b) $20R$
(c) $30R$ (d) $40R$

130 A cylindrical tube of uniform cross-sectional area A is fitted with two air-tight frictionless pistons. The pistons are connected to each other by a metallic wire. Initially the pressure of the gas is p_0 and temperature is T_0. Atmospheric pressure is also p_0. Now, the temperature of the gas is increased to $2T_0$. The tension in the wire will be

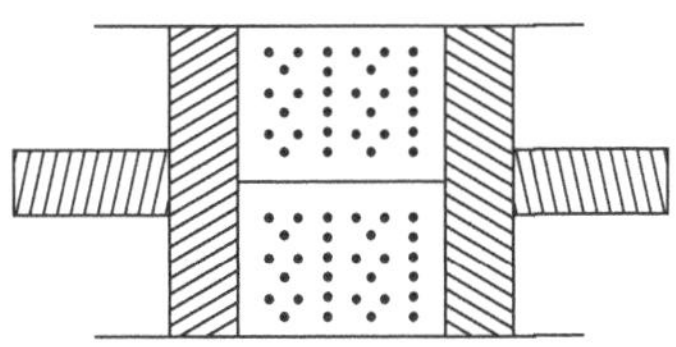

(a) $p_0 A$ (b) $2p_0 A$ (c) $\frac{p_0 A}{2}$ (d) $4p_0 A$

131 A sample of gas follows a process represented by $pV^2 =$ constant. Bulk modulus for this process is B, then which of the following graph is correct?

(a) 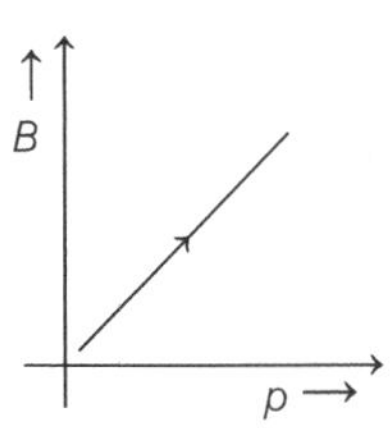

(b) 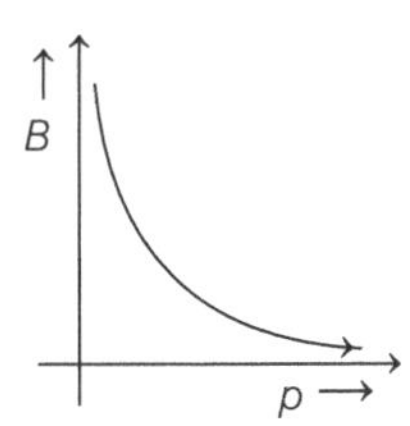

(c) 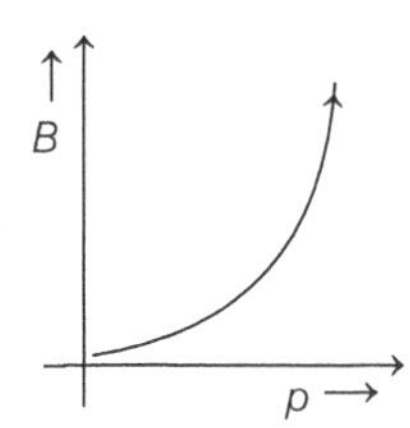

(d) None of these

132 Logarithms of readings of pressure and volume for an ideal gas were plotted on a graph as shown in figure. By measuring the gradient, it can be shown that the gas may be

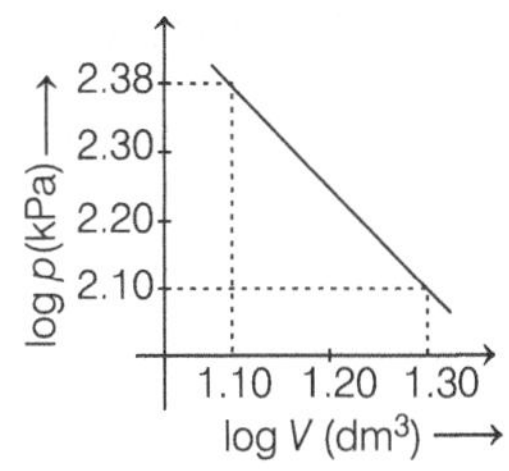

(a) monoatomic and undergoing an adiabatic change
(b) monoatomic and undergoing an isothermal change
(c) diatomic and undergoing an adiabatic change
(d) triatomic and undergoing an isothermal change

133 A monoatomic ideal gas undergoes a cyclic process *ABCDEA* as shown in the *p*-*V* diagram. (*BC* is isothermal and $p_3V_1 < p_1V_3$). Then, which of the following curve is correctly converted?

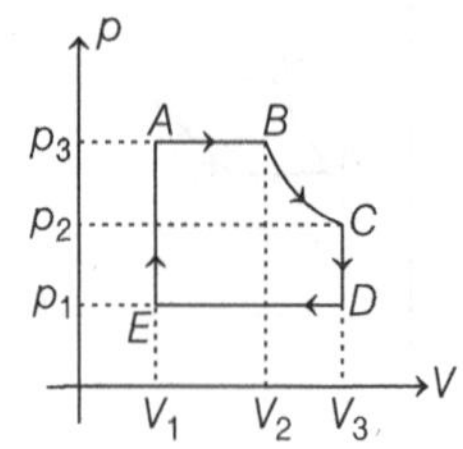

(a) 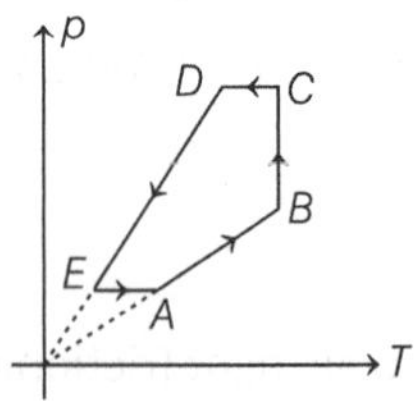

(b) 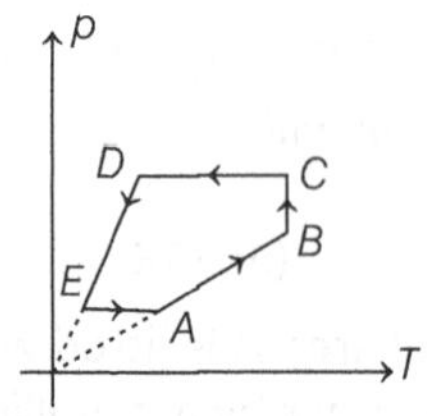

(c) 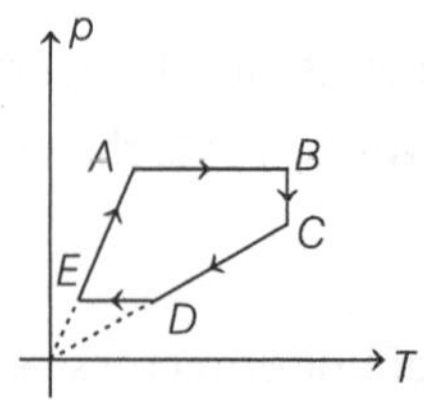

(d) None of these

134 Which of the following graphs correctly shows the change in internal energy of an ideal gas with pressure for the isobaric, isochoric and isothermal processes?

(a) 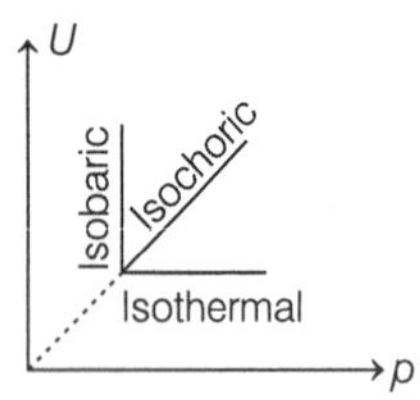

(b) 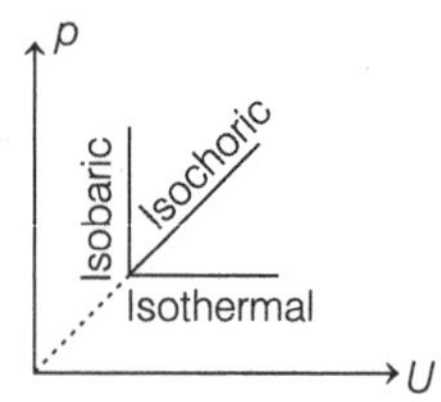

(c) 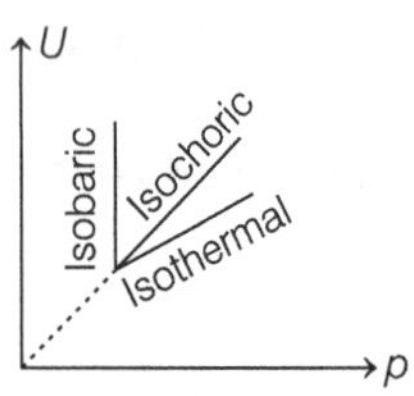

(d) 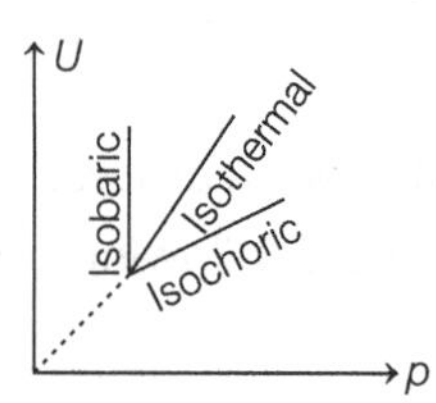

135 A gas is undergoing an adiabatic process. From the details given in the graph, find the value of adiabatic constant.

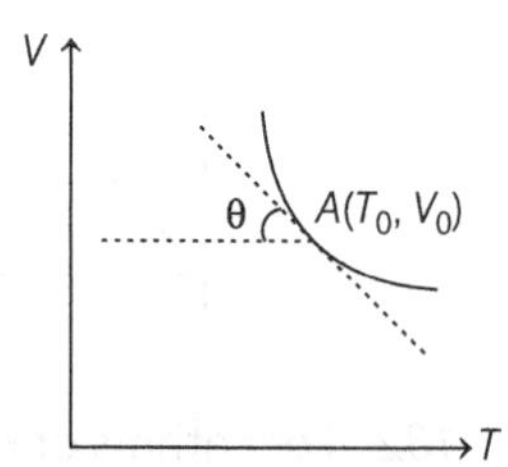

(a) $\dfrac{V_0}{T_0 \tan\theta} + 1$

(b) $\dfrac{V_0 \tan\theta}{T_0} + 1$

(c) $\dfrac{V_0 \tan^2\theta}{T_0} + 1$

(d) $\dfrac{V_0}{T_0} + \tan\theta$

136 The temperature of the two outer surfaces of a composite slab consisting of two materials having coefficients of thermal conductivity *K* and 2*K* and thickness *x* and 3*x* respectively are T_2 and T_1 ($T_2 > T_1$). The rate of heat transfer through the slab, in a steady state is $\left[\dfrac{A(T_2 - T_1)K}{x}\right] f$, with *f* equal to

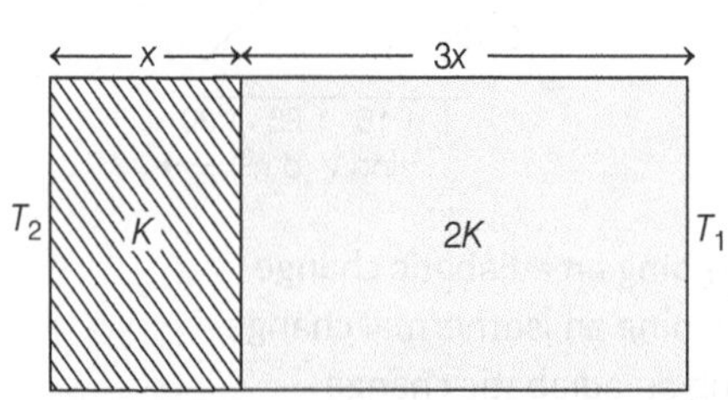

(a) 1
(b) 1/2
(c) 2/5
(d) 1/5

137 A gas is contained in a vessel of volume V_0 at a pressure p_0. If the gas is to be pumped out by a suction pump of stroke volume V, then number of moles of gas remained in the vessel after two strokes is (where, R is gas constant, T is constant temperature of gas)

(a) $\dfrac{p_0V_0^3}{RT(V_0+V)^2}$ (b) $\dfrac{p_0V_0^2}{RT(V_0+V)^2}$ (c) $\dfrac{p_0V_0^3}{RT(V_0+V)^3}$ (d) $\dfrac{p_0V_0^2}{RT(V_0+V)^3}$

138 Two moles of helium gas are taken along the path $ABCD$ (as shown in figure). The work done by the gas is

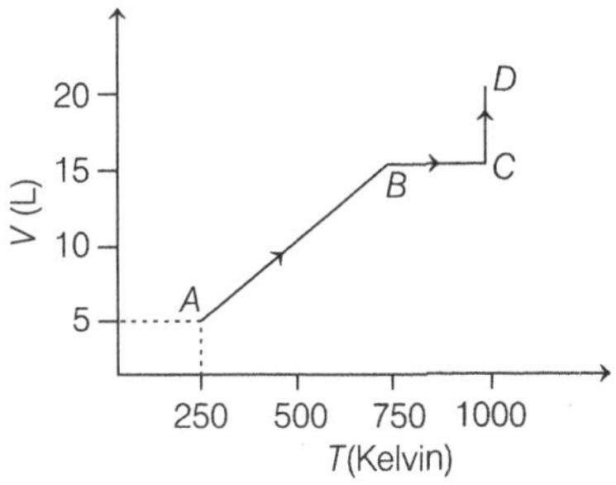

(a) $2000R\left(1+\ln\dfrac{4}{3}\right)$ (b) $5000R\,(3+\ln 4)$ (c) $5000R\left(2+\ln\dfrac{16}{9}\right)$ (d) $1000R\left(1+\ln\dfrac{16}{9}\right)$

139 One mole of a gas is subjected to two processes AB and BC, one after the other as shown in the figure. BC is represented by pV^n = constant. Choose correct option,
(where, T = temperature, W = work done by gas, V = volume and U = internal energy).

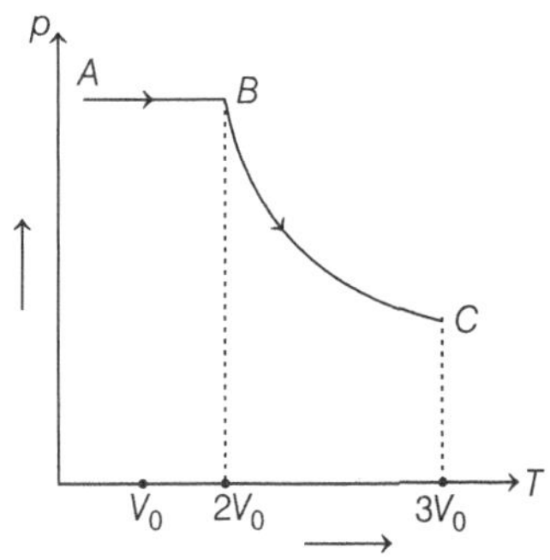

(a) $T_A = T_B = T_C$ (b) $V_A < V_B$, $p_B < p_C$ (c) $W_{AB} < W_{BC}$ (d) $U_A < U_B$

140 In the following p-V diagram, two adiabatics cut two isothermals at temperatures T_1 and T_2 as shown in figure. The value of $\dfrac{V_a}{V_d}$ will be

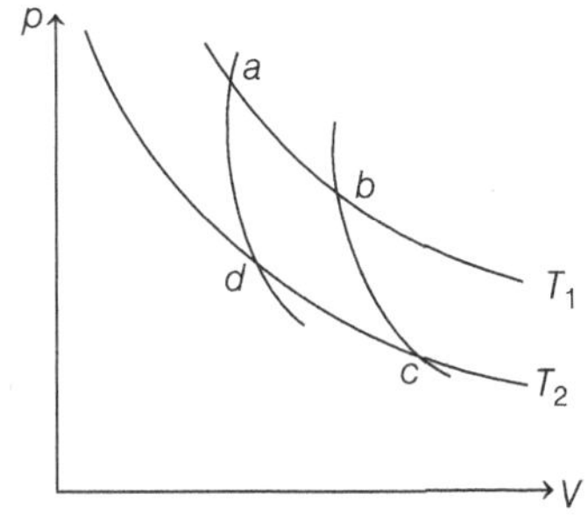

(a) $\dfrac{V_b}{V_c}$ (b) $\dfrac{V_c}{V_b}$ (c) $\dfrac{V_d}{V_a}$ (d) $V_b\,V_c$

141 Two different rods A and B are kept as shown in figure.

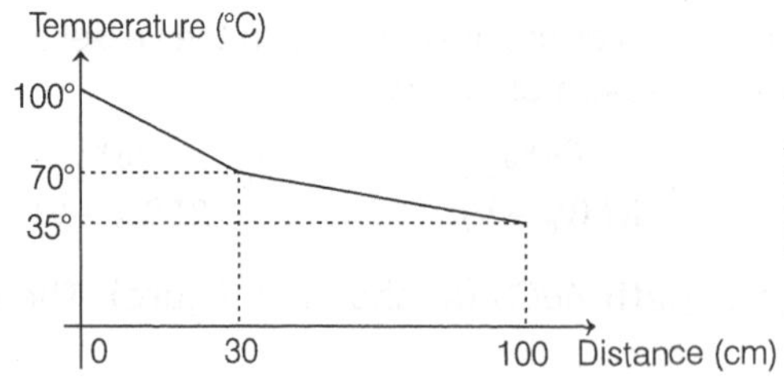

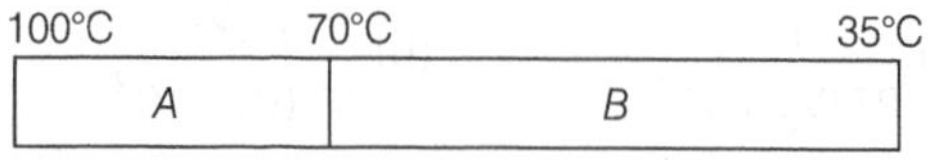

The variation of temperature of different cross-sections is plotted in a graph as shown in figure. The ratio of thermal conductivities of A and B is

(a) 2 (b) 0.5 (c) 1 (d) 2/3

142 An ideal diatomic gas undergoes a thermodynamic process, where pressure (p) as a function of volume (V) is given by $p = p_0 e^{\frac{3V_0}{V}}$. The average rotational kinetic energy of the gas molecules at volume $V = \frac{V_0}{2}$ is

(a) $\frac{p_0 V_0}{2} e^6$ (b) $p_0 V_0$ (c) $\frac{3p_0 V_0}{4} e^6$ (d) $\frac{5}{4} p_0 V_0 e^6$

143 Two elastic rods are joined between fixed supports as shown in the figure. Condition for no change in the lengths of individual rods with the increase of temperature is (α_1, α_2 = linear expansion coefficient, A_1, A_2 = area of rods, Y_1, Y_2 = Young's modulus)

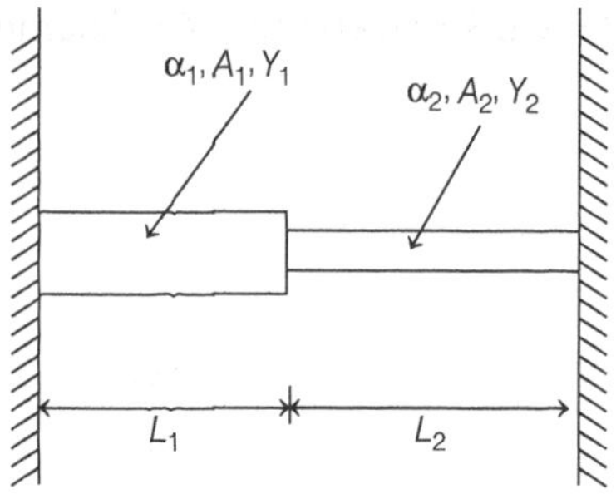

(a) $\frac{A_1}{A_2} = \frac{\alpha_1 Y_1}{\alpha_2 Y_2}$ (b) $\frac{A_1}{A_2} = \frac{L_1 \alpha_1 Y_1}{L_2 \alpha_2 Y_2}$ (c) $\frac{A_1}{A_2} = \frac{L_2 \alpha_2 Y_2}{L_1 \alpha_1 Y_1}$ (d) $\frac{A_1}{A_2} = \frac{\alpha_2 Y_2}{\alpha_1 Y_1}$

144 Two identical rectangular rods of metal are welded end-to-end in series between temperatures 0°C and 100°C and 10 J of heat is conducted (in steady state process) through the rod in 2.00 min. If 5 such rods are taken and joined as shown in figure maintaining the same temperature difference between A and B, then the time in which 20 J heat will flow through the rods is

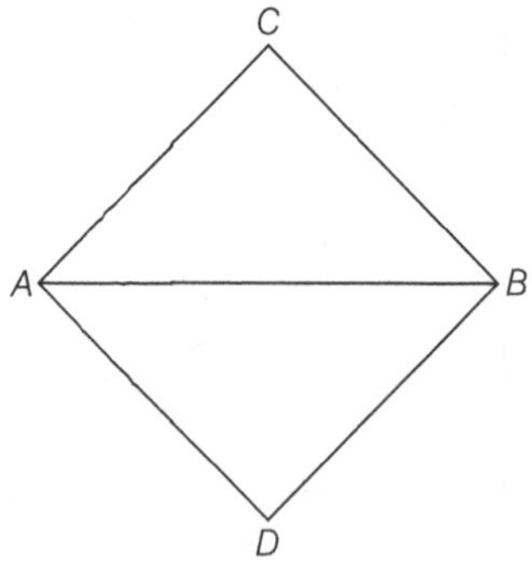

(a) 30 s (b) 2 min (c) 1 min (d) 20 s

145 A cylindrical piston of mass M slides smoothly inside a long cylinder closed at one end, enclosing a certain mass of gas. The cylinder is kept with its axis horizontal. If the piston in disturbed from its equilibrium position, it oscillates simple harmonically. The period of oscillation will be (Assume T = constant)

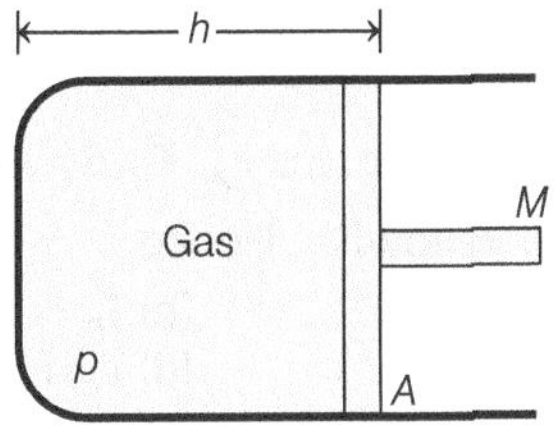

(a) $T = 2\pi\sqrt{\left(\frac{Mh}{pA}\right)}$ (b) $\frac{1}{2\pi}\sqrt{\frac{MA}{ph}}$ (c) $T = 2\pi\sqrt{\left(\frac{M}{pAh}\right)}$ (d) $T = 2\pi\sqrt{MphA}$

146 An ideal gas is taken through a process, where $dQ = dU - dW$. The specific heat of the ideal gas in this process is (dQ = heat supplied to the gas, dU = change in internal energy, dW = work done by gas)

(a) C_V (b) $C_V + \frac{3}{2}R$

(c) $C_V - R$ (d) the process is impossible

147 Certain perfect gas is found to obey pV^n = constant during adiabatic process. The volume expansion coefficient at temperature T is

(a) $\frac{1-n}{T}$ (b) $\frac{1}{(1-n)T}$ (c) $\frac{n}{T}$ (d) $\frac{1}{nT}$

148 The relation between internal energy U, pressure p and volume V of a gas in an adiabatic process is $U = a + bpV$, where a and b are positive constants. The value of ratio of molar specific heats (γ) is

(a) $\frac{a}{b}$ (b) $\frac{b+1}{b}$ (c) $\frac{a+1}{a}$ (d) $\frac{b}{a}$

LEVEL 03 Previous Years Questions

149 A cup of coffee cools from 90°C to 80°C in t minute, when the room temperature is 20°C. The time taken by cup of coffee to cool from 80°C to 60°C in same room at 20°C, is **[NEET 2021]**

(a) $\frac{13}{10}t$ (b) $\frac{13}{5}t$ (c) $\frac{10}{13}t$ (d) $\frac{5}{13}t$

150 Match Column I with Column II and choose the correct match from the given choices. **[NEET 2021]**

	Column I		Column II
A.	Root mean square of gas molecules	1.	$\frac{1}{3}nm\bar{v}^2$
B.	Pressure exerted by ideal gas	2.	$\sqrt{\frac{3RT}{M}}$
C.	Average kinetic energy of a molecule	3.	$\frac{5}{2}RT$
D.	Total internal energy of 1 mole of a diatomic gas	4.	$\frac{3}{2}k_BT$

Codes

	A	B	C	D
(a)	3	1	4	2
(b)	2	3	4	1
(c)	2	1	4	3
(d)	3	2	1	4

151 Three stars *A*,*B*,*C* have surface temperatures T_A, T_B, T_C, respectively. Star *A* appears bluish, start *B* appears reddish and star *C* yellowish. Hence, **[NEET 2020]**

(a) $T_A > T_B > T_C$ (b) $T_B > T_C > T_A$
(c) $T_C > T_B > T_A$ (d) $T_A > T_C > T_B$

152 The *p*-*V* diagram for an ideal gas in a piston cylinder assembly undergoing a thermodynamic process is shown in the figure. The process is **[NEET 2020]**

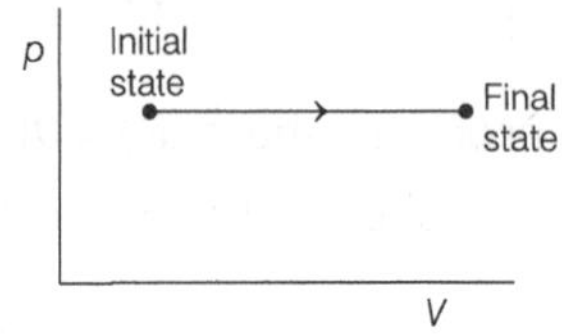

(a) adiabatic (b) isochoric
(c) isobaric (d) isothermal

153 The mean free path *l* for a gas molecule depends upon diameter, *d* of the molecule as **[NEET 2020]**

(a) $l \propto \frac{1}{d^2}$ (b) $l \propto d$ (c) $l \propto d^2$ (d) $l \propto \frac{1}{d}$

154 An ideal gas equation can be written as $p = \frac{\rho RT}{M_0}$, where ρ and M_0 are respectively, **[NEET 2020]**

(a) mass density, mass of the gas (b) number density, molar mass
(c) mass density, molar mass (d) number density, mass of the gas

155 The efficiency of a Carnot engine depends upon **[NEET 2020]**

(a) the temperature of the sink only (b) the temperatures of the source and sink
(c) the volume of the cylinder of the engine (d) the temperature of the source only

156 The quantities of heat required to raise the temperature of two solid copper spheres of radii r_1 and r_2 ($r_1 = 1.5\, r_2$) through 1 K are in the ratio **[NEET 2020]**

(a) $\frac{9}{4}$ (b) $\frac{3}{2}$ (c) $\frac{5}{3}$ (d) $\frac{27}{8}$

157 Two cylinders *A* and *B* of equal capacity are connected to each other *via* a stop cock. A contains an ideal gas at standard temperature and pressure. *B* is completely evacuated. The entire system is thermally insulated. The stop cock is suddenly opened. The process is **[NEET 2020]**

(a) adiabatic (b) isochoric
(c) isobaric (d) isothermal

158 The mean free path λ for a gas, with molecular diameter *d* and number density *n* can be expressed as **[NEET 2020]**

(a) $\frac{1}{\sqrt{2}\, n\pi d^2}$ (b) $\frac{1}{\sqrt{2}\, n^2 \pi d^2}$ (c) $\frac{1}{\sqrt{2}\, n^2 \pi^2 d^2}$ (d) $\frac{1}{\sqrt{2}\, n\pi d}$

159 A cylinder contains hydrogen gas at pressure of 249 kPa and temperature 27°C. Its density is (Take, $R = 8.3\ \text{J mol}^{-1}\ \text{K}^{-1}$) **[NEET 2020]**

(a) 0.2 kg/m^3 (b) 0.1 kg/m^3
(c) 0.02 kg/m^3 (d) 0.5 kg/m^3

160 The average thermal energy for a monoatomic gas is (where, k_B is Boltzmann constant and T is absolute temperature.) **[NEET 2020]**

(a) $\frac{3}{2}k_BT$ (b) $\frac{5}{2}k_BT$ (c) $\frac{7}{2}k_BT$ (d) $\frac{1}{2}k_BT$

161 A copper rod of 88 cm and an aluminium rod of unknown length have their increase in length independent of increase in temperature. The length of aluminium rod is **[NEET 2019]**

(a) 113.9 cm (b) 88 cm
(c) 68 cm (d) 6.8 cm

162 An object kept in a large room having air temperature of 25°C takes 12 min to cool from 80°C to 70°C. The time taken to cool for the same object from 70°C to 60°C would be nearly **[NEET 2019]**

(a) 10 min (b) 12 min
(c) 20 min (d) 15 min

163 A deep rectangular pond of surface area A, containing water (density = ρ, specific heat capacity = s), is located in a region, where the outside air temperature is a steady value at the -26°C. The thickness of the frozen ice layer in this pond, at a certain instant is x.

Taking the thermal conductivity of ice as K and its specific latent heat of fusion as L, the rate of increase of the thickness of ice layer, at this instant would be given by **[NEET 2019]**

(a) $26K/\rho r(L-4s)$ (b) $26K/(\rho x^2 - L)$
(c) $26K/(\rho xL)$ (d) $26K/\rho r(L+4s)$

164 1 g of water of volume 1 cm^3 at 100°C is converted into steam at same temperature under normal atmospheric pressure ($\simeq 1\times10^5$ Pa). The volume of steam formed equals 1671 cm^3. If the specific latent heat of vaporisation of water is 2256 J/g, the change in internal energy is **[NEET 2019]**

(a) 2423 J (b) 2089 J
(c) 167 J (d) 2256 J

165 In which of the following processes, heat is neither absorbed nor released by a system? **[NEET 2019]**

(a) Adiabatic (b) Isobaric
(c) Isochoric (d) Isothermal

166 Increase in temperature of a gas filled in a container would lead to **[NEET 2019]**

(a) increase in its kinetic energy (b) decrease in its pressure
(c) decrease in intermolecular distance (d) increase in its mass

167 The value of $\gamma\left(=\frac{C_p}{C_V}\right)$, for hydrogen, helium and another ideal diatomic gas X (whose molecules are not rigid but have an additional vibrational mode), are respectively equal to **[NEET 2019]**

(a) $\frac{7}{5},\frac{5}{3},\frac{9}{7}$ (b) $\frac{5}{3},\frac{7}{5},\frac{9}{7}$
(c) $\frac{5}{3},\frac{7}{5}\frac{7}{5}$ (d) $\frac{7}{5},\frac{5}{3},\frac{7}{5}$

168 The power radiated by a black body is P and it radiates maximum energy at wavelength, λ_0. If the temperature of the black body is now changed, so that it radiates maximum energy at wavelength $\frac{3}{4}\lambda_0$, the power radiated by it becomes nP. The value of n is **[NEET 2018]**

(a) $\frac{256}{81}$ (b) $\frac{4}{3}$
(c) $\frac{3}{4}$ (d) $\frac{81}{256}$

169 A sample of 0.1 g of water at 100°C and normal pressure (1.013×10^5 Nm^{-2}) requires 54 cal of heat energy to convert to steam at 100°C. If the volume of the steam produced is 167.1 cc, the change in internal energy of the sample, is **[NEET 2018]**

(a) 42.2 J (b) 208.7 J (c) 104.3 J (d) 84.5 J

170 The volume (V) of a monoatomic gas varies with its temperature (T), as shown in the graph. The ratio of work done by the gas, to the heat absorbed by it, when it undergoes a change from state A to state B, is **[NEET 2018]**

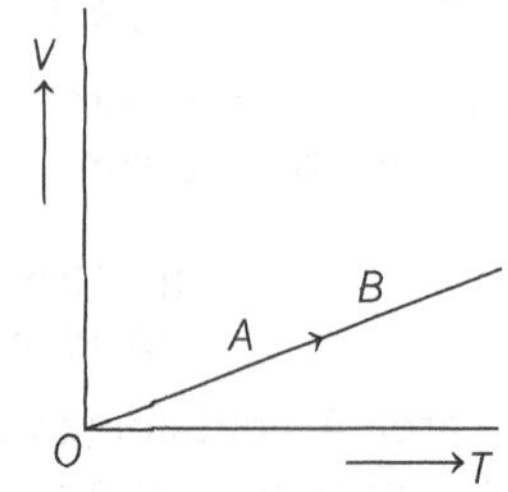

(a) $\frac{1}{3}$ (b) $\frac{2}{3}$ (c) $\frac{2}{5}$ (d) $\frac{2}{7}$

171 The efficiency of an ideal heat engine working between the freezing point and boiling point of water, is **[NEET 2018]**

(a) 6.25% (b) 20% (c) 26.8% (d) 12.5%

172 At what temperature will the rms speed of oxygen molecules become just sufficient for escaping from the earth's atmosphere? (Take, mass of oxygen molecule, $m = 2.76 \times 10^{-26}$ kg, Boltzmann's constant, $k_B = 1.38 \times 10^{-23}$ JK^{-1}.) **[NEET 2018]**

(a) 5.016×10^4 K (b) 8.326×10^4 K
(c) 2.508×10^4 K (d) 1.254×10^4 K

173 Two rods A and B of different materials are welded together as shown in figure. Their thermal conductivities are K_1 and K_2. The thermal conductivity of the composite rod will be **[NEET 2017]**

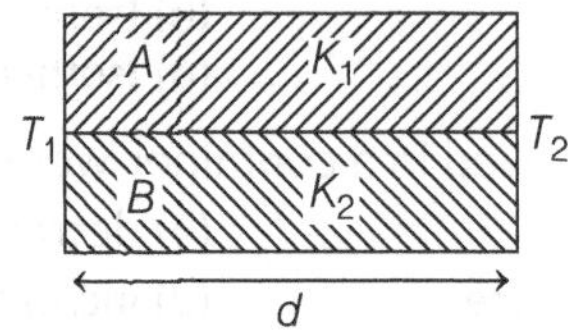

(a) $\frac{K_1 + K_2}{2}$ (b) $\frac{3(K_1 + K_2)}{2}$ (c) $K_1 + K_2$ (d) $2(K_1 + K_2)$

174 A spherical black body with a radius of 12 cm radiates 450 W power at 500 K. If the radius were halved and the temperature doubled, the power radiated in watt would be **[NEET 2017]**

(a) 225 (b) 450
(c) 1000 (d) 1800

175 Thermodynamic processes are indicated in the following diagram **[NEET 2017]**

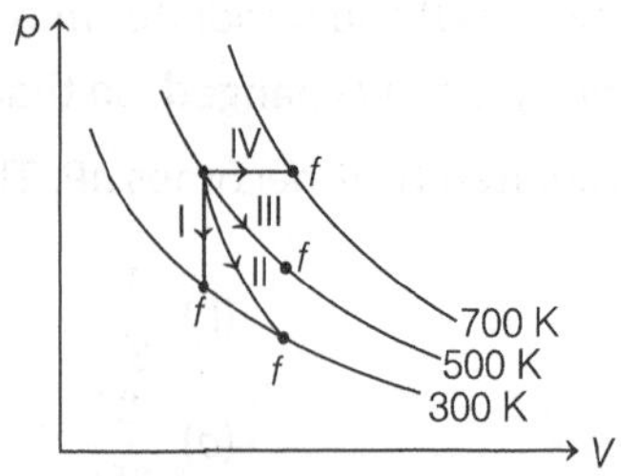

Match the following :

	Column I		Column II
P.	Process I	a.	Adiabatic
Q.	Process II	b.	Isobaric
R.	Process III	c.	Isochoric
S.	Process IV	d.	Isothermal

(a) $P \rightarrow a, Q \rightarrow c, R \rightarrow d, S \rightarrow b$ (b) $P \rightarrow c, Q \rightarrow a, R \rightarrow d, S \rightarrow b$
(c) $P \rightarrow c, Q \rightarrow d, R \rightarrow b, S \rightarrow a$ (d) $P \rightarrow d, Q \rightarrow b, R \rightarrow a, S \rightarrow c$

176 A Carnot engine having an efficiency of $\frac{1}{10}$ as heat engine, is used as a refrigerator. If the work done on the system is 10 J, the amount of energy absorbed from the reservoir at lower temperature is **[NEET 2017]**
(a) 1 J (b) 90 J
(c) 99 J (d) 100 J

177 A gas mixture consists of 2 moles of O_2 and 4 moles of Ar at temperature T. Neglecting all vibrational modes, the total internal energy of the system is **[NEET 2017]**
(a) $4RT$ (b) $15RT$
(c) $9RT$ (d) $11RT$

178 Coefficient of linear expansion of brass and steel rods are α_1 and α_2. Lengths of brass and steel rods are l_1 and l_2, respectively. If $(l_2 - l_1)$ is maintained same at all temperatures, which one of the following relations holds good ? **[NEET 2016]**
(a) $\alpha_1 l_2^2 = \alpha_2 l_1^2$ (b) $\alpha_1^2 l_2 = \alpha_2^2 l_1$
(c) $\alpha_1 l_1 = \alpha_2 l_2$ (d) $\alpha_1 l_2 = \alpha_2 l_1$

179 A piece of ice falls from a height h, so that it melts completely. Only one-quarter of the heat produced is absorbed by the ice and all energy of ice gets converted into heat during its fall. The value of h is (Take, latent heat of ice is 3.4×10^5 J/kg and $g = 10$ N/kg) **[NEET 2016]**
(a) 544 km (b) 136 km
(c) 68 km (d) 34 km

180 Two identical bodies are made of a material, for which the heat capacity increases with temperature. One of these is at 100°C, while the other one is at 0°C. If the two bodies are brought into contact, then assuming no heat loss, the final common temperature is **[NEET 2016]**
(a) 50°C (b) more than 50°C
(c) less than 50°C but greater than 0°C (d) 0°C

181 A black body is at a temperature of 5760 K. The energy of radiation emitted by the body at wavelength 250 nm is U_1, at wavelength 500 nm is U_2 and that at 1000 nm is U_3. Wien's constant, $b = 2.88 \times 10^6$ nmK. Which of the following is correct ? **[NEET 2016]**
(a) $U_3 = 0$ (b) $U_1 > U_2$
(c) $U_2 > U_1$ (d) $U_1 = 0$

182 A body cools from a temperature $3T$ to $2T$ in 10 min. The room temperature is T. Assume that Newton's law of cooling is applicable. The temperature of the body at the end of next 10 min will be **[NEET 2016]**
(a) $\frac{7}{4}T$ (b) $\frac{3}{2}T$
(c) $\frac{4}{3}T$ (d) T

183 A gas is compressed isothermally to half its initial volume. The same gas is compressed separately through an adiabatic process until its volume is again reduced to half. Then, **[NEET 2016]**

(a) compressing the gas through adiabatic process will require more work to be done
(b) compressing the gas isothermally or adiabatically will require the same amount of work
(c) which of the case (whether compression through isothermal or through adiabatic process) requires more work will depend upon the atomicity of the gas
(d) compressing the gas isothermally will require more work to be done

184 Figure below shows two paths that may be taken by a gas to go from a state A to a state C. **[NEET 2016]**

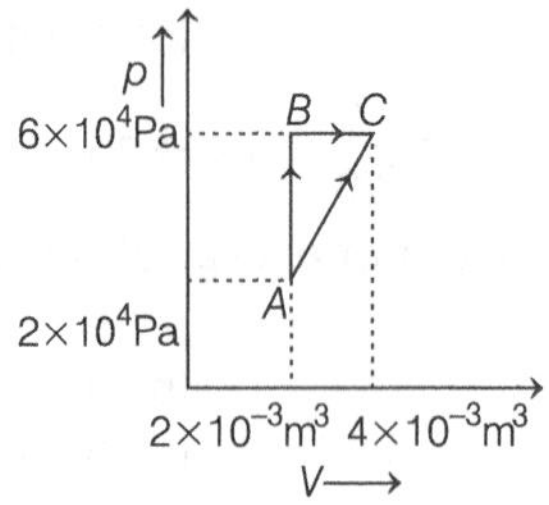

In process AB, 400 J of heat is added to the system and in process BC, 100 J of heat is added to the system. The heat absorbed by the system in the process AC will be

(a) 380 J
(b) 500 J
(c) 460 J
(d) 300 J

185 A refrigerator works between 4°C and 30°C. It is required to remove 600 cal of heat every second in order to keep the temperature of the refrigerated space constant. The power required is (Take, 1 cal $= 4.2$ J) **[NEET 2016]**

(a) 23.65 W
(b) 236.5 W
(c) 2365 W
(d) 2.365 W

186 The temperature inside a refrigerator is t_2 °C and the room temperature is t_1 °C. The amount of heat delivered to the room for each joule of electrical energy consumed ideally will be **[NEET 2016]**

(a) $\frac{t_1}{t_1 - t_2}$
(b) $\frac{t_1 + 273}{t_1 - t_2}$
(c) $\frac{t_2 + 273}{t_1 - t_2}$
(d) $\frac{t_1 + t_2}{t_1 + 273}$

187 The molecules of a given mass of a gas have rms velocity of 200 ms^{-1} at 27°C and $1.0 \times 10^5 Nm^{-2}$ pressure. When the temperature and pressure of the gas are respectively, 127°C and $0.05 \times 10^5 Nm^{-2}$, the rms velocity of its molecules in ms^{-1} is **[NEET 2016]**

(a) $\frac{400}{\sqrt{3}}$
(b) $\frac{100\sqrt{2}}{3}$
(c) $\frac{100}{3}$
(d) $100\sqrt{2}$

188 A given sample of an ideal gas occupies a volume V at a pressure p and absolute temperature T. The mass of each molecule of the gas is m. Which of the following gives the density of the gas? **[NEET 2016]**

(a) $p/(kT)$
(b) $pm/(kT)$
(c) $p/(kTV)$
(d) mkT

189 One mole of an ideal monatomic gas undergoes a process described by the equation $pV^3 =$ constant. The heat capacity of the gas during this process is **[NEET 2016]**

(a) $\frac{3}{2}R$
(b) $\frac{5}{2}R$
(c) $2R$
(d) R

190 The value of coefficient of volume expansion of glycerin is 5×10^{-4} K^{-1}. The fractional change in the density of glycerin for a rise of 40°C in its temperature is **[CBSE AIPMT 2015]**

(a) 0.015 (b) 0.020
(c) 0.025 (d) 0.010

191 The two ends of a metal rod are maintained at temperatures 100°C and 110°C. The rate of heat flow in the rod is found to be 4.0 J/s. If the ends are maintained at temperatures 200°C and 210°C, the rate of heat flow will be **[CBSE AIPMT 2015]**

(a) 44.0 J/s (b) 16.8 J/s
(c) 8.0 J/s (d) 4.0 J/s

192 One mole of an ideal diatomic gas undergoes a transition from *A* to *B* along a path *AB* as shown in the figure. **[CBSE AIPMT 2015]**

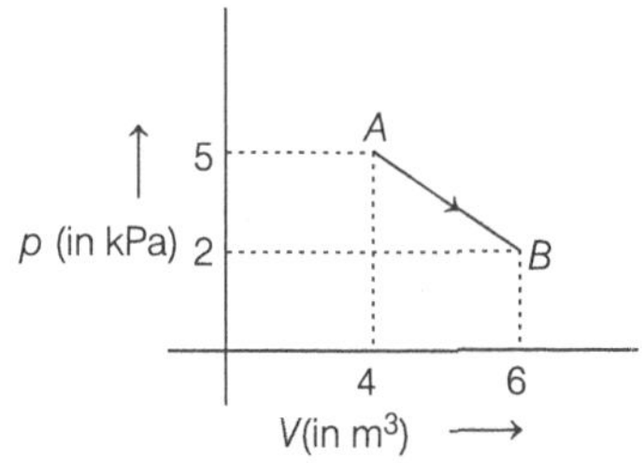

The change in internal energy of the gas during the transition is

(a) 20 kJ (b) – 20 kJ
(c) 20 J (d) – 12 kJ

193 An ideal gas is compressed to half its initial volume by means of several process. Which of the process results in the maximum work done on the gas? **[CBSE AIPMT 2015]**

(a) Adiabatic (b) Isobaric
(c) Isochoric (d) Isothermal

194 A Carnot engine, having an efficiency of $\eta = \frac{1}{10}$ as heat engine, is used as a refrigerator. If the work done on the system is 10 J, the amount of energy absorbed from the reservoir at lower temperature is **[CBSE AIPMT 2015]**

(a) 100 J (b) 99 J
(c) 90 J (d) 1 J

195 The coefficient of performance of a refrigerator is 5. If the temperature inside freezer is – 20°C, the temperature of the surroundings to which it rejects heat is **[CBSE AIPMT 2015]**

(a) 31°C (b) 41°C
(c) 11°C (d) 21°C

196 4.0 g of a gas occupies 22.4 L at NTP. The specific heat capacity of the gas at constant volume is 5.0 JK^{-1} mol^{-1}. If the speed of sound in this gas at NTP is 952 ms^{-1}, then the heat capacity at constant pressure is (Take, gas constant, $R = 8.3$ JK^{-1} mol^{-1}) **[CBSE AIPMT 2015]**

(a) 8.0 JK^{-1} mol^{-1} (b) 7.5 JK^{-1} mol^{-1}
(c) 7.0 JK^{-1} mol^{-1} (d) 8.5 JK^{-1} mol^{-1}

197 Two vessels separately contain two ideal gases *A* and *B* at the same temperature, the pressure of *A* being twice that of *B*. Under such conditions, the density of *A* is found to be 1.5 times the density of *B*. The ratio of molecular weight of *A* and *B* is **[CBSE AIPMT 2015]**

(a) $\frac{2}{3}$ (b) $\frac{3}{4}$
(c) 2 (d) $\frac{1}{2}$

198 The ratio of the specific heats $\frac{C_p}{C_V} = \gamma$ in terms of degrees of freedom (n) is given by **[CBSE AIPMT 2015]**

(a) $\left(1+\frac{1}{n}\right)$ (b) $\left(1+\frac{n}{3}\right)$
(c) $\left(1+\frac{2}{n}\right)$ (d) $\left(1+\frac{n}{2}\right)$

199 Steam at 100°C is passed into 20 g of water at 10°C. When water acquires a temperature of 80°C, the mass of water present will be (Take, specific heat of water = 1 cal $g^{-1}\,°C^{-1}$ and latent heat of steam = 540 cal g^{-1}) **[CBSE AIPMT 2014]**

(a) 24 g (b) 31.5 g
(c) 42.5 g (d) 22.5 g

200 Certain quantity of water cools from 70°C to 60°C in the first 5 min and to 54°C in the next 5 min. The temperature of the surrounding is **[CBSE AIPMT 2014]**

(a) 45°C (b) 20°C
(c) 42°C (d) 10°C

201 A monoatomic gas at a pressure p, having a volume V expands isothermally to a volume $2V$ and then adiabatically to a volume $16V$. The final pressure of the gas is $\left(\text{Take, } \gamma = \frac{5}{3}\right)$ **[CBSE AIPMT 2014]**

(a) $64p$ (b) $32p$
(c) $\frac{p}{64}$ (d) $16p$

202 A thermodynamic system undergoes cyclic process $ABCDA$ as shown in figure. The work done by the system in the cycle is **[CBSE AIPMT 2014]**

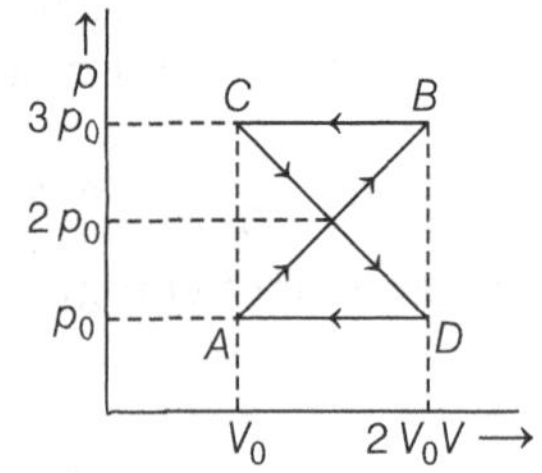

(a) p_0V_0 (b) $2p_0V_0$ (c) $\frac{p_0V_0}{2}$ (d) zero

203 The mean free path of molecules of a gas, (radius r) is inversely proportional to **[CBSE AIPMT 2014]**

(a) r^3 (b) r^2
(c) r (d) $\sqrt{r}$

204 The molar specific heat of an ideal gas at constant pressure and volume are denoted by C_p and C_V, respectively. If $\gamma = \frac{C_p}{C_V}$ and R is the universal gas constant, then C_V is equal to **[CBSE AIPMT 2014]**

(a) $\frac{1+\gamma}{1-\gamma}$ (b) $\frac{R}{(\gamma-1)}$ (c) $\frac{(\gamma-1)}{R}$ (d) γR

205 A piece of iron is heated in a flame. If first becomes dull red, then becomes reddish yellow and finally turns to white hot. The correct explanation for the above observation is possible by using **[NEET 2013]**

(a) Stefan's law (b) Wien's displacement law
(c) Kirchhoff's law (d) Newton's law of cooling

206 A gas is taken through the cycle $A \rightarrow B \rightarrow C \rightarrow A$ as shown. What is the net work done by the gas? **[NEET 2013]**

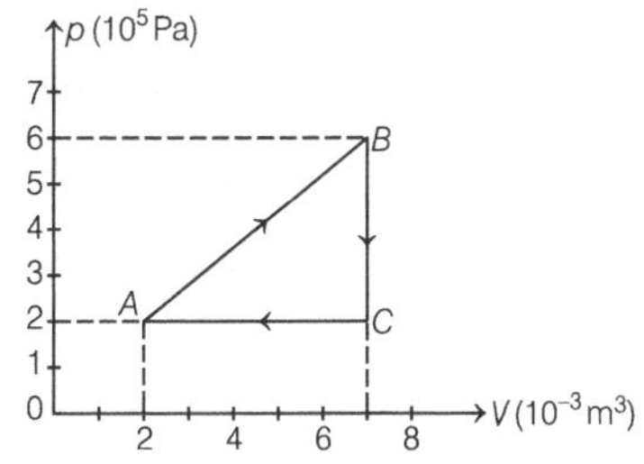

(a) 2000 J (b) 1000 J (c) Zero (d) −2000 J

207 In the given (V-T) diagram, what is the relation between pressures p_1 and p_2? **[NEET 2013]**

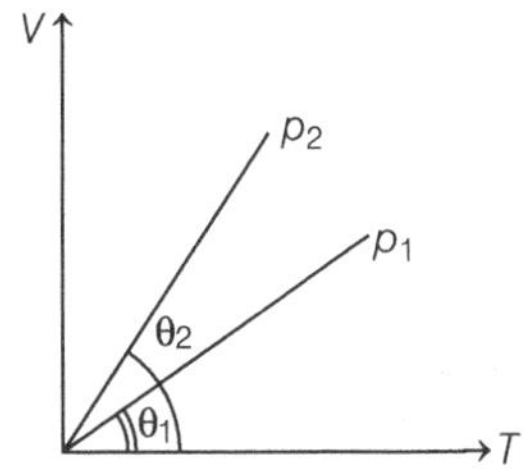

(a) $p_2 = p_1$ (b) $p_2 > p_1$
(c) $p_2 < p_1$ (d) Cannot be predicted

208 During an adiabatic process, the pressure of a gas is found to be proportional to the cube of its temperature. The ratio of $\frac{C_p}{C_V}$ for the gas is **[NEET 2013]**

(a) $\frac{4}{3}$ (b) 2
(c) $\frac{5}{3}$ (d) $\frac{3}{2}$

209 The amount of heat energy required to raise the temperature of 1 g of helium at NTP, from T_1K to T_2K is **[NEET 2013]**

(a) $\frac{3}{8} N_a K_B (T_2 - T_1)$ (b) $\frac{3}{2} N_a K_B (T_2 - T_1)$
(c) $\frac{3}{4} N_a K_B (T_2 - T_1)$ (d) $\frac{3}{4} N_a K_B \left(\frac{T_2}{T_1}\right)$

210 Liquid oxygen at 50 K is heated to 300 K at constant pressure of 1 atm. The rate of heating is constant. Which one of the following graphs represents the variation of temperature with time? **[CBSE AIPMT 2012]**

(a) 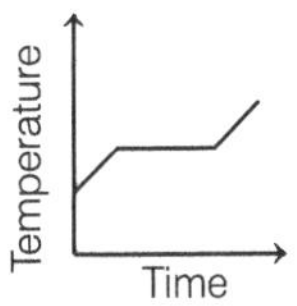

(c) 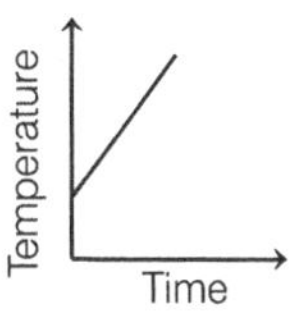

(b) 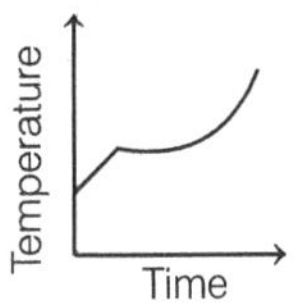

(d) 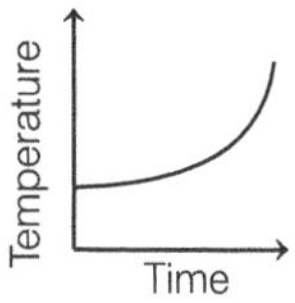

211 If the radius of a star is R and it acts as a black body, what would be the temperature of the star, in which the rate of energy production is Q? **[CBSE AIPMT 2012]**

(σ stands for Stefan's constant.)

(a) $Q/4\pi R^2\sigma$
(b) $(Q/4\pi R^2\sigma)^{-1/2}$
(c) $(4\pi R^2 Q/\sigma)^{1/4}$
(d) $(Q/4\pi R^2\sigma)^{1/4}$

212 A thermodynamic system is taken through the cycle $ABCD$ as shown in figure. Heat rejected by the gas during the cycle is **[CBSE AIPMT 2012]**

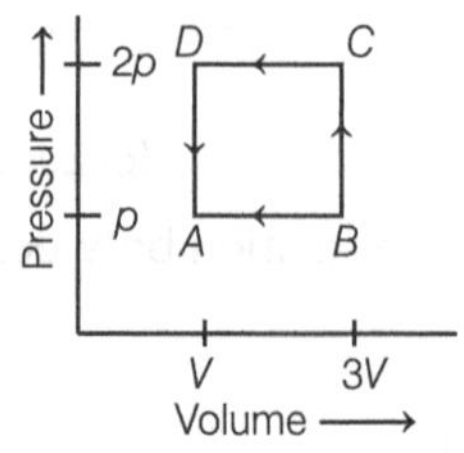

(a) $2\,pV$
(b) $4\,pV$
(c) $\frac{1}{2}pV$
(d) pV

213 One mole of an ideal gas goes from an initial state A to final state B *via* two processes. It first undergoes isothermal expansion from volume V to $3V$ and then its volume is reduced from $3V$ to V at constant pressure. The correct p-V diagram representing the two processes is **[CBSE AIPMT 2012]**

(a)
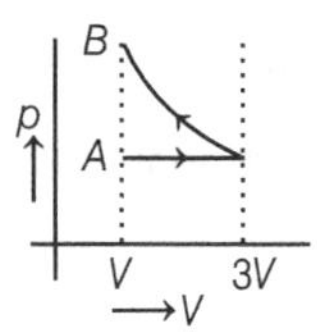

(b)
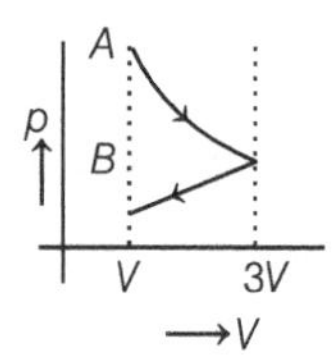

(c)
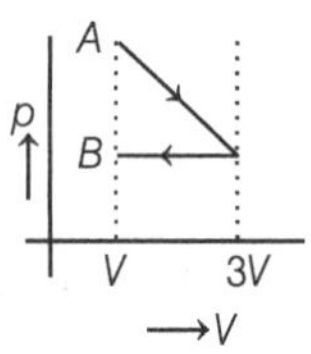

(d)
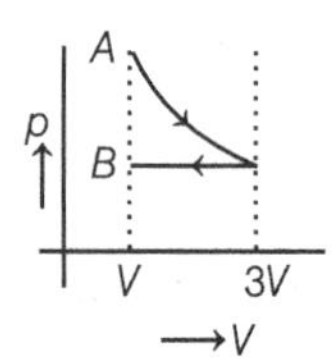

214 When 1 kg of ice at 0°C melts to water at 0°C, the resulting change in its entropy, taking latent heat of ice to be 80 cal/°C, is **[CBSE AIPMT 2011]**

(a) 8×10^4 cal/K
(b) 80 cal/K
(c) 293 cal/K
(d) 273 cal/K

215 During an isothermal expansion, a confined ideal gas does −150 J of work against its surroundings. This implies that **[CBSE AIPMT 2011]**

(a) 300 J of heat has been added to the gas
(b) no heat is transferred because the process is isothermal
(c) 150 J of heat has been added to the gas
(d) 150 J of heat has been removed from the gas

216 A cylindrical metallic rod in thermal contact with two reservoirs of heat at its two ends conducts an amount of heat Q in time t. The metallic rod is melted and the material is formed into a rod of half the radius of the original rod. What is the amount of heat conducted by the new rod when placed in thermal contact with the two reservoirs in time t? **[CBSE AIPMT 2010]**

(a) $Q/4$
(b) $Q/16$
(c) $2Q$
(d) $Q/2$

217 If ΔU and ΔW represent the increase in internal energy and work done by the system respectively in a thermodynamical process, which of the following is true? [CBSE AIPMT 2010]
(a) $\Delta U = -\Delta W$, in an adiabatic process
(b) $\Delta U = \Delta W$, in an isothermal process
(c) $\Delta U = \Delta W$, in an adiabatic process
(d) $\Delta U = -\Delta W$, in an isothermal process

ANSWER KEY

Level 1 Objective Questions

1 (a)	**2** (a)	**3** (c)	**4** (d)	**5** (a)	**6** (d)	**7** (a)	**8** (c)	**9** (c)	**10** (b)
11 (c)	**12** (b)	**13** (b)	**14** (c)	**15** (a)	**16** (a)	**17** (d)	**18** (b)	**19** (d)	**20** (a)
21 (d)	**22** (b)	**23** (c)	**24** (a)	**25** (c)	**26** (a)	**27** (d)	**28** (d)	**29** (a)	**30** (b)
31 (c)	**32** (c)	**33** (c)	**34** (c)	**35** (b)	**36** (b)	**37** (a)	**38** (b)	**39** (c)	**40** (a)
41 (c)	**42** (c)	**43** (b)	**44** (d)	**45** (a)	**46** (c)	**47** (c)	**48** (b)	**49** (d)	**50** (a)
51 (d)	**52** (c)	**53** (a)	**54** (c)	**55** (a)	**56** (b)	**57** (c)	**58** (c)	**59** (c)	**60** (b)
61 (b)	**62** (c)	**63** (b)	**64** (b)	**65** (b)	**66** (c)	**67** (b)	**68** (a)	**69** (c)	**70** (b)

Level 2 Objective Questions

71 (c)	**72** (c)	**73** (d)	**74** (c)	**75** (b)	**76** (d)	**77** (c)	**78** (a)	**79** (b)	**80** (c)
81 (b)	**82** (c)	**83** (d)	**84** (d)	**85** (c)	**86** (a)	**87** (d)	**88** (a)	**89** (c)	**90** (c)
91 (d)	**92** (a)	**93** (d)	**94** (b)	**95** (a)	**96** (c)	**97** (b)	**98** (a)	**99** (b)	**100** (a)
101 (d)	**102** (a)	**103** (c)	**104** (d)	**105** (a)	**106** (c)	**107** (d)	**108** (c)	**109** (a)	**110** (b)
111 (d)	**112** (b)	**113** (b)	**114** (c)	**115** (a)	**116** (c)	**117** (d)	**118** (a)	**119** (a)	**120** (b)
121 (b)	**122** (c)	**123** (a)	**124** (c)	**125** (c)	**126** (b)	**127** (b)	**128** (c)	**129** (b)	**130** (a)
131 (a)	**132** (c)	**133** (c)	**134** (a)	**135** (a)	**136** (c)	**137** (a)	**138** (d)	**139** (d)	**140** (a)
141 (b)	**142** (a)	**143** (d)	**144** (a)	**145** (a)	**146** (a)	**147** (b)	**148** (b)		

Level 3 Previous Years Questions

149 (b)	**150** (c)	**151** (d)	**152** (c)	**153** (a)	**154** (a)	**155** (b)	**156** (d)	**157** (a)	**158** (a)
159 (a)	**160** (a)	**161** (c)	**162** (d)	**163** (c)	**164** (b)	**165** (a)	**166** (a)	**167** (a)	**168** (a)
169 (b)	**170** (c)	**171** (c)	**172** (b)	**173** (a)	**174** (d)	**175** (b)	**176** (b)	**177** (d)	**178** (c)
179 (b)	**180** (b)	**181** (c)	**182** (b)	**183** (a)	**184** (c)	**185** (b)	**186** (b)	**187** (a)	**188** (b)
189 (d)	**190** (b)	**191** (d)	**192** (b)	**193** (a)	**194** (c)	**195** (a)	**196** (a)	**197** (b)	**198** (c)
199 (d)	**200** (a)	**201** (c)	**202** (d)	**203** (b)	**204** (b)	**205** (b)	**206** (b)	**207** (c)	**208** (d)
209 (a)	**210** (a)	**211** (d)	**212** (a)	**213** (d)	**214** (c)	**215** (c)	**216** (b)	**217** (a)	

Solutions

LEVEL 01 Objective Questions

1. Let θ be the final temperature.

Heat taken by ice = Heat given by water

$$m_1L_1 + mc_1\theta = m_2L_2\theta_2$$

$$5 \times 80 + 5 \times 1(\theta - 0) = 20 \times 1 \times (40 - \theta)$$

$$\Rightarrow \quad 25\theta = 400$$

$$\Rightarrow \quad \theta = 16°\,C$$

2. Amount of heat given to copper = Amount of heat given to water

$$m_c c_c T_1 = m_w c_w T_2$$

$$50 \times 0.42 \times 10 = 10 \times 4.2 \times T_2$$

$$\Rightarrow \quad T_2 = 5°C$$

3. From scale conversion,

$$\frac{X - 15°}{75° - 15°} = \frac{Y - 25°}{125° - 25°}$$

We have given, $X = 60°$

$$\frac{60° - 15°}{75° - 15°} = \frac{Y - 25°}{125° - 25°}$$

$$\frac{45°}{60°} = \frac{Y - 25°}{100°}$$

$$\Rightarrow \quad Y = 100°$$

4. Heat given = Heat taken

$$mC\,(2T - T_f) = \left(\frac{m}{2}\right)(2C)(T_f - T)$$

Solving this equation, we get

$$T_f = \frac{3T}{2}$$

5. Let the mass of steam condensed be m.

So, $\quad Q_1 = m \times L + m \times (100 - 90) \times 1 = 550\,m$

$\quad Q_2 = 22 \times (90 - 20) \times 1 = 1540$ cal

$\therefore \quad m = \dfrac{1540}{550} = 2.8$ g

So, total mass = 22 + 2.8 = 24.8 g

6. Let θ be the final temperature.

Heat given = Heat taken

$$110 \times 1(\theta - 10°) + 10(\theta - 20°) = 220 \times 1\,(70° - \theta)$$

$$\theta = 50°C$$

7. Heat taken = Heat given

$$ms\,(50 - 30) = (40 - m)\,s\,(80 - 50)$$

Solving this equation, we get

$$m = 24 \text{ kg} \quad \text{and} \quad 40 - m = 16 \text{ kg}$$

8. Heat gained by ice = Heat lost by water

$$(10 \times 80) + 10 \times 0.5\,[0 - (-20)] = 10 \times 1 \times (10 - 0) + 80 \times 10$$

$$900 \text{ cal} = 900 \text{ cal}$$

So, at equilibrium, the calorimeter will contain 10 g of ice and 10 g of water.

9. Heat required for a minute = $50 \times 0.6 \times 50 = 1500$ cal

From graph, it is clear that boiling point of water is 200°C.

10. $Q = mc\Delta T$

$$1000 \text{ J} = \frac{100}{18} \times 75 \times \Delta T$$

$$\Rightarrow \quad \Delta T = 2.4 \text{ K}$$

11. $x \times 540 = y \times 80 + y \times 100$

$$\therefore \quad \frac{y}{x} = 3$$

12. Thermal expansion is just like photographic expansion. So, a, b, r_1, r_2 and d all will increase.

13. We have given,

Initial temperature, $T_1 = 27°C$

Length of brass wire at T_1, $L = 1.8$ m

Final temperature, $T_2 = -39°C$

Diameter of wire, $d = 2 \text{ mm} = 2 \times 10^{-3}$ m

Tension in the wire = F

Change in length, $\Delta L = \alpha L(T_2 - T_1)$

Now, $Y = \dfrac{F \times L}{A \times \Delta L}$

$$F = \frac{Y \times A \times \Delta L}{L} = \frac{Y \times \pi(d/2)^2 \times \alpha L(T_2 - T_1)}{L}$$

$$= \frac{Y \times \pi d^2}{4} \times \alpha(T_2 - T_1)$$

$$= \frac{0.91 \times 10^{11} \times 3.14 \times (2 \times 10^{-3})^2}{4} \times 2 \times 10^{-5}(-39 - 27)$$

$$= 3.8 \times 10^2 \text{ N}$$

14. Net change in length $= \Delta L_1 + \Delta L_2$

$$= L_1\alpha_1\Delta T + L_2\alpha_2\Delta T$$

$$\Delta L = (L_1 + L_2)\alpha\Delta T$$

$$= \Delta L_1 + \Delta L_2$$

$$(L_1 + L_2)\,\alpha\,\Delta T = L_1\alpha_1\Delta T + L_2\alpha_2\Delta T$$

$$\Rightarrow \quad \alpha = \frac{L_1\alpha_1 + L_2\alpha_2}{L_1 + L_2} = \frac{L_1\alpha_1 + L_2\alpha_2}{L}$$

15. $\Delta L_S = L_S\alpha_S\Delta t$

$\Delta L_C = L_C\alpha_C\Delta t$

We have given, $L_S - L_C = 10$ cm

$$\Rightarrow \quad L_C = (L_S - 10) \text{ cm} \qquad \text{...(i)}$$

$$\therefore \quad L_S\alpha_S\Delta t = L_C\alpha_C\Delta t$$

$\Rightarrow \quad \dfrac{L_S}{L_C} = \dfrac{\alpha_C}{\alpha_S} = \dfrac{3}{2} \Rightarrow L_S = \dfrac{3}{2}L_C$ ...(ii)

From Eqs. (i) and (ii), we get

$$L_S = \frac{3}{2}(L_S - 10)$$

$$L_S = 30 \text{ cm}$$

$$L_C = (30 - 10) = 20 \text{ cm}$$

So, the lengths of steel and copper are 30 cm and 20 cm.

16. As the volume of air inside flask is same.

So, expansion of flask must be equal to the mercury

$$V_g \gamma_g t = V_m \gamma_m t \Rightarrow V_m = 1000 \frac{(9 \times 10^{-6})}{1.8 \times 10^{-4}}$$

$$= 50 \text{ cm}^3$$

17. If the temperature of whole rod is same, it means that heat is being conducted from one end to another infinitely fast.

So, the conductivity is infinite.

18. In the steady state, heat flow is constant.

$$\frac{2kA(T - T_A)}{t - 0} = \frac{kA(T_B - T)}{2t - t}$$

$\Rightarrow \quad 3T = T_B + 2T_A$ ...(i)

Now, the temperature difference,

$T_B - T_A = 36$ ...(ii)

From Eqs. (i) and (ii), we get

$$T - T_A = 12°\text{C}$$

19.

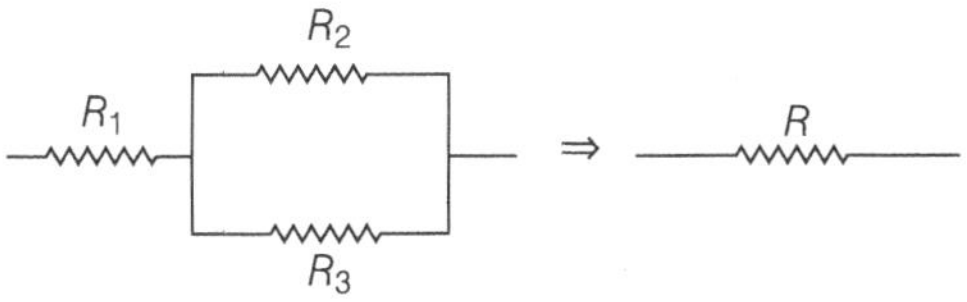

$$R = R_1 + \frac{1}{\frac{1}{R_2} + \frac{1}{R_3}}$$

$$\therefore \quad \frac{3d/2}{KA} = \frac{d}{K_1A} + \frac{1}{\frac{K_2A/2}{d/2} + \frac{K_3A/2}{d/2}}$$

$$\therefore \quad \frac{3}{2K} = \frac{1}{K_1} + \frac{1}{K_2 + K_3}$$

or $$K = \frac{3}{2}\left(\frac{1}{K_1} + \frac{1}{K_2 + K_3}\right)^{-1}$$

20.

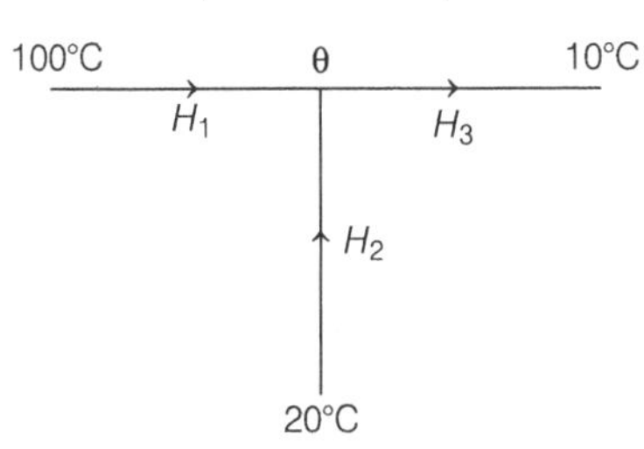

Heat current, $H = \dfrac{\text{Temperature difference}}{\text{Thermal resistance}}$

and thermal resistance $\propto$ length

$$H_1 + H_2 = H_3$$

$$\therefore \quad \frac{100° - \theta}{R/2} + \frac{20° - \theta}{R} = \frac{\theta - 10°}{R/2}$$

Solving this equation, we get

$$\theta = 48°\text{C}$$

21. Rate of flow of heat $= \dfrac{kA\Delta T}{L} \propto \dfrac{A}{L}$

$$\therefore \quad \frac{R_1}{R_2} = \frac{A_1}{A_2} \times \frac{L_2}{L_1} = \left(\frac{1}{2}\right)^2 \times \frac{1}{2} = \frac{1}{8}$$

$$R_1 : R_2 = 1 : 8$$

22. Highly polished mirror like surfaces are good reflectors but not good radiators.

23. Energy radiated per second is called power and expression of power is given by,

$$E = e_r \sigma T^4 A \quad \text{(where, } A = \text{surface area)}$$

$$= e_r \sigma T^4 (4\pi) \left(\frac{d}{2}\right)^2$$

$\therefore \quad E \propto T^4 d^2$ ...(i)

According to question, if temperature becomes $2T$ and diameter becomes d, then using Eq. (i), new radiated energy will be

$$E_{new} \propto (2T)^4 \cdot \left(\frac{d}{4}\right)^2$$

$$E_{new} \propto E$$

24. From Wien's displacement law,

$$\lambda_1 T_1 = \lambda_2 T_2$$

$$\Rightarrow \quad \frac{T_2}{T_1} = \frac{\lambda_1}{\lambda_2} = \frac{4\lambda_0}{3\lambda_0} = \frac{4}{3}$$

Now, the energy radiated per unit area,

$$E = \sigma T^4 \Rightarrow E \propto T^4$$

$$\therefore \quad \frac{E_2}{E_1} = \frac{T_2^4}{T_1^4} = \left(\frac{4}{3}\right)^4 = \frac{256}{81}$$

25. 30% means $e_r = 0.3$

$$P = e_r \sigma T^4 A = e_r \sigma T^4 (4\pi R^2)$$

Power P is same. Therefore, $e_r R^2$ should be same.

$$\therefore \quad 0.3\left(\frac{2.3}{2}\right)^2 = (1)\, R^2$$

$$\Rightarrow \quad R = 0.629 \text{ cm}$$

26. For isothermal process,

$$p_x V = p_A \frac{V}{2}$$

$$p_A = 2p_x$$

For adiabatic process, $p_x V^\gamma = p_B \left(\frac{V}{2}\right)^\gamma$

$$p_B = 2^\gamma p_x$$

Since, $\gamma > 1$, final pressure is higher in adiabatic process.

27. $$\frac{p_1 V_1}{T_1} = \frac{p_2 V_2}{T_2}$$

$$\Rightarrow \quad V_2 = \left(\frac{p_1}{p_2} \times \frac{T_2}{T_1}\right) \times V_1 = \frac{4}{2} \times \frac{270}{300} \times 1500$$

$$= 2700 \text{ m}^3$$

28. At low pressure and high temperature, intermolecular force between gas molecules is insufficient, which is the basic postulate of an ideal gas.

29. Number of moles, $n = \frac{pV}{RT}$

$$= \frac{1.3 \times 10^5 \times 7 \times 10^{-3}}{8.314 \times 273} \approx 0.4 \text{ mol}$$

So, the number of molecules contained in a vessel

$$= 0.4 \times 6.022 \times 10^{23} = 2.4 \times 10^{23}$$

30. Suddenly compression means adiabatic process.

$$TV^{\gamma - 1} = \text{constant}, \quad \gamma = 1.5$$

$$\therefore \quad T\sqrt{V} = \text{constant}$$

$$\text{or} \quad T_f = \sqrt{\frac{V_i}{V_f}} T_i$$

$$= 3T_i = 3(400) = 1200 \text{ K or } 927°\text{C}$$

31. $pV = nRT = \frac{m}{m_0} RT$

$$pV = C$$

So, C depends on the mass of the gas and the temperature.

32. Applying ideal gas equation in the initial condition,

$$pV = \frac{M}{M_0} RT \quad \text{...(i)}$$

When $2M$ is added,

$$p' \frac{V}{3} = \frac{3M}{M_0} \frac{RT}{3} \quad \text{...(ii)}$$

Now, dividing Eq. (ii) by Eq. (i), we get

$$p' = 3p$$

33. Initially, $p_1 = p$

$$V' = V + V = 2V$$

Finally, $V_2 = V$

As T = constant,

$$p_1 V_1 = p_2 V_2$$

$$\Rightarrow \quad p_2 = \frac{p \times 2V}{V} = 2p$$

34. For isothermal process, pV = constant

Initial $p_1 V_1 = 12$

Final $pV = 12$

C process start and end on the same iosthermal.

35. $\rho_A = \frac{p_0 m}{RT_0} \propto \frac{p}{T}$

At point B, $\rho_B = \frac{3p_0(m)}{R(2T_0)} = \frac{3}{2}\rho_0$

36. V = constant

$$\therefore \quad T \propto p$$

Percentage change in pressure is small.

So, percentage change in temperature will also be 0.5%.

$\therefore \quad \Delta T = 0.5\%$ of initial temperature T_i or $2 = \frac{0.5}{100} T_i$

$$\therefore \quad T_i = 400 \text{ K} \quad \text{or} \quad 127°\text{C}$$

37. Let T be the temperature of final mixture.

$$U = U_1 + U_2$$

$$\frac{f}{2}(n_1 + n_2)RT = \frac{f}{2}(n_1)RT_0 + \frac{f}{2}(n_2)R(2T_0)$$

$$(2 + 4)T = 2T_0 + 8T_0$$

$$\Rightarrow \quad T = \frac{5}{3} T_0$$

38. Let the initial pressure and temperature be p_1 and T_1.

So, final pressure, $p_2 = 1.004\, p_1$

Final temperature, $T_2 = T_1 + 1$

Now, $\frac{p_1}{T_1} = \frac{p_2}{T_2} = \frac{1.004\, p_1}{T_1 + 1}$

We get, $T_1 = 250$ °C

39. $n = \frac{pV}{Kt}$

$\frac{pV}{Kt}$ gives the number of molecules of the gas.

40. $W = p\Delta V = 50 \times (4 - 10) = -300$ J

Now, from first law of thermodynamics,

$$\Delta U = \Delta Q - W$$

$$= 100 - (-300)$$

$$= 400 \text{ J}$$

So, internal energy increased by 400 J.

41. $C_p = 7.2$ cal/mol°C

$$R = 8.3 \text{ J/mol-K} = 2.0 \text{ cal/mol°C}$$

$$C_p - C_V = R$$

$$\therefore \quad C_V = C_p - R = 5.2 \text{ cal/mol-°C}$$

Now, process is isochoric. Hence,

$$Q = nC_V \Delta T$$

$$= 5 \times 5.2 \times 10$$

$$= 260 \text{ cal}$$

42. The difference between C_p and C_V must be 2.

So, option (c) is most reliable $= (7 - 5) = 2$

43. $R = \frac{2}{3}C_V$

We know, $\gamma - 1 = \frac{R}{C_V}$

$\Rightarrow \quad R = C_V(\gamma - 1)$

So, $\frac{2}{3}C_V = C_V(\gamma - 1)$

$\Rightarrow \quad \gamma = \frac{5}{3}$

Now, $p^{1-\gamma}T^{\gamma} =$ constant k

On substituting $\gamma = \frac{5}{3}$, we get

$$p^{1-5/3}T^{5/3} = k$$

$$\Rightarrow \quad p^{-2/3} = k \cdot T^{-5/3}$$

$$\Rightarrow \quad p^{-2/3} \propto T^{-5/3} \Rightarrow p \propto T^{\frac{-5}{3} \times \frac{-3}{2}}$$

$$p \propto T^{5/2}$$

44. $R = \frac{2}{5}C_V$

$\therefore \quad C_V = \frac{5R}{2}$

So, the degree of freedom of the gas is 5.

45. Number of molecules per unit volume,

$$n = \frac{p}{kT}$$

46. We know at constant pressure,

$$dQ = nC_p dT$$
$$dW = RdT = (C_p - C_V)dT$$
$$dU = C_V dT$$

$\therefore \quad dQ : dW : dU = C_p dT : (C_p - C_V)\, dT : C_V dT$

$$= \frac{C_p}{C_V} : \left(\frac{C_p - C_V}{C_V}\right) : \frac{C_V}{C_V}$$

$$= \gamma : (\gamma - 1) : 1$$

47. Molar heat capacity, $C = \frac{Q}{\Delta t}$

$$\Delta U = \frac{5R\Delta t}{2} \Rightarrow \Delta t = \frac{2\Delta U}{5R}$$

So, $\quad C = \frac{5QR}{2\Delta U} = \frac{10R}{3}$

$$\frac{Q}{\Delta U} = \frac{4}{3} \qquad \text{...(i)}$$

Now, from first law of thermodynamics,

$$\Delta U = \Delta Q - \Delta W \qquad \text{...(ii)}$$

From Eqs. (i) and (ii), we get

$$W = \frac{Q}{4}$$

48. $dQ = nC_p dT$

$$dW = n(C_p - C_V)\, dT$$

So, ratio of heat absorbed to the work done by the gas,

$$\frac{dQ}{dW} = \frac{nC_p dT}{n(C_p - C_V)dT}$$

$$= \frac{1}{1 - \frac{C_V}{C_p}} = \frac{1}{1 - \frac{3}{5}} \qquad \left[\because \frac{C_V}{C_p} = \frac{3}{5}\right]$$

$$= \frac{5}{2}$$

49. Monoatomic — $f_1 = 3$

Diatomic — $f_2 = 5$

Non-linear polyatomic — $f_3 = 6$

So, $\quad f = \frac{f_1 n_1 + f_2 n_2 + f_3 n_3}{n_1 + n_2 + n_3}$

$$= \frac{3 \times 5 + 5 \times 1 + 6 \times 5}{5 + 1 + 5} = \frac{50}{11}$$

50. Work done by the area under the curve which is negative.

So, energy is lost by the gas.

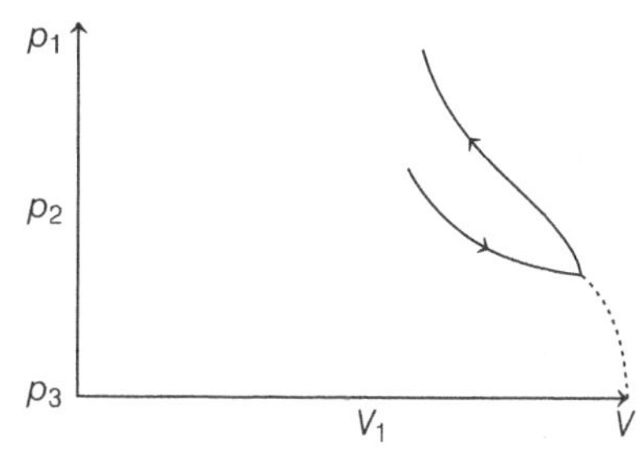

$\therefore \quad p_3 > p_1$ and $W < 0$

51. We know that, the slope of adiabatic process is Y times of the isothermal process.

From graph, the slope of XY is more, so the process is adiabatic and compression is taking place, as volume is decreasing.

YZ process is isothermal and expansion is taking place. As the volume is increasing.

In ZX, process is isobaric, so at constant pressure, volume is decreasing. Thus, compression takes place.

52. At equilibrium in the flasks, pressure must be same, using ideal gas equation,

$$\frac{n_1 T_1}{V_1} = \frac{n_2 T_2}{V_2} \text{ (R and p are constants)}$$

$$\frac{m'_1 \times 100}{2V_2} = \frac{m_2 \times 200}{V_2}$$

$$m' = \frac{m}{4}$$

53. We have given, $p = kV$

$$pV^{-1} = k$$

We know, $pV^r = k$

So, $r = -1$

Now, the molar heat capacity,

$$C = \frac{R}{5/3 - 1} + \frac{R}{1-(-1)}$$

$$= \frac{3R}{2} + \frac{R}{2} = 2R$$

So, heat supplied, $\Delta Q = nC\Delta T$

$$= (1)(2R)(2T_0 - T_0) = 2RT_0$$

54. We know, $\gamma = \frac{C_p}{C_V}$

$$\gamma = 1 + \frac{R}{C_V} \Rightarrow C_V = \frac{R}{\gamma - 1}$$

From Charles' law, $V \propto T$

So, T changes to $2T$.

$\therefore$ Internal energy, $\Delta U = nC_V(\Delta T) = nC_V(2T - T)$

$$= nC_V T = \frac{nRT}{\gamma - 1} = \frac{pV}{\gamma - 1}$$

55. 1 cal = 4.2 J

2000 cal = 2000 × 4.2 = 8400 J

$$\text{Coefficient of performance} = \frac{\text{Heat absorbed}}{\text{Work done}}$$

$$4 = \frac{8400}{W} \Rightarrow W = \frac{8400}{4} = 2100 \text{ J}$$

56. $\beta = \frac{T_2}{T_1 - T_2}$

$$5 = \frac{253}{T_1 - 253} \quad [\because T_2 = -20°C + 273 = 253 \text{ K}]$$

$$5T_1 - 1265 = 253$$

$$\Rightarrow T_1 = \frac{1518}{5} = 303.6 \text{ K}$$

$$\Rightarrow T_1 = 303.6 - 273 = 31°C$$

57. $\eta = 1 - \frac{T_{hot}}{T_{cold}}$

At $T_{cold} = \infty$, $\eta = 1$

So, the source at infinite temperature is not attainable. Thus, the efficiency of heat engine is always less than 1.

58. $$\eta_1 = 1 - \frac{T_2}{T_1} = \frac{1}{6} \quad \text{...(i)}$$

$$n_2 = 1 - \frac{(T_2 - 62)}{T_1} = \frac{1}{3} \quad \text{...(ii)}$$

From Eqs. (i) and (ii), we get

$$T_1 = 372 \text{ K}$$

So, $1 - \frac{T_2}{372} = \frac{1}{6}$

$$\Rightarrow T_2 = 310 \text{ K}$$

59. $v_{rms} \propto \sqrt{T}$

Temperature is same.

$$\therefore V_1 = V_2 = V$$

60. $H = mC_{p_w}\Delta T_1 + mL + mC_{p_i}\Delta T_2$

$$= m(C_{p_w}\Delta T_1 + L + C_{p_i}\Delta T_2)$$

$$= 100[1 \times (25 - 0) + 80 + 0.5 \times \{0 - (-10)\}]$$

$$= 100[25 + 80 + 0.5 \times 10]$$

$$= 11 \text{ kcal}$$

Now, the heat removed in 1 hr 50 min = 11×10^3 cal

$$110 \text{ min} = 11 \times 10^3 \text{ cal}$$

$$1 \text{ min} = \frac{11 \times 10^3}{110} = 100 \text{ cal}$$

61. $\beta = \frac{1 - \eta}{\eta}$

$$= \frac{1 - 1/10}{1/10} = \frac{9/10}{1/10} = 9$$

Now, $\beta = \frac{Q}{W}$

$$Q = 9 \times 10 = 90 \text{ J}$$

62. $\eta = 1 - \frac{T_C}{T_h} = 1 - \frac{400}{800} = 0.5$

Also, $\eta = \frac{W}{Q_h}$

$$0.5 = \frac{750}{Q_h}$$

$$\Rightarrow Q_h = 1500 = \frac{1500}{4.2} \text{ cal/s}$$

$$= 357.14 \text{ cal/s}$$

So, the amount of heat absorbed in calories is 357.14 cal/s and the efficiency of heat engine is 50%.

63. $V \propto \frac{1}{\sqrt{m_{gas}}}$ = momentum, $p \propto \sqrt{m_{gas}}$

So, ratio of momentum $= \sqrt{\frac{m_{H_2}}{m_{O_2}}} = \sqrt{\frac{2}{32}} = \frac{1}{4}$

64. $v_{rms} = \sqrt{\frac{3RT}{m}} \propto \sqrt{\frac{T}{m}}$

Now, when the temperature is doubled and m becomes $\frac{m}{2}$.

$$\therefore v_{rms} = \sqrt{\frac{2T(2)}{m}} = 2v$$

So, the new mean square speed of atom is $2v$.

65. 10 g means 5 g mol.

Average translational kinetic energy of 1 mole is $\frac{3}{2}RT$.

$$\therefore K_T = 5\left(\frac{3}{2}RT\right)$$

$$= \frac{15}{2} \times 8.31 \times 300$$

$$= 18697 \text{ J} \approx 18700 \text{ J}$$

66. $v_{rms} = \sqrt{\frac{3RT}{m}}$

v_{rms} is independent of the quantity of gas. So, v_{rms} remains same.

67. Thermal energy for diatomic gases, $U = \frac{5}{2} nRT$

We know, $pV = nRT$

$$\therefore \quad U = \frac{5}{2} pV = \frac{5}{2} \times 8 \times 10^4 \times \frac{1}{4}$$
$$= 5 \times 10^4 \text{ J}$$

68. From Dalton's law of partial pressure,

$$p = p_1 + p_2 + p_3$$
$$= \frac{nRT}{V} + \frac{nRT}{V} + \frac{nRT}{V} = \frac{3nRT}{V}$$

Here, $n = \frac{1}{2}, T = 273 + 27 = 300 \text{ K}, V = 5 \times 10^{-3} \text{m}^3$

$$\therefore \quad p = \frac{3 \times 8.31 \times 300}{2 \times 5 \times 10^{-2}} = 7.5 \times 10^5 \text{ Nm}^{-2}$$

69. Internal energy of n moles of an ideal gas at temperature T is given as,

$$U = \frac{f}{2} nRT$$

$$\because \quad U_1 = U_2$$
$$f_1 n_1 T_1 = f_2 n_2 T_2$$
$$\Rightarrow \quad \frac{n_1}{n_2} = \frac{f_2 T_2}{f_1 T_1} = \frac{3 \times 10}{5 \times 6} = 1$$

70. $v_{rms} \propto \sqrt{T}$ and $T \propto pV$

$$\therefore \quad v_{rms} \propto \sqrt{pV} \quad \text{...(i)}$$

According to question, if volume of gas increases 2 times and pressure increases 8 times, then using Eq. (i),

$$\Rightarrow \quad v'_{rms} \propto \sqrt{(8p)(2V)}$$
$$\Rightarrow \quad v'_{rms} \propto \sqrt{16pV}$$
$$\Rightarrow \quad v'_{rms} \propto 4v_{rms}$$

LEVEL 02 Objective Questions

71. Let θ be the temperature on X scale corresponding to $50°Y$ on Y-scale,

$$\frac{\theta - (-125°)}{375° - (-125°)} = \frac{50° - (-70°)}{-30° - (-70°)}$$
$$\Rightarrow \quad \theta = 1375° X$$

72. For the temperature scale to show the same reading, $A = B$

$$\frac{A - 42}{110} = \frac{A - 72}{220}$$
$$2(A - 42) = A - 72$$
$$2A - 84 = A - 72$$
$$\Rightarrow \quad A = 12°\text{C}$$

73. $\alpha_1 = \frac{\Delta l}{10 \Delta T} = \frac{1 \times 10^{-3}}{100} = 10^{-5}$

$$\alpha_2 = \frac{2 \times 10^{-3}}{100} = 2 \times 10^{-5}$$
$$\alpha_3 = \frac{3 \times 10^{-3}}{1.5 \times 50} = 4 \times 10^{-5}$$
$$\alpha_4 = \frac{4 \times 10^{-3}}{2.5 \times 20} = 8 \times 10^{-5}$$

So, A_4 has the greatest coefficient of the linear expansion.

74. $T = 2\pi \sqrt{\frac{L}{g}}$

$$\therefore \quad \frac{\Delta T}{T} = \frac{1}{2} \frac{\Delta L}{L} = \frac{1}{2} \alpha \Delta T$$
$$= \frac{1}{2} \times 10^{-5} \times 20$$
$$= 10^{-4}$$
$$\Delta T = T \times 10^{-9}$$
$$= 24 \times 60 \times 60 \times 10^{-4}$$
$$= 8.64 \text{ s}$$

75. Fractional loss of time per second $= \frac{1}{2} \alpha \Delta T$

$$\frac{1}{2} \alpha (T_0 - 15) \times 24 \text{ h} = 5 \quad \text{...(i)}$$
$$\frac{1}{2} \alpha (30 - T_0) \times 24 \text{ h} = 10 \quad \text{...(ii)}$$

On solving Eqs. (i) and (ii), we get

$$T_0 = 20°\text{C}$$

76. $Y = \frac{FL}{A \Delta L}$ ...(i)

$$\frac{\Delta L}{L} = \alpha \Delta T \quad \text{...(ii)}$$

From Eqs. (i) and (ii), we get

$$F = YA\alpha\Delta T$$
$$Y_1 A \alpha_1 \Delta T = Y_2 A \alpha_2 \Delta T$$
$$\Rightarrow \quad \frac{\alpha_1}{\alpha_2} = \frac{Y_2}{Y_1}$$

77. $I = \frac{1}{3} ml^2$

$$I \propto l^2$$
$$\therefore \quad \frac{\Delta I}{I} = 2\left(\frac{\Delta l}{l}\right)$$
$$\text{or} \quad \Delta I = 2I\left(\frac{\Delta l}{l}\right) = 2I\alpha\Delta T$$

78. $\Delta Q = \int msdT$

$$= m\int_0^{20} aT^3 dT = ma\int_0^{20} \frac{T^4}{4} = 4 \times 10^4 \, ma$$

79. Thermal capacity (S) = Mass × Specific heat

$$\therefore \quad \frac{S_A}{S_B}=\frac{m_A s_A}{m_B s_B}=\frac{(4/3)\pi r_A^2 \rho_A s_A}{(4/3)\pi r_B^2 \rho_B s_B}$$

$$=\left(\frac{r_A}{r_B}\right)^3 \frac{\rho_A s_A}{\rho_B s_B}=\left(\frac{1}{2}\right)^3 \times \frac{2}{1}\times\frac{1}{3}$$

$$=\frac{1}{12}$$

80. According to the Dalton's law of partial pressure, the total pressure exerted by a mixture of gases which do not interact with each other is equal to the sum of the partial pressure which each would exert.

$$p=p_1+p_2+p_3$$

81. Let the final temperature be T.

$$m_1c_1(T-T_1)+m_2c_2(T-T_2)+m_3c_3(T-T_3)=0$$

$$T=\frac{m_1c_1T_1+m_2c_2T_2+m_3c_3T_3}{m_1c_1+m_2c_2+m_3c_3}$$

82. From the given graph, melting point for $A = 60°C$ and melting point for $B = 20°C$

Time taken by A for fusion = (6 − 2) = 4 min

Time taken by B for fusion = (6.5 − 4) = 2.5 min

$$\therefore \quad \frac{H_A}{H_B}=\frac{6\times4\times60}{6\times2.5\times60}=\frac{8}{5}$$

83. Heat gain = Heat loss

$$ML=mc\theta \Rightarrow M=\frac{mc\theta}{L}$$

84. We have given, $\lambda_1 = 1987$ nm and $\lambda_2 = 2980$ nm

Now, from Wien's displacement law,

$$T_P\times1987=b$$

$$T_Q\times2980=b$$

$$\frac{T_P}{T_Q}=\frac{1987}{2980}$$

Area under the curve, $E\propto T^4$

$$\frac{E_P}{E_Q}=\left(\frac{T_P}{T_Q}\right)^4=\left(\frac{1987}{2980}\right)^4\approx\frac{16}{81}$$

85. $\frac{Q}{t}=\frac{kA(T_1-T_2)\,t}{L}$ ...(i)

Area of cross-section of new rod,

$$A'=\pi\left(\frac{R}{3}\right)^2=\frac{\pi R^2}{9}=\frac{A}{9}$$

Since, the volume is constant, so the new length of rod,

$$AL=A'L' \Rightarrow L'=L\frac{A}{A'}=9L$$

Now, $Q'=\frac{k(A/9)(T_1-T_2)\,t}{9L}$

$$=\frac{1}{81}\frac{kA(T_1-T_2)\,t}{L}=\frac{Q}{81}$$ [using Eq.(i)]

86. $\frac{1}{2}\left(\frac{1}{2}mv^2\right)=ms\Delta\theta$

$$\therefore \quad \Delta\theta=\frac{v^2}{4s}=\frac{(200)^2}{4\times125}=80\ °C$$

87. $pV=nRT$

Here, n = number of gram moles

$$=\frac{5}{28}$$

$$\therefore \quad pV=\frac{5}{28}RT$$

88. $p\propto T^3$

$$pV=nRT$$

$$p\propto T^3$$

$$p\propto(pV)^3$$

$$p^2V^3=\text{constant}$$

$$pV^{3/2}=\text{constant}$$

$$\gamma=\frac{3}{2}\ \text{expansion.}$$

89.

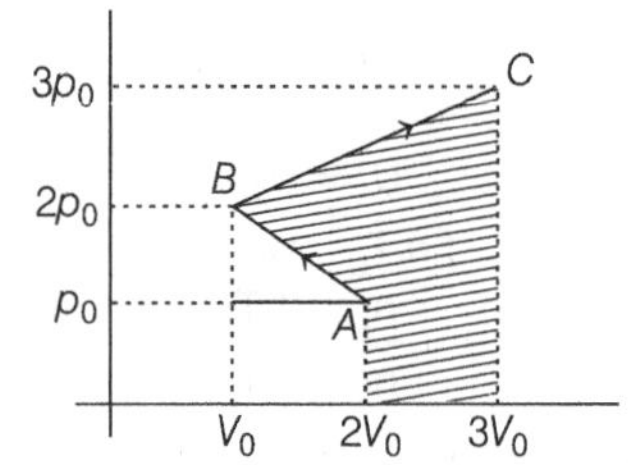

Work done by the gas = Hatched area

$$=\frac{1}{2}(2p_0+3p_0)(3V_0-V_0)-\frac{1}{2}(2p_0+p_0)(2V_0-V_0)$$

$$=5\,p_0V_0-\frac{3}{2}p_0V_0$$

$$=\frac{7}{2}p_0V_0$$

90. From graph, AB is the isobaric process.

So, work done during the isobaric expansion = area under AB

$$=2\times10^2\times(3-1)$$

$$=400\ \text{J}$$

91. Since, all the three thermodynamical process represents a cyclic process.

We know, internal energy is a state function.

So, $\Delta U=0$

92. AB is isochoric process.

$$\therefore \quad Q_{AB}=nC_V\Delta T=n\left(\frac{5}{2}R\right)(T_f-T_i)$$

$$=\frac{5(p_fV_f-p_iV_i)}{2}=\frac{5\,(2p_0V_0-p_0V_0)}{2}$$

$$=2.5\,p_0V_0$$

93. In all isochoric processes, $W = 0$.

In all isobaric processes, $W = p\Delta V$

$$W_1 = 1\,(0.6 - 0.8) \times 10^3 = -200\text{ J}$$
$$W_3 = 2\,(0.4 - 0.6) \times 10^3 = -400\text{ J}$$
$$W_5 = 3\,(0.2 - 0.4) \times 10^3 = -600\text{ J}$$
$$\therefore \quad W_{\text{total}} = -1200\text{ J}$$

94. Work done on the system $= p\Delta V$

ΔV is constant, so $\Delta W \propto p$

$$p_B > p_A,\ \Delta W_2 > \Delta W_1$$

95. From the graph, area under the path A is more and in path C is least.

So, $W_A > W_B > W_C > W_D$

Now, from first law of thermodynamics, $Q = \Delta U + W$

As the ΔU is a state function, so it is constant for all paths.

Thus, $Q_A > Q_B > Q_C > Q_D$

Now, for paths A and D,

$$Q_A = \Delta U + W_A \text{ and } Q_D = \Delta U + W_D$$

So, $Q_A - Q_D = W_A - W_D$

96. From first law of thermodynamics,

$$\Delta U = \Delta Q - p\Delta V$$
$$300 = 400 - (0.5) \times \Delta V$$
$$\Rightarrow \quad \Delta V = 2\text{ L}$$

97. In AB, process is isochoric,

$$\Delta U_{AB} = \Delta Q_{AB} = 400\text{ J} \qquad [\because \Delta W = 0]$$

In BC, process is isobaric,

$$\Delta U_{BC} = \Delta Q_{BC} - \Delta W_{BC} = 100 - p\Delta V$$
$$= 100 - 6 \times 10^4 (4 \times 10^{-3} - 2 \times 10^{-3})$$
$$= -20\text{ J}$$

Now, the total change in internal energy,

$$\Delta U_{AC} = 400 - 20 = 380\text{ J}$$

So, in process AC, heat added is,

$$\Delta Q_{AC} = \Delta U_{AC} + \Delta W_{AC}$$
$$= 380 + \frac{1}{2}(6 \times 10^4 + 2 \times 10^4) \times 2 \times 10^{-3} \qquad [\because \Delta W_{AC} = \text{area}]$$
$$= 460\text{ J}$$

98. Process $A \rightarrow B$ - Isobaric

Process $B \rightarrow C$ - Isothermal

Process $C \rightarrow D$ - Isochoric

Process $D \rightarrow A$ - Isothermal

In curve (a), same process is represented.

99. From graph, $p \propto U$

We know, internal energy U is a function of temperature only.

So, $p \propto T$

$$p \propto \frac{pV}{nR}$$
$$V = \text{constant}$$

100. We know, $pm = \rho RT$

$\rho \propto$ Density of gas

In the given pressure - volume graph, volume is increasing, so density must decrease.

So, graph (a) represents the pressure - density (p - ρ) diagram.

101. We have given,

$$pT = \text{constant}$$
$$pT = k$$
$$\Rightarrow \quad p = \frac{k}{T} \qquad \text{...(i)}$$

Now, substituting p in ideal gas equation,

$$pV = RT \qquad \text{...(ii)}$$

From Eqs. (i) and (ii), we get

$$\Rightarrow \quad V = \frac{RT^2}{k}$$

Now, differentiating both sides, we get

$$dV = \frac{R}{k} \times 2T \times dT \qquad \text{...(iii)}$$

Now, $dW = pdV$

$$dW = \frac{k}{T} \times \frac{R}{k} \times 2T \times dT \qquad \text{[using Eqs. (iii) and (i)]}$$
$$dW = 2RdT$$

From first law of thermodynamics,

$$dQ = dU + dW$$
$$CdT = C_V dT + 2RdT$$
$$C = C_V + 2R$$
$$\therefore \quad -C_V = 2R - C$$
$$-C_V = 2R - 4.5\,R$$
$$C_V = 2.5\,R = \frac{5}{2}R$$

So, the degrees of freedom of molecules is 5.

102. $\Delta U = U_f - U_i = nC_V T_f - nC_V T_i$

$$= n\left(\frac{5}{2}RT_f\right) - n\left(\frac{5}{2}RT_i\right)$$
$$= \frac{5}{2}(p_f V_f - p_i V_i) \qquad (\because pV = nRT)$$
$$= \frac{5}{2}(12 - 20)\text{ kJ}$$
$$= -20\text{ kJ}$$

103. The average energy of gas per degree of freedom $= \frac{1}{2}kT$

So, mean energy of gas with n degree of freedom $= \frac{n}{2}kT$

104. $\gamma = \frac{C_p}{C_V}$

$$\gamma - 1 = \frac{C_p - C_V}{C_V} = \frac{R}{C}$$
$$C_V = \frac{R}{\gamma - 1} = \frac{R}{\frac{67}{45} - 1} = \frac{45R}{22}$$

Now, $C_p = C_V + R = \frac{67}{22}R$

So, given heat of gas mixture $= nC_p\Delta T$

$$= 0.75 \times \frac{67}{22} \times 2 \times 100 = 456.8 \approx 457 \text{ cal}$$

105. In process $A \to B$, $W_{AB} = 0$

In process $B \to C$, $W_{BC} = p_B(V_C - V_B)$

$$= p_B(V_D - V_A) = 240 \text{ J}$$

Now, from the energy conservation,

$$Q = U + W$$

$$\Rightarrow \quad U = Q - W = (600 + 200) - 240$$

$$= 560 \text{ J}$$

106. In process AB, $Q_{AB} = U + W = nC_V\Delta T + nR\Delta T$

$$= \frac{n \times 3R}{2} \times T_0 + nRT_0 = \frac{5}{2}nRT_0$$

In process BC, $W_{BC} = nRT\ln\frac{V_2}{V_1}$

$$= nR \times 2T_0 \ln\frac{2V_0}{2V_0} = nRT_0 \ln 2$$

Now, $\frac{Q_{AB}}{W_{BC}} = \frac{5/2\, nRT_0}{2nRT_0 \ln 2} = \frac{5}{4 \ln 2}$

107. Work done in adiabatic process is given as,

$$W = \frac{-1}{\gamma - 1}(p_fV_f - p_iV_i)$$

$$6R = \frac{-1}{5/3 - 1}R(T_f - T_i)$$

$$\Rightarrow \quad T_f - T_i = -4$$

$$\Rightarrow \quad T_f = (T - 4) \text{ K}$$

108. Work done during process ABC,

$$\Delta W = \text{Area under } p\text{ - }V \text{ graph}$$

$$= \frac{1}{2}5(p)(V) + \frac{1}{2}(4p)(4V) = 10.5\, pV$$

Now, the internal energy, $\Delta U = \frac{3}{2}nR\Delta T$

$$= \frac{3}{2}nR(T_C - T_A)$$

$$= \frac{3}{2}(6pV - 2pV) = \frac{3}{2} \times 4pV$$

$$= 6pV$$

$$\therefore \quad \Delta Q = \Delta U + \Delta W$$

$$= 6\, pV + 10.5\, pV$$

$$= 16.5\, pV$$

109. $\Delta U = Q_{ABC} - W_{ABC} = 10$ J

Note : U or ΔU is state function. So, we can directly find ΔU from $Q_{ABC} - W_{ABC}$.

110. As the gas is enclosed in a closed vessel, so volume remains constant.

$$\therefore \quad W = 0$$

$$Q = \Delta U$$

Initial internal energy of a gas $= \frac{5}{2}NRT$

Now, when the n moles get dissociated, the vessel contains $(N - n)$ moles of diatomic and $2n$ moles of monoatomic.

Its internal energy, $U_2 = \frac{5}{2}(N - n)RT + \frac{3}{2}(2n)RT$

$$= \frac{nRT}{2} + \frac{5}{2}NRT$$

So, $\Delta U = \left(\frac{nRT}{2} + \frac{5NRT}{2} - \frac{5NRT}{2}\right)$

$$\Delta U = \Delta Q = \frac{nRT}{2}$$

111. For adiabatic process,

$$pV^\gamma = \text{constant}$$

$$\ln (pV)^\gamma = \ln k$$

$$\ln p + \gamma \ln V = \ln k$$

$$\ln p = -\gamma \ln V + \ln k$$

So, slope $= -\gamma$

From graph, $m_A < m_B$

$$\gamma_A > \gamma_B$$

So, the gas B is monoatomic and A is diatomic.

112. In process AB,

$$W_{AB} = nR\Delta T = 1 \times R(2T - T) = RT$$

In process BC, $W_{BC} = nRT\ln\frac{p_i}{p_f}$

$$= 1 \times R \times 2T\ln\frac{2p}{p} = 2RT \ln 2$$

Similarly, $W_{CD} = 1 \times R \times (T - 2T) = -RT$

and $\quad W_{DA} = 1 \times R \times T \ln\frac{P}{2P} = -RT \ln 2$

$$\therefore \quad W_{\text{Total}} = RT \ln 2$$

113. In isobaric process,

$$\frac{Q}{W} = \frac{Q}{Q - \Delta U} = \frac{1}{1 - \frac{\Delta U}{Q}}$$

$$= \frac{1}{1 - \frac{nC_V\Delta T}{nC_p\Delta T}} = \frac{1}{1 - \frac{1}{\gamma}} \quad \text{...(i)}$$

$$= \frac{5}{2} \text{ (given)} \quad \text{...(ii)}$$

From Eqs. (i) and (ii), we get

$$r = \frac{5}{3}$$

Hence, gas is monoatomic and process is isobaric.

$$p = \text{constant}$$

$$\therefore \quad V \propto T$$

or V - T graph is a straight line passing through origin.

114. $T \propto pV$

$$(pV)_1 = 4\,(pV)_2$$

$$\therefore \quad T_1 = 4T_2$$

$$v_{rms} \propto \sqrt{T}$$

$$\because \quad \frac{(v_{rms})_{T_1}}{(v_{rms})_{T_2}} = \frac{2}{1}$$

115. $T = \dfrac{pV}{nR}$

$$T_f - T_i = \frac{p_f V_f}{nR} - \frac{p_i V_i}{nR}$$

$$= \frac{\dfrac{p_0}{\left[1+\left(\dfrac{2V_0}{3V_0}\right)^2\right]}(2V_0) - \dfrac{p_0(V_0)}{\left[1+\left(\dfrac{2V_0}{V_0}\right)^2\right]}}{nR} = \frac{4p_0V_0}{5nR}$$

Putting $n = R$, we get

$$T_f - T_i = \frac{4p_0V_0}{5R}$$

116. $p_1V = n_1RT_1$

$$p_2V = n_2RT_2$$

When the vessels are joined,

$$p(2V) = (n_1 + n_2)\,RT$$

$$\therefore \quad \frac{p}{T} = \frac{(n_1+n_2)R}{2V} = \frac{1}{2}\left(\frac{p_1}{T_1} + \frac{p_2}{T_2}\right)$$

$$= \frac{1}{2}\left(\frac{p_1T_2 + p_2T_1}{T_1T_2}\right)$$

117. From the law of equipartition of energy, the energy associated with each degree of the freedom

$$= \frac{1}{2}kT$$

Now, internal energy for n molecules of a gas,

$$U = nf\frac{1}{2}kT$$

$$\Rightarrow \quad n = \frac{2U}{fkT}$$

n represents the number of molecules in the gas.

118. $\eta = 1 - \dfrac{T_{cold}}{T_{hot}} \times 100$

$$0.5 = 1 - \frac{280}{T_{hot}}$$

$$T_{hot} = 560\text{ K}$$

Now, when $\eta = 60\%$

$$0.6 = 1 - \frac{280}{T'_{hot}}$$

$$T'_{hot} = 700\text{ K}$$

So, the temperature of the reservoir is increased by

$$= (700 - 560) = 140\text{ K}$$

119. $\eta = 1 - \dfrac{T_2}{T_1} = \dfrac{T_1 - T_2}{T_1}$

$$\eta_1 = \frac{473 - 273}{473} = \frac{200}{473}$$

$$\eta_2 = \frac{273 - 73}{273} = \frac{200}{273}$$

$$\therefore \quad \eta_1/\eta_2 = \frac{273}{473} = 0.577$$

120. We know, $C = C_V + \dfrac{R}{1-\gamma}$

$$V \propto T^4$$

$$V \propto (pV)^4$$

$$p^4V^3 = \text{constant}$$

$$pV^{3/4} = \text{constant}$$

Now, comparing with $pV^{\gamma} = \text{constant}$

$$\gamma = \frac{3}{4}$$

$$\text{So, } C = C_V + \frac{R}{1-\gamma} = 3R + \frac{R}{1-3/4} = 7R$$

121. Let the mass of ice be m.

Heat given by the burner in first 50 min,

$$\frac{d\theta}{dt} = \frac{mL}{t_1} = \frac{m \times 80 \times 4.2 \times 10^3}{50} \qquad \text{...(i)}$$

Heat given by burner in 50 to 60 min,

$$\frac{d\theta}{dt} = \frac{(m+5) \times (4.2 \times 10^3) \times 2°\text{C}}{10} \qquad \text{...(ii)}$$

From Eqs. (i) and (ii), we get

$$m = \frac{5}{7}\text{ kg}$$

122. $p = \text{constant}$

$$\therefore \quad \Delta U = 2.5p\Delta V \quad \text{and} \quad W = p\Delta V$$

$$\therefore \quad Q = W + \Delta U = 3.5\,p\Delta V$$

$$= 3.5 \times 1.01 \times 10^5 \times (2.1) \times 10^{-3}$$

$$\approx 350\text{ J}$$

123. Heat gained by the ice = Heat released by steam

$$(20)(0.5)(10) + (20)(1)(100) + (20)(80) = m\,(540)$$

$$\Rightarrow \quad m = \frac{185}{27}\text{ g}$$

124. Given, $pT^2 = \text{constant}$

$$\left(\frac{nRT}{V}\right)T^2 = \text{constant}$$

$$T^3V^{-1} = \text{constant}$$

Differentiating with respect to pressure, we get

$$(T^3)(-1)\,V^{-2}dV + (V^{-1})\,3T^2dT = 0$$

$$(3T^2)\,V^{-1}dT = (V^{-2})T^3dV$$

$$\Rightarrow \quad \left(\frac{dV}{V}\right)\frac{1}{dT} = \frac{3T^2}{T^3} = \frac{3}{T}$$

We know, coefficient of volume expansion,

$$\gamma = \left(\frac{dV}{V}\right)\frac{1}{dT} \Rightarrow \gamma = \frac{3}{T}$$

125. For an adiabatic process, pV^γ = constant

Taking its derivative,

$$dpV^\gamma + \gamma pV^{\gamma-1}dV = 0$$

$$dpV^\gamma = -\gamma pV^{\gamma-1}dV$$

$$\frac{dp}{p} = -\gamma\frac{V^{\gamma-1}}{V^\gamma}dV$$

$$\frac{\Delta p}{p} = -\gamma\frac{\Delta V}{V}$$

126. Internal energy of a gas in state 1

$$= nC_V dT = \frac{3}{2}nRT = \frac{3}{2}pV = \frac{3}{2}\times 2\times 1 = 3\text{ J}$$

Internal energy at state 4

$$= \frac{3}{2}pV = \frac{3}{2}\times 2\times 5 = 15\text{ J}$$

So,
$$\Delta Q = \Delta U + \Delta W = (3-15) + p\Delta V = -12 + 2(1-5) = -20\text{ J}$$

127. Applying ideal gas equation,

$$\frac{p_1V}{n_1T_1} = \frac{p_2V}{n_2T_2}$$

$$\Rightarrow \quad n_2 = \frac{p_2}{p_1}\frac{T_1}{T_2}n_1 = \frac{1}{2}\times\frac{(273+57)}{(273+27)} = \frac{11}{20}n_1$$

So, mass of the gas leaked

$$= \left(1-\frac{11}{20}\right)\times 28 = \frac{9}{20}\times 28 = \frac{63}{5}\text{ g}$$

128. Number of moles, $n = 6$

AB — isochoric process, $\Delta W_{AB} = 0$

BC — isobaric process, $\Delta W_{BC} = nR\Delta T = nR(T_C - T_B) = 6R(2200-800) = 8400\,R$

CD — isochoric process, $\Delta W_{CD} = 0$

DA — isobaric process, $\Delta W_{DA} = nR\Delta T = nR(600-1200) = -3600\,R$

$$\Delta W_{\text{cyclic}} = \Delta W_{AB} + \Delta W_{BC} + \Delta W_{CD} + \Delta W_{DA} = 8400\,R - 3600\,R = 4800\,R = 4800\times 8.3 \approx 40\text{ kJ}$$

129. $dW = pdV = \frac{RT}{V}dV$

Given, $V = kT^{2/3}$

$$dV = k^{2/3}T^{-1/3}dT$$

$$\because \quad \frac{dV}{V} = \frac{k\frac{2k}{3}T^{-1/3}dT}{kT^{2/3}}$$

$$\frac{dV}{V} = \frac{2}{3}\frac{dT}{T} \quad \text{...(i)}$$

Now, $W = \int_{T_1}^{T_2} RT\frac{dV}{V}$

$$= \int_{T_1}^{T_2} RT\frac{2}{3}\frac{dT}{T} = \frac{2}{3}R(T_2 - T_1) \quad \text{[using Eq. (i)]}$$

$$= \frac{2R}{3}\times 30 = 20R$$

130. Volume, V = constant

$\therefore \quad p \propto T$

When temperature is doubled, pressure will also become two times or $2p_0$.

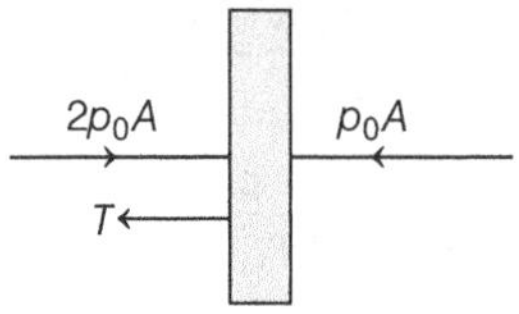

$$T + p_0A = 2p_0A$$

$$\therefore \quad T = p_0A$$

131. In the process, pV^x = constant

Bulk modulus, $B = xp$

Here, $x = 2$

$\therefore \quad B = 2p$

or $\quad B \propto p$

132. From graph, slope, $m = \frac{2.10 - 2.38}{1.3 - 1.1} = -1.4$

Now, $\log p = m\log V + C$

$$\log pV^m = \text{constant}$$

$$pV^{1.4} = \text{constant}$$

$$pV^\gamma = \text{constant}$$

$$\gamma = 1.4 = K = \frac{7}{5}$$

So, gas is diatomic and undergoing an adiabatic change.

133. AB : p = constant

$$T \propto V$$

V is increasing. Therefore, T will also increase.

BC : $T =$ constant

$\therefore \quad p \propto \frac{1}{V}$

V is increasing. Therefore, p will decrease.

CD : $V =$ constant

$\therefore \quad T \propto p$

p is decreasing. So, T will also decrease.

DE : $p =$ constant

$\therefore \quad T \propto p$

p is decreasing. So, T will also decrease.

EA process is just like **CD** process.

134. In isobaric process, p is constant.

In isochoric process, V is constant.

In isothermal process, T is constant.

Internal energy is a function of temperature only.

So, the graph (a) shows the change in internal energy of an ideal gas.

135. In adiabatic process,

$$TV^{\gamma - 1} = \text{constant}$$

or $$VT^{\frac{1}{\gamma - 1}} = \text{constant}$$

Differentiating this equation, we get

$$\frac{dV}{dT} = \text{Slope} = \frac{1}{\gamma - 1}\left(\frac{V}{T}\right)$$

At point A,

$$\text{Slope} = \tan\theta = \frac{1}{\gamma - 1}\left(\frac{V_0}{T_0}\right)$$

$$\therefore \quad \gamma = \frac{V_0}{T_0 \tan\theta} + 1$$

136. Rate of heat transfer,

$$H = \frac{\text{Temperature difference}}{\text{Net thermal resistance}}$$

$$= \frac{T_2 - T_1}{\frac{x}{KA} + \frac{3x}{2KA}} = \frac{2}{5}\left[\frac{(T_2 - T_1)KA}{x}\right]$$

137.

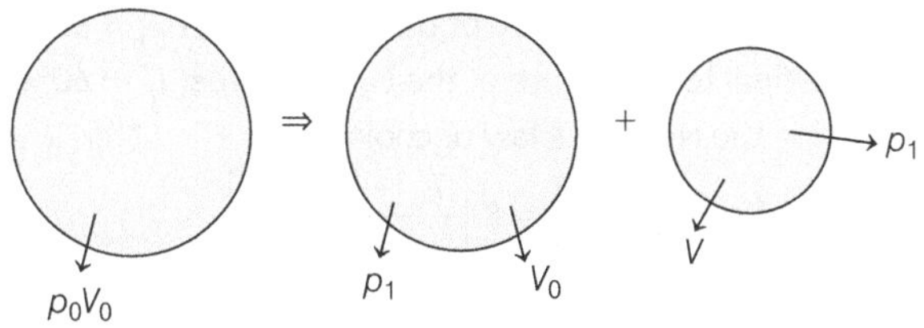

Process is isothermal ($pV =$ constant)

$$\therefore \quad p_0V_0 = p_1(V_0 + V)$$

or $$p_1 = \frac{p_0V_0}{V_0 + V}$$

Similarly, $$p_2 = \frac{p_1V_0}{V_0 + V} = p_0\left(\frac{V_0}{V_0 + V}\right)^2$$

Now, $$n = \frac{p_2V_0}{RT} = \frac{p_0V_0^3}{RT(V_0 + V)^2}$$

138. Line AB passes through origin.

$$p = \text{constant}$$

or $$W_{AB} = nR\Delta T$$

$$= 2(R)(500)$$

$$= 1000\,R$$

BC is isochoric.

$$\therefore \quad W_{BC} = 0$$

CD is isothermal.

$$\therefore \quad W_{CD} = nRT \ln\frac{V_f}{V_i}$$

$$= (2)(R)(1000)\ln\frac{20}{15}$$

$$= 2000\,R \ln\frac{4}{3}$$

Total work done,

$$W_{\text{net}} = 1000\,R\left(\ln\frac{16}{9} + 1\right)$$

139. $T_A < T_B$

$\therefore \quad U_A < U_B$, as U is a function of only T.

140. We know for the adiabatic process, $T_1V_1^{\gamma - 1} =$ constant

For bc curve, $T_1V_b^{\gamma - 1} = T_2V_c^{\gamma - 1}$

$$\frac{T_2}{T_1} = \left(\frac{V_b}{V_c}\right)^{\gamma - 1} \quad \text{...(i)}$$

For ad curve, $T_1V_a^{\gamma - 1} = T_2V_d^{\gamma - 1}$

$$\frac{T_2}{T_1} = \left(\frac{V_a}{V_d}\right)^{\gamma - 1} \quad \text{...(ii)}$$

From Eqs. (i) and (ii), we get

$$\frac{V_b}{V_c} = \frac{V_a}{V_d}$$

141. The graph of temperature *versus* distance gives the thermal conductivity as slope.

$$\text{Slope, } m_1 = \frac{100 - 70}{-30} = \frac{30}{-30} = -1$$

$$\text{Slope, } m_2 = \frac{70 - 35}{30 - 100} = \frac{35}{-70} = \frac{-1}{2}$$

$$\therefore \quad \text{Ratio of thermal conductivity} = \frac{m_2}{m_1} = \frac{1}{2} = 0.5$$

142. Rotational degree of freedom in diatomic gas is 2.

$$\therefore \quad K_R = 2\left[\frac{n}{2}RT\right]$$

$$= nRT = pV$$

$$= (p_0 e^{\frac{3V_0}{V}})(V)$$

Substituting $V = \frac{V_0}{2}$, we get

$$K_R = \frac{p_0 V_0 e^6}{2}$$

143. Tension in the two rods must be same.

So, $A_1 Y_1 \alpha_1 \Delta Q = A_2 Y_2 \alpha_2 \Delta Q$

$$A_1 Y_1 \alpha_1 = A_2 Y_2 \alpha_2$$

$$\therefore \quad \frac{A_1}{A_2} = \frac{\alpha_2 Y_2}{\alpha_1 Y_1}$$

144. Thermal resistance,

$R_i = 2R$ (two identical rods in series)

and $R_f = \frac{R}{2}$

Resistance has decreased four times $\left(2R \text{ to } \frac{R}{2}\right)$.

Therefore, heat current will be four times. But total heat flown is double (10 J to 20 J). So, time taken will be half.

145. In isothermal process,

$$\frac{dp}{dV} = -\frac{p}{V}$$

$$\therefore \quad dp = -\left(\frac{p}{V}\right) dV$$

$$= -\left(\frac{p}{V}\right)(Ax)$$

Restoring force,

$$F = (dp)\, A = -\left(\frac{p}{Ah}\right) A^2 x$$

$$= -\left(\frac{pA}{h}\right) x$$

$$= -kx$$

where, $k = \frac{pA}{h}$

$$\therefore \quad T = 2\pi \sqrt{\frac{M}{k}} = 2\pi \sqrt{\frac{Mh}{pA}}$$

146. $dQ = dU - dW$

From first law of thermodynamics,

$$dQ = dU + dW$$

$\therefore \quad dQ - dW = dU$

$$dW = 0$$

So, $dQ = dU$

$$C = C_V$$

147. $pV^n = k$

$$\frac{nRT}{V} V^n = k$$

or $\quad TV^{n-1} = k$

Now, differentiating both sides,

$$T(n-1)\, V^{n-2}\, dV + V^{n-1} dT = 0$$

$$T(n-1)\, dV + V dT = 0$$

We get,

$$\frac{dV}{V dt} = \frac{1}{(1-n)T}$$

148. $U = a + bpV$ ...(i)

In adiabatic process,

$$dU = -\,dW = \frac{nR}{\gamma - 1}(T_2 - T_1)$$

$$= \frac{nR}{\gamma - 1}(dT)$$

$$\int dU = \frac{nR}{\gamma - 1} \int dT$$

$$U = \left(\frac{nR}{\gamma - 1}\right) T + a = \frac{pV}{\gamma - 1} + a \quad \text{...(ii)}$$

Now, comparing Eqs. (i) and (ii), we get

$$b = \frac{1}{\gamma - 1}, \; \gamma = \frac{b+1}{b}$$

LEVEL 03 Previous Years Questions

149. In first conditions,

Given, the initial temperature of the cup of coffee,

$$T_i = 90°C$$

The final temperature of the cup of coffee, $T_f = 80°C$

The time taken to drop the temperature 90°C to 80°C is t.

The temperature of the surrounding, $T_0 = 20°C$

Using the Newton's law of cooling,

$$\text{rate of cooling} = \frac{dT}{dt} = K\left[\frac{T_i + T_f}{2} - T_0\right]$$

Substituting the values in the above equation, we get

$$\frac{90 - 80}{t} = K\left[\frac{90 + 80}{2} - 20\right]$$

$$\Rightarrow \quad \frac{10}{t} = K[65] \quad \text{...(i)}$$

In second condition,

The initial temperature of the cup of coffee, $T_i' = 80°C$

The final temperature of the cup of coffee, $T_f' = 60°C$.

Using the Newton's law of cooling,

$$\text{rate of cooling} = \frac{dT}{dt} = K\left[\frac{T'_i + T'_f}{2} - T_0\right]$$

Substituting the values in the above equation, we get

$$\frac{80 - 60}{t_1} = K\left[\frac{60 + 80}{2} - 20\right]$$

$$\Rightarrow \quad \frac{20}{t_1} = K\,[50] \quad \text{...(ii)}$$

Solving Eqs. (i) and (ii), we get

$$\frac{t_1}{t} = \frac{65 \times 2}{50} = \frac{13}{5}$$

or $$t_1 = \frac{13}{5} t$$

150. We know that, the rms speed of the gas molecules,

$$v_{rms} = \sqrt{3RT/M}$$

Here, T is the temperature of the gas,
R is the universal gas constant,
M is the molar mass of the gas.
Pressure exerted by an ideal gas,

$$p = \frac{1}{3} nm\bar{v}^2$$

Here, n is the number of moles, m is the mass of the gas, $\bar{v}$ is the average speed of the gas molecules.
The average kinetic energy of a molecule,

$$KE_{av} = \frac{3}{2} k_B T$$

Here, k_B is the Boltzmann's constant,
T is the temperature of the gas.
For diatomic gas, degree of freedom, $f = 5$
As, total internal energy of 1 mole of diatomic gas,

$$\Delta U = \frac{nfRT}{2} \Rightarrow \Delta U = \frac{1\,(5)RT}{2} \Rightarrow \Delta U = \frac{5RT}{2}$$

The correct match is A $\rightarrow$ 2, B $\rightarrow$ 1, C $\rightarrow$ 4 and D $\rightarrow$ 3.

151. According to the Wien's displacement law,

$$\lambda = \frac{b}{T}$$

i.e. $$\lambda \propto \frac{1}{T} \quad \text{... (i)}$$

We know that,

$$\lambda_{bluish} < \lambda_{yellowish} < \lambda_{reddish}$$

Hence, using Eq. (i), we have

$$T_A > T_C > T_B$$

152. p-V diagram for an ideal gas is given as

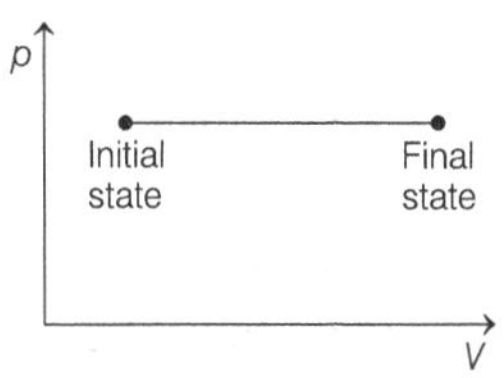

From the p-V diagram, it is clear that pressure of ideal gas in the piston cylinder is constant during thermodynamic process. Hence, this process is isobaric.

153. The mean free path l for a gas molecule is given as

$$l = \frac{1}{\sqrt{2}\pi nd^2} \Rightarrow l \propto \frac{1}{d^2}$$

where, d = diameter of molecule of gas.

154. Ideal gas equation is given as

$$p = \frac{\rho RT}{M_0} \Rightarrow p \cdot \frac{M_0}{\rho} = RT \Rightarrow pV = RT$$

where, $V = \frac{M_0}{\rho}$

Hence, ρ and M_0 are mass density and mass of gas, respectively.

155. Efficiency of Carnot engine is given as

$$\eta = 1 - \frac{T_2}{T_1}$$

where, T_2 = temperature of sink
and T_1 = temperature of source.
Hence, η depends upon the temperature of source and sink both.

156. Since, heat required, $Q = mc\Delta T$

$$= \left(\frac{4}{3}\pi r^3 \cdot \rho\right) c\Delta T \qquad [\because m = V_{sphere}\,\rho]$$

Since, π, ρ, c and T are constants.

$$\Rightarrow \quad Q \propto r^3$$

or $$\frac{Q_1}{Q_2} = \frac{r_1^3}{r_2^3}$$

$$= \left(\frac{r_1}{r_2}\right)^3 = \left(\frac{1.5 r_2}{r_2}\right)^3 = \frac{27}{8}$$

157. Since, the entire system is thermally insulated. Thus, their will be no transfer of heat between the system and the surrounding. Hence, when the stop cock is suddenly opened, a sudden process of expansion will take place with no heat transfer. Thus, the given process will be adiabatic, as it is a process that occurs without transferring heat or mass between a thermodynamic system and its surroundings.
Hence, correct option is (a).

158. The mean free path λ for a gas, with molecular diameter d and number density n is given by the relation

$$\lambda = \frac{1}{\sqrt{2}n\pi d^2}$$

159. Given, pressure, $p = 249$ kPa $= 249 \times 10^3$ Pa

Temperature, $T = 27°C$

$$= (273 + 27)\,K = 300\,K$$

Density, $\rho = ?$

As from ideal gas equation,

$$pV = nRT$$

$$\Rightarrow \quad pV = \frac{m}{M}RT \qquad \left[\text{as } n = \frac{m}{M}\right]$$

$$\Rightarrow \quad pVM = mRT$$

$$\Rightarrow \quad pM = \frac{m}{V}RT = \rho RT \qquad \left[\text{as } \frac{m}{V} = \rho\right]$$

$$\Rightarrow \quad \rho = \frac{pM}{RT}$$

$$= \frac{249 \times 10^3 \times 2 \times 10^{-3}}{8.3 \times 300}$$

[$\because$ for hydrogen gas, $M = 2\,g = 2 \times 10^{-3}$ kg]

$$\Rightarrow \quad \rho = 0.2\,kg/m^3$$

160. The average thermal energy of a system with degree of freedom f is equals to its average energy, which is given as

$$= \frac{f}{2} k_B \cdot T$$

For monoatomic gas, $f = 3$

$\therefore$ Average thermal energy $= \frac{3}{2} k_B \cdot T$

161. Due to change in temperature, the thermal strain produced in a rod of length L is given by

$$\frac{\Delta L}{L} = \alpha\, \Delta T$$

$$\Rightarrow \quad \Delta L = L\, \alpha\, \Delta T$$

where, L = original length of rod

and α = coefficient of linear expansion of solid rod.

As the change in length (ΔL) of the given two rods of copper and aluminium are independent of temperature change. So, difference in lengths is same at all temperature.

$$\Rightarrow \quad (l_{Cu} - l_{Al}) = (l'_{Cu} - l'_{Al})$$

$$\Rightarrow (l_{Cu} - l_{Al}) = [l_{Cu}(1 + \alpha_{Cu}\Delta T) - l_{Al}(1 + \alpha_{Al}\, \Delta T)]$$

$$\Rightarrow \quad l_{Cu} \cdot \alpha_{Cu} \cdot \Delta T = l_{Al} \cdot \alpha_{Al} \cdot \Delta T$$

$$L_{Cu}\alpha_{Cu} = L_{Al}\alpha_{Al} \quad \text{... (i)}$$

Here, $\alpha_{Cu} = 1.7 \times 10^{-5}\, K^{-1}$

$\alpha_{Al} = 2.2 \times 10^{-5}\, K^{-1}$

and $L_{Cu} = 88$ cm

Substituting the given values in Eq. (i), we get

$$L_{Al} = \frac{L_{Cu}\alpha_{Cu}}{\alpha_{Al}}$$

$$= \frac{1.7 \times 10^{-5} \times 88}{2.2 \times 10^{-5}}$$

$$\simeq 68\, cm$$

162. Key Idea From Newton's law of cooling, the time taken t by a body to cool from T_1 to T_2 when placed in a medium of temperature T_0 can be calculated from relation,

$$\frac{T_1 - T_2}{t} = \frac{1}{K}\left(\frac{T_1 + T_2}{2} - T_0\right)$$

When the object cool from 80°C to 70°C in 12 min, then from Newton's law of cooling,

$$\frac{80 - 70}{12} = \frac{1}{K}\left(\frac{80 + 70}{2} - 25\right) \quad [\because T_0 = 25°\, C]$$

$$\frac{5}{6} = \frac{1}{K}(50) \quad \text{...(i)}$$

Similarly, when object cool from 70°C to 60°C, we get

$$\frac{70 - 60}{t} = \frac{1}{K}\left(\frac{70 + 60}{2} - 25\right)$$

$$\frac{10}{t} = \frac{1}{K} 40 \quad \text{...(ii)}$$

Dividing Eq. (i) by Eq. (ii), we get

$$\frac{5}{6} \times \frac{t}{10} = \frac{50}{40} \Rightarrow t = \frac{5}{4} \times 12$$

$$= 15 \text{ min}$$

163. Key Idea If area of cross-section of a surface is not uniform or if the steady state condition is not reached, the heat flow equation can be applied to a thin layer of material perpendicular to direction of heat flow.

The rate of heat flow by conduction for growth of ice is given by,

$$\frac{d\theta}{dt} = \frac{KA(\theta_0 - \theta_1)]}{x}$$

where, $d\theta = \rho A dx L, \theta_0 = 0$ and $\theta_1 = -\theta$

Given, $\theta_0 = 0°\, C, \theta_1 = -26°\, C$

The rate of increase of thickness can be calculated from equation

$$\frac{d\theta}{dt} = \frac{KA(\theta_0 - \theta_1)}{x}$$

$$\Rightarrow \quad \frac{\rho A dx L}{dt} = \frac{KA(\theta_0 - \theta_1)}{x}$$

$$\Rightarrow \quad \frac{dx}{dt} = \frac{KA(\theta_0 - \theta_1)}{\rho A x L}$$

$$= \frac{K[0 - (-26)]}{\rho x L} = \frac{26K}{\rho x L}$$

164. Given, mass of water, $m = 1$ g

Volume of 1 g of water = $1\, cm^3 = 10^{-6}\, m^3$

Volume of 1 g of steam = $1671\, cm^3$

$= 1671 \times 10^{-6}\, m^3$

Pressure, $p = 1 \times 10^5$ Pa

Latent heat of vaporization of water,

$L = 2256$ J/g

Change in volume,

$$\Delta V = (1671 - 1) \times 10^{-6}\, m^3$$

$$= 1670 \times 10^{-6}\, m^3 \quad \text{...(i)}$$

Heat supplied,

$$\Delta Q = mL = 1 \times 2256 = 2256\, J \quad \text{...(ii)}$$

As the steam expands, so the work done in expansion is

$$\Delta W = p\Delta V$$

$$= 1 \times 10^5 \times 1670 \times 10^{-6} \quad \text{[from Eq. (i)]}$$

$$= 167\, J \quad \text{...(iii)}$$

According to the first law of thermodynamics,

$$\Delta Q = \Delta U + \Delta W$$

$$\Rightarrow \quad \Delta U = \Delta Q - \Delta W$$
$$= 2256 - 167 \quad \text{[from Eqs. (ii) and (iii)]}$$
$$= 2089 \text{ J}$$

165. In an adiabatic process, the system is completely insulated from the surroundings. Thus, heat is neither absorbed nor released by the system to the surroundings. So, $\Delta Q = 0$.

166. As the temperature of gas in the container is increased, the kinetic energy also increases. This is because the average kinetic energy of a gas is given by

$$\text{KE} = \frac{f}{2}nRT \quad \text{... (i)}$$

where, f = degree of freedom,
n = number of moles of gas molecules,
R = universal gas constant and
T = absolute temperature of the gas.
From Eq. (i),

$$\text{KE} \propto T$$

option (b) is incorrect as increase in temperature will lead to increase in pressure as $p \propto T$. Other options (c) and (d) are also incorrect as molecular distance increases while mass remains the same for increase in the temperature.

167. The poisson's ratio,

$$\gamma = \frac{C_P}{C_V} \quad \text{...(i)}$$

where, C_p = molar heat capacity constant pressure
and C_V = molar heat capacity at constant volume
Also, $C_p = C_V + R$ (from Mayer's relation)
$C_V = \frac{f}{2}R$ (where, f = degree of freedom)

$$\Rightarrow \quad C_p = \left(\frac{f}{2} + 1\right)R$$

So, Eq. (i) becomes,

$$\Rightarrow \quad \gamma = 1 + \frac{2}{f}$$

For hydrogen gas, which is diatomic, the degree of freedom is 5 (3 translational, 2 rotational).

$$\therefore \quad \gamma = 1 + \frac{2}{5} = \frac{7}{5}$$

For helium gas, which is monoatomic, the degree of freedom is 3 (3 translational only).

$$\therefore \quad \gamma = 1 + \frac{2}{3} = \frac{5}{3}$$

The diatomic gas X also have vibrational motion, so degree of freedom is 7 (3 translational, 2 rotational and 2 vibrational).

$$\therefore \quad \gamma = 1 + \frac{2}{7} = \frac{9}{7}$$

168. According to the Wien's law,

$$\lambda_{max} \propto \frac{1}{T}$$

i.e. $\lambda_{max}T$ = constant
where, λ_{max} is the maximum wavelength of the radiation emitted at temperature T.

$$\therefore \quad \lambda_{max\,1}T_1 = \lambda_{max\,2}T_2$$

$$\text{or} \quad \frac{T_1}{T_2} = \frac{\lambda_{max\,2}}{\lambda_{max\,1}} \quad \text{...(i)}$$

Here, $\lambda_{max\,1} = \lambda_0$
and $\lambda_{max\,2} = \frac{3}{4}\lambda_0$

Substituting the above values in Eq. (i), we get

$$\frac{T_1}{T_2} = \frac{\frac{3}{4}\lambda_0}{\lambda_0} = \frac{3}{4}$$

$$\text{or} \quad \frac{T_1}{T_2} = \frac{3}{4} \quad \text{...(ii)}$$

As we know that, from Stefan's law, the power radiated by a body at temperature T is given as

$$P = \sigma AeT^4$$

$$\text{i.e.} \quad P \propto T^4$$

($\because$ the quantity σAe is constant for a body)

$$\Rightarrow \quad \frac{P_1}{P_2} = \frac{T_1^4}{T_2^4} = \left(\frac{T_1}{T_2}\right)^4$$

From Eq. (i), we get

$$\frac{P_1}{P_2} = \left(\frac{3}{4}\right)^4 = \frac{81}{256}$$

Given, $P_1 = P$ and $P_2 = nP$

$$\Rightarrow \quad \frac{P_1}{P_2} = \frac{P}{nP} = \frac{81}{256}$$

$$\text{or} \quad n = \frac{256}{81}$$

169. According to the question,
heat spent during the conversion of sample of water at 100°C to steam is,

$$\Delta Q = 54 \text{ cal} = 54 \times 4.18 \text{ J}$$
$$= 225.72 \text{ J}$$

Normal pressure, $p = 1.013 \times 10^5 \text{ Nm}^{-2}$

Net work done during the conversion would be given as

$$\Delta W = p\Delta V$$
$$= p[V_{steam} - V_{water}]$$

Here, $V_{steam} = 167.1 \text{ cc} = 167.1 \times 10^{-6} \text{ m}^3$
$V_{water} = 0.1 \text{ g} = 0.1 \text{ cc} = 0.1 \times 10^{-6} \text{ m}^3$

$$\therefore \quad \Delta W = 1.013 \times 10^5 [(167.1 - 0.1) \times 10^{-6}]$$
$$= 1.013 \times 167 \times 10^{-1}$$
$$= 16.917 \text{ J}$$

Now, by the first law of thermodynamics,

$$\Delta Q = \Delta U + \Delta W$$

where, ΔU is the change in internal energy of the sample.

$$\Rightarrow \quad \Delta U = \Delta Q - \Delta W$$

Substituting the values in the above equation, we get

$$\Delta U = 225.72 - 16.917 = 208.7 \text{ J}$$

170. According to the given graph,

Volume (V) $\propto$ Temperature (T)

or $$\frac{V}{T} = \text{constant}$$

Thus, the process is isobaric.

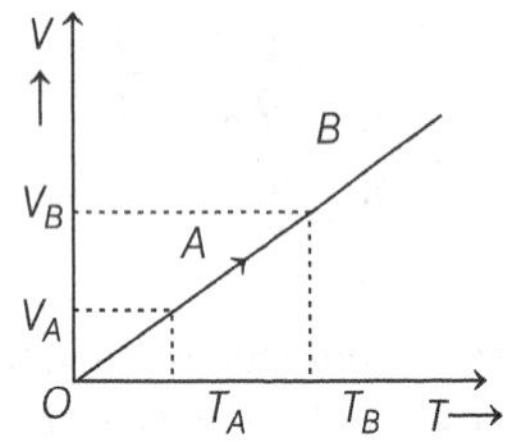

$\therefore$ Work done, $\Delta W = p\Delta V$

$$= nR\Delta T = nR(T_B - T_A) \quad \text{...(i)}$$

Heat absorbed, $\Delta Q = nC_p\Delta T$

$$= nC_p(T_B - T_A) \quad \text{...(ii)}$$

As, $C_p = \dfrac{\gamma R}{\gamma - 1}$, $\left(\text{where, } \gamma = 1 + \dfrac{2}{f}\right)$

For a monoatomic gas, $f = 3$

$$\Rightarrow \quad C_p = \left(R + \frac{3}{2}R\right) = \frac{5}{2}R$$

Substituting the value of C_p in Eq. (ii), we get

$$\Delta Q = n\left(\frac{5}{2}R\right)(T_B - T_A)$$

Hence, $$\frac{\Delta W}{\Delta Q} = \frac{nR(T_B - T_A)}{n\left(\frac{5}{2}R\right)(T_B - T_A)} = \frac{2}{5}$$

171. Efficiency of an ideal heat engine is given as

$$\eta = 1 - \frac{T_2}{T_1}$$

where, T_1 is the temperature of the source and T_2 is the temperature of the sink.

Here, $T_1 = 100 + 273 = 373$ K

$T_2 = 0 + 273 = 273$ K

$$\Rightarrow \quad \eta = 1 - \frac{273}{373} = \frac{373 - 273}{373} = \frac{100}{373} = 0.268$$

$$\therefore \quad \eta\,\% = 0.268 \times 100 = 26.8\,\%$$

172. Key Concept The minimum velocity with which the body must be projected vertically upwards, so that it could escape from the earth's atmosphere, is its escape velocity (v_e).

As, $$v_e = \sqrt{2gR}$$

Substituting the value of g (9.8 ms^{-2}) and radius of earth ($R = 6.4 \times 10^6$ m), we get

$$v_e = \sqrt{2 \times 9.8 \times 6.4 \times 10^6} \cong 11.2 \text{ km s}^{-1} = 11200 \text{ ms}^{-1}$$

Let the temperature of molecule be T when it attains v_e.

According to the question,

$$v_{\text{rms}} = v_e$$

where, v_{rms} is the rms speed of the oxygen molecule.

$$\Rightarrow \quad \sqrt{\frac{3k_BT}{m_{O_2}}} = 11.2 \times 10^3$$

or $$T = \frac{(11.2 \times 10^3)^2 (m_{O_2})}{(3k_B)}$$

Substituting the given values, i.e.

$$K_B = 1.38 \times 10^{-23} \text{ JK}^{-1} \text{ and}$$

$$m_{O_2} = m = 2.76 \times 10^{-26} \text{ kg}$$

We get, $$T = \frac{(11.2 \times 10^3)^2 (2.76 \times 10^{-26})}{(3 \times 1.38 \times 10^{-23})} = 8.326 \times 10^4 \text{ K}$$

173. In parallel arrangement of n rods,

Equivalent thermal conductivity is given by

$$K_{eq} = \frac{K_1A_1 + K_2A_2 + \ldots + K_nA_n}{A_1 + A_2 + \ldots + A_n}$$

If rods are of same area, then

$$K_{eq} = \frac{K_1 + K_2 + \ldots + K_n}{n}$$

Now, in the question, it is not given that rods are of same area. But we can judge that from given diagram.

$\therefore$ Equivalent thermal conductivity of the system of two rods

$$\Rightarrow \quad K_{eq} = \frac{K_1 + K_2}{2}$$

174. Radiated power of a black body,

$$P = \sigma AT^4$$

where, A = surface area of the body,

T = temperature of the body

and σ = Stefan's constant

When radius of the sphere is halved, new area,

$$A' = \frac{A}{4}$$

$\therefore$ Power radiated,

$$P' = \sigma\left(\frac{A}{4}\right)(2T)^4 = \frac{16}{4}\cdot(\sigma A T^4)$$
$$= 4P = 4 \times 450$$
$$= 1800 \text{ W}$$

175. In isochoric process, the curve is parallel to y-axis because volume is constant.Isobaric is parallel to x-axis because pressure is constant. Along the curve, it will be isothermal because temperature is constant.

So, $P \to c$, $Q \to a$, $R \to d$, $S \to b$

176. Consider schematic diagram for a Carnot engine as shown below.

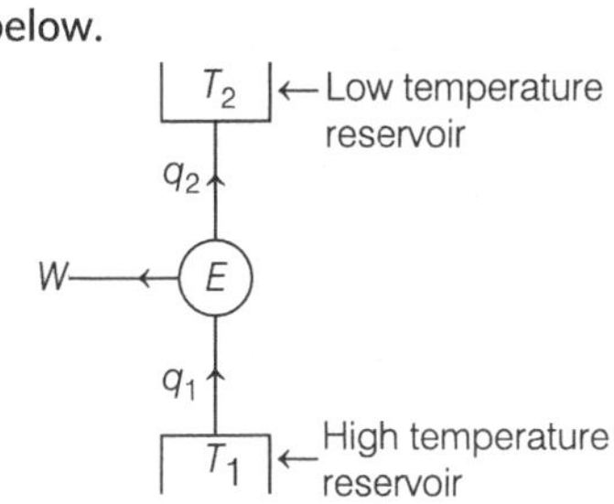

In case of engine,

$$\text{engine efficiency} = \frac{\text{work}}{\text{heat absorbed}} = \frac{W}{q_1}$$

$$\therefore \quad \frac{W}{q_1} = \frac{1}{10}$$

$$\Rightarrow \quad \frac{10 \text{ J}}{q_1} = \frac{1}{10}$$

or $\quad q_1 = 100 \text{ J}$

When this engine is reversed, it takes in work W and heat q_2 from cold reservoir and ejects 100 J of heat to hot reservoir.

$$\therefore \quad W + q_2 = q_1$$
$$\Rightarrow \quad 10 + q_2 = 100$$

or $\quad q_2 = 90 \text{ J}$

177. Total internal energy of system = Internal energy of oxygen molecules + Internal energy of argon molecules

$$= \frac{f_1}{2} n_1 RT + \frac{f_2}{2} n_2 RT$$
$$= \frac{5}{2} \times 2RT + \frac{3}{2} \times 4RT$$
$$= 11RT$$

178. According to question,

Coefficient of linear expansion of brass
$$= \alpha_1$$
Coefficient of linear expansion of steel
$$= \alpha_2$$

Length of brass and steel rods are l_1 and l_2, respectively.

As given difference increase in length $(l_2' - l_1')$ is same for all temperature.

So, $\quad l_2' - l_1' = l_2 - l_1$

$$\Rightarrow l_2(1 + \alpha_2 \Delta t) - l_1(1 + \alpha_1 \Delta t) = l_2 - l_1$$
$$\Rightarrow \quad l_2 \alpha_2 = l_1 \alpha_1$$

179. According to question, as conservation of energy, energy gained by the ice during its fall from height h is given by,

$$E = mgh$$

As given, only one-quarter of its energy is absorbed by the ice.

So, $\quad \frac{mgh}{4} = mL_f$

$$\Rightarrow \quad h = \frac{mL_f \times 4}{mg} = \frac{L_f \times 4}{g}$$
$$= \frac{3.4 \times 10^5 \times 4}{10}$$
$$= 13.6 \times 10^4 = 136000 \text{ m}$$
$$= 136 \text{ km}$$

180. Heat lost by Ist body = heat gained by IInd body. Body at 100°C temperature has greater heat capacity than body at 0°C. So, final temperature will be closer to 100°C. So, $T_c > 50°\text{C}$.

181. Given, temperature, $T_1 = 5760$ K

Since, it is given that energy of radiation emitted by the body at wavelength 250 nm in U_1, at wavelength 500 nm is U_2 and that at 1000 nm is U_3.

$\because$ According to the Wien's law, we get

$$\lambda_m T = b$$

where, b = Wien's constant $= 2.88 \times 10^6$ nmK

$$\Rightarrow \quad \lambda_m = \frac{b}{T} \quad \Rightarrow \quad \lambda_m = \frac{2.88 \times 10^6 \text{ nmK}}{5760 \text{ K}}$$

$$\Rightarrow \quad \lambda_m = 500 \text{ nm}$$

$\because \lambda_m$ = wavelength corresponding to maximum energy, so, $U_2 > U_1$.

182. Using Newton's law of cooling we have, for first 10 min,

$$\frac{3T - 2T}{10} = K\left(\frac{3T + 2T}{2} - T\right)$$

$$\Rightarrow \quad \frac{T}{10} = K\left(\frac{3}{2}T\right) \quad \text{...(i)}$$

After next 10 min, let temperature of body is T_1, then

$$\frac{2T - T_1}{10} = K\left(\frac{2T + T_1}{2} - T\right) \quad \text{...(ii)}$$

Dividing Eq. (ii) by Eq. (i), we get

$$\frac{2T - T_1}{T} = \frac{\frac{2T + T_1}{2} - T}{\frac{3}{2}T}$$

$$\Rightarrow \quad 3(2T - T_1) = T_1$$

$$\Rightarrow \quad T_1 = \frac{3}{2}T$$

183. The solution of this question can be understood by plotting a p-V graph for the compression of a gas isothermally and adiabatically simultaneously to half of its initial volume. i.e.

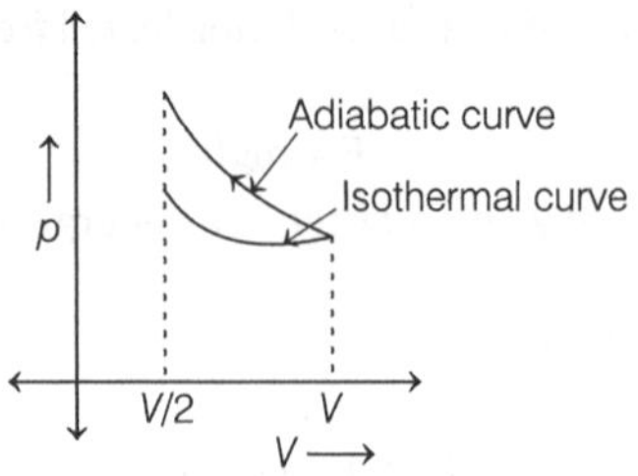

Since, the isothermal curve is less steeper than the adiabatic curve. So, area under the p-V curve for adiabatic process has more magnitude than isothermal curve. Hence, work done in adiabatic process will be more than in isothermal process.

184. Since, initial and final points are same

So, $\Delta U_{A \to B \to C} = \Delta U_{A \to C}$ ...(i)

Also $A \to B$ is isochoric process

So $dW_{A \to B} = 0$ and $dQ = dU + dW$

So, $dQ_{A \to B} = dU_{A \to B} = 400$ J

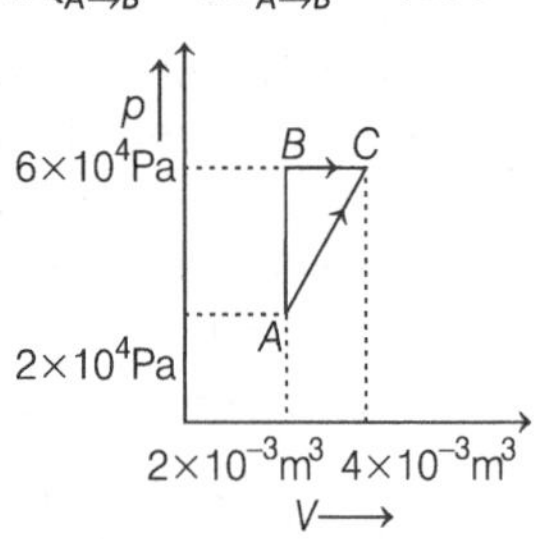

Next $B \to C$ is isobaric process

So, $dQ_{B \to C} = dU_{B \to C} + dW_{B \to C}$

$= dU_{B \to C} + p\Delta V_{B \to C}$

$\Rightarrow \quad 100 = dU_{B \to C} + 6 \times 10^4 (2 \times 10^{-3})$

$\Rightarrow \quad dU_{B \to C} = 100 - 120$

$= -20$ J

From Eq. (i),

$\because \quad \Delta U_{A \to B \to C} = \Delta U_{A \to C}$

$\Rightarrow \Delta U_{A \to B} + \Delta U_{B \to C} = dQ_{A \to C} - dW_{A \to C}$

$\Rightarrow 400 + (-20) = dQ_{A \to C} - (p\Delta V_A + \text{Area of } \Delta ABC)$

$$\Rightarrow dQ_{A \to C} = 380 + \begin{pmatrix} 2 \times 10^4 \times 2 \times 10^{-3} \\ + \frac{1}{2} \times 2 \times 10^{-3} \times 4 \times 10^4 \end{pmatrix}$$

$= 380 + (40 + 40)$

$dQ_{A \to C} = 460$ J

185. Given, temperature of source,

$T = 30°\text{C} = (30 + 273)\, k$

$\Rightarrow \quad T_1 = 303$ K

Temperature of sink, $T_2 = 4°\text{C} = (4 + 273)k$

$T_2 = 277$ K

As, we know that

$$\frac{Q_1}{Q_2} = \frac{T_1}{T_2}$$

$$\Rightarrow \quad \frac{Q_2 + W}{Q_2} = \frac{T_1}{T_2} \qquad (\because W = Q_1 - Q_2)$$

where, Q_2 is the amount of heat drawn from the sink (at T_2), W is work done on working substance, Q_1 is amount of heat rejected to source (at room temperature T_1).

$\Rightarrow \quad WT_2 + T_2Q_2 = T_1Q_2$

$\Rightarrow \quad WT_2 = T_1Q_2 - T_2Q_2$

$\Rightarrow \quad WT_2 = Q_2 (T_1 - T_2)$

$$\Rightarrow \quad W = Q_2 \left(\frac{T_1}{T_2} - 1 \right)$$

$$\Rightarrow \quad W = 600 \times 4.2 \times \left(\frac{303}{277} - 1 \right)$$

$$W = 600 \times 4.2 \times \left(\frac{26}{277} \right)$$

$W = 236.5$ J

$$\text{Power} = \frac{\text{Work done}}{\text{Time}} = \frac{W}{t} = \frac{236.5}{1}$$

$= 236.5$ W

186. For a refrigerator, we know that

$$\frac{Q_1}{W} = \frac{Q_1}{Q_1 - Q_2} = \frac{T_1}{T_1 - T_2}$$

where,

Q_1 = amount of heat delivered to the room,

W = electrical energy consumed,

T_1 = room temperature = $t_1 + 273$,

T_2 = temperature of sink = $t_2 + 273$.

$$\therefore \quad \frac{Q_1}{1} = \frac{t_1 + 273}{t_1 + 273 - (t_2 + 273)}$$

$$\Rightarrow \quad Q_1 = \frac{t_1 + 273}{t_1 - t_2}$$

187. It is given that

$v_{rms} = 200 \text{ ms}^{-1}, T_1 = 300 \text{ K}, p_1 = 10^5 \text{ N/m}^2$

$T_2 = 400 \text{ K}, p_2 = 0.05 \times 10^5 \text{ N/m}^2$

As, rms velocity of gas molecules,

$$\because \quad v_{rms} \propto \sqrt{T} \qquad \left(\because v_{rms} = \sqrt{\frac{3RT}{m}} \right)$$

For two different cases,

$$\Rightarrow \quad \frac{(v_{rms})_1}{(v_{rms})_2} = \sqrt{\frac{T_1}{T_2}}$$

$$\Rightarrow \quad \frac{200}{(v_{rms})_2} = \sqrt{\frac{300}{400}} = \sqrt{\frac{3}{4}}$$

$$\Rightarrow \quad (v_{rms})_2 = \frac{2}{\sqrt{3}} \times 200$$

$$= \frac{400}{\sqrt{3}} \text{ ms}^{-1}$$

188. As we know that,

Pressure, $p = \frac{1}{3} \cdot \frac{nm}{V} v_{rms}^2$

where, nm = mass of the gas, V = volume of the gas

$\therefore \quad \frac{mn}{V}$ = density of the gas. Thus,

$$p = \frac{1}{2} p v_{rms}^2 = \frac{1}{3} \rho \frac{3RT}{M_0} = \frac{\rho RT}{M_0} \qquad \left(\because v_{rms} = \sqrt{\frac{3RT}{M_0}}\right)$$

$$\rho = \frac{pM_0}{RT} = \frac{pmN_A}{k N_A T} \qquad [\because R = N_A k \text{ and } M_0 = mN_A]$$

$$\rho = \frac{pm}{kT}$$

189. As we know that, for polytropic process of index (α), specific heat capacity

$$= C_V + \frac{R}{1-\alpha}$$

$\because$ Process, pV^3 = constant

$$\Rightarrow \quad \alpha = 3$$

$$\therefore \quad C = C_V + \frac{R}{1-\alpha} = \frac{fR}{2} + \frac{R}{1-3} \qquad \text{...(i)}$$

$$\text{where, } C_V = \frac{fR}{2} = \frac{3R}{2} \qquad \text{...(ii)}$$

For monatomic gas, $f = 3 = \frac{3R}{2}$

From Eqs. (i) and (ii), we get

$$\Rightarrow \quad C = \frac{3R}{2} - \frac{R}{2} = R$$

190. Given, the value of coefficient of volume expansion of glycerin is $5 \times 10^{-4} \text{ K}^{-1}$.

As, original density of glycerin,

$$\rho = \rho_0 (1 + Y\Delta T)$$

$$\Rightarrow \quad \rho - \rho_0 = \rho_0 Y \Delta T$$

Thus, fractional change in the density of glycerin for a rise of 40°C in its temperature,

$$\frac{\rho - \rho_0}{\rho_0} = Y\Delta T = 5 \times 10^{-4} \times 40$$

$$= 200 \times 10^{-4} = 0.020$$

191. Here, $\Delta T_1 = 110°\text{C} - 100°\text{C} = 10°\text{ C}$

$$\frac{dQ_1}{dt} = 4 \text{ J/s}$$

$$\Rightarrow \quad \Delta T_2 = 210°\text{C} - 200°\text{C} = 10°\text{ C}$$

$$\frac{dQ_2}{dt} = ?$$

As the rate of heat flow is directly proportional to the temperature difference and the temperature difference in both the cases is same, i.e. 10°C. So, the same rate of heat will flow in the second case.

Hence, $\frac{dQ_2}{dt} = 4$ J/s

192. For a diatomic gas,

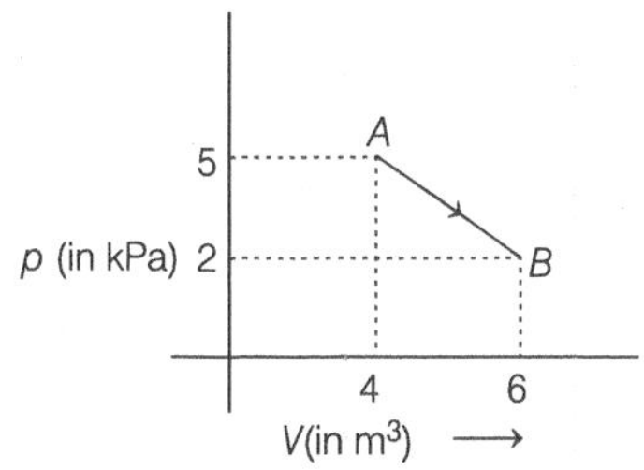

$$C_V = \frac{5}{2} R$$

The change in internal energy of gas in the transition from A to B is

$$\Delta U = nC_V dT = n\left(\frac{5R}{2}\right)(T_B - T_A)$$

$$= nR\frac{5}{2}\left(\frac{p_B V_B}{nR} - \frac{p_A V_A}{nR}\right)$$

$$= \frac{5}{2}(2 \times 10^3 \times 6 - 5 \times 10^3 \times 4)$$

$$= \frac{5}{2} \times (-8 \times 10^3) = -\frac{4 \times 10^4}{2} = -20 \text{ kJ}$$

193. Given, ideal gas is compressed to half its initial volume i.e.

$$V_0 = \frac{V}{2}$$

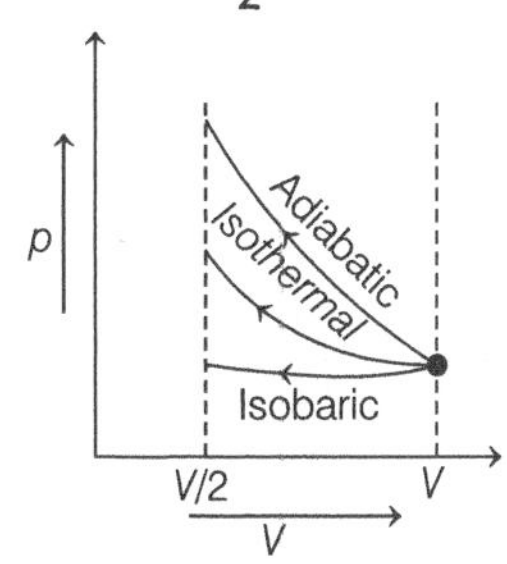

The isochoric process is one in which volume is kept constant, meaning that work done by the system will be zero. i.e. $W_{isochoric} = 0$

As we know, work done on the gas = Area under curve, i.e.

$$W_{adiabatic} > W_{isothermal} > W_{isobaric}$$

194. As, $Q_1 + W = Q_2$

Given, $\eta = \frac{1}{10}$

Now, using $\eta = 1 - \frac{T_1}{T_2}$

So, $\frac{1}{10} = 1 - \frac{T_1}{T_2} \Rightarrow \frac{T_1}{T_2} = \frac{9}{10}$

Now $\quad \dfrac{Q_1}{Q_2} = \dfrac{T_1}{T_2} \quad \Rightarrow \dfrac{Q_1}{Q_1 + W} = \dfrac{9}{10}$

$\Rightarrow \quad 10Q_1 = 9Q_1 + 9W$

$\Rightarrow \quad Q_1 = 9W = 9 \times 10 = 90 \text{ J}$

195. Key Concept Coefficient of performance (β) of a refrigerator is defined as the ratio of quantity of heat removed per cycle (Q_2) to the work done on the working substance per cycle to remove this heat.

Given, coefficient of performance of a refrigerator,

$$\beta = 5$$

Temperature of surface, i.e. inside freezer,

$$T_2 = -20°\text{C} = -20 + 273 = 253\text{K}$$

Temperature of surrounding, i.e. heat rejected outside $T_1 = ?$

So, $\quad \beta = \dfrac{T_2}{T_1 - T_2}$

$\Rightarrow \quad 5 = \dfrac{253}{T_1 - 253}$

$\Rightarrow \quad 5T_1 - 1265 = 253$

$\Rightarrow \quad 5T_1 = 1518$

$$T_1 = \frac{1518}{5} = 303.6 \text{ K}$$

$$T_1 = 303.6 - 273 = 31° \text{ C}$$

196. Given, $M = 4$ g, $V = 22.4$ L, $C_V = 5 \text{ JK}^{-1}\text{mol}^{-1}$

$v_{\text{sound}} = 952$ m/s, $C_p = ?$

As, velocity of sound, $v_{\text{sound}} = \sqrt{\dfrac{\gamma pV}{M}}$

$\Rightarrow \quad \gamma = \dfrac{M}{pV} v^2_{\text{sound}} = \dfrac{C_p}{C_V}$

So, heat capacity at constant pressure,

$$C_p = C_V \left[\frac{M}{pV}\right] v^2_{\text{sound}}$$

$$= 5\left[\frac{4 \times 10^{-3}}{10^5 \times 22.4 \times 10^{-3}}\right](952)^2$$

$$= \frac{20}{22.4} \times (952)^2 \times 10^{-5}$$

$$= 809200 \times 10^{-5}$$

$$= 8.09 \text{ J/mol-K}$$

197.

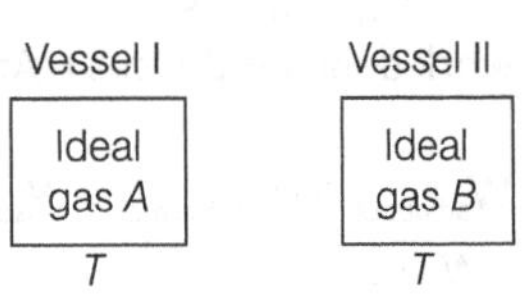

$\rho_A = 1.5\,\rho_B \qquad \rho_B$

$p_A = 2p_B \qquad p_B$

According to ideal gas equation, we have

Pressure, $p = \dfrac{\rho RT}{M}$, where M is molecular weight of ideal gas.

Such that, $\quad \dfrac{p}{\rho} = \dfrac{RT}{M} \Rightarrow M = \dfrac{\rho RT}{P}$

As temperature is same for both gases,

So, $\quad M \propto \dfrac{\rho}{p}$

$\Rightarrow \quad \dfrac{M_A}{M_B} = \dfrac{\rho_A}{\rho_B} \times \dfrac{p_B}{p_A}$

$$= 1.5 \times \frac{1}{2} = 0.75 = \frac{3}{4}$$

198. The specific heat of gas at constant volume in terms of degree of freedom n is

$$C_V = \frac{n}{2}R$$

Also, $\quad C_p - C_V = R$

So, $\quad C_p = \dfrac{n}{2}R + R = R\left(1 + \dfrac{n}{2}\right)$

Now, $\quad \gamma = \dfrac{C_p}{C_V} = \dfrac{R\left(1+\dfrac{n}{2}\right)}{\dfrac{n}{2}R} = \dfrac{2}{n} + 1$

199. Concept Apply principle of calorimetry.

According to principle of calorimetry,

Heat lost by steam = Heat gained by water

Let m' be the amount of steam that converts into water.

$m' \times L + m' s \Delta T = ms\,\Delta t \qquad \begin{bmatrix} s = \text{specific heat of water} \\ L = \text{latent heat of water} \end{bmatrix}$

$$m' \times 540 + m' \times 1 \times (100 - 80) = 20 \times 1 \times (80 - 10)$$

$$m' = \frac{20 \times 70}{560} = 2.5 \text{ g}$$

Now, net mass of water

$$= 20 + 2.5 = 22.5 \text{ g}$$

200. Concept Apply Newton's law of cooling.

Let the temperature of the surrounding be $t°$C.

For first case,

$$\frac{(70 - 60)}{5 \text{ min}} = K(65°\text{C} - t°\text{C})$$

(65° is average of 70°C and 60°C)

$$\frac{10}{5\text{min}} = K(65°\text{C} - t°\text{C}) \qquad \text{...(i)}$$

For second case,

$$\frac{(60-54)}{5\text{min}} = K(57 - t) \qquad \text{...(ii)}$$

(57°C is average of 60°C and 54°C)

From Eqs. (i) and (ii),

$$\frac{10}{6} = \frac{(65 - t)}{(57 - t)}$$

So, $\quad t = 45°$ C

201. For isothermal expansion,

$$pV = p' \times 2V \quad [\because V' = 2V]$$

$$p' = \frac{p}{2}$$

For adiabatic expansion,

$$pV^{\gamma} = \text{constant}$$

$$\Rightarrow \quad p'V^{\gamma} = p''V''^{\gamma}$$

$$\Rightarrow \quad \frac{p}{2}(2V)^{5/3} = p''(16V)^{5/3}$$

$$\Rightarrow \quad p'' = \frac{p}{2}\left[\frac{2V}{16V}\right]^{5/3} = \frac{p}{2}\left(\frac{1}{8}\right)^{5/3}$$

$$= \frac{p}{2}\left(\frac{1}{32}\right) = p/64$$

202. Work done in the cyclic process = Area bounded by the closed configuration

= Area of closed configuration

$$= \frac{1}{2} \times V_0 \times p_0 - \frac{1}{2} \times V_0 \times p_0 = 0 \text{ (zero)}$$

203. Mean free path (l) is given by

$$l = \frac{1}{\sqrt{2} n \pi d^2}$$

$$\Rightarrow \quad l \propto \frac{1}{d^2}$$

So, $$l \propto \frac{1}{r^2} \quad [r = \text{radius of the molecules of gas}]$$

204. As we know that,

$$C_p - C_V = R$$

$$C_p = R + C_V$$

and $$\frac{C_p}{C_V} = \gamma \quad \text{(given)}$$

So, $$\frac{R + C_V}{C_V} = \gamma$$

$$\Rightarrow \quad \gamma C_V = R + C_V$$

$$\Rightarrow \quad \gamma C_V - C_V = R$$

$$\Rightarrow \quad C_V = \frac{R}{\gamma - 1}$$

205. Equation of Wien's displacement law is given by

$$\lambda_m T = \text{constant}$$

206. According to the question,

Net work done = Area enclosed in p-V curve i.e. ΔABC

Area of ΔABC

$$= \frac{1}{2} \times 5 \times 10^{-3} \times 4 \times 10^5 \text{ J}$$

$$= 10^3 \text{ J} = 1000 \text{ J}$$

207. According to the question,

$$\text{Slope of the graph} \propto \frac{1}{\text{Pressure } (p)}$$

So, $p_2 < p_1$

208. According to the question,

$$p \propto T^3 \quad \text{...(i)} \quad \begin{pmatrix} p = \text{pressure,} \\ T = \text{temperature} \end{pmatrix}$$

and we know that,

$$pV = nRT$$

and $$pV \propto T \quad \text{...(ii)}$$

So, putting Eq. (ii) in Eq. (i), we get

$$p \propto (pV)^3$$

$$\Rightarrow \quad p^2V^3 = \text{constant}$$

$$\Rightarrow \quad pV^{3/2} = \text{constant} \quad \text{...(iii)}$$

Comparing Eq. (iii) with $pV^{\gamma} = \text{constant}$.

We have, $\gamma = 3/2$.

209. We know that, $Q = \frac{F}{2} nR\,\Delta T$

$\therefore$ Amount of heat required,

$$Q = \frac{3}{2} \times \frac{1}{4} \times k_B N_a \Delta T$$

$$= \frac{3}{8} N_a k_B (T_2 - T_1)$$

210. Graph (a) shows the variation of temperature with time. At first temperature will increase then there will be state change from liquid to gas.

211. From Stefan' law,

$$E = \sigma T^4$$

So, the rate of energy production

$$Q = E \times A$$

$$Q = \sigma T^4 \times 4\pi R^2$$

Temperature of star

$$T = \left(\frac{Q}{4\pi R^2 \sigma}\right)^{1/4}$$

212. For given cyclic process,

$$\Delta U = 0$$

$$\Rightarrow \quad Q = W$$

Also, W = – area enclosed by the curve

$$= -AB \times AD$$

$$= -(2p - p)(3V - V)$$

$$= -p \times 2V$$

$\therefore$ Heat rejected $= 2pV$

213. According to question, firstly gas expands from volume V to $3V$ and after this volume is reduced from $3V$ to V at constant pressure.

In isothermal expansion, p-V curve is rectangular hyperbola.

214. Change in entropy is given by

$$\Delta S = \frac{ml}{T} = \frac{1000 \times 80}{273}$$

$$= 293 \text{ cal K}^{-1}$$

215. Concept Apply first law of thermodynamics.

According to first law of thermodynamics

$$\Delta U = \Delta Q + \Delta W$$

For isothermal process, $\Delta U = 0$

So, $\Delta Q = -\Delta W$

Given, $\Delta W = -150$ J

So, $\Delta Q = +150$ J

When Q is positive, the heat is added to the gas.

216. In steady state the amount of heat flowing from one face to the other face in time t is given by

$Q = \frac{KA(\theta_1 - \theta_2)t}{l}$, where K is coefficient of thermal conductivity of material of rod

$$\Rightarrow \quad \frac{Q}{t} \propto \frac{A}{l} \propto \frac{r^2}{l} \quad \text{...(i)}$$

As the metallic rod is melted and the material is formed into a rod of half the radius

$$V_1 = V_2$$

$$\pi r_1^2 l_1 = \pi r_2^2 l_2 \quad \left[\because r_2 = \frac{r_1}{2}\right]$$

$$\Rightarrow \quad l_1 = \frac{l_2}{4} \quad \text{...(ii)}$$

Now, from Eqs. (i) and (ii)

$$\frac{Q_1}{Q_2} = \frac{r_1^2}{l_1} \times \frac{l_2}{r_2^2} = \frac{r_1^2}{l_1} \times \frac{4l_1}{(r_1/2)^2}$$

$$\Rightarrow \quad Q_1 = 16\, Q_2$$

$$\Rightarrow \quad Q_2 = \frac{Q_1}{16}$$

217. From first law of thermodynamics

$$\Delta Q = \Delta U + \Delta W$$

For adiabatic process, $\Delta Q = 0$

$\because \quad \Delta Q = 0$

So, $\quad \Delta U = -\Delta W$